Quantum Chemistry

Fourth Edition

IRA N. LEVINE

Chemistry Department
Brooklyn College
City University of New York
Brooklyn, New York

PRENTICE HALL, Englewood Cliffs, New Jersey 07632

Library of Congress Cataloging-in-Publication Data

Levine, Ira N.
    Quantum chemistry / Ira N. Levine.—4th ed.
        p.   cm.
    Includes bibliographical references and index.
    ISBN 0-205-12770-3
    1. Quantum chemistry.   I. Title.
QD462.L48   1991
541.2'8—dc20
                        90-20993
                        CIP

Editorial/production supervision: *Nancy Forsyth/Peter Petraitis*
Editorial/production service: *Raeia Maes*
Cover design: *Linda Dickinson*
Manufacturing buyer: *Linda Cox/Megan Cochran*

© 1991 by Prentice-Hall, Inc.
A Simon & Schuster Company
Englewood Cliffs, New Jersey 07632

Printed in the United States of America

10 9 8   7   6   5

ISBN   0-205-12770-3

PRENTICE-HALL INTERNATIONAL (UK) LIMITED, *London*
PRENTICE-HALL OF AUSTRALIA PTY. LIMITED, *Sydney*
PRENTICE-HALL CANADA INC., *Toronto*
PRENTICE-HALL HISPANOAMERICANA, S.A., *Mexico*
PRENTICE-HALL OF INDIA PRIVATE LIMITED, *New Delhi*
PRENTICE-HALL OF JAPAN, INC., *Tokyo*
SIMON & SCHUSTER ASIA PTE. LTD., *Singapore*
EDITORA PRENTICE-HALL DO BRASIL, LTDA., *Rio de Janeiro*

To my students: Ricardo Alkins, Salvatore Atzeni, Abe Auerbach, Andrew Auerbach, Joseph Barbuto, David Baron, Sene Bauman, Howard Becker, Michael Beitchman, Anna Berne, Kamal Bharucha, Susan Bienenfeld, Mark Blackman, Toby Block, Allen Bloom, Diza Braksmayer, Paul Brumer, Lynn Caporale, Richard Carter, Shih-ching Chang, Ching-hong Chen, Huifen Chen, Kangmin Chen, Guang-Yu Cheng, Yu-Chi Cheng, Jeonghwan Cho, Ting-Yi Chu, Joseph Cincotta, Robert Curran, Joseph D'Amore, Ronald Davy, Aly Dominique, Barry DuRon, Myron Elgart, Anna Eng, Stephen Engel, Quianping Fang, Larry Filler, Seymour Fishman, Donald Franceschetti, Mark Freilich, Michael Freshwater, Tobi Eisenstein Fried, Joel Friedman, Kenneth Friedman, Aryeh Frimer, Mark Froimowitz, Paul Gallant, Mark Gold, Stephen Goldman, Neil Goodman, Roy Goodman, Isaac Gorbaty, Steven Greenberg, Michael Gross, Zhijie Gu, Runyu Han, Sheila Handler, Warren Hirsch, Richard Hom, Kuo-zong Hong, Fu-juan Hsu, Jong-chin Hwan, Leonard Itzkowitz, Mark Johnson, Kirby Juengst, Abraham Karkowsky, Spiros Kassomenakis, Michael Kittay, Colette Knight, Barry Kohn, David Kurnit, Athanasios Ladas, Alan Lambowitz, Yedidyah Langsam, Surin Laosooksathit, Chi-Yin Lee, Stephen Lemont, Elliot Lerner, Israel Liebersohn, Joel Liebman, Steven Lipp, James Liubicich, Rachel Loftoa, Dennis Lynch, Tom McDonough, Pietro Mangiaracina, Louis Maresca, Allen Marks, Ira Michaels, Paul Mogolesko, Safrudin Mustopa, Irving Nadler, Stuart Nagourney, Harold Nelson, Wen-Hui Pan, Padmanabhan Parakat, Albert Pierre-Louis, Eli Pines, Jerry Polesuk, Arlene Gallanter Pollin, James Pollin, Lahanda Punyasena, Robert Richman, Richard Rigg, Bruce Rosenberg, Martin Rosenberg, Robert Rundberg, Edward Sachs, David Schaeffer, Gary Schneier, Neil Schweid, Judith Rosenkranz Selwyn, Gunnar Senum, Steven Shaya, Allen Sheffron, Wu-mian Shen, Yuan Shi, Lawrence Shore, Alvin Silverstein, Barry Siskind, Jerome Solomon, Henry Sperling, Charles Stimler, Helen Sussman, David Trauber, King-hung Tse, Sammy Wainhaus, Alan Waldman, Zheng Wang, Robert Washington, Janet Weaver, William Wihlborg, Shiming Wo, Jinan Wu, Xiaowen Wu, Ming Min Xia, Wei-Guo Xia, Xiaoming Ye, Ching-Chun Yiu, Xue-yi Yuan, Ken Zaner, Juin-tao Zhang, Li Li Zhou, Shan Zhou, Yun Zhou.

Contents

8 THE VARIATION METHOD   189

9 PERTURBATION THEORY   221

10 ELECTRON SPIN AND THE PAULI PRINCIPLE   258

11 MANY-ELECTRON ATOMS   281

Preface

This book is intended for first-year graduate and advanced undergraduate courses in quantum chemistry.

To help students learn the material, the following improvements were made in the fourth edition:

- Worked-out examples have been added to the text.
- Summaries have been added to the first 14 chapters.
- The number of problems has been increased.
- The answers section was expanded to include answers to nearly all numerical problems.
- Important terms are printed in boldface italic type when first defined.
- Equations that should be memorized are marked with an asterisk.
- A systematic listing of theorems has been given in Chapter 7.
- The perturbation theory derivations in Chapter 9 have been simplified.
- Throughout, the text has been revised to clarify and update it.

New material in the fourth edition includes:

- Møller–Plesset perturbation theory (Section 15.12)
- The coupled-cluster method (Section 15.13)
- Density-functional theory (Section 15.14)
- Relativistic effects (Section 15.15)
- The CASSCF and MRCI methods (Section 13.21)
- The AM1 and PM3 methods (Sections 16.5 and 17.1)
- The basis-set superposition error and the counterpoise correction (Section 17.1)
- Size consistency (Section 15.11)
- Nuclear motion in diatomic molecules (Section 13.2)
- Gauss–Jordan elimination (Section 8.4)

The following topics have been substantially expanded:

- Comparisons of methods (Section 17.1)
- The molecular mechanics method (Section 16.6)
- Configuration interaction (Sections 13.21 and 15.11)
- Basis sets (Section 15.4)
- The Hartree–Fock method (Section 13.16)

Matrices are now introduced earlier (Sections 7.10 and 8.6), but the book is written so that matrices can be omitted entirely if time does not allow their inclusion. None of the derivations rely on matrices.

I have tried to make explanations clear and complete, without glossing over difficult or subtle points. Derivations are given with enough detail to make them easy to follow, and resort to the frustrating phrase "it can be shown that" is avoided wherever possible. The aim is to give students a solid understanding of the physical and mathematical aspects of quantum mechanics and molecular electronic structure. The book is designed to be useful to students in all branches of chemistry, not just future quantum chemists. However, the presentation is such that those who do go on in quantum chemistry will have a good foundation and will not be hampered by misconceptions.

An obstacle faced by many chemistry students in learning quantum mechanics is their unfamiliarity with much of the required mathematics. In this text I have included detailed treatments of operators, differential equations, simultaneous linear equations, and other needed topics. Rather than putting all the mathematics in an introductory chapter or a series of appendices, I have integrated the mathematics with the physics and chemistry. Immediate application of the mathematics to solving a quantum-mechanical problem will make the mathematics more meaningful to students than would separate study of the mathematics. I have also kept in mind the limited physics background of many chemistry students by reviewing topics in physics.

This book has benefited from the reviews and suggestions of Leland Allen, N. Colin Baird, Uldis Blukis, James Bolton, Melvyn Feinberg, Gordon A. Gallup, David Goldberg, Hans Jaffé, Harry King, Peter Kollman, Joel Liebman, Frank Meeks, Robert Metzger, William Palke, Kenneth Sando, Harrison Shull, James J. P. Stewart, and Arieh Warshel. Portions of the fourth edition were reviewed by Donald Chesnut, Warren Hehre, Neil Kestner, Gary Pfeiffer, Russell Pitzer, Richard Stratt, and Michael Zerner. I wish to thank these people and several anonymous reviewers.

I would appreciate receiving any suggestions that readers may have for improving the book.

I. N. L.

1

The Schrödinger Equation

1.1 QUANTUM CHEMISTRY

In the late seventeenth century, Isaac Newton discovered classical mechanics, the laws of motion of macroscopic objects. In the early twentieth century, physicists found that classical mechanics does not correctly describe the behavior of very small particles such as the electrons and nuclei of atoms and molecules; the behavior of such particles is described by a set of laws called *quantum mechanics*.

Quantum chemistry applies quantum mechanics to problems in chemistry. The influence of quantum chemistry is felt in all branches of chemistry. Physical chemists use quantum mechanics to calculate (with the aid of statistical mechanics) thermodynamic properties (for example, entropy, heat capacity) of gases; to interpret molecular spectra, thereby allowing experimental determination of molecular properties (for example, bond lengths and bond angles, dipole moments, barriers to internal rotation, energy differences between conformational isomers); to calculate molecular properties theoretically; to calculate properties of transition states in chemical reactions, thereby allowing estimation of rate constants; to understand intermolecular forces; and to deal with bonding in solids.

Organic chemists use quantum mechanics to estimate the relative stabilities of molecules, to calculate properties of reaction intermediates, to investigate the mechanisms of chemical reactions, to predict aromaticity of compounds, and to analyze NMR spectra.

Analytical chemists use spectroscopic methods extensively. The frequencies and intensities of lines in a spectrum can be properly understood and interpreted only through use of quantum mechanics.

Inorganic chemists use ligand field theory, an approximate quantum-mechanical method, to predict and explain the properties of transition-metal complex ions.

Although the large size of biologically important molecules makes quantum-mechanical calculations on them extremely difficult, biochemists are beginning to benefit from quantum-mechanical studies of conformations of biological molecules, enzyme–substrate binding, and solvation of biological molecules.

1.2   HISTORICAL BACKGROUND OF QUANTUM MECHANICS

The development of quantum mechanics began in 1900 with Planck's study of the light emitted by heated solids, so we start by discussing the nature of light.

In 1801, Thomas Young gave convincing experimental evidence for the wave nature of light by showing that light exhibited diffraction and interference when passed through two adjacent pinholes. (See *Halliday and Resnick*, Chapter 45. References with the authors' names italicized are listed in the Bibliography.)

About 1860, James Clerk Maxwell developed four equations, known as Maxwell's equations, which unified the laws of electricity and magnetism. Maxwell's equations predicted that an accelerated electric charge would radiate energy in the form of electromagnetic waves consisting of oscillating electric and magnetic fields. The speed predicted by Maxwell's equations for these waves turned out to be the same as the experimentally measured speed of light. Maxwell concluded that light is an electromagnetic wave.

In 1888, Heinrich Hertz detected radio waves produced by accelerated electric charges in a spark, as predicted by Maxwell's equations. This convinced physicists that light is indeed an electromagnetic wave.

All electromagnetic waves travel at speed $c = 2.998 \times 10^{10}$ cm/s in vacuum. The frequency ν and wavelength λ of a wave are related by

$$\lambda \nu = c \qquad (1.1)^*$$

(An equation with an asterisk after its number should be memorized.) Various conventional labels are applied to electromagnetic waves depending on their frequency. In order of increasing frequency one has radio waves, microwaves, infrared radiation, visible light, ultraviolet radiation, X-rays, and gamma rays. We shall use the term *light* to denote any kind of electromagnetic radiation.

In the late 1800s, physicists measured the intensity of light at various frequencies emitted by a heated blackbody at a fixed temperature. A *blackbody* is an object that absorbs all light falling on it. A good approximation to a blackbody is a cavity with a tiny hole. When physicists used statistical mechanics and the electromagnetic-wave model of light to predict the intensity-versus-frequency curve for emitted blackbody radiation, they found a result in complete disagreement with the high-frequency portion of the experimental curves.

In 1900, Max Planck developed a theory that gave excellent agreement with the observed blackbody-radiation curves. Planck assumed that the atoms of the blackbody could emit light energy only in amounts given by $h\nu$, where ν is the radiation's frequency and h is a proportionality constant, called *Planck's constant*. The value $h = 6.6 \times 10^{-34}$ J·s gave curves that agreed with the experimental blackbody curves. Planck's work marks the beginning of quantum mechanics.

Planck's hypothesis that only certain quantities of light energy can be emitted (that the emission is *quantized*) was in direct contradiction to all previous ideas of physics. The energy of a wave is related to its amplitude, and the amplitude varies continuously from zero on up. Moreover, according to Newtonian mechanics, the energy of a material body can vary continuously. Hence physicists expected the energy of an atom to vary continuously. If one expects the energies of atoms and of electromagnetic waves to vary continuously, one also expects the electromagnetic-radiation energy emitted by atoms to vary continuously. However, only with the hypothesis of quantized energy emission does one obtain the correct blackbody-radiation curves.

The second application of energy quantization was to the photoelectric effect. In the *photoelectric effect*, light shining on a metal causes emission of electrons. The energy of a wave is proportional to its intensity and is not related to its frequency, so the electromagnetic-wave picture of light leads one to expect that the kinetic energy of an emitted photoelectron would increase as the light intensity increases but would not change as the light frequency changes. Instead, one observes that the kinetic energy of an emitted electron is independent of the light's intensity but increases as the light's frequency increases.

In 1905, Albert Einstein showed that these observations could be explained by viewing light as composed of particlelike entities (called **photons**), with each photon having an energy

$$E_{\text{photon}} = h\nu \qquad (1.2)^*$$

When an electron in the metal absorbs a photon, part of the absorbed photon energy is used to overcome the forces holding the electron in the metal, and the remainder appears as kinetic energy of the electron after it has left the metal. Conservation of energy gives

$$h\nu = \Phi + \tfrac{1}{2}mv^2$$

where Φ is the minimum energy needed by an electron to escape the metal (the metal's *work function*), and $\tfrac{1}{2}mv^2$ is the maximum kinetic energy of an emitted electron. An increase in the light's frequency ν increases the photon energy and hence increases the kinetic energy of the emitted electron. An increase in light intensity at fixed frequency increases the rate at which photons strike the metal and hence increases the rate of emission of electrons, but does not change the kinetic energy of each emitted electron.

The photoelectric effect shows that light can exhibit particlelike behavior in addition to the wavelike behavior it shows in diffraction experiments.

Now let us consider the structure of matter.

In the late nineteenth century, investigations of electric discharge tubes and natural radioactivity showed that atoms and molecules are composed of charged particles. Electrons have a negative charge. The proton has a positive charge equal in magnitude but opposite in sign to the electron charge and is 1836 times as heavy as the electron. The third constituent of atoms, the neutron (discovered in 1932), is uncharged and slightly heavier than the proton.

Starting in 1909, Rutherford, Geiger, and Marsden carried out a series of experiments in which they passed a beam of alpha particles through a thin metal foil and observed the deflections of the particles by allowing them to fall on a fluorescent screen. Alpha particles are positively charged helium nuclei obtained from natural radioactive decay. Rutherford observed that most of the alpha particles passed through the foil essentially undeflected, but, surprisingly, a few underwent large deflections, some being deflected backward. To get large deflections, one needs a very close approach between the charges, so that the Coulombic repulsive force is great. If the positive charge were spread throughout the atom (as J. J. Thomson had proposed in 1904), once the high-energy alpha particle penetrated the atom, the repulsive force would fall off, becoming zero at the center of the atom, according to classical electrostatics. Hence Rutherford concluded that such large deflections could occur only if the positive charge were concentrated in a tiny, heavy nucleus.

An atom contains a tiny (10^{-13} to 10^{-12} cm radius), heavy nucleus consisting of neutrons and Z protons, where Z is the atomic number. Outside the nucleus there are Z electrons. The charged particles interact according to Coulomb's law. (The nucleons are held together in the nucleus by strong, short-range nuclear forces, which will not concern us.) The radius of an atom is about one *angstrom* ($1 \text{ Å} \equiv 10^{-8} \text{ cm} = 10^{-10} \text{ m}$), as shown, for example, by results from the kinetic theory of gases. Molecules have more than one nucleus.

The chemical properties of atoms and molecules are determined by their electronic structure, and so the question arises as to the nature of the motions and energies of the electrons. Since the nucleus is much more massive than the electron, we expect the motion of the nucleus to be slight compared with the electrons' motions.

In 1911, Rutherford proposed his planetary model of the atom in which the electrons revolved about the nucleus in various orbits, just as the planets revolve about the sun. However, there is a fundamental difficulty with this model. According to classical electromagnetic theory, an accelerated charged particle radiates energy in the form of electromagnetic (light) waves. An electron circling the nucleus at constant speed is being accelerated, since the direction of its velocity vector is continually changing. Hence the electrons in the Rutherford model should continually lose energy by radiation and therefore would spiral in toward the nucleus. Thus, according to classical (nineteenth-century) physics, the Rutherford atom is unstable and would collapse.

A possible way out of this difficulty was proposed by Niels Bohr in 1913, when he applied the concept of quantization of energy to the hydrogen atom. Bohr assumed that the energy of the electron in a hydrogen atom was quantized, with the electron constrained to move only on one of a number of allowed circles. When an electron makes a transition from one Bohr orbit to another, a photon of light whose frequency ν satisfies

$$E_{\text{upper}} - E_{\text{lower}} = h\nu \qquad (1.3)^*$$

is absorbed or emitted, where E_{upper} and E_{lower} are the energies of the upper and lower states (conservation of energy). With the assumption that an electron making a transition from a free (ionized) state to one of the bound orbits emits a photon whose frequency is an integral multiple of one-half the classical frequency of revolution of the electron in the bound orbit, Bohr used Newtonian mechanics to derive a formula for the hydrogen-atom energy levels. Using (1.3), he obtained agreement with the observed hydrogen spectrum. However, attempts to fit the helium spectrum using the Bohr theory failed. Moreover, the theory could not account for chemical bonds in molecules.

The basic difficulty in the Bohr model arises from the use of classical Newtonian mechanics to describe the electronic motions in atoms. The evidence of atomic spectra, which show discrete frequencies, indicates that only certain energies of motion are allowed; the electronic energy is quantized. However, Newtonian mechanics allows a continuous range of energies. Quantization does occur in wave motion; for example, the fundamental and overtone frequencies of a violin string. Hence Louis de Broglie suggested in 1923 that the motion of electrons might have a wave aspect; that an electron of mass m and speed v would have a wavelength

$$\lambda = \frac{h}{mv} = \frac{h}{p} \qquad (1.4)$$

associated with it, where p is the linear momentum. De Broglie arrived at Eq. (1.4) by reasoning in analogy with photons. The energy of any particle (including a photon) can be expressed, according to Einstein's special theory of relativity, as $E = mc^2$, where c is the speed of light and m is the particle's relativistic mass (not its rest mass). Using $E_{\text{photon}} = h\nu$, we get $mc^2 = h\nu = hc/\lambda$ and $\lambda = h/mc = h/p$ for a photon traveling at speed c. Equation (1.4) is the corresponding equation for an electron.

In 1927, Davisson and Germer experimentally confirmed de Broglie's hypothesis by reflecting electrons from metals and observing diffraction effects. In 1932, Stern observed the same effects with helium atoms and hydrogen molecules, thus verifying that the wave effects are not peculiar to electrons, but result from some general law of motion for microscopic particles.

Thus electrons behave in some respects like particles and in other respects like waves. We are faced with the apparently contradictory "wave–particle duality" of matter (and of light). How can an electron be both a particle, which is a localized entity, and a wave, which is nonlocalized? The answer is that an electron is neither a wave nor a particle, but something else. An accurate pictorial description of an electron's behavior is impossible using the wave or particle concept of classical physics. The concepts of classical physics have been developed from experience in the macroscopic world and do not provide a proper description of the microscopic world. Evolution has shaped the human brain to allow it to understand and deal effectively with macroscopic phenomena. The human nervous system was not developed to deal with phenomena at the atomic and molecular level, so it is not surprising if we cannot fully understand such phenomena.

Although both photons and electrons show an apparent duality, they are not the same kinds of entities. Photons always travel at speed c and have zero rest mass; electrons always have $v < c$ and a nonzero rest mass. Photons must always be treated relativistically, but electrons whose speed is not too high can be treated nonrelativistically.

1.3 THE UNCERTAINTY PRINCIPLE

Let us consider what effect the wave–particle duality has on attempts to measure simultaneously the x coordinate and the x component of linear momentum of a microscopic particle. We start with a beam of particles with momentum p, traveling in the y direction, and we let the beam fall on a narrow slit. Behind this slit is a photographic plate. See Fig. 1.1.

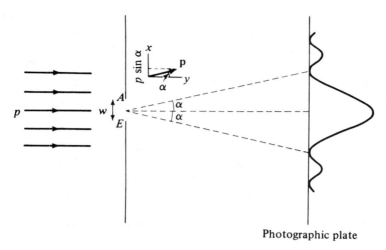

Figure 1.1 Diffraction of electrons by a slit.

Particles that pass through the slit of width w have an uncertainty w in their x coordinate at the time of going through the slit. Calling this spread in x values Δx, we have $\Delta x = w$.

Since microscopic particles have wave properties, they are diffracted by the slit producing (as would a light beam) a diffraction pattern on the plate. The height of the graph in Fig. 1.1 is a measure of the number of particles reaching a given point. The diffraction pattern indicates that when the particles were diffracted by the slit, their direction of motion was changed so that part of their momentum was transferred to the x direction. The x component of momentum is given by the projection of the momentum vector in the x direction. A particle deflected upward by an angle α has an x component of momentum $p \sin \alpha$. A particle deflected downward by an angle α has an x component of momentum $-p \sin \alpha$. Since most of the particles undergo deflections in the range $-\alpha$ to α, where α is the angle to the first minimum in the diffraction pattern, we shall take one-half the spread of momentum values in the central diffraction peak as a measure of the uncertainty Δp_x in the x component of momentum: $\Delta p_x = p \sin \alpha$.

Hence at the slit, where the measurement is made,

$$\Delta x \, \Delta p_x = pw \sin \alpha \qquad (1.5)$$

The angle α at which the first diffraction minimum occurs is readily calculated. The condition for the first minimum is that the difference in the distances traveled by particles passing through the slit at its upper edge and particles passing through the center of the slit be equal to $\frac{1}{2}\lambda$, where λ is the wavelength of the associated wave. Waves originating from the top of the slit are then exactly out of phase with waves originating from the center of the slit and they cancel each other. Waves originating from a point in the slit at a distance d below the slit midpoint cancel with waves originating at a distance d below the top of the slit. Drawing AC in Fig. 1.2 so that $AD = CD$, we have the difference in path length as BC. The distance from the slit to the screen is large compared with

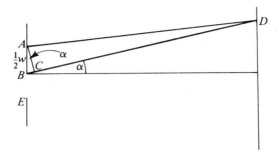

Figure 1.2 Calculation of first diffraction minimum.

the slit width. Hence AD and BD are nearly parallel. This makes the angle ACB essentially a right angle, and so angle $BAC = \alpha$. The path difference BC is then $\frac{1}{2} w \sin \alpha$. Setting BC equal to $\frac{1}{2}\lambda$, we have $w \sin \alpha = \lambda$, and Eq. (1.5) becomes $\Delta x \, \Delta p_x = p\lambda$. The wavelength λ is given by the de Broglie relation $\lambda = h/p$, so $\Delta x \, \Delta p_x = h$. Since the uncertainties have not been precisely defined, the equality sign is not really justified; instead we write

$$\Delta x \, \Delta p_x \approx h \qquad\qquad (1.6)$$

indicating that the product of the uncertainties in x and p_x is of the order of magnitude of Planck's constant. In Section 5.1 we will give a precise statistical definition of the uncertainties and a precise inequality to replace (1.6).

Although we have demonstrated (1.6) for only one experimental setup, its validity is general. No matter what attempts are made, the wave–particle duality of microscopic "particles" imposes a limit on our ability to measure simultaneously the position and momentum of such particles. The more precisely we determine the position, the less accurate is our determination of momentum. (In Fig. 1.1, $\sin \alpha = \lambda/w$, so narrowing the slit increases the spread of the diffraction pattern.) This limitation is the ***uncertainty principle*** and was discovered in 1927 by Werner Heisenberg.

Because of the wave–particle duality, the act of measurement introduces an uncontrollable disturbance in the system being measured. We started with particles having a precise value of p_x (zero); by imposing the slit, we measured the x coordinate of the particles to an accuracy w, but this measurement introduced an uncertainty into the p_x values of the particles. The measurement changed the state of the system.

1.4 THE TIME-DEPENDENT SCHRÖDINGER EQUATION

Classical mechanics applies only to macroscopic particles. For microscopic "particles" we require a new form of mechanics, which we will call ***quantum mechanics***. We now consider some of the contrasts between classical and quantum mechanics. For simplicity we deal with a one-particle, one-dimensional system.

In classical mechanics the motion of the particle is governed by Newton's second law:

$$F = ma = m \frac{d^2x}{dt^2} \qquad (1.7)^*$$

where F is the force acting on the particle, m is its mass, and t is the time; a is the acceleration, given by $a = dv/dt = (d/dt)(dx/dt) = d^2x/dt^2$, where v is the velocity. Equation (1.7) contains the second derivative of the coordinate x with respect to time. To solve it, we must carry out two integrations. This introduces two arbitrary constants c_1 and c_2 into the solution, and

$$x = g(t, c_1, c_2) \qquad (1.8)$$

where g is some function of time. We now ask: What information must we possess at a given time t_0 to be able to predict the future motion of the particle? If we know that at t_0 the particle is at point x_0, we have

$$x_0 = g(t_0, c_1, c_2) \qquad (1.9)$$

Since we have two constants to determine, we need more information. Differentiating (1.8), we have

$$\frac{dx}{dt} = v = \frac{d}{dt}\, g(t, c_1, c_2)$$

If we also know that at time t_0 the particle has velocity v_0, then we have the additional relation

$$v_0 = \frac{d}{dt}\, g(t, c_1, c_2) \bigg|_{t=t_0} \qquad (1.10)$$

We may then use (1.9) and (1.10) to solve for c_1 and c_2 in terms of x_0 and v_0. Knowing c_1 and c_2, we can use Eq. (1.8) to predict the exact future motion of the particle.

As an example of Eqs. (1.7) to (1.10), consider the vertical motion of a particle in the earth's gravitational field. Let the x axis point upward. The force on the particle is downward and is $F = -mg$, where g is the gravitational acceleration constant. Newton's second law (1.7) is $-mg = m\, d^2x/dt^2$, so $d^2x/dt^2 = -g$. A single integration gives $dx/dt = -gt + c_1$. The arbitrary constant c_1 can be found if we know that at time t_0 the particle had velocity v_0. Since $v = dx/dt$, we have $v_0 = -gt_0 + c_1$ and $c_1 = v_0 + gt_0$. Hence $dx/dt = -gt + gt_0 + v_0$. Integration of this equation gives $x = -\frac{1}{2}gt^2 + (gt_0 + v_0)t + c_2$. If we know that at time t_0 the particle had position x_0, then $x_0 = -\frac{1}{2}gt_0^2 + (gt_0 + v_0)t_0 + c_2$ and $c_2 = x_0 - \frac{1}{2}gt_0^2 - v_0t_0$. The expression for x as a function of time becomes $x = -\frac{1}{2}gt^2 + (gt_0 + v_0)t + x_0 - \frac{1}{2}gt_0^2 - v_0t_0$ or $x = x_0 - \frac{1}{2}g(t - t_0)^2 + v_0(t - t_0)$. Knowing x_0 and v_0 at time t_0, we can predict the future position of the particle.

The classical-mechanical potential energy V of a particle moving in one dimension is defined to satisfy

$$\partial V(x, t)/\partial x = -F(x, t) \qquad (1.11)^*$$

For example, for a particle moving in the earth's gravitational field, $\partial V/\partial x = -F = mg$ and integration gives $V = mgx + c$, where c is an arbitrary constant.

We are free to set the zero level of potential energy wherever we please; choosing $c = 0$, we have $V = mgx$ as the potential-energy function.

The word *state* in classical mechanics means a specification of the position and velocity of each particle of the system at some instant of time, plus specification of the forces acting on the particles. According to Newton's second law, given the state of a system at any time, its future state and future motions are exactly determined, as shown by Eqs. (1.8)–(1.10). The impressive success of Newton's laws in explaining planetary motions led many philosophers to use Newton's laws as an argument for philosophical determinism. The mathematician and astronomer Pierre de Laplace (1749–1827) assumed that the universe consisted of nothing but particles that obeyed Newton's laws. Therefore, given the state of the universe at some instant, the future motion of everything in the universe was completely determined. A super-being able to know the state of the universe at any instant could, in principle, calculate all future motions.

Although classical mechanics is deterministic, it was realized in the 1970s that many classical-mechanical systems (for example, a pendulum oscillating under the influence of gravity, friction, and a periodically varying driving force) show chaotic behavior for certain ranges of the systems' parameters. In a chaotic system, the motion is extraordinarily sensitive to the initial values of the particles' positions and velocities and to the forces acting, and two initial states that differ by an experimentally undetectable amount will eventually lead to very different future behavior of the system. (For example, a physicist constructed a double pendulum for which the gravitational attraction of a raindrop one mile away is sufficient to greatly change the motion after two minutes of oscillations; see J. Gleick, *Chaos*, Viking, New York, 1987, p. 230.) Thus, because the accuracy with which one can measure the initial state is limited, prediction of the long-term behavior of a chaotic classical-mechanical system is, in practice, impossible, even though the system obeys deterministic equations. Computer calculations indicate that the motion of the planet Pluto may be chaotic [G. J. Sussman and J. Wisdom, *Science*, **241**, 433 (1988); *Scientific American*, Oct. 1988, p. 20].

Given exact knowledge of the present state of a classical-mechanical system, we can predict its future state. However, the Heisenberg uncertainty principle shows that we cannot determine simultaneously the exact position and velocity of a microscopic particle, so the very knowledge required by classical mechanics for predicting the future motions of a system cannot be obtained. We must be content in quantum mechanics with something less than complete prediction of the exact future motion.

Our approach to quantum mechanics will be to *postulate* the basic principles and then use these postulates to deduce experimentally testable consequences such as the energy levels of atoms. To describe the *state* of a system in quantum mechanics, we postulate the existence of a function of the coordinates called the *wave function* or *state function* Ψ. Since the state will, in general, change with time, Ψ is also a function of time. For a one-particle, one-dimensional system, we have $\Psi = \Psi(x, t)$. The wave function contains all possible information about a system, so instead of speaking of "the state described by the wave function Ψ," we simply say "the state Ψ." Newton's second law tells us how to find the future state of a classical-mechanical system from knowledge of its present state. To find the future state of a quantum-mechanical system from knowledge of its present state, we want an equation that tells us how the wave function changes with time. For a one-particle, one-dimensional system, this equation is postulated to be

$$-\frac{\hbar}{i}\frac{\partial \Psi(x, t)}{\partial t} = -\frac{\hbar^2}{2m}\frac{\partial^2 \Psi(x, t)}{\partial x^2} + V(x, t)\Psi(x, t) \tag{1.12}$$

where the constant $\hbar$ (**h-bar**) is defined as

$$\hbar \equiv \frac{h}{2\pi} \tag{1.13}*$$

The concept of the wave function and the equation governing its change with time were discovered in 1926 by the Austrian physicist Erwin Schrödinger (1887–1961). In this equation, known as the **time-dependent Schrödinger equation** (or the **Schrödinger wave equation**), $i = \sqrt{-1}$, m is the mass of the particle, and $V(x, t)$ is the potential-energy function of the system.

The time-dependent Schrödinger equation contains the first derivative of the wave function with respect to time and allows us to calculate the future wave function (state) at any time, if we know the wave function at time t_0.

The wave function contains all the information we can possibly know about the system it describes. What information does Ψ give us about the result of a measurement of the x coordinate of the particle? We cannot expect Ψ to involve the definite specification of position that the state of a classical-mechanical system does. The correct answer to this question was provided by Max Born shortly after Schrödinger discovered the Schrödinger equation. Born postulated that

$$|\Psi(x, t)|^2 \, dx \tag{1.14}*$$

gives the *probability* at time t of finding the particle in the region of the x axis lying between x and $x + dx$. In (1.14) the bars denote the absolute value and dx is an infinitesimal length on the x axis. The function $|\Psi(x, t)|^2$ is the **probability density** for finding the particle at various places on the x axis. (A review of probability is given in Section 1.6.) For example, suppose that at some particular time t_0 the particle is in a state characterized by the wave function ae^{-bx^2}, where a and b are real constants. If we measure the particle's position at time t_0, we might get any value of x, because the probability density $a^2e^{-2bx^2}$ is nonzero everywhere. Values of x in the region around $x = 0$ are more likely to be found than other values, since $|\Psi|^2$ is a maximum at the origin.

To make a precise connection between $|\Psi|^2$ and experimental measurements, we would take many identical noninteracting systems, each of which was in the same state Ψ. We would then measure the particle's position in each system. If we had n systems and made n measurements, and if dn_x denotes the number of measurements for which we found the particle between x and $x + dx$, then dn_x/n is the probability for finding the particle between x and $x + dx$. Thus

$$\frac{dn_x}{n} = |\Psi|^2 \, dx$$

and a graph of $(1/n) \, dn_x/dx$ versus x gives the probability density $|\Psi|^2$. It might be thought that we could find the probability-density function by taking one system that was in the state Ψ and repeatedly carrying out measurements of the particle's position. This procedure will not do because the process of measurement generally changes the state of a system. We saw an example of this in the discussion of the uncertainty principle (Section 1.3).

Quantum mechanics is basically *statistical* in nature. Knowing the state, we cannot predict the result of a position measurement with certainty; we can only predict the *probabilities* of various possible results. The Bohr theory of the hydrogen atom specified the precise path of the electron and is therefore not a correct quantum-mechanical picture.

Quantum mechanics does not say that an electron is distributed over a large region of space as a wave is distributed. Rather, it is the probability patterns (wave functions) used to describe the electron's motion that behave like waves and satisfy a wave equation.

The reader might ask how the wave function gives us information on other properties (for example, momentum) besides the position. We postpone discussion on this point until later chapters.

The postulates of, say, thermodynamics (the first, second, and third laws of thermodynamics) are stated in terms of macroscopic experience and hence are fairly readily understood. The postulates of quantum mechanics are stated in terms of the microscopic world and appear quite abstract. You should not expect to have a full understanding of the postulates of quantum mechanics at first reading. As we treat various examples, understanding of the postulates will increase.

It may bother the reader that we wrote down the Schrödinger equation without any attempt to prove its plausibility. By using analogies between geometrical optics and classical mechanics on the one hand, and wave optics and quantum mechanics on the other hand, one can show the plausibility of the Schrödinger equation. Geometrical optics is an approximation to wave optics, valid when the wavelength of the light is much less than the size of the apparatus. (Recall its use in treating lenses and mirrors.) Likewise, classical mechanics is an approximation to wave mechanics, valid when the particle's wavelength is much less than the size of the apparatus. One can make a plausible guess as to how to get the proper equation for quantum mechanics from classical mechanics based on the known relation between the equations of geometrical and wave optics. Since most chemists are not particularly familiar with optics, these arguments have been omitted. In any case, such analogies can only make the Schrödinger equation seem *plausible*; they cannot be used to *derive* or *prove* this equation. The Schrödinger equation is a *postulate* of the theory, to be tested by agreement of its predictions with experiment. (Details of the reasoning that led Schrödinger to his equation are given in *Jammer*, Section 5.3.)

Quantum mechanics provides the law of motion for microscopic particles. Experimentally, macroscopic objects obey classical mechanics. Hence for quantum mechanics to be a valid theory, it should reduce to classical mechanics as we make the transition from microscopic to macroscopic particles. Quantum effects are associated with the de Broglie wavelength $\lambda = h/mv$. Since h is very small, the de Broglie wavelength of macroscopic objects is essentially zero. Thus, in the limit $\lambda \rightarrow 0$, we expect the time-dependent Schrödinger equation to reduce to Newton's second law. We can prove this to be so (see Problem 7.50).

A similar situation holds in the relation between special relativity and classical mechanics. In the limit $v/c \rightarrow 0$, where c is the velocity of light, special relativity reduces to classical mechanics. The form of quantum mechanics that we will develop will be nonrelativistic. A complete integration of relativity with quantum mechanics has not been achieved.

Historically, quantum mechanics was first formulated in 1925 by Heisenberg, Born, and Jordan using matrices, several months before Schrödinger's 1926 formulation using differential equations. Schrödinger proved that the Heisenberg formulation (called *matrix mechanics*) is equivalent to the Schrödinger formulation (called *wave mechanics*).

1.5 THE TIME-INDEPENDENT SCHRÖDINGER EQUATION

The time-dependent Schrödinger equation (1.12) is rather formidable looking. Fortunately, for many applications of quantum mechanics to chemistry, it is not necessary to deal with this equation; instead, the simpler time-independent Schrödinger equation is used. We now derive the time-independent from the time-dependent Schrödinger equation for the one-particle, one-dimensional case.

We begin by restricting ourselves to the special case where the potential energy V is not a function of time but depends only on x. This will be true if the system experiences no time-dependent external forces. The time-dependent Schrödinger equation reads

$$-\frac{\hbar}{i}\frac{\partial \Psi(x, t)}{\partial t} = -\frac{\hbar^2}{2m}\frac{\partial^2 \Psi(x, t)}{\partial x^2} + V(x)\Psi(x, t) \tag{1.15}$$

We now restrict ourselves to looking for those solutions of (1.15) that can be written as the product of a function of time and a function of x:

$$\Psi(x, t) = f(t)\psi(x) \tag{1.16}*$$

Capital psi is used for the time-dependent wave function and lowercase psi for the factor that depends only on the coordinate x. States corresponding to wave functions of the form (1.16) possess certain properties (to be discussed shortly) that make them of great interest. [Not all solutions of (1.15) have the form of (1.16); see Problem 3.38.] Taking partial derivatives of (1.16), we have

$$\frac{\partial \Psi(x, t)}{\partial t} = \frac{df(t)}{dt}\psi(x), \qquad \frac{\partial^2 \Psi(x, t)}{\partial x^2} = f(t)\frac{d^2\psi(x)}{dx^2}$$

Substitution into (1.15) gives

$$-\frac{\hbar}{i}\frac{df(t)}{dt}\psi(x) = -\frac{\hbar^2}{2m}f(t)\frac{d^2\psi(x)}{dx^2} + V(x)f(t)\psi(x)$$

$$-\frac{\hbar}{i}\frac{1}{f(t)}\frac{df(t)}{dt} = -\frac{\hbar^2}{2m}\frac{1}{\psi(x)}\frac{d^2\psi(x)}{dx^2} + V(x) \tag{1.17}$$

where we divided by $f\psi$. In general, we expect the quantity to which each side of

(1.17) is equal to be a certain function of x and t. However, the right side of (1.17) does not depend on t, so the function to which each side of (1.17) is equal must be independent of t. The left side of (1.17) is independent of x, so this function must also be independent of x. Since the function is independent of both variables, x and t, it must be a constant. We call this constant E.

Equating the left side of (1.17) to E, we get

$$\frac{df(t)}{f(t)} = -\frac{iE}{\hbar}\, dt$$

Integrating both sides of this equation with respect to t, we have

$$\ln f(t) = -iEt/\hbar + C$$

where C is an arbitrary constant of integration. Hence

$$f(t) = e^C e^{-iEt/\hbar} = A e^{-iEt/\hbar}$$

where the arbitrary constant A has replaced e^C. Since A can be included as a factor in the function $\psi(x)$ that multiplies $f(t)$ in (1.16), A can be omitted from $f(t)$. Thus

$$f(t) = e^{-iEt/\hbar}$$

Equating the right side of (1.17) to E, we have

$$-\frac{\hbar^2}{2m}\frac{d^2\psi(x)}{dx^2} + V(x)\psi(x) = E\psi(x) \qquad \textbf{(1.18)}^*$$

Equation (1.18) is the ***time-independent Schrödinger equation*** for a single particle of mass m moving in one dimension. [Schrödinger actually developed the time-independent equation before the time-dependent equation. The relevant papers are E. Schrödinger, *Ann. Physik*, **79**, 361, 489 (1926); **80**, 437 (1926); **81**, 109 (1926).]

What is the significance of the constant E? Since E occurs as $[E - V(x)]$ in (1.18), E has the same dimensions as V, so E has the dimensions of energy. In fact, we postulate that E is the energy of the system. (This is a special case of a more general postulate to be discussed in a later chapter.) Thus, for cases where the potential energy is a function of x only, there exist wave functions of the form

$$\Psi(x, t) = e^{-iEt/\hbar}\psi(x) \qquad (1.19)$$

and these wave functions correspond to states of constant energy E. Much of our attention in the next few chapters will be devoted to finding the solutions of (1.18) for various systems.

The wave function in (1.19) is complex, but the quantity that is experimentally observable is the probability density $|\Psi(x, t)|^2$. The square of the absolute value of a complex quantity is given by the product of the quantity with its complex conjugate, the complex conjugate being formed by replacing i with $-i$ wherever it occurs. (See Section 1.7.) Thus

$$|\Psi|^2 = \Psi^*\Psi \qquad \textbf{(1.20)}^*$$

where the star denotes the complex conjugate. For the wave function (1.19), we have

$$|\Psi(x, t)|^2 = [e^{-iEt/\hbar}\psi(x)]^* e^{-iEt/\hbar}\psi(x)$$

$$= e^{iEt/\hbar}\psi^*(x)e^{-iEt/\hbar}\psi(x)$$

$$= e^0\psi^*(x)\psi(x) = \psi^*(x)\psi(x)$$

$$|\Psi(x, t)|^2 = |\psi(x)|^2 \tag{1.21}$$

In deriving (1.21), we assumed that E is a real number, so that $E = E^*$. This fact will be proved in Section 7.2.

Hence for states of the form (1.19), the probability density is given by $|\psi(x)|^2$ and does not change with time. Such states are called **stationary states**. Since the physically significant quantity is $|\Psi(x, t)|^2$, and since for stationary states $|\Psi(x, t)|^2 = |\psi(x)|^2$, the function $\psi(x)$ is often called the **wave function**, although the complete wave function of a stationary state is obtained by multiplying $\psi(x)$ by $e^{-iEt/\hbar}$. The term stationary state should not mislead the reader into thinking that a particle in a stationary state is at rest. What is stationary is the probability density $|\Psi|^2$, not the particle itself.

We will be concerned mostly with states of constant energy (stationary states) and hence will usually deal with the time-independent Schrödinger equation (1.18). For simplicity we will refer to this equation as "the Schrödinger equation." Note that the Schrödinger equation contains *two* unknowns, the allowed energies E and the allowed wave functions ψ. To solve for two unknowns, we need to impose additional conditions (called boundary conditions) on ψ besides requiring that it satisfy (1.18); the boundary conditions determine the allowed energies, since it turns out that only certain values of E allow ψ to meet the boundary conditions. This will become clearer when we discuss specific examples in later chapters.

1.6 PROBABILITY

Probability plays a fundamental role in quantum mechanics. In this section we review the mathematics of probability.

There has been much controversy about the proper definition of probability. One definition is the following: If an experiment has n equally probable outcomes, m of which are favorable to the occurrence of a certain event A, then the probability that A occurs is m/n. Note that this definition is circular, since it specifies equally *probable* outcomes when *probability* is what we are attempting to define. It is simply assumed that we can recognize equally probable outcomes. An alternative definition is based on actually performing the experiment many times. Suppose that we perform the experiment N times and that in M of these trials the event A occurs. The probability of A occurring is then defined as

$$\lim_{N\to\infty} \frac{M}{N}$$

Thus, if we toss a coin repeatedly, the fraction of heads will approach $1/2$ as we increase the number of tosses.

For example, suppose we pick a card at random from a deck and ask for the probability of drawing a heart. There are 52 cards and hence 52 equally probable outcomes. Since there are 13 hearts, there are 13 favorable outcomes. Hence $m/n = 13/52 = 1/4$. The probability for drawing a heart is $1/4$.

Sometimes we ask for the probability of two related events both occurring. For example, we may ask for the probability of drawing two hearts from a 52-card deck, assuming we do not replace the first card after it is drawn. There are 52 possible outcomes of the first draw, and for each of these possibilities there are 51 possible second draws. We have $52 \cdot 51$ possible outcomes. Since there are 13 hearts, there are $13 \cdot 12$ different ways to draw two hearts. The desired probability is $13 \cdot 12/52 \cdot 51 = 1/17$. This calculation illustrates the theorem: The probability that two events A and B both occur is the probability that A occurs, multiplied by the conditional probability that B then occurs, calculated with the assumption that A occurred. Thus, if A is the probability of drawing a heart on the first draw, the probability of A is $13/52$. The probability of drawing a heart on the second draw, given that the first draw yielded a heart, is $12/51$ since there remain 12 hearts in the deck. The probability of drawing two hearts is then $(13/52)(12/51) = 1/17$, as found previously.

In quantum mechanics we must deal with probabilities involving a continuous variable, for example, the x coordinate. It does not make much sense to talk about the probability of a particle being found *at* a particular point such as $x = 0.5000 \ldots$, since there are an infinite number of points on the x axis, and for any finite number of measurements we make, the probability of getting *exactly* $0.5000 \ldots$ is vanishingly small. Instead we talk of the probability of finding the particle in a small interval of the x axis lying between x and $x + dx$, dx being an infinitesimal element of length. This probability will naturally be proportional to the length of the small interval, dx, and will vary for different regions of the x axis. Hence the probability that the particle will be found between x and $x + dx$ is equal to $g(x)\, dx$, where $g(x)$ is some function that tells how the probability varies over the x axis. The function $g(x)$ is called the ***probability density***, since it is a probability per unit length. Since probabilities are real, nonnegative numbers, $g(x)$ must be a real function that is everywhere nonnegative. The wave function Ψ can take on negative and complex values and is not a probability density. Quantum mechanics postulates that the probability density is given by $|\Psi|^2$ [Eq. (1.14)].

What is the probability that the particle lies in some finite region of space $a \leq x \leq b$? To find this probability, we sum up the probabilities $|\Psi|^2\, dx$ of finding the particle in all the infinitesimal regions lying between a and b. This is just the definition of the definite integral

$$\int_a^b |\Psi|^2\, dx = \Pr(a \leq x \leq b) \qquad (1.22)^*$$

where Pr denotes a probability. A probability of 1 represents certainty. Since it is certain that the particle is somewhere on the x axis, we have the requirement

$$\int_{-\infty}^{\infty} |\Psi|^2\, dx = 1 \qquad (1.23)^*$$

When Ψ satisfies (1.23), it is said to be **normalized**. For a stationary state, $|\Psi|^2 = |\psi|^2$ and $\int_{-\infty}^{\infty} |\psi|^2 \, dx = 1$.

EXAMPLE   A one-particle, one-dimensional system has $\Psi = a^{-1/2} e^{-|x|/a}$ at $t = 0$, where $a = 1.0000$ nm (1 nm $= 10^{-9}$ m). At $t = 0$, the particle's position is measured. (a) Find the probability that the measured value lies between $x = 1.5000$ nm and $x = 1.5001$ nm. (b) Find the probability that the measured value is between $x = 0$ and $x = 2$ nm. (c) Verify that Ψ is normalized.

(a) In this tiny interval, x changes by only 0.0001 nm, and Ψ goes from $e^{-1.5000}$ nm$^{-1/2} = 0.22313$ nm$^{-1/2}$ to $e^{-1.5001}$ nm$^{-1/2} = 0.22311$ nm$^{-1/2}$, so Ψ is nearly constant in this interval, and it is a very good approximation to consider this interval as infinitesimal. The desired probability is given by (1.14) as

$$|\Psi|^2 \, dx = a^{-1} e^{-2|x|/a} \, dx = (1 \text{ nm})^{-1} e^{-2(1.5 \text{ nm})/(1 \text{ nm})} (0.0001 \text{ nm}) = 4.979 \times 10^{-6}$$

(See also Problem 1.8.)

(b) Use of Eq. (1.22) and $|x| = x$ for $x \geq 0$ gives

$$\text{Pr}(0 \leq x \leq 2 \text{ nm}) = \int_0^{2 \text{ nm}} |\Psi|^2 \, dx = a^{-1} \int_0^{2 \text{ nm}} e^{-2x/a} \, dx$$

$$= -\tfrac{1}{2} e^{-2x/a} \big|_0^{2 \text{ nm}} = -\tfrac{1}{2}(e^{-4} - 1) = 0.4908$$

(c) Use of $|x| = -x$ for $x \leq 0$, $|x| = x$ for $x \geq 0$, and $\int_{-\infty}^{\infty} f(x) \, dx = \int_{-\infty}^{0} f(x) \, dx + \int_{0}^{\infty} f(x) \, dx$ gives

$$\int_{-\infty}^{\infty} |\Psi|^2 \, dx = a^{-1} \int_{-\infty}^{0} e^{2x/a} \, dx + a^{-1} \int_{0}^{\infty} e^{-2x/a} \, dx$$

$$= a^{-1}(\tfrac{1}{2} a e^{2x/a} \big|_{-\infty}^{0}) + a^{-1}(-\tfrac{1}{2} a e^{-2x/a} \big|_{0}^{\infty}) = \tfrac{1}{2} + \tfrac{1}{2} = 1$$

1.7 COMPLEX NUMBERS

We have seen that the wave function can be complex, so we now review some properties of complex numbers.

If i is $\sqrt{-1}$, then we may write a complex number z as $z = x + iy$, where x and y are real numbers; x and y are called the real and imaginary parts of z: $x = \text{Re}(z)$, $y = \text{Im}(z)$. A convenient representation of z is as a point in the complex plane (Fig. 1.3), where the real part of z is plotted on the horizontal axis and the imaginary part on the vertical axis. This diagram immediately suggests defining two quantities that characterize the complex number z: the distance r of the point z from the origin is called the **absolute value** or modulus of z and is denoted by $|z|$; the angle θ that the radius vector to the point z makes with the positive horizontal axis is called the **phase** or argument of z. We have

$$|z| = r = (x^2 + y^2)^{1/2}, \qquad \tan \theta = y/x \tag{1.24}$$

$$x = r \cos \theta, \qquad y = r \sin \theta$$

So we may write $z = x + iy$ as

$$z = r \cos \theta + ir \sin \theta = re^{i\theta} \tag{1.25}$$

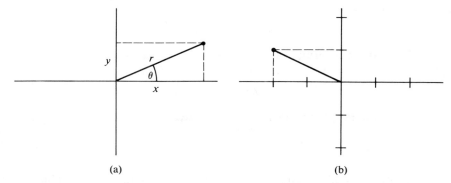

(a)   (b)

Figure 1.3 (a) Plot of a complex number $z = x + iy$. (b) Plot of the number $-2 + i$.

since (Problem 4.3)

$$e^{i\theta} = \cos\theta + i\sin\theta \qquad (1.26)^*$$

The angle θ in these equations is in radians.

The **complex conjugate** z^* of the complex number z is defined as

$$z^* \equiv x - iy = re^{-i\theta} \qquad (1.27)^*$$

If z is a real number, its imaginary part is zero. Thus z is real if and only if $z = z^*$. Taking the complex conjugate twice, we get z back again, $(z^*)^* = z$. Forming the product of z and its complex conjugate, we have

$$zz^* = (x + iy)(x - iy) = x^2 + iyx - iyx - i^2y^2$$
$$zz^* = x^2 + y^2 = r^2 = |z|^2 \qquad (1.28)^*$$

For the product and quotient of two complex numbers $z_1 = r_1 e^{i\theta_1}$ and $z_2 = r_2 e^{i\theta_2}$, we have

$$z_1 z_2 = r_1 r_2 e^{i(\theta_1 + \theta_2)}, \qquad \frac{z_1}{z_2} = \frac{r_1}{r_2} e^{i(\theta_1 - \theta_2)} \qquad (1.29)$$

It is easy to prove, either directly from the definition of complex conjugate or from (1.29), that the complex conjugate of a product is the product of the complex conjugates:

$$(z_1 z_2)^* = z_1^* z_2^* \qquad (1.30)^*$$

Likewise,

$$\left(\frac{z_1}{z_2}\right)^* = \frac{z_1^*}{z_2^*}, \qquad (z_1 + z_2)^* = z_1^* + z_2^*, \qquad (z_1 - z_2)^* = z_1^* - z_2^* \qquad (1.31)$$

For the absolute values of products and quotients, it follows from (1.29) that

$$|z_1 z_2| = |z_1|\,|z_2|, \qquad \left|\frac{z_1}{z_2}\right| = \frac{|z_1|}{|z_2|} \qquad (1.32)$$

Therefore, if ψ is a complex wave function, we have

$$|\psi^2| = |\psi|^2 = \psi^*\psi \tag{1.33}$$

We now obtain a formula for the nth roots of the number 1. We may take the phase of the number 1 to be 0 or 2π or 4π, and so on; hence

$$1 = e^{i2\pi k}$$

where k is any integer, zero, negative, or positive. Now consider the number ω, where

$$\omega = e^{i2\pi k/n}$$

n being a positive integer. Using (1.29) n times, we see that $\omega^n = 1$. Thus ω is an nth root of unity. There are n different complex nth roots of unity, and taking n successive values of the integer k gives us all of them:

$$\omega = e^{i2\pi k/n}, \qquad k = 0, 1, 2, \ldots, n-1 \tag{1.34}$$

Any other value of k besides those in (1.34) gives a number whose phase differs by an integral multiple of 2π from one of the numbers in (1.34) and hence is not a different root.

1.8 UNITS

Two different systems of units are currently used in science. In the cgs Gaussian system, the units of length, mass, and time are the centimeter (cm), gram (g), and second (s). Force is measured in dynes (dyn) and energy in ergs. Coulomb's law for the magnitude of the force between two charges Q_1' and Q_2' separated by a distance r in vacuum is given by $F = Q_1'Q_2'/r^2$, where Q_1' and Q_2' are in statcoulombs (statC), also called electrostatic units of charge (esu).

In the International System (SI), the units of length, mass, and time are the meter (m), kilogram (kg), and second (s). Force is measured in newtons (N) and energy in joules (J). Coulomb's law is written as $F = Q_1Q_2/4\pi\varepsilon_0 r^2$, where the charges Q_1 and Q_2 are in coulombs (C) and ε_0 is a constant (called the permittivity of vacuum) whose experimental value is 8.854×10^{-12} C^2 N^{-1} m^{-2}. In the International System, charge is not expressible in terms of the mechanical units of meters, kilograms, and seconds. SI units are the officially recommended units of science, but the simple form of Coulomb's law in Gaussian units makes Gaussian units popular in quantum chemistry.

In this book, Coulomb's law is usually written as

$$F = Q_1'Q_2'/r^2 \tag{1.35}*$$

One can think of this equation as being in Gaussian units with Q_1' and Q_2' in statcoulombs, r in centimeters, and F in dynes. Alternatively, one can view Eq. (1.35) as being in SI units, with r in meters, F in newtons, and Q_1' and Q_2' as abbreviations for $Q_1/(4\pi\varepsilon_0)^{1/2}$ and $Q_2/(4\pi\varepsilon_0)^{1/2}$, where Q_1 and Q_2 are the charges in coulombs; we have

$$Q' = Q/(4\pi\varepsilon_0)^{1/2} \tag{1.36}*$$

1.9 SUMMARY

The state of a quantum-mechanical system is described by a state function or wave function Ψ, which is a function of the coordinates of the particles of the system and of the time. The state function changes with time according to the time-dependent Schrödinger equation, which for a one-particle, one-dimensional system is Eq. (1.12). For such a system, the quantity $|\Psi(x, t)|^2 \, dx$ gives the probability that a measurement of the particle's position at time t will find it between x and $x + dx$. The state function is normalized according to $\int_{-\infty}^{\infty} |\Psi|^2 \, dx = 1$. If the system's potential-energy function does not depend on t, then the system can exist in one of a number of stationary states of fixed energy. For a stationary state of a one-particle, one-dimensional system, $\Psi(x, t) = e^{-iEt/\hbar} \psi(x)$, where the time-independent wave function $\psi(x)$ is a solution of the time-independent Schrödinger equation (1.18).

PROBLEMS

1.1 Calculate the de Broglie wavelength of an electron moving at 1/137th the speed of light. (At this speed, the relativistic correction to the mass is negligible.)

1.2 The work function of Na is 2.28 eV, where $1 \, eV = 1.602 \times 10^{-19}$ J. (a) Calculate the maximum kinetic energy of photoelectrons emitted from Na exposed to 200 nm ultraviolet radiation. [1 nanometer (nm) $= 10^{-9}$ m.] (b) Calculate the longest wavelength that will cause the photoelectric effect in Na.

1.3 When J. J. Thomson investigated electrons in cathode-ray tubes, he observed the behavior expected for particles obeying classical mechanics. (a) Electrons are accelerated through a potential difference of 1000 volts and passed through a collimating slit of width 0.100 cm. Calculate the diffraction angle α in Fig. 1.1. Use (6.107). (b) What slit width is needed to give $\alpha = 1.00°$ for 1000-volt electrons?

1.4 In classical mechanics the kinetic energy of a particle is defined as $T \equiv \frac{1}{2}mv^2$. Use results from Section 1.4 to show that, for a particle moving vertically in the earth's gravitational field (with g assumed constant), $T + V = \frac{1}{2}mv_0^2 + mgx_0$, so $T + V$ is constant.

1.5 A certain one-particle, one-dimensional system has $\Psi = ae^{-ibt}e^{-bmx^2/\hbar}$, where a and b are constants and m is the particle's mass. Find the potential-energy function V for this system. *Hint:* Use the time-dependent Schrödinger equation.

1.6 A certain one-particle, one-dimensional system has the potential energy $V = 2c^2\hbar^2 x^2/m$ and is in a stationary state with $\psi(x) = bxe^{-cx^2}$, where b is a constant, $c = 2.00 \, nm^{-2}$, and $m = 1.00 \times 10^{-27}$ g. Find the particle's energy.

1.7 At a certain instant of time, a one-particle, one-dimensional system has $\Psi = (2/b^3)^{1/2}xe^{-|x|/b}$, where $b = 3.000$ nm. If a measurement of x is made at this time in the system, find the probability that the result (a) lies between 0.9000 and 0.9001 nm (treat this interval as infinitesimal); (b) lies between 0 and 2 nm (use the table of integrals in the Appendix, if necessary). (c) For what value of x is the probability density a minimum? (There is no need to use calculus to answer this.) (d) Verify that Ψ is normalized.

1.8 Use Eq. (1.22) to find the answer to part (a) of the example at the end of Section 1.6 and compare it with the approximate answer found in the example.

1.9 A one-particle, one-dimensional system has the state function

$$\Psi = (\sin at)(2/\pi c^2)^{1/4}e^{-x^2/c^2} + (\cos at)(32/\pi c^6)^{1/4}xe^{-x^2/c^2}$$

where a is a constant and $c = 2.000$ Å. If the particle's position is measured at $t = 0$, estimate the probability that the result will lie between 2.000 and 2.001 Å.

1.10 What important probability-density function occurs in (a) the kinetic theory of gases? (b) the analysis of random errors of measurement?

1.11 Which of the following functions meet *all* the requirements of a probability-density function (a and b are positive constants)? (a) e^{iax}; (b) xe^{-bx^2}; (c) e^{-bx^2}.

1.12 If the peak in the mass spectrum of C_2F_6 at mass number 138 is 100 units high, calculate the heights of the peaks at mass numbers 139 and 140. Isotopic abundances: ^{12}C, 98.89%; ^{13}C, 1.11%; ^{19}F, 100%.

1.13 In bridge, each of the four players (A, B, C, D) receives 13 cards. Suppose A and C have 11 of the 13 spades between them. What is the probability that the remaining two spades are distributed so that B and D have one spade apiece?

1.14 Suppose a certain disease is present in 0.50% of the general population. Further, suppose that a certain test for this disease correctly detects the disease in 98 out of 100 people who have it and incorrectly yields a false positive result in 1 out of 100 people who do not have the disease. Find the probability that someone who is selected at random from the general population and tests positive with this test actually has the disease.

1.15 Plot these points in the complex plane: (a) 3; (b) $-i$; (c) $-2 + 3i$.

1.16 Show that $1/i = -i$.

1.17 Simplify (a) i^2; (b) i^3; (c) i^4.

1.18 Find the complex conjugate of (a) -4; (b) $-2i$; (c) $6 + 3i$; (d) $2e^{-i\pi/5}$.

1.19 Find the absolute value and the phase of (a) i; (b) $2e^{i\pi/3}$; (c) $-2e^{i\pi/3}$; (d) $1 - 2i$.

1.20 Write each of the following in the form $re^{i\theta}$: (a) i; (b) -1; (c) $1 - 2i$; (d) $-1 - i$.

1.21 (a) Find the cube roots of 1. (b) Explain why the nth roots of 1 when plotted in the complex plane lie on a circle of radius 1 and are separated by an angle $2\pi/n$ from one another.

1.22 Verify that

$$\sin\theta = \frac{e^{i\theta} - e^{-i\theta}}{2i}, \qquad \cos\theta = \frac{e^{i\theta} + e^{-i\theta}}{2}$$

1.23 Express each of the following units in terms of the fundamental units (cm, g, s) of the Gaussian system: (a) dyne; (b) erg; (c) statcoulomb. Express each of the following units in terms of fundamental SI units (m, kg, s): (d) newton; (e) joule.

1.24 Calculate the force on an alpha particle passing a gold atomic nucleus at a distance of 0.00300 Å. Do the calculation twice, once using SI units and once using Gaussian units.

1.25 True or false? (a) A probability density can never be negative. (b) The state function Ψ can never be negative. (c) The state function Ψ must be a real function. (d) If $z = z^*$, then z must be a real number. (e) $\int_{-\infty}^{\infty} \Psi \, dx = 1$ for a one-particle, one-dimensional system.

2
The Particle in a Box

2.1 DIFFERENTIAL EQUATIONS

Since the Schrödinger equation is a differential equation, we now review the mathematics of differential equations. This section considers only *ordinary* differential equations, which are those with only one independent variable. [A *partial* differential equation has more than one independent variable; an example is the time-dependent Schrödinger equation (1.15).] An ordinary differential equation is a relation involving an independent variable x, a dependent variable $y(x)$, and the first, second, . . . , nth derivatives of y (y', y'', . . . , $y^{(n)}$). An example is

$$y''' + 2x(y')^2 + \sin x \cos y = 3e^x \tag{2.1}$$

The *order* of a differential equation is the order of the highest derivative in the equation. Thus, (2.1) is of third order.

A special kind of differential equation is the *linear differential equation*, which has the form

$$A_n(x)y^{(n)} + A_{n-1}(x)y^{(n-1)} + \cdots + A_0(x)y = g(x) \tag{2.2}$$

where the A's (some of which may be zero) are functions of x only. In (2.2), y and its derivatives appear to the first power. A differential equation that cannot be put in the form (2.2) is *nonlinear*. If $g(x) = 0$ in (2.2), the linear differential equation is said to be *homogeneous*; otherwise it is *inhomogeneous*. The one-dimensional Schrödinger equation (1.18) is a linear homogeneous differential equation of second order.

By dividing by the coefficient of y'', we can put any linear homogeneous second-order differential equation into the form

$$y'' + P(x)y' + Q(x)y = 0 \tag{2.3}$$

Suppose we have two independent functions y_1 and y_2, each of which satisfies (2.3). By independent, we mean that y_2 is not simply a multiple of y_1. Then the general solution of the linear homogeneous differential equation (2.3) is

$$y = c_1y_1 + c_2y_2 \tag{2.4}$$

where c_1 and c_2 are arbitrary constants. This is readily verified by substituting

(2.4) into the left side of (2.3):

$$c_1 y_1'' + c_2 y_2'' + P(x) c_1 y_1' + P(x) c_2 y_2' + Q(x) c_1 y_1 + Q(x) c_2 y_2$$

$$= c_1 [y_1'' + P(x) y_1' + Q(x) y_1] + c_2 [y_2'' + P(x) y_2' + Q(x) y_2]$$

$$= c_1 \cdot 0 + c_2 \cdot 0 = 0 \tag{2.5}$$

where the fact that y_1 and y_2 satisfy (2.3) has been used.

In general, the general solution of a differential equation of nth order has n arbitrary constants. To fix these constants, we may have **boundary conditions**, which are conditions that specify the value of y or various of its derivatives at a point or points. Thus, if y represents the displacement of a vibrating string held fixed at two points, we know y must be zero at these points.

An important case is the linear homogeneous second-order differential equation with *constant coefficients*:

$$y'' + py' + qy = 0 \tag{2.6}$$

where p and q are constants. To solve (2.6), let us tentatively assume a solution of the form $y = e^{sx}$. We are looking for a function whose derivatives when multiplied by constants will cancel out the original function. The exponential function repeats itself when differentiated and is thus the correct choice. Substitution in (2.6) gives

$$s^2 e^{sx} + pse^{sx} + qe^{sx} = 0$$

$$s^2 + ps + q = 0 \tag{2.7*}$$

Equation (2.7) is called the **auxiliary equation**. It is a quadratic equation with two roots s_1 and s_2 that, provided s_1 and s_2 are not equal, give two independent solutions to (2.6). Thus the general solution of (2.6) is

$$y = c_1 e^{s_1 x} + c_2 e^{s_2 x} \tag{2.8*}$$

2.2 PARTICLE IN A ONE-DIMENSIONAL BOX

Having obtained the solution of one kind of differential equation, let us look at a case where we can use this solution to solve the time-independent Schrödinger equation. We consider a particle in a one-dimensional box. By this we mean a particle subjected to a potential-energy function that is infinite everywhere along the x axis except for a line segment of length l, where the potential energy is zero. Such a system may seem physically unreal, but we shall later see that this model can be applied with some success to certain conjugated molecules; see Section 16.2 and Problem 2.11. We put the origin at the left end of the line segment (Fig. 2.1).

We have three regions to consider. In regions I and III, the potential energy V equals infinity and the time-independent Schrödinger equation (1.18) is

$$-\frac{\hbar^2}{2m} \frac{d^2\psi}{dx^2} = (E - \infty)\psi$$

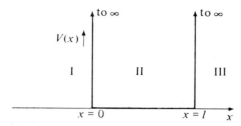

Figure 2.1 Potential energy function $V(x)$ for the particle in a one-dimensional box.

Neglecting E in comparison with ∞, we have

$$\frac{d^2\psi}{dx^2} = \infty\psi, \qquad \psi = \frac{1}{\infty}\frac{d^2\psi}{dx^2}$$

and we conclude that ψ is zero outside the box:

$$\psi_{\mathrm{I}} = 0, \qquad \psi_{\mathrm{III}} = 0 \tag{2.9}$$

For region II, x between zero and l, the potential energy V is zero, and the Schrödinger equation (1.18) becomes

$$\frac{d^2\psi_{\mathrm{II}}}{dx^2} + \frac{2m}{\hbar^2} E\psi_{\mathrm{II}} = 0 \tag{2.10}$$

where m is the mass of the particle and E is its total energy. We recognize (2.10) as a linear homogeneous second-order differential equation with constant coefficients. The auxiliary equation (2.7) gives

$$s^2 + 2mE\hbar^{-2} = 0$$

$$s = \pm(-2mE)^{1/2}\hbar^{-1} \tag{2.11}$$

$$s = \pm i(2mE)^{1/2}/\hbar \tag{2.12}$$

where $i = \sqrt{-1}$. Using (2.8), we have

$$\psi_{\mathrm{II}} = c_1 e^{i(2mE)^{1/2}x/\hbar} + c_2 e^{-i(2mE)^{1/2}x/\hbar} \tag{2.13}$$

Temporarily, let

$$\theta \equiv (2mE)^{1/2}x/\hbar$$

$$\psi_{\mathrm{II}} = c_1 e^{i\theta} + c_2 e^{-i\theta}$$

We have $e^{i\theta} = \cos\theta + i\sin\theta$ [Eq. (1.26)] and $e^{-i\theta} = \cos(-\theta) + i\sin(-\theta) = \cos\theta - i\sin\theta$, since

$$\cos(-\theta) = \cos\theta \quad \text{and} \quad \sin(-\theta) = -\sin\theta \tag{2.14}*$$

Therefore,

$$\psi_{II} = c_1 \cos\theta + ic_1 \sin\theta + c_2 \cos\theta - ic_2 \sin\theta$$
$$= (c_1 + c_2)\cos\theta + (ic_1 - ic_2)\sin\theta$$
$$= A\cos\theta + B\sin\theta$$

where A and B are new arbitrary constants. Hence

$$\psi_{II} = A\cos[\hbar^{-1}(2mE)^{1/2}x] + B\sin[\hbar^{-1}(2mE)^{1/2}x] \qquad (2.15)$$

Now we determine A and B by applying boundary conditions. It seems reasonable to postulate that the wave function will be continuous; that is, it will make no sudden jumps in value. If ψ is to be continuous at the point $x = 0$, then ψ_I and ψ_{II} must approach the same value at $x = 0$:

$$\lim_{x\to 0}\psi_I = \lim_{x\to 0}\psi_{II}$$
$$0 = \lim_{x\to 0}\{A\cos[\hbar^{-1}(2mE)^{1/2}x] + B\sin[\hbar^{-1}(2mE)^{1/2}x]\}$$
$$0 = A$$

since

$$\sin 0 = 0 \quad \text{and} \quad \cos 0 = 1 \qquad \textbf{(2.16)}^*$$

With $A = 0$, Eq. (2.15) becomes

$$\psi_{II} = B\sin[(2\pi/h)(2mE)^{1/2}x] \qquad (2.17)$$

Applying the continuity condition at $x = l$, we get

$$B\sin[(2\pi/h)(2mE)^{1/2}l] = 0 \qquad (2.18)$$

B cannot be zero, because this would make the wave function zero everywhere— we would have an empty box. Therefore,

$$\sin[(2\pi/h)(2mE)^{1/2}l] = 0$$

The zeros of the sine function occur at $0, \pm\pi, \pm 2\pi, \pm 3\pi, \ldots$. Hence

$$(2\pi/h)(2mE)^{1/2}l = \pm n\pi \qquad (2.19)$$

The value $n = 0$ is a special case. From (2.19), $n = 0$ corresponds to $E = 0$. For $E = 0$, the roots (2.12) of the auxiliary equation are equal and (2.13) is not the complete solution of the Schrödinger equation. To find the complete solution, we return to (2.10), which for $E = 0$ reads $d^2\psi_{II}/dx^2 = 0$; integration gives $d\psi_{II}/dx = c$ and $\psi_{II} = cx + d$, where c and d are constants. The boundary condition that $\psi_{II} = 0$ at $x = 0$ gives $d = 0$, and the condition that $\psi_{II} = 0$ at $x = l$ then gives $c = 0$. Thus $\psi_{II} = 0$ for $E = 0$, and $E = 0$ is not an allowed energy value. Hence, $n = 0$ is not allowed.

Solving (2.19) for E, we have

$$E = n^2\frac{h^2}{8ml^2}, \qquad n = 1, 2, 3, \ldots \qquad \textbf{(2.20)}^*$$

Only the energy values (2.20) allow ψ to satisfy the boundary condition of continuity at $x = l$. Application of a boundary condition has forced us to the conclusion that the values of the energy are quantized (Fig. 2.2). This is in striking contrast to the classical result that the particle in the box can have any nonnegative energy. Note that there is a minimum value, greater than zero, for the energy of the particle. The state of lowest energy is called the **ground state**. States with energies higher than the ground-state energy are **excited states**.

EXAMPLE   A particle of mass 2.00×10^{-26} g is in a one-dimensional box of length 4.00 nm. Find the frequency and wavelength of the photon emitted when this particle goes from the $n = 3$ to the $n = 2$ level.

By conservation of energy, the energy $h\nu$ of the emitted photon equals the energy difference between the two stationary states [Eq. (1.3); see also Section 9.10]:

$$h\nu = E_{\text{upper}} - E_{\text{lower}} = n_u^2 h^2/8ml^2 - n_l^2 h^2/8ml^2$$

$$\nu = \frac{(n_u^2 - n_l^2)h}{8ml^2} = \frac{(3^2 - 2^2)(6.626 \times 10^{-34}\ \text{J s})}{8(2.00 \times 10^{-29}\ \text{kg})(4.00 \times 10^{-9}\ \text{m})^2} = 1.29 \times 10^{12}\ \text{s}^{-1}$$

where u and l stand for upper and lower. Use of $\lambda\nu = c$ gives $\lambda = 2.32 \times 10^{-4}$ m.

Substitution of (2.19) into (2.17) gives for the wave function

$$\psi_{\text{II}} = B \sin\left(\frac{n\pi x}{l}\right), \qquad n = 1, 2, 3, \ldots \tag{2.21}$$

The use of the negative sign in front of $n\pi$ does not give us another independent solution. Since $\sin(-\theta) = -\sin\theta$, we would simply get a constant, -1, times the solution with the plus sign.

The constant B in Eq. (2.21) is still arbitrary. To fix its value, we use the

Figure 2.2 Lowest four energy levels for the particle in a one-dimensional box.

normalization requirement, Eqs. (1.23) and (1.21):

$$\int_{-\infty}^{\infty} |\Psi|^2 \, dx = \int_{-\infty}^{\infty} |\psi|^2 \, dx = 1$$

$$\int_{-\infty}^{0} |\psi_I|^2 \, dx + \int_{0}^{l} |\psi_{II}|^2 \, dx + \int_{l}^{\infty} |\psi_{III}|^2 \, dx = 1$$

$$|B|^2 \int_{0}^{l} \sin^2 \left(\frac{n\pi x}{l} \right) dx = 1 = |B|^2 \frac{l}{2} \qquad (2.22)$$

where the integral was evaluated by using $2 \sin^2 t = 1 - \cos 2t$. We have

$$|B| = (2/l)^{1/2}$$

Note that only the absolute value of B has been determined. B could be $-(2/l)^{1/2}$ as well as $(2/l)^{1/2}$. Moreover, B need not be a real number; we could use any complex number with absolute value $(2/l)^{1/2}$. All we can say is that $B = (2/l)^{1/2}e^{i\alpha}$, where α is the phase of B and could be any value in the range 0 to 2π (Section 1.7). Choosing the phase to be zero, we write as the stationary-state wave functions for the particle in a box

$$\psi_{II} = \left(\frac{2}{l} \right)^{1/2} \sin \left(\frac{n\pi x}{l} \right), \qquad n = 1, 2, 3, \ldots \qquad (2.23)^*$$

Graphs of the wave functions and the probability densities are shown in Figs. 2.3 and 2.4.

The wave function is zero at certain points, called **nodes**. For each increase of one in the value of n (n is called a **quantum number**), there is one more node. The existence of nodes in ψ and $|\psi|^2$ may seem surprising. Thus, for $n = 2$, Fig. 2.4 says that there is zero probability of finding the particle in the center of the box at $x = l/2$. How can the particle get from one side of the box to the other

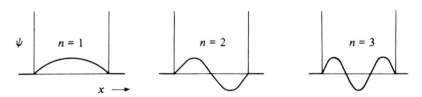

Figure 2.3 Graphs of ψ for the three lowest-energy particle-in-a-box states.

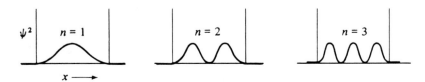

Figure 2.4 Graphs of $|\psi|^2$ for the lowest particle-in-a-box states.

without at any time being found in the center? This apparent paradox arises from attempting to understand the motion of microscopic particles using our everyday experience of the motions of macroscopic particles. However, as stated in Chapter 1, electrons and other microscopic "particles" cannot be fully and correctly described in terms of concepts of classical physics drawn from the macroscopic world.

Figure 2.4 shows that the probability of finding the particle at various places in the box is quite different from the classical result. Classically, a particle of fixed energy in a box bounces back and forth elastically between the two walls, moving at constant speed. Thus it is equally likely to be found at any point in the box. Quantum mechanically, we find a maximum in probability at the center of the box for the lowest energy level. As we go to higher energy levels with more nodes, the maxima and minima of probability come closer together, and the variations in probability along the length of the box ultimately become undetectable. For high quantum numbers, we approach the classical result of uniform probability density. This result, that in the limit of large quantum numbers quantum mechanics goes over into classical mechanics, is known as the *Bohr correspondence principle*. Since Newtonian mechanics is valid for macroscopic bodies (moving at speeds much less than the speed of light), we expect nonrelativistic quantum mechanics to give the same answer as classical mechanics for macroscopic bodies. Because of the extremely small magnitude of Planck's constant, quantization of energy is unobservable for macroscopic bodies. Since the mass of the particle and the length of the box squared appear in the denominator of Eq. (2.20), a macroscopic object in a macroscopic box having a macroscopic energy of motion would have a huge value for n, and hence, according to the correspondence principle, would show classical behavior.

We have a whole set of wave functions, each corresponding to a different value of the energy and characterized by the quantum number n, which may have integral values from 1 up. Let the subscript i denote a particular wave function with the value n_i for its quantum number:

$$\psi_i = \left(\frac{2}{l}\right)^{1/2} \sin\left(\frac{n_i \pi x}{l}\right), \qquad 0 < x < l$$

$$\psi_i = 0, \qquad \text{elsewhere}$$

Since the wave function has been normalized, we have

$$\int_{-\infty}^{\infty} \psi_i^* \psi_j \, dx = 1 \qquad \text{if } i = j \tag{2.24}$$

We now ask for the value of this integral when we use wave functions corresponding to *different* energy levels:

$$\int_{-\infty}^{\infty} \psi_i^* \psi_j \, dx = \int_0^l \left(\frac{2}{l}\right)^{1/2} \sin\left(\frac{n_i \pi x}{l}\right)\left(\frac{2}{l}\right)^{1/2} \sin\left(\frac{n_j \pi x}{l}\right) dx, \qquad n_i \neq n_j$$

Let $t = \pi x/l$:

$$\int_{-\infty}^{\infty} \psi_i^* \psi_j \, dx = \frac{2}{l} \int_0^{\pi} \sin n_i t \sin n_j t \, dt \cdot \frac{l}{\pi} \tag{2.25}$$

The integral may be evaluated with the identity

$$\sin n_i t \sin n_j t = \tfrac{1}{2} \cos (n_i - n_j)t - \tfrac{1}{2} \cos (n_i + n_j)t$$

Hence

$$\int_{-\infty}^{\infty} \psi_i^* \psi_j \, dx = \frac{2}{\pi} \int_0^{\pi} \tfrac{1}{2} \cos (n_i - n_j)t \, dt - \frac{2}{\pi} \int_0^{\pi} \tfrac{1}{2} \cos (n_i + n_j)t \, dt = 0$$

since $\sin m\pi = 0$ for m an integer. We thus have

$$\int_{-\infty}^{\infty} \psi_i^* \psi_j \, dx = 0 , \qquad i \neq j \tag{2.26}$$

When (2.26) holds, we say that ψ_i and ψ_j are **orthogonal** to each other for $i \neq j$. We can combine (2.24) and (2.26) by writing

$$\int_{-\infty}^{\infty} \psi_i^* \psi_j \, dx = \delta_{ij} \tag{2.27}$$

The symbol δ_{ij} is called the **Kronecker delta** (after a mathematician); it equals 1 whenever the two indexes i and j are equal, and equals 0 when i and j are unequal:

$$\delta_{ij} \equiv \begin{cases} 0 & \text{for } i \neq j \\ 1 & \text{for } i = j \end{cases} \tag{2.28}*$$

The property of the wave functions illustrated in (2.27) is called **orthonormality**. We proved orthonormality only for the particle-in-a-box wave functions; we shall prove it more generally in Section 7.2.

A more rigorous way to look at the particle in a box with infinite walls is to first treat the particle in a box with a finite jump in potential energy at the walls and then take the limit as the jump in potential energy becomes infinite. The results, when the limit is taken, will be the same as we have obtained (see Problem 2.14).

2.3 THE FREE PARTICLE IN ONE DIMENSION

By a free particle, we mean a particle subject to no forces whatever. For a free particle, integration of (1.11) shows that the potential energy remains constant no matter what the value of x is. Since the choice of the zero level of energy is arbitrary, we may set $V(x) = 0$. The Schrödinger equation (1.18) becomes

$$\frac{d^2\psi}{dx^2} + \frac{2m}{\hbar^2} E\psi = 0 \tag{2.29}$$

Equation (2.29) is the same as Eq. (2.10) (except for the boundary conditions). Therefore, the general solution of (2.29) is (2.13):

$$\psi = c_1 e^{i(2mE)^{1/2}x/\hbar} + c_2 e^{-i(2mE)^{1/2}x/\hbar} \tag{2.30}$$

What boundary condition might we impose? It seems reasonable to postulate (since $\psi^*\psi\,dx$ represents a probability) that ψ will remain finite as x goes to $\pm\infty$. If the energy E is less than zero, then this boundary condition will be violated, since for $E < 0$ we have

$$i(2mE)^{1/2} = i(-2m|E|)^{1/2} = i \cdot i \cdot (2m|E|)^{1/2} = -(2m|E|)^{1/2}$$

and therefore the first term in (2.30) will become infinite as x approaches minus infinity. Similarly, if E is negative, the second term in (2.30) becomes infinite as x approaches plus infinity. Thus the boundary condition requires

$$E \geqslant 0 \tag{2.31}$$

for the free particle. The wave function is oscillatory and is a linear combination of a sine and a cosine term [Eq. (2.15)]. For the free particle, we do not get quantization of the energy; all nonnegative energies are allowed. Since we set $V = 0$, the energy E is in this case all kinetic energy. If we attempt to evaluate the arbitrary constants c_1 and c_2 by normalization, we will find that the integral $\int_{-\infty}^{\infty} \psi^*(x)\psi(x)\,dx$ is divergent. In other words, the free-particle wave function is not normalizable in the usual sense. This is to be expected on physical grounds, since there is no reason for the probability of finding the free particle to approach zero as x goes to $\pm\infty$.

The free-particle problem represents an unreal situation, since we could not actually have a particle that had no interaction with any other particle in the universe.

2.4 PARTICLE IN A RECTANGULAR WELL

We shall briefly discuss the problem of a particle in a one-dimensional box with walls of finite height (Fig. 2.5a). The potential-energy function is $V = V_0$ for $x < 0$, $V = 0$ for $0 \leqslant x \leqslant l$, and $V = V_0$ for $x > l$. There are two cases to consider, depending on whether the particle's energy E is less than or greater than V_0.

We first consider $E < V_0$. The Schrödinger equation (1.18) in regions I and III is $d^2\psi/dx^2 + (2m/\hbar^2)(E - V_0)\psi = 0$. This is a linear homogeneous differential equation with constant coefficients, and the auxiliary equation (2.7) is $s^2 + (2m/\hbar^2)(E - V_0) = 0$ with roots $s = \pm(2m/\hbar^2)^{1/2}(V_0 - E)^{1/2}$. Therefore,

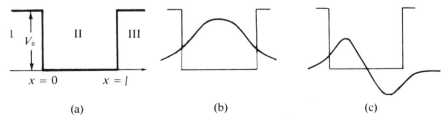

Figure 2.5 (a) Potential energy for a particle in a one-dimensional rectangular well. (b) The ground-state wave function for this potential. (c) The first excited-state wave function.

$$\psi_{\rm I} = C \exp\left[(2m/\hbar^2)^{1/2}(V_0 - E)^{1/2}x\right] + D \exp\left[-(2m/\hbar^2)^{1/2}(V_0 - E)^{1/2}x\right]$$

$$\psi_{\rm III} = F \exp\left[(2m/\hbar^2)^{1/2}(V_0 - E)^{1/2}x\right] + G \exp\left[-(2m/\hbar^2)^{1/2}(V_0 - E)^{1/2}x\right]$$

where C, D, F, and G are constants.

As in Section 2.3, we must prevent $\psi_{\rm I}$ from becoming infinite as $x \rightarrow -\infty$. Since we are assuming $E < V_0$, the quantity $(V_0 - E)^{1/2}$ is a real, positive number, and to keep $\psi_{\rm I}$ finite as $x \rightarrow -\infty$, we must have $D = 0$. Similarly, to keep $\psi_{\rm III}$ finite as $x \rightarrow +\infty$, we must have $F = 0$. Therefore,

$$\psi_{\rm I} = C \exp\left[(2m/\hbar^2)^{1/2}(V_0 - E)^{1/2}x\right], \quad \psi_{\rm III} = G \exp\left[-(2m/\hbar^2)^{1/2}(V_0 - E)^{1/2}x\right]$$

In region II, $V = 0$, the Schrödinger equation is (2.10), and its solution is (2.15):

$$\psi_{\rm II} = A \cos\left[(2m/\hbar^2)^{1/2}E^{1/2}x\right] + B \sin\left[(2m/\hbar^2)^{1/2}E^{1/2}x\right]$$

To complete the problem, we must apply the boundary conditions. As with the particle in a box with infinite walls, we require the wave function to be continuous at $x = 0$ and at $x = l$; so $\psi_{\rm I}(0) = \psi_{\rm II}(0)$ and $\psi_{\rm II}(l) = \psi_{\rm III}(l)$. There are four arbitrary constants in the wave function, so more than these two boundary conditions are needed. As well as requiring ψ to be continuous, we shall require that its derivative $d\psi/dx$ be continuous everywhere. To justify this requirement, we note that if $d\psi/dx$ changed discontinuously at a point then its derivative (its instantaneous rate of change) $d^2\psi/dx^2$ would become infinite at that point. However, for the problem under consideration, the Schrödinger equation $d^2\psi/dx^2 = (2m/\hbar^2)(V - E)\psi$ does not contain anything infinite on the right side, so $d^2\psi/dx^2$ cannot become infinite. [For a more rigorous argument, see D. Branson, *Am. J. Phys.*, **47**, 1000 (1979).] Therefore, $d\psi_{\rm I}/dx = d\psi_{\rm II}/dx$ at $x = 0$ and $d\psi_{\rm II}/dx = d\psi_{\rm III}/dx$ at $x = l$.

From $\psi_{\rm I}(0) = \psi_{\rm II}(0)$, we get $C = A$. From $\psi_{\rm I}'(0) = \psi_{\rm II}'(0)$, we get (Problem 2.13a) $B = (V_0 - E)^{1/2}A/E^{1/2}$. From $\psi_{\rm II}(l) = \psi_{\rm III}(l)$, we get a complicated equation that allows G to be found in terms of A. The constant A is determined by normalization.

Taking $\psi_{\rm II}'(l) = \psi_{\rm III}'(l)$, dividing it by $\psi_{\rm II}(l) = \psi_{\rm III}(l)$, and expressing B in terms of A, we obtain the following equation for the energy levels (Problem 2.13b):

$$\tan\left[(2m/\hbar^2)^{1/2}E^{1/2}l\right] = 2E^{1/2}(V_0 - E)^{1/2}/(2E - V_0) \tag{2.32}$$

Defining the dimensionless constants ε and b as $\varepsilon \equiv E/V_0$ and $b \equiv (2mV_0/\hbar^2)^{1/2}l$, we write (2.32) as

$$\tan(b\varepsilon^{1/2}) = 2\varepsilon^{1/2}(1 - \varepsilon)^{1/2}/(2\varepsilon - 1) \tag{2.33}$$

Only the particular values of E that satisfy (2.32) yield a wave function that is continuous and has a continuous derivative, so the energy levels are quantized for $E < V_0$. To find the allowed energy levels, one can plot the left and right sides of (2.33) on the same graph paper and find the intersection points of the two curves (see *Kauzmann*, p. 191). A detailed study (see *Merzbacher*, Section 6.8) shows that the number of allowed energy levels with $E < V_0$ is N, where N satisfies

$$N - 1 < (8mV_0)^{1/2}l/h \leq N$$

For example, if $V_0 = h^2/ml^2$, then $(8mV_0)^{1/2}l/h = 8^{1/2} = 2.83$ and $N = 3$.

Figure 2.5 shows ψ for the lowest two energy levels. The wave function is oscillatory inside the box and dies off exponentially outside the box. It turns out that the number of nodes increases by one for each higher level.

So far we have considered only states with $E < V_0$. For $E > V_0$, the quantity $(V_0 - E)^{1/2}$ is imaginary, and instead of dying off to zero as x goes to $\pm\infty$, ψ_I and ψ_{III} oscillate (similar to the free-particle ψ). We no longer have any reason to set D in ψ_I and F in ψ_{III} equal to zero, and with these additional constants available to satisfy the boundary conditions on ψ and ψ', one finds that E need not be restricted to obtain properly behaved wave functions. Therefore, all energies above V_0 are allowed.

States with $E < V_0$ are called *bound states*; states with $E > V_0$ are *unbound*.

2.5 TUNNELING

For the particle in a rectangular well (Section 2.4), Fig. 2.5 and the equations for ψ_I and ψ_{III} show that for the bound states there is a nonzero probability of finding the particle in regions I and III, where its total energy E is less than its potential energy $V = V_0$. Classically, this behavior is not allowed. The classical equations $E = T + V$ and $T \geq 0$, where T is the kinetic energy, mean that E cannot be less than V in classical mechanics.

Consider a particle in a one-dimensional box with walls of finite height *and* finite thickness (Fig. 2.6). Classically, the particle cannot escape from the box unless its energy is greater than the potential-energy barrier V_0. However, a quantum-mechanical treatment (which is omitted) shows that there is a finite probability for a particle of total energy less than V_0 to be found outside the box.

The term *tunneling* denotes the penetration of a particle into a classically forbidden region (as in Fig. 2.5) or the passage of a particle through a potential-energy barrier whose height exceeds the particle's energy. Since tunneling is a quantum effect, its probability of occurrence is greater the less classical is the behavior of the particle. Therefore, tunneling is most prevalent with particles of small mass. (Note that the greater the mass m, the more rapidly the functions ψ_I and ψ_{III} of Section 2.4 die away to zero.) Electrons tunnel quite readily; hydrogen atoms tunnel more readily than heavier atoms.

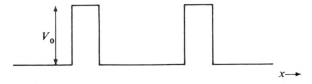

Figure 2.6 Potential energy for a particle in a one-dimensional box of finite height and thickness.

The emission of alpha particles from a radioactive nucleus involves tunneling of the alpha particles through the potential-energy barrier produced by the short-range attractive nuclear forces and the Coulombic repulsive force between the daughter nucleus and the alpha particle. The NH_3 molecule is pyramidal. There is a potential-energy barrier to inversion of the molecule, with the potential-energy maximum occurring at the planar configuration. The hydrogen atoms can tunnel through this barrier, thereby inverting the molecule. In CH_3CH_3 there is a barrier to internal rotation, with a potential-energy maximum at the eclipsed position of the hydrogens. The hydrogens can tunnel through this barrier from one staggered position to the next. Tunneling of electrons is important in oxidation–reduction reactions and in electrode processes. Tunneling usually contributes significantly to the rate of chemical reactions that involve transfer of hydrogen atoms. See R. P. Bell, *The Tunnel Effect in Chemistry*, Chapman & Hall, London, 1980.

The scanning tunneling microscope, invented in 1981, uses the tunneling of electrons through the space between the extremely fine tip of a metal wire and the surface of an electrically conducting solid to produce images of individual atoms on the solid's surface. A small voltage is applied between the solid and the wire, and as the tip is moved across the surface at a height of a few angstroms, the tip height is adjusted to keep the current flow constant. A plot of tip height versus position yields the image of the surface.

2.6 SUMMARY

The general solution to the linear, homogeneous, second-order, constant-coefficients differential equation $y''(x) + py'(x) + qy(x) = 0$ is $y = c_1 e^{s_1 x} + c_2 e^{s_2 x}$, where s_1 and s_2 are the solutions to the auxiliary equation $s^2 + ps + q = 0$.

For a particle in a one-dimensional box ($V = 0$ for $0 \leqslant x \leqslant l$ and $V = \infty$ elsewhere), the stationary-state wave functions and energies are $\psi = (2/l)^{1/2} \sin(n\pi x/l)$ for $0 \leqslant x \leqslant l$, $\psi = 0$ elsewhere, and $E = n^2 h^2/8ml^2$, where $n = 1, 2, 3, \ldots$.

PROBLEMS

2.1 (a) Solve $y''(x) + y'(x) - 2y(x) = 0$. (b) Evaluate the arbitrary constants in the solution if the boundary conditions are $y = 0$ at $x = 0$ and $y' = 1$ at $x = 0$.

2.2 (a) For the case of equal roots of the auxiliary equation, $s_1 = s_2 = s$, we have found only one independent solution of the linear homogeneous second-order differential equation: e^{sx}. Verify that xe^{sx} is the second solution in this case. (b) Solve $y''(x) - 2y'(x) + y(x) = 0$.

2.3 For a macroscopic object of mass 1.0 g moving with speed 1.0 cm/s in a one-dimensional box of length 1.0 cm, find the quantum number n.

2.4 Consider a particle with quantum number n moving in a one-dimensional box of length l. (a) Determine the probability of finding the particle in the left quarter of the box. (b) For what value of n is this probability a maximum? (c) What is the limit of this probability for $n \to \infty$? (d) What principle is illustrated in (c)?

2.5 An extremely crude picture of an electron in an atom or molecule treats it as a particle in a one-dimensional box whose length is on the order of the size of atoms and molecules. (a) For an electron in a box of length 1.0 Å, calculate the separation between the two lowest energy levels. (b) Calculate the wavelength of a photon corresponding to a transition between these two levels. (c) In what portion of the electromagnetic spectrum is this wavelength?

2.6 When a particle of mass 9.1×10^{-28} g in a certain one-dimensional box goes from the $n = 5$ level to the $n = 2$ level, it emits a photon of frequency 6.0×10^{14} s^{-1}. Find the length of the box.

2.7 When an electron in a certain excited energy level in a one-dimensional box of length 2.00 Å makes a transition to the ground state, a photon of wavelength 8.79×10^{-9} m is emitted. Find the quantum number of the initial state.

2.8 For the particle in a one-dimensional box of length l, we could have put the coordinate origin at the center of the box. Find the wave functions and energy levels for this choice of origin.

2.9 (a) Sketch rough graphs of ψ and of ψ^2 for the $n = 4$ and $n = 5$ particle-in-a-box states. (b) Find the slope of the $n = 4$ ψ^2 curve at $x = \frac{1}{2}l$ and check that your curve was drawn with the correct slope.

2.10 Integrals involving trigonometric functions can often be evaluated using the identities of Problem 1.22. Use the complex-exponential form of the sine function to verify Eq. (2.27) for the particle-in-a-box wave functions.

2.11 A crude treatment of the pi electrons of a conjugated molecule regards these electrons as moving in the particle-in-a-box potential of Fig. 2.1, where the box length is somewhat more than the length of the conjugated chain. The Pauli exclusion principle (Chapter 10) allows no more than two electrons to occupy each box level. (These two have opposite spins.) For butadiene $CH_2{=}CHCH{=}CH_2$, take the box length as 7.0 Å and use this model to estimate the wavelength of light absorbed when a pi electron is excited from the highest-occupied to the lowest-vacant box level. The experimental value is 217 nm.

2.12 Write down the time-dependent wave function for a free particle with energy E.

2.13 (a) For the particle in a rectangular well (Section 2.4), verify that $B = (V_0 - E)^{1/2}A/E^{1/2}$. (b) Verify Eq. (2.32).

2.14 For the particle in a rectangular well (Section 2.4), show that in the limit $V_0 \to \infty$ (a) Eq. (2.32) gives $E = n^2h^2/8ml^2$ as in Eq. (2.20); (b) the wave function goes to Eqs. (2.9) and (2.21).

2.15 Sketch ψ for the next-lowest bound level in Fig. 2.5.

2.16 For an electron in a 15.0-eV-deep one-dimensional rectangular well of width 2.00 Å, calculate the number of bound states. Use (6.107).

2.17 For a particle in a one-dimensional rectangular well, (a) must there be at least one bound state? (b) is ψ'' continuous at $x = 0$?

2.18 Find the allowed bound energy levels for Problem 2.16 by using a programmable calculator or a computer to calculate the left side of (2.33) minus the right side (or to calculate each side separately) for values of ε going from 0 to 1 in small steps. [At $\varepsilon = 0$, Eq. (2.33) is satisfied, but this is not an allowed energy level since it turns out to lead to $\psi = 0$ everywhere.]

2.19 For each of these choices of F in Newton's second law (1.7), classify (1.7) as a linear or nonlinear differential equation (a, b, c and k are constants). (a) $F = c$; (b) $F = -kx$; (c) $F = -ax^3$; (d) $F = b \sin ax$; (e) $F = a - kx$. [Classical-mechanical systems that show chaotic behavior (Section 1.4) obey nonlinear differential equations, but not all nonlinear differential equations give rise to chaotic behavior.]

2.20 The energy of most stars results from the fusion of hydrogen nuclei to helium nuclei. The interior of the sun (a typical star) is at 15×10^6 K. At this temperature, virtually no nuclei have sufficient kinetic energy to overcome the electrostatic repulsion between nuclei and approach each other closely enough to undergo a fusion reaction. Therefore, when Eddington proposed in 1920 that nuclear fusion is the source of stellar energy, his idea was rejected. Explain why fusion does occur in stars, despite the above-mentioned apparent difficulty.

2.21 True or false? (a) The particle-in-a-box ground state has quantum number $n = 0$. (b) The particle-in-a-box stationary-state wave functions are discontinuous at certain points. (c) The first derivative of each particle-in-a-box stationary-state wave function is discontinuous at certain points. (d) The maximum probability density for every particle-in-a-box stationary state is at the center of the box. (e) For the particle-in-a-box $n = 2$ stationary state, the probability of finding the particle in the left quarter of the box equals the probability of finding it in the right quarter of the box. (Answer this and the next question without evaluating any integrals.) (f) For the $n = 1$ particle-in-a-box stationary state, the probability of finding the particle in the left third of the box equals the probability of finding it in the middle third.

3

Operators

3.1 OPERATORS

We now develop the theory of quantum mechanics in a more general fashion than previously. We begin by writing the one-particle, one-dimensional time-independent Schrödinger equation (1.18) in the form

$$\left[-\frac{\hbar^2}{2m} \frac{d^2}{dx^2} + V(x) \right] \psi(x) = E\psi(x) \tag{3.1}$$

The entity in brackets in (3.1) is an *operator*. Equation (3.1) suggests that we have an energy operator, which, operating on the wave function, gives us the wave function back again, but multiplied by the allowed values of the energy. We therefore discuss operators.

An *operator* is a rule that transforms a given function into another function. For example, let $\hat{D}$ be the operator that differentiates a function with respect to x. We use a circumflex to indicate an operator. Provided $f(x)$ is differentiable, the result of operating on $f(x)$ with $\hat{D}$ is $\hat{D}f(x) = f'(x)$. For example, $\hat{D}(x^2 + 3e^x) = 2x + 3e^x$. If $\hat{3}$ is the operator that multiplies a function by 3, then $\hat{3}(x^2 + 3e^x) = 3x^2 + 9e^x$. If cos is the operator that takes the cosine of a function, then application of cos to the function $x^2 + 1$ gives $\cos(x^2 + 1)$. In general, if the operator $\hat{A}$ transforms the function $f(x)$ into the function $g(x)$, we write $\hat{A}f(x) = g(x)$.

We define the **sum** and the **difference** of two operators $\hat{A}$ and $\hat{B}$ by the equations

$$(\hat{A} + \hat{B})f(x) \equiv \hat{A}f(x) + \hat{B}f(x) \tag{3.2}*$$

$$(\hat{A} - \hat{B})f(x) \equiv \hat{A}f(x) - \hat{B}f(x)$$

For example, if $\hat{D} \equiv d/dx$, then

$$(\hat{D} + \hat{3})(x^3 - 5) \equiv \hat{D}(x^3 - 5) + 3(x^3 - 5) = 3x^2 + (3x^3 - 15) = 3x^3 + 3x^2 - 15$$

We define the **product** of two operators $\hat{A}$ and $\hat{B}$ by the equation

$$\hat{A}\hat{B}f(x) \equiv \hat{A}[\hat{B}f(x)] \tag{3.3}*$$

35

In other words, we first operate on $f(x)$ with the operator on the right of the operator product, and then we take the resulting function and operate on it with the operator on the left of the operator product. For example,

$$\hat{3}\hat{D}f(x) = \hat{3}[\hat{D}f(x)] = \hat{3}f'(x) = 3f'(x)$$

For this example it would have made no difference in the final result whether we first applied one operator or the other. In general, we cannot assume that $\hat{A}\hat{B}$ and $\hat{B}\hat{A}$ have the same effect. Consider, for example, the operators d/dx and $\hat{x}$:

$$\hat{D}\hat{x}f(x) = \frac{d}{dx}[xf(x)] = f(x) + xf'(x) = (\hat{1} + \hat{x}\hat{D})f(x) \tag{3.4}$$

$$\hat{x}\hat{D}f(x) = \hat{x}\left[\frac{d}{dx}f(x)\right] = xf'(x)$$

Thus $\hat{A}\hat{B}$ and $\hat{B}\hat{A}$ are different operators in this case.

We can develop an *operator algebra* as follows. Two operators $\hat{A}$ and $\hat{B}$ are said to be **equal** if $\hat{A}f = \hat{B}f$ for all functions f. Equal operators produce the same result when they operate on a given function. For example, (3.4) shows that

$$\hat{D}\hat{x} = \hat{1} + \hat{x}\hat{D} \tag{3.5}$$

The operator $\hat{1}$ (multiplication by 1) is the *unit operator*. The operator $\hat{0}$ (multiplication by 0) is the *null operator*. We usually omit the circumflex over operators that are simply multiplication by a constant. We can transfer operators from one side of an operator equation to the other (Problem 3.13); thus (3.5) is equivalent to

$$\hat{D}\hat{x} - \hat{x}\hat{D} - 1 = 0$$

where we have omitted circumflexes over the null and unit operators.

Operators obey the associative law of multiplication:

$$\hat{A}(\hat{B}\hat{C}) = (\hat{A}\hat{B})\hat{C} \tag{3.6}$$

The proof of (3.6) is outlined in Problem 3.5. As an example, let $\hat{A} = d/dx$, $\hat{B} = \hat{x}$, and $\hat{C} = \hat{3}$. Using (3.5), we have

$$(\hat{A}\hat{B}) = \hat{D}\hat{x} = 1 + \hat{x}\hat{D}, \qquad [(\hat{A}\hat{B})\hat{C}]f = (1 + \hat{x}\hat{D})3f = 3f + 3xf'$$

$$(\hat{B}\hat{C}) = 3\hat{x}, \qquad [\hat{A}(\hat{B}\hat{C})]f = \hat{D}(3xf) = 3f + 3xf'$$

A major difference between operator algebra and ordinary algebra is that numbers obey the commutative law of multiplication, but operators do not necessarily do so; $ab = ba$ if a and b are numbers, but $\hat{A}\hat{B}$ and $\hat{B}\hat{A}$ are not necessarily equal operators. We define the **commutator** $[\hat{A}, \hat{B}]$ of the operators $\hat{A}$ and $\hat{B}$ as the operator $\hat{A}\hat{B} - \hat{B}\hat{A}$:

$$[\hat{A}, \hat{B}] \equiv \hat{A}\hat{B} - \hat{B}\hat{A} \tag{3.7}*$$

If $\hat{A}\hat{B} = \hat{B}\hat{A}$, then $[\hat{A}, \hat{B}] = 0$, and we say that $\hat{A}$ and $\hat{B}$ **commute**. If $\hat{A}\hat{B} \neq \hat{B}\hat{A}$, then $\hat{A}$ and $\hat{B}$ do not commute. Note that $[\hat{A}, \hat{B}]f = \hat{A}\hat{B}f - \hat{B}\hat{A}f$. Since the order in which we apply the operators $\hat{3}$ and d/dx makes no difference, we have

$$\left[\hat{3}, \frac{d}{dx} \right] = \hat{3}\frac{d}{dx} - \frac{d}{dx}\hat{3} = 0$$

From Eq. (3.5) we have

$$\left[\frac{d}{dx}, \hat{x} \right] = \hat{D}\hat{x} - \hat{x}\hat{D} = 1 \qquad (3.8)$$

The operators d/dx and $\hat{x}$ do not commute.

EXAMPLE   Find $[z^3, d/dz]$.

To find $[z^3, d/dz]$, we apply this operator to an arbitrary function $g(z)$. Using the commutator definition (3.7) and the definitions of the difference and product of two operators, we have

$$[z^3, d/dz]g = [z^3(d/dz) - (d/dz)z^3]g = z^3(d/dz)g - (d/dz)(z^3 g)$$
$$= z^3 g' - 3z^2 g - z^3 g' = -3z^2 g$$

Deleting the arbitrary function g, we get the operator equation $[z^3, d/dz] = -3z^2$.

The **square** of an operator is defined as the product of the operator with itself: $\hat{A}^2 = \hat{A}\hat{A}$. Let us find the square of the differentiation operator:

$$\hat{D}^2 f(x) = \hat{D}(\hat{D}f) = \hat{D}f' = f''$$
$$\hat{D}^2 = d^2/dx^2$$

As another example, the square of the operator that takes the complex conjugate of a function is equal to the unit operator, since taking the complex conjugate twice gives the original function. The nth power of an operator ($n = 1, 2, 3, \ldots$) is defined to mean applying the operator n times in succession.

It turns out that the operators occurring in quantum mechanics are linear. $\hat{A}$ is a **linear operator** if and only if it has the following two properties:

$$\hat{A}[f(x) + g(x)] = \hat{A}f(x) + \hat{A}g(x) \qquad (3.9)^*$$

$$\hat{A}[cf(x)] = c\hat{A}f(x) \qquad (3.10)^*$$

where f and g are arbitrary functions and c is an arbitrary constant (not necessarily real). Examples of linear operators include $\hat{x}^2$, d/dx, and d^2/dx^2. Some nonlinear operators are cos and $(\ )^2$, where $(\ )^2$ squares the function it acts on.

EXAMPLE   Is d/dx a linear operator? Is $\sqrt{\ }$ a linear operator?

We have

$$(d/dx)[f(x) + g(x)] = df/dx + dg/dx = (d/dx)f(x) + (d/dx)g(x)$$
$$(d/dx)[cf(x)] = c\, df(x)/dx$$

so d/dx obeys (3.9) and (3.10) and is a linear operator. However,

$$\sqrt{f(x) + g(x)} \neq \sqrt{f(x)} + \sqrt{g(x)}$$

so $\sqrt{}$ does not obey (3.9) and is nonlinear.

Equation (2.2) defines the form of a linear differential equation. Using the differentiation operator $\hat{D}$, we can rewrite (2.2) as

$$[A_n(x)\hat{D}^n + A_{n-1}(x)\hat{D}^{n-1} + \cdots + A_0(x)] y(x) = g(x)$$

The operator in brackets in this equation is linear.

Useful identities in linear-operator manipulations are

$$(\hat{A} + \hat{B})\hat{C} = \hat{A}\hat{C} + \hat{B}\hat{C} \tag{3.11}*$$

$$\hat{A}(\hat{B} + \hat{C}) = \hat{A}\hat{B} + \hat{A}\hat{C} \tag{3.12}*$$

EXAMPLE   Prove the distributive law (3.11) for linear operators.

A good way to begin a proof is to first write down what is given and what is to be proved. We are given that $\hat{A}$, $\hat{B}$, and $\hat{C}$ are linear operators. We must prove that $(\hat{A} + \hat{B})\hat{C} = \hat{A}\hat{C} + \hat{B}\hat{C}$.

To prove that the operator $(\hat{A} + \hat{B})\hat{C}$ is equal to the operator $\hat{A}\hat{C} + \hat{B}\hat{C}$, we must prove that these two operators give the same result when applied to an arbitrary function f. Thus we must prove that

$$[(\hat{A} + \hat{B})\hat{C}]f = (\hat{A}\hat{C} + \hat{B}\hat{C})f$$

We start with $[(\hat{A} + \hat{B})\hat{C}]f$. This expression involves the product of the two operators $\hat{A} + \hat{B}$ and $\hat{C}$. The operator-product definition (3.3) with $\hat{A}$ replaced by $\hat{A} + \hat{B}$ and $\hat{B}$ replaced by $\hat{C}$ gives $[(\hat{A} + \hat{B})\hat{C}]f = (\hat{A} + \hat{B})(\hat{C}f)$. The entity $\hat{C}f$ is a function, and use of the definition of the sum $\hat{A} + \hat{B}$ of the two operators $\hat{A}$ and $\hat{B}$ gives $(\hat{A} + \hat{B})(\hat{C}f) = \hat{A}(\hat{C}f) + \hat{B}(\hat{C}f)$. Thus

$$[(\hat{A} + \hat{B})\hat{C}]f = (\hat{A} + \hat{B})(\hat{C}f) = \hat{A}(\hat{C}f) + \hat{B}(\hat{C}f)$$

Use of the operator-product definition (3.3) gives $\hat{A}(\hat{C}f) = \hat{A}\hat{C}f$ and $\hat{B}(\hat{C}f) = \hat{B}\hat{C}f$. Hence

$$[(\hat{A} + \hat{B})\hat{C}]f = \hat{A}\hat{C}f + \hat{B}\hat{C}f \tag{3.13}$$

Use of the operator-sum definition (3.2) with $\hat{A}$ replaced by $\hat{A}\hat{C}$ and $\hat{B}$ replaced by $\hat{B}\hat{C}$ gives $(\hat{A}\hat{C} + \hat{B}\hat{C})f = \hat{A}\hat{C}f + \hat{B}\hat{C}f$, so (3.13) becomes

$$[(\hat{A} + \hat{B})\hat{C}]f = (\hat{A}\hat{C} + \hat{B}\hat{C})f$$

which is what we wanted to prove. Hence $(\hat{A} + \hat{B})\hat{C} = \hat{A}\hat{C} + \hat{B}\hat{C}$.

Note that we did not need to use the linearity of $\hat{A}$, $\hat{B}$, and $\hat{C}$. Hence (3.11) holds for any operators. However, (3.12) holds only if $\hat{A}$ is linear (see Problem 3.14).

EXAMPLE   Find the square of the operator $d/dx + \hat{x}$.

To find the effect of $(d/dx + \hat{x})^2$, we apply this operator to an arbitrary function $f(x)$; letting $\hat{D} \equiv d/dx$, we have

$$(\hat{D} + \hat{x})^2 f(x) = (\hat{D} + \hat{x})[(\hat{D} + x)f] = (\hat{D} + \hat{x})(f' + xf)$$

$$= f'' + f + xf' + xf' + x^2 f = (\hat{D}^2 + 2\hat{x}\hat{D} + \hat{x}^2 + 1)f(x)$$

$$(\hat{D} + \hat{x})^2 = \hat{D}^2 + 2\hat{x}\hat{D} + \hat{x}^2 + 1$$

Let us repeat this calculation, using only operator equations:

$$(\hat{D} + \hat{x})^2 = (\hat{D} + \hat{x})(\hat{D} + \hat{x}) = \hat{D}(\hat{D} + \hat{x}) + \hat{x}(\hat{D} + \hat{x})$$

$$= \hat{D}^2 + \hat{D}\hat{x} + \hat{x}\hat{D} + \hat{x}^2 = \hat{D}^2 + \hat{x}\hat{D} + 1 + \hat{x}\hat{D} + \hat{x}^2$$

$$= \hat{D}^2 + 2x\hat{D} + x^2 + 1$$

where (3.11), (3.12), and (3.5) have been used and the circumflex over the operator "multiplication by x" has been omitted.

3.2 EIGENFUNCTIONS AND EIGENVALUES

Suppose that the effect of operating on some function $f(x)$ with the operator $\hat{A}$ is simply to multiply $f(x)$ by a certain constant k. We then say that $f(x)$ is an *eigenfunction* of $\hat{A}$ with *eigenvalue* k. As part of the definition, we shall require that the eigenfunction $f(x)$ is not identically zero. By this we mean that, although $f(x)$ may vanish at various points, it is not everywhere zero. We have

$$\hat{A}f(x) = kf(x) \tag{3.14}*$$

(*Eigen* is a German word meaning *characteristic*. "Eigenvalue" is a hybrid word; it has been suggested that "characteristicwert" would be just as suitable.) As an example of (3.14), e^{2x} is an eigenfunction of the operator d/dx with eigenvalue 2:

$$(d/dx)e^{2x} = 2e^{2x}$$

EXAMPLE   If $f(x)$ is an eigenfunction of the linear operator $\hat{A}$ and c is any constant, prove that $cf(x)$ is an eigenfunction of $\hat{A}$ with the same eigenvalue as $f(x)$.

A good way to see how to do a proof is to carry out the following steps:

1. Write down the given information and translate this information from words into equations.
2. Write down what is to be proved in the form of an equation or equations.
3. (a) Manipulate the given equations of step 1 so as to transform them to the desired equations of step 2. (b) Alternatively, start with one side of the equation that we want to prove true and use the given equations of step 1 to manipulate this side until it is transformed into the other side of the equation to be proved.

We are given three pieces of information: f is an eigenfunction of $\hat{A}$; $\hat{A}$ is a linear operator; c is a constant. Translating these statements into equations, we have [see Eqs. (3.14), (3.9), and (3.10)]

$$\hat{A}f = kf \tag{3.15}$$

$$\hat{A}(f + g) = \hat{A}f + \hat{A}g \quad \text{and} \quad \hat{A}(bf) = b\hat{A}f \tag{3.16}$$

$$c = \text{a constant}$$

where k and b are constants and f and g are functions.

We want to prove that cf is an eigenfunction of $\hat{A}$ with the same eigenvalue as f, which, written as an equation, is

$$\hat{A}(cf) = k(cf)$$

Using the strategy of step 3(b), we start with the left side $\hat{A}(cf)$ of this last equation and try to show that it equals $k(cf)$. Using the second equation in the linearity definition (3.16), we have $\hat{A}(cf) = c\hat{A}f$. Using the eigenvalue equation (3.15), we have $c\hat{A}f = ckf$. Hence

$$\hat{A}(cf) = c\hat{A}f = ckf = k(cf)$$

which completes the proof.

EXAMPLE   (a) Find the eigenfunctions and eigenvalues of the operator d/dx. (b) If we impose the boundary condition that the eigenfunctions remain finite as $x \rightarrow \pm\infty$, find the eigenvalues.

(a) Equation (3.14) with $\hat{A} = d/dx$ becomes

$$df(x)/dx = kf(x) \tag{3.17}$$

$$df/f = k \, dx$$

$$\ln f = kx + \text{constant}$$

$$f = e^{\text{constant}}e^{kx}$$

$$f = ce^{kx} \tag{3.18}$$

The eigenfunctions of d/dx are given by (3.18). The eigenvalues are k, which can be any number whatever and (3.17) will still be satisfied. The eigenfunctions contain an arbitrary multiplicative constant c. This is true for the eigenfunctions of any linear operator, as was proved in the previous example. Each different value of k in (3.18) gives a different eigenfunction. However, eigenfunctions with the same value of k but different values of c are not independent of each other.

(b) Since k can be complex, we write it as $k = a + ib$, where a and b are real numbers. We then have $f(x) = ce^{ax}e^{ibx}$. The factor e^{ax} goes to infinity as x goes to infinity if a is positive; it goes to infinity as x goes to minus infinity if a is negative. Hence the boundary conditions require that $a = 0$, and the eigenvalues are $k = ib$.

3.3 OPERATORS AND QUANTUM MECHANICS

We now examine the relationship between operators and quantum mechanics. Comparing Eq. (3.1) with (3.14), we see that the Schrödinger equation is an eigenvalue problem. The values of the energy E are the eigenvalues; the eigenfunctions are the wave functions ψ; the operator whose eigenfunctions and eigenvalues are desired is $-(\hbar^2/2m)\,d^2/dx^2 + V(x)$. This operator is called the *Hamiltonian operator* for the system.

Sir William Rowan Hamilton (1805–1865) devised an alternative form of Newton's equations of motion involving a function H, the Hamiltonian function for the system. For a system where the potential energy is a function of the coordinates only, the total energy remains constant with time; that is, E is conserved. We shall restrict ourselves to such conservative systems. For conservative systems, the classical-mechanical *Hamiltonian function* turns out to be simply the total energy expressed in terms of coordinates and conjugate momenta. For Cartesian coordinates x, y, z, the *conjugate momenta* are the components of linear momentum in the x, y, and z directions: p_x, p_y, and p_z:

$$p_x \equiv mv_x\,, \qquad p_y \equiv mv_y\,, \qquad p_z \equiv mv_z \qquad (3.19)^*$$

where v_x, v_y, and v_z are the components of the particle's velocity in the x, y, and z directions.

Let us find the classical-mechanical Hamiltonian function for a particle of mass m moving in one dimension and subject to a potential energy $V(x)$. The Hamiltonian function is equal to the energy, which is composed of kinetic and potential energies. The familiar form of the kinetic energy, $\frac{1}{2}mv_x^2$, will not do, however, since we must express the Hamiltonian as a function of coordinates and momenta, not velocities. Since $v_x = p_x/m$, the form of the kinetic energy we want is $p_x^2/2m$. The Hamiltonian function is

$$H = \frac{p_x^2}{2m} + V(x) \qquad (3.20)$$

The time-independent Schrödinger equation (3.1) indicates that, corresponding to the Hamiltonian function (3.20), we have a quantum-mechanical operator

$$-\frac{\hbar^2}{2m}\frac{d^2}{dx^2} + V(x)$$

whose eigenvalues are the possible values of the system's energy. This correspondence between physical quantities in classical mechanics and operators in quantum mechanics is general. It is a fundamental postulate of quantum mechanics that to every physical property (for example, the energy, the x coordinate, the momentum) there corresponds a quantum-mechanical operator. We further postulate that the operator corresponding to the property B is obtained by writing down the classical-mechanical expression for B as a function of Cartesian coordinates and corresponding momenta and then making the following replacements. Each Cartesian coordinate q is replaced by the operator multiplication by that coordinate:

$$\hat{q} = q \cdot$$

Each Cartesian component of linear momentum p_q is replaced by the operator

$$\hat{p}_q = \frac{\hbar}{i} \frac{\partial}{\partial q} = -i\hbar \frac{\partial}{\partial q}$$

where $i = \sqrt{-1}$ and $\partial/\partial q$ is the operator for the partial derivative with respect to the coordinate q. Note that $1/i = i/i^2 = i/(-1) = -i$.

Consider some examples. The operator corresponding to the x coordinate is multiplication by x:

$$\hat{x} = x \cdot \qquad\qquad\qquad (3.21)^*$$

Also,

$$\hat{y} = y \cdot \qquad \text{and} \qquad \hat{z} = z \cdot \qquad\qquad (3.22)^*$$

The operators for the components of linear momentum are

$$\hat{p}_x = \frac{\hbar}{i} \frac{\partial}{\partial x}, \qquad \hat{p}_y = \frac{\hbar}{i} \frac{\partial}{\partial y}, \qquad \hat{p}_z = \frac{\hbar}{i} \frac{\partial}{\partial z} \qquad (3.23)^*$$

The operator corresponding to p_x^2 is

$$\hat{p}_x^2 = \left(\frac{\hbar}{i} \frac{\partial}{\partial x}\right)^2 = \frac{\hbar}{i} \frac{\partial}{\partial x} \frac{\hbar}{i} \frac{\partial}{\partial x} = -\hbar^2 \frac{\partial^2}{\partial x^2} \qquad (3.24)$$

with similar expressions for $\hat{p}_y^2$ and $\hat{p}_z^2$.

Now consider the potential-energy and kinetic-energy operators in one dimension. Suppose we had a system with the potential-energy function $V(x) = ax^2$, where a is a constant. Replacing x with $x \cdot$, we see that the potential-energy operator is simply multiplication by ax^2:

$$\hat{V}(x) = ax^2 \cdot$$

In general, we have for any potential-energy function

$$\hat{V}(x) = V(x) \cdot \qquad\qquad\qquad (3.25)^*$$

The classical-mechanical expression for the kinetic energy T in (3.20) is

$$T = p_x^2/2m \qquad\qquad\qquad (3.26)^*$$

Replacing p_x by the corresponding operator, we have

$$\hat{T} = -\frac{\hbar^2}{2m} \frac{\partial^2}{\partial x^2} = -\frac{\hbar^2}{2m} \frac{d^2}{dx^2} \qquad (3.27)$$

where (3.24) has been used, and the partial derivative becomes an ordinary derivative in one dimension. The classical-mechanical Hamiltonian (3.20) is

$$H = T + V = p_x^2/2m + V(x) \qquad\qquad (3.28)$$

The corresponding quantum-mechanical Hamiltonian (or energy) operator is

$$\hat{H} = \hat{T} + \hat{V} = -\frac{\hbar^2}{2m}\frac{d^2}{dx^2} + V(x) \qquad (3.29)^*$$

which agrees with the operator in the Schrödinger equation (3.1). Note that all these operators are linear.

How are the quantum-mechanical operators related to the corresponding properties of a system? Each such operator has its own set of eigenfunctions and eigenvalues. Let $\hat{B}$ be the quantum-mechanical operator that corresponds to the physical property B. Letting f_i and b_i symbolize the eigenfunctions and eigenvalues of $\hat{B}$, we have [Eq. (3.14)]

$$\hat{B}f_i = b_i f_i, \qquad i = 1, 2, 3, \ldots \qquad (3.30)$$

The operator $\hat{B}$ has many eigenfunctions and eigenvalues, and the subscript i is used to indicate this. $\hat{B}$ is usually a differential operator, and (3.30) is a differential equation whose solutions give the eigenfunctions and eigenvalues. Quantum mechanics postulates that (no matter what the state function of the system happens to be) *a measurement of the property B must yield one of the eigenvalues b_i of the operator $\hat{B}$*. For example, the only values that can be found for the energy of a system are the eigenvalues of the energy (Hamiltonian) operator $\hat{H}$. Using ψ_i to symbolize the eigenfunctions of $\hat{H}$, we have as the eigenvalue equation (3.30)

$$\hat{H}\psi_i = E_i\psi_i \qquad (3.31)^*$$

Using the Hamiltonian (3.29), we obtain for a one-dimensional, one-particle system

$$\left[-\frac{\hbar^2}{2m}\frac{d^2}{dx^2} + V(x) \right]\psi_i = E_i\psi_i \qquad (3.32)$$

which is the time-independent Schrödinger equation, (3.1). Thus our postulates about operators are consistent with our previous work. We shall later provide further justification of the choice (3.23) for the momentum operator by showing that in the limiting transition to classical mechanics this choice yields $p_x = m(dx/dt)$, as it should. (See Problem 7.50.)

In Chapter 1 we postulated that the state of a quantum-mechanical system is specified by a state function $\Psi(x, t)$, which contains all the information we can know about the system. How does Ψ give us information about the property B? We postulate that *if Ψ is an eigenfunction of $\hat{B}$ with eigenvalue b_k, then a measurement of B is certain to yield the value b_k*. Consider, for example, the energy. The eigenfunctions of the energy operator are the solutions $\psi(x)$ of the time-independent Schrödinger equation (3.32). Suppose the system is in a stationary state with state function [Eq. (1.19)]

$$\Psi(x, t) = e^{-iEt/\hbar}\psi(x) \qquad (3.33)$$

Is $\Psi(x, t)$ an eigenfunction of the energy operator $\hat{H}$? We have

$$\hat{H}\Psi(x, t) = \hat{H}e^{-iEt/\hbar}\psi(x)$$

$\hat{H}$ contains no derivatives with respect to time and therefore does not affect the exponential factor $e^{-iEt/\hbar}$. We have

$$\hat{H}\Psi(x, t) = e^{-iEt/\hbar}\hat{H}\psi(x) = Ee^{-iEt/\hbar}\psi(x) = E\Psi(x, t)$$

$$\hat{H}\Psi = E\Psi \qquad (3.34)$$

where (3.31) was used. Hence, for a stationary state, $\Psi(x, t)$ is an eigenfunction of $\hat{H}$, and we are certain to obtain the value E when we measure the energy.

As an example of another property, consider momentum. The eigenfunctions g of $\hat{p}_x$ are found by solving

$$\hat{p}_x g = kg$$

$$\frac{\hbar}{i}\frac{dg}{dx} = kg \qquad (3.35)$$

We find (Problem 3.21)

$$g = Ae^{ikx/\hbar} \qquad (3.36)$$

where A is an arbitrary constant. To keep g finite for large $|x|$, the eigenvalues k must be real. Thus the eigenvalues of $\hat{p}_x$ are all the real numbers

$$-\infty < k < \infty \qquad (3.37)$$

which is reasonable. Any measurement of p_x must yield one of the eigenvalues (3.37) of p_x. Each different value of k in (3.36) gives a different eigenfunction g. It might seem surprising that the operator for the physical property momentum involves the imaginary number i. Actually, the presence of i in $\hat{p}_x$ ensures that the eigenvalues k are real; recall that the eigenvalues of d/dx are imaginary (Section 3.2).

Comparing the free-particle wave function (2.30) with the eigenfunctions (3.36) of $\hat{p}_x$, we note the following physical interpretation: The first term in (2.30) corresponds to positive momentum and represents motion in the $+x$ direction, while the second term in (2.30) corresponds to negative momentum and represents motion in the $-x$ direction.

Now consider the momentum of a particle in a box. The state function for a particle in a stationary state in a one-dimensional box is [Eqs. (3.33), (2.20), and (2.23)]

$$\Psi(x, t) = e^{-iEt/\hbar}\left(\frac{2}{l}\right)^{1/2}\sin\left(\frac{n\pi x}{l}\right) \qquad (3.38)$$

where $E = n^2h^2/8ml^2$. Do we have a definite value of p_x? That is, is $\Psi(x, t)$ an eigenfunction of $\hat{p}_x$? Looking at the eigenfunctions of $\hat{p}_x$, we see that there is no numerical value of the real constant k that will make the exponential function in (3.36) become a sine function, as in (3.38). Hence Ψ is not an eigenfunction of $\hat{p}_x$. We can verify this directly; we have

$$\hat{p}_x\Psi = \frac{\hbar}{i}\frac{\partial}{\partial x}e^{-iEt/\hbar}\left(\frac{2}{l}\right)^{1/2}\sin\left(\frac{n\pi x}{l}\right) = \frac{n\pi\hbar}{il}e^{-iEt/\hbar}\left(\frac{2}{l}\right)^{1/2}\cos\left(\frac{n\pi x}{l}\right)$$

Since $\hat{p}_x\Psi \neq$ constant $\cdot \Psi$, the state function Ψ is not an eigenfunction of $\hat{p}_x$.

Note that the system's state function Ψ need not be an eigenfunction f_i of the operator $\hat{B}$ in (3.30) that corresponds to the physical property B of the system. Thus, the particle-in-a-box stationary-state wave functions are not eigenfunctions of $\hat{p}_x$. Despite this, we still must get one of the eigenvalues (3.37) of $\hat{p}_x$ when we measure p_x for a particle-in-a-box stationary state.

Are the particle-in-a-box stationary-state wave functions eigenfunctions of $\hat{p}_x^2$? We have [Eq. (3.24)]

$$\hat{p}_x^2 \Psi = -\hbar^2 \frac{\partial^2}{\partial x^2} e^{-iEt/\hbar} \left(\frac{2}{l}\right)^{1/2} \sin\left(\frac{n\pi x}{l}\right) = \frac{n^2\pi^2\hbar^2}{l^2} e^{-iEt/\hbar} \left(\frac{2}{l}\right)^{1/2} \sin\left(\frac{n\pi x}{l}\right)$$

$$\hat{p}_x^2 \Psi = \frac{n^2 h^2}{4l^2} \Psi \tag{3.39}$$

Hence a measurement of p_x^2 will always give the result $n^2 h^2/4l^2$ when the particle is in the stationary state with quantum number n. This should come as no surprise: The potential energy in the box is zero, and the Hamiltonian is

$$\hat{H} = \hat{T} + \hat{V} = \hat{T} = \hat{p}_x^2/2m$$

We then have [Eq. (3.34)]

$$\hat{H}\Psi = E\Psi = \frac{\hat{p}_x^2}{2m} \Psi$$

$$\hat{p}_x^2 \Psi = 2mE\Psi = 2m \frac{n^2 h^2}{8ml^2} \Psi = \frac{n^2 h^2}{4l^2} \Psi \tag{3.40}$$

in agreement with (3.39). The only possible value for p_x^2 is

$$p_x^2 = n^2 h^2/4l^2 \tag{3.41}$$

Equation (3.41) suggests that a measurement of p_x would necessarily yield one of the two values $\pm\frac{1}{2}nh/l$, corresponding to the particle moving to the right or to the left in the box. This plausible suggestion is not accurate. An analysis using the methods of Chapter 7 shows that there is a high probability that the measured value will be close to one of the two values $\pm\frac{1}{2}nh/l$, but that any value consistent with (3.37) can result from a measurement of p_x for the particle in a box; see Problem 7.33.

We postulated that a measurement of the property B must give a result that is one of the eigenvalues of the operator $\hat{B}$. If the state function Ψ happens to be an eigenfunction of $\hat{B}$ with eigenvalue b, we are certain to get b when we measure B. Suppose, however, that Ψ is not one of the eigenfunctions of $\hat{B}$. What then? We still assert that *we will get one of the eigenvalues of $\hat{B}$ when we measure B, but we cannot predict which eigenvalue will be obtained*. We shall see in Chapter 7 that the probabilities for obtaining the various eigenvalues of $\hat{B}$ can be predicted.

EXAMPLE   The energy of a particle of mass m in a one-dimensional box of length l is measured. What are the possible values that can result from the measurement if (a) at the time the measurement begins, the particle's state function is $\Psi = (30/l^5)^{1/2} x(l - x)$ for $0 \leqslant x \leqslant l$; (b) at the time the measurement begins, $\Psi = (2/l)^{1/2} \sin(3\pi x/l)$ for $0 \leqslant x \leqslant l$?

(a) The possible outcomes of a measurement of the property E are the eigenvalues of the system's energy (Hamiltonian) operator $\hat{H}$. Therefore, the measured value must be one of the numbers $n^2 h^2 / 8ml^2$, where $n = 1, 2, 3, \ldots$. Since Ψ is not one of the eigenfunctions $(2/l)^{1/2} \sin(n\pi x/l)$ [Eq. (2.23)] of $\hat{H}$, we cannot predict which one of these eigenvalues will be obtained for this nonstationary state. (b) Since Ψ is an eigenfunction of $\hat{H}$ with eigenvalue $3^2 h^2 / 8ml^2$ [Eq. (2.20)], the measurement must give $9h^2 / 8ml^2$.

3.4 THE THREE-DIMENSIONAL MANY-PARTICLE SCHRÖDINGER EQUATION

Up to now we have restricted ourselves to one-dimensional, one-particle systems. The operator formalism developed in the last section enables us to extend our work to three-dimensional many-particle systems. The time-dependent Schrödinger equation for the time development of the state function is postulated to have the form of Eq. (1.12):

$$i\hbar \, \frac{\partial \Psi}{\partial t} = \hat{H}\Psi \tag{3.42}*$$

The time-independent Schrödinger equation for the energy eigenfunctions and eigenvalues is

$$\hat{H}\psi = E\psi \tag{3.43}*$$

which is obtained from (3.42) by taking the potential energy as independent of time and applying the separation-of-variables procedure used to obtain (1.18) from (1.12).

For a one-particle, three-dimensional system, the classical-mechanical Hamiltonian is

$$H = T + V = \frac{1}{2m} \, (p_x^2 + p_y^2 + p_z^2) + V(x, y, z) \tag{3.44}$$

Introducing the quantum-mechanical operators [Eq. (3.24)], we have for the Hamiltonian operator

$$\hat{H} = - \frac{\hbar^2}{2m} \left(\frac{\partial^2}{\partial x^2} + \frac{\partial^2}{\partial y^2} + \frac{\partial^2}{\partial z^2} \right) + V(x, y, z) \tag{3.45}$$

The operator in parentheses in (3.45) is called the **Laplacian operator** ∇^2 (read as "del squared"):

$$\nabla^2 \equiv \frac{\partial^2}{\partial x^2} + \frac{\partial^2}{\partial y^2} + \frac{\partial^2}{\partial z^2} \tag{3.46}*$$

The one-particle, three-dimensional time-independent Schrödinger equation is then

$$- \frac{\hbar^2}{2m} \, \nabla^2 \psi + V\psi = E\psi \tag{3.47}$$

Now consider a three-dimensional system with n particles. Let particle i

have mass m_i and coordinates (x_i, y_i, z_i), where $i = 1, 2, 3, \ldots, n$. The kinetic energy is the sum of the kinetic energies of the individual particles:

$$T = \frac{1}{2m_1}(p_{x_1}^2 + p_{y_1}^2 + p_{z_1}^2) + \frac{1}{2m_2}(p_{x_2}^2 + p_{y_2}^2 + p_{z_2}^2) + \cdots + \frac{1}{2m_n}(p_{x_n}^2 + p_{y_n}^2 + p_{z_n}^2)$$

where p_{x_i} is the x component of the linear momentum of particle i, and so on. The kinetic-energy operator is

$$\hat{T} = -\frac{\hbar^2}{2m_1}\left(\frac{\partial^2}{\partial x_1^2} + \frac{\partial^2}{\partial y_1^2} + \frac{\partial^2}{\partial z_1^2}\right) - \cdots - \frac{\hbar^2}{2m_n}\left(\frac{\partial^2}{\partial x_n^2} + \frac{\partial^2}{\partial y_n^2} + \frac{\partial^2}{\partial z_n^2}\right)$$

$$\hat{T} = -\sum_{i=1}^{n} \frac{\hbar^2}{2m_i} \nabla_i^2 \qquad (3.48)^*$$

$$\nabla_i^2 \equiv \frac{\partial^2}{\partial x_i^2} + \frac{\partial^2}{\partial y_i^2} + \frac{\partial^2}{\partial z_i^2} \qquad (3.49)^*$$

We shall usually restrict ourselves to cases where the potential energy depends only on the $3n$ coordinates:

$$V = V(x_1, y_1, z_1, \ldots, x_n, y_n, z_n)$$

The Hamiltonian operator for an n-particle three-dimensional system is then

$$\hat{H} = -\sum_{i=1}^{n} \frac{\hbar^2}{2m_i} \nabla_i^2 + V(x_1, \ldots, z_n) \qquad (3.50)^*$$

and the time-independent Schrödinger equation is

$$\left[-\sum_{i=1}^{n} \frac{\hbar^2}{2m_i} \nabla_i^2 + V(x_1, \ldots, z_n)\right]\psi = E\psi \qquad (3.51)$$

where the time-independent wave function is a function of the $3n$ coordinates of the n particles:

$$\psi = \psi(x_1, y_1, z_1, \ldots, x_n, y_n, z_n) \qquad (3.52)$$

The Schrödinger equation (3.51) is a linear partial differential equation.

As an example, consider a system of two particles interacting so that the potential energy is inversely proportional to the distance between them, with c being the proportionality constant. The Schrödinger equation (3.51) becomes

$$\left[-\frac{\hbar^2}{2m_1}\left(\frac{\partial^2}{\partial x_1^2} + \frac{\partial^2}{\partial y_1^2} + \frac{\partial^2}{\partial z_1^2}\right) - \frac{\hbar^2}{2m_2}\left(\frac{\partial^2}{\partial x_2^2} + \frac{\partial^2}{\partial y_2^2} + \frac{\partial^2}{\partial z_2^2}\right)\right.$$
$$\left. + \frac{c}{[(x_1 - x_2)^2 + (y_1 - y_2)^2 + (z_1 - z_2)^2]^{1/2}}\right]\psi = E\psi \qquad (3.53)$$

$$\psi = \psi(x_1, y_1, z_1, x_2, y_2, z_2)$$

Although (3.53) looks formidable, we shall solve it in Chapter 6.

For a one-particle, one-dimensional system, the Born postulate [Eq. (1.14)] states that $|\Psi(x', t)|^2 \, dx$ is the probability of observing the particle between x' and

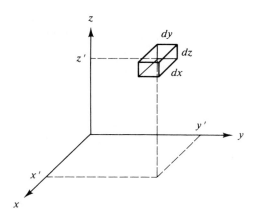

Figure 3.1 An infinitesimal box-shaped region located at x', y', z'.

$x' + dx$ at time t, where x' is a particular value of x. We extend this postulate as follows. *For a three-dimensional, one-particle system, the quantity*

$$|\Psi(x', y', z', t)|^2 \, dx \, dy \, dz \qquad \qquad (3.54)^*$$

is the probability of finding the particle in the infinitesimal region of space with the x coordinate lying between x' and x' + dx, the y coordinate lying between y' and y' + dy, and the z coordinate lying between z' and z' + dz (Figure 3.1). Since the total probability of finding the particle is 1, the normalization condition is

$$\int_{-\infty}^{\infty} \int_{-\infty}^{\infty} \int_{-\infty}^{\infty} |\Psi(x, y, z, t)|^2 \, dx \, dy \, dz = 1 \qquad \qquad (3.55)$$

For a three-dimensional *n-particle system, we postulate that*

$$|\Psi(x_1', y_1', z_1', x_2', y_2', z_2', \ldots, x_n', y_n', z_n', t)|^2 \, dx_1 \, dy_1 \, dz_1$$
$$\times \, dx_2 \, dy_2 \, dz_2 \cdots dx_n \, dy_n \, dz_n \qquad (3.56)^*$$

is the probability at time t of simultaneously finding particle 1 in the infinitesimal rectangular box-shaped region at (x_1', y_1', z_1') with edges dx_1, dy_1, dz_1, particle 2 in the infinitesimal box-shaped region at (x_2', y_2', z_2') with edges dx_2, dy_2, dz_2, $\ldots$, and particle n in the infinitesimal box-shaped region at (x_n', y_n', z_n') with edges dx_n, dy_n, dz_n. The total probability of finding all the particles is 1, and the normalization condition is

$$\int_{-\infty}^{\infty} \int_{-\infty}^{\infty} \int_{-\infty}^{\infty} \cdots \int_{-\infty}^{\infty} \int_{-\infty}^{\infty} \int_{-\infty}^{\infty} |\Psi|^2 \, dx_1 \, dy_1 \, dz_1 \cdots dx_n \, dy_n \, dz_n = 1 \quad (3.57)$$

It is customary in quantum mechanics to denote integration over the full range of all the coordinates of a system by $\int dq$ or $\int d\tau$. A shorthand way of writing (3.55) or (3.57) is

$$\int |\Psi|^2 \, d\tau = 1 \qquad \qquad (3.58)^*$$

Although (3.58) may look like an indefinite integral, it is understood to be a definite integral; the integration variables and their ranges are understood from the context.

For a stationary state, $|\Psi|^2 = |\psi|^2$, and

$$\int |\psi|^2 \, d\tau = 1 \qquad (3.59)^*$$

3.5   THE PARTICLE IN A THREE-DIMENSIONAL BOX

For the present, we confine ourselves to one-particle problems. In this section we consider the three-dimensional case of the problem solved in Section 2.2, the particle in a box.

There are many possible shapes for a three-dimensional box. The box we consider is a rectangular parallelepiped with edges of length a, b, and c. We choose our coordinate system so that one corner of the box lies at the origin and the box lies in the first octant of space. Within the box, the potential energy is zero; outside the box, it is infinite:

$$V(x, y, z) = 0, \qquad \text{in the region} \quad \begin{cases} 0 < x < a \\ 0 < y < b \\ 0 < z < c \end{cases} \qquad (3.60)$$

$$V = \infty \qquad \text{elsewhere}$$

Since the probability for the particle to have infinite energy is zero, the wave function must be zero outside the box. Within the box, the potential-energy operator is zero and the Schrödinger equation (3.47) is

$$-\frac{\hbar^2}{2m} \left(\frac{\partial^2 \psi}{\partial x^2} + \frac{\partial^2 \psi}{\partial y^2} + \frac{\partial^2 \psi}{\partial z^2} \right) = E\psi \qquad (3.61)$$

To solve (3.61), we start by assuming that the solution can be written as the product of a function of x alone times a function of y alone times a function of z alone:

$$\psi(x, y, z) = f(x)g(y)h(z) \qquad (3.62)$$

It might be thought that this assumption throws away solutions that are not of the form (3.62). However, it can be shown that, if we can find solutions of the form (3.62) that satisfy the boundary conditions, then there are no other solutions of the Schrödinger equation that will satisfy the boundary conditions. The method we are using to solve (3.62) is called *separation of variables*.

From (3.62), we find

$$\frac{\partial^2 \psi}{\partial x^2} = f''(x)g(y)h(z), \qquad \frac{\partial^2 \psi}{\partial y^2} = f(x)g''(y)h(z), \qquad \frac{\partial^2 \psi}{\partial z^2} = f(x)g(y)h''(z)$$

$$(3.63)$$

Substitution of (3.62) and (3.63) into (3.61) gives

$$-(\hbar^2/2m)f''gh - (\hbar^2/2m)fg''h - (\hbar^2/2m)fgh'' - Efgh = 0 \qquad (3.64)$$

Division of this equation by fgh gives

$$-\frac{\hbar^2 f''}{2mf} - \frac{\hbar^2 g''}{2mg} - \frac{\hbar^2 h''}{2mh} - E = 0 \tag{3.65}$$

$$-\frac{\hbar^2 f''(x)}{2mf(x)} = \frac{\hbar^2 g''(y)}{2mg(y)} + \frac{\hbar^2 h''(z)}{2mh(z)} + E \tag{3.66}$$

Let us define E_x as equal to the left side of (3.66):

$$E_x \equiv -\hbar^2 f''(x)/2mf(x) \tag{3.67}$$

The definition (3.67) shows that E_x is independent of y and z. Equation (3.66) shows that E_x equals $\hbar^2 g''(y)/2mg(y) + \hbar^2 h''(z)/2mh(z) + E$; therefore, E_x must be independent of x. Being independent of x, y, and z, the quantity E_x must be a constant.

Similar to (3.67), we define E_y and E_z by

$$E_y \equiv -\hbar^2 g''(y)/2mg(y) , \qquad E_z \equiv -\hbar^2 h''(z)/2mh(z) \tag{3.68}$$

Since x, y, and z occur symmetrically in (3.65), the same reasoning that showed E_x to be a constant shows that E_y and E_z are constants. Substitution of the definitions (3.67) and (3.68) into (3.65) gives

$$E_x + E_y + E_z = E \tag{3.69}$$

Equations (3.67) and (3.68) are

$$\frac{d^2 f(x)}{dx^2} + \frac{2m}{\hbar^2} E_x f(x) = 0 \tag{3.70}$$

$$\frac{d^2 g(y)}{dy^2} + \frac{2m}{\hbar^2} E_y g(y) = 0 , \qquad \frac{d^2 h(z)}{dz^2} + \frac{2m}{\hbar^2} E_z h(z) = 0 \tag{3.71}$$

We have converted the partial differential equation in three variables into three ordinary differential equations. What are the boundary conditions on (3.70)? Since the wave function vanishes outside the box, continuity of ψ requires that it vanish on the walls of the box. In particular, ψ must be zero on the wall of the box lying in the yz plane, where $x = 0$, and it must be zero on the parallel wall of the box, where $x = a$. Therefore,

$$f(0) = 0 , \qquad f(a) = 0$$

Now compare Eq. (3.70) with Eq. (2.10) in Section 2.2, dealing with the particle in a one-dimensional box. The equations are the same in form, with E_x in (3.70) corresponding to E in (2.10). Are the boundary conditions the same? Yes, except that we have $x = a$ instead of $x = l$ as the second point where the independent variable vanishes. Thus we can use the work in Section 2.2 to write down as the solution [see Eqs. (2.23) and (2.20)]

$$f(x) = \left(\frac{2}{a}\right)^{1/2} \sin\left(\frac{n_x \pi x}{a}\right)$$

$$E_x = \frac{n_x^2 h^2}{8ma^2} , \qquad n_x = 1, 2, 3, \ldots$$

The same reasoning applied to the y and z equations gives

$$g(y) = \left(\frac{2}{b}\right)^{1/2} \sin\left(\frac{n_y \pi y}{b}\right), \qquad h(z) = \left(\frac{2}{c}\right)^{1/2} \sin\left(\frac{n_z \pi z}{c}\right)$$

$$E_y = \frac{n_y^2 h^2}{8mb^2}, \qquad n_y = 1, 2, 3, \ldots \quad \text{and} \quad E_z = \frac{n_z^2 h^2}{8mc^2}, \qquad n_z = 1, 2, 3, \ldots$$

From (3.69), we have for the energy

$$E = \frac{h^2}{8m}\left(\frac{n_x^2}{a^2} + \frac{n_y^2}{b^2} + \frac{n_z^2}{c^2}\right) \tag{3.72}$$

From (3.62), the wave function in the box is

$$\psi(x, y, z) = \left(\frac{8}{abc}\right)^{1/2} \sin\left(\frac{n_x \pi x}{a}\right) \sin\left(\frac{n_y \pi y}{b}\right) \sin\left(\frac{n_z \pi z}{c}\right) \tag{3.73}$$

The wave function has three quantum numbers, n_x, n_y, n_z; we can attribute this to the three-dimensional nature of the problem. The three quantum numbers vary independently of one another.

Since the x, y, and z factors in the wave function are each independently normalized, the wave function is normalized:

$$\int_{-\infty}^{\infty}\int_{-\infty}^{\infty}\int_{-\infty}^{\infty} |\psi|^2 \, dx \, dy \, dz = \int_0^a |f(x)|^2 \, dx \int_0^b |g(y)|^2 \, dy \int_0^c |h(z)|^2 \, dz = 1$$

where we used (Problem 3.29)

$$\int\int\int F(x)G(y)H(z) \, dx \, dy \, dz = \int F(x) \, dx \int G(y) \, dy \int H(z) \, dz \qquad \textbf{(3.74)*}$$

Suppose that $a = b = c$. We then have a cube. The energy levels are then

$$E = (h^2/8ma^2)(n_x^2 + n_y^2 + n_z^2) \tag{3.75}$$

Let us tabulate some of the allowed energies of a particle confined to a cube with infinitely strong walls:

$n_x n_y n_z$	111	211	121	112	122	212	221	113	131	311	222
$E(8ma^2/h^2)$	3	6	6	6	9	9	9	11	11	11	12

Observe that states with different quantum numbers may have the same energy. For example, the states ψ_{211}, ψ_{121}, and ψ_{112} (where the subscripts give the quantum numbers) all have the same energy. However, Eq. (3.73) shows that these three sets of quantum numbers give three different, independent wave functions and therefore do represent different states of the system. When two or more independent wave functions correspond to states with the same energy eigenvalue, the eigenvalue is said to be **degenerate**. The **degree of degeneracy** (or, simply, the **degeneracy**) of an energy level is the number of states that have that energy. Thus the second-lowest energy level of the particle in a cube is threefold degenerate. We got the degeneracy when we made the edges of the box equal; degeneracy is usually related to the symmetry of the system. Note that the wave functions ψ_{211}, ψ_{121}, and ψ_{112} can be transformed into one another by rotating the cubic box. Usually, one does not find degeneracy in one-dimensional problems.

In the statistical-mechanical evaluation of the molecular partition function of an ideal gas, the translational energy levels of each gas molecule are taken to be the levels of a particle in a three-dimensional rectangular box; see *Levine, Physical Chemistry*, Sections 22.6 and 22.7.

3.6 DEGENERACY

Let us prove an important theorem about the wave functions of an n-fold degenerate energy level. We have n independent wave functions $\psi_1, \psi_2, \ldots, \psi_n$. Let W be the energy of the degenerate level:

$$\hat{H}\psi_1 = W\psi_1, \quad \hat{H}\psi_2 = W\psi_2, \quad \ldots, \quad \hat{H}\psi_n = W\psi_n \tag{3.76}$$

We wish to prove that any linear combination

$$\phi \equiv c_1\psi_1 + c_2\psi_2 + \cdots + c_n\psi_n$$

of the n wave functions of the *degenerate* level is an eigenfunction of the Hamiltonian with eigenvalue W. We must show $\hat{H}\phi = W\phi$ or

$$\hat{H}(c_1\psi_1 + c_2\psi_2 + \cdots + c_n\psi_n) = W(c_1\psi_1 + c_2\psi_2 + \cdots + c_n\psi_n) \tag{3.77}$$

Since $\hat{H}$ is a linear operator, we can apply Eq. (3.9) $n - 1$ times to the left side of (3.77) to obtain

$$\hat{H}(c_1\psi_1 + c_2\psi_2 + \cdots + c_n\psi_n) = \hat{H}(c_1\psi_1) + \hat{H}(c_2\psi_2) + \cdots + \hat{H}(c_n\psi_n)$$

Use of Eqs. (3.10) and (3.76) gives

$$\hat{H}(c_1\psi_1 + c_2\psi_2 + \cdots + c_n\psi_n) = c_1\hat{H}\psi_1 + c_2\hat{H}\psi_2 + \cdots + c_n\hat{H}\psi_n$$

$$= c_1 W\psi_1 + c_2 W\psi_2 + \cdots + c_n W\psi_n$$

$$\hat{H}(c_1\psi_1 + c_2\psi_2 + \cdots + c_n\psi_n) = W(c_1\psi_1 + c_2\psi_2 + \cdots + c_n\psi_n) \tag{3.78}$$

which completes the proof.

As an example, the stationary-state wave functions ψ_{211}, ψ_{121}, and ψ_{112} for the particle in a cubic box are degenerate, and the linear combination $c_1\psi_{211} + c_2\psi_{121} + c_3\psi_{112}$ is an eigenfunction of the particle-in-a-cubic-box Hamiltonian with eigenvalue $6h^2/8ma^2$, the same eigenvalue as for each of ψ_{211}, ψ_{121}, and ψ_{112}.

Note that the linear combination $c_1\psi_1 + c_2\psi_2$ is not an eigenfunction of $\hat{H}$ if ψ_1 and ψ_2 correspond to different energy eigenvalues ($\hat{H}\psi_1 = E_1\psi_1$ and $\hat{H}\psi_2 = E_2\psi_2$ with $E_1 \neq E_2$).

Since any linear combination of the wave functions corresponding to a degenerate energy level is an eigenfunction of $\hat{H}$ with the same eigenvalue, we can construct an infinite number of different wave functions for any degenerate energy level. Actually, we are only interested in eigenfunctions that are linearly independent. The n functions $f_1, \ldots, f_n$ are said to be ***linearly independent*** if the equation $c_1 f_1 + \cdots + c_n f_n = 0$ can only be satisfied with all the constants

$c_1, \ldots, c_n$ equal to zero. This means that no member of the set of functions can be expressed as a linear combination of the remaining members. For example, the functions $f_1 = 3x$, $f_2 = 5x^2 - x$, $f_3 = x^2$ are not linearly independent, since $f_2 = 5f_3 - \frac{1}{3}f_1$. The functions $g_1 = 1$, $g_2 = x$, $g_3 = x^2$ are linearly independent, since none of them can be written as a linear combination of the other two. The **degree of degeneracy** of an energy level is equal to the number of linearly independent wave functions corresponding to that value of the energy.

3.7 AVERAGE VALUES

It was pointed out in Section 3.3 that, when the state function Ψ is not an eigenfunction of the operator $\hat{B}$, a measurement of B will give one of a number of possible values (the eigenvalues of $\hat{B}$). We now consider the average value of the property B for a system whose state is Ψ.

To determine the average value of B experimentally, we take many identical, noninteracting systems each in the same state Ψ and we measure B in each system. The **average value** of B (symbolized by $\langle B \rangle$) is defined as the arithmetic mean of the observed values $b_1, b_2, \ldots, b_N$:

$$\langle B \rangle = \frac{\sum_{j=1}^{N} b_j}{N} \tag{3.79}$$

where N, the number of systems, is extremely large.

Instead of summing over the observed values of B, we can sum over all the possible values of B, multiplying each possible value by the number of times it is observed, to obtain the equivalent expression

$$\langle B \rangle = \frac{\sum_{b} n_b b}{N} \tag{3.80}$$

where n_b is the number of times the value b is observed. An example will make this clear. Suppose a class of nine students takes a quiz that has five questions and that the students receive these grades: 0, 20, 20, 60, 60, 80, 80, 80, 100. Calculating the average grade according to (3.79), we have

$$\frac{1}{N} \sum_{j=1}^{N} b_j = \frac{0 + 20 + 20 + 60 + 60 + 80 + 80 + 80 + 100}{9} = 56$$

To calculate the average grade according to (3.80), we sum over the possible grades: 0, 20, 40, 60, 80, 100. We have

$$\frac{1}{N} \sum_{b} n_b b = \frac{1(0) + 2(20) + 0(40) + 2(60) + 3(80) + 1(100)}{9} = 56$$

Equation (3.80) can be written as

$$\langle B \rangle = \sum_{b} \left(\frac{n_b}{N} \right) b$$

Since N is extremely large, n_b/N is the probability P_b of observing the value b, and

$$\langle B \rangle = \sum_b P_b b \tag{3.81}$$

Now consider the average value of the x coordinate for a one-particle, one-dimensional system in the state $\Psi(x, t)$. The x coordinate takes on a continuous range of values, and the probability of observing the particle between x and $x + dx$ is $|\Psi|^2\, dx$. The summation over the infinitesimal probabilities is equivalent to an integration, and (3.81) becomes

$$\langle x \rangle = \int_{-\infty}^{\infty} x |\Psi(x, t)|^2\, dx \tag{3.82}$$

For the one-particle, three-dimensional case, the probability of finding the particle in the volume element at point (x, y, z) with edges dx, dy, dz is

$$|\Psi(x, y, z, t)|^2\, dx\, dy\, dz \tag{3.83}$$

If we want the probability that the particle is between x and $x + dx$, we must integrate (3.83) over all possible values of y and z, since the particle can have any values for its y and z coordinates while its x coordinate lies between x and $x + dx$. Hence, in the three-dimensional case (3.82) becomes

$$\langle x \rangle = \int_{-\infty}^{\infty} \left[\int_{-\infty}^{\infty} \int_{-\infty}^{\infty} |\Psi(x, y, z, t)|^2\, dy\, dz \right] x\, dx$$

$$\langle x \rangle = \int_{-\infty}^{\infty} \int_{-\infty}^{\infty} \int_{-\infty}^{\infty} |\Psi(x, y, z, t)|^2 x\, dx\, dy\, dz \tag{3.84}$$

Now consider the average value of some physical property $B(x, y, z)$ that is a function of the particle's coordinates. An example is the potential energy $V(x, y, z)$. The same reasoning that gave Eq. (3.84) yields

$$\langle B(x, y, z) \rangle = \int_{-\infty}^{\infty} \int_{-\infty}^{\infty} \int_{-\infty}^{\infty} |\Psi(x, y, z, t)|^2 B(x, y, z)\, dx\, dy\, dz \tag{3.85}$$

$$\langle B(x, y, z) \rangle = \int_{-\infty}^{\infty} \int_{-\infty}^{\infty} \int_{-\infty}^{\infty} \Psi^* B \Psi\, dx\, dy\, dz \tag{3.86}$$

The form (3.86) might seem like a bit of whimsy, since it is no different from (3.85). In a moment we shall see its significance.

In general, the property B depends on both coordinates *and* momenta:

$$B = B(x, y, z, p_x, p_y, p_z)$$

for the one-particle three-dimensional case. How do we find the average value of B? We *postulate* that $\langle B \rangle$ for a system in state Ψ is

$$\langle B \rangle = \int_{-\infty}^{\infty} \int_{-\infty}^{\infty} \int_{-\infty}^{\infty} \Psi^* B\left(x, y, z, \frac{\hbar}{i}\frac{\partial}{\partial x}, \frac{\hbar}{i}\frac{\partial}{\partial y}, \frac{\hbar}{i}\frac{\partial}{\partial z}\right) \Psi\, dx\, dy\, dz$$

$$\langle B \rangle = \int_{-\infty}^{\infty} \int_{-\infty}^{\infty} \int_{-\infty}^{\infty} \Psi^* \hat{B} \Psi\, dx\, dy\, dz \tag{3.87}$$

where $\hat{B}$ is the quantum-mechanical operator for the property B. [Later we shall provide some justification for this postulate by using (3.87) to show that the time-dependent Schrödinger equation reduces to Newton's second law in the transition from quantum to classical mechanics; see Problem 7.50.] For the n-particle case, we postulate that

$$\langle B \rangle = \int \Psi^* \hat{B} \Psi \, d\tau \qquad \text{(3.88)}^*$$

where $\int d\tau$ indicates integration over the full range of the $3n$ coordinates. The state function in (3.88) must be normalized, since we took $\Psi^*\Psi$ as the probability density. It is important to have the operator properly sandwiched between Ψ^* and Ψ. The quantities $\hat{B}\Psi^*\Psi$ and $\Psi^*\Psi\hat{B}$ are not the same as $\Psi^*\hat{B}\Psi$, unless B is a function of coordinates only. In $\int \Psi^* \hat{B} \Psi \, d\tau$, one first operates on Ψ with $\hat{B}$ to produce a new function $\hat{B}\Psi$, which is then multiplied by Ψ^*; one then integrates over all space to produce a number, which is $\langle B \rangle$.

For a stationary state, we have

$$\Psi^* \hat{B} \Psi = e^{iEt/\hbar} \psi^* \hat{B} e^{-iEt/\hbar} \psi = e^0 \psi^* \hat{B} \psi = \psi^* \hat{B} \psi$$

since $\hat{B}$ contains no time derivatives and does not affect the time factor in Ψ. Hence, for a stationary state,

$$\langle B \rangle = \int \psi^* \hat{B} \psi \, d\tau \qquad (3.89)$$

Thus, if $\hat{B}$ is time independent, then $\langle B \rangle$ is time independent in a stationary state.

Consider the special case where Ψ is an eigenfunction of $\hat{B}$; $\hat{B}\Psi = k\Psi$. Equation (3.88) becomes

$$\langle B \rangle = \int \Psi^* \hat{B} \Psi \, d\tau = \int \Psi^* k \Psi \, d\tau = k \int \Psi^* \Psi \, d\tau = k$$

since Ψ is normalized. This result is reasonable, since k is the only possible value we can find for B when we make a measurement.

From Eq. (3.88), it readily follows that the average value of a sum is the sum of the average values:

$$\langle B + C \rangle = \langle B \rangle + \langle C \rangle \qquad (3.90)$$

where B and C are any two properties. However, it is not necessarily true that the average value of a product is the product of the average values:

$$\langle BC \rangle \neq \langle B \rangle \langle C \rangle$$

The term *expectation value* is often used instead of average value. The expectation value is not necessarily one of the possible values we might observe; for example, the average number of children born to an American woman in her lifetime (as calculated from 1988 fertility rates for all age groups) is 1.94.

EXAMPLE Find $\langle x \rangle$ and $\langle p_x \rangle$ for the ground stationary state of a particle in a three-dimensional box.

Substitution of the stationary-state wave function $\psi = f(x)g(y)h(z)$ [Eq. (3.62)] into the average-value postulate (3.89) gives

$$\langle x \rangle = \int \psi^* \hat{x}\psi \, d\tau = \int_0^c \int_0^b \int_0^a f^* g^* h^* xfgh \, dx \, dy \, dz$$

since $\psi = 0$ outside the box. Use of (3.74) gives

$$\langle x \rangle = \int_0^a x|f(x)|^2 \, dx \int_0^b |g(y)|^2 \, dy \int_0^c |h(z)|^2 \, dz = \int_0^a x|f(x)|^2 \, dx$$

since $g(y)$ and $h(z)$ are each normalized. For the ground state, $n_x = 1$ and $f(x) = (2/a)^{1/2} \sin(\pi x/a)$. Hence

$$\langle x \rangle = \frac{2}{a} \int_0^a x \sin^2\left(\frac{\pi x}{a}\right) dx = \frac{a}{2} \tag{3.91}$$

where the Appendix integral (A.3) was used. A glance at Fig. 2.4 shows that (3.91) is reasonable.

Also,

$$\langle p_x \rangle = \int \psi^* \hat{p}_x \psi \, d\tau = \int_0^c \int_0^b \int_0^a f^* g^* h^* \frac{\hbar}{i} \frac{\partial}{\partial x}[f(x)g(y)h(z)] \, dx \, dy \, dz$$

$$\langle p_x \rangle = \frac{\hbar}{i} \int_0^a f^*(x)f'(x) \, dx \int_0^b |g(y)|^2 \, dy \int_0^c |h(z)|^2 \, dz$$

$$\langle p_x \rangle = \frac{\hbar}{i} \int_0^a f(x)f'(x) \, dx = \frac{\hbar}{2i} f^2(x)\Big|_0^a = 0 \tag{3.92}$$

where the boundary conditions $f(0) = 0$ and $f(a) = 0$ were used. The result (3.92) is reasonable since the particle is equally likely to be headed in the $+x$ or $-x$ direction.

3.8 REQUIREMENTS FOR AN ACCEPTABLE WAVE FUNCTION

In solving the particle in a box, we required ψ to be continuous. We now discuss other requirements the wave function must satisfy.

Since $\psi^*\psi \, d\tau$ is a probability, we want to be able to normalize the wave function by choosing a suitable **normalization constant** as a multiplier of the wave function. However, we can do this only if the integral over all space $\int \psi^*\psi \, d\tau$ exists. If this integral exists, ψ is said to be **quadratically integrable.** Thus we generally demand that ψ be quadratically integrable. The important exception is a particle that is not bound. Thus the wave functions for the unbound states of the particle in a well (Section 2.4) and for a free particle are not quadratically integrable.

Since $\psi^*\psi$ is the probability density, it must be single valued. It would be embarrassing if our theory gave two different values for the probability of finding

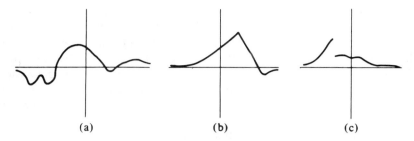

Figure 3.2 Function (a) is continuous and its first derivative is continuous. Function (b) is continuous, but its first derivative has a discontinuity. Function (c) is discontinuous.

a particle at a certain point. If we demand that ψ be single valued, then surely $\psi^*\psi$ will be single valued. It is possible to have ψ multivalued [for example, $\psi(q) = -1, +1, i$] and still have $\psi^*\psi$ single valued. We will, however, demand single valuedness for ψ.

In addition to demanding that ψ be continuous, we usually also require that all the partial derivatives $\partial\psi/\partial x$, $\partial\psi/\partial y$, and so on, be continuous. (See Fig. 3.2.) Referring back to Section 2.2, however, we note that for the particle in a box there is a discontinuity in the derivative of the wave function at the walls of the box; ψ and $d\psi/dx$ are zero everywhere outside the box, but from Eq. (2.23) we see that $d\psi/dx$ does not become zero at the walls. The discontinuity in ψ' is due to the infinite jump in potential energy at the walls of the box. For a box with walls of finite height, ψ' is continuous at the walls (Section 2.4).

In line with the requirement of quadratic integrability, it is sometimes stated that the wave function must be finite everywhere, including infinity. However, this is usually a much stronger requirement than quadratic integrability, and, in fact, it turns out that some of the relativistic wave functions for the hydrogen atom are infinite at the origin, but are quadratically integrable. Occasionally, one encounters nonrelativistic wave functions that are infinite at the origin. Thus the fundamental requirement is quadratic integrability, rather than finiteness.

We require that the eigenfunctions of any operator representing a physical quantity meet the above requirements. A function meeting these requirements is said to be **well behaved**.

3.9 SUMMARY

An operator is a rule that transforms one function into another function. The sum and product of operators are defined by $(\hat{A} + \hat{B})f(x) \equiv \hat{A}f(x) + \hat{B}f(x)$ and $\hat{A}\hat{B}f(x) \equiv \hat{A}[\hat{B}f(x)]$. The commutator of two operators is $[\hat{A}, \hat{B}] \equiv \hat{A}\hat{B} - \hat{B}\hat{A}$. The operators in quantum mechanics are linear, meaning that they satisfy $\hat{A}[f(x) + g(x)] = \hat{A}f(x) + \hat{A}g(x)$ and $\hat{A}[cf(x)] = c\hat{A}f(x)$. The eigenfunctions F_i and eigenvalues b_i of the operator $\hat{B}$ obey $\hat{B}F_i = b_iF_i$.

So far, the following quantum-mechanical postulates have been introduced:

(a) The state of a system is described by a function Ψ (the state function or wave function) of the coordinates and the time. Ψ is single valued, continuous, and (except for unbound states) quadratically integrable.

(b) To each physical property B of a system, there corresponds an operator $\hat{B}$. This operator is found by taking the classical-mechanical expression for the property in terms of Cartesian coordinates and momenta and replacing each coordinate x by $x \cdot$ and each momentum component p_x by $(\hbar/i)\partial/\partial x$.

(c) The only possible values that can result from measurements of the property B are the eigenvalues b_i of the equation $\hat{B}g_i = b_i g_i$, where the eigenfunctions g_i are required to be well behaved.

(d) The average value of the property B is given by $\langle B \rangle = \int \Psi^* \hat{B} \Psi \, d\tau$, where Ψ is the system's state function.

(e) The state function of an undisturbed system changes with time according to $-(\hbar/i)(\partial\Psi/\partial t) = \hat{H}\Psi$, where $\hat{H}$ is the Hamiltonian operator (the energy operator) of the system.

(f) For a three-dimensional n-particle system, the quantity (3.56) is the probability of finding the system's particles in the infinitesimal regions of space listed after (3.56).

The Hamiltonian operator for an n-particle, three-dimensional system is $\hat{H} = -\sum_{i=1}^{n} (\hbar^2/2m_i)\nabla_i^2 + V$, where $\nabla_i^2 \equiv \partial^2/\partial x_i^2 + \partial^2/\partial y_i^2 + \partial^2/\partial z_i^2$. The time-independent Schrödinger equation is $\hat{H}\psi = E\psi$.

The stationary-state wave functions and energy levels of a particle in a three-dimensional rectangular box were found by the use of separation of variables.

The degree of degeneracy of an energy level is the number of linearly independent wave functions that correspond to that energy value. Any linear combination of wave functions of a degenerate level with energy W is an eigenfunction of $\hat{H}$ with eigenvalue W.

PROBLEMS

3.1 If $g = \hat{A}f$, find g for each of these choices of $\hat{A}$ and f. (a) $\hat{A} = d/dx$ and $f = \cos(x^2 + 1)$; (b) $\hat{A} = \hat{5}$ and $f = \sin x$; (c) $\hat{A} = (\quad)^2$ and $f = \sin x$; (d) $\hat{A} = \exp$ and $f = \ln x$; (e) $\hat{A} = d^2/dx^2$ and $f = \ln 3x$; (f) $\hat{A} = d^2/dx^2 + 3x \, d/dx$ and $f = 4x^3$.

3.2 State whether each of the following entities is an operator or a function: (a) $\hat{A}\hat{B}$; (b) $\hat{A}f(x)$; (c) $\hat{B}\hat{A}f(x)$; (d) $[\hat{B}, \hat{A}]$; (e) $f(x)\hat{A}$; (f) $f(x)\hat{A}\hat{B}g(x)$.

3.3 Prove that $\hat{A} + \hat{B} = \hat{B} + \hat{A}$.

3.4 Let $\hat{D} = d/dx$. Verify that $(\hat{D} + x)(\hat{D} - x) = \hat{D}^2 - x^2 - 1$.

3.5 By repeated application of the definition of the product of two operators show that

$$[(\hat{A}\hat{B})\hat{C}]f = \hat{A}[\hat{B}(\hat{C}f)]$$

$$[\hat{A}(\hat{B}\hat{C})]f = \hat{A}[\hat{B}(\hat{C}f)]$$

3.6 (a) Show that $(\hat{A} + \hat{B})^2 = (\hat{B} + \hat{A})^2$ for any two operators (linear or nonlinear). (b) Under what conditions is $(\hat{A} + \hat{B})^2$ equal to $\hat{A}^2 + 2\hat{A}\hat{B} + \hat{B}^2$?

3.7 What do you suppose we mean by the zeroth power of an operator?

3.8 Classify these operators as linear or nonlinear: (a) $3x^2 d^2/dx^2$; (b) $(\ )^2$; (c) $\int dx$; (d) exp; (e) $\Sigma_{x=1}^n$.

3.9 (a) Give an example of an operator that satisfies Eq. (3.9) but does not satisfy (3.10). (b) Give an example of an operator that satisfies (3.10) but not (3.9).

3.10 Prove that the product of two linear operators is a linear operator.

3.11 Verify the commutator identity $[\hat{A}, \hat{B}] = -[\hat{B}, \hat{A}]$.

3.12 Evaluate (a) $[\sin z, d/dz]$; (b) $[d^2/dx^2, ax^2 + bx + c]$, where a, b, and c are constants; (c) $[d/dx, d^2/dx^2]$.

3.13 Prove that if $\hat{A} + \hat{B} = \hat{C}$ then $\hat{A} = \hat{C} - \hat{B}$.

3.14 Prove that Eq. (3.12) holds for linear operators.

3.15 (a) If $\hat{A}$ is linear, show that

$$\hat{A}(bf + cg) = b\hat{A}f + c\hat{A}g \qquad (3.93)$$

where b and c are arbitrary constants and f and g are arbitrary functions. (b) If (3.93) is true, show that $\hat{A}$ is linear.

3.16 The *Laplace transform operator* $\hat{L}$ is defined by

$$\hat{L}f(x) = \int_0^\infty e^{-px} f(x)\, dx$$

(a) Is $\hat{L}$ linear? (b) Evaluate $\hat{L}(1)$. (c) Evaluate $\hat{L}e^{ax}$, assuming that $p > a$.

3.17 We define the *translation operator* $\hat{T}_h$ by

$$\hat{T}_h f(x) = f(x + h)$$

(a) Is $\hat{T}_h$ a linear operator? (b) Evaluate $(\hat{T}_1^2 - 3\hat{T}_1 + 2)x^2$.

3.18 We define the operator $e^{\hat{A}}$ by the equation

$$e^{\hat{A}} = \hat{1} + \hat{A} + \frac{\hat{A}^2}{2!} + \frac{\hat{A}^3}{3!} + \cdots = \sum_{k=0}^\infty \frac{\hat{A}^k}{k!}$$

Show that $e^{\hat{D}} = \hat{T}_1$, where $\hat{D} = d/dx$ and $\hat{T}_1$ is defined in Problem 3.17.

3.19 Which of the following functions are eigenfunctions of d^2/dx^2? (a) e^x; (b) x^2; (c) $\sin x$; (d) $3 \cos x$; (e) $\sin x + \cos x$. Give the eigenvalue for each eigenfunction.

3.20 Find the eigenfunctions of

$$-\frac{\hbar^2}{2m} \frac{d^2}{dx^2}$$

If the eigenfunctions are to remain finite for $x \to \pm\infty$, what are the allowed eigenvalues?

3.21 Fill in the details leading to (3.36) and (3.37) as the eigenfunctions and eigenvalues of $\hat{p}_x$.

3.22 Give the quantum-mechanical operators for the following physical quantities. (a) p_y^3; (b) $xp_y - yp_x$; (c) $(xp_y - yp_x)^2$.

3.23 Evaluate the commutators (a) $[\hat{x}, \hat{p}_x]$; (b) $[\hat{x}, \hat{p}_x^2]$; (c) $[\hat{x}, \hat{p}_y]$; (d) $[\hat{x}, \hat{V}(x, y, z)]$; (e) $[\hat{x}, \hat{H}]$, where the Hamiltonian operator is given by Eq. (3.45); (f) $[\hat{x}\hat{y}\hat{z}, \hat{p}_x^2]$.

3.24 Write the expression for the probability of finding particle number 1 with its x coordinate between 0 and 2 for (a) a one-particle, one-dimensional system; (b) a one-particle, three-dimensional system; (c) a two-particle, three-dimensional system.

3.25 If ψ is a normalized wave function, what are its SI units for (a) the one-particle, one-dimensional case; (b) the one-particle, three-dimensional case; (c) the n-particle, three-dimensional case?

3.26 An electron in a three-dimensional rectangular box with dimensions of 5.00, 3.00, and 6.00 Å makes a radiative transition from the lowest-lying excited state to the ground state. Calculate the frequency of the photon emitted.

3.27 An electron is in the ground state in a three-dimensional box with V given by (3.60), with $a = 1.00\,\text{nm}$, $b = 2.00\,\text{nm}$, and $c = 5.00\,\text{nm}$. Find the probability that a measurement of the electron's position will find it in the region defined by (a) $0 \leqslant x \leqslant 0.40\,\text{nm}$, $1.50\,\text{nm} \leqslant y \leqslant 2.00\,\text{nm}$, $2.00\,\text{nm} \leqslant z \leqslant 3.00\,\text{nm}$; (b) $0 \leqslant x \leqslant 0.40\,\text{nm}$, $0 \leqslant y \leqslant 2.00\,\text{nm}$, $0 \leqslant z \leqslant 5.00\,\text{nm}$. (c) What is the probability that a measurement of the position of this electron will find it with its x coordinate between 0 and 0.40 nm?

3.28 The stationary-state wave functions of a particle in a three-dimensional rectangular box are eigenfunctions of which of these operators? (a) $\hat{p}_x$; (b) $\hat{p}_x^2$; (c) $\hat{p}_z^2$; (d) $\hat{x}$. For each of these operators that $\psi_{n_x n_y n_z}$ is an eigenfunction of, state in terms of n_x, n_y, and n_z what value will be observed when the corresponding property is measured.

3.29 Prove the multiple-integral identity (3.74).

3.30 Explain how degeneracy can occur for a particle in a rectangular box with $a \neq b \neq c$.

3.31 Solve the one-particle, three-dimensional time-independent Schrödinger equation for the free particle.

3.32 If ψ is an unnormalized wave function and N is a constant such that $N\psi$ is normalized, express $|N|$ in terms of ψ.

3.33 The terms *state* and *energy level* are not synonymous in quantum mechanics. For the particle in a cubic box, consider the energy range $E < 15h^2/8ma^2$. (a) How many states lie in this range? (b) How many energy levels lie in this range?

3.34 For the particle in a cubic box, what is the degree of degeneracy of the energy levels with the following values of $8ma^2E/h^2$? (a) 12; (b) 14; (c) 27.

3.35 Which of the following are sets of linearly independent functions? (a) x, x^2, x^6; (b) 8, x, x^2, $3x^2 - 1$; (c) $\sin x$, $\cos x$; (d) $\sin z$, $\cos z$, $\tan z$; (e) $\sin x$, $\cos x$, e^{ix}; (f) $\sin^2 x$, $\cos^2 x$, 1; (g) $\sin^2 x$, $\cos^2 y$, 1.

3.36 For the particle confined to a box with dimensions a, b, and c, find the following values for the state with quantum numbers n_x, n_y, n_z. (a) $\langle x \rangle$; (b) $\langle y \rangle$, $\langle z \rangle$. Use symmetry considerations and the answer to part a. (c) $\langle p_x \rangle$; (d) $\langle x^2 \rangle$. Is $\langle x^2 \rangle = \langle x \rangle^2$? Is $\langle xy \rangle = \langle x \rangle \langle y \rangle$?

3.37 Which of the following functions, when multiplied by a normalization constant, would be acceptable one-dimensional wave functions for a bound particle? (a and b are positive constants.) (a) e^{-ax}; (b) e^{-bx^2}; (c) xe^{-bx^2}; (d) ie^{-bx^2}; (e) $f(x) = e^{-bx^2}$ for $x < 0$, $f(x) = 2e^{-bx^2}$ for $x \geqslant 0$.

3.38 Show that if Ψ_1 and Ψ_2 satisfy the time-dependent Schrödinger equation then $c_1\Psi_1 + c_2\Psi_2$ satisfies this equation, where c_1 and c_2 are constants.

3.39 True or false? (a) If g is an eigenfunction of the linear operator $\hat{B}$, then cg is an eigenfunction of $\hat{B}$, where c is an arbitrary constant. (b) If we measure the property B when the system's state function is not an eigenfunction of $\hat{B}$, then we can get a result that is not an eigenvalue of $\hat{B}$. (c) If f_1 and f_2 are eigenfunctions of $\hat{B}$, then $c_1 f_1 + c_2 f_2$ must be an eigenfunction of $\hat{B}$, where c_1 and c_2 are constants. (d) The state function Ψ must be an eigenfunction of each operator $\hat{B}$ that represents a physical property of the system. (e) A linear combination of two solutions to the time-independent Schrödinger equation must be a solution of this equation. (f) The system's state function Ψ must be an eigenfunction of $\hat{H}$.

3.40 (a) Write a computer program that will find all sets of positive integers n_x, n_y, n_z for which $n_x^2 + n_y^2 + n_z^2 \leqslant 60$, will print the sets in order of increasing $n_x^2 + n_y^2 + n_z^2$, and will print the $n_x^2 + n_y^2 + n_z^2$ values. (b) What is the degeneracy of the particle-in-a-cubic-box level with $n_x^2 + n_y^2 + n_z^2 = 54$?

4

The Harmonic Oscillator

4.1 POWER-SERIES SOLUTION OF DIFFERENTIAL EQUATIONS

So far we have considered only cases where the potential energy $V(x)$ is a constant. This makes the Schrödinger equation a second-order linear homogeneous differential equation with *constant* coefficients, which we know how to solve. However, we want to deal with cases in which V varies with x. A useful approach in this case is to attempt a power-series solution of the Schrödinger equation.

To illustrate the method, consider the differential equation

$$y''(x) + c^2 y(x) = 0 \tag{4.1}$$

where c^2 is a real, positive number. Of course, this differential equation has *constant* coefficients, but we can solve it with the power-series method if we want. Let us first obtain the solution by using the auxiliary equation, which is $s^2 + c^2 = 0$. We find $s = \pm ic$. Recalling the work in Section 2.2 [Eqs. (2.10) and (4.1) are the same], we get trigonometric solutions when the roots of the auxiliary equation are pure imaginary:

$$y = A \cos(cx) + B \sin(cx) \tag{4.2}$$

where A and B are the constants of integration. A different form of (4.2) is sometimes useful, namely

$$y = D \sin(cx + e) \tag{4.3}$$

where D and e are arbitrary constants. Using the formula for the sine of the sum of two angles, we can show that (4.3) is equivalent to (4.2).

Now let us solve (4.1) using the power-series method. We start by assuming that the solution can be expanded in a Taylor series (see Problem 4.1) about $x = 0$; that is, we assume that

$$y(x) = \sum_{n=0}^{\infty} a_n x^n = a_0 + a_1 x + a_2 x^2 + a_3 x^3 + \cdots \tag{4.4}$$

where the a's are coefficients to be determined so as to satisfy (4.1). Differentiating (4.4), we have

$$y'(x) = a_1 + 2a_2 x + 3a_3 x^2 + \cdots = \sum_{n=1}^{\infty} n a_n x^{n-1} \tag{4.5}$$

where we have assumed that term-by-term differentiation is valid for the series. (This is not always true for infinite series.) For y'', we have

$$y''(x) = 2a_2 + 3(2)a_3x + \cdots = \sum_{n=2}^{\infty} n(n-1)a_nx^{n-2} \qquad (4.6)$$

Substituting (4.4) and (4.6) into (4.1), we get

$$\sum_{n=2}^{\infty} n(n-1)a_nx^{n-2} + \sum_{n=0}^{\infty} c^2a_nx^n = 0 \qquad (4.7)$$

We want to combine the two sums in (4.7). Provided certain conditions are met, we can add two infinite series term by term to get their sum:

$$\sum_{j=0}^{\infty} b_jx^j + \sum_{j=0}^{\infty} c_jx^j = \sum_{j=0}^{\infty} (b_j + c_j)x^j \qquad (4.8)$$

To apply (4.8) to the two sums in (4.7), we want the summation limits in each sum to be the same and the powers of x to be the same. We therefore make a change of summation index in the first sum in (4.7), defining k as $k \equiv n - 2$. The limits $n = 2$ to ∞ correspond to $k = 0$ to ∞, and use of $n = k + 2$ gives

$$\sum_{n=2}^{\infty} n(n-1)a_nx^{n-2} = \sum_{k=0}^{\infty} (k+2)(k+1)a_{k+2}x^k \qquad (4.9)$$

$$\sum_{n=2}^{\infty} n(n-1)a_nx^{n-2} = \sum_{n=0}^{\infty} (n+2)(n+1)a_{n+2}x^n \qquad (4.10)$$

Equation (4.10) follows from (4.9) because the summation index is a ***dummy variable***; it makes no difference what letter we use to denote this variable. An example will make this clear. Consider the sums

$$\sum_{i=1}^{3} c_ix^i \quad \text{and} \quad \sum_{j=1}^{3} c_jx^j \qquad (4.11)$$

Because only the dummy variables in the two sums differ, the sums are equal. This is easy to see if we write them out:

$$\sum_{i=1}^{3} c_ix^i = c_1x + c_2x^2 + c_3x^3 \quad \text{and} \quad \sum_{j=1}^{3} c_jx^j = c_1x + c_2x^2 + c_3x^3$$

In going from (4.9) to (4.10), we simply changed the symbol denoting the summation index from k to n.

The integration variable in a definite integral is also a dummy variable, since the value of a definite integral is unaffected by what letter we use for this variable:

$$\int_a^b f(x)\,dx = \int_a^b f(t)\,dt \qquad (4.12)$$

Using (4.10) in (4.7), we find, after applying (4.8), that

$$\sum_{n=0}^{\infty} [(n+2)(n+1)a_{n+2} + c^2a_n]x^n = 0 \qquad (4.13)$$

If (4.13) is to be true for all values of x, then the coefficient of each power of x must vanish. To see this, consider the equation

$$\sum_{j=0}^{\infty} b_j x^j = 0 \qquad (4.14)$$

Putting $x = 0$ in (4.14) shows that $b_0 = 0$. Taking the first derivative of (4.14) with respect to x and then putting $x = 0$ shows that $b_1 = 0$. Taking the nth derivative and putting $x = 0$ gives $b_n = 0$. Thus, from (4.13), we have

$$(n + 2)(n + 1)a_{n+2} + c^2 a_n = 0 \qquad (4.15)$$

$$a_{n+2} = -\frac{c^2}{(n + 1)(n + 2)} a_n \qquad (4.16)$$

An equation like (4.16) is called a **recursion relation**. Using (4.16), if we know the value of a_0, we can find $a_2, a_4, a_6, \ldots$. If we know a_1, we can find $a_3, a_5, a_7, \ldots$. Since there is no restriction on the values of a_0 and a_1, they are arbitrary constants, which we denote by A and Bc:

$$a_0 = A, \qquad a_1 = Bc \qquad (4.17)$$

Using (4.16), we find for the coefficients

$$a_0 = A, \qquad a_2 = -\frac{c^2 A}{1 \cdot 2}, \qquad a_4 = \frac{c^4 A}{4 \cdot 3 \cdot 2 \cdot 1}, \qquad a_6 = -\frac{c^6 A}{6!}, \ldots$$

$$a_{2k} = (-1)^k \frac{c^{2k} A}{(2k)!}, \qquad k = 0, 1, 2, 3, \ldots \qquad (4.18)$$

$$a_1 = Bc, \qquad a_3 = -\frac{c^3 B}{2 \cdot 3}, \qquad a_5 = \frac{c^5 B}{5 \cdot 4 \cdot 3 \cdot 2}, \qquad a_7 = -\frac{c^7 B}{7!}, \ldots$$

$$a_{2k+1} = (-1)^k \frac{c^{2k+1} B}{(2k + 1)!}, \qquad k = 0, 1, 2, \ldots \qquad (4.19)$$

From (4.4), (4.18), and (4.19), we have

$$y = \sum_{n=0}^{\infty} a_n x^n = \sum_{n=0,2,4,\ldots}^{\infty} a_n x^n + \sum_{n=1,3,5,\ldots}^{\infty} a_n x^n$$

$$y = A \sum_{k=0}^{\infty} (-1)^k \frac{c^{2k} x^{2k}}{(2k)!} + B \sum_{k=0}^{\infty} (-1)^k \frac{c^{2k+1} x^{2k+1}}{(2k + 1)!} \qquad (4.20)$$

The two series in (4.20) are the Taylor series for $\cos(cx)$ and $\sin(cx)$ (Problem 4.2). Hence, in agreement with (4.2), we have

$$y = A \cos(cx) + B \sin(cx) \qquad (4.21)$$

4.2 THE ONE-DIMENSIONAL HARMONIC OSCILLATOR

In this section we will increase our quantum-mechanical repertoire by solving the Schrödinger equation for the one-dimensional harmonic oscillator. This system is important as a model for molecular vibrations.

Classical-Mechanical Treatment. Before looking at the wave mechanics of the harmonic oscillator, we review the classical treatment. We have a single particle of mass m attracted toward the origin by a force proportional to the particle's displacement from the origin:

$$F_x = -kx \qquad (4.22)$$

The proportionality constant k is called the **force constant**. F_x is the x component of the force on the particle. This is also the total force in this one-dimensional problem. Equation (4.22) is obeyed by a particle attached to a spring, provided the spring is not stretched greatly from its equilibrium position.

Newton's second law, $F = ma$, gives

$$-kx = m\,\frac{d^2x}{dt^2} \qquad (4.23)$$

where t is the time. Equation (4.23) is the same as Eq. (4.1) with $c^2 = k/m$; hence the solution is [Eq. (4.3)]

$$x = A\sin\left(2\pi\nu t + b\right) \qquad (4.24)$$

where A (the **amplitude** of the vibration) and b are the integration constants, and the **vibration frequency** ν is

$$\nu = \frac{1}{2\pi}\left(\frac{k}{m}\right)^{1/2} \qquad \mathbf{(4.25)^*}$$

Now consider the energy. The potential energy V is related to the components of force in the three-dimensional case by

$$F_x = -\frac{\partial V}{\partial x}, \qquad F_y = -\frac{\partial V}{\partial y}, \qquad F_z = -\frac{\partial V}{\partial z} \qquad \mathbf{(4.26)^*}$$

Equation (4.26) is the definition of potential energy. Since we have a one-dimensional problem, we have [Eq. (1.11)]

$$F_x = -\frac{dV}{dx} = -kx \qquad (4.27)$$

Integration of (4.27) gives $V = \int kx\,dx = \frac{1}{2}kx^2 + C$, where C is a constant. The potential energy always has an arbitrary additive constant. Choosing $C = 0$, we have [Eq. (4.25)]

$$V = \tfrac{1}{2}kx^2 \qquad \mathbf{(4.28)^*}$$

$$V = 2\pi^2\nu^2 mx^2 \qquad (4.29)$$

The graph of $V(x)$ is a parabola (Fig. 4.4). The kinetic energy T is

$$T = \tfrac{1}{2}m(dx/dt)^2 \qquad (4.30)$$

and can be found by differentiating (4.24) with respect to x. Adding T and V, one finds for the total energy (Problem 4.4)

$$E = T + V = \tfrac{1}{2}kA^2 = 2\pi^2\nu^2 mA^2 \qquad (4.31)$$

where the identity $\sin^2\theta + \cos^2\theta = 1$ was used.

Quantum-Mechanical Treatment. The harmonic-oscillator Hamiltonian operator is

$$\hat{H} = \hat{T} + \hat{V} = -\frac{\hbar^2}{2m}\frac{d^2}{dx^2} + 2\pi^2\nu^2 mx^2 = -\frac{\hbar^2}{2m}\left(\frac{d^2}{dx^2} - \alpha^2 x^2\right) \qquad (4.32)$$

where, to save time in writing, we defined α as

$$\alpha \equiv 2\pi\nu m/\hbar \qquad (4.33)$$

The Schrödinger equation $\hat{H}\psi = E\psi$ reads, after multiplication by $2m/\hbar^2$,

$$\frac{d^2\psi}{dx^2} + (2mE\hbar^{-2} - \alpha^2 x^2)\psi = 0 \qquad (4.34)$$

We might now attempt a power-series solution of (4.34) using the methods of the last section. If we do now try a power series for ψ of the form (4.4), we shall find that it leads to a three-term recursion relation, which is much more difficult to deal with than a two-term recursion relation like Eq. (4.15). We therefore attempt to modify the form of (4.34) so as to get a two-term recursion relation when we try a series solution. A substitution that will achieve this purpose is (see Problem 4.20) $f(x) \equiv e^{\alpha x^2/2}\psi(x)$. Thus

$$\psi = e^{-\alpha x^2/2}f(x) \qquad (4.35)$$

This equation is simply the definition of a new function $f(x)$ that replaces $\psi(x)$ as the unknown function to be solved for. (We can make any substitution we please in a differential equation.) Differentiating (4.35) twice, we have

$$\psi'' = e^{-\alpha x^2/2}(f'' - 2\alpha x f' - \alpha f + \alpha^2 x^2 f) \qquad (4.36)$$

Substituting (4.35) and (4.36) into (4.34), we find

$$f''(x) - 2\alpha x f'(x) + (2mE\hbar^{-2} - \alpha)f(x) = 0 \qquad (4.37)$$

Now we try a series solution for $f(x)$:

$$f(x) = \sum_{n=0}^{\infty} c_n x^n \qquad (4.38)$$

Assuming the validity of term-by-term differentiation of (4.38), we get

$$f'(x) = \sum_{n=1}^{\infty} nc_n x^{n-1} = \sum_{n=0}^{\infty} nc_n x^{n-1} \qquad (4.39)$$

[The first term in the second sum in (4.39) is zero.]   Also,

$$f''(x) = \sum_{n=2}^{\infty} n(n-1)c_n x^{n-2} = \sum_{j=0}^{\infty} (j+2)(j+1)c_{j+2}x^j$$

$$= \sum_{n=0}^{\infty} (n+2)(n+1)c_{n+2}x^n \qquad (4.40)$$

where we made the substitution $j = n - 2$ and then changed the summation index from j to n. [Compare Eqs. (4.9) and (4.10).] Substitution into (4.37) gives

$$\sum_{n=0}^{\infty} (n+2)(n+1)c_{n+2}x^n - 2\alpha \sum_{n=0}^{\infty} nc_n x^n + (2mE\hbar^{-2} - \alpha) \sum_{n=0}^{\infty} c_n x^n = 0$$

$$\sum_{n=0}^{\infty} [(n+2)(n+1)c_{n+2} - 2\alpha nc_n + (2mE\hbar^{-2} - \alpha)c_n]x^n = 0 \qquad (4.41)$$

Setting the coefficient of x^n equal to zero [for the same reason as in Eq. (4.13)], we have

$$c_{n+2} = \frac{\alpha + 2\alpha n - 2mE\hbar^{-2}}{(n+1)(n+2)} c_n \qquad (4.42)$$

which is the desired two-term recursion relation. Equation (4.42) has the same form as (4.16), in that knowing c_n we can calculate c_{n+2}; we thus have two arbitrary constants: c_0 and c_1. If we set c_1 equal to zero, then we will have as a solution a power series containing only even powers of x, multiplied by the exponential factor:

$$\psi = e^{-\alpha x^2/2}f(x) = e^{-\alpha x^2/2} \sum_{n=0,2,4,\ldots}^{\infty} c_n x^n = e^{-\alpha x^2/2} \sum_{l=0}^{\infty} c_{2l}x^{2l} \qquad (4.43)$$

If we set c_0 equal to zero, we get another independent solution:

$$\psi = e^{-\alpha x^2/2} \sum_{n=1,3,\ldots}^{\infty} c_n x^n = e^{-\alpha x^2/2} \sum_{l=0}^{\infty} c_{2l+1}x^{2l+1} \qquad (4.44)$$

The general solution of the Schrödinger equation is a linear combination of these two independent solutions [recall Eq. (2.4)]:

$$\psi = Ae^{-\alpha x^2/2} \sum_{l=0}^{\infty} c_{2l+1}x^{2l+1} + Be^{-\alpha x^2/2} \sum_{l=0}^{\infty} c_{2l}x^{2l} \qquad (4.45)$$

where A and B are arbitrary constants.

We now must see if the boundary conditions on the wave function lead to any restrictions on the solution. To see how the two infinite series behave for large x, we examine the ratio of successive coefficients in each series. The ratio of the coefficient of x^{2l+2} to that of x^{2l} in the second series is [set $n = 2l$ in Eq. (4.42)]

$$\frac{c_{2l+2}}{c_{2l}} = \frac{\alpha + 4\alpha l - 2mE\hbar^{-2}}{(2l+1)(2l+2)} \qquad (4.46)$$

Assuming that for large values of x the later terms in the series are the dominant ones, we look at the ratio (4.46) for large values of l:

$$\frac{c_{2l+2}}{c_{2l}} \sim \frac{4\alpha l}{(2l)(2l)} = \frac{\alpha}{l}, \qquad \text{for } l \text{ large} \qquad (4.47)$$

Setting $n = 2l + 1$ in (4.42), we find that for large l the ratio of successive coefficients in the first series is also α/l. Now consider the power-series expansion for the function $e^{\alpha x^2}$. Using (Problem 4.3)

$$e^z = \sum_{n=0}^{\infty} \frac{z^n}{n!} = 1 + z + \frac{z^2}{2!} + \cdots \qquad (4.48)$$

we get

$$e^{\alpha x^2} = 1 + \alpha x^2 + \cdots + \frac{\alpha^l x^{2l}}{l!} + \frac{\alpha^{l+1} x^{2l+2}}{(l+1)!} + \cdots \tag{4.49}$$

The ratio of the coefficients of x^{2l+2} and x^{2l} in this series is

$$\frac{\alpha^{l+1}}{(l+1)!} \div \frac{\alpha^l}{l!} = \frac{\alpha}{l+1} \sim \frac{\alpha}{l}, \qquad \text{for large } l$$

Thus the ratio of successive coefficients in each of the infinite series in the solution (4.45) is the same as in the series for $e^{\alpha x^2}$ for large l. We conclude that, for large x, each series goes as $e^{\alpha x^2}$. [This is not a rigorous proof. A proper mathematical derivation is given in H. A. Buchdahl, *Am. J. Phys.*, **42**, 47 (1974); see also M. Bowen and J. Coster, *Am. J. Phys.*, **48**, 307 (1980).]

If each series goes as $e^{\alpha x^2}$, then (4.45) shows that ψ will behave as $e^{\alpha x^2/2}$ for large x. The wave function will become infinite as x goes to infinity and will not be quadratically integrable. If we could somehow break off the series after a finite number of terms, then the factor $e^{-\alpha x^2/2}$ would ensure that ψ went to zero as x became infinite. (Using l'Hospital's rule, it is easy to show that $x^p e^{-\alpha x^2/2}$ goes to zero as $x \rightarrow \infty$, where p is any finite power.) To have one of the series break off after a finite number of terms, the coefficient of c_n in the recursion relation (4.42) must become zero for some value of n, say for $n = v$. This makes $c_{v+2}, c_{v+4}, \ldots$ all equal to zero, and one of the series in (4.45) will have a finite number of terms. In the recursion relation (4.42), there is one quantity whose value is not yet fixed, but can be adjusted to make the coefficient of c_v vanish; this quantity is the energy E. Setting the coefficient of c_v equal to zero in (4.42) and using (4.33), we get

$$\alpha + 2\alpha v - 2mE\hbar^{-2} = 0$$

$$2mE\hbar^{-2} = (2v+1)2\pi\nu m\hbar^{-1}$$

$$E = (v + \tfrac{1}{2})h\nu, \qquad v = 0, 1, 2, \ldots \tag{4.50}*$$

The harmonic-oscillator stationary-state energy levels (4.50) are equally spaced (Fig. 4.1). Do not confuse the quantum number v (vee) with the vibrational frequency ν (nu).

Substitution of (4.50) into the recursion relation (4.42) gives

$$c_{n+2} = \frac{2\alpha(n - v)}{(n+1)(n+2)} c_n \tag{4.51}$$

By quantizing the energy according to (4.50), we have made one of the series break off after a finite number of terms. To get rid of the other infinite series in (4.45), we must set the arbitrary constant that multiplies it equal to zero. This leaves us with a wave function that is $e^{-\alpha x^2/2}$ times a finite power series containing only even or only odd powers of x, depending on whether v is even or odd, respectively. The highest power of x in this power series is x^v, since we chose E to make $c_{v+2}, c_{v+4}, \ldots$ all vanish. The wave functions (4.45) are thus

$$\psi_v = \begin{cases} e^{-\alpha x^2/2}(c_0 + c_2 x^2 + \cdots + c_v x^v), & \text{for } v \text{ even} \\ e^{-\alpha x^2/2}(c_1 x + c_3 x^3 + \cdots + c_v x^v), & \text{for } v \text{ odd} \end{cases} \tag{4.52}$$

Figure 4.1 Lowest five energy levels for the harmonic oscillator.

where the arbitrary constants A and B in (4.45) can be absorbed into c_1 and c_0, respectively, and can therefore be omitted. The coefficients after c_0 and c_1 are found from the recursion relation (4.51). Since the quantum number v occurs in the recursion relation, we obtain a different set of coefficients c_i for each different v.

As in the particle in a box, it is the boundary conditions that force us to quantize the energy. For values of E that differ from (4.50), ψ is not quadratically integrable. For example, Fig. 4.2 plots ψ of Eq. (4.43) for the values $E/h\nu = 0.499, 0.500$, and 0.501, where the recursion relation (4.42) is used to calculate the coefficients c_n (see also Problem 4.23).

The harmonic-oscillator ground-state energy is nonzero; this energy, $\frac{1}{2}h\nu$, is called the **zero-point energy**. This would be the vibrational energy of a harmonic oscillator in a collection of harmonic oscillators at a temperature of absolute zero. The zero-point energy can be understood from the uncertainty principle. If the lowest state had an energy of zero, both its potential and kinetic energies (which are nonnegative) would have to be zero. Zero kinetic energy would mean that the momentum was exactly zero, and Δp_x would be zero. Zero potential energy would mean that the particle was always located at the origin, and Δx would be zero. But we cannot have both Δx and Δp_x equal to zero. Hence the necessity for a nonzero ground-state energy. Similar considerations apply for the particle in a box.

Even and Odd Functions.   Before considering the wave functions in detail, we define even and odd functions. If $f(x)$ satisfies

$$f(-x) = f(x) \qquad \text{(4.53)}^*$$

then f is an **even function** of x. Thus x^2 and e^{-bx^2} are both even functions of x since $(-x)^2 = x^2$ and $e^{-b(-x)^2} = e^{-bx^2}$. The graph of an even function is symmetric about

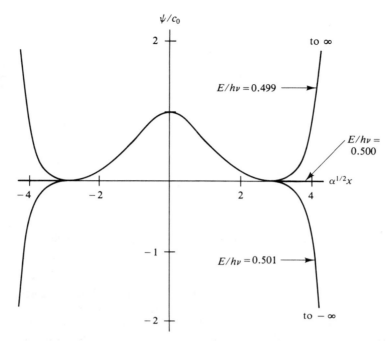

Figure 4.2 Plots of the harmonic-oscillator Schrödinger-equation solution containing only even powers of x for $E = 0.499h\nu$, $E = 0.500h\nu$, and $E = 0.501h\nu$. (In the region around $x = 0$, the three curves nearly coincide.)

the y axis (for example, see Fig. 4.3a); hence

$$\int_{-a}^{+a} f(x)\, dx = 2 \int_{0}^{a} f(x)\, dx , \qquad \text{for } f(x) \text{ even} \qquad \textbf{(4.54)}^*$$

If $g(x)$ satisfies

$$g(-x) = -g(x) \qquad\qquad\qquad\qquad \textbf{(4.55)}^*$$

then g is an ***odd function*** of x. Examples are x, $1/x$, and xe^{x^2}. Setting $x = 0$ in (4.55), we see that an odd function must be zero at $x = 0$, provided $g(0)$ is defined and single valued. The graph of an odd function has the general appearance of Fig. 4.3b. Because positive contributions on one side of the y axis are canceled by corresponding negative contributions on the other side, we have

$$\int_{-a}^{+a} g(x)\, dx = 0 , \qquad \text{for } g(x) \text{ odd} \qquad \textbf{(4.56)}^*$$

It is easy to show that the product of two even functions or of two odd functions is an even function, while the product of an even and an odd function is an odd function.

The Harmonic-Oscillator Wave Functions. The exponential factor $e^{-\alpha x^2/2}$ in (4.52) is an even function of x. If v is an even number, the polynomial factor contains only even powers of x, which makes ψ_v an even function. If v is odd, the polynomial factor contains only odd powers of x, and ψ_v, being the product of an even function and an odd function, is an odd function. Each harmonic-oscillator stationary state ψ is either an even or odd function according to whether the quantum number v is even or odd. In Section 7.5, we shall see that when the potential energy V is an even function the wave functions of nondegenerate levels must be either even or odd functions.

We now find the explicit forms of the wave functions of the lowest three levels. For the $v = 0$ ground state, Eq. (4.52) gives

$$\psi_0 = c_0 e^{-\alpha x^2/2} \tag{4.57}$$

where the subscript on ψ gives the value of v. We fix c_0 by normalization:

$$1 = \int_{-\infty}^{\infty} |c_0|^2 e^{-\alpha x^2} \, dx = 2|c_0|^2 \int_0^{\infty} e^{-\alpha x^2} \, dx$$

where Eq. (4.54) has been used. Using the integral (A.8) in the Appendix, we find

$$|c_0| = (\alpha/\pi)^{1/4} \tag{4.58}$$

$$\psi_0 = (\alpha/\pi)^{1/4} e^{-\alpha x^2/2} \tag{4.59}$$

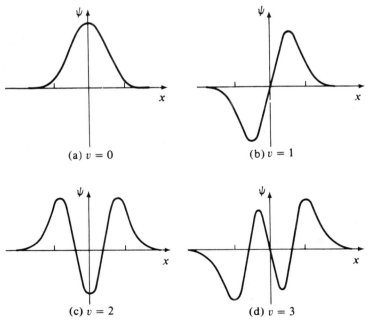

(a) $v = 0$ (b) $v = 1$

(c) $v = 2$ (d) $v = 3$

Figure 4.3 Harmonic-oscillator wave functions. The same scale is used for all graphs. The points marked on the x axes are for $\alpha^{1/2}x = \pm 2$.

if we choose the phase of the normalization constant to be zero. The wave function (4.59) is a Gaussian function (Fig. 4.3a).

For the $v = 1$ state, Eq. (4.52) gives

$$\psi_1 = c_1 x e^{-\alpha x^2/2} \tag{4.60}$$

After normalization using the integral in Eq. (A.9), we have

$$\psi_1 = (4\alpha^3/\pi)^{1/4} x e^{-\alpha x^2/2} \tag{4.61}$$

ψ_1 is graphed in Fig. 4.3b.

For $v = 2$, Eq. (4.52) gives

$$\psi_2 = (c_0 + c_2 x^2) e^{-\alpha x^2/2} \tag{4.62}$$

recursion gives you the constants The recursion relation (4.51) with $v = 2$ gives

$$c_2 = \frac{2\alpha(-2)}{1 \cdot 2} c_0 = -2\alpha c_0 \tag{4.63}$$

Hence

$$\psi_2 = c_0(1 - 2\alpha x^2) e^{-\alpha x^2/2} \tag{4.64}$$

Evaluating c_0 by normalization, we find (Problem 4.10)

$$\psi_2 = (\alpha/4\pi)^{1/4}(2\alpha x^2 - 1) e^{-\alpha x^2/2} \tag{4.65}$$

The number of nodes in the wave function equals the quantum number v. It can be proved (see *Messiah*, pages 109–110) that in a one-dimensional problem the number of nodes interior to the boundary points is zero for the ground-state ψ and increases by one for each successive excited state. The boundary points for the harmonic oscillator are $\pm\infty$.

The polynomial factors in the harmonic-oscillator wave functions are well known in mathematics and are called *Hermite polynomials*, after a French mathematician. (See Problem 4.18.)

According to the quantum-mechanical solution, there is some probability of finding the particle at any point on the x axis (except at the nodes). Classically, the particle is confined to the region where the potential energy does not exceed the total energy; this is the region from $-a$ to $+a$ in Fig. 4.4. It might seem that, by saying the particle can be found outside the classically allowed region, we are allowing it to have negative kinetic energy. Actually, there is no paradox in the quantum-mechanical view. To verify that the particle is in the **classically forbidden region** (where $V > E$), we must measure its position. This measurement changes the state of the system (Sections 1.3 and 1.4); the interaction of the oscillator with the measuring apparatus transfers sufficient energy to the oscillator for it to be in the classically forbidden region. An accurate measurement of x introduces a large uncertainty in the momentum and hence in the kinetic energy. Penetration of classically forbidden regions was previously discussed in Sections 2.4 and 2.5.

Note from Fig. 4.3 that ψ oscillates in the classically allowed region [which is $-(2v + 1)^{1/2} \leq \alpha^{1/2} x \leq (2v + 1)^{1/2}$; Problem 4.22] and decreases exponentially to zero in the classically forbidden region. We previously saw this behavior for the particle in a rectangular well (Section 2.4).

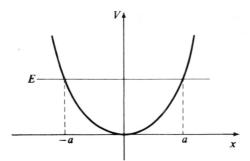

Figure 4.4 The classically allowed ($|x| \leq a$) and forbidden ($|x| > a$) regions for the harmonic oscillator.

Figure 4.3 shows that, as we go to higher-energy states of the harmonic oscillator, ψ and $|\psi|^2$ tend to have maxima farther and farther from the origin. Since $V = \frac{1}{2}kx^2$ increases as we go farther from the origin, the average potential energy $\langle V \rangle = \int_{-\infty}^{\infty} |\psi|^2 V \, dx$ increases as the quantum number increases. The average kinetic energy is $\langle T \rangle = -(\hbar^2/2m) \int_{-\infty}^{\infty} \psi^* \psi'' \, dx$. Integration by parts gives (Problem 7.4b) $\langle T \rangle = (\hbar^2/2m) \int_{-\infty}^{\infty} |d\psi/dx|^2 \, dx$. The higher number of nodes in states with higher quantum number produces a faster rate of change of ψ, so $\langle T \rangle$ increases as the quantum number increases.

4.3 VIBRATION OF MOLECULES

We shall see in Section 13.1 that to an excellent approximation one can treat separately the motions of the electrons and the motions of the nuclei of a molecule. (This is due to the much heavier mass of the nuclei.) One first imagines the nuclei to be held stationary and solves a Schrödinger equation for the electronic energy U. (U also includes the energy of nuclear repulsion.) For a diatomic (two-atom) molecule, the electronic energy U depends on the distance R between the nuclei, $U = U(R)$, and the U versus R curve has the typical appearance of Fig. 13.1.

After finding $U(R)$, one then solves a Schrödinger equation for nuclear motion, using $U(R)$ as the potential energy for nuclear motion. For a diatomic molecule, the nuclear Schrödinger equation is a two-particle equation. We shall see in Section 6.3 that, when the potential energy of a two-particle system depends only on the distance between the particles, the energy of the system is the sum of (a) the kinetic energy of translational motion of the entire system through space and (b) the energy of internal motion of the particles relative to each other. The classical expression for the two-particle internal-motion energy turns out to be the sum of the potential energy of interaction between the particles and the kinetic energy of a hypothetical particle whose mass is $m_1 m_2/(m_1 + m_2)$ (where m_1 and m_2 are the masses of the two particles) and whose coordinates are the coordinates of one particle relative to the other. The quantity $m_1 m_2/(m_1 + m_2)$ is called the **reduced mass** μ.

The internal motion of a diatomic molecule consists of *vibration*, corresponding to a change in the distance R between the two nuclei, and *rotation*, corresponding to a change in the spatial orientation of the line joining the nuclei. To a good approximation, one can usually treat the vibrational and rotational motions separately. The rotational energy levels are found in Section 6.4. Here we consider the vibrational levels.

The Schrödinger equation for the vibration of a diatomic molecule has a kinetic-energy operator for the hypothetical particle of mass $\mu = m_1 m_2 / (m_1 + m_2)$ and a potential-energy term given by $U(R)$. If we place the origin to coincide with the minimum point of the U curve in Fig. 13.1 and take the zero of potential energy at the energy of this minimum point, then the lower portion of the $U(R)$ curve will nearly coincide with the potential-energy curve of a harmonic oscillator with the appropriate force constant k (see Fig. 4.5 and Problem 4.28). The minimum in the $U(R)$ curve occurs at the *equilibrium distance* R_e between the nuclei. In Fig. 4.5, x is the deviation of the internuclear distance from its equilibrium value: $x \equiv R - R_e$.

The harmonic-oscillator force constant k in Eq. (4.28) is obtained as $k = d^2 V/dx^2$, and the harmonic-oscillator curve essentially coincides with the $U(R)$ curve at $R = R_e$, so the molecular force constant is $k = d^2 U/dR^2|_{R=R_e}$ (see also Problem 4.28). Differences in nuclear mass have virtually no effect on the electronic-energy curve $U(R)$, so different isotopic species of the same molecule have essentially the same force constant k.

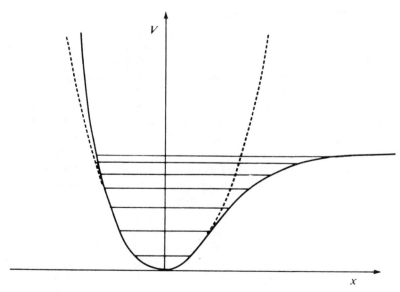

Figure 4.5 Potential energy for vibration of a diatomic molecule (solid curve) and for a harmonic oscillator (dashed curve). Also shown are the bound-state vibrational energy levels for the diatomic molecule. In contrast to the harmonic oscillator, a diatomic molecule has only a finite number of bound vibrational levels.

We expect, therefore, that a reasonable approximation to the vibrational energy levels E_{vib} of a diatomic molecule would be the harmonic-oscillator vibrational energy levels; Eqs. (4.50) and (4.25) give

$$E_{\text{vib}} \approx (v + \tfrac{1}{2})h\nu_e, \qquad v = 0, 1, 2, \ldots \qquad \textbf{(4.66)}^*$$

$$\nu_e = \frac{1}{2\pi}\left(\frac{k}{\mu}\right)^{1/2}, \qquad \mu = \frac{m_1 m_2}{m_1 + m_2}, \qquad k = \frac{d^2 U}{dR^2}\bigg|_{R=R_e} \qquad \textbf{(4.67)}^*$$

ν_e is called the **equilibrium** (or **harmonic**) **vibrational frequency**. This approximation is best for the lower vibrational levels. As v increases, the nuclei spend more and more time in regions far from their equilibrium separation. For such regions the potential energy deviates substantially from that of a harmonic oscillator and the harmonic-oscillator approximation is poor. Instead of being equally spaced, one finds that the vibrational levels of a diatomic molecule come closer and closer together as v increases (Fig. 4.5). Eventually, the vibrational energy is sufficiently large to dissociate the diatomic molecule into atoms that are not bound to each other. A more accurate expression for the molecular vibrational energy that allows for the anharmonicity of the vibration is

$$E_{\text{vib}} = (v + \tfrac{1}{2})h\nu_e - (v + \tfrac{1}{2})^2 h\nu_e x_e \qquad (4.68)$$

where the *anharmonicity constant* $\nu_e x_e$ is positive in nearly all cases.

Using the time-dependent Schrödinger equation, one finds (see Section 9.10) that the most probable vibrational transitions when a diatomic molecule is exposed to electromagnetic radiation are those where v changes by ± 1; furthermore, for absorption or emission of electromagnetic radiation to occur, the vibration must change the molecule's dipole moment. Hence homonuclear diatomics (H_2, N_2, Cl_2, and the like) cannot undergo transitions between vibrational levels by absorption or emission of radiation. (Such transitions can occur during intermolecular collisions.) The relation $E_{\text{upper}} - E_{\text{lower}} = h\nu$, the approximate equation (4.66), and the **selection rule** $\Delta v = 1$ for absorption of radiation show that a heteronuclear diatomic molecule whose vibrational frequency is ν_e will most strongly absorb light of frequency ν_{light} given approximately by

$$\nu_{\text{light}} = (E_2 - E_1)/h \approx [(v_2 + \tfrac{1}{2})h\nu_e - (v_1 + \tfrac{1}{2})h\nu_e]/h = (v_2 - v_1)\nu_e = \nu_e \qquad (4.69)$$

The values of k and μ in (4.67) for diatomic molecules are such that ν_{light} usually falls in the infrared region of the spectrum. Transitions with $\Delta v = 2, 3, \ldots$ also occur, but these (called *overtones*) are much weaker than the $\Delta v = 1$ absorption.

Use of the more accurate equation (4.68) gives (Problem 4.25)

$$\nu_{\text{light}} = \nu_e - 2\nu_e x_e(v_1 + 1) \qquad (4.70)$$

where v_1 is the quantum number of the lower level and $\Delta v = 1$.

The relative population of two molecular energy levels is given by the **Boltzmann distribution law** (see any physical chemistry text) as

$$\frac{N_i}{N_j} = \frac{g_i}{g_j}\, e^{-(E_i - E_j)/kT} \qquad \textbf{(4.71)}^*$$

where energy levels i and j have energies E_i and E_j and degeneracies g_i and g_j and are populated by N_i and N_j molecules, and where k is Boltzmann's constant and T the absolute temperature. For a nondegenerate level, $g_i = 1$.

The magnitude of $\nu = (1/2\pi)(k/\mu)^{1/2}$ is such that for light diatomics (for example, H_2, HCl, CO) only the $v = 0$ vibrational level is significantly populated at room temperature. For heavy diatomics (for example, I_2), there is significant room-temperature population of one or more excited vibrational levels.

The vibrational absorption spectrum of a polar diatomic molecule consists of a $v = 0 \rightarrow 1$ band, much weaker overtone bands ($v = 0 \rightarrow 2$, $0 \rightarrow 3, \ldots$), and, if there is significant population of $v > 0$ levels, *hot bands* such as $v = 1 \rightarrow 2$, $2 \rightarrow 3$. Each band corresponding to a particular vibrational transition consists of several closely spaced lines; each such line corresponds to a different change in rotational state simultaneous with the change in vibrational state; each line is the result of a vibration–rotation transition.

The SI unit for spectroscopic frequencies is the **hertz** (Hz), defined by $1 \, \text{Hz} \equiv 1 \, \text{s}^{-1}$. Multiples such as the megahertz (MHz) equal to 10^6 Hz and the gigahertz (GHz) equal to 10^9 Hz are often used. Infrared absorption lines are usually specified by giving their **wave number** $\tilde{\nu}$, defined as

$$\tilde{\nu} \equiv 1/\lambda = \nu/c \tag{4.72}$$

where λ is the wavelength in vacuum.

In the harmonic-oscillator approximation, the quantum-mechanical energy levels of a polyatomic molecule turn out to be $E_{\text{vib}} = \Sigma_i \, (v_i + \frac{1}{2})h\nu_i$, where the ν_i's are the frequencies of the normal modes of vibration of the molecule and v_i is the vibrational quantum number of the ith normal mode. Each v_i takes on the values $0, 1, 2, \ldots$ independently of the values of the other vibrational quantum numbers. A linear molecule with n atoms has $3n - 5$ normal modes; a nonlinear molecule has $3n - 6$ normal modes. (See *Levine, Molecular Spectroscopy*, Chapter 6 for details.)

To calculate the reduced mass μ in (4.67), one needs the masses of isotopic species. Some relative isotopic masses are listed in Table A.3 in the Appendix.

EXAMPLE   The strongest infrared band of $^{12}\text{C}^{16}\text{O}$ occurs at $\tilde{\nu} = 2143 \, \text{cm}^{-1}$. Find the force constant of $^{12}\text{C}^{16}\text{O}$. State any approximation made.

The strongest infrared band corresponds to the $v = 0 \rightarrow 1$ transition. We approximate the molecular vibration as that of a harmonic oscillator. From (4.69), the equilibrium molecular vibrational frequency is approximately

$$\nu_e \approx \nu_{\text{light}} = \tilde{\nu}c = (2143 \, \text{cm}^{-1})(2.9979 \times 10^{10} \, \text{cm/s}) = 6.424 \times 10^{13} \, \text{s}^{-1}$$

To relate k to ν_e in (4.67), we need the reduced mass $\mu = m_1 m_2/(m_1 + m_2)$. One mole of ^{12}C has a mass of 12 g and contains Avogadro's number of atoms. Hence the mass of one atom of ^{12}C is $(12 \, \text{g})/(6.02214 \times 10^{23})$. The reduced mass and force constant are

$$\mu = \frac{12(15.9949) \, \text{g}}{27.9949} \frac{1}{6.02214 \times 10^{23}} = 1.1385 \times 10^{-23} \, \text{g}$$

$$k = 4\pi^2 \nu_e^2 \mu = 4\pi^2 (6.424 \times 10^{13} \, \text{s}^{-1})^2 (1.1385 \times 10^{-23} \, \text{g})$$

$$= 1.855 \times 10^6 \, \text{dyn/cm} = 1855 \, \text{N/m}$$

4.4 SUMMARY

The one-dimensional harmonic oscillator has $V = \frac{1}{2}kx^2$. Its stationary-state energies are $E = (v + \frac{1}{2})h\nu$, where the vibrational frequency is $\nu = (1/2\pi)(k/m)^{1/2}$ and the quantum number v is $v = 0, 1, 2, \ldots$. The eigenfunctions are even or odd functions and are given by (4.52). An even function satisfies $f(-x) = f(x)$. An odd function satisfies $f(-x) = -f(x)$. If f is even, then $\int_{-a}^{a} f(x)\,dx = 2\int_{0}^{a} f(x)\,dx$. If f is odd, then $\int_{-a}^{a} f(x)\,dx = 0$. The vibrational energy of a diatomic molecule can be roughly approximated by the harmonic-oscillator energies with $\nu = (1/2\pi)(k/\mu)^{1/2}$, where the reduced mass is $\mu = m_1 m_2 / (m_1 + m_2)$.

PROBLEMS

4.1 Provided certain conditions are met, we can expand the function $f(x)$ in an infinite power series about the point $x = a$:

$$f(x) = \sum_{n=0}^{\infty} c_n (x - a)^n \tag{4.73}$$

Differentiate (4.73) m times, and then set $x = a$ to show that $c_n = f^{(n)}(a)/n!$, thus giving the familiar **Taylor series**:

$$f(x) = \sum_{n=0}^{\infty} \frac{f^{(n)}(a)}{n!} (x - a)^n \tag{4.74*}$$

4.2 (a) Use (4.74) to derive the first few terms in the Taylor-series expansion about $x = 0$ for the function $\sin x$, and infer the general formula. (b) Differentiate the Taylor series in (a) to obtain the Taylor series for $\cos x$.

4.3 (a) Obtain the Taylor-series expansion about $x = 0$ for e^x. (b) Use the Taylor series (about $x = 0$) of $\sin x$, $\cos x$, and e^x to verify that $e^{i\theta} = \cos\theta + i\sin\theta$ [Eq. (1.26)].

4.4 Derive (4.31) for E of a classical oscillator.

4.5 (a) Obtain the recursion relation for the coefficients c_n in the power-series solution of $(1 - x^2)y''(x) - 2xy'(x) + 3y(x) = 0$. (b) Express c_4 in terms of c_0 and c_5 in terms of c_1.

4.6 Which of the following are even functions? are odd functions? (a) $\sin x$; (b) $\cos x$; (c) $\tan x$; (d) e^x; (e) 13; (f) $x\cosh x$; (g) $2 - 2x$; (h) $(3 + x)(3 - x)$.

4.7 Prove the statements made after Eq. (4.56) about products of even and odd functions.

4.8 What single-valued function is both odd and even?

4.9 (a) If $f(x)$ is an even function that is everywhere differentiable, prove that $f'(x)$ is an odd function. Do not assume that $f(x)$ can be expanded in a Taylor series. (b) Prove that the derivative of an everywhere-differentiable odd function is an even function. (c) If $f(x)$ is an even function that is differentiable at the origin, find $f'(0)$.

4.10 Verify the normalization factors for the $v = 1$ and $v = 2$ harmonic-oscillator wave functions.

4.11 For the ground state of the one-dimensional harmonic oscillator, find the average value of the kinetic energy and of the potential energy; verify that $\langle T \rangle = \langle V \rangle$ in this case.

4.12 Use the recursion relation (4.51) to find the $v = 3$ normalized harmonic-oscillator wave function.

4.13 Find ψ/c_0 for the $v = 4$ harmonic-oscillator wave function.

4.14 Point out the similarities and differences between the one-dimensional particle-in-a-box and the harmonic-oscillator wave functions and energies.

4.15 For the $v = 1$ harmonic-oscillator state, find the most likely position(s) of the particle.

4.16 Draw rough graphs of ψ and of ψ^2 for the $v = 5$ state of the one-dimensional harmonic oscillator without finding the explicit formula for ψ.

4.17 (a) The three-dimensional harmonic oscillator has the potential-energy function

$$V = \tfrac{1}{2}k_x x^2 + \tfrac{1}{2}k_y y^2 + \tfrac{1}{2}k_z z^2$$

where the k's are three force constants. Find the energy eigenvalues by solving the Schrödinger equation. (b) If $k_x = k_y = k_z$, find the degree of degeneracy of each of the four lowest energy levels.

4.18 The *Hermite polynomials* are defined by

$$H_n(z) = (-1)^n e^{z^2} \frac{d^n e^{-z^2}}{dz^n}$$

(a) Verify that

$$H_0 = 1, \qquad H_1 = 2z, \qquad H_2 = 4z^2 - 2, \qquad H_3 = 8z^3 - 12z$$

(b) The Hermite polynomials obey the relation (*Pauling and Wilson*, pages 77–79)

$$zH_n(z) = nH_{n-1}(z) + \tfrac{1}{2}H_{n+1}(z)$$

Verify this identity for $n = 0$, 1, and 2. (c) The normalized harmonic-oscillator wave functions can be written as (*Pauling and Wilson*, pages 79–80)

$$\psi_v(x) = (2^v v!)^{-1/2} \left(\frac{\alpha}{\pi}\right)^{1/4} e^{-\alpha x^2/2} H_v(\alpha^{1/2}x) \tag{4.75}$$

Verify (4.75) for the three lowest states.

4.19 Find $\langle x \rangle$ for the harmonic-oscillator state with quantum number v.

4.20 When a second-order linear homogeneous differential equation is written in the form (2.3), any point at which $P(x)$ or $Q(x)$ becomes infinite is called a *singular point* or *singularity*. In solving a differential equation by the power-series method, one can often find the proper substitution to give a two-term recursion relation by examining the differential equation near its singularities. For the harmonic-oscillator Schrödinger equation (4.34), the singularities are at $x = \pm\infty$. To check whether $x = \infty$ is a singular point, one substitutes $z = 1/x$ and examines the coefficients at $z = 0$. Verify that $\exp(-\alpha x^2/2)$ is an approximate solution of (4.34) for very large $|x|$.

4.21 Find the eigenvalues and eigenfunctions of $\hat{H}$ for a one-dimensional system with $V(x) = \infty$ for $x < 0$, $V(x) = \tfrac{1}{2}kx^2$ for $x \geqslant 0$.

4.22 For the harmonic-oscillator state with quantum number v, what range of the x coordinate is allowed classically?

4.23 (a) Write a computer program that uses the recursion relation (4.42) to calculate ψ/c_0 of (4.43) versus $\alpha^{1/2}x$ for $\alpha^{1/2}x$ values from 0 to 6 in increments of 0.5 for specified values of $mE\hbar^{-2}/\alpha = E/h\nu$. Include a test to stop adding terms in the infinite series when the last calculated term is sufficiently small. (b) Run the program for $E/h\nu = 0.499$, 0.5, and 0.501 to verify Fig. 4.1.

4.24 (a) The infrared absorption spectrum of $^1H^{35}Cl$ has its strongest band at 8.65×10^{13} Hz. Calculate the force constant of the bond in this molecule. (b) Find the approximate zero-point vibrational energy of $^1H^{35}Cl$. (c) Predict the frequency of the strongest infrared band of $^2H^{35}Cl$.

4.25 (a) Verify (4.70). (b) Find the corresponding equation for the $v = 0 \rightarrow v_2$ transition.

4.26 The $v = 0 \rightarrow 1$ and $v = 0 \rightarrow 2$ bands of $^1H^{35}Cl$ occur at 2885.98 cm^{-1} and 5667.98 cm^{-1}. (a) Calculate ν_e/c and $\nu_e x_e/c$ for this molecule. (b) Predict the wave number of the $v = 0 \rightarrow 3$ band of $^1H^{35}Cl$.

4.27 (a) The $v = 0 \rightarrow 1$ band of LiH occurs at 1359 cm^{-1}. Calculate the ratio of the $v = 1$ to $v = 0$ populations at 25°C and at 200°C. (b) Do the same as in (a) for ICl, whose strongest infrared band occurs at 381 cm^{-1}.

4.28 Show that if one expands $U(R)$ in Fig. 4.5 in a Taylor series about $R = R_e$ and neglects terms containing $(R - R_e)^3$ and higher powers (these terms are small for R near R_e) then one obtains a harmonic-oscillator potential with $k = d^2U/dR^2|_{R=R_e}$.

4.29 The *Morse function* $U(R) = D_e[1 - e^{-a(R-R_e)}]^2$ is often used to approximate the $U(R)$ curve of a diatomic molecule, where the molecule's equilibrium dissociation energy D_e is $D_e \equiv U(\infty) - U(R_e)$. (a) Verify that this equation for D_e is satisfied by the Morse function. (b) Show that $a = (k_e/2D_e)^{1/2}$.

4.30 (a) Show that if k_i and f_i are eigenvalues and eigenfunctions of $\hat{A}$ then ck_i and f_i are eigenvalues and eigenfunctions of $c\hat{A}$. (b) Give an operator whose eigenvalues are $\frac{1}{2}, \frac{3}{2}, \frac{5}{2}, \dots$. (c) Give an operator whose eigenvalues are 1, 2, 3, $\dots$.

4.31 (a) A certain system in a certain stationary state has $\psi = Ne^{-ax^4}$. (N is the normalization constant.) Find the system's potential-energy function $V(x)$ and its energy E. *Hint*: The zero level of energy is arbitrary, so choose $V(0) = 0$. (b) Sketch $V(x)$. (c) Is this the ground-state ψ? Explain.

4.32 Show that adding a constant C to the potential energy leaves the stationary-state wave functions unchanged and simply adds C to the energy eigenvalues.

4.33 The one-dimensional double-well potential has $V = \infty$ for $x < -a$, $V = 0$ for $-a \le x \le -b$, $V = V_0$ for $-b < x < b$, $V = 0$ for $b \le x \le a$, and $V = \infty$ for $x > a$ (where a, b, and V_0 are positive constants). (a) Sketch V. (b) Use the general properties of wave functions discussed in Chapters 2 and 4 to sketch ψ for the lowest four bound states for a finite value of V_0. (c) Do the same as in (b) for $V_0 = \infty$.

5

Angular Momentum

5.1 SIMULTANEOUS MEASUREMENT OF SEVERAL PROPERTIES

In this chapter we discuss angular momentum, and in the next chapter we show that for the stationary states of the hydrogen atom the magnitude of the electron's angular momentum is constant. As a preliminary, we consider what criterion we can use to decide which properties of a system can be simultaneously assigned definite values.

In Section 3.3, we postulated that if the state function Ψ is an eigenfunction of the operator $\hat{A}$ with eigenvalue s, then a measurement of the physical property A is certain to yield the result s. If Ψ is simultaneously an eigenfunction of the two operators $\hat{A}$ and $\hat{B}$, that is, if $\hat{A}\Psi = s\Psi$ and $\hat{B}\Psi = t\Psi$, then we can simultaneously assign definite values to the physical quantities A and B. When will it be possible for Ψ to be simultaneously an eigenfunction of two different operators? In Chapter 7, we shall prove the following two theorems. First, a necessary condition for the existence of a complete set of simultaneous eigenfunctions of two operators is that the operators commute with each other. (The word *complete* is used here in a certain technical sense, which we won't worry about until Chapter 7.) Conversely, if $\hat{A}$ and $\hat{B}$ are two commuting operators that correspond to physical quantities, then there exists a complete set of functions that are eigenfunctions of both $\hat{A}$ and $\hat{B}$. Thus, if $[\hat{A}, \hat{B}] = 0$, then Ψ can be an eigenfunction of both $\hat{A}$ and $\hat{B}$.

Recall that the commutator of $\hat{A}$ and $\hat{B}$ is defined as $[\hat{A}, \hat{B}] \equiv \hat{A}\hat{B} - \hat{B}\hat{A}$ [Eq. (3.7)]. The following commutator identities are helpful in evaluating commutators; these identities are easily proved by writing out the commutators in detail (Problem 5.1):

$$[\hat{A}, \hat{B}] = -[\hat{B}, \hat{A}] \tag{5.1}*$$

$$[\hat{A}, \hat{A}^n] = 0, \qquad n = 1, 2, 3, \ldots \tag{5.2}*$$

$$[k\hat{A}, \hat{B}] = [\hat{A}, k\hat{B}] = k[\hat{A}, \hat{B}] \tag{5.3}*$$

$$[\hat{A}, \hat{B} + \hat{C}] = [\hat{A}, \hat{B}] + [\hat{A}, \hat{C}], \qquad [\hat{A} + \hat{B}, \hat{C}] = [\hat{A}, \hat{C}] + [\hat{B}, \hat{C}] \tag{5.4}*$$

$$[\hat{A}, \hat{B}\hat{C}] = [\hat{A}, \hat{B}]\hat{C} + \hat{B}[\hat{A}, \hat{C}], \qquad [\hat{A}\hat{B}, \hat{C}] = [\hat{A}, \hat{C}]\hat{B} + \hat{A}[\hat{B}, \hat{C}] \tag{5.5}*$$

where k is a constant and the operators are assumed to be linear.

EXAMPLE   Starting from Eq. (3.8)

$$[\partial/\partial x, x] = 1$$

use the commutator identities (5.1)–(5.5) to find (a) $[\hat{x}, \hat{p}_x]$; (b) $[\hat{x}, \hat{p}_x^2]$; (c) $[\hat{x}, \hat{H}]$ and $[\hat{p}_x, \hat{H}]$ for a one-particle, three-dimensional system.

(a) Use of (5.3), (5.1), and $[\partial/\partial x, x] = 1$ gives

$$[\hat{x}, \hat{p}_x] = \left[x, \frac{\hbar}{i} \frac{\partial}{\partial x} \right] = \frac{\hbar}{i} \left[x, \frac{\partial}{\partial x} \right] = -\frac{\hbar}{i} \left[\frac{\partial}{\partial x}, x \right] = -\frac{\hbar}{i}$$

$$[\hat{x}, \hat{p}_x] = i\hbar \tag{5.6}$$

(b) Use of (5.5) and (5.6) gives

$$[\hat{x}, \hat{p}_x^2] = [\hat{x}, \hat{p}_x]\hat{p}_x + \hat{p}_x[\hat{x}, \hat{p}_x] = i\hbar \cdot \frac{\hbar}{i} \frac{\partial}{\partial x} + \frac{\hbar}{i} \frac{\partial}{\partial x} \cdot i\hbar$$

$$[\hat{x}, \hat{p}_x^2] = 2\hbar^2 \frac{\partial}{\partial x} \tag{5.7}$$

(c) Use of (5.4), (5.3), and (5.7) gives

$$[\hat{x}, \hat{H}] = [\hat{x}, \hat{T} + \hat{V}] = [\hat{x}, \hat{T}] + [\hat{x}, \hat{V}(x, y, z)] = [\hat{x}, \hat{T}]$$

$$= [x, (1/2m)(\hat{p}_x^2 + \hat{p}_y^2 + \hat{p}_z^2)]$$

$$= (1/2m)[\hat{x}, \hat{p}_x^2] + (1/2m)[\hat{x}, \hat{p}_y^2] + (1/2m)[\hat{x}, \hat{p}_z^2]$$

$$= \frac{1}{2m} \cdot 2\hbar^2 \frac{\partial}{\partial x} + 0 + 0$$

$$[\hat{x}, \hat{H}] = \frac{\hbar^2}{m} \frac{\partial}{\partial x} = \frac{i\hbar}{m} \hat{p}_x \tag{5.8}$$

It is left as a problem (Problem 5.2) to show that

$$[\hat{p}_x, \hat{H}] = \frac{\hbar}{i} \frac{\partial V(x, y, z)}{\partial x} \tag{5.9}$$

The above commutators have important physical consequences. Since $[\hat{x}, \hat{p}_x] \neq 0$, we cannot expect the state function to be simultaneously an eigenfunction of $\hat{x}$ and of $\hat{p}_x$. Hence we cannot simultaneously assign definite values to x and p_x, in agreement with the uncertainty principle. Since $\hat{x}$ and $\hat{H}$ do not commute, we cannot expect to assign definite values to the energy and the x coordinate at the same time. A stationary state (which has a definite energy) shows a spread of possible values for x, the probabilities for observing various values of x being given by the Born postulate.

For a state function Ψ that is not an eigenfunction of $\hat{A}$, we get various possible outcomes when we measure A in identical systems. We want some measure of the spread or dispersion in the set of observed values A_i. If $\langle A \rangle$ is the average of these values, then the deviation of each measurement from the average is $A_i - \langle A \rangle$. If we averaged all the deviations, we would get zero, since positive and negative deviations would cancel. Hence to make all deviations positive, we square them. The average of the squares of the deviations is called the *variance* of A, symbolized in statistics by σ_A^2 and in quantum mechanics by $(\Delta A)^2$:

$$(\Delta A)^2 \equiv \sigma_A^2 \equiv \langle (A - \langle A \rangle)^2 \rangle = \int \Psi^* (\hat{A} - \langle A \rangle)^2 \Psi \, d\tau \qquad (5.10)$$

where the average-value expression (3.88) was used. The definition (5.10) is equivalent to (Problem 5.4)

$$(\Delta A)^2 = \langle A^2 \rangle - \langle A \rangle^2 \qquad (5.11)$$

The positive square root of the variance is called the **standard deviation,** σ_A or ΔA. The standard deviation is the most commonly used measure of spread, and we shall take it as the measure of the "uncertainty" in the property A.

For the product of the standard deviations of two properties of a quantum-mechanical system whose state function is Ψ, one can show that (see Problem 7.52)

$$\Delta A \, \Delta B \geq \frac{1}{2} \left| \int \Psi^* [\hat{A}, \hat{B}] \Psi \, d\tau \right| \qquad (5.12)$$

If $\hat{A}$ and $\hat{B}$ commute, then the integral in (5.12) is zero, and we have the possibility of having ΔA and ΔB both zero, in agreement with the previous discussion.

As an example of (5.12), we find, using (5.6) and $|z_1 z_2| = |z_1| \, |z_2|$ [Eq. (1.32)],

$$\Delta x \, \Delta p_x \geq \frac{1}{2} \left| \int \Psi^* [\hat{x}, \hat{p}_x] \Psi \, d\tau \right| = \frac{1}{2} \left| \int \Psi^* i\hbar \Psi \, d\tau \right| = \frac{1}{2} \hbar |i| \left| \int \Psi^* \Psi \, d\tau \right|$$

$$\Delta x \, \Delta p_x \geq \tfrac{1}{2} \hbar \qquad (5.13)$$

Equation (5.13) is the quantitative statement of the **Heisenberg uncertainty principle**.

EXAMPLE   For the ground state of the particle in a three-dimensional box, use the following results of Eqs. (3.91), (3.92), (3.39), the equation following (3.89), and Problem 3.36

$$\langle x \rangle = a/2 , \qquad \langle x^2 \rangle = a^2 (1/3 - 1/2\pi^2)$$

$$\langle p_x \rangle = 0 , \qquad \langle p_x^2 \rangle = h^2/4a^2$$

to check that the uncertainty principle (5.13) is obeyed.

We have

$$(\Delta x)^2 = \langle x^2 \rangle - \langle x \rangle^2 = a^2 (1/3 - 1/2\pi^2) - a^2/4 = a^2 (\pi^2 - 6)/12\pi^2$$

$$\Delta x = a(\pi^2 - 6)^{1/2}/(12)^{1/2}\pi$$

$$(\Delta p_x)^2 = \langle p_x^2 \rangle - \langle p_x \rangle^2 = h^2/4a^2 , \qquad \Delta p_x = h/2a$$

$$\Delta x \, \Delta p_x = \frac{h}{2\pi} \left(\frac{\pi^2 - 6}{12} \right)^{1/2} = 0.568\hbar > \tfrac{1}{2}\hbar$$

There is also an uncertainty relation involving energy and time:

$$\Delta E \, \Delta t \geq \tfrac{1}{2}\hbar \qquad \tag{5.14}$$

Some texts state that (5.14) is derived from (5.12) by taking $i\hbar\,\partial/\partial t$ as the energy operator and multiplication by t as the time operator. However, the energy operator is the Hamiltonian $\hat{H}$, and not $i\hbar\,\partial/\partial t$. Moreover, time is not an observable but is a parameter in quantum mechanics. Hence there is no quantum-mechanical time operator. (The noun **observable** in quantum mechanics means a physically measurable property of a system.) Equation (5.14) must be derived by a special treatment, which we omit. (See *Ballentine*, Section 12-3.) The derivation of (5.14) shows that Δt is to be interpreted as the lifetime of the state whose energy is uncertain by ΔE. It is often stated that Δt in (5.14) is the duration of the energy measurement; however, Aharonov and Bohm have shown that "energy can be measured reproducibly in an arbitrarily short time." [See Y. Aharonov and D. Bohm, *Phys. Rev.*, **122**, 1649 (1961); **134**, B1417 (1964).]

Now consider the possibility of simultaneously assigning definite values to *three* physical quantities: A, B, and C. Suppose

$$[\hat{A}, \hat{B}] = 0 \tag{5.15}$$

$$[\hat{A}, \hat{C}] = 0 \tag{5.16}$$

Is this sufficient to ensure that there exist simultaneous eigenfunctions of all three operators? Equation (5.15) ensures that we can construct a common set of eigenfunctions for $\hat{A}$ and $\hat{B}$; Eq. (5.16) ensures that we can construct a common set of eigenfunctions for $\hat{A}$ and $\hat{C}$. If these two sets of eigenfunctions are the same, then we will have a common set of eigenfunctions for all three operators. Hence we ask: Is the set of eigenfunctions of the linear operator $\hat{A}$ uniquely determined (apart from arbitrary multiplicative constants)? The answer is, in general, no. If there is more than one independent eigenfunction corresponding to an eigenvalue of $\hat{A}$ (that is, degeneracy), then any linear combination of the eigenfunctions of the degenerate eigenvalue is an eigenfunction of $\hat{A}$ (Section 3.6). It might well be that the proper linear combinations needed to give eigenfunctions of $\hat{B}$ would differ from the linear combinations that give eigenfunctions of $\hat{C}$. It turns out that, if we are to have a common complete set of eigenfunctions of all three operators, we require that $[\hat{B}, \hat{C}] = 0$, in addition to (5.15) and (5.16). *To have a complete set of functions that are simultaneous eigenfunctions of several operators, each operator must commute with every other operator.*

5.2 VECTORS

In the next section we shall solve the eigenvalue problem for angular momentum, which is a vector property. We therefore first review vectors.

Physical properties (for example, mass, length, energy) that are completely specified by their magnitude are called *scalars*. Physical properties (for example, force, velocity, momentum) that require specification of both magnitude and

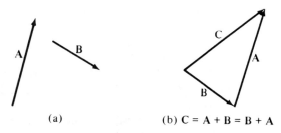

(a)   (b) C = A + B = B + A

Figure 5.1 Addition of two vectors.

direction are called **vectors**. A vector is represented by a directed line segment whose length and direction give the magnitude and direction of the property. We use boldface type for vectors.

The sum of two vectors **A** and **B** is defined by the following procedure: Slide the first vector so that its tail touches the head of the second vector, keeping the direction of the first vector fixed. Then draw a new vector from the tail of the second vector to the head of the first vector. See Fig. 5.1. The product of a vector and a scalar, $c\mathbf{A}$, is defined as a vector of length $|c|$ times the length of **A** with the same direction as **A** if c is positive, or the opposite direction to **A** if c is negative.

To obtain an algebraic (as well as geometric) way of representing vectors, we set up Cartesian coordinates in space. We draw a vector of unit length directed along the positive x axis and call it **i**. (No connection with $i = \sqrt{-1}$.) Unit vectors in the positive y and z directions are called **j** and **k** (Fig. 5.2). To represent any vector **A** in terms of the three unit vectors, we first slide **A** so that its tail is at the origin, preserving its direction during this process. We then find the projections of **A** on the x, y, and z axes: A_x, A_y, and A_z. From the definition of vector addition, it follows that (Fig. 5.2)

$$\mathbf{A} = A_x\mathbf{i} + A_y\mathbf{j} + A_z\mathbf{k} \qquad (5.17)^*$$

To specify **A**, it is sufficient to specify its three components: (A_x, A_y, A_z). We

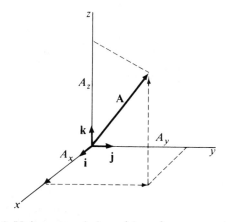

Figure 5.2 Unit vectors **i**, **j**, and **k** and components of **A**.

can therefore define a vector in three-dimensional space as an ordered set of three numbers. The advantage of this definition is that it can be extended to more than three dimensions. Thus a vector in five-dimensional "space" is an ordered set of five numbers. Such abstract vector spaces (which can be of infinite dimension) are used in advanced formulations of quantum mechanics.

Two vectors **A** and **B** are equal if and only if all the corresponding components are equal: $A_x = B_x$, $A_y = B_y$, $A_z = B_z$. Hence a vector equation is equivalent to three scalar equations.

To add two vectors analytically, we add corresponding components:

$$\mathbf{A} + \mathbf{B} = A_x\mathbf{i} + A_y\mathbf{j} + A_z\mathbf{k} + B_x\mathbf{i} + B_y\mathbf{j} + B_z\mathbf{k}$$

$$\mathbf{A} + \mathbf{B} = (A_x + B_x)\mathbf{i} + (A_y + B_y)\mathbf{j} + (A_z + B_z)\mathbf{k} \tag{5.18}*$$

Also, if c is a scalar, then

$$c\mathbf{A} = cA_x\mathbf{i} + cA_y\mathbf{j} + cA_z\mathbf{k} \tag{5.19}*$$

The **magnitude** A of a vector **A** is its length and is therefore a scalar. Often the notation $|\mathbf{A}|$ is used for the magnitude of **A**.

The **dot product** or *scalar product* $\mathbf{A} \cdot \mathbf{B}$ of two vectors is defined by

$$\mathbf{A} \cdot \mathbf{B} = |\mathbf{A}|\,|\mathbf{B}|\cos\theta = \mathbf{B} \cdot \mathbf{A} \tag{5.20}*$$

where θ is the angle between the vectors. The dot product, being the product of three scalars, is a scalar. Note that $|\mathbf{A}|\cos\theta$ is the projection of **A** on **B**. From the definition of vector addition, it follows that the projection of the vector $\mathbf{A} + \mathbf{B}$ on some vector **C** is the sum of the projections of **A** and of **B** on **C**. Hence

$$(\mathbf{A} + \mathbf{B}) \cdot \mathbf{C} = \mathbf{A} \cdot \mathbf{C} + \mathbf{B} \cdot \mathbf{C} \tag{5.21}$$

Since the three unit vectors **i**, **j**, and **k** are each of unit length and are mutually perpendicular, we have

$$\mathbf{i} \cdot \mathbf{i} = \mathbf{j} \cdot \mathbf{j} = \mathbf{k} \cdot \mathbf{k} = \cos 0 = 1\,, \qquad \mathbf{i} \cdot \mathbf{j} = \mathbf{j} \cdot \mathbf{k} = \mathbf{k} \cdot \mathbf{i} = \cos(\pi/2) = 0 \tag{5.22}$$

We can use (5.22) and the distributive law (5.21) to get an important formula for the dot product of two vectors:

$$\mathbf{A} \cdot \mathbf{B} = (A_x\mathbf{i} + A_y\mathbf{j} + A_z\mathbf{k}) \cdot (B_x\mathbf{i} + B_y\mathbf{j} + B_z\mathbf{k})$$

$$\mathbf{A} \cdot \mathbf{B} = A_xB_x + A_yB_y + A_zB_z \tag{5.23}*$$

where six of the nine terms in the dot product are zero.

Consider the dot product of a vector with itself. From (5.20) we have

$$\mathbf{A} \cdot \mathbf{A} = |\mathbf{A}|^2 \tag{5.24}*$$

Using (5.23), we therefore have

$$|\mathbf{A}| = (A_x^2 + A_y^2 + A_z^2)^{1/2} \tag{5.25}*$$

For three-dimensional vectors, there is another type of product. The **cross product** or *vector product* $\mathbf{A} \times \mathbf{B}$ is a vector whose magnitude is

$$|\mathbf{A} \times \mathbf{B}| = |\mathbf{A}|\,|\mathbf{B}|\sin\theta \tag{5.26}$$

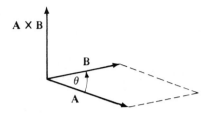

Figure 5.3 Cross product of two vectors.

whose line segment is perpendicular to the plane defined by **A** and **B**, and whose direction is such that **A**, **B**, and **A** × **B** form a right-handed system (just as the x, y, and z axes form a right-handed system). (See Fig. 5.3.) From the definition it follows that

$$\mathbf{B} \times \mathbf{A} = -\mathbf{A} \times \mathbf{B} \tag{5.27}$$

Also, it can be shown that (*Taylor and Mann*, pages 281–282)

$$\mathbf{A} \times (\mathbf{B} + \mathbf{C}) = \mathbf{A} \times \mathbf{B} + \mathbf{A} \times \mathbf{C} \tag{5.28}$$

For the three unit vectors, we have

$$\mathbf{i} \times \mathbf{i} = \mathbf{j} \times \mathbf{j} = \mathbf{k} \times \mathbf{k} = \sin 0 = 0$$

$$\mathbf{i} \times \mathbf{j} = \mathbf{k}, \quad \mathbf{j} \times \mathbf{i} = -\mathbf{k}, \quad \mathbf{j} \times \mathbf{k} = \mathbf{i}, \quad \mathbf{k} \times \mathbf{j} = -\mathbf{i}, \quad \mathbf{k} \times \mathbf{i} = \mathbf{j}, \quad \mathbf{i} \times \mathbf{k} = -\mathbf{j}$$

Using these equations and the distributive property (5.28), we find

$$\mathbf{A} \times \mathbf{B} = (A_x \mathbf{i} + A_y \mathbf{j} + A_z \mathbf{k}) \times (B_x \mathbf{i} + B_y \mathbf{j} + B_z \mathbf{k})$$

$$\mathbf{A} \times \mathbf{B} = (A_y B_z - A_z B_y)\mathbf{i} + (A_z B_x - A_x B_z)\mathbf{j} + (A_x B_y - A_y B_x)\mathbf{k}$$

As a mnemonic device, we can express the cross product as a determinant (see Section 8.3):

$$\mathbf{A} \times \mathbf{B} = \begin{vmatrix} \mathbf{i} & \mathbf{j} & \mathbf{k} \\ A_x & A_y & A_z \\ B_x & B_y & B_z \end{vmatrix} = \mathbf{i} \begin{vmatrix} A_y & A_z \\ B_y & B_z \end{vmatrix} - \mathbf{j} \begin{vmatrix} A_x & A_z \\ B_x & B_z \end{vmatrix} + \mathbf{k} \begin{vmatrix} A_x & A_y \\ B_x & B_y \end{vmatrix} \tag{5.29}$$

We define the vector operator *del* as

$$\boldsymbol{\nabla} \equiv \mathbf{i}\, \frac{\partial}{\partial x} + \mathbf{j}\, \frac{\partial}{\partial y} + \mathbf{k}\, \frac{\partial}{\partial z} \tag{5.30}$$

From Eq. (3.23), the operator for the linear-momentum vector is $\hat{\mathbf{p}} = -i\hbar\boldsymbol{\nabla}$.

The ***gradient*** of a function $g(x, y, z)$ is defined as the result of operating on the function with del:

$$\text{grad } g(x, y, z) \equiv \boldsymbol{\nabla} g(x, y, z) \equiv \mathbf{i}\, \frac{\partial g}{\partial x} + \mathbf{j}\, \frac{\partial g}{\partial y} + \mathbf{k}\, \frac{\partial g}{\partial z} \tag{5.31}$$

The gradient of a scalar function is a vector function. The vector $\boldsymbol{\nabla} g(x, y, z)$ represents the spatial rate of change of the function g; the x component of $\boldsymbol{\nabla} g$ is the rate of change of g with respect to x, and so on. It can be shown that the vector $\boldsymbol{\nabla} g$ points in the direction in which the rate of change of g is greatest. From

Eq. (4.26), we have

$$\mathbf{F} = -\nabla V(x, y, z) = -\mathbf{i}\frac{\partial V}{\partial x} - \mathbf{j}\frac{\partial V}{\partial y} - \mathbf{k}\frac{\partial V}{\partial z} \qquad (5.32)$$

For example, the force field of a point charge is radially directed, and the rate of change of potential energy of another charge in this field is greatest in the radial direction.

Finally, suppose that the components of a vector are each functions of some parameter t; $A_x = A_x(t)$, $A_y = A_y(t)$, $A_z = A_z(t)$. We define the derivative of the vector with respect to t as

$$\frac{d\mathbf{A}}{dt} = \mathbf{i}\frac{dA_x}{dt} + \mathbf{j}\frac{dA_y}{dt} + \mathbf{k}\frac{dA_z}{dt}$$

5.3 ANGULAR MOMENTUM OF A ONE-PARTICLE SYSTEM

In Section 3.3 we found the eigenfunctions and eigenvalues for the linear-momentum operator $\hat{p}_x$. In this section we consider the same problem for the angular momentum of a particle. Angular momentum is extremely important in the quantum mechanics of atomic structure. We begin by reviewing the classical mechanics of angular momentum.

Classical Mechanics of One-Particle Angular Momentum. Consider a moving particle of mass m. We set up a Cartesian coordinate system that is fixed in space. Let $\mathbf{r}$ be the vector from the origin to the instantaneous position of the particle. We have

$$\mathbf{r} = \mathbf{i}x + \mathbf{j}y + \mathbf{k}z \qquad (5.33)$$

where x, y, and z are the particle's coordinates at a given instant. These coordinates are functions of time, and defining the velocity vector $\mathbf{v}$ as the time derivative of the position vector, we have (Section 5.2)

$$\mathbf{v} \equiv \frac{d\mathbf{r}}{dt} = \mathbf{i}\frac{dx}{dt} + \mathbf{j}\frac{dy}{dt} + \mathbf{k}\frac{dz}{dt} \qquad (5.34)$$

$$v_x = dx/dt, \qquad v_y = dy/dt, \qquad v_z = dz/dt$$

We define the particle's *linear momentum* vector $\mathbf{p}$ by

$$\mathbf{p} \equiv m\mathbf{v} \qquad (5.35)^*$$

$$p_x = mv_x, \qquad p_y = mv_y, \qquad p_z = mv_z \qquad (5.36)$$

The particle's *angular momentum* $\mathbf{L}$ with respect to the coordinate origin is defined in classical mechanics as

$$\mathbf{L} \equiv \mathbf{r} \times \mathbf{p} \qquad (5.37)^*$$

$$\mathbf{L} = \begin{vmatrix} \mathbf{i} & \mathbf{j} & \mathbf{k} \\ x & y & z \\ p_x & p_y & p_z \end{vmatrix} \qquad (5.38)$$

$$L_x = yp_z - zp_y, \qquad L_y = zp_x - xp_z, \qquad L_z = xp_y - yp_x \qquad (5.39)$$

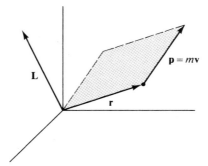

Figure 5.4 $\mathbf{L} = \mathbf{r} \times \mathbf{p}$.

where (5.29) was used. L_x, L_y, and L_z are the components of $\mathbf{L}$ along the x, y, and z axes. The angular-momentum vector $\mathbf{L}$ is perpendicular to the plane defined by the particle's position vector $\mathbf{r}$ and its velocity $\mathbf{v}$ (Fig. 5.4).

The *torque* $\boldsymbol{\tau}$ acting on a particle is defined as the cross product of $\mathbf{r}$ and the force $\mathbf{F}$ acting on the particle: $\boldsymbol{\tau} \equiv \mathbf{r} \times \mathbf{F}$. It is readily shown that (*Halliday and Resnick*, Section 12-3) $\boldsymbol{\tau} = d\mathbf{L}/dt$. When there is no torque acting on the particle, the rate of change of its angular momentum is zero; that is, its angular momentum is constant (or conserved). For a planet orbiting the sun, the gravitational force is radially directed; since the cross product of two parallel vectors is zero, there is no torque on the planet and its angular momentum is conserved.

One-Particle Orbital-Angular-Momentum Operators. Now for the quantum-mechanical treatment. In quantum mechanics, there are two kinds of angular momenta: *orbital angular momentum* results from the motion of a particle through space, and is the analog of the classical-mechanical quantity $\mathbf{L}$; *spin angular momentum* (Chapter 10) is an intrinsic property of many microscopic particles and has no classical-mechanical analog. We are now considering only orbital angular momentum. We get the quantum-mechanical operators for the components of orbital angular momentum of a particle by replacing the coordinates and momenta in the classical equations (5.39) by their corresponding operators [Eqs. (3.21)–(3.23)]. We find

$$\hat{L}_x = -i\hbar\left(y\,\frac{\partial}{\partial z} - z\,\frac{\partial}{\partial y}\right) \tag{5.40}$$

$$\hat{L}_y = -i\hbar\left(z\,\frac{\partial}{\partial x} - x\,\frac{\partial}{\partial z}\right) \tag{5.41}$$

$$\hat{L}_z = -i\hbar\left(x\,\frac{\partial}{\partial y} - y\,\frac{\partial}{\partial x}\right) \tag{5.42}$$

(Since $\hat{y}\hat{p}_z = \hat{p}_z\hat{y}$, and so on, we do not run into any problems of noncommutativity in constructing these operators.) Using

$$\hat{L}^2 = |\hat{\mathbf{L}}|^2 = \hat{\mathbf{L}}\cdot\hat{\mathbf{L}} = \hat{L}_x^2 + \hat{L}_y^2 + \hat{L}_z^2 \tag{5.43}$$

we can construct the operator for the square of the angular-momentum magnitude from the operators in (5.40)–(5.42).

Since the commutation relations determine which physical quantities can be simultaneously assigned definite values, we investigate these relations for angular momentum. Operating on some function $f(x, y, z)$ with $\hat{L}_y$, we have

$$\hat{L}_y f = -i\hbar \left(z \frac{\partial f}{\partial x} - x \frac{\partial f}{\partial z} \right)$$

Operating on this last equation with $\hat{L}_x$, we have

$$\hat{L}_x \hat{L}_y f = -\hbar^2 \left(y \frac{\partial f}{\partial x} + yz \frac{\partial^2 f}{\partial z\, \partial x} - yx \frac{\partial^2 f}{\partial z^2} - z^2 \frac{\partial^2 f}{\partial y\, \partial x} + zx \frac{\partial^2 f}{\partial y\, \partial z} \right) \quad (5.44)$$

Similarly,

$$\hat{L}_x f = -i\hbar \left(y \frac{\partial f}{\partial z} - z \frac{\partial f}{\partial y} \right)$$

$$\hat{L}_y \hat{L}_x f = -\hbar^2 \left(zy \frac{\partial^2 f}{\partial x\, \partial z} - z^2 \frac{\partial^2 f}{\partial x\, \partial y} - xy \frac{\partial^2 f}{\partial z^2} + x \frac{\partial f}{\partial y} + xz \frac{\partial^2 f}{\partial z\, \partial y} \right) \quad (5.45)$$

Subtracting (5.45) from (5.44), we have

$$\hat{L}_x \hat{L}_y f - \hat{L}_y \hat{L}_x f = -\hbar^2 \left(y \frac{\partial f}{\partial x} - x \frac{\partial f}{\partial y} \right)$$

$$[\hat{L}_x, \hat{L}_y] = i\hbar \hat{L}_z \quad (5.46)$$

where we have used relations such as

$$\frac{\partial^2 f}{\partial z\, \partial x} = \frac{\partial^2 f}{\partial x\, \partial z} \quad \mathbf{(5.47)^*}$$

which are true for well-behaved functions. We could use the same procedure to find $[\hat{L}_y, \hat{L}_z]$ and $[\hat{L}_z, \hat{L}_x]$, but we can save time by noting a certain kind of symmetry in (5.40)–(5.42). By a *cyclic permutation* of x, y, and z, we mean replacing x by y, replacing y by z, and replacing z by x. If we carry out a cyclic permutation in $\hat{L}_x$, we get $\hat{L}_y$; a cyclic permutation in $\hat{L}_y$ gives $\hat{L}_z$; and $\hat{L}_z$ is transformed into $\hat{L}_x$ by a cyclic permutation. Hence, by carrying out two successive cyclic permutations on (5.46), we have

$$[\hat{L}_y, \hat{L}_z] = i\hbar \hat{L}_x, \qquad [\hat{L}_z, \hat{L}_x] = i\hbar \hat{L}_y \quad (5.48)$$

Now we evaluate the commutators of $\hat{L}^2$ with each of its components, using commutator identities of Section 5.1.

$$[\hat{L}^2, \hat{L}_x] = [\hat{L}_x^2 + \hat{L}_y^2 + \hat{L}_z^2, \hat{L}_x]$$

$$= [\hat{L}_x^2, \hat{L}_x] + [\hat{L}_y^2, \hat{L}_x] + [\hat{L}_z^2, \hat{L}_x]$$

$$= [\hat{L}_y^2, \hat{L}_x] + [\hat{L}_z^2, \hat{L}_x]$$

$$= [\hat{L}_y, \hat{L}_x]\hat{L}_y + \hat{L}_y[\hat{L}_y, \hat{L}_x] + [\hat{L}_z, \hat{L}_x]\hat{L}_z + \hat{L}_z[\hat{L}_z, \hat{L}_x]$$

$$= -i\hbar \hat{L}_z \hat{L}_y - i\hbar \hat{L}_y \hat{L}_z + i\hbar \hat{L}_y \hat{L}_z + i\hbar \hat{L}_z \hat{L}_y$$

$$[\hat{L}^2, \hat{L}_x] = 0 \quad \mathbf{(5.49)^*}$$

Since a cyclic permutation of x, y, and z leaves $\hat{L}^2 = \hat{L}_x^2 + \hat{L}_y^2 + \hat{L}_z^2$ unchanged, if we carry out two such permutations on (5.49), we get

$$[\hat{L}^2, \hat{L}_y] = 0, \qquad [\hat{L}^2, \hat{L}_z] = 0 \qquad \textbf{(5.50)}*$$

To which of the quantities L^2, L_x, L_y, L_z can we assign definite values simultaneously? Because $\hat{L}^2$ commutes with each of its components, we can specify an exact value for L^2 and any *one* component. However, because no two components of $\hat{L}$ commute with each other, we cannot specify more than one component simultaneously. (There is one exception to this statement, which will be discussed shortly.) It is traditional to take L_z as the component of angular momentum that will be specified along with L^2. Note that in specifying $L^2 = |\mathbf{L}|^2$ we are not specifying the vector $\mathbf{L}$, only its magnitude. A complete specification of $\mathbf{L}$ requires simultaneous specification of each of its three components, which we usually cannot do. In classical mechanics when angular momentum is conserved, each of its three components has a definite value. In quantum mechanics when angular momentum is conserved, only its magnitude and one of its components are specifiable.

We could now attempt to find the eigenvalues and common eigenfunctions of $\hat{L}^2$ and $\hat{L}_z$ by using the forms for these operators in Cartesian coordinates. However, we would find that the partial differential equations obtained would not be separable. For this reason we carry out a transformation to **spherical polar coordinates** (Fig. 5.5). The coordinate r is the distance from the origin to the point (x, y, z). The angle θ is the angle the vector $\mathbf{r}$ makes with the positive z axis. The angle that the projection of $\mathbf{r}$ in the xy plane makes with the positive x axis is ϕ. (Mathematics texts often interchange θ and ϕ.) A little trigonometry gives

$$x = r \sin\theta \cos\phi, \qquad y = r \sin\theta \sin\phi, \qquad z = r \cos\theta \qquad (5.51)$$

$$r^2 = x^2 + y^2 + z^2, \qquad \cos\theta = \frac{z}{(x^2 + y^2 + z^2)^{1/2}}, \qquad \tan\phi = y/x \qquad (5.52)$$

To transform the angular-momentum operators to spherical polar coordinates, we must transform $\partial/\partial x$, $\partial/\partial y$, and $\partial/\partial z$ into these coordinates. [This transformation may be skimmed if desired. Begin reading again after Eq. (5.64).]

To perform this transformation, we use the *chain rule*. Suppose we have a function of r, θ, and ϕ: $f(r, \theta, \phi)$. If we carry out a change of independent variables by

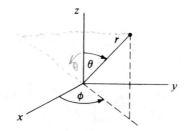

Figure 5.5 Spherical polar coordinates.

substituting

$$r = r(x, y, z), \qquad \theta = \theta(x, y, z), \qquad \phi = \phi(x, y, z)$$

into f, we transform it into a function of x, y, and z:

$$f[r(x, y, z), \theta(x, y, z), \phi(x, y, z)] = g(x, y, z)$$

For example, suppose that $f(r, \theta, \phi) = 3r \cos \theta + 2 \tan^2 \phi$. Using (5.52), we have $g(x, y, z) = 3z + 2y^2 x^{-2}$.

The chain rule tells us how the partial derivatives of $g(x, y, z)$ are related to those of $f(r, \theta, \phi)$. In fact,

$$\left(\frac{\partial g}{\partial x} \right)_{y,z} = \left(\frac{\partial f}{\partial r} \right)_{\theta,\phi} \left(\frac{\partial r}{\partial x} \right)_{y,z} + \left(\frac{\partial f}{\partial \theta} \right)_{r,\phi} \left(\frac{\partial \theta}{\partial x} \right)_{y,z} + \left(\frac{\partial f}{\partial \phi} \right)_{r,\theta} \left(\frac{\partial \phi}{\partial x} \right)_{y,z} \qquad (5.53)$$

$$\left(\frac{\partial g}{\partial y} \right)_{x,z} = \left(\frac{\partial f}{\partial r} \right)_{\theta,\phi} \left(\frac{\partial r}{\partial y} \right)_{x,z} + \left(\frac{\partial f}{\partial \theta} \right)_{r,\phi} \left(\frac{\partial \theta}{\partial y} \right)_{x,z} + \left(\frac{\partial f}{\partial \phi} \right)_{r,\theta} \left(\frac{\partial \phi}{\partial y} \right)_{x,z} \qquad (5.54)$$

$$\left(\frac{\partial g}{\partial z} \right)_{x,y} = \left(\frac{\partial f}{\partial r} \right)_{\theta,\phi} \left(\frac{\partial r}{\partial z} \right)_{x,y} + \left(\frac{\partial f}{\partial \theta} \right)_{r,\phi} \left(\frac{\partial \theta}{\partial z} \right)_{x,y} + \left(\frac{\partial f}{\partial \phi} \right)_{r,\theta} \left(\frac{\partial \phi}{\partial z} \right)_{x,y} \qquad (5.55)$$

To convert these equations to operator equations, we delete f and g. However, it would not do to write, for example,

$$\frac{\partial}{\partial r} \left(\frac{\partial r}{\partial x} \right)_{y,z}$$

for the first term on the right side of the operator equation corresponding to (5.53), because this would imply that $\partial / \partial r$ was to operate on

$$\left[\left(\frac{\partial r}{\partial x} \right)_{y,z} f \right]$$

whereas, according to (5.53), it should operate only on f. For the operator equation corresponding to (5.53), we thus write

$$\frac{\partial}{\partial x} = \left(\frac{\partial r}{\partial x} \right)_{y,z} \frac{\partial}{\partial r} + \left(\frac{\partial \theta}{\partial x} \right)_{y,z} \frac{\partial}{\partial \theta} + \left(\frac{\partial \phi}{\partial x} \right)_{y,z} \frac{\partial}{\partial \phi} \qquad (5.56)$$

with similar equations for $\partial / \partial y$ and $\partial / \partial z$. The task now is to evaluate the partial derivatives such as $(\partial r / \partial x)_{y,z}$. Taking the partial derivative of the first equation in (5.52) with respect to x at constant y and z, we have

$$2r \left(\frac{\partial r}{\partial x} \right)_{y,z} = 2x = 2r \sin \theta \cos \phi$$

$$\left(\frac{\partial r}{\partial x} \right)_{y,z} = \sin \theta \cos \phi \qquad (5.57)$$

Differentiating $r^2 = x^2 + y^2 + z^2$ with respect to y and with respect to z, we find

$$\left(\frac{\partial r}{\partial y} \right)_{x,z} = \sin \theta \sin \phi, \qquad \left(\frac{\partial r}{\partial z} \right)_{x,y} = \cos \theta \qquad (5.58)$$

From the second equation in (5.52), we find

$$-\sin \theta \left(\frac{\partial \theta}{\partial x} \right)_{y,z} = -\frac{xz}{r^3}$$

$$\left(\frac{\partial \theta}{\partial x} \right)_{y,z} = \frac{\cos \theta \cos \phi}{r} \qquad (5.59)$$

Also,

$$\left(\frac{\partial \theta}{\partial y}\right)_{x,z} = \frac{\cos \theta \sin \phi}{r} \ , \qquad \left(\frac{\partial \theta}{\partial z}\right)_{x,y} = -\frac{\sin \theta}{r} \tag{5.60}$$

From $\tan \phi = y/x$, we find

$$\left(\frac{\partial \phi}{\partial x}\right)_{y,z} = -\frac{\sin \phi}{r \sin \theta} \ , \qquad \left(\frac{\partial \phi}{\partial y}\right)_{x,z} = \frac{\cos \phi}{r \sin \theta} \ , \qquad \left(\frac{\partial \phi}{\partial z}\right)_{x,y} = 0 \tag{5.61}$$

Substituting (5.57), (5.59), and (5.61) into (5.56), we find

$$\frac{\partial}{\partial x} = \sin \theta \cos \phi \frac{\partial}{\partial r} + \frac{\cos \theta \cos \phi}{r} \frac{\partial}{\partial \theta} - \frac{\sin \phi}{r \sin \theta} \frac{\partial}{\partial \phi} \tag{5.62}$$

Similarly,

$$\frac{\partial}{\partial y} = \sin \theta \sin \phi \frac{\partial}{\partial r} + \frac{\cos \theta \sin \phi}{r} \frac{\partial}{\partial \theta} + \frac{\cos \phi}{r \sin \theta} \frac{\partial}{\partial \phi} \tag{5.63}$$

$$\frac{\partial}{\partial z} = \cos \theta \frac{\partial}{\partial r} - \frac{\sin \theta}{r} \frac{\partial}{\partial \theta} \tag{5.64}$$

At long last, we are ready to express the angular-momentum components in spherical polar coordinates. Substituting (5.51), (5.63), and (5.64) into (5.40), we have

$$\hat{L}_x = -i\hbar \left[r \sin \theta \sin \phi \left(\cos \theta \frac{\partial}{\partial r} - \frac{\sin \theta}{r} \frac{\partial}{\partial \theta} \right) \right.$$

$$\left. - r \cos \theta \left(\sin \theta \sin \phi \frac{\partial}{\partial r} + \frac{\cos \theta \sin \phi}{r} \frac{\partial}{\partial \theta} + \frac{\cos \phi}{r \sin \theta} \frac{\partial}{\partial \phi} \right) \right]$$

$$\hat{L}_x = i\hbar \left(\sin \phi \frac{\partial}{\partial \theta} + \cot \theta \cos \phi \frac{\partial}{\partial \phi} \right) \tag{5.65}$$

Also, we find

$$\hat{L}_y = -i\hbar \left(\cos \phi \frac{\partial}{\partial \theta} - \cot \theta \sin \phi \frac{\partial}{\partial \phi} \right) \tag{5.66}$$

$$\hat{L}_z = -i\hbar \frac{\partial}{\partial \phi} \tag{5.67}$$

By squaring each of $\hat{L}_x$, $\hat{L}_y$, and $\hat{L}_z$ and then adding their squares, we can construct $\hat{L}^2$ [Eq. (5.43)]. The result is (Problem 5.15)

$$\hat{L}^2 = -\hbar^2 \left(\frac{\partial^2}{\partial \theta^2} + \cot \theta \frac{\partial}{\partial \theta} + \frac{1}{\sin^2 \theta} \frac{\partial^2}{\partial \phi^2} \right) \tag{5.68}$$

Although the angular-momentum operators depend on all three Cartesian coordinates, x, y, and z, they involve only the two spherical polar coordinates θ and ϕ.

One-Particle Orbital-Angular-Momentum Eigenfunctions and Eigenvalues.

We now find the common eigenfunctions of $\hat{L}^2$ and $\hat{L}_z$, which we denote by Y. Since these operators involve only θ and ϕ, Y is a function of these two coordinates: $Y = Y(\theta, \phi)$. (Of course, since the operators are linear, we can multiply Y by an arbitrary function of r and still have an eigenfunction of $\hat{L}^2$ and $\hat{L}_z$.) We must solve

$$\hat{L}_z Y(\theta, \phi) = bY(\theta, \phi) \tag{5.69}$$

$$\hat{L}^2 Y(\theta, \phi) = cY(\theta, \phi) \tag{5.70}$$

where b and c are the eigenvalues of $\hat{L}_z$ and $\hat{L}^2$.

Using the $\hat{L}_z$ operator, we have

$$-i\hbar \frac{\partial}{\partial \phi} Y(\theta, \phi) = bY(\theta, \phi) \tag{5.71}$$

Since the operator in (5.71) does not involve θ, we try a separation of variables, writing

$$Y(\theta, \phi) = S(\theta) T(\phi) \tag{5.72}$$

Equation (5.71) becomes

$$-i\hbar \frac{\partial}{\partial \phi} [S(\theta) T(\phi)] = bS(\theta) T(\phi)$$

$$-i\hbar S(\theta) \frac{dT(\phi)}{d\phi} = bS(\theta) T(\phi)$$

$$\frac{dT(\phi)}{T(\phi)} = \frac{ib}{\hbar} d\phi$$

$$T(\phi) = Ae^{ib\phi/\hbar} \tag{5.73}$$

where A is an arbitrary constant.

Is T suitable as an eigenfunction? The answer is no, since it is not, in general, a single-valued function. If we add 2π to ϕ, we will still be at the same point in space, and hence we want no change in T when this is done. For T to be single valued, we have the restriction :

$$T(\phi + 2\pi) = T(\phi)$$

$$Ae^{ib\phi/\hbar} e^{ib2\pi/\hbar} = Ae^{ib\phi/\hbar}$$

$$e^{ib2\pi/\hbar} = 1 \tag{5.74}$$

To satisfy $e^{i\alpha} = \cos \alpha + i \sin \alpha = 1$, we must have $\alpha = 2\pi m$, where

$$m = 0, \pm 1, \pm 2, \pm \cdots$$

Therefore, (5.74) gives

$$2\pi b/\hbar = 2\pi m$$

$$b = m\hbar, \qquad m = \cdots -2, -1, 0, 1, 2, \ldots \tag{5.75}$$

and (5.73) becomes

$$T(\phi) = Ae^{im\phi}, \qquad m = 0, \pm 1, \pm 2, \ldots \tag{5.76}$$

The eigenvalues for the z component of angular momentum are quantized.

We fix A by normalizing T. First let us consider normalizing some function F of r, θ, and ϕ. The ranges of the independent variables are (see Fig. 5.5)

normalization ranges for spherical polar coords:

$$0 \leqslant r \leqslant \infty , \qquad 0 \leqslant \theta \leqslant \pi , \qquad 0 \leqslant \phi \leqslant 2\pi \qquad (5.77)^*$$

The infinitesimal volume element in spherical polar coordinates is (*Taylor and Mann*, page 417)

$$d\tau = r^2 \sin \theta \, dr \, d\theta \, d\phi \qquad (5.78)^*$$

The quantity (5.78) is the volume of an infinitesimal region of space for which the spherical polar coordinates lie in the ranges r to $r + dr$, θ to $\theta + d\theta$, and ϕ to $\phi + d\phi$. The normalization condition for F in spherical polar coordinates is therefore

$$\int_0^\infty \left[\int_0^\pi \left[\int_0^{2\pi} |F^2(r, \theta, \phi)| \, d\phi \right] \sin \theta \, d\theta \right] r^2 \, dr = 1 \qquad (5.79)$$

If F happens to have the form

$$F(r, \theta, \phi) = R(r)S(\theta)T(\phi)$$

then use of the integral identity (3.74) gives for (5.79)

$$\int_0^\infty |R^2(r)| r^2 \, dr \int_0^\pi |S^2(\theta)| \sin \theta \, d\theta \int_0^{2\pi} |T^2(\phi)| \, d\phi = 1$$

and it is <u>convenient to normalize each factor of F separately</u>:

$$\int_0^\infty |R^2| r^2 \, dr = 1 , \qquad \int_0^\pi |S^2| \sin \theta \, d\theta = 1 , \qquad \int_0^{2\pi} |T^2| \, d\phi = 1 \qquad (5.80)$$

(We did the same thing for the wave function of the particle in a three-dimensional box.) Therefore,

$$\int_0^{2\pi} (Ae^{im\phi})^* Ae^{im\phi} \, d\phi = 1 = |A|^2 \int_0^{2\pi} d\phi$$

$$|A| = (2\pi)^{-1/2}$$

$$T(\phi) = \frac{1}{\sqrt{2\pi}} e^{im\phi} , \qquad m = 0, \pm 1, \pm 2, \ldots \qquad (5.81)$$

We now solve $\hat{L}^2 Y = cY$ [Eq. (5.70)] for the eigenvalues c of $\hat{L}^2$. Using (5.68) for $\hat{L}^2$, (5.72) for Y, and (5.81), we have

$$-\hbar^2 \left(\frac{\partial^2}{\partial \theta^2} + \cot \theta \frac{\partial}{\partial \theta} + \frac{1}{\sin^2 \theta} \frac{\partial^2}{\partial \phi^2} \right) \left(S(\theta) \frac{1}{\sqrt{2\pi}} e^{im\phi} \right) = cS(\theta) \frac{1}{\sqrt{2\pi}} e^{im\phi}$$

$$\frac{d^2 S}{d\theta^2} + \cot \theta \frac{dS}{d\theta} - \frac{m^2}{\sin^2 \theta} S = - \frac{c}{\hbar^2} S \qquad (5.82)$$

To solve (5.82), we carry out some tedious manipulations, which may be skimmed if desired. Begin reading again at Eq. (5.91). First, for convenience, we change the independent variable by making the substitution

$$w = \cos \theta \qquad (5.83)$$

This transforms S into some new function of w:

$$S(\theta) = G(w) \tag{5.84}$$

The chain rule gives

$$\frac{dS}{d\theta} = \frac{dG}{dw}\frac{dw}{d\theta} = -\sin\theta\,\frac{dG}{dw} = -(1-w^2)^{1/2}\frac{dG}{dw} \tag{5.85}$$

To calculate $d^2S/d\theta^2$, we use some operator algebra:

$$\frac{d}{d\theta} = -(1-w^2)^{1/2}\frac{d}{dw}$$

$$\frac{d^2}{d\theta^2} = (1-w^2)^{1/2}\frac{d}{dw}(1-w^2)^{1/2}\frac{d}{dw}$$

$$\frac{d^2}{d\theta^2} = (1-w^2)\frac{d^2}{dw^2} + (1-w^2)^{1/2}(\tfrac{1}{2})(1-w^2)^{-1/2}(-2w)\frac{d}{dw}$$

$$\frac{d^2S}{d\theta^2} = (1-w^2)\frac{d^2G}{dw^2} - w\frac{dG}{dw} \tag{5.86}$$

Using (5.86), (5.85), and $\cot\theta = \cos\theta/\sin\theta = w/(1-w^2)^{1/2}$, we find that (5.82) becomes

$$(1-w^2)\frac{d^2G}{dw^2} - 2w\frac{dG}{dw} + \left[\frac{c}{\hbar^2} - \frac{m^2}{(1-w^2)}\right]G(w) = 0 \tag{5.87}$$

The range of w is $-1 \le w \le 1$.

To get a two-term recursion relation when we try a power-series solution, we make the following change of dependent variable:

$$G(w) = (1-w^2)^{|m|/2}H(w) \tag{5.88}$$

Differentiating (5.88), we evaluate G' and G'', and (5.87) becomes, after we divide by $(1-w^2)^{|m|/2}$,

$$(1-w^2)H'' - 2(|m|+1)wH' + [c\hbar^{-2} - |m|(|m|+1)]H = 0 \tag{5.89}$$

We now try a power series for H:

$$H(w) = \sum_{j=0}^{\infty} a_j w^j \tag{5.90}$$

Differentiating [compare Eqs. (4.38)–(4.40)], we have

$$H'(w) = \sum_{j=0}^{\infty} ja_j w^{j-1}$$

$$H''(w) = \sum_{j=0}^{\infty} j(j-1)a_j w^{j-2} = \sum_{j=0}^{\infty} (j+2)(j+1)a_{j+2}w^j$$

Substitution of these power series into (5.89) yields, after combining sums,

$$\sum_{j=0}^{\infty}\left[(j+2)(j+1)a_{j+2} + \left(-j^2 - j - 2|m|j + \frac{c}{\hbar^2} - |m|^2 - |m|\right)a_j\right]w^j = 0$$

Setting the coefficient of w^j equal to zero, we have the recursion relation

$$a_{j+2} = \frac{[(j+|m|)(j+|m|+1) - c/\hbar^2]}{(j+1)(j+2)}a_j \tag{5.91}$$

Just as in the harmonic-oscillator case, the general solution of (5.89) is an arbitrary linear combination of a series of even powers (whose coefficients are determined by a_0) and a series of odd powers (whose coefficients are determined by a_1). It can be shown that the infinite series defined by the recursion relation (5.91) does not give well-behaved eigenfunctions. [Many texts point out that the infinite series diverges at $w = \pm 1$. However, this is not sufficient cause to reject the infinite series, since the eigenfunctions might be quadratically integrable, even though infinite at two points. For a careful discussion, see M. Whippman, *Am. J. Phys.*, **34**, 656 (1966).] Hence, as in the harmonic-oscillator case, we must cause one of the series to break off, its last term being $a_k w^k$. We eliminate the other series by setting a_0 or a_1 equal to zero, depending on whether k is odd or even.

Setting the coefficient of a_k in (5.91) equal to zero, we have

$$c = \hbar^2(k + |m|)(k + |m| + 1), \qquad k = 0, 1, 2, \ldots \qquad (5.92)$$

Since $|m|$ takes on the values $0, 1, 2, \ldots$, the quantity $k + |m|$ takes on the values $0, 1, 2, \ldots$. We therefore define the quantum number l as

$$l \equiv k + |m| \qquad (5.93)$$

and the allowed eigenvalues for the square of the magnitude of angular momentum are

$$c = l(l + 1)\hbar^2, \qquad l = 0, 1, 2, \ldots \qquad (5.94)$$

The magnitude of the orbital angular momentum of a particle is

$$|\mathbf{L}| = [l(l + 1)]^{1/2}\hbar \qquad (5.95)$$

From (5.93), it follows that $|m| \leq l$. The possible values for m are thus

$$m = -l, -l + 1, -l + 2, \ldots, -1, 0, 1, \ldots, l - 2, l - 1, l \qquad (5.96)$$

Let us examine the angular-momentum eigenfunctions. From (5.83), (5.84), (5.88), (5.90), and (5.93), the theta factor in the eigenfunctions is

$$S_{l,m}(\theta) = \sin^{|m|}(\theta) \sum_{\substack{j=1, 3, \ldots \\ \text{or } j=0, 2, \ldots}}^{l-|m|} a_j \cos^j \theta \qquad (5.97)$$

where the sum is over even or odd values of j, depending on whether $l - |m|$ is even or odd. The coefficients a_j satisfy the recursion relation (5.91), which, using (5.94), becomes

$$a_{j+2} = \frac{[(j + |m|)(j + |m| + 1) - l(l + 1)]}{(j + 1)(j + 2)} a_j \qquad (5.98)$$

The $\hat{L}^2$ and $\hat{L}_z$ eigenfunctions are given by Eqs. (5.72) and (5.81) as

$$Y_l^m(\theta, \phi) = S_{l,m}(\theta) T(\phi) = \frac{1}{\sqrt{2\pi}} S_{l,m}(\theta) e^{im\phi} \qquad \textbf{(5.99)}^*$$

EXAMPLE   Find $Y_l^m(\theta, \phi)$ and the $\hat{L}^2$ and $\hat{L}_z$ eigenvalues for (a) $l = 0$; (b) $l = 1$.
(a) For $l = 0$, Eq. (5.96) gives $m = 0$, and (5.97) becomes

$$S_{0,0}(\theta) = a_0 \qquad (5.100)$$

The normalization condition (5.80) gives

$$\int_0^\pi |a_0^2| \sin \theta \; d\theta = 1 = 2|a_0^2|$$

$$|a_0| = 2^{-1/2}$$

Equation (5.99) gives

$$Y_0^0(\theta, \phi) = \frac{1}{\sqrt{4\pi}} \qquad (5.101)$$

[Obviously, (5.101) is an eigenfunction of the operators $\hat{L}^2$, $\hat{L}_x$, $\hat{L}_y$, and $\hat{L}_z$, Eqs. (5.65)–(5.68).] For $l = 0$, there is no angular dependence in the eigenfunction; we say that the eigenfunctions are *spherically symmetric* for $l = 0$.
For $l = 0$ and $m = 0$, Eqs. (5.69), (5.70), (5.75), and (5.94) give the $\hat{L}^2$ eigenvalue as $c = 0$ and the $\hat{L}_z$ eigenvalue as $b = 0$.
(b) For $l = 1$, the possible values for m in (5.96) are -1, 0, and 1. For $|m| = 1$, (5.97) gives

$$S_{1,\pm 1}(\theta) = a_0 \sin \theta \qquad (5.102)$$

Note that a_0 in (5.102) is not necessarily the same as a_0 in (5.100). Normalization gives

$$1 = |a_0^2| \int_0^\pi \sin^2 \theta \sin \theta \; d\theta = |a_0^2| \int_{-1}^1 (1 - w^2) \; dw$$

$$|a_0| = \sqrt{3}/2$$

where the substitution $w = \cos \theta$ was made. Thus $S_{1,\pm 1} = (3^{1/2}/2) \sin \theta$ and (5.99) gives

$$Y_1^1 = (3/8\pi)^{1/2} \sin \theta \; e^{i\phi} , \qquad Y_1^{-1} = (3/8\pi)^{1/2} \sin \theta \; e^{-i\phi} \qquad (5.103)$$

For $l = 1$ and $m = 0$, (5.97) gives $S_{1,0} = a_1 \cos \theta$. Normalizing, we find $S_{1,0} = (3/2)^{1/2} \cos \theta$. Hence $Y_1^0 = (3/4\pi)^{1/2} \cos \theta$.
For $l = 1$, (5.94) gives the $\hat{L}^2$ eigenvalues as $2\hbar^2$; for $m = -1$, 0, and 1, (5.75) gives the $\hat{L}_z$ eigenvalues as $-\hbar$, 0, and $\hbar$, respectively.

The functions $S_{l,m}(\theta)$ are well known in mathematics, and are *associated Legendre functions* multiplied by a normalization constant. The associated Legendre functions $P_l^{|m|}(w)$ are defined by

$$P_l^{|m|}(w) \equiv \frac{1}{2^l l!} (1 - w^2)^{|m|/2} \frac{d^{l+|m|}}{dw^{l+|m|}} (w^2 - 1)^l , \qquad l = 0, 1, 2, \ldots \qquad (5.104)$$

These functions are related to the *Legendre polynomials* $P_l(w)$, which are defined by

$$P_l(w) \equiv \frac{1}{2^l l!} \frac{d^l}{dw^l} (w^2 - 1)^l , \qquad l = 0, 1, 2, \ldots \tag{5.105}$$

From the definitions, we have

$$P_l^{|m|}(w) = (1 - w^2)^{|m|/2} \frac{d^{|m|}}{dw^{|m|}} P_l(w) , \qquad P_l^0(w) = P_l(w) \tag{5.106}$$

The first few Legendre polynomials are

$$\begin{aligned} P_0(w) &= 1 & P_2(w) &= \tfrac{1}{2}(3w^2 - 1) \\ P_1(w) &= w & P_3(w) &= \tfrac{1}{2}(5w^3 - 3w) \end{aligned} \tag{5.107}$$

Some associated Legendre functions are

$$\begin{aligned} P_0^0(w) &= 1 & P_2^0(w) &= \tfrac{1}{2}(3w^2 - 1) \\ P_1^0(w) &= w & P_2^1(w) &= 3w(1 - w^2)^{1/2} \\ P_1^1(w) &= (1 - w^2)^{1/2} & P_2^2(w) &= 3 - 3w^2 \end{aligned} \tag{5.108}$$

It can be shown that (*Pauling and Wilson*, page 129)

$$S_{l,m}(\theta) = \left[\frac{(2l + 1)}{2} \frac{(l - |m|)!}{(l + |m|)!} \right]^{1/2} P_l^{|m|}(\cos \theta) \tag{5.109}$$

Equations (5.109) and (5.104) give the explicit formula for the normalized theta factor in the angular-momentum eigenfunctions. Using (5.109), we construct Table 5.1, which gives the theta factor in the angular-momentum eigenfunctions.

The eigenfunctions of $\hat{L}^2$ and $\hat{L}_z$ are called **spherical harmonics** (or *surface harmonics*) and are given by Eqs. (5.109) and (5.99) as

$$Y_l^m(\theta, \phi) = \left[\frac{2l + 1}{4\pi} \frac{(l - |m|)!}{(l + |m|)!} \right]^{1/2} P_l^{|m|}(\cos \theta) e^{im\phi} \tag{5.110}$$

The phase of the normalization constant of the spherical harmonics is arbitrary

TABLE 5.1 $S_{l,m}(\theta)$

$l = 0$:	$S_{0,0}$	$= \tfrac{1}{2}\sqrt{2}$
$l = 1$:	$S_{1,0}$	$= \tfrac{1}{2}\sqrt{6} \cos \theta$
	$S_{1,\pm 1}$	$= \tfrac{1}{2}\sqrt{3} \sin \theta$
$l = 2$:	$S_{2,0}$	$= \tfrac{1}{4}\sqrt{10}(3 \cos^2 \theta - 1)$
	$S_{2,\pm 1}$	$= \tfrac{1}{2}\sqrt{15} \sin \theta \cos \theta$
	$S_{2,\pm 2}$	$= \tfrac{1}{4}\sqrt{15} \sin^2 \theta$
$l = 3$:	$S_{3,0}$	$= \tfrac{3}{4}\sqrt{14}(\tfrac{5}{3} \cos^3 \theta - \cos \theta)$
	$S_{3,\pm 1}$	$= \tfrac{1}{8}\sqrt{42} \sin \theta (5 \cos^2 \theta - 1)$
	$S_{3,\pm 2}$	$= \tfrac{1}{4}\sqrt{105} \sin^2 \theta \cos \theta$
	$S_{3,\pm 3}$	$= \tfrac{1}{8}\sqrt{70} \sin^3 \theta$

(see Section 1.7); many texts use a different phase convention than in (5.110). Thus Y_l^m differs from text to text by a minus sign.

Summarizing our results, we see that the one-particle orbital angular-momentum eigenfunctions and eigenvalues are [Eqs. (5.69), (5.70), (5.75), and (5.94)]

$$\hat{L}^2 Y_l^m(\theta, \phi) = l(l+1)\hbar^2 Y_l^m(\theta, \phi) , \qquad l = 0, 1, 2, \ldots \qquad \textbf{(5.111)}^*$$

$$\hat{L}_z Y_l^m(\theta, \phi) = m\hbar Y_l^m(\theta, \phi) , \qquad m = -l, -l+1, \ldots, l-1, l \qquad \textbf{(5.112)}^*$$

where the eigenfunctions are given by (5.110). Often the symbol m_l is used instead of m for the L_z quantum number.

Since $l \geq |m|$, the magnitude $[l(l+1)]^{1/2}\hbar$ of the orbital angular momentum $\mathbf{L}$ is greater than the magnitude $|m|\hbar$ of its z component L_z, except for $l = 0$. If it were possible to have the angular-momentum magnitude equal to its z component, this would mean that the x and y components were zero, and we would have specified all three components of $\mathbf{L}$. However, since the components of angular momentum do not commute with each other, we cannot do this. The one exception is when l is zero. In this case, $|\mathbf{L}|^2 = L_x^2 + L_y^2 + L_z^2$ has zero for its eigenvalue, and it must be true that all three components L_x, L_y, and L_z have zero eigenvalues. From Eq. (5.12), the uncertainties in angular-momentum components satisfy

$$\Delta L_x \, \Delta L_y \geq \frac{1}{2} \left| \int \Psi^* [\hat{L}_x, \hat{L}_y] \Psi \, d\tau \right| = \frac{\hbar}{2} \left| \int \Psi^* \hat{L}_z \Psi \, d\tau \right| \qquad (5.113)$$

and two similar equations obtained by cyclic permutation. When the eigenvalues of $\hat{L}_z$, $\hat{L}_x$, and $\hat{L}_y$ are zero, $\hat{L}_x \Psi = 0$, $\hat{L}_y \Psi = 0$, $\hat{L}_z \Psi = 0$, the right-hand sides of (5.113) and the two similar equations are zero, and having $\Delta L_x = \Delta L_y = \Delta L_z = 0$ is permitted. But what about the statement in Section 5.1 that to have simultaneous eigenfunctions of two operators the operators must commute? The answer is that this theorem refers to the possibility of having the whole set of eigenfunctions of one operator be eigenfunctions of the other operator. Thus, even though $\hat{L}_x$ and $\hat{L}_z$ do not commute, it is possible to have *some* of the eigenfunctions of $\hat{L}_z$ (those with $l = 0 = m$) be eigenfunctions of $\hat{L}_x$. However, it is impossible to have *all* the $\hat{L}_z$ eigenfunctions also be eigenfunctions of $\hat{L}_x$.

Since we cannot specify L_x and L_y, the vector $\mathbf{L}$ can lie anywhere on the surface of a cone whose axis is the z axis, whose altitude is $m\hbar$, and whose slant height is $\sqrt{l(l+1)}\hbar$ (Fig. 5.6). The possible orientations of $\mathbf{L}$ with respect to the z axis for the case $l = 1$ are shown in Fig. 5.7. For each eigenvalue of $\hat{L}^2$, there are $2l + 1$ different eigenfunctions Y_l^m, corresponding to the $2l + 1$ values of m. We say that the $\hat{L}^2$ eigenvalues are $(2l + 1)$-fold degenerate. (The term **degeneracy** is applicable to the eigenvalues of any operator, not just the Hamiltonian.)

Of course, there is nothing special about the z axis; all directions of space are equivalent. If we had chosen to specify L^2 and L_x (rather than L_z), we would have gotten the same eigenvalues for L_x as we found for L_z. However, it is easier to solve the $\hat{L}_z$ eigenvalue equation because $\hat{L}_z$ has a simple form in spherical polar coordinates, which involve the angle of rotation ϕ about the z axis.

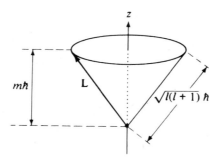

Figure 5.6 Orientation of **L**.

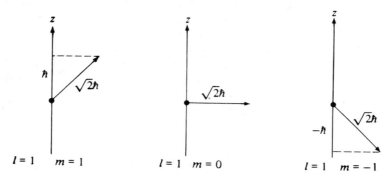

Figure 5.7 Orientations of **L** for $l = 1$.

5.4 THE LADDER-OPERATOR METHOD FOR ANGULAR MOMENTUM

We obtained the eigenvalues of $\hat{L}^2$ and $\hat{L}_z$ by expressing these orbital angular-momentum operators as differential operators and solving the resulting differential equations. We now demonstrate that it is possible to solve for these eigenvalues using only the operator commutation relations. The work in this section applies to any operators that satisfy the angular-momentum commutation relations. In particular, it applies to spin angular momentum (Chapter 10), as well as orbital angular momentum. The operator manipulations that follow are abstract and may be difficult to fully grasp at first reading. However, the process of finding the eigenvalues of operators without using their explicit forms has an elegance that will reward the persistent student.

We used the letter L for orbital angular momentum. Here we will use the letter M to indicate that we are dealing with any kind of angular momentum. We have three linear operators $\hat{M}_x$, $\hat{M}_y$, and $\hat{M}_z$, and all we know about them is that they obey the commutation relations [similar to (5.46) and (5.48)]

$$[\hat{M}_x, \hat{M}_y] = i\hbar \hat{M}_z , \qquad [\hat{M}_y, \hat{M}_z] = i\hbar \hat{M}_x , \qquad [\hat{M}_z, \hat{M}_x] = i\hbar \hat{M}_y \qquad (5.114)$$

We define the operator $\hat{M}^2$ as

$$\hat{M}^2 = \hat{M}_x^2 + \hat{M}_y^2 + \hat{M}_z^2 \tag{5.115}$$

Our problem is to find the eigenvalues of $\hat{M}^2$ and $\hat{M}_z$.

We begin by evaluating the commutators of $\hat{M}^2$ with its components, using Eqs. (5.114) and (5.115). The work is identical with that used to derive Eqs. (5.49) and (5.50), and we have

$$[\hat{M}^2, \hat{M}_x] = [\hat{M}^2, \hat{M}_y] = [\hat{M}^2, \hat{M}_z] = 0 \tag{5.116}$$

Hence we can have simultaneous eigenfunctions of $\hat{M}^2$ and $\hat{M}_z$.

Next we define two new operators, the **raising operator** $\hat{M}_+$ and the **lowering operator** $\hat{M}_-$:

$$\hat{M}_+ \equiv \hat{M}_x + i\hat{M}_y \tag{5.117}$$

$$\hat{M}_- \equiv \hat{M}_x - i\hat{M}_y \tag{5.118}$$

These are examples of **ladder operators**. The reason for the terminology will become clear shortly. Let us investigate their properties. We have

$$\hat{M}_+\hat{M}_- = (\hat{M}_x + i\hat{M}_y)(\hat{M}_x - i\hat{M}_y) = \hat{M}_x(\hat{M}_x - i\hat{M}_y) + i\hat{M}_y(\hat{M}_x - i\hat{M}_y)$$

$$= \hat{M}_x^2 - i\hat{M}_x\hat{M}_y + i\hat{M}_y\hat{M}_x + \hat{M}_y^2 = \hat{M}^2 - \hat{M}_z^2 + i[\hat{M}_y, \hat{M}_x]$$

$$\hat{M}_+\hat{M}_- = \hat{M}^2 - \hat{M}_z^2 + \hbar\hat{M}_z \tag{5.119}$$

Similarly, we find

$$\hat{M}_-\hat{M}_+ = \hat{M}^2 - \hat{M}_z^2 - \hbar\hat{M}_z \tag{5.120}$$

For the commutators of these operators with $\hat{M}_z$, we have

$$[\hat{M}_+, \hat{M}_z] = [\hat{M}_x + i\hat{M}_y, \hat{M}_z] = [\hat{M}_x, \hat{M}_z] + i[\hat{M}_y, \hat{M}_z] = -i\hbar\hat{M}_y - \hbar\hat{M}_x$$

$$[\hat{M}_+, \hat{M}_z] = -\hbar\hat{M}_+$$

$$\hat{M}_+\hat{M}_z = \hat{M}_z\hat{M}_+ - \hbar\hat{M}_+ \tag{5.121}$$

where (5.114) was used. Similarly, we find

$$\hat{M}_-\hat{M}_z = \hat{M}_z\hat{M}_- + \hbar\hat{M}_- \tag{5.122}$$

Using Y for the common eigenfunctions of $\hat{M}^2$ and $\hat{M}_z$, we have

$$\hat{M}^2 Y = cY \tag{5.123}$$

$$\hat{M}_z Y = bY \tag{5.124}$$

where c and b are the eigenvalues. Operating on Eq. (5.124) with $\hat{M}_+$, we get

$$\hat{M}_+\hat{M}_z Y = \hat{M}_+ bY$$

Using Eq. (5.121) and the fact that $\hat{M}_+$ is linear, we have

$$(\hat{M}_z\hat{M}_+ - \hbar\hat{M}_+)Y = b\hat{M}_+ Y$$

$$\hat{M}_z(\hat{M}_+ Y) = (b + \hbar)(\hat{M}_+ Y) \tag{5.125}$$

This last equation says that the function $\hat{M}_+ Y$ is an eigenfunction of $\hat{M}_z$ with eigenvalue $b + \hbar$. In other words, operating on the eigenfunction Y with the raising operator $\hat{M}_+$ converts Y into another eigenfunction of $\hat{M}_z$ with eigenvalue $\hbar$ higher than the eigenvalue of Y. If we now apply the raising operator to (5.125) and use (5.121) again, we find similarly

$$\hat{M}_z(\hat{M}_+^2 Y) = (b + 2\hbar)(\hat{M}_+^2 Y)$$

Repeated application of the raising operator gives

$$\hat{M}_z(\hat{M}_+^k Y) = (b + k\hbar)(\hat{M}_+^k Y), \qquad k = 0, 1, 2, \ldots \tag{5.126}$$

If we operate on (5.124) with the lowering operator and apply (5.122), we find in the same manner

$$\hat{M}_z(\hat{M}_- Y) = (b - \hbar)(\hat{M}_- Y) \tag{5.127}$$

$$\hat{M}_z(\hat{M}_-^k Y) = (b - k\hbar)(\hat{M}_-^k Y) \tag{5.128}$$

Thus by using the raising and lowering operators on the eigenfunction with eigenvalue b, we generate a ladder of eigenvalues, the difference from step to step being $\hbar$:

$$\cdots \quad b - 2\hbar, \quad b - \hbar, \quad b, \quad b + \hbar, \quad b + 2\hbar, \quad \cdots$$

The functions $\hat{M}_\pm^k Y$ are eigenfunctions of $\hat{M}_z$ with eigenvalues $b \pm k\hbar$ [Eqs. (5.126) and (5.128)]. We now show that these functions are also eigenfunctions of $\hat{M}^2$, all with the *same* eigenvalue c:

$$\hat{M}_z \hat{M}_\pm^k Y = (b \pm k\hbar) \hat{M}_\pm^k Y \tag{5.129}$$

$$\hat{M}^2 \hat{M}_\pm^k Y = c \hat{M}_\pm^k Y, \qquad k = 0, 1, 2, \ldots \tag{5.130}$$

To prove (5.130), we first show that $\hat{M}^2$ commutes with $\hat{M}_+$ and $\hat{M}_-$:

$$[\hat{M}^2, \hat{M}_\pm] = [\hat{M}^2, \hat{M}_x \pm i\hat{M}_y] = [\hat{M}^2, \hat{M}_x] \pm i[\hat{M}^2, \hat{M}_y] = 0 \pm 0 = 0$$

We also have

$$[\hat{M}^2, \hat{M}_\pm^2] = [\hat{M}^2, \hat{M}_\pm]\hat{M}_\pm + \hat{M}_\pm[\hat{M}^2, \hat{M}_\pm] = 0 + 0 = 0$$

and it follows by induction that

$$[\hat{M}^2, \hat{M}_\pm^k] = 0 \quad \text{or} \quad \hat{M}^2 \hat{M}_\pm^k = \hat{M}_\pm^k \hat{M}^2, \qquad k = 0, 1, 2, \ldots \tag{5.131}$$

If we operate on (5.123) with $\hat{M}_\pm^k$ and use (5.131), we get

$$\hat{M}_\pm^k \hat{M}^2 Y = \hat{M}_\pm^k c Y$$

$$\hat{M}^2(\hat{M}_\pm^k Y) = c(\hat{M}_\pm^k Y) \tag{5.132}$$

which is what we wanted to prove.

Next we show that the set of eigenvalues of $\hat{M}_z$ generated using the ladder operators must be bounded. For the particular eigenfunction Y with $\hat{M}_z$ eigenvalue b, we have

$$\hat{M}_z Y = bY$$

and for the set of eigenfunctions and eigenvalues generated by the ladder operators, we have

$$\hat{M}_z Y_k = b_k Y_k \tag{5.133}$$

where

$$Y_k = \hat{M}_\pm^k Y \tag{5.134}$$

$$b_k = b \pm k\hbar \tag{5.135}$$

(Application of $\hat{M}_+$ or $\hat{M}_-$ destroys the normalization of Y, so Y_k is not normalized. For the normalization constant, see Problem 10.19.)

Operating on (5.133) with $\hat{M}_z$, we have

$$\hat{M}_z^2 Y_k = b_k \hat{M}_z Y_k$$

$$\hat{M}_z^2 Y_k = b_k^2 Y_k \tag{5.136}$$

Now subtract (5.136) from (5.130), and use (5.134) and (5.115):

$$\hat{M}^2 Y_k - \hat{M}_z^2 Y_k = c Y_k - b_k^2 Y_k$$

$$(\hat{M}_x^2 + \hat{M}_y^2) Y_k = (c - b_k^2) Y_k \tag{5.137}$$

The operator $\hat{M}_x^2 + \hat{M}_y^2$ corresponds to a nonnegative physical quantity and hence has nonnegative eigenvalues. Therefore, (5.137) implies that $c - b_k^2 \geqslant 0$ and $c^{1/2} \geqslant |b_k|$. Thus

$$c^{1/2} \geqslant b_k \geqslant -c^{1/2}, \qquad k = 0, \pm 1, \pm 2, \ldots \tag{5.138}$$

Since c remains constant as k varies, (5.138) shows that the set of eigenvalues b_k is bounded above and below. Let b_{max} and b_{min} denote the maximum and minimum values of b_k. Y_{max} and Y_{min} will be the corresponding eigenfunctions:

$$\hat{M}_z Y_{max} = b_{max} Y_{max} \tag{5.139}$$

$$\hat{M}_z Y_{min} = b_{min} Y_{min} \tag{5.140}$$

Now operate on (5.139) with the raising operator and use (5.121):

$$\hat{M}_+ \hat{M}_z Y_{max} = b_{max} \hat{M}_+ Y_{max}$$

$$\hat{M}_z (\hat{M}_+ Y_{max}) = (b_{max} + \hbar)(\hat{M}_+ Y_{max}) \tag{5.141}$$

This last equation seems to contradict the statement that b_{max} is the largest eigenvalue of $\hat{M}_z$, since it says that $\hat{M}_+ Y_{max}$ is an eigenfunction of $\hat{M}_z$ with eigenvalue $b_{max} + \hbar$. The only way out of this contradiction is to have the function $\hat{M}_+ Y_{max}$ vanish. (We always reject zero as an eigenfunction on physical grounds.) Thus

$$\hat{M}_+ Y_{max} = 0 \tag{5.142}$$

Operating on (5.142) with the lowering operator and using (5.120), (5.139), and

(5.123), we have

$$\hat{M}_- \hat{M}_+ Y_{max} = 0$$

$$(\hat{M}^2 - \hat{M}_z^2 - \hbar \hat{M}_z) Y_{max} = 0$$

$$(c - b_{max}^2 - \hbar b_{max}) Y_{max} = 0$$

$$c - b_{max}^2 - \hbar b_{max} = 0$$

$$c = b_{max}^2 + \hbar b_{max} \tag{5.143}$$

A similar argument shows that

$$\hat{M}_- Y_{min} = 0 \tag{5.144}$$

and by applying the raising operator to this equation and using (5.119), we find

$$c = b_{min}^2 - \hbar b_{min}$$

Subtracting this last equation from (5.143), we have

$$b_{max}^2 + \hbar b_{max} + (\hbar b_{min} - b_{min}^2) = 0$$

This is a quadratic equation in the unknown b_{max}, and using the usual formula (it still works in quantum mechanics), we find

$$b_{max} = -b_{min}, \qquad b_{max} = b_{min} - \hbar$$

The second root is rejected, since it says that b_{max} is less than b_{min}. So

$$b_{min} = -b_{max} \tag{5.145}$$

Moreover, from (5.135) we know that b_{max} and b_{min} differ by an integral multiple of $\hbar$:

$$b_{max} - b_{min} = n\hbar, \qquad n = 0, 1, 2, \ldots \tag{5.146}$$

Substituting (5.145) in (5.146), we have for the $\hat{M}_z$ eigenvalues

$$b_{max} = \tfrac{1}{2} n\hbar$$

$$b_{max} = j\hbar, \qquad j = 0, \tfrac{1}{2}, 1, \tfrac{3}{2}, 2, \ldots \tag{5.147}$$

$$b_{min} = -j\hbar$$

$$b = -j\hbar, (-j+1)\hbar, (-j+2)\hbar, \ldots, (j-2)\hbar, (j-1)\hbar, j\hbar \tag{5.148}$$

and from (5.143) we find as the $\hat{M}^2$ eigenvalues

$$c = j(j+1)\hbar^2, \qquad j = 0, \tfrac{1}{2}, 1, \tfrac{3}{2}, \ldots \tag{5.149}$$

Thus

$$\hat{M}^2 Y = j(j+1)\hbar^2 Y, \qquad j = 0, \tfrac{1}{2}, 1, \tfrac{3}{2}, 2, \ldots \tag{5.150}$$

$$\hat{M}_z Y = m_j \hbar Y, \qquad m_j = -j, -j+1, \ldots, j-1, j \tag{5.151}$$

We have found the eigenvalues of $\hat{M}^2$ and $\hat{M}_z$ using just the commutation relations. However, comparison of (5.150) and (5.151) with (5.111) and (5.112) shows that in addition to integral values for the angular-momentum quantum number ($l = 0, 1, 2, \ldots$) we now also have the possibility for half-integral values ($j = 0, \frac{1}{2}, 1, \frac{3}{2}, \ldots$). This perhaps suggests that there might be another kind of angular momentum besides orbital angular momentum. In Chapter 10 we shall see that spin angular momentum can have half-integral, as well as integral, quantum numbers. For orbital angular momentum, the boundary condition of single valuedness of the $T(\phi)$ eigenfunctions [see the equation following (5.73)] eliminates the half-integral values of the angular-momentum quantum numbers. [Not everyone accepts single valuedness as a valid boundary condition on wave functions, and many other reasons have been given for rejecting half-integral orbital-angular-momentum quantum numbers; see C. G. Gray, *Am. J. Phys.*, **37**, 559 (1969); M. L. Whippman, *Am. J. Phys.*, **34**, 656 (1966).]

The ladder-operator method can be used to solve other eigenvalue problems; see Problem 5.30.

5.5 SUMMARY

For a complete set of eigenfunctions to be simultaneously eigenfunctions of several operators, each operator must commute with every other operator.

The standard deviation ΔA measures the uncertainty in a quantum-mechanical property A, where $(\Delta A)^2 = \langle A^2 \rangle - \langle A \rangle^2$. For the properties x and p_x, we have $\Delta x \, \Delta p_x \geq \frac{1}{2}\hbar$.

The vector **B** can be written as $\mathbf{B} = B_x \mathbf{i} + B_y \mathbf{j} + B_z \mathbf{k}$, where **i**, **j**, and **k** are unit vectors along the x, y, and z axes, and B_x, B_y, and B_z are the components of **B**. The magnitude of **B** is $|\mathbf{B}| = (B_x^2 + B_y^2 + B_z^2)^{1/2}$. The dot product of two vectors that make an angle θ with each other is $\mathbf{B} \cdot \mathbf{A} = B_x A_x + B_y A_y + B_z A_z = |\mathbf{B}| \, |\mathbf{A}| \cos \theta$. The cross product is given by (5.29).

The classical-mechanical definition of orbital angular momentum is $\mathbf{L} \equiv \mathbf{r} \times \mathbf{p}$. The operator $\hat{L}^2$ commutes with $\hat{L}_x$, $\hat{L}_y$, and $\hat{L}_z$, but $\hat{L}_x$, $\hat{L}_y$, and $\hat{L}_z$ do not commute with one another. When expressed in spherical polar coordinates, the operators $\hat{L}^2$, $\hat{L}_x$, $\hat{L}_y$, and $\hat{L}_z$ depend only on the angles θ (the angle between the z axis and **r**) and ϕ (the angle between the projection of **r** in the xy plane and the x axis) and not on the radial coordinate r.

The ranges of the spherical polar coordinates are 0 to π for θ, 0 to 2π for ϕ, and 0 to ∞ for r. The volume element in these coordinates is $d\tau = r^2 \sin \theta \, dr \, d\theta \, d\phi$.

The common eigenfunctions and eigenvalues of $\hat{L}^2$ and $\hat{L}_z$ are given by $\hat{L}^2 Y_l^m = l(l+1)\hbar^2 Y_l^m$ and $\hat{L}_z Y_l^m = m\hbar Y_l^m$, where the angular-momentum quantum numbers are $l = 0, 1, 2, \ldots$, and $m = -l, -l+1, \ldots, l-1, l$, and the functions $Y_l^m(\theta, \phi)$ are spherical harmonics.

For operators $\hat{M}_x$, $\hat{M}_y$, $\hat{M}_z$, and $\hat{M}^2 = \hat{M}_x^2 + \hat{M}_y^2 + \hat{M}_z^2$ that obey the angular-momentum commutation relations, use of the ladder operators $\hat{M}_+ = \hat{M}_x + i\hat{M}_y$ and $\hat{M}_- = \hat{M}_x - i\hat{M}_y$ gives the possible $\hat{M}^2$ eigenvalues as $j(j+1)\hbar^2$, where j can be integral or half-integral, and gives the $\hat{M}_z$ eigenvalues as $m_j \hbar$, where m_j ranges from $-j$ to j in integral steps.

PROBLEMS

5.1 Verify the commutator identities (5.1)–(5.5).

5.2 Verify Eq. (5.9) for $[\hat{p}_x, \hat{H}]$.

5.3 Find $[\hat{x}, \hat{p}_x^3]$ starting from (5.7) for $[\hat{x}, \hat{p}_x^2]$.

5.4 Derive $(\Delta A)^2 = \langle A^2 \rangle - \langle A \rangle^2$ [Eq. (5.11)].

5.5 Show that the standard deviation ΔA is 0 when Ψ is an eigenfunction of $\hat{A}$.

5.6 For the ground state of the one-dimensional harmonic oscillator, compute the standard deviations Δx and Δp_x, and check that the uncertainty principle is obeyed. Use the results of Problem 4.11 to save time.

5.7 At a certain instant of time, a particle in a one-dimensional box of length l (Fig. 2.1) is in a nonstationary state with $\Psi = (105/l^7)^{1/2}x^2(l - x)$ inside the box. For this state, find Δx and Δp_x and verify that the uncertainty principle $\Delta x \, \Delta p_x \geq \frac{1}{2}\hbar$ is obeyed.

5.8 Let w be the variable defined as the number of heads that show when two coins are tossed simultaneously. Find $\langle w \rangle$ and σ_w. [*Hint:* Use (3.81) and (5.11).]

5.9 Let **A** have the components $(3, -2, 6)$; let **B** have the components $(-1, 4, 4)$. Find $|\mathbf{A}|$, $|\mathbf{B}|$, $\mathbf{A} + \mathbf{B}$, $\mathbf{A} - \mathbf{B}$, $\mathbf{A} \cdot \mathbf{B}$, $\mathbf{A} \times \mathbf{B}$. Find the angle between **A** and **B**.

5.10 Use the vector dot product to find the obtuse angle between two diagonals of a cube. What is the chemical significance of this angle?

5.11 Show that

$$\nabla^2[f(x, y, z)g(x, y, z)] = g\nabla^2 f + 2\nabla f \cdot \nabla g + f\nabla^2 g$$

5.12 Let $f = 2x^2 - 5xyz + z^2 - 1$. Find **grad** f. Find $\nabla^2 f$.

5.13 The *divergence* of a vector function **A** is a scalar function defined by

$$\text{div } \mathbf{A} \equiv \nabla \cdot \mathbf{A} = \left(\mathbf{i}\frac{\partial}{\partial x} + \mathbf{j}\frac{\partial}{\partial y} + \mathbf{k}\frac{\partial}{\partial z} \right) \cdot (A_x\mathbf{i} + A_y\mathbf{j} + A_z\mathbf{k})$$

$$\text{div } \mathbf{A} \equiv \frac{\partial A_x}{\partial x} + \frac{\partial A_y}{\partial y} + \frac{\partial A_z}{\partial z}$$

(a) Verify that div $[\mathbf{grad}\, g(x, y, z)] \equiv \nabla \cdot \nabla g = \partial^2 g/\partial x^2 + \partial^2 g/\partial y^2 + \partial^2 g/\partial z^2$. This is the origin of the notation ∇^2 for $\partial^2/\partial x^2 + \partial^2/\partial y^2 + \partial^2/\partial z^2$. (b) Find $\nabla \cdot \mathbf{r}$, where $\mathbf{r} = \mathbf{i}x + \mathbf{j}y + \mathbf{k}z$.

5.14 The *curl* of a vector function **A** is defined by $\mathbf{curl\, A} \equiv \nabla \times \mathbf{A}$. Prove that **curl grad** $g(x, y, z) = 0$ for all well-behaved functions g.

5.15 Derive Eq. (5.68) for $\hat{L}^2$ from Eqs. (5.65)–(5.67).

5.16 (a) Show that the three commutation relations (5.46) and (5.48) are equivalent to the single relation $\hat{\mathbf{L}} \times \hat{\mathbf{L}} = i\hbar\hat{\mathbf{L}}$. (b) Find $[\hat{L}_x^2, \hat{L}_y]$.

5.17 Consider the following incorrect derivation. Differentiation of x in (5.51) gives $\partial x/\partial r = \sin\theta\cos\phi$. Then, since $\partial r/\partial x = 1/(\partial x/\partial r)$, we have $(\partial r/\partial x)_{y,z} = 1/(\sin\theta\cos\phi)$ (?). But this result disagrees with (5.57). Find the error in this reasoning.

5.18 Find the spherical polar coordinates for points with the following (x, y, z) coordinates: (a) $(1, 2, 0)$; (b) $(-1, 0, 3)$; (c) $(3, 1, -2)$; (d) $(-1, -1, -1)$.

5.19 Find the (x, y, z) coordinates of the points with the following spherical polar coordinates: (a) $r = 1$, $\theta = \pi/2$, $\phi = \pi$; (b) $r = 2$, $\theta = \pi/4$, $\phi = 0$.

5.20 Give the shape of a surface on which (a) r is constant; (b) θ is constant; (c) ϕ is constant.

5.21 By integrating the spherical-polar-coordinates differential volume element $d\tau$ over appropriate limits, verify the formula $\frac{4}{3}\pi R^3$ for the volume of a sphere of radius R.

5.22 Derive the formula for $S_{2,0}$ (Table 5.1) in two ways: (a) by using (5.109); (b) by using the recursion relation and normalization.

5.23 Calculate the possible angles between **L** and the z axis for $l = 2$.

5.24 Show that the spherical harmonics are eigenfunctions of the operator $\hat{L}_x^2 + \hat{L}_y^2$. (The proof is short.) What are the eigenvalues?

5.25 (a) If we measure L_z of a particle that has angular-momentum quantum number $l = 2$, what are the possible outcomes of the measurement? (b) If we measure L_z of a particle whose state function is an eigenfunction of $\hat{L}^2$ with eigenvalue $12\hbar^2$, what are the possible outcomes of the measurement?

5.26 At a certain instant of time t', a particle has the state function $\Psi = Ne^{-ar^2}Y_2^1(\theta, \phi)$, where N and a are constants. (a) If L^2 of this particle were to be measured at time t', what would be the outcome? Give a numerical answer. (b) If L_z of this particle were to be measured at t', what would be the outcome? Give a numerical answer.

5.27 Use the recursion relation (5.98) and normalization to find (a) Y_3^0; (b) Y_3^1.

5.28 Complete this equation: $\hat{L}_z^3 Y_l^m = ?$

5.29 Apply the lowering operator $\hat{L}_-$ three times in succession to $Y_1^1(\theta, \phi)$ and verify that we obtain functions that are proportional to Y_1^0, Y_1^{-1}, and zero.

5.30 The one-dimensional harmonic-oscillator Hamiltonian is

$$\hat{H} = \frac{\hat{p}_x^2}{2m} + 2\pi^2\nu^2 m\hat{x}^2$$

The raising and lowering operators for this problem are defined as

$$\hat{A}_+ \equiv \frac{1}{(2m)^{1/2}}[\hat{p}_x + 2\pi i\nu m\hat{x}], \qquad \hat{A}_- \equiv \frac{1}{(2m)^{1/2}}[\hat{p}_x - 2\pi i\nu m\hat{x}]$$

Show that

$$\hat{A}_+\hat{A}_- = \hat{H} - \tfrac{1}{2}h\nu, \qquad \hat{A}_-\hat{A}_+ = \hat{H} + \tfrac{1}{2}h\nu$$

$$[\hat{A}_+, \hat{A}_-] = -h\nu$$

$$[\hat{H}, \hat{A}_+] = h\nu\hat{A}_+, \qquad [\hat{H}, \hat{A}_-] = -h\nu\hat{A}_-$$

Show that $\hat{A}_+$ and $\hat{A}_-$ are indeed ladder operators and that the eigenvalues are spaced at intervals of $h\nu$. Since both the kinetic energy and the potential energy are nonnegative, we expect the energy eigenvalues to be nonnegative. Hence there must be a state of minimum energy. Operate on the wave function for this state first with $\hat{A}_-$ and then with $\hat{A}_+$ and show that the lowest-energy eigenvalue is $\tfrac{1}{2}h\nu$. Finally, conclude that

$$E = (n + \tfrac{1}{2})h\nu, \qquad n = 0, 1, 2, \ldots$$

(See also Problem 7.54.)

6

The Hydrogen Atom

6.1 THE ONE-PARTICLE CENTRAL-FORCE PROBLEM

Before tackling the hydrogen atom, we shall consider the more general problem of a single particle moving under a central force. The results of this section will apply to any central-force problem, for example, the hydrogen atom (Section 6.5), the isotropic three-dimensional harmonic oscillator (Problem 6.1).

A *central force* is one derived from a potential-energy function that is spherically symmetric, which means that it is a function only of the distance of the particle from the origin: $V = V(r)$. The relation between force and potential energy is given by (5.32) as

$$\mathbf{F} = -\boldsymbol{\nabla}V(x, y, z) = -\mathbf{i}(\partial V/\partial x) - \mathbf{j}(\partial V/\partial y) - \mathbf{k}(\partial V/\partial z) \qquad (6.1)$$

The partial derivatives in (6.1) can be found by the chain rule [Eqs. (5.53)–(5.55)]. Since V in this case is a function of r only, we have

$$\left(\frac{\partial V}{\partial \theta}\right)_{r,\phi} = \left(\frac{\partial V}{\partial \phi}\right)_{r,\theta} = 0$$

Hence

$$\left(\frac{\partial V}{\partial x}\right)_{y,z} = \frac{dV}{dr}\left(\frac{\partial r}{\partial x}\right)_{y,z} = \frac{x}{r}\frac{dV}{dr} \qquad (6.2)$$

$$\left(\frac{\partial V}{\partial y}\right)_{x,z} = \frac{y}{r}\frac{dV}{dr}, \qquad \left(\frac{\partial V}{\partial z}\right)_{x,y} = \frac{z}{r}\frac{dV}{dr} \qquad (6.3)$$

where Eqs. (5.57) and (5.58) have been used. Equation (6.1) becomes

$$\mathbf{F} = -\frac{1}{r}\frac{dV}{dr}(x\mathbf{i} + y\mathbf{j} + z\mathbf{k}) = -\frac{dV(r)}{dr}\frac{\mathbf{r}}{r} \qquad (6.4)$$

where (5.33) for $\mathbf{r}$ was used. The quantity $\mathbf{r}/r$ in (6.4) is a unit vector in the radial direction. A central force is radially directed.

Now we consider the quantum mechanics of a single particle subject to a central force. The Hamiltonian operator is

$$\hat{H} = \hat{T} + \hat{V} = -(\hbar^2/2m)\nabla^2 + V(r) \qquad (6.5)$$

where $\nabla^2 \equiv \partial^2/\partial x^2 + \partial^2/\partial y^2 + \partial^2/\partial z^2$ [Eq. (3.46)]. Since V is spherically symmetric, we shall work in spherical polar coordinates. Hence we want to transform the Laplacian operator to these coordinates. We already have the forms of the operators $\partial/\partial x$, $\partial/\partial y$, and $\partial/\partial z$ in these coordinates [Eqs. (5.62)–(5.64)], and by squaring each of these operators and then adding their squares, we get the Laplacian. This calculation is left as an exercise. The result is (Problem 6.3)

$$\nabla^2 = \frac{\partial^2}{\partial r^2} + \frac{2}{r}\frac{\partial}{\partial r} + \frac{1}{r^2}\frac{\partial^2}{\partial \theta^2} + \frac{1}{r^2}\cot\theta\,\frac{\partial}{\partial \theta} + \frac{1}{r^2\sin^2\theta}\frac{\partial^2}{\partial \phi^2} \tag{6.6}$$

Looking back to (5.68), which gives the operator for the square of the magnitude of the orbital angular momentum of a single particle, $\hat{L}^2$, we see that

$$\nabla^2 = \frac{\partial^2}{\partial r^2} + \frac{2}{r}\frac{\partial}{\partial r} - \frac{1}{r^2\hbar^2}\hat{L}^2 \tag{6.7}$$

The Hamiltonian (6.5) becomes

$$\hat{H} = -\frac{\hbar^2}{2m}\left(\frac{\partial^2}{\partial r^2} + \frac{2}{r}\frac{\partial}{\partial r}\right) + \frac{1}{2mr^2}\hat{L}^2 + V(r) \tag{6.8}$$

In classical mechanics a particle subject to a central force has its angular momentum conserved (Section 5.3). In quantum mechanics we might ask whether we can have states with definite values for both the energy and the angular momentum. To have the set of eigenfunctions of $\hat{H}$ also be eigenfunctions of $\hat{L}^2$, the commutator $[\hat{H}, \hat{L}^2]$ must vanish. We have

$$[\hat{H}, \hat{L}^2] = [\hat{T}, \hat{L}^2] + [\hat{V}, \hat{L}^2]$$

$$[\hat{T}, \hat{L}^2] = \left[-\frac{\hbar^2}{2m}\left(\frac{\partial^2}{\partial r^2} + \frac{2}{r}\frac{\partial}{\partial r}\right) + \frac{1}{2mr^2}\hat{L}^2, \hat{L}^2\right]$$

$$[\hat{T}, \hat{L}^2] = -\frac{\hbar^2}{2m}\left[\frac{\partial^2}{\partial r^2} + \frac{2}{r}\frac{\partial}{\partial r}, \hat{L}^2\right] + \frac{1}{2m}\left[\frac{1}{r^2}\hat{L}^2, \hat{L}^2\right] \tag{6.9}$$

Recall that $\hat{L}^2$ involves only θ and ϕ and not r [Eq. (5.68)]. Hence it commutes with any operator that involves only r. [To reach this conclusion, we must use relations like (5.47) with x and z replaced by r and θ.] Thus the first commutator in (6.9) is zero. Moreover, since any operator commutes with itself, the second commutator in (6.9) is zero. Therefore, $[\hat{T}, \hat{L}^2] = 0$. Also, since $\hat{L}^2$ does not involve r and V is a function of r only, we have $[\hat{V}, \hat{L}^2] = 0$. Therefore,

$$[\hat{H}, \hat{L}^2] = 0, \qquad \text{if } V = V(r) \tag{6.10}$$

The Hamiltonian commutes with $\hat{L}^2$ when the potential-energy function is independent of θ and ϕ.

Now consider the operator $\hat{L}_z = -i\hbar\,\partial/\partial\phi$ [Eq. (5.67)]. Since $\hat{L}_z$ does not involve r and since it commutes with $\hat{L}^2$ [Eq. (5.50)], it follows that $\hat{L}_z$ commutes with the Hamiltonian (6.8):

$$[\hat{H}, \hat{L}_z] = 0, \qquad \text{if } V = V(r) \tag{6.11}$$

We can therefore have a set of simultaneous eigenfunctions of $\hat{H}$, $\hat{L}^2$, and $\hat{L}_z$ for the central-force problem. Let ψ denote these common eigenfunctions:

$$\hat{H}\psi = E\psi \tag{6.12}*$$

$$\hat{L}^2\psi = l(l+1)\hbar^2\psi, \qquad l = 0, 1, 2, \ldots \tag{6.13}*$$

$$\hat{L}_z\psi = m\hbar\psi, \qquad m = -l, -l+1, \ldots, l \tag{6.14}*$$

where Eqs. (5.111) and (5.112) have been used.

Using (6.8) and (6.13), we have for the Schrödinger equation (6.12)

$$\left[-\frac{\hbar^2}{2m}\left(\frac{\partial^2}{\partial r^2} + \frac{2}{r}\frac{\partial}{\partial r}\right) + \frac{1}{2mr^2}\hat{L}^2 + V(r)\right]\psi = E\psi$$

$$\left[-\frac{\hbar^2}{2m}\left(\frac{\partial^2}{\partial r^2} + \frac{2}{r}\frac{\partial}{\partial r}\right) + \frac{l(l+1)\hbar^2}{2mr^2} + V(r)\right]\psi = E\psi \tag{6.15}$$

The eigenfunctions of $\hat{L}^2$ are the spherical harmonics $Y_l^m(\theta, \phi)$, and since $\hat{L}^2$ does not involve r, we can multiply Y_l^m by an arbitrary function of r and still have eigenfunctions of $\hat{L}^2$ and $\hat{L}_z$. Therefore,

$$\psi = R(r)Y_l^m(\theta, \phi) \tag{6.16}*$$

Using (6.16) in (6.15), we then divide both sides by Y_l^m to obtain an ordinary differential equation for the unknown function $R(r)$:

$$-\frac{\hbar^2}{2m}\left(R'' + \frac{2}{r}R'\right) + \frac{l(l+1)\hbar^2}{2mr^2}R + V(r)R = ER(r) \tag{6.17}$$

We have shown that, *for any one-particle problem with a spherically symmetric potential-energy function $V(r)$, the wave function is $\psi = R(r)Y_l^m(\theta, \phi)$, where the radial factor $R(r)$ satisfies (6.17)*. By using a specific form for $V(r)$ in (6.17), we can solve it for a particular problem.

6.2 NONINTERACTING PARTICLES AND SEPARATION OF VARIABLES

Up to this point, we have solved only one-particle quantum-mechanical problems. The hydrogen atom is a two-particle system, and as a preliminary to dealing with the H atom, we first consider a simpler case, that of two noninteracting particles.

Suppose that a system is composed of the noninteracting particles 1 and 2. Let q_1 symbolize the coordinates (x_1, y_1, z_1) of particle 1 and let q_2 symbolize the coordinates (x_2, y_2, z_2) of particle 2. Because the particles exert no forces on each other, the classical-mechanical energy of the system is the sum of the energies of the two particles: $E = E_1 + E_2 = T_1 + V_1 + T_2 + V_2$, and the classical Hamiltonian is the sum of Hamiltonians for each particle: $H = H_1 + H_2$. Therefore, the Hamiltonian operator is

$$\hat{H} = \hat{H}_1 + \hat{H}_2$$

where $\hat{H}_1$ involves only the coordinates q_1 and the momentum operators $\hat{p}_1$ that correspond to q_1. The Schrödinger equation for the system is

$$(\hat{H}_1 + \hat{H}_2)\psi(q_1, q_2) = E\psi(q_1, q_2) \tag{6.18}$$

We attempt a solution of (6.18) by separation of variables, setting

$$\psi(q_1, q_2) = G_1(q_1)G_2(q_2) \tag{6.19}$$

We have

$$\hat{H}_1 G_1(q_1)G_2(q_2) + \hat{H}_2 G_1(q_1)G_2(q_2) = EG_1(q_1)G_2(q_2) \tag{6.20}$$

Since $\hat{H}_1$ involves only the coordinate and momentum operators of particle 1, we have

$$\hat{H}_1[G_1(q_1)G_2(q_2)] = G_2(q_2)\hat{H}_1 G_1(q_1) \tag{6.21}$$

since, as far as $\hat{H}_1$ is concerned, G_2 is a constant. Using (6.21) and a similar equation for $\hat{H}_2$, we find that Eq. (6.20) becomes

$$G_2(q_2)\hat{H}_1 G_1(q_1) + G_1(q_1)\hat{H}_2 G_2(q_2) = EG_1(q_1)G_2(q_2)$$

$$\frac{\hat{H}_1 G_1(q_1)}{G_1(q_1)} + \frac{\hat{H}_2 G_2(q_2)}{G_2(q_2)} = E \tag{6.22}$$

Now, by the same arguments used in connection with Eq. (3.65), we conclude that each term on the left in (6.22) must be a constant. Using E_1 and E_2 to denote these constants, we have

$$\frac{\hat{H}_1 G_1(q_1)}{G_1(q_1)} = E_1 , \qquad \frac{\hat{H}_2 G_2(q_2)}{G_2(q_2)} = E_2$$

$$E = E_1 + E_2 \tag{6.23}$$

In other words, when the system is composed of two noninteracting particles, we can reduce the two-particle problem to two separate one-particle problems by solving

$$\hat{H}_1 G_1(q_1) = E_1 G_1(q_1) , \qquad \hat{H}_2 G_2(q_2) = E_2 G_2(q_2) \tag{6.24}$$

which are separate Schrödinger equations for each particle.

This result is easily generalized to any number of noninteracting particles. For n such particles, we have

$$\hat{H} = \hat{H}_1 + \hat{H}_2 + \cdots + \hat{H}_n$$

$$\psi(q_1, q_2, \ldots, q_n) = G_1(q_1)G_2(q_2)\cdots G_n(q_n) \tag{6.25}*$$

$$E = E_1 + E_2 + \cdots + E_n \tag{6.26}*$$

$$\hat{H}_i G_i = E_i G_i , \qquad i = 1, 2, \ldots, n \tag{6.27}*$$

For a system of noninteracting particles, the energy is the sum of the individual energies of each particle and the wave function is the product of wave functions for each particle; the wave function of particle i is found by solving a Schrödinger equation for particle i using the Hamiltonian $\hat{H}_i$.

These results also apply to a single particle whose Hamiltonian is the sum of separate terms for each coordinate:

$$\hat{H} = \hat{H}_x(\hat{x}, \hat{p}_x) + \hat{H}_y(\hat{y}, \hat{p}_y) + \hat{H}_z(\hat{z}, \hat{p}_z)$$

In this case, we conclude that the wave functions and energies are

$$\psi(x, y, z) = F(x)G(y)K(z), \qquad E = E_x + E_y + E_z$$

$$\hat{H}_x F(x) = E_x F(x), \qquad \hat{H}_y G(y) = E_y G(y), \qquad \hat{H}_z K(z) = E_z K(z)$$

Examples include the particle in a three-dimensional box (Section 3.5), the three-dimensional free particle (Problem 3.31), and the three-dimensional harmonic oscillator (Problem 4.17).

6.3 REDUCTION OF THE TWO-PARTICLE PROBLEM TO A ONE-PARTICLE PROBLEM

The hydrogen atom contains two particles, the proton and the electron. We now show how we can frequently reduce a two-particle problem to a one-particle problem.

Consider the classical-mechanical treatment of two interacting particles of masses m_1 and m_2. We specify their positions by the radius vectors $\mathbf{r}_1$ and $\mathbf{r}_2$ drawn from the origin of a Cartesian coordinate system (Fig. 6.1). Particles 1 and 2 have coordinates (x_1, y_1, z_1) and (x_2, y_2, z_2). We draw the vector $\mathbf{r} = \mathbf{r}_2 - \mathbf{r}_1$ from particle 1 to 2 and denote the components of $\mathbf{r}$ by x, y, and z:

$$x = x_2 - x_1, \qquad y = y_2 - y_1, \qquad z = z_2 - z_1 \qquad \textbf{(6.28)}^*$$

The coordinates x, y, and z are called the ***relative*** or ***internal coordinates***.

We now draw the vector $\mathbf{R}$ from the origin to the system's center of mass, point C, and denote the coordinates of C by X, Y, and Z:

$$\mathbf{R} = \mathbf{i}X + \mathbf{j}Y + \mathbf{k}Z \qquad (6.29)$$

The definition of the center of mass of this two-particle system gives

$$X = \frac{m_1 x_1 + m_2 x_2}{m_1 + m_2}, \qquad Y = \frac{m_1 y_1 + m_2 y_2}{m_1 + m_2}, \qquad Z = \frac{m_1 z_1 + m_2 z_2}{m_1 + m_2} \qquad (6.30)$$

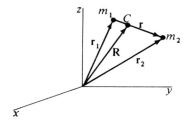

Figure 6.1 A two-particle system with center of mass at C.

These three equations are equivalent to the vector equation

$$\mathbf{R} = \frac{m_1\mathbf{r}_1 + m_2\mathbf{r}_2}{m_1 + m_2} \tag{6.31}$$

We also have

$$\mathbf{r} = \mathbf{r}_2 - \mathbf{r}_1 \tag{6.32}$$

We regard (6.31) and (6.32) as simultaneous linear equations in the two unknowns $\mathbf{r}_1$ and $\mathbf{r}_2$ and solve for them to get

$$\mathbf{r}_1 = \mathbf{R} - \frac{m_2}{m_1 + m_2}\,\mathbf{r}\,, \qquad \mathbf{r}_2 = \mathbf{R} + \frac{m_1}{m_1 + m_2}\,\mathbf{r} \tag{6.33}$$

Equations (6.31) and (6.32) represent a transformation of coordinates from $x_1,\ y_1,\ z_1,\ x_2,\ y_2,\ z_2$ to $X,\ Y,\ Z,\ x,\ y,\ z$. Consider what happens to the Hamiltonian under this transformation. Let an overhead dot indicate differentiation with respect to time. The velocity of particle 1 is [Eq. (5.34)] $\mathbf{v}_1 = d\mathbf{r}_1/dt = \dot{\mathbf{r}}_1$. The kinetic energy is the sum of the kinetic energies of the two particles:

$$T = \tfrac{1}{2}m_1|\dot{\mathbf{r}}_1|^2 + \tfrac{1}{2}m_2|\dot{\mathbf{r}}_2|^2 \tag{6.34}$$

Introducing the time derivatives of Eqs. (6.33) into (6.34), we have

$$T = \tfrac{1}{2}m_1\left(\dot{\mathbf{R}} - \frac{m_2}{m_1 + m_2}\,\dot{\mathbf{r}}\right)\cdot\left(\dot{\mathbf{R}} - \frac{m_2}{m_1 + m_2}\,\dot{\mathbf{r}}\right)$$

$$+ \tfrac{1}{2}m_2\left(\dot{\mathbf{R}} + \frac{m_1}{m_1 + m_2}\,\dot{\mathbf{r}}\right)\cdot\left(\dot{\mathbf{R}} + \frac{m_1}{m_1 + m_2}\,\dot{\mathbf{r}}\right)$$

where $|\mathbf{A}|^2 = \mathbf{A}\cdot\mathbf{A}$ [Eq. (5.24)] has been used. Using the distributive law for the dot products, we find, after simplifying,

$$T = \tfrac{1}{2}(m_1 + m_2)|\dot{\mathbf{R}}|^2 + \frac{1}{2}\frac{m_1 m_2}{(m_1 + m_2)}\,|\dot{\mathbf{r}}|^2 \tag{6.35}$$

Let M be the total mass of the system:

$$M \equiv m_1 + m_2 \tag{6.36}$$

We define the **reduced mass** μ of the two-particle system as

$$\mu \equiv \frac{m_1 m_2}{m_1 + m_2} \tag{6.37}*$$

Then

$$T = \tfrac{1}{2}M|\dot{\mathbf{R}}|^2 + \tfrac{1}{2}\mu|\dot{\mathbf{r}}|^2 \tag{6.38}$$

The first term in (6.38) is the kinetic energy due to translational motion of the whole system of mass M. *Translational motion* is motion in which each particle undergoes the same displacement. The quantity $\tfrac{1}{2}M|\dot{\mathbf{R}}|^2$ would be the kinetic energy of a hypothetical particle of mass M located at the center of mass. The second term in (6.38) is the kinetic energy of internal (relative) motion of the two particles. This internal motion is of two types. The distance r between the two particles can change (vibration), and the direction of the **r** vector can change (rotation). Note that $|\dot{\mathbf{r}}| = |d\mathbf{r}/dt| \neq d|\mathbf{r}|/dt$.

Corresponding to the original coordinates x_1, y_1, z_1, x_2, y_2, z_2, we had six linear momenta:

$$p_{x_1} = m_1\dot{x}_1, \quad \ldots, \quad p_{z_2} = m_2\dot{z}_2 \tag{6.39}$$

Comparing Eqs. (6.34) and (6.38), we define the six linear momenta for the new coordinates X, Y, Z, x, y, z as

$$p_X \equiv M\dot{X}, \qquad p_Y \equiv M\dot{Y}, \qquad p_Z \equiv M\dot{Z}$$

$$p_x \equiv \mu\dot{x}, \qquad p_y \equiv \mu\dot{y}, \qquad p_z \equiv \mu\dot{z}$$

We define two new momentum vectors as

$$\mathbf{p}_M \equiv \mathbf{i}M\dot{X} + \mathbf{j}M\dot{Y} + \mathbf{k}M\dot{Z} \quad \text{and} \quad \mathbf{p}_\mu \equiv \mathbf{i}\mu\dot{x} + \mathbf{j}\mu\dot{y} + \mathbf{k}\mu\dot{z}$$

Introducing these momenta into (6.38), we have

$$T = \frac{|\mathbf{p}_M|^2}{2M} + \frac{|\mathbf{p}_\mu|^2}{2\mu} \tag{6.40}$$

Now consider the potential energy. We make the restriction that V is a function *only* of the relative coordinates x, y, and z of the two particles:

$$V = V(x, y, z) \tag{6.41}$$

An example of (6.41) is two charged particles interacting according to Coulomb's law [see Eq. (3.53)]. With this restriction on V, the Hamiltonian function is

$$H = \frac{p_M^2}{2M} + \left[\frac{p_\mu^2}{2\mu} + V(x, y, z)\right] \tag{6.42}$$

Now suppose we had a system composed of a particle of mass M subject to no forces and a particle of mass μ subject to the potential-energy function $V(x, y, z)$, and further suppose that there was no interaction between these particles. If (X, Y, Z) are the coordinates of the particle of mass M, and (x, y, z) are the coordinates of the particle of mass μ, what is the Hamiltonian of this hypothetical system? Clearly, it is identical with (6.42).

The Hamiltonian (6.42) can be viewed as the sum of the Hamiltonians $p_M^2/2M$ and $[p_\mu^2/2\mu + V(x, y, z)]$ of two hypothetical noninteracting particles of mass M and μ. Therefore, the results of Section 6.2 show that the system's quantum-mechanical energy is the sum of energies of the two hypothetical particles [Eq. (6.23)]: $E = E_M + E_\mu$. From Eqs. (6.24) and (6.42), the translational energy E_M is found by solving the Schrödinger equation $(\hat{p}_M^2/2M)\psi_M = E_M\psi_M$. This is the Schrödinger equation for a free particle of mass M, so its possible eigenvalues are all nonnegative numbers: $E_M \geq 0$ [Eq. (2.31)]. From (6.24) and (6.42), the energy E_μ is found by solving the Schrödinger equation

$$\left[\frac{\hat{p}_\mu^2}{2\mu} + V(x, y, z)\right]\psi_\mu(x, y, z) = E_\mu\psi_\mu(x, y, z) \tag{6.43}$$

We have thus separated the problem of two particles interacting according to a potential-energy function $V(x, y, z)$ that depends on only the relative coordinates x, y, z into two separate one-particle problems: (1) the translational motion of the entire system of mass M, which simply adds a nonnegative constant energy E_M to the system's energy, and (2) the relative or internal motion, which is dealt with by solving the Schrödinger equation (6.43) for a hypothetical particle of mass μ whose coordinates are the relative coordinates x, y, z and that moves subject to the potential energy $V(x, y, z)$.

For example, for the hydrogen atom, which is composed of an electron (e) and a proton (p), the atom's total energy is $E = E_M + E_\mu$, where E_M is the translational energy of motion through space of the entire atom of mass $M = m_e + m_p$, and where E_μ is found by solving (6.43) with $\mu = m_e m_p / (m_e + m_p)$ and V being the Coulomb's law potential energy of interaction of the electron and proton; see Section 6.5.

6.4 THE TWO-PARTICLE RIGID ROTOR

Although we are now equipped to solve the Schrödinger equation for the hydrogen atom, we will first solve a simpler problem: that of the two-particle rigid rotor. By this we mean a two-particle system with the particles held at a fixed distance from each other by a rigid massless rod of length d. For this problem the vector $\mathbf{r}$ in Fig. 6.1 has the constant magnitude $|\mathbf{r}| = d$. Therefore (see Section 6.3), the kinetic energy of internal motion is wholly rotational energy. The energy of the rotor is wholly kinetic, and

$$V = 0 \tag{6.44}$$

Equation (6.44) is a special case of Eq. (6.41), and we may therefore use the results of the last section to separate off the translational motion of the system as a whole. We will concern ourselves only with the rotational energy. The Hamiltonian for the rotation is given by the terms in brackets in (6.43) as

$$\hat{H} = \frac{\hat{p}_\mu^2}{2\mu} = -\frac{\hbar^2}{2\mu} \nabla^2, \qquad \mu = \frac{m_1 m_2}{m_1 + m_2} \tag{6.45}$$

where m_1 and m_2 are the masses of the two particles. The coordinates of the fictitious particle are the relative coordinates of m_1 and m_2, as given by Eq. (6.28).

Instead of the relative Cartesian coordinates, x, y, z, it will prove more fruitful to use the relative spherical polar coordinates, r, θ, ϕ. The r coordinate is equal to the magnitude of the $\mathbf{r}$ vector in Fig. 6.1, and since m_1 and m_2 are constrained to remain a fixed distance apart, we have $r = d$. Thus the problem is equivalent to a particle of mass μ constrained to move on the surface of a sphere of radius d. Because the radial coordinate is constant, the wave function will be a function of θ and ϕ only. Hence the first two terms of the Laplacian operator in (6.8) will give zero when operating on the wave function and may be omitted. Looking at things in a slightly different way, we note that the operators in (6.8) that involve r derivatives correspond to the kinetic energy of radial motion, and since there is no radial motion, the r derivatives are omitted from the Hamiltonian.

Since $V = 0$ is a special case of $V = V(r)$, the results of Section 6.1 tell us that the eigenfunctions are given by (6.16) with the r factor omitted:

$$\psi = Y_J^m(\theta, \phi) \tag{6.46}$$

where J rather than l is used for the rotational angular-momentum quantum number.

The Hamiltonian operator is given by Eq. (6.8) with the r derivatives omitted and $V(r) = 0$; thus

$$\hat{H} = (2\mu d^2)^{-1} \hat{L}^2$$

Use of (6.13) gives

$$\hat{H}\psi = E\psi$$

$$(2\mu d^2)^{-1} \hat{L}^2 Y_J^m(\theta, \phi) = E Y_J^m(\theta, \phi)$$

$$(2\mu d^2)^{-1} J(J + 1)\hbar^2 Y_J^m(\theta, \phi) = E Y_J^m(\theta, \phi)$$

$$E = \frac{J(J + 1)\hbar^2}{2\mu d^2}, \qquad J = 0, 1, 2, \ldots \tag{6.47}$$

The **moment of inertia** I of a system of n particles about some particular axis in space is defined as

$$I \equiv \sum_{i=1}^{n} m_i \rho_i^2 \tag{6.48}$$

where m_i is the mass of the ith particle and ρ_i is the perpendicular distance from this particle to the axis. The value of I depends on the choice of axis. For the two-particle rigid rotor, we choose our axis to be a line that passes through the center of mass and is perpendicular to the line joining m_1 and m_2 (Fig. 6.2). If we place the rotor so that the center of mass, point C, lies at the origin of a Cartesian coordinate system and the line joining m_1 and m_2 lies on the x axis, then C will have the coordinates $(0, 0, 0)$, m_1 will have the coordinates $(-\rho_1, 0, 0)$, and m_2

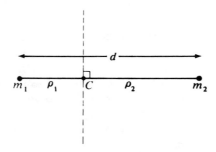

Figure 6.2 Axis (dashed line) for calculating the moment of inertia of a two-particle rigid rotor. C is the center of mass.

will have the coordinates $(\rho_2, 0, 0)$. Using these coordinates in (6.30), we find

$$m_1\rho_1 = m_2\rho_2 \tag{6.49}$$

The moment of inertia of the rotor about the axis we have chosen is

$$I = m_1\rho_1^2 + m_2\rho_2^2 \tag{6.50}$$

Using (6.49), we transform Eq. (6.50) to (see Problem 6.6)

$$I = \mu d^2 \tag{6.51}*$$

where $\mu \equiv m_1 m_2/(m_1 + m_2)$ is the reduced mass of the system and $d \equiv \rho_1 + \rho_2$ is the distance between m_1 and m_2. The allowed energy levels (6.47) of the two-particle rigid rotor are

$$E = \frac{J(J+1)\hbar^2}{2I}, \qquad J = 0, 1, 2, \ldots \tag{6.52}*$$

The lowest level is $E = 0$, so there is no zero-point rotational energy. Having zero rotational energy and therefore zero angular momentum for the rotor does not violate the uncertainty principle; recall the discussion following Eq. (5.112). Note that E increases as $J^2 + J$, so the spacing between adjacent rotational levels increases as J increases.

Are the rotor energy levels (6.52) degenerate? The energy depends on J only, but the wave function (6.46) depends on J and m, where $m\hbar$ is the z component of the rotor's angular momentum. For each value of J, there are $2J + 1$ values of m, ranging from $-J$ to J. Hence the levels are $(2J + 1)$-fold degenerate. The states of a degenerate level have different orientations of the angular-momentum vector of the rotor about a space-fixed axis.

The angles θ and ϕ in the wave function (6.46) are relative coordinates of the two point masses. If we set up a Cartesian coordinate system with the origin at the rotor's center of mass, θ and ϕ will be as shown in Fig. 6.3. This coordinate system undergoes the same translational motion as the rotor's center of mass but does not rotate in space.

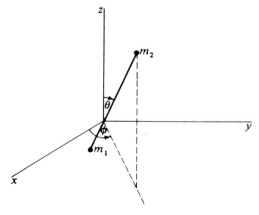

Figure 6.3 Coordinate system for the two-particle rigid rotor.

The rotational angular momentum $[J(J + 1)\hbar^2]^{1/2}$ is the angular momentum of the two particles with respect to an origin at the system's center of mass C.

The rotational levels of a diatomic molecule can be well approximated by the two-particle rigid-rotor energies (6.52). It is found (*Levine, Molecular Spectroscopy*, Section 4.4) that when a diatomic molecule absorbs or emits radiation the allowed pure-rotational transitions are

$$\Delta J = \pm 1 \tag{6.53}$$

In addition, a molecule must have a nonzero dipole moment in order to show a pure-rotational spectrum. A *pure-rotational transition* is one where only the rotational quantum number changes. [Vibration–rotation transitions (Section 4.3) involve changes in both vibrational and rotational quantum numbers.] The spacing between adjacent low-lying rotational levels is significantly less than that between adjacent vibrational levels, and the pure-rotational spectrum falls in the microwave (or the far-infrared) region. The frequencies of the pure-rotational spectral lines of a diatomic molecule are then (approximately)

$$\nu = \frac{E(J + 1) - E(J)}{h} = \frac{[(J + 1)(J + 2) - J(J + 1)]h}{8\pi^2 I} = 2(J + 1)B \tag{6.54}$$

$$B \equiv h/8\pi^2 I, \qquad J = 0, 1, 2, \ldots \tag{6.55}$$

B is called the **rotational constant** of the molecule.

The spacings between the diatomic rotational levels (6.52) for low and moderate values of J are generally less than or of the same order of magnitude as kT at room temperature, so the Boltzmann distribution law (4.71) shows that many rotational levels are significantly populated at room temperature. Absorption of radiation by diatomic molecules having $J = 0$ (the $J = 0 \rightarrow 1$ transition) gives a line at the frequency $2B$; absorption by molecules having $J = 1$ (the $J = 1 \rightarrow 2$ transition) gives a line at $4B$; absorption by $J = 2$ molecules gives a line at $6B$; and so on. See Fig. 6.4.

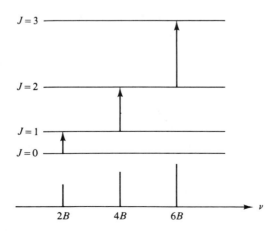

Figure 6.4 Two-particle rigid-rotor absorption transitions.

Measurement of the rotational absorption frequencies allows B to be found. From B, we get the molecule's moment of inertia I, and from I we get the bond distance d. The value of d found is an average over the $v = 0$ vibrational motion of the molecule. Because of the asymmetry of the potential-energy curve in Figs. 4.5 and 13.1, d is very slightly longer than the equilibrium bond length R_e in Fig. 13.1.

As noted in Section 4.3, isotopic species such as $^1H^{35}Cl$ and $^1H^{37}Cl$ have virtually the same electronic energy curve $U(R)$ and so have virtually the same equilibrium bond distance. However, the different isotopic masses produce different moments of inertia and hence different rotational absorption frequencies.

For the rotational energies of polyatomic molecules, see *Levine, Molecular Spectroscopy*, Chapter 5.

EXAMPLE   The lowest-frequency pure-rotational absorption line of $^{12}C^{32}S$ occurs at 48991.0 MHz. Find the bond distance in $^{12}C^{32}S$.

The lowest-frequency rotational absorption is the $0 \rightarrow 1$ line. From (6.54), the $J \rightarrow J + 1$ transition occurs at $\nu = 2(J + 1)B$, so the lowest-frequency absorption is at $\nu = 2B$. Hence $B \equiv h/8\pi^2 I = \nu/2$ and $I = h/4\pi^2\nu$. Since $I = \mu d^2$, we have $d = (h/4\pi^2\nu\mu)^{1/2}$. Table A.3 gives

$$\mu = \frac{m_1 m_2}{m_1 + m_2} = \frac{12(31.97207)}{12 + 31.97207} \frac{1}{6.02214 \times 10^{23}} \text{ g} = 1.44885 \times 10^{-23} \text{ g}$$

Hence

$$d = \frac{1}{2\pi}\left(\frac{h}{\nu\mu}\right)^{1/2} = \frac{1}{2\pi}\left[\frac{6.62608 \times 10^{-34} \text{ J s}}{(48991.0 \times 10^6 \text{ s}^{-1})(1.44885 \times 10^{-26} \text{ kg})}\right]^{1/2}$$

$$= 1.5377 \times 10^{-10} \text{ m} = 1.5377 \text{ Å}$$

6.5   THE HYDROGEN ATOM

The hydrogen atom consists of a proton and an electron. If e represents the charge on the proton ($e = +1.6 \times 10^{-19}$ C), then the electron's charge is $-e$.

A few scientists have speculated that the proton and electron charges might not be exactly equal in magnitude. Experiments show that the magnitudes of the electron and proton charges are equal to within one part in 10^{21}. See J. G. King, *Phys. Rev. Lett.*, **5**, 562 (1960); H. F. Dylla and J. G. King, *Phys. Rev. A*, **7**, 1224 (1973).

We shall assume the electron and proton to be point masses whose interaction is given by Coulomb's law. In discussing atoms and molecules, we shall usually be considering isolated systems, ignoring interatomic and intermolecular interactions.

Instead of treating just the hydrogen atom, we consider a slightly more general problem: the **hydrogenlike atom**. By this we mean a system consisting of one electron and a nucleus of charge Ze. For $Z = 1$, we have the hydrogen atom;

for $Z = 2$, the He^+ ion; for $Z = 3$, the Li^{2+} ion; and so on. The hydrogenlike atom is the single most important system in quantum chemistry. An exact solution of the Schrödinger equation for atoms with more than one electron cannot be obtained because of the interelectronic repulsions. If, as a first approximation, we ignore these repulsions, then the electrons can be treated independently. (See Section 6.2.) The atomic wave function will be approximated by a product of one-electron functions, which will be hydrogenlike wave functions. A one-electron wave function is called an *orbital*. An orbital for an electron in an atom is called an *atomic orbital*. We shall use atomic orbitals to construct approximate wave functions for many-electron atoms. Orbitals are also used to construct approximate wave functions for molecules.

For the hydrogenlike atom, let (x, y, z) be the coordinates of the electron relative to the nucleus, and let $\mathbf{r} = \mathbf{i}x + \mathbf{j}y + \mathbf{k}z$. The Coulomb's law force on the electron in the hydrogenlike atom is [see Eq. (1.35)]

$$\mathbf{F} = -\frac{Ze'^2}{r^2}\frac{\mathbf{r}}{r} \tag{6.56}$$

where $\mathbf{r}/r$ is a unit vector in the $\mathbf{r}$ direction. The minus sign indicates an attractive force. The quantity e' can be viewed either as the proton charge in statcoulombs or as $e' \equiv e/(4\pi\varepsilon_0)^{1/2}$, where e is the proton charge in coulombs. (See Section 1.8.)

> The possibility of small deviations from Coulomb's law has been considered. Experiments have shown that if the Coulomb's law force is written as being proportional to r^{-2+s}, then $|s| < 10^{-15}$. It can be shown that a deviation from Coulomb's law would imply a nonzero rest mass for the photon; see A. S. Goldhaber and M. M. Nieto, *Rev. Mod. Phys.*, **43**, 277 (1971). There is no evidence for a nonzero photon rest mass, and data indicate that any such mass must be less than 10^{-48} g; L. Davis et al., *Phys. Rev. Lett.*, **35**, 1402 (1975).

The force in (6.56) is central, and comparison with Eq. (6.4) gives $dV(r)/dr = Ze'^2/r^2$. Integration gives

$$V = Ze'^2 \int \frac{dr}{r^2} = -\frac{Ze'^2}{r} \tag{6.57}$$

where the integration constant has been taken as 0 to make $V = 0$ at infinite separation between the charges. For any two charges Q_1 and Q_2 separated by distance r_{12}, Eq. (6.57) becomes

$$V = \frac{Q_1'Q_2'}{r_{12}} \tag{6.58}*$$

Since the potential energy of this two-particle system depends only on the relative coordinates of the particles, we can apply the results of Section 6.3 to reduce the problem to two one-particle problems. The translational motion of the atom as a whole simply adds some constant to the total energy, and we shall not concern ourselves with it. To deal with the internal motion of the system, we

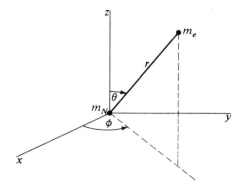

Figure 6.5 Relative spherical polar coordinates.

introduce a fictitious particle of mass

$$\mu = \frac{m_e m_N}{m_e + m_N} \tag{6.59}$$

where m_e and m_N are the electronic and nuclear masses. The particle of reduced mass μ moves subject to the potential-energy function (6.57), and its coordinates (r, θ, ϕ) are the spherical polar coordinates of one particle relative to the other (Fig. 6.5).

The Hamiltonian for the internal motion is [Eq. (6.43)]

$$\hat{H} = -\frac{\hbar^2}{2\mu} \nabla^2 - \frac{Ze'^2}{r} \tag{6.60}$$

Since V is a function of the r coordinate only, we have a one-particle central-force problem, and we may apply the results of Section 6.1. Using Eqs. (6.16) and (6.17), we have for the wave function

$$\psi(r, \theta, \phi) = R(r) Y_l^m(\theta, \phi), \qquad l = 0, 1, 2, \ldots, \qquad |m| \le l \tag{6.61}$$

where Y_l^m is a spherical harmonic, and the radial function $R(r)$ satisfies

$$-\frac{\hbar^2}{2\mu} \left(R'' + \frac{2}{r} R' \right) + \frac{l(l+1)\hbar^2}{2\mu r^2} R - \frac{Ze'^2}{r} R = ER(r) \tag{6.62}$$

To save time in writing, we define the constant a as

$$a \equiv \hbar^2 / \mu e'^2 \tag{6.63}$$

and (6.62) becomes

$$R'' + \frac{2}{r} R' + \left[\frac{2E}{ae'^2} + \frac{2Z}{ar} - \frac{l(l+1)}{r^2} \right] R = 0 \tag{6.64}$$

Solution of the Radial Equation. We could now try a power-series solution of (6.64), but we would get a three-term rather than a two-term recursion relation. We therefore seek a substitution that will lead to a two-term recursion relation. It turns out that the proper substitution can be found by examining the behavior of the solution for large values of r. For large r, (6.64) becomes

$$R'' + \frac{2E}{ae'^2} R = 0, \qquad r \text{ large} \tag{6.65}$$

which may be solved using the auxiliary equation (2.7). The solutions are

$$\exp\left[\pm(-2E/ae'^2)^{1/2}r\right] \tag{6.66}$$

Suppose that E is positive. The quantity under the square-root sign in (6.66) is negative, and the factor multiplying r is imaginary:

$$R(r) \sim e^{\pm i\sqrt{2\mu E}\,r/\hbar}, \qquad E \geqslant 0 \tag{6.67}$$

where (6.63) has been used. The symbol $\sim$ in (6.67) indicates that we are giving the behavior of $R(r)$ for large values of r; this is called the *asymptotic* behavior of the function. Note the resemblance of (6.67) to Eq. (2.30), the free-particle wave function. Equation (6.67) does not give the complete radial factor in the wave function for positive energies. Further study (*Bethe and Salpeter*, pages 21–24) shows that the radial function for $E \geqslant 0$ remains finite for all values of r, no matter what the value of E. Thus, just as for the free particle, all nonnegative energies of the hydrogen atom are allowed. Physically, these eigenfunctions correspond to states in which the electron is not bound to the nucleus; that is, the atom is ionized. (A classical-mechanical analogy is a comet moving in a hyperbolic orbit about the sun. The comet is not bound and makes but one visit to the solar system.) Since we get continuous rather than discrete allowed values for $E \geqslant 0$, the positive-energy eigenfunctions are called *continuum eigenfunctions*. The angular part of a continuum wave function is a spherical harmonic. Like the free-particle wave functions, the continuum eigenfunctions are not normalizable in the usual sense.

We now consider the *bound* states of the hydrogen atom, with $E < 0$. In this case, the quantity under the square-root sign in (6.66) is positive. Since we want the wave functions to remain finite as r goes to infinity, we prefer the minus sign in (6.66), and in order to get a two-term recursion relation, we make the substitution

$$R(r) = e^{-Cr}K(r) \tag{6.68}$$

$$C \equiv \left(-\frac{2E}{ae'^2}\right)^{1/2} \tag{6.69}$$

where e in (6.68) stands for the base of natural logarithms, and not the proton charge. Use of the substitution (6.68) will guarantee nothing about the behavior of the wave function for large r. The differential equation we obtain from this substitution will still have two linearly independent solutions. We can make any substitution we please in a differential equation; in fact, we could make the substitution $R(r) = e^{+Cr}J(r)$ and still wind up with the correct eigenfunctions and eigenvalues. The relation between J and K would naturally be $J(r) = e^{-2Cr}K(r)$.

Proceeding with (6.68), we evaluate R' and R'', substitute into (6.64), multiply through by r^2e^{Cr}, and use (6.69) to obtain the following differential equation for $K(r)$:

$$r^2 K'' + (2r - 2Cr^2)K' + [(2Za^{-1} - 2C)r - l(l+1)]K = 0 \qquad (6.70)$$

We could now substitute a power series of the form

$$K = \sum_{k=0}^{\infty} c_k r^k \qquad (6.71)$$

for K; if we did we would find that, in general, the first few coefficients in (6.71) are zero. If c_s is the first nonzero coefficient, (6.71) can be written as

$$K = \sum_{k=s}^{\infty} c_k r^k, \qquad c_s \neq 0 \qquad (6.72)$$

Letting $j \equiv k - s$, and then defining b_j as $b_j \equiv c_{j+s}$, we have

$$K = \sum_{j=0}^{\infty} c_{j+s} r^{j+s} = r^s \sum_{j=0}^{\infty} b_j r^j, \qquad b_0 \neq 0 \qquad (6.73)$$

(Although the various substitutions we are making might seem arbitrary, they are standard procedure in solving differential equations by power series.) In (6.73), s is an integer whose value is to be determined by substitution into the differential equation. We thus set

$$K(r) = r^s M(r) \qquad (6.74)$$

$$M(r) = \sum_{j=0}^{\infty} b_j r^j, \qquad b_0 \neq 0 \qquad (6.75)$$

Evaluating K' and K'' from (6.74) and substituting into (6.70), we get

$$r^2 M'' + [(2s+2)r - 2Cr^2]M' + [s^2 + s + (2Za^{-1} - 2C - 2Cs)r - l(l+1)]M = 0$$
$$(6.76)$$

To determine s, let us look at (6.76) for $r = 0$. From (6.75), we have

$$M(0) = b_0, \qquad M'(0) = b_1, \qquad M''(0) = 2b_2 \qquad (6.77)$$

Using (6.77) in (6.76), we find for $r = 0$

$$b_0(s^2 + s - l^2 - l) = 0 \qquad (6.78)$$

Since b_0 is not zero, the terms in parentheses must vanish: $s^2 + s - l^2 - l = 0$. This is a quadratic equation in the unknown s, with the roots

$$s = l, \qquad s = -l - 1 \qquad (6.79)$$

These roots correspond to the two linearly independent solutions of the differential equation. Let us examine them from the standpoint of proper behavior of the wave function. From Eqs. (6.68), (6.74), and (6.75), we have

$$R(r) = e^{-Cr} r^s \sum_{j=0}^{\infty} b_j r^j \qquad (6.80)$$

Since $e^{-Cr} = 1 - Cr + \cdots$, the function $R(r)$ behaves for small r as $b_0 r^s$. For the root $s = l$, $R(r)$ behaves properly at the origin. However, for $s = -l - 1$, $R(r)$ is proportional to

$$\frac{1}{r^{l+1}} \tag{6.81}$$

for small r. Since $l = 0, 1, 2, \ldots$, the root $s = -l - 1$ makes the radial factor in the wave function infinite at the origin. Many texts take this as sufficient reason for rejecting this root. However, this is not a good argument, since for the *relativistic* hydrogen atom, the $l = 0$ eigenfunctions are infinite at $r = 0$. Let us therefore look at (6.81) from the standpoint of quadratic integrability, since we certainly require the bound-state eigenfunctions to be normalizable.

The normalization integral [Eq. (5.80)] for the radial functions that behave like (6.81) looks like

$$\int_0 |R|^2 r^2 \, dr \approx \int_0 \frac{1}{r^{2l}} \, dr$$

for small r. The behavior of the integral at the lower limit of integration is

$$\left. \frac{1}{r^{2l-1}} \right|_{r=0} \tag{6.82}$$

For $l = 1, 2, 3, \ldots$, (6.82) is infinite, and the normalization integral is infinite. Hence we must reject the root $s = -l - 1$ for $l \geqslant 1$. However, for $l = 0$, (6.82) is finite, and there is no trouble with quadratic integrability. Thus there is a solution to the radial equation that behaves as r^{-1} for small r and is quadratically integrable.

Further study of this solution shows that it corresponds to an energy value that the experimental hydrogen-atom spectrum shows does not exist. Thus the r^{-1} solution must be rejected, but there is some dispute over the reason for doing so. We shall state the two reasons most often given, omitting the detailed arguments. One view is that the 1/r solution satisfies the Schrödinger equation everywhere in space except at the origin and hence must be rejected [*Dirac*, page 156; B. H. Armstrong and E. A. Power, *Am. J. Phys.*, **31**, 262 (1963)]. A second view is that the $1/r$ solution must be rejected because the Hamiltonian operator is not Hermitian with respect to it (*Merzbacher*, Section 10.5). (In Chapter 7 we shall define Hermitian operators and show that quantum-mechanical operators are required to be Hermitian.)

Taking the first root in (6.79), we have for the radial factor (6.80)

$$R(r) = e^{-Cr} r^l M(r) \tag{6.83}$$

With $s = l$, Eq. (6.76) becomes

$$rM'' + (2l + 2 - 2Cr)M' + (2Za^{-1} - 2C - 2Cl)M = 0 \tag{6.84}$$

From (6.75), we have

$$M(r) = \sum_{j=0}^{\infty} b_j r^j \tag{6.85}$$

$$M' = \sum_{j=0}^{\infty} j b_j r^{j-1} = \sum_{j=1}^{\infty} j b_j r^{j-1} = \sum_{k=0}^{\infty} (k+1) b_{k+1} r^k = \sum_{j=0}^{\infty} (j+1) b_{j+1} r^j$$

$$M'' = \sum_{j=0}^{\infty} j(j-1) b_j r^{j-2} = \sum_{j=1}^{\infty} j(j-1) b_j r^{j-2} = \sum_{k=0}^{\infty} (k+1) k b_{k+1} r^{k-1}$$

$$= \sum_{j=0}^{\infty} (j+1) j b_{j+1} r^{j-1}$$

Substituting these expressions in (6.84) and combining sums, we get

$$\sum_{j=0}^{\infty} \left[j(j+1) b_{j+1} + 2(l+1)(j+1) b_{j+1} + \left(\frac{2Z}{a} - 2C - 2Cl - 2Cj \right) b_j \right] r^j = 0$$

Setting the coefficient of r^j equal to zero, we get the recursion relation

$$b_{j+1} = \frac{(2C + 2Cl + 2Cj - 2Za^{-1})}{j(j+1) + 2(l+1)(j+1)} b_j \tag{6.86}$$

We now must examine the behavior of the infinite series (6.85) for large r. Since for large r the behavior of the series is determined by the terms with large j, we examine the ratio b_{j+1}/b_j for large j:

$$\frac{b_{j+1}}{b_j} \sim \frac{2Cj}{j^2} = \frac{2C}{j} , \qquad \text{for } j \text{ large} \tag{6.87}$$

Now consider the power series for e^{2Cr}:

$$e^{2Cr} = 1 + 2Cr + \cdots + \frac{(2C)^j r^j}{j!} + \frac{(2C)^{j+1} r^{j+1}}{(j+1)!} + \cdots \tag{6.88}$$

The ratio of successive powers of r in (6.88) is

$$\frac{(2C)^{j+1}}{(j+1)!} \cdot \frac{j!}{(2C)^j} = \frac{2C}{j+1} \sim \frac{2C}{j} , \qquad \text{for } j \text{ large}$$

which is the same as (6.87) for large j. This suggests that for large r the infinite series (6.85) behaves like e^{2Cr}. For large r, the radial function (6.83) behaves like

$$R(r) \sim e^{-Cr} r^l e^{2Cr} = r^l e^{Cr} \tag{6.89}$$

Therefore, $R(r)$ will become infinite as r goes to infinity and will not be quadratically integrable. The only way to avoid this "infinity catastrophe" (as in the harmonic-oscillator case) is to have the series terminate after a finite number of terms, in which case the e^{-Cr} factor will ensure that the wave function goes to zero as r goes to infinity. Let the last term in the series be $b_k r^k$. Then, to have $b_{k+1}, b_{k+2}, \ldots,$ all vanish, the fraction multiplying b_j in the recursion relation (6.86) must vanish when $j = k$; we have

$$2C(k + l + 1) = 2Za^{-1} , \qquad k = 0, 1, 2, \ldots \tag{6.90}$$

k and l are integers, and we now define a new integer n by

$$n \equiv k + l + 1, \qquad n = 1, 2, 3, \ldots \tag{6.91}$$

From (6.91) the quantum number l must satisfy

$$l \leq n - 1 \tag{6.92}$$

Hence l ranges from 0 to $n - 1$.

Energy Levels. Use of (6.91) in (6.90) gives

$$Cn = Za^{-1} \tag{6.93}$$

Substituting $C \equiv (-2E/ae'^2)^{1/2}$ [Eq. (6.69)] into (6.93) and solving for E, we get

$$E = -\frac{Z^2}{n^2} \left(\frac{e'^2}{2a} \right) = -\frac{Z^2 \mu e'^4}{2n^2 \hbar^2} \tag{6.94}$$

where $a \equiv \hbar^2/\mu e'^2$ [Eq. (6.63)]. These are the bound-state energy levels of the hydrogenlike atom, and they are discrete. Figure 6.6 shows the potential-energy curve [Eq. (6.57)] and some of the allowed energy levels for the hydrogen atom ($Z = 1$). The crosshatching indicates that all positive energies are allowed.

It turns out that all changes in n are allowed in light absorption and emission; the reciprocal wavelengths of hydrogen-atom spectral lines are then

$$\frac{1}{\lambda} = \frac{\nu}{c} = \frac{E_2 - E_1}{hc} = \frac{e'^2}{2ahc} \left(\frac{1}{n_1^2} - \frac{1}{n_2^2} \right) \equiv R_{\mathrm{H}} \left(\frac{1}{n_1^2} - \frac{1}{n_2^2} \right) \tag{6.95}$$

where $R_{\mathrm{H}} = 109677.6 \, \mathrm{cm}^{-1}$ is the *Rydberg constant* for hydrogen.

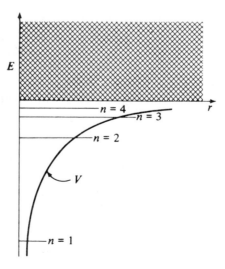

Figure 6.6 Energy levels of the hydrogen atom.

Degeneracy. Are the hydrogen-atom energy levels degenerate? For the bound states, the energy (6.94) depends only on n. However, the wave function (6.61) depends on all three quantum numbers n, l, and m, whose allowed values are [Eqs. (6.91), (6.92), (5.111), and (5.112)]

$$n = 1, 2, 3, \ldots \tag{6.96}*$$

$$l = 0, 1, 2, \ldots, n - 1 \tag{6.97}*$$

$$m = -l, -l + 1, \ldots, 0, \ldots, l - 1, l \tag{6.98}*$$

States with different values of l or m, but the same value of n, have the same energy; we have degeneracy, except for $n = 1$, where l and m must both be 0. For a given value of n, we can have n different values of l. For each of these values of l, we can have $2l + 1$ values of m. The degree of degeneracy of a given bound-state level is then

$$\sum_{l=0}^{n-1} (2l + 1) = \sum_{l=0}^{n-1} (2l) + \sum_{l=0}^{n-1} 1$$

The second sum on the right side of this equation has n terms, each term having the value 1. Therefore, it is equal to n. For the first sum, we have

$$\sum_{l=0}^{n-1} (2l) = 2 \sum_{l=0}^{n-1} l = 2 \sum_{l=1}^{n-1} l$$

The sum of the integers from 1 to k is (Problem 6.9) $\frac{1}{2}k(k + 1)$. Thus the sum of the integers from 1 to $n - 1$ is $\frac{1}{2}(n - 1)n$. We have

$$\sum_{l=0}^{n-1} (2l + 1) = 2 \cdot \tfrac{1}{2}(n - 1)n + n = n^2 \tag{6.99}$$

and the degeneracy of the discrete levels for the hydrogenlike atom is n^2 (spin considerations being omitted). For the continuum levels, it turns out that for a given energy there is no restriction on the maximum value of l; hence these levels are infinity-fold degenerate.

The radial equation for the hydrogen atom can also be solved by the use of ladder operators (also known as *factorization*); see Z. W. Salsburg, *Am. J. Phys.*, **33**, 36 (1965).

6.6   THE BOUND-STATE HYDROGEN-ATOM WAVE FUNCTIONS

The Radial Factor. Using (6.93), we have for the recursion relation (6.86)

$$b_{j+1} = \frac{2Z}{na} \frac{j + l + 1 - n}{(j + 1)(j + 2l + 2)} b_j \tag{6.100}$$

The discussion preceding Eq. (6.91) shows that the highest power of r in the polynomial $M(r) = \sum_j b_j r^j$ [Eq. (6.85)] is $k = n - l - 1$; hence use of $C = Z/na$ [Eq. (6.93)] in $R(r) = e^{-Cr} r^l M(r)$ [Eq. (6.83)] gives the radial factor in the hydrogen-atom ψ as

$$R_{nl}(r) = r^l e^{-Zr/na} \sum_{j=0}^{n-l-1} b_j r^j \tag{6.101}$$

where $a \equiv \hbar^2/\mu e'^2$ [Eq. (6.63)]. The complete hydrogenlike bound-state wave functions are [Eq. (6.61)]

$$\psi_{nlm} = R_{nl}(r) Y_l^m(\theta, \phi) = R_{nl}(r) S_{lm}(\theta) \frac{1}{\sqrt{2\pi}} e^{im\phi} \tag{6.102}$$

where the first few theta functions are given in Table 5.1.

How many nodes does $R(r)$ have? The radial function is zero at $r = \infty$, at $r = 0$ for $l \neq 0$, and at values of r that make $M(r)$ vanish. $M(r)$ is a polynomial of degree $n - l - 1$, and it can be shown that the roots of $M(r) = 0$ are all real and positive. Thus, aside from the origin and infinity, there are $n - l - 1$ nodes in $R(r)$. The nodes of the spherical harmonics are discussed in Problem 6.35.

Ground-State Wave Function and Energy. For the ground state of the hydrogenlike atom, we have $n = 1$, $l = 0$, and $m = 0$. The radial factor (6.101) is

$$R_{10}(r) = b_0 e^{-Zr/a}$$

The constant b_0 is determined by normalization [Eq. (5.80)]:

$$|b_0|^2 \int_0^\infty e^{-2Zr/a} r^2 \, dr = 1$$

Using the Appendix integral (A.7), we find

$$R_{10}(r) = 2\left(\frac{Z}{a}\right)^{3/2} e^{-Zr/a} \tag{6.103}$$

Multiplying by $Y_0^0 = 1/(4\pi)^{1/2}$, we have as the ground-state wave function

$$\psi_{100} = \frac{1}{\pi^{1/2}} \left(\frac{Z}{a}\right)^{3/2} e^{-Zr/a} \tag{6.104}$$

The hydrogen-atom energies and wave functions involve the reduced mass, given by (6.59) as

$$\mu_H = \frac{m_e m_p}{m_e + m_p} = \frac{m_e}{1 + m_e/m_p} = \frac{m_e}{1 + 0.000544617} = 0.9994557 m_e \tag{6.105}$$

where m_p is the proton mass and m_e/m_p was found from Table A.1. The reduced mass is very close to the electron mass. Because of this, some texts use the electron mass instead of the reduced mass in the H atom Schrödinger equation. Physically, this corresponds to assuming that the proton mass is infinite in comparison with the electron mass in (6.105) and that all the internal motion is motion of the electron. The error introduced by using the electron mass for the reduced mass is about 1 part in 2000 for the hydrogen atom. For heavier atoms, the error introduced by assuming an infinitely heavy nucleus is even less than this. Also, for many-electron atoms, the form of the correction for nuclear motion is quite complicated. For these reasons we shall in future assume an infinitely heavy nucleus and simply use the electron mass in writing the Schrödinger equation for atoms.

If we replace the reduced mass of the hydrogen atom by the electron mass, the quantity a defined by (6.63) becomes

$$a_0 \equiv \frac{\hbar^2}{m_e e'^2} = 0.52918 \ \text{Å} \tag{6.106}$$

where the subscript zero indicates use of the electron mass instead of the reduced mass. For historical reasons, a_0 is called the **Bohr radius**; it was the radius of the circle in which the electron moved in the ground state of the hydrogen atom, according to the Bohr theory. Of course, since the ground-state wave function (6.104) is nonzero for all finite values of r, there is some probability of finding the electron at any distance from the nucleus. The electron is certainly not confined to a circle.

A convenient unit for electronic energies is the **electron volt** (eV), defined as the kinetic energy acquired by an electron accelerated through a potential difference of 1 volt (V). Potential difference is defined as energy per unit charge. Since $e = 1.602177 \times 10^{-19}$ C and $1 \ \text{V C} = 1 \ \text{J} = 10^7$ ergs, we have

$$1 \ \text{eV} = 1.602177 \times 10^{-19} \ \text{J} = 1.602177 \times 10^{-12} \ \text{erg} \tag{6.107}$$

EXAMPLE   Calculate the ground-state energy of the hydrogen atom using SI units and convert the result to electron volts.

The H atom ground-state energy is given by (6.94) with $n = 1$, $Z = 1$, and $e' = e/(4\pi\varepsilon_0)^{1/2}$ as $E = -\mu e^4/8h^2\varepsilon_0^2$. Use of (6.105) for μ gives

$$E = -\frac{0.9994557(9.10939 \times 10^{-31} \ \text{kg})(1.602177 \times 10^{-19} \ \text{C})^4}{8(6.62608 \times 10^{-34} \ \text{J s})^2(8.8541878 \times 10^{-12} \ \text{C}^2/\text{N-m}^2)^2}$$

$$= -(2.17868 \times 10^{-18} \ \text{J})[(1 \ \text{eV})/(1.602177 \times 10^{-19} \ \text{J})]$$

$$E = -13.598 \ \text{eV} \tag{6.108}$$

a number worth remembering. This is the minimum energy needed to ionize a ground-state hydrogen atom.

Let us examine a significant property of the ground-state wave function (6.104). We have $r = (x^2 + y^2 + z^2)^{1/2}$. For points on the x axis, where $y = 0$ and $z = 0$, we have $r = (x^2)^{1/2} = |x|$, and

$$\psi_{100}(x, 0, 0) = \pi^{-1/2}(Z/a)^{3/2} e^{-Z|x|/a} \tag{6.109}$$

Figure 6.7 shows how (6.109) varies along the x axis. Although the wave function is continuous at the origin, the slope of the tangent to the curve is positive at the left of the origin but negative at its right. Thus $\partial\psi/\partial x$ is discontinuous at the origin. We say that the wave function has a *cusp* at the origin. The cusp is present because the potential energy $V = -Ze'^2/r$ becomes infinite at the origin. (Recall the discontinuous slope of the particle-in-a-box wave functions at the walls of the box.)

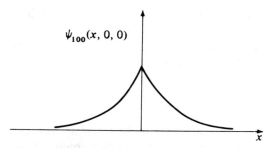

$\psi_{100}(x, 0, 0)$

Figure 6.7 Cusp in the hydrogen-atom ground-state wave function.

We denoted the hydrogen-atom bound-state wave functions by three subscripts that give the values of n, l, and m. We now introduce a slightly different notation, in which the value of l is indicated by a letter:

letter	s	p	d	f	g	h	i	k	$\cdots$
l	0	1	2	3	4	5	6	7	$\cdots$

(6.110)*

The letters s, p, d, f are of spectroscopic origin, standing for sharp, principal, diffuse, fundamental. After these we go alphabetically, except that j is omitted. Preceding the code letter for l, we write the value of n. Thus the ground-state wave function ψ_{100} is called ψ_{1s}.

Wave Functions for $n = 2$.   For $n = 2$, we have the states ψ_{200}, ψ_{21-1}, ψ_{210}, and ψ_{211}. We denote ψ_{200} as ψ_{2s}. To distinguish the three $2p$ functions, we use a subscript giving the m value: $\psi_{2p_{-1}}$, ψ_{2p_0}, $\psi_{2p_{+1}}$. The radial factor in the wave function depends on n and l, but not on m, as can be seen from (6.101). Each of the three $2p$ wave functions thus has the same radial factor. The $2s$ and $2p$ radial factors may be found in the usual manner from (6.101) and (6.100), followed by normalization. The results are given in Table 6.1. Note that the exponential factor

TABLE 6.1   Radial factors in the hydrogenlike-atom wave functions

$$R_{1s} = 2\left(\frac{Z}{a}\right)^{3/2} e^{-Zr/a}$$

$$R_{2s} = \frac{1}{\sqrt{2}}\left(\frac{Z}{a}\right)^{3/2}\left(1 - \frac{Zr}{2a}\right)e^{-Zr/2a}$$

$$R_{2p} = \frac{1}{2\sqrt{6}}\left(\frac{Z}{a}\right)^{5/2} r e^{-Zr/2a}$$

$$R_{3s} = \frac{2}{3\sqrt{3}}\left(\frac{Z}{a}\right)^{3/2}\left(1 - \frac{2Zr}{3a} + \frac{2Z^2r^2}{27a^2}\right)e^{-Zr/3a}$$

$$R_{3p} = \frac{8}{27\sqrt{6}}\left(\frac{Z}{a}\right)^{3/2}\left(\frac{Zr}{a} - \frac{Z^2r^2}{6a^2}\right)e^{-Zr/3a}$$

$$R_{3d} = \frac{4}{81\sqrt{30}}\left(\frac{Z}{a}\right)^{7/2} r^2 e^{-Zr/3a}$$

in the $n = 2$ radial functions is not the same as in the R_{1s} function. The complete wave function is found by multiplying the radial factor by the appropriate spherical harmonic. Using (6.102), Table 6.1, and Table 5.1, we have

$$\psi_{2s} = \frac{1}{\pi^{1/2}} \left(\frac{Z}{2a} \right)^{3/2} \left(1 - \frac{Zr}{2a} \right) e^{-Zr/2a} \tag{6.111}$$

$$\psi_{2p_{-1}} = \frac{1}{8\pi^{1/2}} \left(\frac{Z}{a} \right)^{5/2} re^{-Zr/2a} \sin\theta \, e^{-i\phi} \tag{6.112}$$

$$\psi_{2p_0} = \frac{1}{\pi^{1/2}} \left(\frac{Z}{2a} \right)^{5/2} re^{-Zr/2a} \cos\theta \tag{6.113}$$

$$\psi_{2p_1} = \frac{1}{8\pi^{1/2}} \left(\frac{Z}{a} \right)^{5/2} re^{-Zr/2a} \sin\theta \, e^{i\phi} \tag{6.114}$$

Table 6.1 lists some of the normalized radial factors in the hydrogenlike wave functions. Figure 6.8 graphs some of the radial functions. The r^l factor makes the radial functions zero at $r = 0$, except for s states.

The Radial Distribution Function.   The probability of finding the electron in the region of space where its coordinates lie in the ranges r to $r + dr$, θ to $\theta + d\theta$, and ϕ to $\phi + d\phi$ is [Eq. (5.78)]

$$|\psi|^2 \, d\tau = [R_{nl}(r)]^2 |Y_l^m(\theta, \phi)|^2 r^2 \sin\theta \, dr \, d\theta \, d\phi \tag{6.115}$$

We now ask: What is the probability of the electron having its radial coordinate between r and $r + dr$ with no restriction on the values of θ and ϕ? We are asking for the probability of finding the electron in a thin spherical shell centered at the origin, of inner radius r and outer radius $r + dr$. We must thus add up the probabilities (6.115) for all possible values of θ and ϕ, keeping r fixed. This amounts to integrating (6.115) over θ and ϕ. Hence the probability of finding the electron between r and $r + dr$ is

$$[R_{nl}(r)]^2 r^2 \, dr \int_0^{2\pi} \int_0^{\pi} |Y_l^m(\theta, \phi)|^2 \sin\theta \, d\theta \, d\phi = [R_{nl}(r)]^2 r^2 \, dr \tag{6.116}$$

since the spherical harmonics are normalized:

$$\int_0^{2\pi} \int_0^{\pi} |Y_l^m(\theta, \phi)|^2 \sin\theta \, d\theta \, d\phi = 1 \tag{6.117}*$$

as can be seen from (5.72) and (5.80). The function $R^2(r)r^2$, which determines the probability of finding the electron at a distance r from the nucleus, is called the *radial distribution function*; see Fig. 6.9. Although $R_{1s}(r)$ is not zero at the origin, the $1s$ radial distribution function is zero at $r = 0$ because of the r^2 factor; the volume of the thin spherical shell becomes zero as r goes to zero. The maximum in the radial distribution function for the $1s$ state of hydrogen is at $r = a$.

EXAMPLE   Find the probability that the electron in the ground-state H atom is less than a distance a from the nucleus.

We want the probability that the radial coordinate lies between 0 and a. This

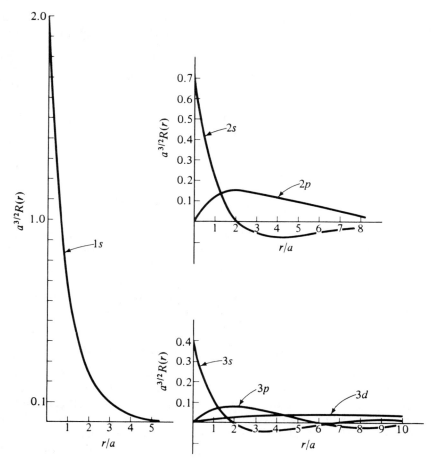

Figure 6.8 Graphs of the radial factor $R_{nl}(r)$ in the hydrogen-atom $(Z = 1)$ wave functions. The same scale is used in all graphs. (In some texts, these functions are not properly drawn to scale.)

is found by taking the infinitesimal probability (6.116) of being between r and $r + dr$ and summing it over the range from 0 to a. This sum of infinitesimal quantities is the definite integral

$$\int_0^a R_{nl}^2 r^2 \, dr = \frac{4}{a^3} \int_0^a e^{-2r/a} r^2 \, dr = \frac{4}{a^3} e^{-2r/a} \left(-\frac{r^2 a}{2} - \frac{2ra^2}{4} - \frac{2a^3}{8} \right) \Big|_0^a$$

$$= 4[e^{-2}(-5/4) - (-1/4)] = 0.323$$

where R_{10} was taken from Table 6.1 and the Appendix integral A.6 was used.

Real Hydrogenlike Functions. The factor $e^{im\phi}$ makes the spherical harmonics complex, except when $m = 0$. Instead of working with complex wave functions such as (6.112) and (6.114), chemists often use real hydrogenlike wave

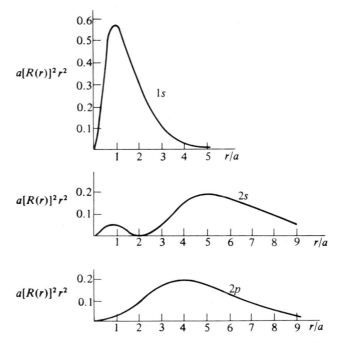

Figure 6.9 Plots of the radial distribution function $[R_{nl}(r)]^2 r^2$ for the hydrogen atom.

functions formed by taking linear combinations of the complex functions. The justification for this procedure is given by the theorem of Section 3.6: Any linear combination of eigenfunctions of a degenerate energy level is an eigenfunction of the Hamiltonian with the same eigenvalue. Since the energy of the hydrogen atom does not depend on m, the $2p_1$ and $2p_{-1}$ states belong to a degenerate energy level. Any linear combination of them is an eigenfunction of the Hamiltonian with the same energy eigenvalue.

One way to combine these two functions to obtain a real function is

$$\psi_{2p_x} \equiv \frac{1}{\sqrt{2}} \left(\psi_{2p_{-1}} + \psi_{2p_1} \right) = \frac{1}{4\sqrt{2\pi}} \left(\frac{Z}{a} \right)^{5/2} r e^{-Zr/2a} \sin\theta \cos\phi \qquad (6.118)$$

where we have used (6.112), (6.114), and $e^{\pm i\phi} = \cos\phi \pm i \sin\phi$. The $1/\sqrt{2}$ factor normalizes ψ_{2p_x}:

$$\int |\psi_{2p_x}|^2 \, d\tau = \frac{1}{2} \left(\int |\psi_{2p_{-1}}|^2 \, d\tau + \int |\psi_{2p_1}|^2 \, d\tau + \int \psi_{2p_{-1}}^* \psi_{2p_1} \, d\tau + \int \psi_{2p_1}^* \psi_{2p_{-1}} \, d\tau \right)$$

$$= \tfrac{1}{2}(1 + 1 + 0 + 0) = 1$$

Here we have used the fact that ψ_{2p_1} and $\psi_{2p_{-1}}$ are normalized and are orthogonal to each other, since

$$\int_0^{2\pi} (e^{-i\phi})^* e^{i\phi} \, d\phi = \int_0^{2\pi} e^{2i\phi} \, d\phi = 0$$

The designation ψ_{2p_x} for (6.118) becomes clearer if we note that (5.51) gives

$$\psi_{2p_x} = \frac{1}{4\sqrt{2\pi}} \left(\frac{Z}{a}\right)^{5/2} x e^{-Zr/2a} \qquad (6.119)$$

A second way of combining the functions is

$$\psi_{2p_y} \equiv \frac{1}{i\sqrt{2}} (\psi_{2p_1} - \psi_{2p_{-1}}) = \frac{1}{4\sqrt{2\pi}} \left(\frac{Z}{a}\right)^{5/2} r \sin\theta \sin\phi \, e^{-Zr/2a} \qquad (6.120)$$

$$\psi_{2p_y} = \frac{1}{4\sqrt{2\pi}} \left(\frac{Z}{a}\right)^{5/2} y e^{-Zr/2a} \qquad (6.121)$$

The function ψ_{2p_0} is real and is often denoted by

$$\psi_{2p_0} = \psi_{2p_z} = \frac{1}{\sqrt{\pi}} \left(\frac{Z}{2a}\right)^{5/2} z e^{-Zr/2a} \qquad (6.122)$$

where capital Z stands for the number of protons in the nucleus and small z is the z coordinate of the electron. We leave it to you to show that ψ_{2p_x}, ψ_{2p_y}, and ψ_{2p_z} are mutually orthogonal. Note that ψ_{2p_z} is zero in the xy plane, positive above this plane, and negative below it.

The functions $\psi_{2p_{-1}}$ and ψ_{2p_1} are eigenfunctions of $\hat{L}^2$ with the *same* eigenvalue: $2\hbar^2$. The reasoning of Section 3.6 shows that the linear combinations (6.118) and (6.120) are also eigenfunctions of $\hat{L}^2$ with eigenvalue $2\hbar^2$. However, $\psi_{2p_{-1}}$ and ψ_{2p_1} are eigenfunctions of $\hat{L}_z$ with *different* eigenvalues: $-\hbar$ and $+\hbar$. Therefore, ψ_{2p_x} and ψ_{2p_y} are not eigenfunctions of $\hat{L}_z$.

We can extend this procedure to construct real wave functions for higher states. Since m ranges from $-l$ to $+l$, for each complex function containing the factor $e^{-i|m|\phi}$, there is a function with the same value of n and l but having the factor $e^{+i|m|\phi}$. Addition and subtraction of these functions gives two real functions, one with the factor $\cos(|m|\phi)$, the other with the factor $\sin(|m|\phi)$. Table 6.2 lists these real wave functions for the hydrogenlike atom. The subscripts on these functions come from similar considerations as for the $2p_x$, $2p_y$, and $2p_z$ functions. For example,

$$\psi_{3d_{x^2-y^2}} \equiv 2^{-1/2}(\psi_{3d_{+2}} + \psi_{3d_{-2}})$$

$$= \frac{1}{81\sqrt{2\pi}} \left(\frac{Z}{a}\right)^{7/2} e^{-Zr/3a} r^2 \sin^2\theta \, (\cos^2\phi - \sin^2\phi)$$

$$= \frac{1}{81\sqrt{2\pi}} \left(\frac{Z}{a}\right)^{7/2} e^{-Zr/3a}(x^2 - y^2)$$

where we have used Table 6.1 for the radial factor, Table 5.1 for the theta factor, and $e^{im\phi}/\sqrt{2\pi}$ for the phi factor.

The real hydrogenlike functions are derived from the complex functions by replacing $e^{im\phi}/(2\pi)^{1/2}$ with $\pi^{-1/2}\sin|m|\phi$ or $\pi^{-1/2}\cos|m|\phi$ for $m \neq 0$; for $m = 0$ the ϕ factor is $1/(2\pi)^{1/2}$ for both real and complex functions.

In dealing with molecules, the real hydrogenlike orbitals are more useful than the complex ones. For example, we shall see in Section 15.6 that the real

TABLE 6.2   Real hydrogenlike wave functions

$$\psi_{1s} = \frac{1}{\pi^{1/2}} \left(\frac{Z}{a}\right)^{3/2} e^{-Zr/a}$$

$$\psi_{2s} = \frac{1}{4(2\pi)^{1/2}} \left(\frac{Z}{a}\right)^{3/2} \left(2 - \frac{Zr}{a}\right) e^{-Zr/2a}$$

$$\psi_{2p_z} = \frac{1}{4(2\pi)^{1/2}} \left(\frac{Z}{a}\right)^{5/2} re^{-Zr/2a} \cos\theta$$

$$\psi_{2p_x} = \frac{1}{4(2\pi)^{1/2}} \left(\frac{Z}{a}\right)^{5/2} re^{-Zr/2a} \sin\theta \cos\phi$$

$$\psi_{2p_y} = \frac{1}{4(2\pi)^{1/2}} \left(\frac{Z}{a}\right)^{5/2} re^{-Zr/2a} \sin\theta \sin\phi$$

$$\psi_{3s} = \frac{1}{81(3\pi)^{1/2}} \left(\frac{Z}{a}\right)^{3/2} \left(27 - 18\frac{Zr}{a} + 2\frac{Z^2 r^2}{a^2}\right) e^{-Zr/3a}$$

$$\psi_{3p_z} = \frac{2^{1/2}}{81\pi^{1/2}} \left(\frac{Z}{a}\right)^{5/2} \left(6 - \frac{Zr}{a}\right) re^{-Zr/3a} \cos\theta$$

$$\psi_{3p_x} = \frac{2^{1/2}}{81\pi^{1/2}} \left(\frac{Z}{a}\right)^{5/2} \left(6 - \frac{Zr}{a}\right) re^{-Zr/3a} \sin\theta \cos\phi$$

$$\psi_{3p_y} = \frac{2^{1/2}}{81\pi^{1/2}} \left(\frac{Z}{a}\right)^{5/2} \left(6 - \frac{Zr}{a}\right) re^{-Zr/3a} \sin\theta \sin\phi$$

$$\psi_{3d_{z^2}} = \frac{1}{81(6\pi)^{1/2}} \left(\frac{Z}{a}\right)^{7/2} r^2 e^{-Zr/3a} (3\cos^2\theta - 1)$$

$$\psi_{3d_{xz}} = \frac{2^{1/2}}{81\pi^{1/2}} \left(\frac{Z}{a}\right)^{7/2} r^2 e^{-Zr/3a} \sin\theta \cos\theta \cos\phi$$

$$\psi_{3d_{yz}} = \frac{2^{1/2}}{81\pi^{1/2}} \left(\frac{Z}{a}\right)^{7/2} r^2 e^{-Zr/3a} \sin\theta \cos\theta \sin\phi$$

$$\psi_{3d_{x^2-y^2}} = \frac{1}{81(2\pi)^{1/2}} \left(\frac{Z}{a}\right)^{7/2} r^2 e^{-Zr/3a} \sin^2\theta \cos 2\phi$$

$$\psi_{3d_{xy}} = \frac{1}{81(2\pi)^{1/2}} \left(\frac{Z}{a}\right)^{7/2} r^2 e^{-Zr/3a} \sin^2\theta \sin 2\phi$$

atomic orbitals $2p_x$, $2p_y$, and $2p_z$ of the oxygen atom have the proper symmetry to be used in constructing a wave function for the H_2O molecule, whereas the complex $2p$ orbitals do not.

6.7   HYDROGENLIKE ORBITALS

We shall refer to the hydrogenlike wave functions as hydrogenlike *orbitals*. These functions have been derived for a one-electron atom, and we cannot expect to use them to get a truly accurate representation of the wave function of a many-electron atom. We shall consider the use of the orbital concept to approximate many-electron atomic wave functions in Chapter 11. For the present we restrict ourselves to one-electron atoms.

There are two fundamentally different ways of depicting orbitals: method I is to draw graphs of the functions; method II is to draw contour surfaces of constant probability density.

First consider drawing graphs (method I). To graph the variation of ψ as a function of the three independent variables r, θ, and ϕ, we need four dimensions. The three-dimensional nature of our world prevents us from drawing such a graph. Instead, we draw graphs of the factors in ψ. Graphing $R(r)$ versus r, we get the curves of Fig. 6.8. Such graphs contain no information on the angular variation of ψ.

Now consider graphs of $S(\theta)$. We have (Table 5.1)

$$S_{0,0} = 1/\sqrt{2}, \qquad S_{1,0} = \tfrac{1}{2}\sqrt{6}\cos\theta$$

We can graph these functions using two-dimensional Cartesian coordinates, plotting S on the vertical axis and θ on the horizontal axis. $S_{0,0}$ gives a horizontal straight line, and $S_{1,0}$ gives a cosine curve. More commonly, S is graphed using plane polar coordinates. The variable θ is the angle with the positive z axis, and $S(\theta)$ is the distance from the origin to the point on the graph. For $S_{0,0}$, we get a circle; for $S_{1,0}$ we obtain two tangent circles (Fig. 6.10). The negative sign on the lower circle of the graph of $S_{1,0}$ indicates that $S_{1,0}$ is negative for $\tfrac{1}{2}\pi < \theta \leqslant \pi$. Strictly speaking, in graphing $\cos\theta$ we only get the upper circle, which is traced out twice; to get two tangent circles, we must graph $|\cos\theta|$.

Instead of graphing the angular factors separately, we can draw a single graph that plots $|S(\theta)T(\phi)|$ as a function of θ and ϕ. We will use spherical polar coordinates, and the distance from the origin to a point on the graph will be $|S(\theta)T(\phi)|$. For an s state, ST is independent of the angles, and we get a sphere of radius $1/(4\pi)^{1/2}$ as the graph. For a p_z state, $ST = \tfrac{1}{2}(3/\pi)^{1/2}\cos\theta$, and the graph of $|ST|$ consists of two spheres with centers on the z axis and tangent at the origin (Fig. 6.11). No doubt Fig. 6.11 is familiar. Some texts say this gives the shape of a p_z orbital, which is wrong; Fig. 6.11 is simply a *graph* of the *angular factor* in a p_z wave function. Graphs of the p_x and p_y angular factors give tangent spheres lying on the x and y axes, respectively. If we graph S^2T^2 in spherical polar

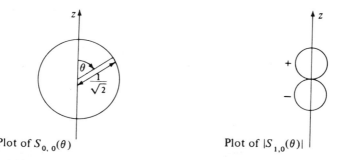

Plot of $S_{0,0}(\theta)$ Plot of $|S_{1,0}(\theta)|$

Figure 6.10. Polar graphs of the θ factors in the s and p_z hydrogen-atom wave functions.

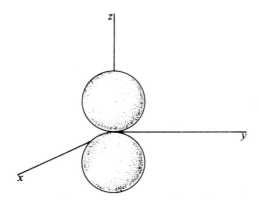

Figure 6.11 Graph of $|Y_1^0(\theta, \phi)|$, the angular factor in a p_z wave function.

coordinates, we get surfaces with the familiar figure-eight cross sections; again, these are graphs and not orbital shapes.

Now consider method II, drawing contour surfaces of constant probability density. No doubt you are familiar with geographical contour maps, which show lines of constant altitude. We shall draw surfaces in space, on each of which the value of $|\psi|^2$, the probability density, is constant. Naturally, if $|\psi|^2$ is constant on a given surface, $|\psi|$ is also constant on that surface; the contour surfaces for $|\psi|^2$ and for $|\psi|$ are identical.

For an s orbital, the wave function depends only on r, and a contour surface is a surface of constant r, that is, a sphere centered at the origin. To pin down the size of an orbital, we take a contour surface within which the probability of finding the electron is, say, 90%; thus we want

$$\int_V |\psi|^2 \, d\tau = 0.90$$

where V is the volume enclosed by the orbital contour surface.

Let us obtain the cross section of the $2p_y$ hydrogenlike orbital in the yz plane. In this plane, $\phi = \pi/2$ (Fig. 6.5), and $\sin \phi = 1$; hence Table 6.2 gives for this orbital in the yz plane

$$|\psi_{2p_y}| = k^{5/2}\pi^{-1/2}re^{-kr}|\sin \theta| \tag{6.123}$$

where $k = Z/2a$. To find the orbital cross section, we use plane polar coordinates to plot (6.123) for a fixed value of ψ; r is the distance from the origin, and θ is the angle with the z axis. The result for a typical contour (Problem 6.38) is shown in Fig. 6.12. Since $ye^{-kr} = y \exp[-k(x^2 + y^2 + z^2)^{1/2}]$, we see that ψ_{2p_y} is a function of y and $(x^2 + z^2)$; hence, on a circle centered on the y axis and parallel to the xz plane, ψ_{2p_y} is constant. Thus a three-dimensional contour surface may be developed by rotating the cross section in Fig. 6.12 about the y axis, giving a pair of distorted ellipsoids. The shape of a real $2p$ orbital is two separated, distorted ellipsoids, and not two tangent spheres.

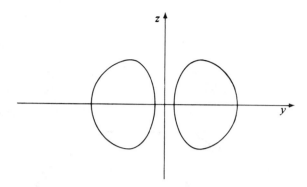

Figure 6.12 Contour of a $2p_y$ orbital.

Now consider the shape of the complex orbitals $\psi_{2p_{\pm 1}}$. We have

$$\psi_{2p_{\pm 1}} = k^{5/2}\pi^{-1/2}re^{-kr}\sin\theta\,e^{\pm i\phi}$$

$$|\psi_{2p_{\pm 1}}| = k^{5/2}\pi^{-1/2}e^{-kr}r|\sin\theta| \qquad (6.124)$$

and the two orbitals have the same shape. Since the right sides of (6.124) and (6.123) are identical, we conclude that Fig. 6.12 also gives the cross section of the $2p_{\pm 1}$ orbitals in the yz plane. Since [Eq. (5.51)]

$$e^{-kr}r|\sin\theta| = \exp\left[-k(x^2+y^2+z^2)^{1/2}\right](x^2+y^2)^{1/2}$$

we see that $|\psi_{2p_{\pm 1}}|$ is a function of z and (x^2+y^2); so we get the three-dimensional orbital shape by rotating Fig. 6.12 about the z axis. This gives a doughnut-shaped surface.

Various hydrogenlike orbital surfaces are shown in Fig. 6.13. The $2s$ orbital has a spherical node, which is not visible; the $3s$ orbital has two such nodes. The $3p_z$ orbital has a spherical node (indicated by a dashed line) and a nodal plane (the xy plane). The $3d_{z^2}$ orbital has two nodal cones. The $3d_{x^2-y^2}$ orbital has two nodal planes. Note that the view shown is not the same for the various orbitals. The relative signs of the wave functions are indicated. The other three real $3d$ orbitals in Table 6.2 have the same shape as the $3d_{x^2-y^2}$ orbital but have different orientations. The $3d_{xy}$ orbital has its lobes lying between the x and y axes and is obtained by rotating the $3d_{x^2-y^2}$ orbital by $45°$ about the z axis. The $3d_{yz}$ and $3d_{xz}$ orbitals have their lobes between the y and z axes and between the x and z axes, respectively.

Figure 6.14 represents the probability density in the yz plane for various orbitals; the number of dots in a given region is proportional to the value of $|\psi|^2$ in that region. Rotation of these diagrams about the vertical (z) axis gives the three-dimensional probability density. The $2s$ orbital has a constant for its angular factor and hence has no angular nodes; for this orbital, $n - l - 1 = 1$, indicating one radial node. The sphere on which $\psi_{2s} = 0$ is evident in Fig. 6.14.

Schrödinger's original interpretation of $|\psi|^2$ was that the electron is "smeared out" into a charge cloud. If we consider an electron passing from one

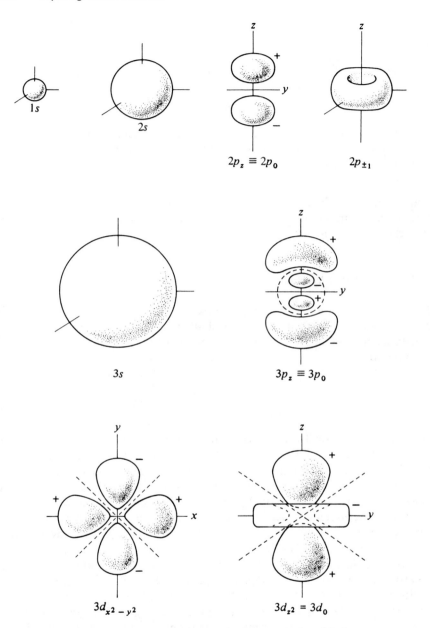

Figure 6.13 Shapes of some hydrogen-atom orbitals.

medium to another, we find that $|\psi|^2$ is nonzero in both mediums. According to the charge-cloud interpretation, this would mean that part of the electron was reflected and part transmitted. However, experimentally one never detects a fraction of an electron; electrons behave as indivisible entities. This difficulty is removed by the Born interpretation, according to which the values of $|\psi|^2$ in the

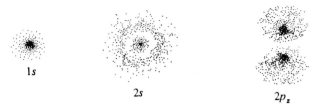

Figure 6.14 Probability densities for some hydrogen-atom states. [For accurate stereo plots, see D. T. Cromer, *J. Chem. Educ.*, **45**, 626 (1968).]

two mediums give the *probabilities* for reflection and transmission. The orbital shapes we have drawn give the regions of space in which the total probability of finding the electron is 90%.

6.8 THE ZEEMAN EFFECT

In 1896, Zeeman observed that application of an external magnetic field caused a splitting of atomic spectral lines. We shall consider this *Zeeman effect* for the hydrogen atom. We begin by reviewing magnetism.

Magnetic fields arise from moving electric charges. A charge Q with velocity **v** gives rise to a magnetic field **B** at point P in space, such that

$$\mathbf{B} = \frac{\mu_0}{4\pi} \frac{Q\mathbf{v} \times \mathbf{r}}{r^3} \tag{6.125}$$

where **r** is the vector from Q to point P and where μ_0 (the *permeability of vacuum*) is defined as $4\pi \times 10^{-7} \, \text{N C}^{-2} \, \text{s}^2$. [Equation (6.125) is valid only for a nonaccelerated charge moving with a speed much less than the speed of light.] The vector **B** is called the *magnetic induction* or *magnetic flux density*. (It was formerly believed that the vector **H** was the fundamental magnetic field vector, so **H** was called the *magnetic field strength*. It is now known that **B** is the fundamental magnetic vector.) Equation (6.125) is in SI units with Q in coulombs and **B** in teslas (T), where $1 \, \text{T} = 1 \, \text{N C}^{-1} \, \text{m}^{-1} \, \text{s}$. Only SI units will be used in this section.

Two electric charges $+Q$ and $-Q$ separated by a small distance b constitute an electric dipole. The *electric dipole moment* is defined as a vector from $-Q$ to $+Q$ with magnitude Qb. For a small planar loop of electric current, it turns out that the magnetic field generated by the moving charges of the current is given by the same mathematical expression as that giving the electric field due to an electric dipole, except that the electric dipole moment is replaced by the *magnetic dipole moment* **μ**; **μ** is a vector of magnitude IA, where I is the current flowing in a loop of area A; the direction of **μ** is perpendicular to the plane of the current loop.

Consider the magnetic (dipole) moment associated with a charge Q moving in a circle of radius r with speed v. The current is the charge flow per unit time. The circumference of the circle is $2\pi r$, and the time for one revolution is $2\pi r/v$.

Hence $I = Qv/2\pi r$. The magnitude of $\boldsymbol{\mu}$ is

$$|\boldsymbol{\mu}| = IA = (Qv/2\pi r)\pi r^2 = Qvr/2 = Qrp/2m \qquad (6.126)$$

where m is the mass of the charged particle and p is its linear momentum. Since the radius vector $\mathbf{r}$ is perpendicular to $\mathbf{p}$, we have

$$\boldsymbol{\mu}_L = \frac{Q\mathbf{r} \times \mathbf{p}}{2m} = \frac{Q}{2m} \mathbf{L} \qquad (6.127)$$

where the definition of orbital angular momentum $\mathbf{L}$ was used and where the subscript on $\boldsymbol{\mu}$ indicates that it arises from the orbital motion of the particle. Although we derived (6.127) for the special case of circular motion, its validity is general. For an electron, $Q = -e$, and the magnetic moment due to its orbital motion is

$$\boldsymbol{\mu}_L = -\frac{e}{2m_e} \mathbf{L} \qquad (6.128)$$

The magnitude of $\mathbf{L}$ is given by (5.95), and the magnitude of the orbital magnetic moment of an electron with orbital-angular-momentum quantum number l is

$$|\boldsymbol{\mu}_L| = \frac{e\hbar}{2m_e} [l(l+1)]^{1/2} = \beta_e [l(l+1)]^{1/2} \qquad (6.129)$$

The constant $e\hbar/2m_e$ is called the *Bohr magneton* β_e:

$$\beta_e = e\hbar/2m_e = 9.274 \times 10^{-24} \text{ J/T} \qquad (6.130)$$

Now consider applying an external magnetic field to the hydrogen atom. The energy of interaction between a magnetic dipole $\boldsymbol{\mu}$ and an external magnetic field $\mathbf{B}$ is [*Halliday and Resnick*, Eq. (33-12)]

$$E_B = -\boldsymbol{\mu} \cdot \mathbf{B} \qquad (6.131)$$

Using Eq. (6.128), we have

$$E_B = \frac{e}{2m_e} \mathbf{L} \cdot \mathbf{B} \qquad (6.132)$$

We take the z axis along the direction of the applied field: $\mathbf{B} = B\mathbf{k}$, where $\mathbf{k}$ is a unit vector in the z direction. We have

$$E_B = \frac{e}{2m_e} B(L_x\mathbf{i} + L_y\mathbf{j} + L_z\mathbf{k}) \cdot \mathbf{k} = \frac{e}{2m_e} BL_z = \frac{\beta_e}{\hbar} BL_z$$

where L_z is the z component of orbital angular momentum. We now replace L_z by the operator $\hat{L}_z$ to give the following additional term in the Hamiltonian, resulting from the external magnetic field:

$$\hat{H}_B = \beta_e B\hbar^{-1}\hat{L}_z \qquad (6.133)$$

The Schrödinger equation for the hydrogen atom in a magnetic field is

$$(\hat{H} + \hat{H}_B)\psi = E\psi \qquad (6.134)$$

where $\hat{H}$ is the hydrogen-atom Hamiltonian in the absence of an external field. We readily verify that the solutions of Eq. (6.134) are the complex hydrogenlike wave functions (6.61):

$$(\hat{H} + \hat{H}_B)R(r)Y_l^m(\theta, \phi) = \hat{H}RY_l^m + \beta_e \hbar^{-1} B \hat{L}_z RY_l^m$$

$$= \left(-\frac{Z^2}{n^2}\frac{e'^2}{2a} + \beta_e Bm\right)RY_l^m \qquad (6.135)$$

where Eqs. (6.94) and (5.112) were used. Thus there is an additional term $\beta_e Bm$ in the energy, and the external magnetic field removes the m degeneracy. For obvious reasons, m is often called the *magnetic quantum number*. Actually, the observed energy shifts do *not* match the predictions of Eq. (6.135), because of the existence of electron spin magnetic moment (Chapter 10 and Section 11.7).

In Chapter 5 we found that in quantum mechanics **L** lies on the surface of a cone. A *classical*-mechanical treatment (*Halliday and Resnick*, Sections 13-2 and 37-7) of the motion of **L** in an applied magnetic field shows that the field exerts a torque on $\boldsymbol{\mu}_L$, causing **L** to revolve about the direction of **B** at a constant frequency given by $|\boldsymbol{\mu}_L|B/2\pi|\mathbf{L}|$, while maintaining a constant angle with **B**. This gyroscopic motion is called *precession*. In quantum mechanics, a complete specification of **L** is impossible; however, one finds that $\langle \mathbf{L} \rangle$ precesses about the field direction (*Dicke and Wittke*, Section 12-3).

6.9 SUMMARY

For a one-particle system with potential energy a function of r only [$V = V(r)$, a central-force problem], the stationary-state wave functions have the form $\psi = R(r)Y_l^m(\theta, \phi)$, where $R(r)$ satisfies the radial equation (6.17) and Y_l^m are the spherical harmonics.

For a system of two noninteracting particles 1 and 2, the Hamiltonian is $\hat{H} = \hat{H}_1 + \hat{H}_2$, and the stationary state wave functions and energies satisfy $\psi = \psi_1(q_1)\psi_2(q_2)$, $E = E_1 + E_2$, where $\hat{H}_1\psi_1 = E_1\psi_1$ and $\hat{H}_2\psi_2 = E_2\psi_2$; q_1 and q_2 stand for the coordinates of particles 1 and 2.

For a system of two interacting particles 1 and 2 with Hamiltonian $\hat{H} = \hat{T}_1 + \hat{T}_2 + \hat{V}$, where V is a function of only the relative coordinates x, y, z of the particles, the energy is the sum of the energies of two hypothetical particles: $E = E_M + E_\mu$. One hypothetical particle has mass $M \equiv m_1 + m_2$; its coordinates are the coordinates of the center of mass, and its energy E_M is that of a free particle. The second particle has mass $\mu \equiv m_1 m_2/(m_1 + m_2)$; its coordinates are the relative coordinates x, y, z, and its energy E_μ is found by solving the Schrödinger equation for internal motion: $[(-\hbar^2/2\mu)\nabla^2 + V]\psi(x, y, z) = E_\mu\psi(x, y, z)$.

The two-particle rigid rotor consists of particles of masses m_1 and m_2 separated by a fixed distance d. Its energy is the sum of the energy E_M of translation and the energy E_μ of rotation. Its stationary-state rotational wave functions are $\psi = Y_J^m(\theta, \phi)$, where θ and ϕ give the orientation of the rotor axis with respect to an origin at the rotor's center of mass, and the quantum numbers are $J = 0, 1, 2, \ldots$ and $m = -J, -J+1, \ldots, J-1, J$. The rotational energy levels are $E_\mu = J(J + 1)\hbar^2/2I$, where $I = \mu d^2$, with $\mu = m_1 m_2/(m_1 + m_2)$. The selection rule for spectroscopic transitions is $\Delta J = \pm 1$.

The hydrogenlike atom has $V = -Ze'^2/r$. With the translational energy separated off, the internal motion is a central-force problem and $\psi = R(r)Y_l^m(\theta, \phi)$. The continuum states have $E \geq 0$ and correspond to an ionized atom. The bound states have the allowed energies $E = -(Z^2/n^2)(e'^2/2a)$, where $a \equiv \hbar^2/\mu e'^2$; the bound-state radial wave function is (6.101). The bound-state quantum numbers are $n = 1, 2, 3, \ldots$; $l = 0, 1, 2, \ldots, n-1$; $m = -l, -l+1, \ldots, l-1, l$.

A one-electron wave function is called an orbital. The shape of an orbital is defined by a contour surface of constant $|\psi|$ that encloses a specified amount of probability.

PROBLEMS

6.1 If the three force constants in Problem 4.17 are all equal, we have a three-dimensional isotropic harmonic oscillator. (a) State why the wave functions for this case can be written as $\psi = F(r)G(\theta, \phi)$. (b) What is the function G? (c) Write a differential equation satisfied by $F(r)$. (d) Use the results found in Problem 4.17 to show that the ground-state wave function does have the form $F(r)G(\theta, \phi)$ and verify that the ground-state $F(r)$ satisfies the differential equation in (c).

6.2 The particle in a spherical box has $V = 0$ for $r \leq b$ and $V = \infty$ for $r > b$. For this system: (a) Explain why $\psi = R(r)f(\theta, \phi)$, where $R(r)$ satisfies (6.17). What is the function $f(\theta, \phi)$? (b) Solve (6.17) for $R(r)$ for the $l = 0$ states. *Hints:* The substitution $R(r) = g(r)/r$ reduces (6.17) to an easily solved equation. Use the boundary condition that ψ is finite at $r = 0$ [see the discussion after Eq. (6.82)] and use a second boundary condition. Show that $\psi = N(\sin kr)/r$ for the $l = 0$ states, where $k \equiv (2mE/\hbar^2)^{1/2}$ and $E = n^2h^2/8mb^2$ with $n = 1, 2, 3, \ldots$. (For $l \neq 0$, the energy-level formula is more complicated.)

6.3 Verify Eq. (6.6) for the Laplacian in spherical polar coordinates.

6.4 For a system of two noninteracting particles of mass 9.0×10^{-26} g and 5.0×10^{-26} g in a one-dimensional box of length 1.00×10^{-8} cm, calculate the energies of the six lowest stationary states.

6.5 The lowest observed microwave absorption frequency of $^{12}C^{16}O$ is 115271 MHz. (a) Compute the bond distance in $^{12}C^{16}O$. (b) Predict the next two lowest microwave absorption frequencies of $^{12}C^{16}O$. (c) Predict the lowest microwave absorption frequency of $^{13}C^{16}O$. (d) For $^{12}C^{16}O$ at 25°C, calculate the ratio of the $J = 1$ population to the $J = 0$ population. Repeat for the $J = 2$ to $J = 0$ ratio. Don't forget degeneracy.

6.6 Verify Eq. (6.51) for I of a two-particle rotor. Begin by multiplying and dividing the right side of (6.50) by $m_1 m_2/(m_1 + m_2)$. Then use (6.49).

6.7 Calculate the ratio of the electrical and gravitational forces between a proton and an electron. Is neglect of the gravitational force justified?

6.8 For the particle in a box with infinitely high walls and for the harmonic oscillator, there are no continuum eigenfunctions, whereas for the hydrogen atom we do have continuum functions. Explain this in terms of the nature of the potential-energy function for each problem.

6.9 To find the degeneracy of the hydrogen-atom energy levels, we used

$$\sum_{k=1}^{n} k = \frac{n(n+1)}{2}$$

ᴕ result by adding corresponding terms of the two series: $1, 2, 3, \ldots, n$ and $1, n - 2, \ldots, 1$.

6.10 (a) Calculate the wavelength and frequency for the spectral line that arises from an $n = 6$ to $n = 3$ transition in the hydrogen atom. (b) Repeat the calculations for He^+; neglect the change in reduced mass from H to He^+.

6.11 Assign each of the following observed vacuum wavelengths to a transition between two hydrogen-atom levels:

6564.7 Å,   4862.7 Å,   4341.7 Å,   4102.9 Å   (Balmer series)

Predict the wavelengths of the next two lines in this series and the wavelength of the series limit. (Balmer was a Swiss mathematician who, in 1885, came up with an empirical formula that fitted lines of the hydrogen spectrum.)

6.12 Each hydrogen-atom line of Problem 6.11 shows a very weak nearby satellite line. Two of the satellites occur at the vacuum wavelengths 6562.9 Å and 4861.4 Å. (a) Explain their origin. (The person who first answered this question got a Nobel Prize.) (b) Calculate the other two satellite wavelengths.

6.13 Calculate the ground-state energy of the hydrogen atom by using Gaussian units in (6.94). Convert the result to electron volts.

6.14 The positron has charge $+e$ and mass equal to the electron mass. Calculate in electron volts the ground-state energy of positronium—an "atom" that consists of a positron and an electron.

6.15 If we were to ignore the interelectronic repulsion in helium, what would be its ground-state energy and wave function? (See Section 6.2.) Compute the percent error in the energy; the experimental He ground-state energy is -79.0 eV.

6.16 For the ground state of the hydrogenlike atom, show that $\langle r \rangle = 3a/2Z$.

6.17 Find $\langle r \rangle$ for the $2p_0$ state of the hydrogenlike atom.

6.18 Find $\langle r^2 \rangle$ for the $2p_1$ state of the hydrogenlike atom.

6.19 For a hydrogenlike atom in a stationary state with quantum numbers n, l, and m, prove that $\langle r \rangle = \int_0^\infty r^3 |R_{nl}|^2 \, dr$.

6.20 Derive the $2s$ and $2p$ radial hydrogenlike functions.

6.21 What is the value of the angular-momentum quantum number l for a t orbital?

6.22 For the ground state of the hydrogenlike atom, find the most probable value of r.

6.23 Where is the probability density a maximum for the hydrogen-atom ground state?

6.24 For the hydrogen-atom ground state, find the probability of finding the electron farther than $2a$ from the nucleus.

6.25 For the H atom ground state, find the probability of finding the electron in the classically forbidden region.

6.26 Find the radius of the sphere defining the $1s$ hydrogen orbital using the 90% probability definition.

6.27 For which hydrogen-atom states is ψ nonzero at the nucleus?

6.28 A stationary-state wave function is an eigenfunction of the Hamiltonian operator $\hat{H} = \hat{T} + \hat{V}$. Students sometimes erroneously believe that ψ is an eigenfunction of $\hat{T}$ and of $\hat{V}$. For the ground state of the hydrogen atom, verify directly that ψ is not an eigenfunction of $\hat{T}$ or of $\hat{V}$, but is an eigenfunction of $(\hat{T} + \hat{V})$. Can you think of a problem we solved where ψ is an eigenfunction of $\hat{T}$ and of $\hat{V}$?

6.29 Show that $\langle T \rangle + \langle V \rangle = E$ for a stationary state.

6.30 Compute $\langle V \rangle$, $\langle T \rangle$, and $\langle T \rangle / \langle V \rangle$ for the hydrogen-atom ground state. Use the result of Problem 6.29 to shorten the work.

6.31 For the hydrogen-atom ground state, use the value of $\langle T \rangle$ from Problem 6.30 to calculate the root-mean-square speed of the electron; find the numerical value of $\langle v^2 \rangle^{1/2}/c$, where c is the speed of light.

6.32 The hydrogenlike wave functions $2p_1$, $2p_0$, and $2p_{-1}$ can be characterized as those $2p$ functions that are eigenfunctions of $\hat{L}_z$. What operators can we use to characterize the functions $2p_x$, $2p_y$, and $2p_z$, and what are the corresponding eigenvalues?

6.33 Given that $\hat{A}f = af$ and $\hat{A}g = bg$, where f and g are functions and a and b are constants, under what condition(s) is the linear combination $c_1 f + c_2 g$ an eigenfunction of the linear operator $\hat{A}$?

6.34 State which of the three operators $\hat{L}^2$, $\hat{L}_z$, and the H-atom $\hat{H}$ each of the following functions is an eigenfunction of: (a) $2p_z$; (b) $2p_x$; (c) $2p_1$.

6.35 For the *real* hydrogenlike functions: (a) What is the shape of the $n - l - 1$ nodal surfaces for which the radial factor is zero? (b) The nodal surfaces for which the ϕ factor vanishes are of the form $\phi = $ constant; thus they are planes perpendicular to the xy plane. How many such planes are there? (Values of ϕ that differ by π are considered to be part of the same plane.) (c) It can be shown that there are $l - m$ surfaces on which the θ factor vanishes. What is the shape of these surfaces? (d) How many nodal surfaces are there for the real hydrogenlike wave functions?

6.36 (a) The *Laguerre polynomials* are defined as

$$L_q(z) \equiv e^z \frac{d^q}{dz^q} (e^{-z} z^q)$$

Verify that

$$L_0 = 1, \quad L_1 = -z + 1, \quad L_2 = z^2 - 4z + 2, \quad L_3 = -z^3 + 9z^2 - 18z + 6$$

(b) The *associated Laguerre polynomials* are usually defined as

$$L_q^s(z) = \frac{d^s}{dz^s} L_q(z)$$

Verify that

$$L_q^0(z) = L_q(z), \quad L_1^1(z) = -1, \quad L_2^1(z) = 2z - 4, \quad L_2^2(z) = 2$$

$$L_3^1(z) = -3z^2 + 18z - 18, \quad L_3^2(z) = -6z + 18, \quad L_3^3(z) = -6$$

(c) The radial factor in the hydrogenlike wave functions can be shown (*Pauling and Wilson*, page 132) to be

$$R_{nl}(r) = -\left[\frac{4Z^3}{n^4 a^3} \frac{(n - l - 1)!}{[(n + l)!]^3} \right]^{1/2} \left(\frac{2Zr}{na} \right)^l e^{-Zr/na} L_{n+l}^{2l+1} \left(\frac{2Zr}{na} \right)$$

Verify this equation for the $1s$, $2s$, and $2p$ states.

6.37 On the x axis the ground-state hydrogen-atom wave function is given by (6.109). Calculate the limit that the derivative of (6.109) approaches (a) as $x \rightarrow 0$ from the left; (b) as $x \rightarrow 0$ from the right.

6.38 Show that the maximum value for ψ_{2p_y} [Eq. (6.123)] is $k^{3/2}\pi^{-1/2}e^{-1}$. Use Eq. (6.123) to plot the $2p_y$ contour for which $\psi = 0.316\psi_{\max}$.

6.39 Sketch rough contours of constant $|\psi|$ for each of the following states of a particle in a two-dimensional square box: $n_x n_y = 11$; 12; 21; 22. What are you reminded of?

6.40 For each of the following systems, give the expression for $d\tau$ and give the limits of each variable in the equation $\int |\psi|^2 \, d\tau = 1$. (a) The particle in a one-dimensional box of length l. (b) The one-dimensional harmonic oscillator. (c) A one-particle, three-dimensional system where Cartesian coordinates are used. (d) The hydrogen atom, using spherical polar coordinates.

6.41 Find the $n = 2$ to $n = 1$ energy-level population ratio for a gas of hydrogen atoms at (a) 25°C; (b) 1000 K; (c) 10000 K.

6.42 Name a quantum-mechanical system for which the spacing between adjacent bound-state energy levels (a) remains constant as E increases; (b) increases as E increases; (c) decreases as E increases.

7

Theorems of
Quantum Mechanics

7.1 INTRODUCTION

The Schrödinger equation for the one-electron atom (Chapter 6) is exactly solvable. However, because of the interelectronic repulsion terms in the Hamiltonian, the Schrödinger equation for many-electron atoms and molecules is not separable in any coordinate system and cannot be solved exactly. Hence we must seek approximate methods of solution. The two main approximation methods, the variation method and perturbation theory, will be developed in Chapters 8 and 9. To derive these methods, we must develop further the theory of quantum mechanics, which is what we will do in this chapter. Although much of the work will be abstract and perhaps not so easy to master, the theorems involved are basic to an understanding of quantum mechanics.

Before starting, we introduce some notations for the integrals we will be using. The definite integral over all space of an operator sandwiched between two functions frequently occurs, and various abbreviations are used:

$$\int f_m^* \hat{A} f_n \, d\tau \equiv \langle f_m | \hat{A} | f_n \rangle \equiv (f_m | \hat{A} | f_n) \equiv \langle m | \hat{A} | n \rangle \qquad (7.1)^*$$

where f_m and f_n are two functions. If it is clear what functions are meant, we can use just the indexes, as indicated in (7.1). The above notation, introduced by Dirac, is called **bracket notation**. Another notation is

$$\int f_m^* \hat{A} f_n \, d\tau \equiv A_{mn} \qquad (7.2)^*$$

The notations A_{mn} and $\langle m | \hat{A} | n \rangle$ imply that we use the complex conjugate of the function whose letter appears first. The integral $\int f_m^* \hat{A} f_n \, d\tau$ is called a **matrix element** of the operator $\hat{A}$. Matrices are rectangular arrays of numbers and obey certain rules of combination (see Section 7.10).

For the integral over all space between two functions, we write

$$\int f_m^* f_n \, d\tau \equiv \langle f_m | f_n \rangle \equiv (f_m, f_n) \equiv \langle m | n \rangle \qquad (7.3)^*$$

147

Since

$$\left[\int f_m^* f_n \, d\tau\right]^* = \int f_n^* f_m \, d\tau$$

we have the identity

$$\langle m|n\rangle^* = \langle n|m\rangle \qquad (7.4)^*$$

In particular, $\langle m|m\rangle^* = \langle m|m\rangle$.

7.2 HERMITIAN OPERATORS

The quantum-mechanical operators that represent physical quantities are linear (Section 3.1). These operators must meet an additional requirement, which we now discuss.

Definition of Hermitian Operators.   Let $\hat{A}$ be the linear operator representing the physical property A. The average value of A is [Eq. (3.88)]

$$\langle A\rangle = \int \Psi^* \hat{A}\Psi \, d\tau \qquad (7.5)$$

where Ψ is the state function of the system. Since the average value of a physical quantity must be a real number, we demand that

$$\langle A\rangle = \langle A\rangle^*$$

$$\int \Psi^* \hat{A}\Psi \, d\tau = \int \Psi(\hat{A}\Psi)^* \, d\tau \qquad (7.6)$$

Equation (7.6) must hold for any function Ψ that can represent a possible state of the system; that is, it must hold for all well-behaved functions Ψ. A linear operator that satisfies (7.6) for all well-behaved functions is called a **Hermitian operator**.

Many texts define a Hermitian operator as a linear operator that satisfies

$$\int f^* \hat{A}g \, d\tau = \int g(\hat{A}f)^* \, d\tau \qquad (7.7)$$

for all well-behaved functions f and g. Note especially that on the left side of (7.7) $\hat{A}$ operates on g, but on the right side $\hat{A}$ operates on f. For the special case $f = g$, (7.7) reduces to (7.6). Equation (7.7) is apparently a more stringent requirement than (7.6), but we shall prove that (7.7) is a consequence of (7.6); hence the two definitions of a Hermitian operator are equivalent.

We begin the proof by setting $\Psi = f + cg$ in (7.6), where c is an arbitrary parameter; this gives

$$\int (f + cg)^* \hat{A}(f + cg) \, d\tau = \int (f + cg)[\hat{A}(f + cg)]^* \, d\tau$$

$$\int (f^* + c^* g^*)\hat{A}f \, d\tau + \int (f^* + c^* g^*)\hat{A}cg \, d\tau$$

$$= \int (f + cg)(\hat{A}f)^* \, d\tau + \int (f + cg)(\hat{A}cg)^* \, d\tau$$

$$\int f^* \hat{A}f \, d\tau + c^* \int g^* \hat{A}f \, d\tau + c \int f^* \hat{A}g \, d\tau + c^*c \int g^* \hat{A}g \, d\tau$$

$$= \int f(\hat{A}f)^* \, d\tau + c \int g(\hat{A}f)^* \, d\tau + c^* \int f(\hat{A}g)^* \, d\tau + cc^* \int g(\hat{A}g)^* \, d\tau$$

By virtue of (7.6), the first terms on each side of this last equation are equal to each other; likewise, the last terms on each side are equal. Hence

$$c^* \int g^* \hat{A}f \, d\tau + c \int f^* \hat{A}g \, d\tau = c \int g(\hat{A}f)^* \, d\tau + c^* \int f(\hat{A}g)^* \, d\tau \qquad (7.8)$$

Setting $c = 1$ in (7.8), we have

$$\int g^* \hat{A}f \, d\tau + \int f^* \hat{A}g \, d\tau = \int g(\hat{A}f)^* \, d\tau + \int f(\hat{A}g)^* \, d\tau \qquad (7.9)$$

Setting $c = i$ in (7.8), we have, after dividing by i,

$$- \int g^* \hat{A}f \, d\tau + \int f^* \hat{A}g \, d\tau = \int g(\hat{A}f)^* \, d\tau - \int f(\hat{A}g)^* \, d\tau \qquad (7.10)$$

We now add (7.9) and (7.10) to get (7.7). This completes the proof.

Therefore, *a Hermitian operator $\hat{A}$ possesses the property that*

$$\int f_i^* \hat{A}f_j \, d\tau = \int f_j(\hat{A}f_i)^* \, d\tau \qquad \textbf{(7.11)}^*$$

where f_i and f_j are arbitrary well-behaved functions. Using the bracket and matrix-element notations, we write

$$\langle f_i | \hat{A} | f_j \rangle = \langle f_j | \hat{A} | f_i \rangle^* \qquad \textbf{(7.12)}^*$$

$$\langle i | \hat{A} | j \rangle = \langle j | \hat{A} | i \rangle^* \qquad (7.13)$$

$$A_{ij} = (A_{ji})^* \qquad \textbf{(7.14)}^*$$

Examples of Hermitian Operators. Let us show that some of the operators we have been using are indeed Hermitian. For simplicity, we shall work in one dimension. To prove that an operator is Hermitian, it suffices to show that it satisfies (7.6) for all well-behaved functions. However, we shall make things a bit harder by proving that (7.11) is satisfied.

First consider the potential-energy operator. The right side of (7.11) is

$$\int_{-\infty}^{\infty} f_j(x)[V(x)f_i(x)]^* \, dx \qquad (7.15)$$

We have $V^* = V$, since the potential energy is a real function. Moreover, the order of the factors in (7.15) is immaterial. Hence

$$\int f_j(Vf_i)^* \, dx = \int f_j V^* f_i^* \, dx = \int f_i^* V f_j \, dx$$

which proves that V is Hermitian.

The operator for the x component of linear momentum is $\hat{p}_x = -i\hbar\, d/dx$ [Eq. (3.23)]. For this operator, the left side of (7.11) is

$$-i\hbar \int_{-\infty}^{\infty} f_i^*(x)\, \frac{df_j(x)}{dx}\, dx$$

Now we use the formula for integration by parts:

$$\int_a^b u(x)\, \frac{dv(x)}{dx}\, dx = u(x)v(x)\Big|_a^b - \int_a^b v(x)\, \frac{du(x)}{dx}\, dx \tag{7.16}$$

Let

$$u(x) \equiv -i\hbar f_i^*(x), \qquad v(x) \equiv f_j(x)$$

Then

$$-i\hbar \int_{-\infty}^{\infty} f_i^* \frac{df_j}{dx}\, dx = -i\hbar f_i^* f_j \Big|_{-\infty}^{\infty} + i\hbar \int_{-\infty}^{\infty} f_j(x)\, \frac{df_i^*(x)}{dx}\, dx \tag{7.17}$$

Because f_i and f_j are well-behaved functions, they vanish at $x = \pm\infty$. (If they didn't vanish at infinity, they wouldn't be quadratically integrable.) Therefore, (7.17) becomes

$$\int_{-\infty}^{\infty} f_i^*\left(-i\hbar \frac{df_j}{dx}\right) dx = \int_{-\infty}^{\infty} f_j\left(-i\hbar \frac{df_i}{dx}\right)^* dx$$

which is the same as (7.11) and proves that $\hat{p}_x$ is Hermitian. The proof that the kinetic-energy operator is Hermitian is left to the reader. The sum of two Hermitian operators can be shown to be Hermitian. Hence the Hamiltonian operator $\hat{H} = \hat{T} + \hat{V}$ is Hermitian.

Theorems about Hermitian Operators. We now prove some important theorems about the eigenvalues and eigenfunctions of Hermitian operators.

Since the eigenvalues of the operator $\hat{A}$ corresponding to the physical quantity A are the possible results of a measurement of A, these eigenvalues should all be real numbers. To prove this, we start with the Hermitian property of $\hat{A}$, as expressed by Eq. (7.11):

$$\int f_i^* \hat{A} f_j\, d\tau = \int f_j(\hat{A}f_i)^*\, d\tau \tag{7.18}$$

which holds for all well-behaved functions f_i and f_j. Let g_i and a_i be the eigenfunctions and eigenvalues of $\hat{A}$:

$$\hat{A}g_i = a_i g_i$$

For the special case that $f_i = g_i$ and $f_j = g_i$, Eq. (7.18) becomes

$$\int g_i^* \hat{A} g_i\, d\tau = \int g_i(\hat{A}g_i)^*\, d\tau$$

Use of $\hat{A}g_i = a_i g_i$ gives

$$a_i \int g_i^* g_i\, d\tau = \int g_i(a_i g_i)^*\, d\tau = a_i^* \int g_i g_i^*\, d\tau$$

$$(a_i - a_i^*) \int |g_i|^2\, d\tau = 0 \tag{7.19}$$

Since the integrand $|g_i|^2$ is never negative, the only way the integral in (7.19) could be zero would be if g_i were zero for all values of the coordinates. However, we always reject $g_i = 0$ as an eigenfunction on physical grounds. Hence the integral in (7.19) cannot be zero. Therefore, $(a_i - a_i^*) = 0$, and $a_i = a_i^*$. We have proved:

Theorem 1. The eigenvalues of a Hermitian operator are real numbers.

To help develop a familiarity with bracket notation, we shall repeat the proof of Theorem 1 using bracket notation. We begin by setting $i = j$ in (7.13):

$$\langle i|\hat{A}|i\rangle = \langle i|\hat{A}|i\rangle^*$$

Choosing the function with index i to be an eigenfunction of $\hat{A}$ and using the eigenvalue equation $\hat{A}_i g_i = a_i g_i$, we have

$$\langle i|a_i|i\rangle = \langle i|a_i|i\rangle^*$$

$$a_i\langle i|i\rangle = a_i^*\langle i|i\rangle^* = a_i^*\langle i|i\rangle$$

$$(a_i - a_i^*)\langle i|i\rangle = 0$$

$$a_i = a_i^*$$

where (7.4) with $m = n$ was used.

We showed that two different particle-in-a-box energy eigenfunctions ψ_i and ψ_j are orthogonal, meaning that $\int_{-\infty}^{\infty} \psi_i^* \psi_j \, dx = 0$ for $i \neq j$ [Eq. (2.26)]. Two functions f_1 and f_2 of the same set of coordinates are said to be ***orthogonal*** if

$$\int f_1^* f_2 \, d\tau = 0 \qquad\qquad (7.20)^*$$

where the integral is a definite integral over the full range of the coordinates. We now prove the general theorem that *the eigenfunctions of a Hermitian operator are, or can be chosen to be, mutually orthogonal.* Given that

$$\hat{B}F = sF, \qquad \hat{B}G = tG \qquad\qquad (7.21)$$

where F and G are two linearly independent eigenfunctions of the Hermitian operator $\hat{B}$, we want to prove that

$$\int F^* G \, d\tau \equiv \langle F|G\rangle = 0$$

We begin with Eq. (7.13), which expresses the Hermitian nature of $\hat{B}$:

$$\langle F|\hat{B}|G\rangle = \langle G|\hat{B}|F\rangle^*$$

Using (7.21), we have

$$\langle F|t|G\rangle = \langle G|s|F\rangle^*$$

$$t\langle F|G\rangle = s^*\langle G|F\rangle^*$$

Since eigenvalues of Hermitian operators are real (Theorem 1), we have $s^* = s$.

Use of $\langle G|F\rangle^* = \langle F|G\rangle$ [Eq. (7.4)] gives

$$t\langle F|G\rangle = s\langle F|G\rangle$$

$$(t - s)\langle F|G\rangle = 0$$

If $s \neq t$, then

$$\langle F|G\rangle = 0 \tag{7.22}$$

We have proved that two eigenfunctions of a Hermitian operator that correspond to *different* eigenvalues are orthogonal. The question now is: Can we have two independent eigenfunctions that have the *same* eigenvalue? The answer is yes. In the case of *degeneracy*, we have the same eigenvalue for more than one independent eigenfunction. Therefore, we can only be certain that two independent eigenfunctions of a Hermitian operator are orthogonal to each other if they do not correspond to a degenerate eigenvalue. We now show that in the case of degeneracy we may *construct* eigenfunctions that will be orthogonal to one another. We shall use the theorem proved in Section 3.6, that any linear combination of eigenfunctions corresponding to a degenerate eigenvalue is an eigenfunction with the same eigenvalue. Let us therefore suppose that F and G are independent eigenfunctions that have the same eigenvalue:

$$\hat{B}F = sF, \qquad \hat{B}G = sG$$

We take linear combinations of F and G to form two new eigenfunctions g_1 and g_2 that will be orthogonal to each other. We choose g_1 equal to F, and for g_2 we write

$$g_2 = G + cF, \qquad g_1 = F$$

The constant c will be chosen to ensure orthogonality. We want

$$\int g_1^* g_2 \, d\tau = 0$$

$$\int F^*(G + cF) \, d\tau = \int F^*G \, d\tau + c \int F^*F \, d\tau = 0$$

Hence choosing

$$c = -\int F^*G \, d\tau \bigg/ \int F^*F \, d\tau \tag{7.23}$$

we have two orthogonal eigenfunctions g_1 and g_2 corresponding to the degenerate eigenvalue. This procedure (called **Schmidt orthogonalization**) can be extended to the case of n-fold degeneracy, to give n linearly independent orthogonal eigenfunctions corresponding to the degenerate eigenvalue.

Thus, although there is no guarantee that the eigenfunctions of a degenerate eigenvalue are orthogonal, we can always *choose* them to be orthogonal, if we desire, by using the Schmidt (or some other) orthogonalization method. In fact, unless stated otherwise, we shall always assume that we have chosen the eigenfunctions to be orthogonal:

$$\int g_i^* g_j \, d\tau = 0, \qquad i \neq j \tag{7.24}$$

where g_i and g_j are independent eigenfunctions of a Hermitian operator. We have proved:

Theorem 2.   Two eigenfunctions of a Hermitian operator $\hat{B}$ that correspond to different eigenvalues are orthogonal; eigenfunctions of $\hat{B}$ that belong to a degenerate eigenvalue can always be chosen to be orthogonal.

An eigenfunction can usually be multiplied by a suitable constant to normalize it, and we shall assume, unless stated otherwise, that all eigenfunctions are normalized:

$$\int g_i^* g_i \, d\tau = 1 \tag{7.25}$$

The exception is where the eigenvalues form a continuum, rather than a discrete set of values; in this case, the eigenfunctions are not quadratically integrable. Examples are the linear-momentum eigenfunctions, the free-particle energy eigenfunctions, and the hydrogen-atom continuum energy eigenfunctions.

Using the Kronecker delta, defined by $\delta_{ij} \equiv 1$ if $i = j$ and $\delta_{ij} \equiv 0$ if $i \neq j$ [Eq. (2.28)], we can combine (7.24) and (7.25) into one equation:

$$\int g_i^* g_j \, d\tau = \langle i | j \rangle = \delta_{ij} \tag{7.26*}$$

where g_i and g_j are eigenfunctions of some Hermitian operator.

As an example, consider the spherical harmonics. We shall prove that

$$\int_0^{2\pi} \int_0^\pi [Y_l^m(\theta, \phi)]^* Y_{l'}^{m'}(\theta, \phi) \sin\theta \, d\theta \, d\phi = \delta_{l,l'} \, \delta_{m,m'} \tag{7.27}$$

where the $\sin\theta$ factor comes from the volume element in spherical polar coordinates, (5.78). The spherical harmonics are eigenfunctions of the Hermitian operator $\hat{L}^2$ [Eq. (5.111)]. Since eigenfunctions of a Hermitian operator belonging to different eigenvalues are orthogonal, we conclude that the integral in (7.27) is zero unless $l = l'$. Similarly, since the Y_l^m functions are eigenfunctions of $\hat{L}_z$ [Eq. (5.112)], we conclude that the integral in (7.27) is zero unless $m = m'$. Also, the multiplicative constant in Y_l^m [Eq. (5.110)] has been chosen so that the spherical harmonics are normalized [Eq. (6.117)]. Therefore, (7.27) is valid.

A proof of the uncertainty principle is outlined in Problem 7.52.

7.3   EXPANSION IN TERMS OF EIGENFUNCTIONS

In the previous section, we proved the orthogonality of the eigenfunctions of a Hermitian operator. We now discuss another important property of these functions; this property allows us to expand an arbitrary well-behaved function in terms of these eigenfunctions.

We have used the Taylor-series expansion (Problem 4.1) of a function as a linear combination of the nonnegative integral powers of $(x - a)$. Can we expand a function as a linear combination of some other set of functions besides 1, $(x - a)$, $(x - a)^2$, . . . ? The answer is yes, as was first shown by Fourier in 1807. A

Fourier series is an expansion of a function as a linear combination of an infinite number of sine and cosine functions. We shall not go into detail about Fourier series, but shall simply look at one example.

Expansion of a Function Using Particle-in-a-Box Wave Functions.

Let us consider expanding a function in terms of the particle-in-a-box, stationary-state wave functions, which are [Eq. (2.23)]

$$\psi_n = \left(\frac{2}{l}\right)^{1/2} \sin\left(\frac{n\pi x}{l}\right), \qquad n = 1, 2, 3, \ldots \tag{7.28}$$

for x between 0 and l. What are our chances for representing an arbitrary function $f(x)$, in the interval $0 \leqslant x \leqslant l$ by a series of the form

$$f(x) = \sum_{n=1}^{\infty} a_n \psi_n = \left(\frac{2}{l}\right)^{1/2} \sum_{n=1}^{\infty} a_n \sin\left(\frac{n\pi x}{l}\right), \qquad 0 \leqslant x \leqslant l \tag{7.29}$$

where the a_n's are constants. Substitution of $x = 0$ and $x = l$ in (7.29) gives the restrictions that $f(0) = 0$ and $f(l) = 0$. In other words, $f(x)$ must satisfy the same boundary conditions as the ψ_n functions. We shall also assume that $f(x)$ is finite, single valued, and continuous, but not necessarily differentiable. With these assumptions it can be shown that the expansion (7.29) is valid. We shall not prove (7.29) but simply illustrate its use to represent a function.

Before we can apply (7.29) to a specific $f(x)$, we must derive an expression for the expansion coefficients a_n. We start by multiplying (7.29) by ψ_m^*:

$$\psi_m^* f(x) = \sum_{n=1}^{\infty} a_n \psi_m^* \psi_n = \left(\frac{2}{l}\right) \sum_{n=1}^{\infty} a_n \sin\left(\frac{n\pi x}{l}\right) \sin\left(\frac{m\pi x}{l}\right)$$

Now we integrate this equation from 0 to l. Assuming the validity of interchanging the integration and the infinite summation, we have

$$\int_0^l \psi_m^* f(x) \, dx = \sum_{n=1}^{\infty} a_n \int_0^l \psi_m^* \psi_n \, dx \tag{7.30}$$

$$= \sum_{n=1}^{\infty} a_n \left(\frac{2}{l}\right) \int_0^l \sin\left(\frac{n\pi x}{l}\right) \sin\left(\frac{m\pi x}{l}\right) dx$$

Recall that we proved the orthonormality of the particle-in-a-box wave functions [Eq. (2.27)]. Therefore, (7.30) becomes

$$\int_0^l \psi_m^* f(x) \, dx = \sum_{n=1}^{\infty} a_n \delta_{mn} \tag{7.31}$$

The type of sum in (7.31) occurs often. Writing it in detail, we have

$$\sum_{n=1}^{\infty} a_n \delta_{mn} = a_1 \delta_{m,1} + a_2 \delta_{m,2} + \cdots + a_m \delta_{m,m} + a_{m+1} \delta_{m,m+1} + \cdots$$

$$= 0 + 0 + \cdots + a_m + 0 + \cdots$$

$$\sum_{n=1}^{\infty} a_n \delta_{mn} = a_m \tag{7.32}$$

Thus since δ_{mn} is zero except when the summation index n is equal to m, all terms but one vanish, and (7.31) becomes

$$a_m = \int_0^l \psi_m^* f(x)\, dx \qquad (7.33)$$

which is the desired expression for the expansion coefficients.

Changing m to n in (7.33) and substituting it into (7.29), we have

$$f(x) = \sum_{n=1}^{\infty} \left[\int_0^l \psi_n^* f(x)\, dx \right] \psi_n(x) \qquad (7.34)$$

This is the desired expression for the expansion of an arbitrary well-behaved function $f(x)$ $(0 \leqslant x \leqslant l)$ as a linear combination of the particle-in-a-box wave functions ψ_n. [Note that the definite integral $\int_0^l \psi_n^* f(x)\, dx$ is a number and not a function of x.]

We now use (7.29) to represent a specific function, the function of Fig. 7.1, which is defined by

$$f(x) = x, \qquad \text{for } 0 \leqslant x \leqslant l/2$$
$$f(x) = l - x, \qquad \text{for } \tfrac{1}{2}l \leqslant x \leqslant l \qquad (7.35)$$

To evaluate the expansion coefficients a_n, we substitute (7.28) and (7.35) into (7.33):

$$a_n = \int_0^l \psi_n^* f(x)\, dx = \left(\frac{2}{l} \right)^{1/2} \int_0^l \sin\left(\frac{n\pi x}{l} \right) f(x)\, dx$$

$$= \left(\frac{2}{l} \right)^{1/2} \int_0^{l/2} x \sin\left(\frac{n\pi x}{l} \right) dx + \left(\frac{2}{l} \right)^{1/2} \int_{l/2}^l (l - x) \sin\left(\frac{n\pi x}{l} \right) dx$$

Using the Appendix integral (A.1), we find

$$a_n = \frac{(2l)^{3/2}}{n^2 \pi^2} \sin\left(\frac{n\pi}{2} \right) \qquad (7.36)$$

Using (7.36) in the expansion (7.29), we have [note that $\sin (n\pi/2)$ vanishes for n

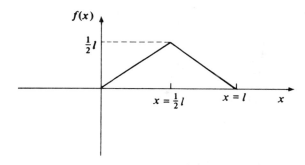

Figure 7.1 Function to be expanded in terms of particle-in-a-box functions.

even and equals $+1$ or -1 for n odd]

$$f(x) = \frac{4l}{\pi^2} \left[\sin\left(\frac{\pi x}{l}\right) - \frac{1}{3^2} \sin\left(\frac{3\pi x}{l}\right) + \frac{1}{5^2} \sin\left(\frac{5\pi x}{l}\right) - \cdots \right]$$

$$f(x) = \frac{4l}{\pi^2} \sum_{n=1}^{\infty} (-1)^{n+1} \sin\left[(2n-1)\frac{\pi x}{l}\right] \frac{1}{(2n-1)^2} \tag{7.37}$$

where $f(x)$ is given by (7.35). Let us check the accuracy of (7.37) at $x = \frac{1}{2}l$. We have

$$f\left(\frac{l}{2}\right) = \frac{4l}{\pi^2} \left(1 + \frac{1}{3^2} + \frac{1}{5^2} + \frac{1}{7^2} + \cdots\right) \tag{7.38}$$

Let us tabulate the right side of (7.38) as a function of the number of terms we take in the infinite series:

Number of terms	1	2	3	4	5	20	100
Right side of (7.38)	$0.405l$	$0.450l$	$0.467l$	$0.475l$	$0.480l$	$0.495l$	$0.499l$

If we take an infinite number of terms, the series should sum to $\frac{1}{2}l$, which is the value of $f(\frac{1}{2}l)$. Assuming the validity of the series, we have the interesting result that the infinite sum in parentheses in (7.38) equals $\pi^2/8$. Figure 7.2 plots $f(x) - \sum_{n=1}^{k} a_n \psi_n$ [where f, a_n, and ψ_n are given by (7.35), (7.36), and (7.28)] for

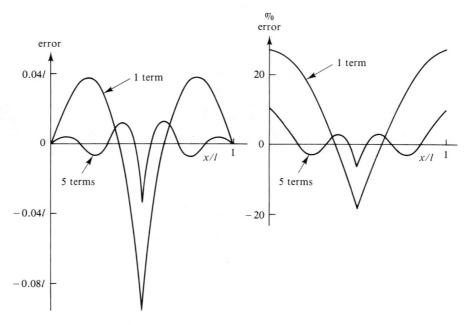

Figure 7.2 Plots of the error and the percent error in the expansion of the function of Fig. 7.1 in terms of particle-in-a-box wave functions when 1 and 5 terms are taken in the expansion.

k values of 1 and 5. As k, the number of terms in the expansion, increases, the series comes closer to $f(x)$, and the difference between f and the series goes to zero.

Expansion of a Function in Terms of Eigenfunctions. We have seen an example of the expansion of a function in terms of a set of functions—the particle-in-a-box energy eigenfunctions. Many different sets of functions can be used to expand an arbitrary function. A set of functions $g_1, g_2, \ldots, g_i, \ldots$ is said to be a *complete set* if any well-behaved function f that obeys the same boundary conditions as the g_i's can be expanded as a linear combination of the g_i's according to

$$f = \sum_i a_i g_i \qquad (7.39)^*$$

where the a_i's are constants. Of course, it is understood that f and the g_i's are all functions of the same set of variables. The limits have been omitted from the sum in (7.39); it is understood that this sum goes over all members of the complete set. By virtue of theorems of Fourier analysis (which we have not proved), the particle-in-a-box energy eigenfunctions can be shown to be a complete set.

We now *postulate* that *the set of eigenfunctions of any Hermitian operator that represents a physical quantity forms a complete set.* (Completeness of the eigenfunctions can be proved in the one-dimensional case and in certain multi-dimensional cases, but must be postulated for most multidimensional systems.) Thus, any well-behaved function that satisfies the same boundary conditions as the set of eigenfunctions can be expanded according to (7.39). Equation (7.29) is an example of (7.39).

The harmonic-oscillator wave functions are given by a Hermite polynomial H_v times an exponential factor (Problem 4.18b). By virtue of the expansion postulate, any well-behaved function $f(x)$ can be expanded as a linear combination of harmonic-oscillator energy eigenfunctions:

$$f(x) = \left(\frac{\alpha}{\pi}\right)^{1/4} \sum_{n=0}^{\infty} \frac{a_n}{(2^n n!)^{1/2}} H_n(\alpha^{1/2} x) e^{-\alpha x^2/2}$$

How about using the hydrogen-atom bound-state wave functions to expand an arbitrary function $f(r, \theta, \phi)$? The answer is that these functions do *not* form a complete set, and we cannot expand f using them. To have a complete set, we must use *all* the eigenfunctions of a particular Hermitian operator. In addition to the bound-state eigenfunctions of the hydrogen-atom Hamiltonian, we have the continuum eigenfunctions, corresponding to ionized states. If we include the continuum eigenfunctions along with the bound-state eigenfunctions, then we have a complete set. (For the particle-in-a-box and the harmonic oscillator, there are no continuum functions.) Equation (7.39) implies an integration over the continuum eigenfunctions, if there are any. Thus, if $\psi_{nlm}(r, \theta, \phi)$ is a bound-state wave function of the hydrogen atom and $\psi_{Elm}(r, \theta, \phi)$ is a continuum eigenfunction, then (7.39) becomes

$$f(r, \theta, \phi) = \sum_{n=1}^{\infty} \sum_{l=0}^{n-1} \sum_{m=-l}^{l} a_{nlm} \psi_{nlm}(r, \theta, \phi) + \sum_{l=0}^{\infty} \sum_{m=-l}^{l} \int_0^{\infty} a_{lm}(E) \psi_{Elm}(r, \theta, \phi)\, dE$$

As another example, consider the eigenfunctions of $\hat{p}_x$ [Eq. (3.36)]:

$$g_k = e^{ikx/\hbar}, \qquad -\infty < k < \infty$$

Here the eigenvalues are all continuous, and the eigenfunction expansion (7.39) of an arbitrary function f becomes

$$f(x) = \int_{-\infty}^{\infty} a(k) e^{ikx/\hbar} \, dk$$

The reader with a good mathematical background may recognize this integral as very nearly the Fourier transform of $a(k)$.

Let us evaluate the expansion coefficients in $f = \Sigma_i a_i g_i$ [Eq. (7.39)], where the g_i functions are the complete set of eigenfunctions of a Hermitian operator. The procedure is the same as used to derive (7.33). We multiply $f = \Sigma_i a_i g_i$ by g_j^* and integrate over all space:

$$g_j^* f = \sum_i a_i g_j^* g_i$$

$$\int g_j^* f \, d\tau = \sum_i a_i \int g_j^* g_i \, d\tau = \sum_i a_i \delta_{ij} = a_j$$

$$a_j = \int g_j^* f \, d\tau \qquad (7.40)$$

where we used the orthonormality of the eigenfunctions of a Hermitian operator: $\int g_j^* g_i \, d\tau = \delta_{ij}$ [Eq. (7.26)]. Substitution of (7.40) for a_i in $f = \Sigma_i a_i g_i$ gives

$$f = \sum_i \left[\int g_i^* f \, d\tau \right] g_i = \sum_i \langle g_i | f \rangle g_i \qquad (7.41)$$

EXAMPLE   Let $F(x) = x(l - x)$ for $0 \leqslant x \leqslant l$ and $F(x) = 0$ elsewhere. Expand F in terms of the particle-in-a-box energy eigenfunctions $\psi_n = (2/l)^{1/2} \sin(n\pi x/l)$ for $0 \leqslant x \leqslant l$.

We begin by noting that $F(0) = 0$ and $F(l) = 0$, so F obeys the same boundary conditions as the ψ_n's and can be expanded using the ψ_n's. The expansion is $F = \Sigma_{n=1}^{\infty} a_n \psi_n$, where $a_n = \int \psi_n^* F \, d\tau$ [Eqs. (7.39) and (7.40)]. We have

$$a_n = \int \psi_n^* F \, d\tau = \left(\frac{2}{l}\right)^{1/2} \int_0^l \left(\sin \frac{n\pi x}{l}\right) x(l - x) \, dx = \frac{2^{3/2} l^{5/2}}{n^3 \pi^3} [1 - (-1)^n]$$

where details of the integral evaluation are left as a problem (Problem 7.11). The expansion $F = \Sigma_{n=1}^{\infty} a_n \psi_n$ is

$$x(l - x) = \frac{4l^2}{\pi^3} \sum_{n=1}^{\infty} \frac{1 - (-1)^n}{n^3} \sin \frac{n\pi x}{l}, \qquad \text{for } 0 \leqslant x \leqslant l$$

A useful theorem is the following:

Theorem 3.   Let the functions $g_1, g_2, \ldots$ be the complete set of eigenfunctions of the Hermitian operator $\hat{A}$, and let the function F be an eigenfunction of $\hat{A}$

with eigenvalue k (that is, $\hat{A}F = kF$); then if F is expanded as $F = \sum_i a_i g_i$, the only nonzero coefficients a_i are those for which g_i has the eigenvalue k. (Because of degeneracy, several g_i's may have the same eigenvalue k.)

Thus in the expansion of F, we include only those eigenfunctions that have the same eigenvalue as F. The proof of Theorem 3 follows at once from $a_j = \int g_j^* F \, d\tau$ [Eq. (7.40)]; if F and g_j correspond to different eigenvalues of the Hermitian operator $\hat{A}$, they will be orthogonal [Eq. (7.22)] and a_j will vanish.

We shall occasionally use a notation (called *ket* notation) in which the function f is denoted by the symbol $|f\rangle$. There doesn't seem to be any point to this notation, but in advanced formulations of quantum mechanics, it takes on a special significance. In ket notation, Eq. (7.41) reads

$$|f\rangle = \sum_i |g_i\rangle \langle g_i|f\rangle = \sum_i |i\rangle \langle i|f\rangle \tag{7.42}$$

Ket notation is conveniently used to specify eigenfunctions by listing their eigenvalues. For example, the hydrogen-atom wave function with quantum numbers n, l, m is denoted by $\psi_{nlm} = |nlm\rangle$.

The contents of Sections 7.2 and 7.3 can be summarized by the statement that *the eigenfunctions of a Hermitian operator form a complete, orthonormal set, and the eigenvalues are real.*

7.4 EIGENFUNCTIONS OF COMMUTING OPERATORS

If the state function Ψ is simultaneously an eigenfunction of the two operators $\hat{A}$ and $\hat{B}$ with eigenvalues a_j and b_j, respectively, then a measurement of the physical property A will yield the result a_j and a measurement of B will yield b_j. Hence the two properties A and B have definite values when Ψ is simultaneously an eigenfunction of $\hat{A}$ and $\hat{B}$.

In Section 5.1, some statements were made about simultaneous eigenfunctions of two operators. We now prove these statements.

First, we show that if there exists a common *complete set* of eigenfunctions for two linear operators then these operators commute. Let $\hat{A}$ and $\hat{B}$ denote two linear operators that have a common complete set of eigenfunctions $g_1, g_2, \ldots$:

$$\hat{A}g_i = a_i g_i \,, \qquad \hat{B}g_i = b_i g_i \tag{7.43}$$

where a_i and b_i are the eigenvalues. We must prove that

$$[\hat{A}, \hat{B}] = \hat{0} \tag{7.44}$$

Equation (7.44) is an operator equation. For two operators to be equal, the results of operating with either of them on an arbitrary well-behaved function f must be the same. Hence we must show that

$$(\hat{A}\hat{B} - \hat{B}\hat{A})f = \hat{0}f = 0$$

where f is an arbitrary function. We begin the proof by expanding f (assuming that it obeys the proper boundary conditions) in terms of the complete set of eigenfunctions g_i:

$$f = \sum_i c_i g_i$$

Operating on each side of this last equation with $\hat{A}\hat{B} - \hat{B}\hat{A}$, we have

$$(\hat{A}\hat{B} - \hat{B}\hat{A})f = (\hat{A}\hat{B} - \hat{B}\hat{A})\sum_i c_i g_i$$

Since $\hat{A}\hat{B}$ and $\hat{B}\hat{A}$ are linear operators (Problem 3.10), we have

$$(\hat{A}\hat{B} - \hat{B}\hat{A})f = \sum_i c_i(\hat{A}\hat{B} - \hat{B}\hat{A})g_i = \sum_i c_i[\hat{A}(\hat{B}g_i) - \hat{B}(\hat{A}g_i)]$$

where the definitions of the sum and the product of operators were used. Use of the eigenvalue equations (7.43) gives

$$(\hat{A}\hat{B} - \hat{B}\hat{A})f = \sum_i c_i[\hat{A}(b_i g_i) - \hat{B}(a_i g_i)] = \sum_i c_i(b_i a_i g_i - a_i b_i g_i) = 0$$

This completes the proof of:

Theorem 4.   If the linear operators $\hat{A}$ and $\hat{B}$ have a common complete set of eigenfunctions, then $\hat{A}$ and $\hat{B}$ commute.

It is sometimes erroneously stated that if there exists a common eigenfunction of $\hat{A}$ and $\hat{B}$ then they commute. We have already seen an example that shows this statement to be false; in Section 5.3 we saw that the spherical harmonic Y_0^0 is an eigenfunction of both $\hat{L}_z$ and $\hat{L}_x$ even though these two operators do not commute. It is instructive to examine the so-called proof that is often given of this erroneous statement. Let g be the common eigenfunction:

$$\hat{A}g = ag , \qquad \hat{B}g = bg$$

We have

$$\hat{A}\hat{B}g = \hat{A}bg = abg \quad \text{and} \quad \hat{B}\hat{A}g = \hat{B}ag = bag = abg$$

$$\hat{A}\hat{B}g = \hat{B}\hat{A}g \tag{7.45}$$

The "proof" is completed by canceling g from each side of (7.45) to get

$$\hat{A}\hat{B} = \hat{B}\hat{A} \, (?) \tag{7.46}$$

It is in going from (7.45) to (7.46) that the error occurs. Just because the two operators $\hat{A}\hat{B}$ and $\hat{B}\hat{A}$ give the same result when acting on the single function g is no reason to conclude that $\hat{A}\hat{B} = \hat{B}\hat{A}$. (For example, d/dx and d^2/dx^2 give the same result when operating on e^x, but d/dx is certainly not equal to d^2/dx^2.) The two operators must give the same result when acting on *every* well-behaved function before we can conclude that they are equal. Thus, even though $\hat{A}$ and $\hat{B}$ do not commute, there might exist one or more common eigenfunctions of $\hat{A}$ and $\hat{B}$. However, we cannot have a common *complete* set of eigenfunctions of two noncommuting operators, as we proved earlier in this section.

We have shown that if there exists a common complete set of eigenfunctions of the linear operators $\hat{A}$ and $\hat{B}$ then they commute. We now prove the converse:

Theorem 5. If the Hermitian operators $\hat{A}$ and $\hat{B}$ commute, we can select a common complete set of eigenfunctions for them.

The proof is as follows. Let the functions g_i and the numbers a_i be the eigenfunctions and eigenvalues of $\hat{A}$:

$$\hat{A}g_i = a_i g_i$$

Operating on both sides of this equation with $\hat{B}$, we have

$$\hat{B}\hat{A}g_i = \hat{B}(a_i g_i)$$

Since $\hat{A}$ and $\hat{B}$ commute and since $\hat{B}$ is linear, we have

$$\hat{A}(\hat{B}g_i) = a_i(\hat{B}g_i) \tag{7.47}$$

This equation states that the function $\hat{B}g_i$ is an eigenfunction of the operator $\hat{A}$ with the same eigenvalue a_i as the eigenfunction g_i. Let us for the moment assume that the eigenvalues of $\hat{A}$ are nondegenerate, so that for any given eigenvalue a_i there exists one and only one linearly independent eigenfunction. If this is so, then the two eigenfunctions g_i and $\hat{B}g_i$, which correspond to the same eigenvalue a_i, must be linearly dependent; that is, one function must be simply a multiple of the other:

$$\hat{B}g_i = k_i g_i \tag{7.48}$$

where k_i is a constant. This equation states that the functions g_i are eigenfunctions of $\hat{B}$, which is what we wanted to prove. In Section 7.3, we postulated that the eigenfunctions of any operator that represents a physical quantity form a complete set. Hence the g_i's form a complete set.

We have just proved the desired theorem for the nondegenerate case, but what about the degenerate case? Let the eigenvalue a_i be n-fold degenerate. We know from Eq. (7.47) that $\hat{B}g_i$ is an eigenfunction of $\hat{A}$ with eigenvalue a_i. Hence, Theorem 3 of Section 7.3 tells us that, if the function $\hat{B}g_i$ is expanded in terms of the complete set of eigenfunctions of $\hat{A}$, then all the expansion coefficients will be zero except those for which the $\hat{A}$ eigenfunction has the eigenvalue a_i. In other words, $\hat{B}g_i$ must be a linear combination of the n linearly independent $\hat{A}$ eigenfunctions that correspond to the eigenvalue a_i:

$$\hat{B}g_i = \sum_{k=1}^{n} c_k g_k, \qquad \text{where } \hat{A}g_k = a_i g_k \quad \text{for } k = 1 \text{ to } n \tag{7.49}$$

where $g_1, \ldots, g_n$ denote those $\hat{A}$ eigenfunctions that have the degenerate eigenvalue a_i. Equation (7.49) shows that g_i is not necessarily an eigenfunction of $\hat{B}$. However, by taking suitable linear combinations of the n linearly independent $\hat{A}$ eigenfunctions corresponding to the degenerate eigenvalue a_i, one can construct a new set of n linearly independent eigenfunctions of $\hat{A}$ that will also be eigenfunctions of $\hat{B}$. Proof of this statement is given in *Merzbacher*, Section 8.5.

Thus, when $\hat{A}$ and $\hat{B}$ commute, it is always possible to *select* a common complete set of eigenfunctions for them. For example, consider the hydrogen atom, where the operators $\hat{L}_z$ and $\hat{H}$ were shown to commute. If we desired, we could take the phi factor in the eigenfunctions of $\hat{H}$ as $\sin m\phi$ and $\cos m\phi$ (Section 6.6). If we did this, we would not have eigenfunctions of $\hat{L}_z$, except for $m = 0$. However, the linear combinations

$$R(r)S(\theta)(\cos m\phi + i \sin m\phi) = RSe^{im\phi}, \qquad m = -l, \ldots, l$$

give us eigenfunctions of $\hat{L}_z$ that are still eigenfunctions of $\hat{H}$ by virtue of the theorem in Section 3.6.

Extension of the above proofs to the case of more than two operators shows that for a set of Hermitian operators $\hat{A}$, $\hat{B}$, $\hat{C}$, . . . there exists a common complete set of eigenfunctions if and only if every pair of operators commutes.

A useful theorem that is related to Theorem 5 is:

Theorem 6. If g_i and g_j are eigenfunctions of the Hermitian operator $\hat{A}$ with different eigenvalues (that is, if $\hat{A}g_i = a_i g_i$ and $\hat{A}g_j = a_j g_j$ with $a_i \neq a_j$), and if $\hat{B}$ is a linear operator that commutes with $\hat{A}$, then

$$\langle g_j | \hat{B} | g_i \rangle = 0, \qquad \text{for } a_j \neq a_i \tag{7.50}$$

The proof of (7.50) is as follows. If the eigenvalue a_i is nondegenerate, then, because $\hat{B}$ commutes with $\hat{A}$, Theorem 5 shows that g_i is also an eigenfunction of $\hat{B}$, with $\hat{B}g_i = k_i g_i$ [Eq. (7.48)]. We have $\langle g_j | \hat{B} | g_i \rangle = \langle g_j | \hat{B}g_i \rangle = k_i \langle g_j | g_i \rangle = 0$, since g_j and g_i are eigenfunctions of the Hermitian operator $\hat{A}$ and correspond to different eigenvalues ($a_j \neq a_i$) and so are orthogonal. If a_i is degenerate, then Eq. (7.49) shows that $\hat{B}g_i$ is a linear combination of the n eigenfunctions of the degenerate eigenvalue a_i, according to $\hat{B}g_i = \sum_{k=1}^{n} c_k g_k$; we then have $\langle g_j | \hat{B}g_i \rangle = \langle g_j | \sum_{k=1}^{n} c_k g_k \rangle = \sum_{k=1}^{n} c_k \langle g_j | g_k \rangle = 0$, because $\langle g_j | g_k \rangle = 0$. The integrals $\langle g_j | g_k \rangle$ are zero because g_j corresponds to the eigenvalue a_j, and each g_k corresponds to the eigenvalue a_i, which differs from a_j.

7.5 PARITY

Certain quantum-mechanical operators have no classical analog. An example is the parity operator. Recall that the harmonic-oscillator wave functions are either even or odd. We shall show how this property is related to the parity operator.

The *parity operator* $\hat{\Pi}$ is defined in terms of its effect on an arbitrary function f:

$$\hat{\Pi}f(x, y, z) = f(-x, -y, -z) \tag{7.51}*$$

The parity operator replaces each Cartesian coordinate with its negative. For example, $\hat{\Pi}(x^2 - ze^{ay}) = x^2 + ze^{-ay}$.

As with any quantum-mechanical operator, we are interested in the eigenvalues c_i and the eigenfunctions g_i of the parity operator:

$$\hat{\Pi}g_i = c_i g_i \tag{7.52}$$

The key to the problem is to calculate the square of $\hat{\Pi}$:

$$\hat{\Pi}^2 f(x, y, z) = \hat{\Pi}[\hat{\Pi}f(x, y, z)] = \hat{\Pi}[f(-x, -y, -z)] = f(x, y, z)$$

Since f is arbitrary, we conclude that $\hat{\Pi}^2$ equals the unit operator:

$$\hat{\Pi}^2 = \hat{1} \tag{7.53}$$

We now operate on (7.52) with $\hat{\Pi}$ to get $\hat{\Pi}\hat{\Pi}g_i = \hat{\Pi}c_ig_i$. Since $\hat{\Pi}$ is linear (Problem 7.20), we have

$$\hat{\Pi}^2g_i = c_i\hat{\Pi}g_i$$

$$\hat{\Pi}^2g_i = c_i^2 g_i \tag{7.54}$$

where the eigenvalue equation (7.52) has been used. Since $\hat{\Pi}^2$ is the unit operator, the left side of (7.54) is simply g_i, and we have

$$g_i = c_i^2 g_i$$

The function g_i cannot be zero everywhere (zero is always rejected as an eigenfunction on physical grounds); hence we can divide by g_i to get $c_i^2 = 1$ and

$$c_i = \pm 1 \tag{7.55}$$

The eigenvalues of $\hat{\Pi}$ are $+1$ and -1. Note that this derivation is applicable to any operator whose square is the unit operator.

What are the eigenfunctions g_i? The eigenvalue equation (7.52) reads

$$\hat{\Pi}g_i(x, y, z) = \pm g_i(x, y, z)$$

$$g_i(-x, -y, -z) = \pm g_i(x, y, z)$$

If the eigenvalue is $+1$, then $g_i(-x, -y, -z) = g_i(x, y, z)$ and g_i is an even function. If the eigenvalue is -1, then g_i is odd: $g_i(-x, -y, -z) = -g_i(x, y, z)$. Hence, *the eigenfunctions of the parity operator $\hat{\Pi}$ are all possible well-behaved even and odd functions.*

When the parity operator commutes with the Hamiltonian operator $\hat{H}$, we can select a common set of eigenfunctions for these operators, as proved in Section 7.4. The eigenfunctions of $\hat{H}$ are the stationary-state wave functions ψ_i. Hence when

$$[\hat{\Pi}, \hat{H}] = 0 \tag{7.56}$$

the wave functions ψ_i can be chosen to be eigenfunctions of $\hat{\Pi}$. We just proved that the eigenfunctions of $\hat{\Pi}$ are either even or odd. Hence, when (7.56) holds, each wave function can be chosen to be either even or odd. Let us find out when the parity and Hamiltonian operators commute.

We have, for a one-particle system,

$$[\hat{H}, \hat{\Pi}] = [\hat{T}, \hat{\Pi}] + [\hat{V}, \hat{\Pi}] = -\frac{\hbar^2}{2m}\left[\frac{\partial^2}{\partial x^2}, \hat{\Pi}\right] - \frac{\hbar^2}{2m}\left[\frac{\partial^2}{\partial y^2}, \hat{\Pi}\right]$$

$$- \frac{\hbar^2}{2m}\left[\frac{\partial^2}{\partial z^2}, \hat{\Pi}\right] + [\hat{V}, \hat{\Pi}] \tag{7.57}$$

Since

$$\hat{\Pi}\left[\frac{\partial^2}{\partial x^2} f(x, y, z)\right] = \frac{\partial}{\partial(-x)} \frac{\partial}{\partial(-x)} f(-x, -y, -z) = \frac{\partial^2}{\partial x^2} f(-x, -y, -z)$$

$$= \frac{\partial^2}{\partial x^2} \hat{\Pi} f(x, y, z)$$

where f is any function, we conclude that

$$\left[\frac{\partial^2}{\partial x^2}, \hat{\Pi}\right] = 0$$

for parity to hold a ...... function (handwritten)

Similar equations hold for the y and z coordinates, and (7.57) becomes

$$[\hat{H}, \hat{\Pi}] = [\hat{V}, \hat{\Pi}]$$

Now

$$\hat{\Pi}[V(x, y, z)f(x, y, z)] = V(-x, -y, -z)f(-x, -y, -z) \qquad (7.58)$$

If the potential energy is an even function, that is, if $V(-x, -y, -z) = V(x, y, z)$, then (7.58) becomes

$$\hat{\Pi}[V(x, y, z)f(x, y, z)] = V(x, y, z)f(-x, -y, -z) = V(x, y, z)\hat{\Pi}f(x, y, z)$$

so $[\hat{V}, \hat{\Pi}] = 0$. Hence, when the potential energy is an even function, the parity operator commutes with the Hamiltonian:

$$[\hat{H}, \hat{\Pi}] = 0, \qquad \text{if } V \text{ is even} \qquad (7.59)$$

These results are easily extended to the n-particle case. For an n-particle system, the parity operator is defined by

$$\hat{\Pi}f(x_1, y_1, z_1, \ldots, x_n, y_n, z_n) = f(-x_1, -y_1, -z_1, \ldots, -x_n, -y_n, -z_n)$$

It is easy to see that (7.56) holds when

$$V(x_1, y_1, z_1, \ldots, x_n, y_n, z_n) = V(-x_1, -y_1, -z_1, \ldots, -x_n, -y_n, -z_n)$$

If V satisfies this equation, V is said to be an even function of the $3n$ coordinates. In summary, we have:

Theorem 7. When the potential energy V is an even function, we can choose the stationary-state wave functions so that each ψ_i is either an even function or an odd function.

A function that is either even or odd is said to be of *definite parity*.

If the energy levels are all nondegenerate (as is usually true in one-dimensional problems), then only one independent wave function corresponds to each energy eigenvalue and there is no element of choice (apart from an arbitrary multiplicative constant) in the wave functions. Thus, for the nondegenerate case, the stationary-state wave functions must be of definite parity when V is an even function. For example, the one-dimensional harmonic oscillator has $V = \frac{1}{2}kx^2$, which is an even function, and we found the wave functions to be of definite parity.

For the degenerate case, we have an element of choice in the wave functions, since an arbitrary linear combination of the functions corresponding to the degenerate level is an eigenfunction of $\hat{H}$. For a degenerate energy level, by taking appropriate linear combinations we can choose wave functions that are of definite parity, but there is no necessity that they be of definite parity.

Parity aids in evaluating integrals. We showed that $\int_{-\infty}^{\infty} f(x)\, dx = 0$ when $f(x)$ is an odd function [Eq. (4.56)]. Let us extend this result to the $3n$-dimensional case. An odd function of $3n$ variables satisfies

$$g(-x_1, -y_1, -z_1, \ldots, -x_n, -y_n, -z_n) = -g(x_1, y_1, z_1, \ldots, x_n, y_n, z_n)$$

If g is an odd function of the $3n$ variables, then

$$\int_{-\infty}^{\infty} \cdots \int_{-\infty}^{\infty} g(x_1, \ldots, z_n)\, dx_1 \cdots dz_n = 0 \qquad (7.60)$$

where the integration is over the $3n$ coordinates. This equation holds because the contribution to the integral from the value of g at $(x_1, y_1, z_1, \ldots, x_n, y_n, z_n)$ is canceled by the contribution from $(-x_1, -y_1, -z_1, \ldots, -x_n, -y_n, -z_n)$.

A more general case is where the integrand is an odd function of some (but not necessarily all) of the variables. Let f be a function such that

$$f(-q_1, -q_2, \ldots, -q_k, q_{k+1}, q_{k+2}, \ldots, q_m)$$
$$= -f(q_1, q_2, \ldots, q_k, q_{k+1}, q_{k+2}, \ldots, q_m) \qquad (7.61)$$

where $1 \le k \le m$. We assert that if f obeys (7.61) then

$$\int_{-\infty}^{\infty} \cdots \int_{-\infty}^{\infty} f(q_1, \ldots, q_m)\, dq_1 \cdots dq_m = 0 \qquad (7.62)$$

The proof of (7.62) is simple. The integral in (7.62) can be written as

$$\int_{-\infty}^{\infty} \cdots \int_{-\infty}^{\infty} \left[\int_{-\infty}^{\infty} \cdots \int_{-\infty}^{\infty} f(q_1, \ldots, q_k, q_{k+1}, \ldots, q_m)\, dq_1 \cdots dq_k \right] dq_{k+1} \cdots dq_m$$
$$(7.63)$$

As far as the multiple integral in brackets is concerned, q_{k+1} through q_m are constants. By virtue of (7.61), the contributions from

$$f(-q_1, \ldots, -q_k, q_{k+1}, \ldots, q_m) \quad \text{and} \quad f(q_1, \ldots, q_k, q_{k+1}, \ldots, q_m)$$

cancel, so that the bracketed integral in (7.63) vanishes and (7.62) follows.

7.6 MEASUREMENT AND THE SUPERPOSITION OF STATES

Quantum mechanics can be regarded as a scheme for calculating the probabilities of the various possible outcomes of a measurement. For example, if we know the state function $\Psi(x, t)$, then the probability that a measurement at time t of the particle's position yields a value between x and $x + dx$ is given by $|\Psi(x, t)|^2\, dx$. We now consider measurement of the general property B; our aim is to find out how to use Ψ to calculate the probabilities for each possible result of a measurement of B. The results of this section, which tell us what information is contained in the state function Ψ, lie at the heart of quantum mechanics.

We shall deal with an n-particle system and use q to symbolize the $3n$ coordinates. We have postulated that the eigenvalues b_i of the operator $\hat{B}$ are the only possible results of a measurement of the property B. Using g_i for the eigenfunctions of $\hat{B}$, we have

$$\hat{B}g_i(q) = b_i g_i(q) \tag{7.64}$$

We postulated in Section 7.3 that the eigenfunctions of any Hermitian operator that represents a physically observable property form a complete set. Since the g_i's form a complete set, we can expand the state function Ψ in terms of the g_i's:

$$\Psi(q, t) = \sum_i c_i(t) g_i(q) \tag{7.65}$$

To allow for the change of Ψ with time, the expansion coefficients c_i vary with time.

Since $|\Psi|^2$ is a probability density, we require that

$$\int \Psi^* \Psi \, d\tau = 1 \tag{7.66}$$

Substituting (7.65) into the normalization condition and using (1.31) and (1.30), we get

$$1 = \int \sum_i c_i^* g_i^* \sum_i c_i g_i \, d\tau = \int \sum_i c_i^* g_i^* \sum_j c_j g_j \, d\tau$$

$$1 = \int \sum_i \sum_j c_i^* c_j g_i^* g_j \, d\tau \tag{7.67}$$

Since the summation indexes in the two sums in (7.67) need not have the same value, different symbols must be used for these two dummy indexes. For example, consider the following product of two sums:

$$\sum_{i=1}^{2} s_i \sum_{i=1}^{2} t_i = (s_1 + s_2)(t_1 + t_2) = s_1 t_1 + s_1 t_2 + s_2 t_1 + s_2 t_2$$

If we carelessly write

$$\sum_{i=1}^{2} s_i \sum_{i=1}^{2} t_i \overset{\text{(wrong)}}{=} \sum_{i=1}^{2} \sum_{i=1}^{2} s_i t_i = \sum_{i=1}^{2} (s_1 t_1 + s_2 t_2) = 2(s_1 t_1 + s_2 t_2)$$

we get the wrong answer. The correct way to write the product is

$$\sum_{i=1}^{2} s_i \sum_{i=1}^{2} t_i = \sum_{i=1}^{2} s_i \sum_{j=1}^{2} t_j = \sum_{i=1}^{2} \sum_{j=1}^{2} s_i t_j = \sum_{i=1}^{2} (s_i t_1 + s_i t_2) = s_1 t_1 + s_1 t_2 + s_2 t_1 + s_2 t_2$$

which gives the right answer.

Assuming the validity of interchanging the infinite summation and the integration in (7.67), we have

$$\sum_i \sum_j c_i^* c_j \int g_i^* g_j \, d\tau = 1$$

Since $\hat{B}$ is Hermitian, its eigenfunctions g_i are orthonormal [Eq. (7.26)]; hence

$$\sum_i \sum_j c_i^* c_j \delta_{ij} = 1$$

$$\sum_i |c_i|^2 = 1 \qquad\qquad (7.68)$$

We shall point out the significance of (7.68) shortly.

Recall the postulate (Section 3.7) that if Ψ is the normalized state function of a system then the average value of the property B is

$$\langle B \rangle = \int \Psi^*(q, t) \hat{B} \Psi(q, t) \, d\tau$$

Using the expansion (7.65) in the average-value expression, we have

$$\langle B \rangle = \int \sum_i c_i^* g_i^* \hat{B} \sum_j c_j g_j \, d\tau = \sum_i \sum_j c_i^* c_j \int g_i^* \hat{B} g_j \, d\tau$$

where the linearity of $\hat{B}$ was used. Use of $\hat{B} g_j = b_j g_j$ [Eq. (7.64)] gives

$$\langle B \rangle = \sum_i \sum_j c_i^* c_j b_j \int g_i^* g_j \, d\tau = \sum_i \sum_j c_i^* c_j b_j \delta_{ij}$$

$$\langle B \rangle = \sum_i |c_i|^2 b_i \qquad\qquad (7.69)$$

How do we interpret (7.69)? We postulated in Section 3.3 that the eigenvalues of an operator are the only possible numbers we can get when we measure the property that the operator represents. In any measurement of B, we get one of the values b_i (assuming there is no experimental error). Now recall Eq. (3.81):

$$\langle B \rangle = \sum_{b_i} P_{b_i} b_i \qquad\qquad (7.70)$$

where P_{b_i} is the probability of getting b_i in a measurement of B. The sum in (7.70) goes over the different eigenvalues b_i, whereas the sum in (7.69) goes over the different eigenfunctions g_i, since the expansion (7.65) is over the g_i's. If there is only one independent eigenfunction for each eigenvalue, then a sum over eigenfunctions is the same as a sum over eigenvalues, and comparison of (7.70) and (7.69) shows that, when there is no degeneracy in the $\hat{B}$ eigenvalues, $|c_i|^2$ is the probability of getting the value b_i in a measurement of the property B. Note that the $|c_i|^2$ values sum to 1, as probabilities should [Eq. (7.68)]. Suppose the eigenvalue b_i is degenerate. From (7.70), P_{b_i} is given by the quantity that multiplies b_i. With degeneracy, more than one term in (7.69) contains b_i, so the probability P_{b_i} of getting b_i in a measurement is found by adding the $|c_i|^2$ values for those eigenfunctions that have the same eigenvalue b_i. We have proved:

Theorem 8. If b_m is a nondegenerate eigenvalue of the operator $\hat{B}$ and g_m is the corresponding normalized eigenfunction ($\hat{B} g_m = b_m g_m$), then, when the property B is measured in a quantum-mechanical system whose state function at the time of the measurement is Ψ, the probability of getting the result b_m is given by $|c_m|^2$, where c_m is the coefficient of g_m in the expansion $\Psi = \sum_i c_i g_i$. If the eigenvalue b_m is degenerate, the probability of obtaining b_m when B is measured is found by adding the $|c_i|^2$ values for those eigenfunctions whose eigenvalue is b_m.

When can the result of a measurement of B be predicted with certainty? We can do this if all the coefficients in the expansion $\Psi = \Sigma_i c_i g_i$ are zero, except one: $c_i = 0$ for all $i \neq k$ and $c_k \neq 0$. For this case, Eq. (7.68) gives $|c_k|^2 = 1$, and we are certain to find the result b_k. In this case, the state function $\Psi = \Sigma_i c_i g_i$ is given by $\Psi = g_k$.

We can thus view the expansion $\Psi = \Sigma_i c_i g_i$ [Eq. (7.65)] as expressing the general state Ψ as a *superposition* of the eigenstates g_i of the operator $\hat{B}$. Each eigenstate g_i corresponds to the value b_i for the property B. The degree to which any eigenfunction g_i occurs in the expansion of Ψ, as measured by $|c_i|^2$, determines the probability of getting the value b_i in a measurement of B.

How do we calculate the expansion coefficients c_i so that we can get the probabilities $|c_i|^2$? We multiply $\Psi = \Sigma_i c_i g_i$ by g_j^*, integrate over all space, and use the orthonormality of the eigenfunctions of the Hermitian operator $\hat{B}$ to get

$$\int g_j^* \Psi \, d\tau = \sum_i c_i \int g_j^* g_i \, d\tau = \sum_i c_i \delta_{ij}$$

$$c_j = \int g_j^* \Psi \, d\tau = \langle g_j | \Psi \rangle \tag{7.71}$$

The probability of finding the nondegenerate eigenvalue b_j in a measurement of B is

$$|c_j|^2 = \left| \int g_j^* \Psi \, d\tau \right|^2 = |\langle g_j | \Psi \rangle|^2 \tag{7.72}$$

where $\hat{B} g_j = b_j g_j$. The quantity $\langle g_j | \Psi \rangle$ is called a ***probability amplitude***.

Thus, if we know the state of the system, as determined by the state function Ψ, we can use (7.72) to predict the probabilities of the various possible outcomes of a measurement of any property B. Determination of the eigenfunctions g_j and eigenvalues b_j of $\hat{B}$ is a mathematical problem.

To determine experimentally the probability of finding g_j when B is measured, we take a very large number n of identical, noninteracting systems, each in the same state Ψ, and measure B in each system. If n_j of the measurements yield b_j, then $P_{b_j} = n_j/n = |\langle g_j | \Psi \rangle|^2$.

We can restate the first part of Theorem 8 as:

Theorem 9. If the property B is measured in a quantum-mechanical system whose state function at the time of the measurement is Ψ, then the probability of observing the nondegenerate $\hat{B}$ eigenvalue b_j is $|\langle g_j | \Psi \rangle|^2$, where g_j is the normalized eigenfunction corresponding to the eigenvalue b_j.

The integral $\langle g_j | \Psi \rangle = \int g_j^* \Psi \, d\tau$ will have a substantial absolute value if the normalized functions g_j and Ψ resemble each other closely and so have similar magnitudes in each region of space. If g_j and Ψ do not resemble each other, then in regions where g_j is large Ψ will be small (and vice versa), so the product $g_j^* \Psi$

will always be small and the absolute value of the integral $\int g_j^* \Psi \, d\tau$ will be small; the probability $|\langle g_j | \Psi \rangle|^2$ of getting b_j will then be small.

EXAMPLE   Suppose that we measure L_z of the electron in a hydrogen atom whose state at the time the measurement begins is the $2p_x$ state. Give the possible outcomes of the measurement and give the probability of each outcome.

From (6.118),

$$\Psi = \psi_{2p_x} = 2^{-1/2} \psi_{2p_1} + 2^{-1/2} \psi_{2p_{-1}}$$

This equation is the expansion of Ψ as a linear combination of $\hat{L}_z$ eigenfunctions. The only nonzero coefficients are for ψ_{2p_1} and $\psi_{2p_{-1}}$, which are eigenfunctions of $\hat{L}_z$ with eigenvalues $\hbar$ and $-\hbar$, respectively. (Recall that the 1 and -1 subscripts give the m quantum number and that the $\hat{L}_z$ eigenvalues are $m\hbar$.) Using Theorem 8, we take the squares of the absolute values of the coefficients in the expansion of Ψ to get the probabilities. Hence the probability for getting $\hbar$ when L_z is measured is $|2^{-1/2}|^2 = 0.5$, and the probability for getting $-\hbar$ is $|2^{-1/2}|^2 = 0.5$.

EXAMPLE   Suppose that the energy E is measured for a particle in a box of length l and that at the time of the measurement the particle is in the nonstationary state $\Psi = 30^{1/2} l^{-5/2} x(l-x)$ for $0 \leqslant x \leqslant l$. Give the possible outcomes of the measurement and give the probability of each possible outcome.

The possible outcomes are given by postulate (c) of Section 3.9 as the eigenvalues of the energy operator $\hat{H}$. The eigenvalues of the particle-in-a-box Hamiltonian are $E = n^2 h^2 / 8ml^2$ $(n = 1, 2, 3, \ldots)$ and these are nondegenerate. The probabilities are found by expanding Ψ in terms of the eigenfunctions ψ_n of $\hat{H}$; $\Psi = \sum_{n=1}^{\infty} c_n \psi_n$, where $\psi_n = (2/l)^{1/2} \sin(n\pi x/l)$. In the example after Eq. (7.41), the function $x(l-x)$ was expanded in terms of the particle-in-a-box energy eigenfunctions. The state function $30^{1/2} l^{-5/2} x(l-x)$ equals $x(l-x)$ multiplied by the normalization constant $30^{1/2} l^{-5/2}$; hence the expansion coefficients c_n are obtained by multiplying the a_n coefficients in the earlier example by $30^{1/2} l^{-5/2}$ to get

$$c_n = \frac{(240)^{1/2}}{n^3 \pi^3} [1 - (-1)^n]$$

The probability P_{E_n} of observing the value $E_n = n^2 h^2 / 8ml^2$ equals $|c_n|^2$:

$$P_{E_n} = \frac{240}{n^6 \pi^6} [1 - (-1)^n]^2 \tag{7.73}$$

The first few probabilities are

n	1	2	3	4	5
E_n	$h^2/8ml^2$	$4h^2/8ml^2$	$9h^2/8ml^2$	$16h^2/8ml^2$	$25h^2/8ml^2$
P_{E_n}	0.998555	0	0.001370	0	0.000064

The very high probability of finding the $n = 1$ energy is related to the fact that the parabolic state function $30^{1/2} l^{-5/2} x(l-x)$ closely resembles the $n = 1$

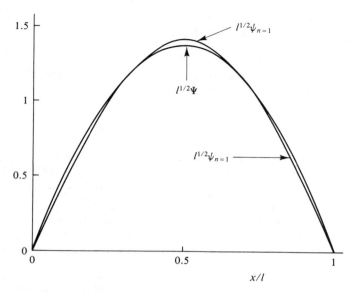

Figure 7.3 Plots of $\Psi = (30)^{1/2}l^{-5/2}x(l - x)$ and the $n = 1$ particle-in-a-box wave function.

particle-in-a-box wave function $(2/l)^{1/2} \sin (\pi x/l)$ (Fig. 7.3). The zero probabilities for $n = 2, 4, 6, \ldots$ are due to the fact that, if the origin is put at the center of the box, the state function $\Psi = 30^{1/2}l^{-5/2}x(l - x)$ is an even function, whereas the $n = 2, 4, 6, \ldots$ functions are odd functions (Fig. 2.3) and so can make no contribution to the expansion of Ψ; the integral $\langle g_n|\Psi \rangle$ vanishes when the integrand is an odd function.

If the property B has a continuous range of eigenvalues (for example, position; Section 7.7), the summation in the expansion (7.65) of Ψ is replaced by an integration over the values of b:

$$\Psi = \int c_b g_b(q) \, db \tag{7.74}$$

and the quantity $|\langle g_b(q)|\Psi \rangle|^2$ is interpreted as a probability density; that is, the probability of finding a value of B between b and $b + db$ for a system in the state Ψ is

$$|\langle g_b(q)|\Psi(q, t)\rangle|^2 \, db \tag{7.75}$$

7.7 POSITION EIGENFUNCTIONS

We have derived the eigenfunctions of the linear-momentum and angular-momentum operators. We now ask: What are the eigenfunctions of the position operator?

We have

$$\hat{x} = x \cdot$$

Denoting the position eigenfunctions by $g_a(x)$, we write

$$x g_a(x) = a g_a(x) \qquad (7.76)$$

where a symbolizes the possible eigenvalues. It follows that

$$(x - a)g_a(x) = 0 \qquad (7.77)$$

We conclude from (7.77) that

$$g_a(x) = 0 \qquad \text{for } x \neq a \qquad (7.78)$$

Moreover, since an eigenfunction that is zero everywhere is unacceptable, we have

$$g_a(x) \neq 0 \qquad \text{for } x = a \qquad (7.79)$$

These conclusions make sense. If the state function is an eigenfunction of $\hat{x}$ with eigenvalue a, $\Psi = g_a(x)$, we know (Section 7.6) that a measurement of x is certain to give the value a; this can only be true if the probability density $|\Psi|^2$ is zero for $x \neq a$, in agreement with (7.78).

Before considering further properties of $g_a(x)$, we define the *Heaviside step function* $H(x)$ by (see Fig. 7.4)

$$H(x) = 1 , \qquad \text{for } x > 0$$
$$H(x) = 0 , \qquad \text{for } x < 0 \qquad (7.80)$$
$$H(x) = \tfrac{1}{2} , \qquad \text{for } x = 0$$

We next define the *Dirac delta function* $\delta(x)$ as the derivative of the Heaviside step function:

$$\delta(x) \equiv dH(x)/dx \qquad (7.81)$$

From (7.80) and (7.81), we have at once (see also Fig. 7.4)

$$\delta(x) = 0 , \qquad \text{for } x \neq 0 \qquad (7.82)$$

Since $H(x)$ makes a sudden jump at $x = 0$, its derivative is infinite at the origin:

$$\delta(x) = \infty , \qquad \text{for } x = 0 \qquad (7.83)$$

We can generalize these equations slightly by setting $x = t - a$ and then changing

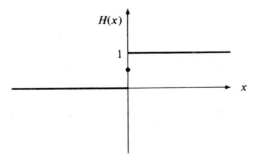

Figure 7.4 The Heaviside step function.

the symbol t to x; Eqs. (7.80) to (7.83) become

$$H(x - a) = 1, \qquad x > a \tag{7.84}$$

$$H(x - a) = 0, \qquad x < a \tag{7.85}$$

$$H(x - a) = \tfrac{1}{2}, \qquad x = a \tag{7.86}$$

$$\delta(x - a) = dH(x - a)/dx \tag{7.87}$$

$$\delta(x - a) = 0, \qquad x \neq a \quad \text{and} \quad \delta(x - a) = \infty, \qquad x = a \tag{7.88}$$

Now consider the following integral:

$$\int_{-\infty}^{\infty} f(x)\delta(x - a)\, dx$$

We evaluate it using integration by parts:

$$\int u\, dv = uv - \int v\, du$$

$$u = f(x), \qquad dv = \delta(x - a)\, dx$$

Using (7.87), we have

$$du = f'(x)\, dx, \qquad v = H(x - a)$$

$$\int_{-\infty}^{\infty} f(x)\delta(x - a)\, dx = f(x)H(x - a)\Big|_{-\infty}^{\infty} - \int_{-\infty}^{\infty} H(x - a)f'(x)\, dx$$

$$\int_{-\infty}^{\infty} f(x)\delta(x - a)\, dx = f(\infty) - \int_{-\infty}^{\infty} H(x - a)f'(x)\, dx \tag{7.89}$$

where (7.84) and (7.85) were used. Since $H(x - a)$ vanishes for $x < a$, Eq. (7.89) becomes

$$\int_{-\infty}^{\infty} f(x)\delta(x - a)\, dx = f(\infty) - \int_{a}^{\infty} H(x - a)f'(x)\, dx$$

$$= f(\infty) - \int_{a}^{\infty} f'(x)\, dx = f(\infty) - f(x)\Big|_{a}^{\infty}$$

$$\int_{-\infty}^{\infty} f(x)\delta(x - a)\, dx = f(a) \tag{7.90}$$

Comparing (7.90) with the equation $\sum_j c_j \delta_{ij} = c_i$, we see that the Dirac delta function plays the same role in an integral that the Kronecker delta plays in a sum. The special case of (7.90) with $f(x) = 1$ is

$$\int_{-\infty}^{\infty} \delta(x - a)\, dx = 1$$

The properties (7.88) of the Dirac delta function agree with the properties (7.78) and (7.79) of the position eigenfunctions $g_a(x)$. We therefore tentatively set

$$g_a(x) = \delta(x - a) \tag{7.91}$$

To verify (7.91), we now show it to be in accord with the Born postulate that $|\Psi(a, t)|^2\, da$ is the probability of observing a value of x between a and $a + da$. According to (7.75), this probability is given by

$$|\langle g_a(x)|\Psi(x, t)\rangle|^2\, da = \left|\int_{-\infty}^{\infty} g_a^*(x)\Psi(x, t)\, dx\right|^2 da \tag{7.92}$$

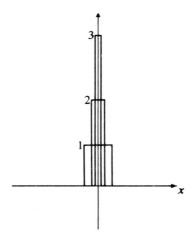

Figure 7.5 Functions that approximate $\delta(x)$ with successively increasing accuracy. The area under each curve is 1.

Using (7.91) and then (7.90), we have for (7.92)

$$\left| \int_{-\infty}^{\infty} \delta(x - a)\Psi(x, t)\, dx \right|^2 da = |\Psi(a, t)|^2\, da$$

which completes the proof.

Since the quantity a in $\delta(x - a)$ can have any real value, the eigenvalues of $\hat{x}$ form a continuum: $-\infty < a < \infty$. As usual for continuum eigenfunctions, $\delta(x - a)$ is not quadratically integrable (Problem 7.36).

Summarizing, the eigenfunctions and eigenvalues of position are

$$\hat{x}\delta(x - a) = a\delta(x - a) \tag{7.93}$$

where a is any real number.

The delta function is badly behaved, and consequently the manipulations we performed are lacking in rigor and would make a mathematician shudder. However, one can put use of the delta function on a rigorous basis by considering it to be the limiting case of a function that becomes successively more peaked at the origin; see Fig. 7.5.

7.8 THE POSTULATES OF QUANTUM MECHANICS

We have introduced the postulates of quantum mechanics as we have gone along. In this section we summarize them.

Postulate 1. The state of a system is described by a function Ψ of the coordinates and the time. This function, called the state function or wave function, contains all the information that can be determined about the system. We further postulate that Ψ is single valued, continuous, and quadratically integrable. For continuum states, the quadratic integrability requirement is omitted.

The designation "wave function" for Ψ is perhaps not the best choice. A physical wave moving in three-dimensional space is a function of the three spatial coordinates and the time. However, for an n-particle system, the function Ψ is a function of $3n$ spatial coordinates and the time. Hence, for a many-particle system, we cannot interpret Ψ as any sort of physical wave. The state function is best thought of as a function from which we can calculate various properties of the system. The nature of the information that Ψ contains is the subject of Postulate 5 and its consequences.

Postulate 2. To every physical observable there corresponds a linear Hermitian operator. To find this operator, write down the classical-mechanical expression for the observable in terms of Cartesian coordinates and corresponding linear-momentum components, and then replace each coordinate x by the operator $x \cdot$ and each momentum component p_x by the operator $-i\hbar \partial/\partial x$.

We saw in Section 7.2 that the restriction to Hermitian operators arises from the requirement that average values of physical quantities be real numbers. The requirement of linearity is closely connected to the superposition of states discussed in Section 7.6. In our derivation of (7.69) for the average value of a property B for a state that was expanded as a superposition of the eigenfunctions of $\hat{B}$, the linearity of $\hat{B}$ played a key role.

When the classical quantity contains a product of a Cartesian coordinate and its conjugate momentum, we run into the problem of noncommutativity in constructing the correct quantum-mechanical operator. Several different rules have been proposed to handle this case. See J. R. Shewell, *Am. J. Phys.*, **27**, 16 (1959); E. H. Kerner and W. G. Sutcliffe, *J. Math. Phys.*, **11**, 391 (1970).

The process of finding quantum-mechanical operators in non-Cartesian coordinates is complicated. See K. Simon, *Am. J. Phys.*, **33**, 60 (1965); G. R. Gruber, *Found. Phys.*, **1**, 227 (1971).

Postulate 3. The only possible values that can result from measurements of the physically observable property B are the eigenvalues b_i in the equation $\hat{B}g_i = b_i g_i$, where $\hat{B}$ is the operator corresponding to the property B. The eigenfunctions g_i are required to be well behaved.

Our main concern is with the energy levels of atoms and molecules. These are given by the eigenvalues of the energy operator, the Hamiltonian $\hat{H}$. The eigenvalue equation for $\hat{H}$, $\hat{H}\psi = E\psi$, is the time-independent Schrödinger equation. However, finding the possible values of any property involves solving an eigenvalue equation.

Postulate 4. If $\hat{B}$ is any linear Hermitian operator that represents a physically observable property, then the eigenfunctions g_i of $\hat{B}$ form a complete set.

This postulate is more a mathematical than a physical postulate. Since there is no mathematical proof (except in various special cases) of the completeness of the eigenfunctions of a linear Hermitian operator, we must assume completeness. Postulate 4 allows us to expand the wave function for any state as a superposition of the orthonormal eigenfunctions of any quantum-mechanical operator:

$$\Psi = \sum_i c_i g_i = \sum_i |g_i\rangle\langle g_i|\Psi\rangle \tag{7.94}$$

Postulate 5. If $\Psi(q, t)$ is the normalized state function of a system at time t, then the average value of a physical observable B at time t is

$$\langle B\rangle = \int \Psi^* \hat{B}\Psi \, d\tau \tag{7.95}^*$$

The definition of the quantum-mechanical average value is given in Section 3.7 and should not be confused with the time average used in classical mechanics.

From Postulates 4 and 5, we showed in Section 7.6 that the probability of observing the nondegenerate eigenvalue b_i in a measurement of B is $P_{b_i} = |\int g_i^* \Psi \, d\tau|^2 = |\langle g_i|\Psi\rangle|^2$, where $\hat{B}g_i = b_i g_i$. If the state function happens to be one of the eigenfunctions of $\hat{B}$, that is, if $\Psi = g_k$, then P_{b_i} becomes $P_{b_i} = |\int g_i^* g_k \, d\tau|^2 = |\delta_{ik}|^2 = \delta_{ik}$, where the orthonormality of the eigenfunctions of the Hermitian operator $\hat{B}$ was used. We are certain to observe the value b_k when $\Psi = g_k$.

Postulate 6. The time development of the state of an undisturbed quantum-mechanical system is given by the Schrödinger time-dependent equation

$$-\frac{\hbar}{i}\frac{\partial\Psi}{\partial t} = \hat{H}\Psi \tag{7.96}^*$$

where $\hat{H}$ is the Hamiltonian (that is, energy) operator of the system.

The time-dependent Schrödinger equation is a first-order differential equation in the time, so that, just as in classical mechanics, the present state of an undisturbed system determines the future state. However, unlike knowledge of the state in classical mechanics, knowledge of the state in quantum mechanics involves a knowledge of only the *probabilities* for various possible outcomes of a measurement. Thus, suppose we have several identical noninteracting systems, each having the same state function $\Psi(t_0)$ at time t_0. If we leave each system to itself, then the state function for each system will change in accord with (7.96). Since each system has the same Hamiltonian, each system will have the same state function $\Psi(t_1)$ at any future time t_1. However, suppose that at time t_2 we measure property B in each system. Although each system has the same state function $\Psi(t_2)$ at the instant the measurement begins, we will not get the same result for each system. Rather, we will get a spread of possible values b_i, where b_i are the eigenvalues of $\hat{B}$. The relative number of times we get each b_i can be calculated from the quantities $|c_i|^2$, where $\Psi(t_2) = \sum_i c_i g_i$, with g_i being the eigenfunctions of $\hat{B}$.

If the Hamiltonian is independent of time, we have the possibility of states of definite energy E. For such states the state function must satisfy

$$\hat{H}\Psi = E\Psi \tag{7.97}$$

and the time-dependent Schrödinger equation becomes

$$-\frac{\hbar}{i}\frac{\partial \Psi}{\partial t} = E\Psi$$

which integrates to $\Psi = Ae^{-iEt/\hbar}$, where A, the integration "constant," is independent of time. The function Ψ depends on the coordinates and the time, so A is some function of the coordinates, which we designate as $\psi(q)$. We have

$$\Psi(q, t) = e^{-iEt/\hbar}\psi(q) \tag{7.98}$$

for a state of constant energy. The function $\psi(q)$ satisfies the time-independent Schrödinger equation

$$\hat{H}\psi(q) = E\psi(q)$$

which follows from (7.97) and (7.98). The factor $\exp(-iEt/\hbar)$ simply indicates a change in the phase of the wave function $\Psi(q, t)$ with time and has no direct physical significance; hence we generally refer to $\psi(q)$ as "the wave function." The Hamiltonian operator plays a unique role in quantum mechanics in that it occurs in the fundamental dynamical equation, the time-dependent Schrödinger equation; the eigenstates of $\hat{H}$ (known as stationary states) have the special property that the probability density $|\Psi|^2$ is independent of time.

The time-dependent Schrödinger equation (7.96) is $(i\hbar\partial/\partial t - \hat{H})\Psi = 0$. Because the operator $i\hbar\partial/\partial t - \hat{H}$ is linear, any linear combination of solutions of the time-dependent Schrödinger equation (7.96) is a solution of (7.96). For example, if the Hamiltonian $\hat{H}$ is independent of time, then there exist stationary-state solutions $\Psi_n = e^{-iE_nt/\hbar}\psi_n(q)$ [Eq. (7.98)] of the time-dependent Schrödinger equation; any linear combination

$$\Psi = \sum_n c_n\psi_n = \sum_n c_n e^{-iE_nt/\hbar}\psi_n(q) \tag{7.99}$$

where the c_n's are time-independent constants, is a solution of the time-dependent Schrödinger equation, although it is not an eigenfunction of $\hat{H}$. Because of the completeness of the eigenfunctions ψ_n, any state function can be written in the form (7.99) if $\hat{H}$ is independent of time. (See also Section 9.9.) The state function (7.99) represents a state that does not have a definite energy; rather, when we make an energy measurement, the probability of getting E_n is $|c_n e^{-iE_nt/\hbar}|^2 = |c_n|^2$.

What determines whether a system is in a stationary state such as (7.98) or a nonstationary state such as (7.99)? The answer is that the past history of the system determines its present state. For example, if we take a system that is in a stationary state and expose it to radiation, the time-dependent Schrödinger equation shows that the radiation causes the state to change to a nonstationary state; see Section 9.9.

You might be wondering about the absence from the list of postulates of the Born postulate that $|\Psi(x, t)|^2\, dx$ is the probability of finding the particle between x and $x + dx$. This postulate is a consequence of Postulate 5, as we now show. Equation (3.81) is $\langle B \rangle = \Sigma_b\, P_b b$, where P_b is the probability of observing the value b in a measurement of the property B that takes on discrete values. The corresponding equation for the continuous variable x is $\langle x \rangle = \int_{-\infty}^{\infty} P(x)x\, dx$, where $P(x)$ is the probability density for observing various values of x. According to Postulate 5, we have $\langle x \rangle = \int_{-\infty}^{\infty} \Psi^* \hat{x} \Psi\, dx = \int_{-\infty}^{\infty} |\Psi|^2 x\, dx$. Comparison of these two expressions for $\langle x \rangle$ shows that $|\Psi|^2$ is the probability density $P(x)$.

Chapter 10 gives two further quantum-mechanical postulates dealing with spin and the Pauli principle.

7.9 MEASUREMENT AND THE INTERPRETATION OF QUANTUM MECHANICS

In quantum mechanics, the state function of a system changes in two ways. [See E. P. Wigner, *Am. J. Phys.*, **31**, 6 (1963).] First, there is the continuous, causal change with time given by the time-dependent Schrödinger equation (7.96). Second, there is the sudden, discontinuous, probabilistic change that occurs when a measurement is made on the system. This kind of change cannot be predicted with certainty, since the result of a measurement cannot be predicted with certainty; only the probabilities (7.72) are predictable. The sudden change in Ψ caused by a measurement is called the *reduction of the wave function. A measurement of the property B that yields the result b_k changes the state function to g_k, the eigenfunction of $\hat{B}$ whose eigenvalue is b_k.* (If b_k is degenerate, Ψ is changed to a linear combination of the eigenfunctions corresponding to b_k.)

Consider an example. Suppose that at time t we measure a particle's position. Let $\Psi(x, t_-)$ be the state function of the particle the instant before the measurement is made (Fig. 7.6a). We further suppose that the result of the

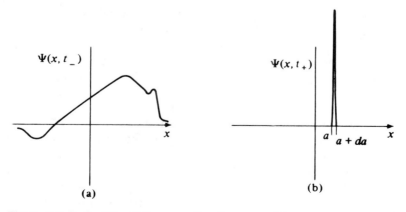

Figure 7.6 Reduction of the wave function caused by a measurement of position.

measurement is that the particle is found to be in the small region of space

$$a < x < a + da \qquad (7.100)$$

We ask: What is the state function $\Psi(x, t_+)$ the instant after the measurement? To answer this question, suppose we were to make a second measurement of position at time t_+. Since t_+ differs from the time t of the first measurement by an infinitesimal amount, we must still find that the particle is confined to the region (7.100). If the particle moved a finite distance in an infinitesimal amount of time, it would have infinite velocity, which is unacceptable. Since $|\Psi(x, t_+)|^2$ is the probability density for finding various values of x, we conclude that $\Psi(x, t_+)$ must be zero outside the region (7.100) and must look something like Fig. 7.6b. Thus the position measurement at time t has reduced Ψ from a function that is spread out over all space to one that is localized in the region (7.100). The change from $\Psi(x, t_-)$ to $\Psi(x, t_+)$ is a probabilistic change.

The measurement process is one of the most controversial areas in quantum mechanics. Just how and at what stage in the measurement process reduction occurs is unclear. Some physicists take the reduction of Ψ as an additional quantum-mechanical postulate, while others claim it is a theorem derivable from the other postulates. Some physicists reject the idea of reduction [see M. Jammer, *The Philosophy of Quantum Mechanics*, Wiley, New York, 1974, Section 11.4; L. E. Ballentine, *Am. J. Phys.*, **55**, 785 (1987)]. Ballentine advocates Einstein's statistical-ensemble interpretation of quantum mechanics, in which the wave function does not describe the state of a single system (as in the orthodox interpretation) but gives a statistical description of a collection of a large number of systems each prepared in the same way (an ensemble); in this interpretation, the need for reduction of the wave function does not occur. [See L. E. Ballentine, *Am. J. Phys.*, **40**, 1763 (1972); *Rev. Mod. Phys.*, **42**, 358 (1970).]

"For the majority of physicists the problem of finding a consistent and plausible quantum theory of measurement is still unsolved. . . . The immense diversity of opinion . . . concerning quantum measurements . . . [is] a reflection of the fundamental disagreement as to the interpretation of quantum mechanics as a whole." (M. Jammer, *The Philosophy of Quantum Mechanics*, pp. 519, 521.)

The probabilistic nature of quantum mechanics has disturbed many eminent physicists, including Einstein, de Broglie, and Schrödinger. These physicists and others have suggested that quantum mechanics may not furnish a complete description of physical reality. Rather, the probabilistic laws of quantum mechanics might be simply a reflection of deterministic laws that operate at a subquantum-mechanical level and that involve "hidden variables." An analogy given by the physicist Bohm is the Brownian motion of a dust particle in air. The particle undergoes random fluctuations of position, and its motion is not completely determined by its position and velocity. Of course, Brownian motion is a result of collisions with the gas molecules and is determined by variables existing on the level of molecular motion. Analogously, the motions of electrons might be determined by hidden variables existing on a subquantum-mechanical level. The orthodox interpretation (often called the Copenhagen interpretation) of quantum mechanics, which was developed by Heisenberg and Bohr, denies the existence of hidden variables and asserts that the laws of quantum mechanics provide a

complete description of physical reality. (Hidden-variables theories are discuss
in F. J. Belinfante, *A Survey of Hidden-Variables Theories*, Pergamon, Elmsford,
N.Y., 1973.)

In 1964, J. S. Bell proved that, in certain experiments involving measurements on two widely separated particles that originally were in the same region of
space, any possible local hidden-variable theory must make predictions that differ
from those that quantum mechanics makes (see *Ballentine*, Chapter 20). In a local
theory, two systems very far from each other act independently of each other.
The results of such experiments agree with quantum-mechanical predictions, thus
providing very strong evidence against all deterministic, local hidden-variable
theories. (The experiments do not rule out nonlocal hidden-variable theories.)
These experiments are described in J. F. Clauser and A. Shimony, *Rep. Prog.
Phys.*, **41**, 1881 (1978); F. M. Pipkin, *Adv. At. Mol. Phys.*, **14**, 281 (1978); B.
Hiley, *New Scientist*, **85**, 746 (1980); A. Aspect et al., *Phys. Rev. Lett.*, **47**, 460
(1981), **49**, 91, 1804 (1982); A. Aspect, in *The Wave-Particle Dualism*, S. Diner et
al. (eds.), Reidel, Hingham, Mass., 1984, pp. 377–390; B. Hiley, *New Scientist*,
97, 17 (1983); A. Shimony, *Scientific American*, Jan. 1988, p. 46.

Further analysis by Bell and others shows that the results of these experiments and the predictions of quantum mechanics are incompatible with a view of
the world in which both realism and locality hold. Realism (also called
objectivity) is the doctrine that external reality exists and has definite properties
independent of whether or not this reality is observed by us. Locality excludes
instantaneous action-at-a-distance and asserts that any influence from one system
to another must travel at a speed that does not exceed the speed of light. Clauser
and Shimony stated that quantum mechanics leads to the "philosophically start-
ling" conclusion that we must either "totally abandon the realistic philosophy of
most working scientists, or dramatically revise our concept of space–time" to
permit "some kind of action-at-a-distance." (J. F. Clauser and A. Shimony, *Rep.
Prog. Phys.*, **41**, 1881; see also B. d'Espagnat, *Scientific American*, Nov. 1979, p.
158.)

Quantum theory predicts and experiments confirm that when measurements
are made on two particles that once interacted but now are separated by an
unlimited distance the results obtained in the measurement on one particle
depend on the results obtained from the measurement on the second particle and
depend on which property of the second particle is measured. Such instantaneous
"spooky actions at a distance" (Einstein's phrase) have led one physicist to
remark that "quantum mechanics is magic" (D. Greenberger, quoted in N. D.
Mermin, *Physics Today*, April 1985, p. 38).

Although the experimental predictions of quantum mechanics are not
arguable, its conceptual interpretation is still the subject of heated debate, with
several different viewpoints being advocated. Excellent bibliographies with com-
mentaries on this subject are B. S. DeWitt and R. N. Graham, *Am. J. Phys.*, **39**,
724 (1971); L. E. Ballentine, *Am. J. Phys.*, **55**, 785 (1987). See also B.
d'Espagnat, *Conceptual Foundations of Quantum Mechanics*, 2nd ed., Benjamin,
Menlo Park, Calif., 1976; M. Jammer, *The Philosophy of Quantum Mechanics*,
Wiley, New York, 1974.

The relation between quantum mechanics and the mind has been the subject of much speculation. Wigner argued that the reduction of the wave function occurs when the result of a measurement enters the consciousness of an observer and thus "the being with consciousness must have a different role in quantum mechanics than the inanimate measuring device." He believes it likely that conscious beings obey different laws of nature than inanimate objects and proposes that scientists look for unusual effects of consciousness acting on matter. [E. P. Wigner, "Remarks on the Mind–Body Question," in *The Scientist Speculates*, I. J. Good, ed., Capricorn, New York, 1965, p. 284; *Proc. Amer. Phil. Soc.*, **113**, 95 (1969); *Found. Phys.*, **1**, 35 (1970).]

Some physicists see parallels between ideas of Eastern mystical thought and certain concepts of quantum mechanics, relativity, and elementary-particle physics. (See G. Zukav, *The Dancing Wu Li Masters: An Overview of the New Physics*, Morrow, New York, 1979.) Physicists in California have formed two groups to explore the relations between the ideas of modern physics, the nature of consciousness, and Eastern mystical thought (*Newsweek*, July 23, 1979, p. 85). The majority of physicists are skeptical of the value of such philosophizing.

7.10 MATRICES

Matrix algebra is a key mathematical tool in doing modern-day quantum-mechanical calculations on molecules. Matrices also furnish a convenient way to formulate much of the theory of quantum mechanics. This section therefore gives an introduction to matrices. Matrix methods will be used in later chapters, but this book is written so that the material on matrices can be omitted if time does not allow this material to be covered.

A *matrix* is a rectangular array of numbers. The numbers that compose a matrix are called the *matrix elements*. Let the matrix A have m rows and n columns, and let a_{ij} ($i = 1, 2, \ldots, m$ and $j = 1, 2, \ldots, n$) denote the element in row i and column j. Then

$$
A = \begin{pmatrix}
a_{11} & a_{12} & \cdots & a_{1n} \\
a_{21} & a_{22} & \cdots & a_{2n} \\
\cdot & \cdot & \cdots & \cdot \\
a_{m1} & a_{m2} & \cdots & a_{mn}
\end{pmatrix}
$$

A is said to be an m by n matrix. Do not confuse A with a determinant (Section 8.3); a matrix need not be square and is not equal to a single number.

A *row matrix* (also called a *row vector*) is a matrix having only one row. A *column matrix* has only one column.

Two matrices R and S are *equal* if they have the same number of rows, the same number of columns, and have corresponding elements equal. If $R = S$, then $r_{ij} = s_{ij}$ for $i = 1, \ldots, m$ and $j = 1, \ldots, n$, where m and n are the dimensions of R and S. A matrix equation is thus equivalent to mn scalar equations.

The *sum* of two matrices A and B is defined as the matrix formed by adding corresponding elements of A and B; the sum is defined only if A and B have the same dimensions. If $\mathbf{C} = \mathbf{A} + \mathbf{B}$, then we have the mn scalar equations $c_{ij} = a_{ij} + b_{ij}$ for $i = 1, \ldots, m$ and $j = 1, \ldots, n$.

$$\text{If} \quad \mathbf{C} = \mathbf{A} + \mathbf{B}, \quad \text{then} \quad c_{ij} = a_{ij} + b_{ij} \qquad (7.101)^*$$

The product of the scalar k and the matrix $\mathbf{A}$ is defined as the matrix formed by multiplying every element of $\mathbf{A}$ by k.

$$\text{If} \quad \mathbf{D} = k\mathbf{A}, \quad \text{then} \quad d_{ij} = ka_{ij} \qquad (7.102)^*$$

If $\mathbf{A}$ is an m by n matrix and $\mathbf{B}$ is an n by p matrix, the **matrix product** $\mathbf{C} = \mathbf{AB}$ is defined to be the m by p matrix whose elements are

$$c_{ij} \equiv a_{i1}b_{1j} + a_{i2}b_{2j} + \cdots + a_{in}b_{nj} = \sum_{k=1}^{n} a_{ik}b_{kj} \qquad (7.103)^*$$

To calculate c_{ij}, we take row i of $\mathbf{A}$ (this row's elements are $a_{i1}, a_{i2}, \ldots, a_{in}$), multiply each element of this row by the corresponding element in column j of $\mathbf{B}$ (this column's elements are $b_{1j}, b_{2j}, \ldots, b_{nj}$), and add the n products. For example, suppose

$$\mathbf{A} = \begin{pmatrix} -1 & 3 & \tfrac{1}{2} \\ 0 & 4 & 1 \end{pmatrix} \quad \text{and} \quad \mathbf{B} = \begin{pmatrix} 1 & 0 & -2 \\ 2 & 5 & 6 \\ -8 & 3 & 10 \end{pmatrix}$$

The number of columns of $\mathbf{A}$ equals the number of rows of $\mathbf{B}$, so the matrix product is defined. The product $\mathbf{C} = \mathbf{AB}$ is a 2 by 3 matrix. The element c_{21} is found from the second row of $\mathbf{A}$ and the first column of $\mathbf{B}$ as follows: $c_{21} = 0(1) + 4(2) + 1(-8) = 0$. Calculation of the remaining elements gives

$$\mathbf{C} = \begin{pmatrix} 1 & 16\tfrac{1}{2} & 25 \\ 0 & 23 & 34 \end{pmatrix}$$

Matrix multiplication is not commutative; the products $\mathbf{AB}$ and $\mathbf{BA}$ need not be equal. (In the preceding example, the product $\mathbf{BA}$ happens to be undefined.)

A matrix with equal numbers of rows and columns is a **square matrix**; the **order** of a square matrix equals the number of rows.

The elements $a_{11}, a_{22}, \ldots, a_{nn}$ of a square matrix of order n lie on its **principal diagonal**. A **diagonal matrix** is a square matrix having zero as the value of each element not on the principal diagonal.

A diagonal matrix whose diagonal elements are each equal to 1 is a **unit matrix**. The (i, j)th element of a unit matrix is the Kronecker delta δ_{ij}; $(\mathbf{I})_{ij} = \delta_{ij}$, where $\mathbf{I}$ is a unit matrix. For example, the unit matrix of order 3 is

$$\begin{pmatrix} 1 & 0 & 0 \\ 0 & 1 & 0 \\ 0 & 0 & 1 \end{pmatrix}$$

Let $\mathbf{B}$ be a square matrix of the same order as a unit matrix $\mathbf{I}$. The (i, j)th element of the product $\mathbf{IB}$ is given by (7.103) as $(\mathbf{IB})_{ij} = \sum_k (\mathbf{I})_{ik}b_{kj} = \sum_k \delta_{ik}b_{kj} = b_{ij}$. Since the (i, j)th elements of $\mathbf{IB}$ and $\mathbf{B}$ are equal for all i and j, we have $\mathbf{IB} = \mathbf{B}$. Similarly, we find $\mathbf{BI} = \mathbf{B}$. Multiplication by a unit matrix has no effect.

Matrices and Quantum Mechanics. In Section 7.1, the integral $\int f_i^* \hat{A} f_j \, d\tau$ was called a matrix element of $\hat{A}$. We now justify this name by showing that such integrals obey the rules of matrix algebra.

Let the functions $f_1, f_2, \ldots$ be a complete, orthonormal set and let the symbol $\{f_i\}$ denote this complete set. The numbers $A_{ij} \equiv \langle f_i | \hat{A} | f_j \rangle \equiv \int f_i^* \hat{A} f_j \, d\tau$ are called *matrix elements of the linear operator $\hat{A}$ in the basis* $\{f_i\}$. The square matrix

$$\mathbf{A} = \begin{pmatrix} A_{11} & A_{12} & \cdots \\ A_{21} & A_{22} & \cdots \\ \cdots\cdots\cdots\cdots \end{pmatrix} = \begin{pmatrix} \langle f_1 | \hat{A} | f_1 \rangle & \langle f_1 | \hat{A} | f_2 \rangle & \cdots \\ \langle f_2 | \hat{A} | f_1 \rangle & \langle f_2 | \hat{A} | f_2 \rangle & \cdots \\ \cdots\cdots\cdots\cdots\cdots\cdots\cdots\cdots \end{pmatrix} \qquad (7.104)$$

is called the *matrix representative* of the linear operator $\hat{A}$ in the $\{f_i\}$ *basis*. Since $\{f_i\}$ usually consists of an infinite number of functions, $\mathbf{A}$ is an infinite-order matrix.

Consider the addition of matrix-element integrals. Suppose $\hat{C} = \hat{A} + \hat{B}$. A typical matrix element of $\hat{C}$ in the $\{f_i\}$ basis is

$$C_{ij} = \langle f_i | \hat{C} | f_j \rangle = \langle f_i | \hat{A} + \hat{B} | f_j \rangle = \int f_i^* (\hat{A} + \hat{B}) f_j \, d\tau$$

$$= \int f_i^* \hat{A} f_j \, d\tau + \int f_i^* \hat{B} f_j \, d\tau = A_{ij} + B_{ij}$$

Thus, if $\hat{C} = \hat{A} + \hat{B}$, then $C_{ij} = A_{ij} + B_{ij}$, which is the rule (7.101) for matrix addition. Hence, if $\hat{C} = \hat{A} + \hat{B}$, then $\mathbf{C} = \mathbf{A} + \mathbf{B}$, where $\mathbf{A}$, $\mathbf{B}$, and $\mathbf{C}$ are the matrix representatives of the operators $\hat{A}$, $\hat{B}$, $\hat{C}$.

Similarly, if $\hat{D} = k\hat{C}$, then we find (Problem 7.44) $D_{ij} = kC_{ij}$, which is the rule for multiplication of a matrix by a scalar.

Finally, suppose that $\hat{R} = \hat{S}\hat{T}$. We have

$$R_{ij} = \int f_i^* \hat{R} f_j \, d\tau = \int f_i^* \hat{S}\hat{T} f_j \, d\tau \qquad (7.105)$$

The function $\hat{T} f_j$ can be expanded in terms of the complete orthonormal set $\{f_i\}$ as [Eq. (7.41)]:

$$\hat{T} f_j = \sum_k c_k f_k = \sum_k \langle f_k | \hat{T} f_j \rangle f_k = \sum_k \langle f_k | \hat{T} | f_j \rangle f_k = \sum_k T_{kj} f_k$$

and R_{ij} becomes

$$R_{ij} = \int f_i^* \hat{S} \sum_k T_{kj} f_k \, d\tau = \sum_k \int f_i^* \hat{S} f_k \, d\tau \, T_{kj} = \sum_k S_{ik} T_{kj} \qquad (7.106)$$

The equation $R_{ij} = \sum_k S_{ik} T_{kj}$ is the rule (7.103) for matrix multiplication. Hence, if $\hat{R} = \hat{S}\hat{T}$, then $\mathbf{R} = \mathbf{ST}$.

We have proved that *the matrix representatives of linear operators in a complete orthonormal basis set obey the same equations that the operators obey.*

Combining Eqs. (7.105) and (7.106), we have the useful sum rule

$$\sum_k \langle i|\hat{S}|k\rangle \langle k|\hat{T}|j\rangle = \langle i|\hat{S}\hat{T}|j\rangle \tag{7.107}$$

Suppose the basis set $\{f_i\}$ is chosen to be the complete, orthonormal set of eigenfunctions g_i of $\hat{A}$, where $\hat{A}g_i = a_i g_i$. Then the matrix element A_{ij} is

$$A_{ij} = \langle g_i|\hat{A}|g_j\rangle = \langle g_i|\hat{A}g_j\rangle = \langle g_i|a_j g_j\rangle = a_j\langle g_i|g_j\rangle = a_j\delta_{ij}$$

The matrix that represents $\hat{A}$ in the basis of orthonormal $\hat{A}$ eigenfunctions is thus a diagonal matrix whose diagonal elements are the eigenvalues of $\hat{A}$. Conversely, one can prove (Problem 7.42) that, when the matrix representative of $\hat{A}$ using a complete orthonormal set is a diagonal matrix, then the basis functions are the eigenfunctions of $\hat{A}$ and the diagonal matrix elements are the eigenvalues of $\hat{A}$.

We have used the complete, orthonormal basis $\{f_i\}$ to represent the operator $\hat{A}$ by the matrix $\mathbf{A}$ of (7.104). The basis $\{f_i\}$ can also be used to represent an arbitrary function u, as follows. We expand u in terms of the complete set $\{f_i\}$, according to $u = \sum_i u_i f_i$, where the expansion coefficients u_i are numbers (not functions) given by Eq. (7.40) as $u_i = \langle f_i|u\rangle$. The set of expansion coefficients $u_1, u_2, \ldots$ is formed into a column matrix (column vector), which we call $\mathbf{u}$, and $\mathbf{u}$ is said to be the ***representative*** of the function u in the $\{f_i\}$ basis. If $\hat{A}u = w$, where w is another function, then we can show (Problem 7.43) that $\mathbf{Au} = \mathbf{w}$, where $\mathbf{A}$, $\mathbf{u}$, and $\mathbf{w}$ are the matrix representatives of $\hat{A}$, u, and w in the $\{f_i\}$ basis. Thus, the effect of the linear operator $\hat{A}$ on an arbitrary function u can be found if the matrix representative $\mathbf{A}$ of $\hat{A}$ is known. Hence, knowing the matrix representative $\mathbf{A}$ is equivalent to knowing what the operator $\hat{A}$ is.

7.11 SUMMARY

A quantum-mechanical operator $\hat{A}$ that represents a physical quantity must be Hermitian, meaning that it satisfies $\int f^*\hat{A}u \, d\tau = \int u(\hat{A}f)^* \, d\tau$ for all well-behaved functions f and u. The eigenvalues of Hermitian operators are real numbers. For a Hermitian operator, eigenfunctions that correspond to different eigenvalues are orthogonal, and eigenfunctions that correspond to a degenerate eigenvalue can be chosen to be orthogonal.

We postulated that the eigenfunctions g_i of any Hermitian operator that represents a physical quantity form a complete set, meaning that any well-behaved function f can be expanded as $f = \sum_k c_k g_k$, where the g_i's are orthonormal and where $c_k = \int g_k^* f \, d\tau$.

If two quantum-mechanical operators commute, they have a complete set of eigenfunctions in common. Two quantum-mechanical operators that have a common complete set of eigenfunctions must commute.

If the potential energy V is an even function, each stationary-state wave function can be chosen to be either even or odd.

If the property B is measured in a system whose state function is Ψ, the probability that the nondegenerate eigenvalue b_i is found is given by $|\langle g_i|\Psi\rangle|^2$, where $\hat{B}g_i = b_i g_i$.

The postulates of quantum mechanics are summarized in Section 7.8.

PROBLEMS

7.1 Is $\langle f_m|\hat{A}|f_n\rangle$ equal to $\langle f_m|\hat{A}f_n\rangle$?

7.2 What operator is shown to be Hermitian by the equation $\langle m|n\rangle = \langle n|m\rangle^*$?

7.3 Let $\hat{A}$ and $\hat{B}$ be Hermitian operators and let c be a real constant. (a) Show that $c\hat{A}$ is Hermitian. (b) Show that $\hat{A} + \hat{B}$ is Hermitian.

7.4 (a) Show that d^2/dx^2 and $\hat{T}_x$ are Hermitian, where $\hat{T}_x = -(\hbar^2/2m)d^2/dx^2$. (b) Show that $\langle T_x\rangle = (\hbar^2/2m)\int |\partial\Psi/\partial x|^2\, d\tau$.

7.5 Which of the following operators are Hermitian: d/dx, $i(d/dx)$, $4d^2/dx^2$, $i(d^2/dx^2)$?

7.6 Let $\hat{A}$ be a Hermitian operator. Show that $\langle A^2\rangle = \int |\hat{A}\psi|^2\, d\tau$ and therefore $\langle A^2\rangle \geq 0$. [This result can be used to derive Eq. (5.138) more rigorously than in the text. Thus, since $\hat{M}^2 - \hat{M}_z^2 = \hat{M}_x^2 + \hat{M}_y^2$, we have $\langle M^2\rangle - \langle M_z^2\rangle = \langle M_x^2\rangle + \langle M_y^2\rangle$. Now $\langle M^2\rangle = c$, $\langle M_z^2\rangle = b_k^2$, and by the theorem of this exercise $\langle M_x^2\rangle \geq 0$, $\langle M_y^2\rangle \geq 0$. Hence $c - b_k^2 \geq 0$.]

7.7 Verify that $\hat{L}_z$ is Hermitian using (a) spherical polar coordinates; (b) Cartesian coordinates.

7.8 Which of the following operators meet the requirements for a quantum-mechanical operator that is to represent a physical quantity: (a) $(\quad)^{1/2}$; (b) d/dx; (c) d^2/dx^2; (d) $i(d/dx)$?

7.9 (a) If $\hat{A}$ and $\hat{B}$ are Hermitian operators, prove that their product $\hat{A}\hat{B}$ is Hermitian if and only if $\hat{A}$ and $\hat{B}$ commute. (b) If $\hat{A}$ and $\hat{B}$ are Hermitian, prove that $\frac{1}{2}(\hat{A}\hat{B} - \hat{B}\hat{A})$ is Hermitian. (c) Is $\hat{x}\hat{p}_x$ Hermitian? (d) Is $\frac{1}{2}(\hat{x}\hat{p}_x + \hat{p}_x\hat{x})$ Hermitian?

7.10 Explain why each of the following integrals must be zero, where the functions are hydrogenlike wave functions: (a) $\langle 2p_1|\hat{L}_z|3p_{-1}\rangle$; (b) $\langle 3p_0|\hat{L}_z|3p_0\rangle$.

7.11 (a) Fill in the details in the evaluation of the expansion coefficients a_n in the example in Section 7.3. (b) Write the $x(l-x)$ expansion of this example for $x = \frac{1}{2}$ and use the first five nonzero terms to approximate π^3. (c) Calculate the percent error in this $x(l-x)$ expansion at $x = l/4$ when the first 1, 3, and 5 nonzero terms are taken.

7.12 (a) Expand the function $f(x) = -1$ for $0 \leq x \leq \frac{1}{2}l$ and $f(x) = 1$ for $\frac{1}{2}l < x \leq l$ in terms of the particle-in-a-box wave functions. (Since f is discontinuous at $\frac{1}{2}l$, the expansion cannot be expected to represent f at this point; since f does not obey the particle-in-a-box boundary conditions of being zero at $x = 0$ and $x = l$, the expansion will not represent f at these points; the expansion will represent f at other points.) (b) Calculate the percent error at $x = l/4$ when the first 1, 3, 5, and 7 nonzero terms are taken in this expansion.

7.13 (a) Show that the hydrogenlike wave functions $2p_x$ and $2p_1$ are not orthogonal. (b) Use the Schmidt procedure to construct linear combinations of $2p_x$ and $2p_1$ that will be orthogonal. Then normalize these functions. Which of the operators $\hat{H}$, $\hat{L}^2$, $\hat{L}_z$ are your linear combinations eigenfunctions of?

7.14 Extend the Schmidt orthogonalization procedure to the case of threefold degeneracy.

7.15 For the hydrogenlike atom, we have $V = -Ze'^2(x^2 + y^2 + z^2)^{-1/2}$ so that the potential energy is an even function of the coordinates. (a) What is the parity of ψ_{2s}? (b) What is the parity of ψ_{2p_x}? (c) Consider $\psi_{2s} + \psi_{2p_x}$. Is it an eigenfunction of $\hat{H}$? Does it have definite parity?

7.16 Let $\hat{\Pi}$ be the parity operator and m be a positive integer. What is $\hat{\Pi}^m$ if m is even? if m is odd?

7.17 Let $\hat{\Pi}$ be the parity operator and let $\psi_i(x)$ be a normalized harmonic-oscillator wave function. We define the matrix elements Π_{ij} as

$$\Pi_{ij} = \int_{-\infty}^{\infty} \psi_i^* \hat{\Pi} \psi_j \, dx$$

Show that $\Pi_{ij} = 0$ for $i \neq j$, and $\Pi_{ii} = \pm 1$.

7.18 Let $\hat{R}$ be a linear operator such that $\hat{R}^n = \hat{1}$, where n is a positive integer and no lower power of $\hat{R}$ equals $\hat{1}$. Find the eigenvalues of $\hat{R}$.

7.19 Use parity to find which of the following integrals must be zero: (a) $\langle 2s|x|2p_x \rangle$; (b) $\langle 2s|x^2|2p_x \rangle$; (c) $\langle 2p_y|x|2p_x \rangle$. The functions in these integrals are hydrogenlike wave functions.

7.20 (a) Show that the parity operator is linear. (b) Show that the parity operator is Hermitian; a proof in one dimension is sufficient.

7.21 (a) In Cartesian coordinates the parity operator $\hat{\Pi}$ corresponds to the transformation of variables $x \rightarrow -x$, $y \rightarrow -y$, $z \rightarrow -z$. Show that in spherical polar coordinates the parity operator corresponds to the transformation $r \rightarrow r$, $\theta \rightarrow \pi - \theta$, $\phi \rightarrow \phi + \pi$. (b) Show that $\hat{\Pi} e^{im\phi} = (-1)^m e^{im\phi}$. (c) Use Eq. (5.97) to show that

$$\hat{\Pi} S_{l,m}(\theta) = (-1)^{l-|m|} S_{l,m}(\theta) = (-1)^{l-m} S_{l,m}(\theta)$$

(d) Combine the results of (b) and (c) to conclude that the spherical harmonic $Y_l^m(\theta, \phi)$ is an even function if l is even and is an odd function if l is odd.

7.22 Since the parity operator is Hermitian (Problem 7.20), two eigenfunctions of $\hat{\Pi}$ that correspond to different eigenvalues must be orthogonal. Show directly that this is so.

7.23 Consider the integral $\langle v_2|x|v_1 \rangle$, where the functions are one-dimensional harmonic-oscillator wave functions with quantum numbers v_2 and v_1. Under what conditions do parity considerations allow us to conclude that this integral must be zero? Might the integral be zero in other cases as well? (This integral is important in discussing radiative transitions.)

7.24 For a hydrogen atom in a p state, the possible outcomes of a measurement of L_z are $-\hbar$, 0, and $\hbar$. For each of the following wave functions, give the probabilities of each of these three results: (a) ψ_{2p_z}; (b) ψ_{2p_y}; (c) ψ_{2p_1}.

7.25 Suppose that at time t' a hydrogen atom is in a nonstationary state with

$$\Psi = 6^{-1/2}(2p_1) - 2^{-1/2}i(2p_0) - 3^{-1/2}(3d_1)$$

If L_z is measured at t', give the possible outcomes and give the probability of each possible outcome.

7.26 If L^2 is measured in a hydrogen atom whose state function is that in Problem 7.25, give the possible outcomes and their probabilities.

7.27 Suppose that a particle in a box of length l is in a nonstationary state with $\Psi = 0$ for $x < l$ and for $x > l$, and

$$\Psi = \tfrac{1}{2} e^{-ih^2 t/8ml^2\hbar} \left(\frac{2}{l}\right)^{1/2} \sin \frac{\pi x}{l} + \tfrac{1}{2}\sqrt{3} e^{i\pi} e^{-ih^2 t/2ml^2\hbar} \left(\frac{2}{l}\right)^{1/2} \sin \frac{2\pi x}{l}$$

for $0 \leq x \leq l$. If the energy is measured at time t, give the possible outcomes and their probabilities.

7.28 Suppose that a particle in a box of length l is in the nonstationary state $\Psi = (105/l^7)^{1/2} x^2 (l - x)$, for $0 \leq x \leq l$, at the time the energy is measured. Give the possible outcomes and their probabilities.

7.29 Use the results of the last example in Section 7.6 to evaluate the sum $\sum_{m=0}^{\infty} [1/(2m+1)^6]$.

7.30 A measurement yields $2^{1/2}\hbar$ for the magnitude of a particle's orbital angular momentum. If L_x is now measured, what are the possible outcomes?

7.31 Suppose that the electron in a hydrogen atom has the state function $\Psi = (27/\pi a^3)^{1/2} e^{-3r/a}$ (where $a = \hbar^2/\mu e'^2$) at the time its energy is measured. Find the probability that the energy value $-e'^2/2a$ is found.

7.32 (a) Combine Eqs. (2.30) and (3.33) to write down Ψ for a free particle in one dimension. (b) Show that this Ψ is a linear combination of two eigenfunctions of $\hat{p}_x$. What are the eigenvalues for these eigenfunctions? (c) If p_x of a free particle in one dimension is measured, give the possible outcomes and give their probabilities.

7.33 (a) Show that, for a particle in a one-dimensional box (Fig. 2.1) of length l, the probability of observing a value of p_x between p and $p + dp$ is

$$\frac{4|N|^2 s^2}{l(s^2 - b^2)^2} [1 - (-1)^n \cos bl] \, dp \tag{7.108}$$

where $s \equiv n\pi l^{-1}$ and $b \equiv p\hbar^{-1}$. The constant N is to be chosen so that the integral from minus infinity to infinity of (7.108) is unity. (b) Evaluate (7.108) for $p = \pm nh/2l$. [At these values of p_x, the denominator of (7.108) is zero and the probability reaches a large but finite value.]

7.34 Find (a) $\int_{-\infty}^{\infty} \delta(x) \, dx$; (b) $\int_{-\infty}^{-1} \delta(x) \, dx$; (c) $\int_{-1}^{1} \delta(x) \, dx$; (d) $\int_{1}^{2} f(x)\delta(x - 3) \, dx$.

7.35 What is the value of $\int_{0}^{\infty} f(x)\delta(x) \, dx$?

7.36 Show that $\int_{-\infty}^{\infty} |\delta(x - a)|^2 \, dx = \infty$.

7.37 The functions in Fig. 7.5 approximate the Dirac delta function. Draw graphs of the corresponding functions that approximate the Heaviside step function with successively increasing accuracy.

7.38 Quantum mechanics postulates that the present state of an undisturbed system determines its future state. Consider the special case of a system with a time-independent Hamiltonian $\hat{H}$. Suppose it is known that at time t_0 the state function is $\Psi(x, t_0)$. Use Eqs. (7.96) and (7.65) to show that

$$\Psi(x, t) = \sum_j \langle \psi_j(x) | \Psi(x, t_0) \rangle \exp\left[-iE_j(t - t_0)/\hbar\right] \psi_j(x)$$

where $\hat{H}\psi_j = E_j\psi_j$.

7.39 For the matrices

$$\mathbf{A} = \begin{pmatrix} 2 & 1 \\ 0 & -3 \end{pmatrix}, \quad \mathbf{B} = \begin{pmatrix} 1 & -1 \\ 4 & 4 \end{pmatrix}$$

find (a) $\mathbf{AB}$; (b) $\mathbf{BA}$; (c) $\mathbf{A} + \mathbf{B}$; (d) $3\mathbf{A}$; (e) $\mathbf{A} - 4\mathbf{B}$.

7.40 Calculate the matrix products $\mathbf{CD}$ and $\mathbf{DC}$, where

$$\mathbf{C} = \begin{pmatrix} 5 \\ 0 \\ -1 \end{pmatrix} \quad \text{and} \quad \mathbf{D} = (i \quad 2 \quad 1)$$

7.41 Find the matrix representative of the unit operator $\hat{1}$ in a complete, orthonormal basis.

7.42 If $\{f_i\}$ is a complete, orthonormal basis such that $\langle f_i | \hat{A} | f_j \rangle = a_i \delta_{ij}$ for all i and j (that is, the matrix representative of $\hat{A}$ is diagonal), prove that the functions $\{f_i\}$ are eigenfunctions of $\hat{A}$ and the a_i's are the eigenvalues of $\hat{A}$. *Hint:* Expand $\hat{A}f_j$ in terms of the set $\{f_i\}$.

7.43 (a) If $\hat{A}$ is a linear operator, $\{f_i\}$ is a complete, orthonormal basis, and u is an arbitrary function, show that $\hat{A}u = \sum_j (\sum_i \langle f_j | \hat{A} | f_i \rangle \langle f_i | u \rangle) f_j$. *Hint:* Expand u in terms of the set $\{f_i\}$, apply $\hat{A}$ to this expansion, and then expand $\hat{A}f_i$. (b) If $\hat{A}u = w$, with $w = \Sigma_j \, w_j f_j$ and $u = \Sigma_i \, u_i f_i$ (where the expansion coefficients w_j and u_i are numbers), show that the equation in (a) says that $w_j = \Sigma_i \, A_{ji} u_i$, and hence that $\mathbf{w} = \mathbf{A}\mathbf{u}$, where $\mathbf{w}$, $\mathbf{A}$, and $\mathbf{u}$ are the representatives of w, $\hat{A}$, and u.

7.44 If $\hat{D} = k\hat{C}$, show that $D_{ij} = kC_{ij}$.

7.45 Write down the $l = 2$ portion of the matrix representative of $\hat{L}_z$ in the $\{Y_l^m\}$ basis.

7.46 Verify the following Kronecker-delta identities: (a) $\delta_{ij} = \delta_{ji}$; (b) $(\delta_{ij})^2 = \delta_{ij}$; (c) $\delta_{ij}\delta_{ik} = \delta_{ij}\delta_{jk}$; (d) $\delta_{i+n,j} = \delta_{i,j-n}$.

7.47 Calculate the uncertainty ΔL_z for these hydrogen-atom stationary states: (a) $2p_z$; (b) $2p_x$.

7.48 For the system of Problem 7.27, (a) use orthonormality [Eq. (2.27)] to verify that Ψ is normalized; (b) find $\langle E \rangle$ at time t; (c) find $\langle x \rangle$ at time t, and find the maximum and minimum values of $\langle x \rangle$ as a function of t.

$\rightarrow$ **7.49** Consider an operator $\hat{A}$ that contains the time as a parameter. We are interested in how the average value of the property A changes with time, and we have

$$\frac{d\langle A \rangle}{dt} = \frac{d}{dt} \int \Psi^* \hat{A} \Psi \, d\tau$$

The definite integral on the right side of this equation is a function of the parameter t, and it is generally a valid mathematical operation to calculate its derivative with respect to t by differentiating the integrand with respect to t:

$$\frac{d\langle A \rangle}{dt} = \int \frac{\partial}{\partial t} [\Psi^* \hat{A} \Psi] \, d\tau = \int \frac{\partial \Psi^*}{\partial t} \hat{A} \Psi \, d\tau + \int \Psi^* \frac{\partial \hat{A}}{\partial t} \Psi \, d\tau + \int \Psi^* \hat{A} \frac{\partial \Psi}{\partial t} \, d\tau$$

Use the time-dependent Schrödinger equation, its complex conjugate, and the Hermitian property of $\hat{H}$ to show that

$$\frac{d\langle A \rangle}{dt} = \int \Psi^* \frac{\partial \hat{A}}{\partial t} \Psi \, d\tau + \frac{i}{\hbar} \int \Psi^* (\hat{H}\hat{A} - \hat{A}\hat{H})\Psi \, d\tau$$

$$\frac{d\langle A \rangle}{dt} = \left\langle \frac{\partial \hat{A}}{\partial t} \right\rangle + \frac{i}{\hbar} \langle [\hat{H}, \hat{A}] \rangle \qquad (7.109)$$

$\rightarrow$ **7.50** Use (7.109) and (5.8) to show that

$$\frac{d}{dt} \langle x \rangle = \frac{\langle p_x \rangle}{m} = \frac{1}{m} \int \Psi^* \frac{\hbar}{i} \frac{\partial \Psi}{\partial x} \, d\tau \qquad (7.110)$$

From (7.110) it follows that

$$\frac{d^2 \langle x \rangle}{dt^2} = \frac{1}{m} \frac{d}{dt} \langle p_x \rangle$$

Use (7.109), (5.9), and (4.26) to show that

$$\langle F_x \rangle = m \frac{d^2 \langle x \rangle}{dt^2} \qquad (7.111)$$

If we consider a classical-mechanical particle, its wave function will be large only in a very small region corresponding to its position, and we may then drop the averages in (7.111) to obtain Newton's second law. Thus classical mechanics is a special case of quantum mechanics. Equation (7.111) is known as *Ehrenfest's theorem*, after the physicist who derived it in 1927.

7.51 In accord with the probability interpretation of $\Psi^*\Psi$, we have normalized the wave function: $\int \Psi^*\Psi \, d\tau = 1$. Assuming that particles are neither created nor destroyed, we expect this normalization to hold for all time; that is, we expect $(d/dt) \int \Psi^*\Psi \, d\tau = 0$. Prove this equation as a special case of (7.109).

7.52 Prove the uncertainty principle (5.12) for any two Hermitian operators $\hat{A}$ and $\hat{B}$, as follows. (a) Define functions f and g as $f = (\hat{A} - \langle A \rangle)\Psi$ and $g = i(\hat{B} - \langle B \rangle)\Psi$. Prove that $\langle f|f \rangle = \langle f|f \rangle^* = (\Delta A)^2$ and $\langle g|g \rangle = (\Delta B)^2$. (b) Now let $I \equiv \langle f + sg | f + sg \rangle$, where s is defined as an arbitrary *real* parameter. The integrand $|f + sg|^2$ of I is everywhere nonnegative and I must therefore be positive, unless it happens that $f = -sg$. We have two possible cases: either (1) $f = -sg$, or (2) $f \neq -sg$. Show directly that for case (1) we have $4\langle f|f \rangle \langle g|g \rangle = [\langle f|g \rangle + \langle g|f \rangle]^2$. For case (2) we have $I = as^2 + bs + c > 0$, where a, b, and c are certain integrals. This inequality means that the equation $as^2 + bs + c = 0$ can have no real roots for s, which (using the quadratic formula) means that $4ac > b^2$. Show that this inequality leads to $4\langle f|f \rangle \langle g|g \rangle > [\langle f|g \rangle + \langle g|f \rangle]^2$. (c) Combining cases (1) and (2), we have shown that $4\langle f|f \rangle \langle g|g \rangle \geq [\langle g|f \rangle^* + \langle f|g \rangle^*]^2$. Show that this leads to (5.12).

7.53 Write a computer program that uses the expansion (7.37) to calculate the function (7.35) at x values of $0, 0.1l, 0.2l, \ldots, l$; have the program do the calculations for 5, 10, 15, and 20 terms taken in the expansion.

7.54 In finding eigenvalues by solving differential equations, quantization occurs only when we apply the condition that the eigenfunctions be well behaved. Ladder operators were used in Section 5.4 to find the eigenvalues of $\hat{M}^2$ and $\hat{M}_z$ and in Problem 5.30 to find the harmonic-oscillator energy eigenvalues. Where was the condition that the eigenfunctions be well behaved used in the derivations in Section 5.4 and Problem 5.30? *Hint:* See Problem 7.6.

7.55 True or false: (a) The state function is always equal to a function of time multiplied by a function of the coordinates. (b) In both classical and quantum mechanics, knowledge of the present state of an isolated system allows its future state to be calculated. (c) The state function is always an eigenfunction of the Hamiltonian. (d) Any linear combination of eigenfunctions of the Hamiltonian is an eigenfunction of the Hamiltonian. (e) If the state function is not an eigenfunction of the operator $\hat{A}$, then a measurement of the property A might give a value that is not one of the eigenvalues of $\hat{A}$. (f) The probability density is independent of time for a stationary state. (g) If two operators do not commute, then they cannot possess any common eigenfunctions. (h) If two operators commute, then every eigenfunction of one must be an eigenfunction of the other. (i) Two linearly independent eigenfunctions of the same Hermitian operator are always orthogonal to each other. (j) If the operator $\hat{B}$ corresponds to a physical property of a quantum-mechanical system, the state function Ψ must be an eigenfunction of $\hat{B}$.

8
The Variation Method

8.1 THE VARIATION THEOREM

We now begin the study of the approximation methods needed to deal with the time-independent Schrödinger equation for systems of interacting particles. This chapter deals with the variation method, which allows us to obtain an approximation to the ground-state energy of a system without solving the Schrödinger equation. The variation method is based on the following theorem:

The Variation Theorem. Given a system whose Hamiltonian operator $\hat{H}$ is time independent and whose lowest-energy eigenvalue is E_1, if ϕ is any normalized, well-behaved function of the coordinates of the system's particles that satisfies the boundary conditions of the problem, then

$$\int \phi^* \hat{H} \phi \, d\tau \geq E_1 , \qquad \phi \text{ normalized} \qquad (8.1)^*$$

The variation theorem allows us to calculate an upper bound for the system's ground-state energy.

To prove (8.1), we expand ϕ in terms of the complete, orthonormal set of eigenfunctions of $\hat{H}$, the stationary-state eigenfunctions ψ_k:

$$\phi = \sum_k a_k \psi_k \qquad (8.2)$$

where

$$\hat{H}\psi_k = E_k \psi_k \qquad (8.3)$$

Note that the expansion (8.2) requires that ϕ obey the same boundary conditions as the ψ_k's. Substitution of (8.2) into the left side of (8.1) gives

$$\int \phi^* \hat{H} \phi \, d\tau = \int \sum_k a_k^* \psi_k^* \hat{H} \sum_j a_j \psi_j \, d\tau = \int \sum_k a_k^* \psi_k^* \sum_j a_j \hat{H} \psi_j \, d\tau$$

Using the eigenvalue equation (8.3) and assuming the validity of interchanging the integration and the infinite summations, we get

$$\int \phi^* \hat{H} \phi \, d\tau = \int \sum_k a_k^* \psi_k^* \sum_j a_j E_j \psi_j \, d\tau = \sum_k \sum_j a_k^* a_j E_j \int \psi_k^* \psi_j \, d\tau = \sum_k \sum_j a_k^* a_j E_j \delta_{kj}$$

where the orthonormality of the eigenfunctions ψ_k was used. We perform the sum over j, and, as usual, the Kronecker delta makes all terms zero except the one with $j = k$, giving

$$\int \phi^* \hat{H} \phi \, d\tau = \sum_k a_k^* a_k E_k = \sum_k |a_k|^2 E_k \tag{8.4}$$

Since E_1 is the lowest-energy eigenvalue of $\hat{H}$, we have $E_k \geq E_1$. Since $|a_k|^2$ is never negative, we can multiply the inequality $E_k \geq E_1$ by $|a_k|^2$ without changing the direction of the inequality sign, and we have $|a_k|^2 E_k \geq |a_k|^2 E_1$. Therefore, $\sum_k |a_k|^2 E_k \geq \sum_k |a_k|^2 E_1$, and use of (8.4) gives

$$\int \phi^* \hat{H} \phi \, d\tau = \sum_k |a_k|^2 E_k \geq \sum_k |a_k|^2 E_1 = E_1 \sum_k |a_k|^2 \tag{8.5}$$

Because ϕ is normalized, we have $\int \phi^* \phi \, d\tau = 1$. Substitution of the expansion (8.2) into the normalization condition gives

$$1 = \int \phi^* \phi \, d\tau = \int \sum_k a_k^* \psi_k^* \sum_j a_j \psi_j \, d\tau = \sum_k \sum_j a_k^* a_j \int \psi_k^* \psi_j \, d\tau$$

$$= \sum_k \sum_j a_k^* a_j \delta_{kj} = \sum_k |a_k|^2 \tag{8.6}$$

[Note that in deriving Eqs. (8.4) and (8.6) we essentially repeated the derivations of Eqs. (7.69) and (7.68), respectively.]

Use of $\sum_k |a_k|^2 = 1$ in (8.5) gives

$$\int \phi^* \hat{H} \phi \, d\tau \geq E_1 \, , \qquad \phi \text{ normalized} \tag{8.7}$$

which is the variation theorem (8.1).

Suppose we have a function ϕ that is not normalized. To apply the variation theorem, we multiply ϕ by a normalization constant N so that $N\phi$ is normalized. Replacing ϕ by $N\phi$ in (8.7), we have

$$|N|^2 \int \phi^* \hat{H} \phi \, d\tau \geq E_1 \tag{8.8}$$

N is determined by $\int (N\phi)^*(N\phi) \, d\tau = |N|^2 \int \phi^* \phi \, d\tau = 1$; thus $|N|^2 = 1/\int \phi^* \phi \, d\tau$ and (8.8) becomes

$$\frac{\int \phi^* \hat{H} \phi \, d\tau}{\int \phi^* \phi \, d\tau} \geq E_1 \tag{8.9*}$$

where ϕ is any well-behaved function (not necessarily normalized) that satisfies the boundary conditions of the problem.

The function ϕ is called a ***trial variation function***, and the integral in (8.1) [or the ratio of integrals in (8.9)] is called the ***variational integral***. To arrive at a good approximation to the ground-state energy E_1, we try many trial variation functions and look for the one that gives the lowest value of the variational integral. From (8.1), the lower the value of the variational integral, the better the approximation we have to E_1. One way to disprove quantum mechanics would be to find a trial variation function that made the variational integral less than E_1 for some system where E_1 is known.

Let ψ_1 be the true ground-state wave function:

$$\hat{H}\psi_1 = E_1\psi_1 \tag{8.10}$$

If we happened to be lucky enough to hit upon a variation function that was equal to ψ_1, then, using (8.10) in (8.1), we see that the variational integral will be equal to E_1. Thus the ground-state wave function gives the minimum value of the variational integral. We therefore expect that the lower the value of the variational integral, the closer the trial variational function will approach the true ground-state wave function. However, it turns out that the variational integral approaches E_1 a lot faster than the trial variation function approaches ψ_1, and it is possible to get a rather good approximation to E_1 using a rather poor ϕ.

In practice, one usually puts several parameters into the trial function ϕ, and then varies these parameters so as to minimize the variational integral. Successful use of the variation method depends on the ability to make a shrewd choice (or guess) for the trial function.

Let us look at some examples of the variation method. Although the real utility of the method is for problems to which we do not know the true solutions, we will here consider problems that are exactly solvable so that we can judge the accuracy of our results.

EXAMPLE   Devise a trial variation function for the particle in a one-dimensional box of length l.

The wave function is zero outside the box and the boundary conditions require that $\psi = 0$ at $x = 0$ and at $x = l$. The variation function ϕ must meet these boundary conditions of being zero at the ends of the box. As noted after Eq. (4.65), the ground-state ψ has no nodes interior to the boundary points, so it is desirable that ϕ have no interior nodes. A simple function that has these properties is the parabolic function

$$\phi = x(l - x), \qquad \text{for } 0 \leq x \leq l \tag{8.11}$$

and $\phi = 0$ outside the box. Since we have not normalized ϕ, we use Eq. (8.9). Inside the box the Hamiltonian is $-(\hbar^2/2m)\, d^2/dx^2$. For the numerator and denominator of (8.9), we have

$$\int \phi^* \hat{H}\phi \, d\tau = -\frac{\hbar^2}{2m} \int_0^l (lx - x^2) \frac{d^2}{dx^2} (lx - x^2) \, dx = -\frac{\hbar^2}{m} \int_0^l (x^2 - lx)\, dx$$

$$\int \phi^* \hat{H}\phi \, d\tau = \frac{\hbar^2 l^3}{6m} \tag{8.12}$$

$$\int \phi^* \phi \, d\tau = \int_0^l x^2 (l - x)^2 \, dx = \frac{l^5}{30}$$

Substituting in (8.9), we get

$$\frac{5h^2}{4\pi^2 l^2 m} \geq E_1$$

From Eq. (2.20), E_1 is $h^2/8ml^2$ and the percent error is

$$\frac{(5/4\pi^2) - (1/8)}{1/8} \times 100\% = 1.3\%$$

Since $\int |\phi|^2 \, d\tau = l^5/30$, the normalized form of (8.11) is $(30/l^5)^{1/2}x(l - x)$. Figure 7.3 shows that this function rather closely resembles the true ground-state particle-in-a-box wave function.

The preceding example did not have a parameter in the trial function. The next example does.

EXAMPLE   For the one-dimensional harmonic oscillator, devise a variation function with a parameter and find the optimum value of the parameter.

The variation function ϕ must be quadratically integrable and so must go to zero as x goes to $\pm\infty$. The function e^{-x} has the proper behavior at $+\infty$ but becomes infinite at $-\infty$. The function e^{-x^2} has the proper behavior at $\pm\infty$. However, it is dimensionally unsatisfactory, since the power to which we raise e must be dimensionless. This can be seen from the Taylor series $e^z = 1 + z + z^2/2! + \cdots$ [Eq. (4.48)]. Since all the terms in this series must have the same dimensions, z must have the same dimensions as 1; that is, z in e^z must be dimensionless. Hence we modify e^{-x^2} to e^{-cx^2}, where c has units of length^{-2}. We shall take c as a variational parameter. The true ground-state ψ must have no nodes. Also, since $V = \frac{1}{2}kx^2$ is an even function, the ground-state ψ must have definite parity and must be an even function, since an odd function has a node at the origin. The trial function e^{-cx^2} has the desired properties of having no nodes and of being an even function.

Use of (4.32) for $\hat{H}$ and Appendix integrals gives (Problem 8.2)

$$\int \phi^* \hat{H} \phi \, d\tau = -\frac{\hbar^2}{2m} \int_{-\infty}^{\infty} e^{-cx^2} \frac{d^2 e^{-cx^2}}{dx^2} \, dx + 2\pi^2 \nu^2 m \int_{-\infty}^{\infty} x^2 e^{-2cx^2} \, dx$$

$$= (\hbar^2/m)(\pi/8)^{1/2}c^{1/2} + \nu^2 m(\pi^5/8)^{1/2}c^{-3/2} \tag{8.13}$$

$$\int \phi^* \phi \, d\tau = \int_{-\infty}^{\infty} e^{-2cx^2} \, dx = 2 \int_0^{\infty} e^{-2cx^2} \, dx = (\pi/2)^{1/2}c^{-1/2}$$

The variational integral W is

$$W \equiv \frac{\int \phi^* \hat{H} \phi \, d\tau}{\int \phi^* \phi \, d\tau} = (\hbar^2/2m)c + (\pi^2/2)\nu^2 m c^{-1} \tag{8.14}$$

We now vary c to minimize the variational integral (8.14). A necessary condition that W be minimized is that

$$dW/dc = 0 = (\hbar^2/2m) - (\pi^2/2)\nu^2 m c^{-2} \tag{8.15}$$

$$c = \pm \pi \nu m/\hbar$$

The negative root $c = -\pi \nu m/\hbar$ is rejected, since it would make $\phi \; (= e^{-cx^2})$ not quadratically integrable. Substitution of $c = \pi \nu m/\hbar$ into (8.14) gives $W = \frac{1}{2}h\nu$. This is the exact ground-state harmonic-oscillator energy and corresponds to the fact that with $c = \pi \nu m/\hbar$ the variation function ϕ is the same (except for being unnormalized) as the harmonic-oscillator ground-state wave function (4.59) and (4.33).

For the normalized harmonic-oscillator variation function $\phi = (2c/\pi)^{1/4} \times e^{-cx^2}$, a large value of c makes ϕ fall off very rapidly from its maximum value at $x = 0$. This makes the probability density large only near $x = 0$. The potential energy $V = \frac{1}{2}kx^2$ is low near $x = 0$, so a large c means a low $\langle V \rangle = \langle \phi | V | \phi \rangle$. [Note also that $\langle V \rangle$ equals the second term on the right side of (8.14).] However, because a large c makes ϕ fall off very rapidly from its maximum, it makes $|d\phi/dx|$ large in the region near $x = 0$. From Problem 7.4b, a large $|d\phi/dx|$ means a large value of $\langle T \rangle$ [which equals the first term on the right side of (8.14)]. The optimum value of c minimizes the sum $\langle T \rangle + \langle V \rangle = W$. In atoms and molecules, the true wave function is a compromise between the tendency to minimize $\langle V \rangle$ by confining the electrons to regions of low V (near the nuclei) and the tendency to minimize $\langle T \rangle$ by allowing the electron probability density to spread out over a large region.

8.2 EXTENSION OF THE VARIATION METHOD

The variation method as presented in the last section has two major limitations. First, it provides information about only the *ground*-state energy and wave function. Second, it provides only an upper bound to the ground-state energy. We now discuss some extensions of the variation method. (See also Section 8.5.)

Consider how we might extend the variation method to obtain an estimate for the energy of the first excited state. We number the stationary states of the system 1, 2, 3, . . . in order of increasing energy:

$$E_1 \le E_2 \le E_3 \le \cdots$$

We showed that for a normalized variational function ϕ [Eqs. (8.4) and (8.6)]

$$\int \phi^* \hat{H} \phi \, d\tau = \sum_{k=1}^{\infty} |a_k|^2 E_k \quad \text{and} \quad \int \phi^* \phi \, d\tau = \sum_{k=1}^{\infty} |a_k|^2 = 1$$

where the a_k's are the expansion coefficients in $\phi = \sum_k a_k \psi_k$ [Eq. (8.2)]. We have $a_k = \langle \psi_k | \phi \rangle$ [Eq. (7.40)]. Let us restrict ourselves to normalized functions ϕ that are orthogonal to the true ground-state wave function ψ_1. Then we have $a_1 = \langle \psi_1 | \phi \rangle = 0$ and

$$\int \phi^* \hat{H} \phi \, d\tau = \sum_{k=2}^{\infty} |a_k|^2 E_k \quad \text{and} \quad \int \phi^* \phi \, d\tau = \sum_{k=2}^{\infty} |a_k|^2 = 1 \qquad (8.16)$$

For $k \ge 2$, we have $E_k \ge E_2$ and $|a_k|^2 E_k \ge |a_k|^2 E_2$. Hence

$$\sum_{k=2}^{\infty} |a_k|^2 E_k \ge \sum_{k=2}^{\infty} |a_k|^2 E_2 = E_2 \sum_{k=2}^{\infty} |a_k|^2 = E_2 \qquad (8.17)$$

Combining (8.16) and (8.17), we have the desired result:

$$\int \phi^* \hat{H} \phi \, d\tau \ge E_2, \quad \text{if} \quad \int \psi_1^* \phi \, d\tau = 0 \quad \text{and} \quad \int \phi^* \phi \, d\tau = 1 \qquad (8.18)$$

The inequality (8.18) provides a way to obtain an upper bound to the energy E_2 of the first excited state. However, the restriction $\langle \psi_1 | \phi \rangle = 0$ makes this method troublesome to apply.

For certain systems, it is possible to be sure that $\langle \psi_1 | \phi \rangle = 0$ even though we do not know the true ground-state wave function. An example is a one-dimensional problem for which V is an even function of x. In this case the ground-state wave function is always an even function, while the first excited-state wave function is odd. (All the wave functions must be of definite parity. The ground-state wave function is nodeless, and, since an odd function vanishes at the origin, the ground-state wave function must be even. The first excited-state wave function has one node and must be odd.) Therefore, for odd trial functions, it must be true that $\langle \psi_1 | \phi \rangle = 0$; the even function ψ_1 times the odd function ϕ gives an odd integrand whose integral from $-\infty$ to ∞ is zero.

Another example is a particle moving in a central field (Section 6.1). The form of the potential energy might be such that we could not solve for the radial factor $R(r)$ in the eigenfunction. However, we know that the angular factor is a spherical harmonic [Eq. (6.16)] and that spherical harmonics with different values of l are orthogonal. Thus we can get an upper bound to the energy of the lowest state with any given angular momentum l by using the factor Y_l^m in the trial function. This result depends on the extension of (8.18) to higher excited states:

$$\frac{\int \phi^* \hat{H} \phi \, d\tau}{\int \phi^* \phi \, d\tau} \geq E_{k+1}, \quad \text{if} \quad \int \psi_1^* \phi \, d\tau = \int \psi_2^* \phi \, d\tau = \cdots = \int \psi_k^* \phi \, d\tau = 0$$

(8.19)

8.3 DETERMINANTS

We shall shortly consider a kind of variational function that gives rise to an equation involving a determinant. Therefore, we now review the properties of determinants.

A **determinant** is a square array of n^2 quantities (called **elements**); the value of the determinant is calculated from its elements in a manner to be given shortly. The number n is the **order** of the determinant. Using a_{ij} to represent a typical element, we write the nth-order determinant as

$$\det (a_{ij}) = \begin{vmatrix} a_{11} & a_{12} & a_{13} & \cdots & a_{1n} \\ a_{21} & a_{22} & a_{23} & \cdots & a_{2n} \\ a_{31} & a_{32} & a_{33} & \cdots & a_{3n} \\ \cdot & \cdot & \cdot & \cdots & \cdot \\ \cdot & \cdot & \cdot & \cdots & \cdot \\ \cdot & \cdot & \cdot & \cdots & \cdot \\ a_{n1} & a_{n2} & a_{n3} & \cdots & a_{nn} \end{vmatrix}$$

(8.20)

The vertical lines in (8.20) have nothing to do with absolute value. Before considering how the value of the nth-order determinant is defined, we consider determinants of first, second, and third orders.

A first-order determinant has one element, and its value is simply the value of that element. Thus

$$|a_{11}| = a_{11} \qquad (8.21)$$

where the vertical lines indicate a determinant and not absolute value.

A second-order determinant has four elements, and its value is defined by

$$\begin{vmatrix} a_{11} & a_{12} \\ a_{21} & a_{22} \end{vmatrix} = a_{11}a_{22} - a_{12}a_{21} \qquad \mathbf{(8.22)^*}$$

The value of a third-order determinant is defined by

$$\begin{vmatrix} a_{11} & a_{12} & a_{13} \\ a_{21} & a_{22} & a_{23} \\ a_{31} & a_{32} & a_{33} \end{vmatrix} = a_{11}\begin{vmatrix} a_{22} & a_{23} \\ a_{32} & a_{33} \end{vmatrix} - a_{12}\begin{vmatrix} a_{21} & a_{23} \\ a_{31} & a_{33} \end{vmatrix} + a_{13}\begin{vmatrix} a_{21} & a_{22} \\ a_{31} & a_{32} \end{vmatrix} \qquad \mathbf{(8.23)^*}$$

$$= a_{11}a_{22}a_{33} - a_{11}a_{32}a_{23} - a_{12}a_{21}a_{33} + a_{12}a_{31}a_{23}$$

$$+ a_{13}a_{21}a_{32} - a_{13}a_{31}a_{22} \qquad (8.24)$$

A third-order determinant is evaluated by writing down the elements of the top row with alternating plus and minus signs and then multiplying each element by a certain second-order determinant; the second-order determinant that multiplies a given element is found by crossing out the row and column of the third-order determinant in which that element appears. The $(n-1)$-order determinant obtained by striking out the ith row and the jth column of the nth-order determinant is called the ***minor*** of the element a_{ij}. We define the ***cofactor*** of a_{ij} as the minor of a_{ij} times the factor $(-1)^{i+j}$. Thus (8.23) states that the third-order determinant is evaluated by multiplying each element of the top row by its cofactor and adding up the three products. [Note that (8.22) conforms to this evaluation by means of cofactors, since the cofactor of a_{11} in (8.22) is a_{22}, and the cofactor of a_{12} is $-a_{21}$.] A numerical example is

$$\begin{vmatrix} 5 & 10 & 2 \\ 0.1 & 3 & 1 \\ 0 & 4 & 4 \end{vmatrix} = 5\begin{vmatrix} 3 & 1 \\ 4 & 4 \end{vmatrix} - 10\begin{vmatrix} 0.1 & 1 \\ 0 & 4 \end{vmatrix} + 2\begin{vmatrix} 0.1 & 3 \\ 0 & 4 \end{vmatrix}$$

$$= 5(8) - 10(0.4) + 2(0.4) = 36.8$$

Denoting the minor of a_{ij} by M_{ij} and the cofactor of a_{ij} by C_{ij}, we have

$$C_{ij} = (-1)^{i+j}M_{ij} \qquad (8.25)$$

The expansion (8.23) of the third-order determinant can be written as

$$\det(a_{ij}) = \begin{vmatrix} a_{11} & a_{12} & a_{13} \\ a_{21} & a_{22} & a_{23} \\ a_{31} & a_{32} & a_{33} \end{vmatrix} = a_{11}C_{11} + a_{12}C_{12} + a_{13}C_{13} \qquad (8.26)$$

A third-order determinant can be expanded using the elements of any row and the corresponding cofactors. For example, using the second row to expand the third-order determinant, we have

$$\det{(a_{ij})} = a_{21}C_{21} + a_{22}C_{22} + a_{23}C_{23} \tag{8.27}$$

$$\det{(a_{ij})} = -a_{21}\begin{vmatrix} a_{12} & a_{13} \\ a_{32} & a_{33} \end{vmatrix} + a_{22}\begin{vmatrix} a_{11} & a_{13} \\ a_{31} & a_{33} \end{vmatrix} - a_{23}\begin{vmatrix} a_{11} & a_{12} \\ a_{31} & a_{32} \end{vmatrix} \tag{8.28}$$

and expansion of the second-order determinants shows that (8.28) is equal to (8.24). We may also use the elements of any column and the corresponding cofactors to expand the determinant, as can be readily verified. Thus for the third-order determinant, we can write

$$\det{(a_{ij})} = a_{k1}C_{k1} + a_{k2}C_{k2} + a_{k3}C_{k3} = \sum_{l=1}^{3} a_{kl}C_{kl}, \qquad k = 1 \text{ or } 2 \text{ or } 3$$

$$\det{(a_{ij})} = a_{1k}C_{1k} + a_{2k}C_{2k} + a_{3k}C_{3k} = \sum_{l=1}^{3} a_{lk}C_{lk}, \qquad k = 1 \text{ or } 2 \text{ or } 3$$

The first expansion uses one of the rows, while the second uses one of the columns.

We now define determinants of higher order by an analogous row (or column) expansion. Thus a fourth-order determinant is defined by $\det{(a_{ij})} = \sum_{l=1}^{4} a_{kl}C_{kl} = \sum_{l=1}^{4} a_{lk}C_{lk}$ with $k = 1$ or 2 or 3 or 4. For an nth-order determinant,

$$\det{(a_{ij})} = \sum_{l=1}^{n} a_{kl}C_{kl} = \sum_{l=1}^{n} a_{lk}C_{lk}, \qquad k = 1 \text{ or } 2 \text{ or } \ldots \text{ or } n \tag{8.29}$$

Some theorems on determinants are (for proofs, see *Sokolnikoff and Redheffer*, pp. 702–707):

   I. If every element of a row (or column) of a determinant is zero, the value of the determinant is zero.
  II. Interchanging any two rows (or columns) multiplies the value of a determinant by -1.
 III. If any two rows (or columns) of a determinant are identical, the determinant has the value zero.
  IV. Multiplication of each element of any one row (or any one column) by some constant k multiplies the value of the determinant by k.
   V. Addition to each element of one row of the same constant multiple of the corresponding element of another row leaves the value of the determinant unchanged. (This theorem also applies to the addition of a multiple of one column to another column.)
  VI. The interchange of all corresponding rows and columns leaves the value of the determinant unchanged.

EXAMPLE   Use Theorem V to evaluate the determinant

$$B = \begin{vmatrix} 1 & 2 & 3 & 4 \\ 4 & 1 & 2 & 3 \\ 3 & 4 & 1 & 2 \\ 2 & 3 & 4 & 1 \end{vmatrix} \tag{8.30}$$

Addition of -2 times the elements of row one to the corresponding elements of row four changes row four to $0, -1, -2, -7$. Then, addition of -3 times row one to row three and -4 times row one to row two gives

$$B = \begin{vmatrix} 1 & 2 & 3 & 4 \\ 0 & -7 & -10 & -13 \\ 0 & -2 & -8 & -10 \\ 0 & -1 & -2 & -7 \end{vmatrix} = 1 \begin{vmatrix} -7 & -10 & -13 \\ -2 & -8 & -10 \\ -1 & -2 & -7 \end{vmatrix}$$

where we expanded B in terms of elements of the first column. Subtracting twice row three from row two and seven times row three from row one, we have

$$B = \begin{vmatrix} 0 & 4 & 36 \\ 0 & -4 & 4 \\ -1 & -2 & -7 \end{vmatrix} = (-1) \begin{vmatrix} 4 & 36 \\ -4 & 4 \end{vmatrix} = -(16 + 144) = -160 \qquad (8.31)$$

The diagonal of a determinant that runs from the top left to the lower right is the **principal diagonal**. A determinant all of whose elements are zero except those on the principal diagonal is said to be in **diagonal form**; for such a determinant:

$$\begin{vmatrix} a_{11} & 0 & 0 & \cdots & 0 \\ 0 & a_{22} & 0 & \cdots & 0 \\ 0 & 0 & a_{33} & \cdots & 0 \\ . & . & . & \cdots & . \\ 0 & 0 & 0 & \cdots & a_{nn} \end{vmatrix} = a_{11} \begin{vmatrix} a_{22} & 0 & \cdots & 0 \\ 0 & a_{33} & \cdots & 0 \\ . & . & \cdots & . \\ 0 & 0 & \cdots & a_{nn} \end{vmatrix}$$

$$= a_{11}a_{22} \begin{vmatrix} a_{33} & 0 & \cdots & 0 \\ 0 & a_{44} & \cdots & 0 \\ . & . & \cdots & . \\ 0 & 0 & \cdots & a_{nn} \end{vmatrix}$$

$$= \cdots = a_{11}a_{22}a_{33} \cdots a_{nn} \qquad (8.32)$$

A diagonal determinant is equal to the product of its diagonal elements.

A determinant whose only nonzero elements occur in square blocks centered about the principal diagonal is in **block-diagonal form**. If we regard each square block as a determinant, then a block-diagonal determinant is equal to the product of the blocks. For example,

$$\begin{vmatrix} a & b & 0 & 0 & 0 & 0 \\ c & d & 0 & 0 & 0 & 0 \\ 0 & 0 & e & 0 & 0 & 0 \\ 0 & 0 & 0 & f & g & h \\ 0 & 0 & 0 & i & j & k \\ 0 & 0 & 0 & l & m & n \end{vmatrix} = \begin{vmatrix} a & b \\ c & d \end{vmatrix} (e) \begin{vmatrix} f & g & h \\ i & j & k \\ l & m & n \end{vmatrix} \qquad (8.33)$$

The dashed lines outline the blocks. The proof of (8.33) is simple. Expanding in terms of elements of the top row, we have

$$
\begin{vmatrix} a & b & 0 & \cdots & 0 \\ c & d & 0 & \cdots & 0 \\ 0 & 0 & e & \cdots & \\ \cdot & \cdot & \cdot & \cdots & \cdot \\ \cdot & \cdot & \cdot & \cdots & \cdot \\ \cdot & \cdot & \cdot & \cdots & \cdot \\ 0 & 0 & \cdot & \cdots & \cdot \end{vmatrix} = a \begin{vmatrix} d & 0 & \cdots & 0 \\ 0 & e & \cdots & \cdot \\ \cdot & \cdot & \cdots & \cdot \\ \cdot & \cdot & \cdots & \cdot \\ 0 & \cdot & \cdots & \cdot \end{vmatrix} - b \begin{vmatrix} c & 0 & \cdots & 0 \\ 0 & e & \cdots & \cdot \\ \cdot & \cdot & \cdots & \cdot \\ \cdot & \cdot & \cdots & \cdot \\ 0 & \cdot & \cdots & \cdot \end{vmatrix}
$$

and expanding the two determinants on the right side of this equation in terms of elements of the top rows, we get

$$
ad \begin{vmatrix} e & \cdot & \cdot & \cdot \\ \cdot & \cdot & \cdot & \cdot \\ \cdot & \cdot & \cdot & \cdot \\ \cdot & \cdot & \cdot & \cdot \end{vmatrix} - bc \begin{vmatrix} e & \cdot & \cdot & \cdot \\ \cdot & \cdot & \cdot & \cdot \\ \cdot & \cdot & \cdot & \cdot \\ \cdot & \cdot & \cdot & \cdot \end{vmatrix} = \begin{vmatrix} a & b \\ c & d \end{vmatrix} \begin{vmatrix} e & \cdot & \cdot & \cdot \\ \cdot & \cdot & \cdot & \cdot \\ \cdot & \cdot & \cdot & \cdot \\ \cdot & \cdot & \cdot & \cdot \end{vmatrix} \tag{8.34}
$$

A similar expansion of the second determinant on the right side of (8.34) completes the proof.

8.4 SIMULTANEOUS LINEAR EQUATIONS

To deal with the kind of variation function discussed in the next section, we need to know about simultaneous linear equations.

Consider the following system of n linear equations in n unknowns:

$$
\begin{aligned}
a_{11}x_1 + a_{12}x_2 + \cdots + a_{1n}x_n &= b_1 \\
a_{21}x_1 + a_{22}x_2 + \cdots + a_{2n}x_n &= b_2 \\
&\cdots \\
a_{n1}x_1 + a_{n2}x_2 + \cdots + a_{nn}x_n &= b_n
\end{aligned} \tag{8.35}
$$

where the a's and b's are known constants and $x_1, x_2, \ldots, x_n$ are the unknowns. If at least one of the b's is not zero, we have a system of **inhomogeneous** linear equations. Such a system can be solved by Cramer's rule. (For a proof of Cramer's rule, see *Sokolnikoff and Redheffer*, p. 708.) Let $\det(a_{ij})$ be the determinant of the coefficients of the unknowns in (8.35). *Cramer's rule* states that x_k $(k = 1, 2, \ldots, n)$ is given by

$$
x_k = \frac{\begin{vmatrix} a_{11} & a_{12} & \cdots & a_{1,k-1} & b_1 & a_{1,k+1} & \cdots & a_{1n} \\ a_{21} & a_{22} & \cdots & a_{2,k-1} & b_2 & a_{2,k+1} & \cdots & a_{2n} \\ \cdot & \cdot & \cdots & \cdot & \cdot & \cdot & \cdots & \cdot \\ a_{n1} & a_{n2} & \cdots & a_{n,k-1} & b_n & a_{n,k+1} & \cdots & a_{nn} \end{vmatrix}}{\det(a_{ij})}, \quad k = 1, 2, \ldots, n \tag{8.36}
$$

where $\det(a_{ij})$ is given by (8.20) and the numerator is the determinant obtained by replacing the kth column of $\det(a_{ij})$ with the elements $b_1, b_2, \ldots, b_n$.

Although Cramer's rule is of theoretical significance, it should not be used for numerical calculations, since successive elimination of unknowns is much more efficient.

A widely used successive-elimination procedure is **Gaussian elimination**, which proceeds as follows: Divide the first equation in (8.35) by the coefficient a_{11} of x_1, thereby making the coefficient of x_1 equal to 1 in this equation. Then subtract a_{21} times the first equation from the second equation, subtract a_{31} times the first equation from the third equation, ..., and subtract a_{n1} times the first equation from the nth equation. This eliminates x_1 from all equations but the first. Now divide the second equation by the coefficient of x_2; then subtract appropriate multiples of the second equation from the 3rd, 4th, ..., nth equations, so as to eliminate x_2 from all equations but the first and second. Continue in this manner. Ultimately, equation n will contain only x_n, equation $n-1$ only x_{n-1} and x_n, and so on. The value of x_n found from equation n is substituted into equation $n-1$ to give x_{n-1}; the values of x_n and x_{n-1} are substituted into equation $n-2$ to give x_{n-2}; and so on. If at any stage a coefficient we want to divide by happens to be zero, the equation with the zero coefficient is exchanged with a later equation that has a nonzero coefficient in the desired position. (The Gaussian elimination procedure also gives an efficient way to evaluate a determinant; see Problem 8.20.)

A related method is **Gauss–Jordan elimination**, which proceeds the same way as Gaussian elimination, except that instead of eliminating x_2 from equations $3, 4, \ldots, n$, we eliminate x_2 from equations $1, 3, 4, \ldots, n$ by subtracting appropriate multiples of the second equation from equations $1, 3, 4, \ldots, n$; instead of eliminating x_3 from equations $4, 5, \ldots, n$, we eliminate x_3 from equations $1, 2, 4, 5, \ldots, n$; and so on. At the end of Gauss–Jordan elimination, equation 1 contains only x_1, equation 2 contains only $x_2, \ldots$, equation n contains only x_n. Gauss–Jordan elimination requires more computation than Gaussian elimination.

If all the b's in (8.35) are zero, we have a system of **linear homogeneous equations**:

$$a_{11}x_1 + a_{12}x_2 + \cdots + a_{1n}x_n = 0$$
$$a_{21}x_1 + a_{22}x_2 + \cdots + a_{2n}x_n = 0$$
$$\cdots \cdots \cdots \cdots \cdots \cdots \cdots \cdots \cdots \tag{8.37}$$
$$a_{n1}x_1 + a_{n2}x_2 + \cdots + a_{nn}x_n = 0$$

One obvious solution of (8.37) is $x_1 = x_2 = \cdots = x_n = 0$, which is called the **trivial solution**. If the determinant of the coefficients in (8.37) is not equal to zero, $\det(a_{ij}) \neq 0$, then we can use Cramer's rule (8.36) to solve for the unknowns, and we find $x_k = 0$, $k = 1, 2, \ldots, n$, since the determinant in the numerator of (8.36) has a column all of whose elements are zero. Thus, when $\det(a_{ij}) \neq 0$, the only solution is the trivial solution, which is of no interest. For there to be a nontrivial solution of a system of n linear homogeneous equations in n unknowns, the determinant of the coefficients must be zero. Also, this condition can be shown to be sufficient to ensure the existence of a nontrivial solution (see T. L. Wade, *The Algebra of Vectors and Matrices*, Addison-Wesley, Reading, Mass., 1951, p. 146). We thus have the extremely important theorem:

A system of n linear homogeneous equations in n unknowns has a nontrivial solution if and only if the determinant of the coefficients is zero.

Suppose that $\det(a_{ij}) = 0$, so that (8.37) has a nontrivial solution. How do we find it? With $\det(a_{ij}) = 0$, Cramer's rule (8.36) gives $x_k = 0/0$, $k = 1, \ldots, n$, which is indeterminate. Thus Cramer's rule is of no immediate help. We also observe that, if $x_1 = d_1$, $x_2 = d_2, \ldots, x_n = d_n$ is a solution of (8.37), then so is $x_1 = cd_1$, $x_2 = cd_2, \ldots, x_n = cd_n$, where c is an arbitrary constant. This is easily seen, since

$$a_{11}cd_1 + a_{12}cd_2 + \cdots + a_{1n}cd_n = c(a_{11}d_1 + a_{12}d_2 + \cdots + a_{1n}d_n) = c \cdot 0 = 0$$

and so on. Therefore, the solution to the linear homogeneous system of equations will contain an arbitrary constant, and we cannot determine a unique value for each unknown. To solve (8.37), we therefore assign an arbitrary value to any one of the unknowns, say x_n; we set $x_n = c$, where c is an arbitrary constant. Having assigned a value to x_n, we transfer the last term in each of the equations of (8.37) to the right side to get

$$
\begin{aligned}
a_{11}x_1 \quad + a_{12}x_2 \quad + \cdots + a_{1,n-1}x_{n-1} \quad &= -a_{1,n}c \\
a_{21}x_1 \quad + a_{22}x_2 \quad + \cdots + a_{2,n-1}x_{n-1} \quad &= -a_{2,n}c \\
\cdot \quad \cdot \qquad \cdot \quad \cdot \qquad\qquad \cdot \quad \cdot \qquad\quad \cdot \quad \cdot \quad &\qquad\qquad (8.38) \\
a_{n-1,1}x_1 + a_{n-1,2}x_2 + \cdots + a_{n-1,n-1}x_{n-1} &= -a_{n-1,n}c \\
a_{n1}x_1 \quad + a_{n2}x_2 \quad + \cdots + a_{n,n-1}x_{n-1} \quad &= -a_{nn}c
\end{aligned}
$$

We now have n equations in $n - 1$ unknowns, which is one more equation than we need. We therefore discard any one of the equations of (8.38), say the last one. This gives a system of $n - 1$ linear *in*homogeneous equations in $n - 1$ unknowns. We could then apply Cramer's rule (8.36) to solve for $x_1, x_2, \ldots, x_{n-1}$. Since the constants on the right side of the equations in (8.38) all contain the factor c, Theorem IV in Section 8.3 shows that all the unknowns contain this arbitrary constant as a factor. The form of the solution is therefore

$$x_1 = ce_1, \quad x_2 = ce_2, \quad \ldots, \quad x_{n-1} = ce_{n-1}, \quad x_n = c \qquad (8.39)$$

where $e_1, \ldots, e_{n-1}$ are numbers and c is an arbitrary constant.

The procedure just outlined fails if the determinant of the inhomogeneous system of $n - 1$ equations in $n - 1$ unknowns [(8.38) with the last equation omitted] happens to be zero. Cramer's rule then has a zero in the denominator and is of no use. We could attempt to get around this difficulty by initially assigning the arbitrary value to another of the unknowns rather than to x_n. We could also try discarding some other equation of (8.38), rather than the last one. What we are looking for, really, is a nonvanishing determinant of order $n - 1$, formed from the determinant of the coefficients of the system (8.37) by striking out a row and a column. If such a determinant exists, then by the procedure given, with the right choice of the equation to be discarded and the right choice of the unknown to be assigned an arbitrary value, we can solve the system and will obtain solutions of the form (8.39). If no such determinant exists, we must assign arbitrary values to two of the unknowns and attempt to proceed from there. Thus the solution to (8.37) might contain two (or even more) arbitrary constants.

An efficient way to solve a system of linear homogeneous equations is to perform Gauss–Jordan elimination on the equations. If only the trivial solution exists, the final set of equations obtained will be $x_1 = 0$, $x_2 = 0$, ..., $x_n = 0$. If a nontrivial solution exists, at least one equation will be reduced to the form $0 = 0$; if m equations of the form $0 = 0$ are obtained, we assign arbitrary constants to m of the unknowns and express the remaining unknowns in terms of these m unknowns.

EXAMPLE   Solve

$$3x_1 + 4x_2 + x_3 = 0$$
$$x_1 + 3x_2 - 2x_3 = 0$$
$$x_1 - 2x_2 + 5x_3 = 0$$

In doing Gaussian or Gauss–Jordan elimination on a set of n inhomogeneous or homogeneous equations, we can eliminate needless writing by omitting the variables $x_1, \ldots, x_n$ and writing down only the n-row, $(n + 1)$-column array of coefficients and constant terms (including any zero coefficients); we then produce the next array by operating on the numbers of each row as if that row were the equation it represents.

To eliminate one set of divisions, we interchange the first and second equations so that we start off with $a_{11} = 1$. Detaching the coefficients and proceeding with Gauss–Jordan elimination, we have

$$
\begin{array}{cccc}
1 & 3 & -2 & 0 \\
3 & 4 & 1 & 0 \\
1 & -2 & 5 & 0
\end{array}
\rightarrow
\begin{array}{cccc}
1 & 3 & -2 & 0 \\
0 & -5 & 7 & 0 \\
0 & -5 & 7 & 0
\end{array}
\rightarrow
\begin{array}{cccc}
1 & 3 & -2 & 0 \\
0 & 1 & -\frac{7}{5} & 0 \\
0 & -5 & 7 & 0
\end{array}
\rightarrow
\begin{array}{cccc}
1 & 0 & \frac{11}{5} & 0 \\
0 & 1 & -\frac{7}{5} & 0 \\
0 & 0 & 0 & 0
\end{array}
$$

The first array is the original set of equations with the first and second equations interchanged. To eliminate x_1 from the second and third equations, we subtract 3 times row one from row two and 1 times row one from row three, thereby producing the second array. Division of row two by -5 produces the third array. To eliminate x_2 from the first and third equations, we subtract 3 times row two from row one and -5 times row two from row three, thereby producing the fourth array. Because the fourth array has the x_3 coefficient in row three equal to zero, we cannot use row three to eliminate x_3 from rows one and two (as would be the last step in the Gauss–Jordan algorithm). Discarding the last equation, which reads $0 = 0$, we assign $x_3 = k$, where k is an arbitrary constant. The first and second equations in the last array read $x_1 + \frac{11}{5}x_3 = 0$ and $x_2 - \frac{7}{5}x_3 = 0$, or $x_1 = -\frac{11}{5}x_3$, $x_2 = \frac{7}{5}x_3$. Hence the general solution is $x_1 = -\frac{11}{5}k$, $x_2 = \frac{7}{5}k$, $x_3 = k$. For those allergic to fractions, we define a new arbitrary constant s as $s \equiv \frac{1}{5}k$ and write $x_1 = -11s$, $x_2 = 7s$, $x_3 = 5s$.

8.5 LINEAR VARIATION FUNCTIONS

A special kind of variation function widely used in the study of molecules is the linear variation function. A *linear variation function* is a linear combination of n linearly independent functions $f_1, f_2, \ldots, f_n$:

$$\phi = c_1 f_1 + c_2 f_2 + \cdots + c_n f_n = \sum_{j=1}^{n} c_j f_j \tag{8.40}$$

where ϕ is the trial variation function and the coefficients c_j are parameters to be determined by minimizing the variational integral. The functions f_j must satisfy the boundary conditions of the problem. We shall restrict ourselves to *real* ϕ so that the c_j's and f_j's are all real.

We now apply the variation theorem (8.9). For the real linear variation function, we have

$$\int \phi^* \phi \, d\tau = \int \sum_{j=1}^{n} c_j f_j \sum_{k=1}^{n} c_k f_k \, d\tau = \sum_{j=1}^{n} \sum_{k=1}^{n} c_j c_k \int f_j f_k \, d\tau$$

We define the *overlap integral* S_{jk} as

$$S_{jk} \equiv \int f_j^* f_k \, d\tau \tag{8.41}*$$

We have

$$\int \phi^* \phi \, d\tau = \sum_{j=1}^{n} \sum_{k=1}^{n} c_j c_k S_{jk} \tag{8.42}$$

Note that S_{jk} is not necessarily equal to δ_{jk}, since there is no reason to suppose that the functions f_j are mutually orthogonal. They are not necessarily the eigenfunctions of any operator. The numerator of (8.9) is

$$\int \phi^* \hat{H} \phi \, d\tau = \int \sum_{j=1}^{n} c_j f_j \hat{H} \sum_{k=1}^{n} c_k f_k \, d\tau = \sum_{j=1}^{n} \sum_{k=1}^{n} c_j c_k \int f_j \hat{H} f_k \, d\tau$$

and using the abbreviation

$$H_{jk} \equiv \int f_j^* \hat{H} f_k \, d\tau \tag{8.43}*$$

we write

$$\int \phi^* \hat{H} \phi \, d\tau = \sum_{j=1}^{n} \sum_{k=1}^{n} c_j c_k H_{jk} \tag{8.44}$$

The variational integral W is

$$W \equiv \frac{\int \phi^* \hat{H} \phi \, d\tau}{\int \phi^* \phi \, d\tau} = \frac{\sum_{j=1}^{n} \sum_{k=1}^{n} c_j c_k H_{jk}}{\sum_{j=1}^{n} \sum_{k=1}^{n} c_j c_k S_{jk}} \tag{8.45}$$

$$W \sum_{j=1}^{n} \sum_{k=1}^{n} c_j c_k S_{jk} = \sum_{j=1}^{n} \sum_{k=1}^{n} c_j c_k H_{jk} \tag{8.46}$$

We now minimize W so as to approach as closely as we can to E_1 ($W \geqslant E_1$). The variational integral W is a function of the n independent variables $c_1, c_2, \ldots, c_n$:

$$W = W(c_1, c_2, \ldots, c_n)$$

A necessary condition for a minimum in a function W of several variables is that its partial derivatives with respect to each of the variables must be zero at the minimum point:

$$\frac{\partial W}{\partial c_i} = 0, \qquad i = 1, 2, \ldots, n \qquad (8.47)$$

We now differentiate (8.46) partially with respect to each c_i to obtain n equations:

$$\frac{\partial W}{\partial c_i} \sum_{j=1}^{n} \sum_{k=1}^{n} c_j c_k S_{jk} + W \frac{\partial}{\partial c_i} \sum_{j=1}^{n} \sum_{k=1}^{n} c_j c_k S_{jk}$$

$$= \frac{\partial}{\partial c_i} \sum_{j=1}^{n} \sum_{k=1}^{n} c_j c_k H_{jk}, \qquad i = 1, 2, \ldots, n \qquad (8.48)$$

Now

$$\frac{\partial}{\partial c_i} \sum_{j=1}^{n} \sum_{k=1}^{n} c_j c_k S_{jk} = \sum_{j=1}^{n} \sum_{k=1}^{n} \left[\frac{\partial}{\partial c_i} (c_j c_k) \right] S_{jk} = \sum_{j=1}^{n} \sum_{k=1}^{n} \left(c_k \frac{\partial c_j}{\partial c_i} + c_j \frac{\partial c_k}{\partial c_i} \right) S_{jk}$$

The c_j's are independent variables, and therefore

$$\frac{\partial c_j}{\partial c_i} = 0 \;\; \text{if } j \neq i \qquad \frac{\partial c_j}{\partial c_i} = 1 \;\; \text{if } j = i$$

$$\frac{\partial c_j}{\partial c_i} = \delta_{ij}$$

We then have

$$\frac{\partial}{\partial c_i} \sum_{j=1}^{n} \sum_{k=1}^{n} c_j c_k S_{jk} = \sum_{k=1}^{n} \sum_{j=1}^{n} c_k \delta_{ij} S_{jk} + \sum_{j=1}^{n} \sum_{k=1}^{n} c_j \delta_{ik} S_{jk}$$

$$= \sum_{k=1}^{n} c_k S_{ik} + \sum_{j=1}^{n} c_j S_{ji} \qquad (8.49)$$

where we have evaluated one of the sums in each double summation using Eq. (7.32). Now, according to Eq. (7.4), we have

$$S_{ji} = S_{ij}^* = S_{ij} \qquad (8.50)$$

where the last equality follows because we are dealing with real functions. Hence,

$$\frac{\partial}{\partial c_i} \sum_{j=1}^{n} \sum_{k=1}^{n} c_j c_k S_{jk} = \sum_{k=1}^{n} c_k S_{ik} + \sum_{j=1}^{n} c_j S_{ij} = \sum_{k=1}^{n} c_k S_{ik} + \sum_{k=1}^{n} c_k S_{ik}$$

$$\frac{\partial}{\partial c_i} \sum_{j=1}^{n} \sum_{k=1}^{n} c_j c_k S_{jk} = 2 \sum_{k=1}^{n} c_k S_{ik} \qquad (8.51)$$

where the fact that j is a dummy variable was used.

By replacing S_{jk} by H_{jk} in each of these manipulations, we get

$$\frac{\partial}{\partial c_i} \sum_{j=1}^{n} \sum_{k=1}^{n} c_j c_k H_{jk} = 2 \sum_{k=1}^{n} c_k H_{ik} \qquad (8.52)$$

This result depends on the fact that

$$H_{ji} = H_{ij}^* = H_{ij} \tag{8.53}$$

which is true because $\hat{H}$ is a Hermitian operator, and we are dealing with real functions and a real Hamiltonian.

Substitution of Eqs. (8.47), (8.51), and (8.52) into (8.48) gives

$$2W \sum_{k=1}^{n} c_k S_{ik} = 2 \sum_{k=1}^{n} c_k H_{ik} , \qquad i = 1, 2, \ldots, n$$

$$\sum_{k=1}^{n} [(H_{ik} - S_{ik}W)c_k] = 0 , \qquad i = 1, 2, \ldots, n \tag{8.54}$$

Equation (8.54) is a set of n simultaneous, linear, homogeneous equations in the n unknowns $c_1, c_2, \ldots, c_n$ [the coefficients in the linear variation function (8.40)]. For example, for $n = 2$, (8.54) gives

$$(H_{11} - S_{11}W)c_1 + (H_{12} - S_{12}W)c_2 = 0$$
$$(H_{21} - S_{21}W)c_1 + (H_{22} - S_{22}W)c_2 = 0 \tag{8.55}$$

For the general case of n functions $f_1, \ldots, f_n$, (8.54) is

$$(H_{11} - S_{11}W)c_1 + (H_{12} - S_{12}W)c_2 + \cdots + (H_{1n} - S_{1n}W)c_n = 0$$
$$(H_{21} - S_{21}W)c_1 + (H_{22} - S_{22}W)c_2 + \cdots + (H_{2n} - S_{2n}W)c_n = 0$$
$$\cdots\cdots\cdots\cdots\cdots\cdots\cdots\cdots\cdots\cdots\cdots\cdots\cdots\cdots\cdots \tag{8.56}$$
$$(H_{n1} - S_{n1}W)c_1 + (H_{n2} - S_{n2}W)c_2 + \cdots + (H_{nn} - S_{nn}W)c_n = 0$$

From the theorem of the last section, for there to be a solution to the system of linear homogeneous equations (8.56) besides the trivial solution $0 = c_1 = c_2 = \cdots = c_n$ (which would make the variation function ϕ zero), the determinant of the coefficients must vanish. For $n = 2$, we have

$$\begin{vmatrix} H_{11} - S_{11}W & H_{12} - S_{12}W \\ H_{21} - S_{21}W & H_{22} - S_{22}W \end{vmatrix} = 0 \tag{8.57}$$

and for the general case

$$\det (H_{ij} - S_{ij}W) = 0 \tag{8.58}^*$$

$$\begin{vmatrix} H_{11} - S_{11}W & H_{12} - S_{12}W & \cdots & H_{1n} - S_{1n}W \\ H_{21} - S_{21}W & H_{22} - S_{22}W & \cdots & H_{2n} - S_{2n}W \\ \cdot & \cdot & \cdots & \cdot \\ \cdot & \cdot & \cdots & \cdot \\ \cdot & \cdot & \cdots & \cdot \\ H_{n1} - S_{n1}W & H_{n2} - S_{n2}W & \cdots & H_{nn} - S_{nn}W \end{vmatrix} = 0 \tag{8.59}$$

Expansion of the determinant in (8.59) gives an algebraic equation of degree n in the unknown W. This algebraic equation has n roots, which can be shown to be real. Arranging these roots in order of increasing value, we denote them as

$$W_1 \leqslant W_2 \leqslant \cdots \leqslant W_n \tag{8.60}$$

If we number the states of the system in order of increasing energy, we have

$$E_1 \le E_2 \le \cdots \le E_n \le E_{n+1} \le \cdots \tag{8.61}$$

where the E's denote the true energies of various states. From the variation theorem, we know that

$$E_1 \le W_1$$

Moreover, it can be proved that [J. K. L. MacDonald, *Phys. Rev.*, **43**, 830 (1933); R. H. Young, *Int. J. Quantum Chem.*, **6**, 596 (1972)]

$$E_2 \le W_2, \ E_3 \le W_3, \ \ldots, \ E_n \le W_n \tag{8.62}$$

Thus, the linear variation method provides upper bounds to the energies of the lowest n states of the system. We use the roots $W_1, W_2, \ldots, W_n$ as approximations to the energies of the lowest states. If approximations to the energies of more states are wanted, we add more functions f_k to the trial function ϕ. The addition of more functions f_k can be shown to increase (or cause no change in) the accuracy of the previously calculated energies. If the functions f_k in $\phi = \Sigma_k c_k f_k$ form a complete set, then we will obtain the true wave functions of the system. Unfortunately, to have a complete set, we usually need an infinite number of functions.

Quantum chemists may use dozens, hundreds, thousands, or even millions of terms in linear variation functions so as to get accurate results for molecules. Obviously, a computer is essential for this work. The most efficient way to solve (8.59) (which is called the *secular equation*) and the associated linear equations (8.56) is by matrix methods (Section 8.6).

To obtain an approximation to the wave function of the ground state, we take the lowest root W_1 of the secular equation and substitute it in the set of equations (8.56); we then solve this set of equations for the coefficients $c_1^{(1)}, c_2^{(1)}, \ldots, c_n^{(1)}$, where the superscript $^{(1)}$ was added to indicate that these coefficients correspond to W_1. [As noted in the previous section, we can determine only the ratios of the coefficients; we solve for $c_2^{(1)}, \ldots, c_n^{(1)}$ in terms of $c_1^{(1)}$, and then determine $c_1^{(1)}$ by normalization.] Having found the $c_k^{(1)}$'s, we take $\phi_1 = \Sigma_k c_k^{(1)} f_k$ as an approximate ground-state wave function. Use of higher roots of (8.59) in (8.56) gives approximations to excited-state wave functions. These approximate wave functions can be shown to be orthogonal (Problem 8.31).

Solution of (8.59) and (8.56) is simplified by having as many of the integrals equal to zero as possible. We can make some of the off-diagonal H_{ij}'s vanish by choosing the functions f_k as eigenfunctions of some operator $\hat{A}$ that commutes with $\hat{H}$. If f_i and f_j correspond to different eigenvalues of $\hat{A}$, then H_{ij} vanishes (Theorem 6 of Section 7.4). If the functions f_k are orthonormal, the off-diagonal S_{ij}'s vanish ($S_{ij} = \delta_{ij}$). If the initially chosen f_k's are not orthogonal, we can use the Schmidt (or some other) procedure to find n linear combinations of these f_k's that are orthogonal and then use the orthogonalized functions.

Equations (8.56) and (8.59) are also valid when the restriction that the variation function be real is removed (Problem 8.30).

EXAMPLE Add functions to the function $x(l - x)$ of the first example of Section 8.1 to form a linear variation function for the particle in a one-dimensional box of length l and find approximate energies and wave functions for the lowest four states.

In the trial function $\phi = \sum_{k=1}^{n} c_k f_k$, we take $f_1 = x(l - x)$. Since we want approximations to the lowest four states, n must be at least 4. There are an infinite number of possible well-behaved functions that could be used for f_2, f_3, and f_4. The function $x^2(l - x)^2$ obeys the boundary conditions of vanishing at $x = 0$ and $x = l$ and leads to simple integrals, so we take $f_2 = x^2(l - x)^2$.

If the origin is placed at the center of the box, the potential energy (Fig. 2.1) is an even function, and, as noted in Section 8.2, the wave functions alternate between being even and odd functions (see also Fig. 2.3). (Throughout this example, the terms even and odd will refer to having the origin at the box's center.) The functions $f_1 = x(l - x)$ and $f_2 = x^2(l - x)^2$ are both even functions (see Problem 8.26). If we were to take $\phi = c_1 x(l - x) + c_2 x^2(l - x)^2$, we would end up with upper bounds to the energies of the lowest two states with even wave functions (the $n = 1$ and $n = 3$ states) and would get approximate wave functions for these two states. Since we also want to approximate the $n = 2$ and $n = 4$ states, we must add in two functions that are odd. An odd function must vanish at the origin [as noted after Eq. (4.55)], so we need functions that vanish at the box midpoint $x = \frac{1}{2}l$, as well as at $x = 0$ and l. A simple function with these properties is $f_3 = x(l - x)(\frac{1}{2}l - x)$. To get f_4, we shall multiply f_2 by $(\frac{1}{2}l - x)$. Thus we take $\phi = \sum_{k=1}^{4} c_k f_k$ with

$$f_1 = x(l - x), \quad f_2 = x^2(l - x)^2, \quad f_3 = x(l - x)(\tfrac{1}{2}l - x), \quad f_4 = x^2(l - x)^2(\tfrac{1}{2}l - x)$$
(8.63)

Because f_1 and f_2 are even, while f_3 and f_4 are odd, many integrals will vanish. Thus

$$S_{13} = S_{31} = 0, \quad S_{14} = S_{41} = 0, \quad S_{23} = S_{32} = 0, \quad S_{24} = S_{42} = 0 \quad (8.64)$$

because the integrand in each of these overlap integrals is an odd function with respect to the origin at the box center. The functions f_1, f_2, f_3, f_4 are eigenfunctions of the parity operator $\hat{\Pi}$ (Section 7.5) with the even functions f_1 and f_2 having parity eigenvalue $+1$ and f_3 and f_4 having eigenvalue -1. The operator $\hat{\Pi}$ commutes with $\hat{H}$ (since V is an even function), so by Theorem 6 of Section 7.4, H_{ij} vanishes if f_i is an odd function and f_j is even, or vice versa. Thus

$$H_{13} = H_{31} = 0, \quad H_{14} = H_{41} = 0, \quad H_{23} = H_{32} = 0, \quad H_{24} = H_{42} = 0 \quad (8.65)$$

From (8.64) and (8.65), the $n = 4$ secular equation (8.59) becomes

$$\begin{vmatrix} H_{11} - S_{11}W & H_{12} - S_{12}W & 0 & 0 \\ H_{21} - S_{21}W & H_{22} - S_{22}W & 0 & 0 \\ 0 & 0 & H_{33} - S_{33}W & H_{34} - S_{34}W \\ 0 & 0 & H_{43} - S_{43}W & H_{44} - S_{44}W \end{vmatrix} = 0 \quad (8.66)$$

The secular determinant is in block-diagonal form and so is equal to the product of its blocks [Eq. (8.33)]:

$$\begin{vmatrix} H_{11} - S_{11}W & H_{12} - S_{12}W \\ H_{21} - S_{21}W & H_{22} - S_{22}W \end{vmatrix} \times \begin{vmatrix} H_{33} - S_{33}W & H_{34} - S_{34}W \\ H_{43} - S_{43}W & H_{44} - S_{44}W \end{vmatrix} = 0$$

The four roots of this equation are found from the equations

$$\begin{vmatrix} H_{11} - S_{11}W & H_{12} - S_{12}W \\ H_{21} - S_{21}W & H_{22} - S_{22}W \end{vmatrix} = 0 \tag{8.67}$$

$$\begin{vmatrix} H_{33} - S_{33}W & H_{34} - S_{34}W \\ H_{43} - S_{43}W & H_{44} - S_{44}W \end{vmatrix} = 0 \tag{8.68}$$

Let the roots of (8.67) (which are approximations to the $n = 1$ and $n = 3$ energies) be W_1 and W_3 and let the roots of (8.68) be W_2 and W_4. After solving the secular equation for the W's, we substitute them one at a time into the set of equations (8.56) to find the coefficients c_k in the variation function. From the secular equation (8.66), the set of equations (8.56) with the root W_1 is

$$\left. \begin{matrix} (H_{11} - S_{11}W_1)c_1^{(1)} + (H_{12} - S_{12}W_1)c_2^{(1)} & = 0 \\ (H_{21} - S_{21}W_1)c_1^{(1)} + (H_{22} - S_{22}W_1)c_2^{(1)} & = 0 \end{matrix} \right\} \tag{8.69a}$$

$$\left. \begin{matrix} (H_{33} - S_{33}W_1)c_3^{(1)} + (H_{34} - S_{34}W_1)c_4^{(1)} = 0 \\ (H_{43} - S_{43}W_1)c_3^{(1)} + (H_{44} - S_{44}W_1)c_4^{(1)} = 0 \end{matrix} \right\} \tag{8.69b}$$

Because W_1 is a root of (8.67), the set of equations (8.69a) has the determinant of its coefficients [which is the determinant in (8.67)] equal to zero; hence (8.69a) has a nontrivial solution for $c_1^{(1)}$ and $c_2^{(1)}$. However, W_1 is not a root of (8.68), so the determinant of the coefficients of the set of equations (8.69b) is nonzero; hence, (8.69b) has only the trivial solution $c_3^{(1)} = c_4^{(1)} = 0$. The trial function ϕ_1 corresponding to the root W_1 thus has the form $\phi_1 = \sum_{k=1}^{4} c_k^{(1)} f_k = c_1^{(1)} f_1 + c_2^{(1)} f_2$. The same reasoning shows that ϕ_3 is a linear combination of f_1 and f_2, while ϕ_2 and ϕ_4 are each linear combinations of f_3 and f_4:

$$\begin{matrix} \phi_1 = c_1^{(1)} f_1 + c_2^{(1)} f_2, & \phi_3 = c_1^{(3)} f_1 + c_2^{(3)} f_2 \\ \phi_2 = c_3^{(2)} f_3 + c_4^{(2)} f_4, & \phi_4 = c_3^{(4)} f_3 + c_4^{(4)} f_4 \end{matrix} \tag{8.70}$$

The even wave functions ψ_1 and ψ_3 are approximated by linear combinations of the even functions f_1 and f_2; the odd functions ψ_2 and ψ_4 are approximated by linear combinations of the odd functions f_3 and f_4.

When the secular equation is in block-diagonal form, it factors into two or more smaller secular equations, and the set of simultaneous equations (8.56) breaks up into two or more smaller sets of equations.

We now must evaluate the H_{ij} and S_{ij} integrals so as to solve (8.67) and (8.68) for W_1, W_2, W_3, and W_4. We have

$$H_{11} = \langle f_1 | \hat{H} | f_1 \rangle = \int_0^l x(l-x) \left(\frac{-\hbar^2}{2m} \right) \frac{d^2}{dx^2} [x(l-x)] \, dx = \frac{\hbar^2 l^3}{6m}$$

$$S_{11} = \langle f_1 | f_1 \rangle = \int_0^l x^2 (l-x)^2 \, dx = \frac{l^5}{30}$$

where (8.12) and the equation following it were used. Evaluating the remaining integrals using the functions (8.63) and the relations (8.50) and (8.53), we find (Problem 8.27)

$$H_{12} = H_{21} = \langle f_2|\hat{H}|f_1\rangle = \hbar^2 l^5/30m , \quad H_{22} = \hbar^2 l^7/105m$$

$$H_{33} = \hbar^2 l^5/40m , \quad H_{44} = \hbar^2 l^9/1260m , \quad H_{34} = H_{43} = \hbar^2 l^7/280m$$

$$S_{12} = S_{21} = \langle f_1|f_2\rangle = l^7/140 , \quad S_{22} = l^9/630$$

$$S_{33} = l^7/840 , \quad S_{44} = l^{11}/27720 , \quad S_{34} = S_{43} = l^9/5040$$

Equation (8.67) becomes

$$\begin{vmatrix} \dfrac{\hbar^2 l^3}{6m} - \dfrac{l^5}{30}W & \dfrac{\hbar^2 l^5}{30m} - \dfrac{l^7}{140}W \\[3mm] \dfrac{\hbar^2 l^5}{30m} - \dfrac{l^7}{140}W & \dfrac{\hbar^2 l^7}{105m} - \dfrac{l^9}{630}W \end{vmatrix} = 0 \qquad (8.71)$$

Using Theorem IV of Section 8.3, we eliminate the fractions by multiplying row 1 of the determinant by $420m/l^3$, row 2 by $1260m/l^5$, and the right side of (8.71) by both factors, to get

$$\begin{vmatrix} 70\hbar^2 - 14ml^2W & 14\hbar^2 l^2 - 3ml^4W \\[2mm] 42\hbar^2 - 9ml^2W & 12\hbar^2 l^2 - 2ml^4W \end{vmatrix} = 0 \qquad (8.72)$$

$$m^2 l^4 W^2 - 56ml^2\hbar^2 W + 252\hbar^4 = 0$$

$$W = (\hbar^2/ml^2)(28 \pm \sqrt{532}) = 0.1250018\hbar^2/ml^2 , \quad 1.293495\hbar^2/ml^2$$

Substitution of the integrals into (8.68) leads to the roots (Problem 8.28)

$$W = (\hbar^2/ml^2)(60 \pm \sqrt{1620}) = 0.5002930\hbar^2/ml^2 , \quad 2.5393425\hbar^2/ml^2 \qquad (8.73)$$

The approximate values $(ml^2/\hbar^2)W = 0.1250018$, 0.5002930, 1.293495, and 2.5393425 may be compared with the exact values [Eq. (2.20)] $(ml^2/\hbar^2)E = 0.125$, 0.5, 1.125, and 2 for the four lowest states. The percent errors are 0.0014%, 0.059%, 15.0%, and 27.0% for $n = 1$, 2, 3, and 4, respectively. We did great for $n = 1$ and 2; lousy for $n = 3$ and 4.

We now find the approximate wave functions corresponding to these W's. Substitution of $W_1 = 0.1250018\hbar^2/ml^2$ into the set of equations (8.69a) corresponding to (8.72) gives (after division by h^2)

$$0.023095c_1^{(1)} - 0.020381c_2^{(1)}l^2 = 0$$
$$-0.061144c_1^{(1)} + 0.053960c_2^{(1)}l^2 = 0 \qquad (8.74)$$

where, for example, the first coefficient is found from

$$70\hbar^2 - 14ml^2W = 70h^2/4\pi^2 - 14(0.1250018)h^2 = 0.023095h^2$$

To solve the homogeneous equations (8.74), we follow the procedure given at the end of Section 8.4. We discard the second equation of (8.74), transfer the $c_2^{(1)}$ term to the right side, and solve for the coefficient ratio; we get

$$c_1^{(1)} = k , \quad c_2^{(1)} = 1.133k/l^2$$

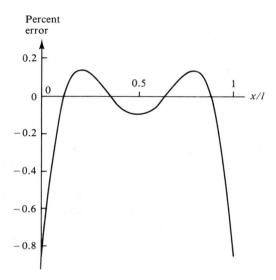

Figure 8.1 Percent deviation of the linear variation function (8.75) from the true ground-state particle-in-a-box wave function.

where k is a constant. We find k from the normalization condition:

$$\langle \phi_1 | \phi_1 \rangle = 1 = \langle kf_1 + 1.133kf_2/l^2 | kf_1 + 1.133kf_2/l^2 \rangle$$
$$= k^2(\langle f_1 | f_1 \rangle + 2.266\langle f_1 | f_2 \rangle /l^2 + 1.284\langle f_2 | f_2 \rangle /l^4)$$
$$= k^2(S_{11} + 2.266S_{12}/l^2 + 1.284S_{22}/l^4) = 0.05156k^2l^5$$

where the previously found values of the overlap integrals were used. Hence $k = 4.404/l^{5/2}$ and

$$\phi_1 = c_1^{(1)} f_1 + c_2^{(1)} f_2 = 4.404f_1/l^{5/2} + 4.990f_2/l^{9/2}$$
$$\phi_1 = l^{-1/2}[4.404(x/l)(1 - x/l) + 4.990(x/l)^2(1 - x/l)^2] \qquad (8.75)$$

where (8.63) was used. Figure 8.1 shows the percent departure of ϕ_1 from the true wave function ψ_1 along the box.

Using W_2, W_3, and W_4 in turn in (8.56), we find the following normalized linear variation functions (Problem 8.29), where $X \equiv x/l$:

$$\phi_2 = l^{-1/2}[16.78X(1 - X)(\tfrac{1}{2} - X) + 71.85X^2(1 - X)^2(\tfrac{1}{2} - X)]$$
$$\phi_3 = l^{-1/2}[28.65X(1 - X) - 132.7X^2(1 - X)^2] \qquad (8.76)$$
$$\phi_4 = l^{-1/2}[98.99X(1 - X)(\tfrac{1}{2} - X) - 572.3X^2(1 - X)^2(\tfrac{1}{2} - X)]$$

8.6 MATRICES, EIGENVALUES, AND EIGENVECTORS

Matrices were introduced in 1857 by the mathematician and lawyer Arthur Cayley as a shorthand way of dealing with simultaneous linear equations and linear transformations from one set of variables to another. The set of linear inhomogeneous equations (8.35) can be written as the matrix equation

$$
\begin{pmatrix}
a_{11} & a_{12} & \cdots & a_{1n} \\
a_{21} & a_{22} & \cdots & a_{2n} \\
\multicolumn{4}{c}{\dotfill} \\
a_{n1} & a_{n2} & \cdots & a_{n1}
\end{pmatrix}
\begin{pmatrix}
x_1 \\
x_2 \\
\vdots \\
x_n
\end{pmatrix}
=
\begin{pmatrix}
b_1 \\
b_2 \\
\vdots \\
b_n
\end{pmatrix}
\tag{8.77}
$$

$$
\mathbf{Ax} = \mathbf{b} \tag{8.78}
$$

where $\mathbf{A}$ is the coefficient matrix and $\mathbf{x}$ and $\mathbf{b}$ are column matrices. The equivalence of (8.35) and (8.77) is readily verified using the matrix-multiplication rule (7.103).

The *determinant* of a square matrix $\mathbf{A}$ is the determinant whose elements are the same as the elements of $\mathbf{A}$. If $\det \mathbf{A} \neq 0$, the matrix $\mathbf{A}$ is said to be *nonsingular*.

The *inverse* of a square matrix $\mathbf{A}$ of order n is the square matrix whose product with $\mathbf{A}$ is the unit matrix of order n. Denoting the inverse of $\mathbf{A}$ by $\mathbf{A}^{-1}$, we have

$$
\mathbf{AA}^{-1} = \mathbf{A}^{-1}\mathbf{A} = \mathbf{I} \tag{8.79}*
$$

One can prove that $\mathbf{A}^{-1}$ exists if and only if $\det \mathbf{A} \neq 0$. (For efficient methods of computing $\mathbf{A}^{-1}$, see *Press et al.*, Secs. 2.3 and 2.4; *Meck*, Sec. 6-2; *Shoup*, Sec. 3.3; Problem 8.39.)

If $\det \mathbf{A} \neq 0$ for the coefficient matrix $\mathbf{A}$ in (8.77), then we can multiply each side of (8.78) by $\mathbf{A}^{-1}$ on the left to get $\mathbf{A}^{-1}(\mathbf{Ax}) = \mathbf{A}^{-1}\mathbf{b}$. Matrix multiplication can be shown to be associative, meaning that $\mathbf{A}(\mathbf{BC}) = (\mathbf{AB})\mathbf{C}$. Hence $\mathbf{A}^{-1}(\mathbf{Ax}) = (\mathbf{A}^{-1}\mathbf{A})\mathbf{x} = \mathbf{Ix} = \mathbf{x}$. Thus, left multiplication of (8.78) by $\mathbf{A}^{-1}$ gives $\mathbf{x} = \mathbf{A}^{-1}\mathbf{b}$ as the solution for the unknowns.

The linear variation method is the most commonly used method to find approximate molecular wave functions, and matrix algebra gives the most computationally efficient method to solve the equations of the linear variation method. If the functions $f_1, \ldots, f_n$ in the linear variation function $\phi = \sum_{k=1}^{n} c_k f_k$ are made to be orthonormal, then $S_{ij} \equiv \int f_i^* f_j \, d\tau = \delta_{ij}$, and the homogeneous set of equations (8.56) for the coefficients c_k that minimize the variational integral becomes

$$
\begin{aligned}
H_{11}c_1 + H_{12}c_2 + \cdots + H_{1n}c_n &= Wc_1 \\
H_{21}c_1 + H_{22}c_2 + \cdots + H_{2n}c_n &= Wc_2 \\
&\ \vdots \\
H_{n1}c_1 + H_{n2}c_2 + \cdots + H_{nn}c_n &= Wc_n
\end{aligned}
\tag{8.80a}
$$

$$
\begin{pmatrix}
H_{11} & H_{12} & \cdots & H_{1n} \\
H_{21} & H_{22} & \cdots & H_{2n} \\
\multicolumn{4}{c}{\dotfill} \\
H_{n1} & H_{n2} & \cdots & H_{nn}
\end{pmatrix}
\begin{pmatrix}
c_1 \\
c_2 \\
\vdots \\
c_n
\end{pmatrix}
= W
\begin{pmatrix}
c_1 \\
c_2 \\
\vdots \\
c_n
\end{pmatrix}
\tag{8.80b}
$$

$$
\mathbf{Hc} = W\mathbf{c} \tag{8.80c}
$$

where $\mathbf{H}$ is the square matrix whose elements are $H_{ij} = \langle f_i | \hat{H} | f_j \rangle$ and $\mathbf{c}$ is the column vector of coefficients $c_1, \ldots, c_n$. In (8.80c), $\mathbf{H}$ is a known matrix and $\mathbf{c}$ and W are unknowns to be solved for.

If

$$\mathbf{Ac} = \lambda\mathbf{c} \qquad (8.81)^*$$

where $\mathbf{A}$ is a square matrix, $\mathbf{c}$ is a column vector with at least one nonzero element, and λ is a scalar, then $\mathbf{c}$ is said to be an *eigenvector* (or characteristic vector) of $\mathbf{A}$ and λ is an *eigenvalue* (or characteristic value) of $\mathbf{A}$.

Comparison of (8.81) with (8.80c) shows that solving the linear variation problem with $S_{ij} = \delta_{ij}$ amounts to finding the eigenvalues and eigenvectors of the matrix $\mathbf{H}$. The matrix eigenvalue equation $\mathbf{Hc} = W\mathbf{c}$ is equivalent to the set of homogeneous equations (8.56), which has a nontrivial solution for the c's if and only if $\det(H_{ij} - \delta_{ij}W) = 0$ [Eq. (8.58) with $S_{ij} = \delta_{ij}$]. For a general square matrix $\mathbf{A}$ of order n, the corresponding equation satisfied by the eigenvalues is

$$\det(A_{ij} - \delta_{ij}\lambda) = 0 \qquad (8.82)^*$$

This equation (called the *characteristic equation*) has n roots for λ (some of which may be equal), so a square matrix of order n has n eigenvalues and n independent eigenvectors.

If $\mathbf{c}$ is an eigenvector of $\mathbf{A}$, then clearly $k\mathbf{c}$ is also an eigenvector of $\mathbf{A}$, where k is any constant. If k is chosen so that

$$\sum_{i=1}^{n} |c_i|^2 = 1 \qquad (8.83)^*$$

then the column vector $\mathbf{c}$ is said to be *normalized*. Two column vectors $\mathbf{b}$ and $\mathbf{c}$ that each have n elements are said to be *orthogonal* if

$$\sum_{i=1}^{n} b_i^* c_i = 0 \qquad (8.84)^*$$

Let us denote the n eigenvalues and the corresponding eigenvectors of $\mathbf{H}$ in (8.80) by $W_1, W_2, \ldots, W_n$ and $\mathbf{c}^{(1)}, \mathbf{c}^{(2)}, \ldots, \mathbf{c}^{(n)}$, so that

$$\mathbf{Hc}^{(i)} = W_i \mathbf{c}^{(i)}, \qquad \text{for } i = 1, 2, \ldots, n \qquad (8.85)$$

where $\mathbf{c}^{(i)}$ is a column vector whose elements are $c_1^{(i)}, \ldots, c_n^{(i)}$. Furthermore, let $\mathbf{C}$ be the square matrix whose columns are the eigenvectors of $\mathbf{H}$, and let $\mathbf{W}$ be the diagonal matrix whose diagonal elements are the eigenvalues of $\mathbf{H}$:

$$\mathbf{C} = \begin{pmatrix} c_1^{(1)} & c_1^{(2)} & \cdots & c_1^{(n)} \\ c_2^{(1)} & c_2^{(2)} & \cdots & c_2^{(n)} \\ \cdots\cdots\cdots\cdots\cdots \\ c_n^{(1)} & c_n^{(2)} & \cdots & c_n^{(n)} \end{pmatrix} \qquad \mathbf{W} = \begin{pmatrix} W_1 & 0 & \cdots & 0 \\ 0 & W_2 & \cdots & 0 \\ \cdots\cdots\cdots\cdots \\ 0 & 0 & \cdots & W_n \end{pmatrix} \qquad (8.86)$$

The set of n eigenvalue equations (8.85) can be written as the single equation:

$$\mathbf{HC} = \mathbf{CW} \qquad (8.87)$$

To verify the matrix equation (8.87), we show that each element $(\mathbf{HC})_{ij}$ of the matrix $\mathbf{HC}$ equals the corresponding element $(\mathbf{CW})_{ij}$ of $\mathbf{CW}$. The matrix-multiplication rule (7.103) gives $(\mathbf{HC})_{ij} = \sum_k H_{ik}(\mathbf{C})_{kj} = \sum_k H_{ik}c_k^{(j)}$. Consider the eigenvalue equation $\mathbf{Hc}^{(j)} = W_j\mathbf{c}^{(j)}$ [Eq. (8.85)]. $\mathbf{Hc}^{(j)}$ and $W_j\mathbf{c}^{(j)}$ are column matrices.

Using (7.103) to equate the elements in row i of each of these column matrices, we have $\sum_k H_{ik} c_k^{(j)} = W_j c_i^{(j)}$. Then

$$(\mathbf{HC})_{ij} = \sum_k H_{ik} c_k^{(j)} = W_j c_i^{(j)}$$

$$(\mathbf{CW})_{ij} = \sum_k (\mathbf{C})_{ik} (\mathbf{W})_{kj} = \sum_k c_i^{(k)} \delta_{kj} W_k = c_i^{(j)} W_j$$

Hence $(\mathbf{HC})_{ij} = (\mathbf{CW})_{ij}$ and (8.87) is proved.

Provided $\mathbf{C}$ has an inverse (see below), we can multiply each side of (8.87) by $\mathbf{C}^{-1}$ on the left to get $\mathbf{C}^{-1}\mathbf{HC} = \mathbf{C}^{-1}(\mathbf{CW})$. [Since matrix multiplication is not commutative, when we multiply each side of $\mathbf{HC} = \mathbf{CW}$ by $\mathbf{C}^{-1}$, we must put the factor $\mathbf{C}^{-1}$ on the left of $\mathbf{HC}$ and on the left of $\mathbf{CW}$ (or on the right of $\mathbf{HC}$ and the right of $\mathbf{CW}$).] We have $\mathbf{C}^{-1}\mathbf{HC} = \mathbf{C}^{-1}(\mathbf{CW}) = (\mathbf{C}^{-1}\mathbf{C})\mathbf{W} = \mathbf{IW} = \mathbf{W}$:

$$\mathbf{C}^{-1}\mathbf{HC} = \mathbf{W} \tag{8.88}$$

To simplify (8.88), we must learn more about matrices.

A square matrix $\mathbf{B}$ is a *symmetric* matrix if all its elements satisfy $b_{ij} = b_{ji}$. The elements of a symmetric matrix are symmetric about the principal diagonal; for example, $b_{12} = b_{21}$. A square matrix $\mathbf{D}$ is a *Hermitian matrix* if all its elements satisfy $d_{ij} = d_{ji}^*$. For example, if

$$\mathbf{M} = \begin{pmatrix} 2 & 5 & 0 \\ 5 & i & 2i \\ 0 & 2i & 4 \end{pmatrix}, \qquad \mathbf{N} = \begin{pmatrix} 6 & 1+2i & 8 \\ 1-2i & -1 & -i \\ 8 & i & 0 \end{pmatrix} \tag{8.89}$$

then $\mathbf{M}$ is symmetric and $\mathbf{N}$ is Hermitian. (Note that the diagonal elements of a Hermitian matrix must be real; $d_{ii} = d_{ii}^*$.) A *real matrix* is one whose elements are all real numbers. A real Hermitian matrix is a symmetric matrix.

The *transpose* $\tilde{\mathbf{A}}$ of the matrix $\mathbf{A}$ is the matrix formed by interchanging rows and columns of $\mathbf{A}$ so that column 1 becomes row 1, column 2 becomes row 2, and so on. The elements $\tilde{a}_{ij}$ of $\tilde{\mathbf{A}}$ are related to the elements of $\mathbf{A}$ by $\tilde{a}_{ij} = a_{ji}$. For a square matrix, the transpose is found by reflecting the elements about the principal diagonal. A symmetric matrix is equal to its transpose. Thus, for the matrix $\mathbf{M}$ in (8.89), we have $\tilde{\mathbf{M}} = \mathbf{M}$.

The *complex conjugate* $\mathbf{A}^*$ of $\mathbf{A}$ is the matrix formed by taking the complex conjugate of each element of $\mathbf{A}$. The *conjugate transpose* $\mathbf{A}^\dagger$ of the matrix $\mathbf{A}$ is formed by taking the transpose of $\mathbf{A}^*$; thus $\mathbf{A}^\dagger = \tilde{\mathbf{A}}^*$ and

$$a_{ij}^\dagger = a_{ji}^* \tag{8.90}$$

(Physicists call $\mathbf{A}^\dagger$ the *adjoint* of $\mathbf{A}$, a name that is used by mathematicians to refer to an entirely different matrix.) An example is

$$\mathbf{B} = \begin{pmatrix} 2 & 3+i \\ 0 & 4i \end{pmatrix}, \qquad \tilde{\mathbf{B}} = \begin{pmatrix} 2 & 0 \\ 3+i & 4i \end{pmatrix}, \qquad \mathbf{B}^\dagger = \begin{pmatrix} 2 & 0 \\ 3-i & -4i \end{pmatrix}$$

An *orthogonal matrix* is one whose inverse is equal to its transpose:

$$\mathbf{A}^{-1} = \tilde{\mathbf{A}}, \qquad \text{if } \mathbf{A} \text{ is orthogonal} \tag{8.91}$$

A *unitary matrix* is one whose inverse is equal to its conjugate transpose:

$$\mathbf{U}^{-1} = \mathbf{U}^{\dagger}, \qquad \text{if } \mathbf{U} \text{ is unitary} \tag{8.92}$$

From the definition (8.92), we have $\mathbf{U}^{\dagger}\mathbf{U} = \mathbf{I}$ if $\mathbf{U}$ is unitary. By equating $(\mathbf{U}^{\dagger}\mathbf{U})_{ij}$ to $(\mathbf{I})_{ij}$, we find (Problem 8.33)

$$\sum_{k} u_{ki}^{*} u_{kj} = \delta_{ij} \tag{8.93}$$

for columns i and j of a unitary matrix. Thus the columns of a unitary matrix (viewed as column vectors) are orthogonal and normalized (orthonormal), as defined by (8.84) and (8.83). Conversely, if (8.93) is true for all columns, then $\mathbf{U}$ is a unitary matrix. If $\mathbf{U}$ is unitary and real, then $\mathbf{U}^{\dagger} = \tilde{\mathbf{U}}$, and $\mathbf{U}$ is an orthogonal matrix.

One can prove that two eigenvectors of a Hermitian matrix $\mathbf{H}$ that correspond to different eigenvalues are orthogonal (see *Levine, Molecular Spectroscopy*, Sec. 2.2). For eigenvectors of $\mathbf{H}$ that correspond to the same eigenvalue, one can take linear combinations of them that will be orthogonal eigenvectors of $\mathbf{H}$. Moreover, the elements of an eigenvector can be multiplied by a constant to normalize the eigenvector. Hence, the eigenvectors of a Hermitian matrix can be chosen to be orthonormal. If the eigenvectors are chosen to be orthonormal, then the eigenvector matrix $\mathbf{C}$ in (8.86) is a unitary matrix, and $\mathbf{C}^{-1} = \mathbf{C}^{\dagger}$; (8.88) then becomes

$$\mathbf{C}^{\dagger}\mathbf{H}\mathbf{C} = \mathbf{W}, \qquad \text{if } \mathbf{H} \text{ is Hermitian} \tag{8.94}$$

For the common case that $\mathbf{H}$ is real as well as Hermitian (that is, $\mathbf{H}$ is real and symmetric), the c's in (8.80a) are real (since W and the H_{ij}'s are real) and $\mathbf{C}$ is real as well as unitary; that is, $\mathbf{C}$ is orthogonal, with $\mathbf{C}^{-1} = \tilde{\mathbf{C}}$; Eq. (8.94) becomes

$$\tilde{\mathbf{C}}\mathbf{H}\mathbf{C} = \mathbf{W}, \qquad \text{if } \mathbf{H} \text{ is real and symmetric} \tag{8.95}$$

To find the eigenvalues and eigenvectors of a Hermitian matrix of order n, we can use either of the following procedures: (1) Solve the algebraic equation $\det(H_{ij} - \delta_{ij}W) = 0$ [Eq. (8.82)] for the eigenvalues $W_1, \ldots, W_n$. Then substitute each W_k into the set of algebraic equations (8.80a) and solve for the elements $c_1^{(k)}, \ldots, c_n^{(k)}$ of the kth eigenvector. (2) Search for a unitary matrix $\mathbf{C}$ such that $\mathbf{C}^{\dagger}\mathbf{H}\mathbf{C}$ is a diagonal matrix. The diagonal elements of $\mathbf{C}^{\dagger}\mathbf{H}\mathbf{C}$ are the eigenvalues of $\mathbf{H}$, and the columns of $\mathbf{C}$ are the orthonormal eigenvectors of $\mathbf{H}$. For the large matrices that occur in quantum chemistry, procedure (2) (called *matrix diagonalization*) is computationally much faster than (1).

A systematic way to diagonalize a real symmetric matrix $\mathbf{H}$ is: Construct the orthogonal matrix $\mathbf{O}_1$ such that the matrix $\mathbf{H}_1 \equiv \tilde{\mathbf{O}}_1\mathbf{H}\mathbf{O}_1$ has zero in place of the largest off-diagonal element of $\mathbf{H}$; then construct $\mathbf{O}_2$ such that $\mathbf{H}_2 \equiv \tilde{\mathbf{O}}_2\mathbf{H}_1\mathbf{O}_2 = \tilde{\mathbf{O}}_2\tilde{\mathbf{O}}_1\mathbf{H}\mathbf{O}_1\mathbf{O}_2$ has zero in place of the largest off-diagonal element of $\mathbf{H}_1$; and so on. The process is repeated until all the off-diagonal elements become negligibly small. The eigenvalues are then the diagonal elements of the transformed matrix $\cdots \tilde{\mathbf{O}}_3\tilde{\mathbf{O}}_2\tilde{\mathbf{O}}_1\mathbf{H}\mathbf{O}_1\mathbf{O}_2\mathbf{O}_3 \cdots$, and the eigenvector matrix is the product $\mathbf{O}_1\mathbf{O}_2\mathbf{O}_3 \cdots$. Details and computer programs for this procedure (the Jacobi method) and for the faster Givens and Householder matrix-diagonalization procedures are given in

Lowe, Appendixes 9 and 10; *Press et al.*, Chapter 11; *Acton*, Chapters 8 and 13; *Shoup*, Chapter 4.

We have discussed matrix diagonalization in the context of the linear variation method. However, finding the eigenvalues a_k and eigenfunctions g_k of any Hermitian operator $\hat{A}$ ($\hat{A}g_k = a_k g_k$) can be formulated as a matrix-diagonalization problem. If we choose a complete, orthonormal basis set $\{f_i\}$ and expand the eigenfunctions as $g_k = \sum_i c_i^{(k)} f_i$, then (Problem 8.40) the eigenvalues of the matrix $\mathbf{A}$ whose elements are $a_{ij} = \langle f_i | \hat{A} | f_j \rangle$ are the eigenvalues of the operator $\hat{A}$, and the elements $c_i^{(k)}$ of the eigenvectors $\mathbf{c}^{(k)}$ of $\mathbf{A}$ give the coefficients in the expansions of the eigenfunctions g_k.

The material of this section further emphasizes the correspondence between linear operators and matrices and the correspondence between functions and column vectors (Section 7.10).

EXAMPLE   Find the eigenvalues and normalized eigenvectors of the Hermitian matrix

$$\mathbf{A} = \begin{pmatrix} 3 & 2i \\ -2i & 0 \end{pmatrix}$$

by solving algebraic equations. Then verify that $\mathbf{C}^\dagger \mathbf{A} \mathbf{C}$ is diagonal, where $\mathbf{C}$ is the eigenvector matrix.

The characteristic equation (8.82) for the eigenvalues λ is det $(a_{ij} - \delta_{ij}\lambda) = 0$, which becomes

$$\begin{vmatrix} 3 - \lambda & 2i \\ -2i & -\lambda \end{vmatrix} = 0$$

$$\lambda^2 - 3\lambda - 4 = 0$$

$$\lambda_1 = 4, \qquad \lambda_2 = -1$$

For the root $\lambda_1 = 4$, the set of simultaneous equations (8.80a) (with H and W replaced by A and λ, respectively) is

$$(3 - \lambda_1)c_1^{(1)} + 2ic_2^{(1)} = 0$$

$$-2ic_1^{(1)} - \lambda_1 c_2^{(1)} = 0$$

or

$$-c_1^{(1)} + 2ic_2^{(1)} = 0$$

$$-2ic_1^{(1)} - 4c_2^{(1)} = 0$$

Discarding either one of these equations, we find

$$c_1^{(1)} = 2ic_2^{(1)}$$

Normalization gives

$$1 = |c_1^{(1)}|^2 + |c_2^{(1)}|^2 = 4|c_2^{(1)}|^2 + |c_2^{(1)}|^2$$

$$|c_2^{(1)}| = 1/\sqrt{5}, \qquad c_2^{(1)} = 1/\sqrt{5}$$

$$c_1^{(1)} = 2ic_2^{(1)} = 2i/\sqrt{5}$$

where the phase of $c_2^{(1)}$ was chosen to be zero.

Similarly, we find for $\lambda_2 = -1$ (Problem 8.37)

$$c_1^{(2)} = -i/\sqrt{5}, \qquad c_2^{(2)} = 2/\sqrt{5}$$

The normalized eigenvectors are then

$$\mathbf{c}^{(1)} = \begin{pmatrix} 2i/\sqrt{5} \\ 1/\sqrt{5} \end{pmatrix}, \qquad \mathbf{c}^{(2)} = \begin{pmatrix} -i/\sqrt{5} \\ 2/\sqrt{5} \end{pmatrix}$$

Because the eigenvalues λ_1 and λ_2 of the Hermitian matrix $\mathbf{A}$ differ, $\mathbf{c}^{(1)}$ and $\mathbf{c}^{(2)}$ are orthogonal (as the reader should verify). Also, $\mathbf{c}^{(1)}$ and $\mathbf{c}^{(2)}$ are normalized. Therefore, $\mathbf{C}$ is unitary and $\mathbf{C}^{-1} = \mathbf{C}^{\dagger}$. Forming $\mathbf{C}$ and its conjugate transpose, we have

$$\mathbf{C}^{-1}\mathbf{A}\mathbf{C} = \mathbf{C}^{\dagger}\mathbf{A}\mathbf{C} = \begin{pmatrix} -2i/\sqrt{5} & 1/\sqrt{5} \\ i/\sqrt{5} & 2/\sqrt{5} \end{pmatrix} \begin{pmatrix} 3 & 2i \\ -2i & 0 \end{pmatrix} \begin{pmatrix} 2i/\sqrt{5} & -i/\sqrt{5} \\ 1/\sqrt{5} & 2/\sqrt{5} \end{pmatrix}$$

$$= \begin{pmatrix} -2i/\sqrt{5} & 1/\sqrt{5} \\ i/\sqrt{5} & 2/\sqrt{5} \end{pmatrix} \begin{pmatrix} 8i/\sqrt{5} & i/\sqrt{5} \\ 4/\sqrt{5} & -2/\sqrt{5} \end{pmatrix} = \begin{pmatrix} 4 & 0 \\ 0 & -1 \end{pmatrix}$$

which is the diagonal matrix of eigenvectors.

8.7 SUMMARY

The variation theorem states that for a system with time-independent Hamiltonian $\hat{H}$ and ground-state energy E_1, we have $\int \phi^* \hat{H} \phi \, d\tau / \int \phi^* \phi \, d\tau \geq E_1$, where ϕ is any well-behaved function that obeys the boundary conditions of the problem. The variation theorem allows us to obtain approximations for the ground-state energy and wave function.

The mathematics of determinants was reviewed in Section 8.3. A set of n linear homogeneous equations in n unknowns has a nontrivial solution if and only if the determinant of the coefficients is equal to zero.

For the linear variation function $\phi = \sum_{i=1}^{n} c_i f_i$, variation of the coefficients c_i to minimize the variational integral W leads to the secular equation $\det(H_{ij} - S_{ij}W) = 0$, whose roots $W_1, \ldots, W_n$ are upper bounds for the n lowest energy eigenvalues; here, $H_{ij} \equiv \langle f_i | \hat{H} | f_j \rangle$ and $S_{ij} \equiv \langle f_i | f_j \rangle$. Substitution of $W_1, \ldots, W_n$ one at a time into the simultaneous homogeneous equations (8.56) allows the coefficients c_i that correspond to each W to be found.

PROBLEMS

8.1 Apply the variation function $\phi = e^{-cr}$ to the hydrogen atom; choose the parameter c to minimize the variational integral, and calculate the percent error in the ground-state energy.

8.2 Verify Eq. (8.13) for $\langle \phi | \hat{H} | \phi \rangle$.

8.3 If the normalized variation function $\phi = (3/l^3)^{1/2} x$ for $0 \leq x \leq l$ is applied to the particle-in-a-one-dimensional-box problem, one finds that the variational integral equals zero, which is *less* than the true ground-state energy. What is wrong?

8.4 For a particle in a three-dimensional box with sides of length a, b, c, write down the variation function that is the three-dimensional extension of the variation function $\phi = x(l - x)$ used in Section 8.1 for the particle in a one-dimensional box. Use the integrals in (8.12) and the equation following it to evaluate the variational integral for the three-dimensional case; find the percent error in the ground-state energy.

8.5 (a) For a one-particle, one-dimensional system with potential energy

$$V = b \quad \text{for } \tfrac{1}{4}l \leqslant x \leqslant \tfrac{3}{4}l, \qquad V = 0 \quad \text{for } 0 \leqslant x \leqslant \tfrac{1}{4}l \text{ and } \tfrac{3}{4}l \leqslant x \leqslant l$$

and $V = \infty$ elsewhere (where b is a constant), use the trial variation function $\phi_1 = (2/l)^{1/2} \sin(\pi x/l)$ for $0 \leqslant x \leqslant l$ to estimate the ground-state energy for $b = \hbar^2/ml^2$ and compare with the true ground-state energy $E = 5.750345\hbar^2/ml^2$. To save time in evaluating integrals, note that $\langle \phi_1 | \hat{H} | \phi_1 \rangle = \langle \phi_1 | \hat{T} | \phi_1 \rangle + \langle \phi_1 | V | \phi_1 \rangle$ and explain why $\langle \phi_1 | \hat{T} | \phi_1 \rangle$ equals the particle-in-a-box ground-state energy $h^2/8ml^2$. (b) For this system, use the variation function $\phi_2 = x(l - x)$. To save time, note that $\langle \phi_2 | \hat{T} | \phi_2 \rangle$ is given by Eq. (8.12). (Why?)

8.6 (a) A particle in a spherical box of radius b has $V = 0$ for $0 \leqslant r \leqslant b$ and $V = \infty$ for $r > b$. Use the trial function $\phi = b - r$ for $0 \leqslant r \leqslant b$ and $\phi = 0$ for $r > b$ to estimate the ground-state energy and compare with the true value $h^2/8mb^2$ (Problem 6.2). (b) Devise another simple variation function that obeys the boundary conditions for this problem and work out the percent error in the ground-state energy given by your function.

8.7 A one-dimensional quartic oscillator has $V = cx^4$, where c is a constant. Devise a variation function with a parameter for this problem, and find the optimum value of the parameter to minimize the variational integral and estimate the ground-state energy.

8.8 For a particle in a box of length l, use the variation function $\phi = x^k(l - x)^k$ for $0 \leqslant x \leqslant l$. You will need the integral

$$\int_0^l x^s(l - x)^t \, dx = l^{s+t+1} \frac{\Gamma(s + 1)\Gamma(t + 1)}{\Gamma(s + t + 2)}$$

where the gamma function obeys the relation $\Gamma(z + 1) = z\Gamma(z)$. The definition of the gamma function $\Gamma(z)$ need not concern you, since the gamma functions will ultimately cancel. (a) Show that the variational integral equals $(\hbar^2/ml^2)(4k^2 + k)/(2k - 1)$. (b) Find the optimum value of k and calculate the percent error in the ground-state energy for this k.

8.9 (a) Apply the variation function $\phi = 1/(a^2 + x^2)$ to the one-dimensional harmonic oscillator. Choose a to minimize the variational integral and find the percent error in the ground-state energy. Some useful integrals are

$$\int_0^\infty \frac{1}{(x^2 + a^2)^2} \, dx = \frac{\pi}{4a^3} \qquad \int_0^\infty \frac{1}{(x^2 + a^2)^3} \, dx = \frac{3\pi}{16a^5}$$

$$\int_0^\infty \frac{x^2}{(x^2 + a^2)^2} \, dx = \frac{\pi}{4a} \qquad \int_0^\infty \frac{x^2}{(x^2 + a^2)^4} \, dx = \frac{\pi}{32a^5}$$

(b) These integrals are functions of the parameter a, and the above equations have the form $\int_0^\infty f(x, a) \, dx = g(a)$. We have $(d/da) \int_0^\infty f(x, a) \, dx = dg/da$. Provided the integrand is well behaved, we can interchange the differentiation and integration, which gives $\int_0^\infty [\partial f(x, a)/\partial a] \, dx = dg/da$. By differentiating the first integral equation in part (a) with respect to a, derive the second integral equation.

8.10 In 1971 a paper was published that applied the normalized variation function $N \exp(-br^2/a_0^2 - cr/a_0)$ to the hydrogen atom and stated that minimization of the variational integral with respect to the parameters b and c yielded an energy 0.7% above the true ground-state energy for infinite nuclear mass. Without doing any calculations, state why this result must be in error.

8.11 (a) Apply the function of Eq. (7.35) as a trial variation function for the ground state of the particle in a box. Note that $f''(x)$ is infinite at $x = \frac{1}{2}l$ because of the discontinuity in $f'(x)$ at this point. Therefore, in evaluating $\int f^*\hat{H}f\,dx$, we run into difficulty in evaluating the integral of ff''. One way around this problem is to first show that

$$\int_0^l ff''\,dx = -\int_0^l (f')^2\,dx = -\int_0^{l/2} (f')^2\,dx - \int_{l/2}^l (f')^2\,dx \qquad (8.96)$$

for any function obeying the boundary conditions. Then, using the expression on the right of (8.96), we can calculate the variational integral. Prove (8.96), and then calculate the percent error in the ground-state energy, using this triangular function. Note that there is no parameter in this trial function. [If you are ambitious, try this alternative procedure: Note that $f'(x)$ involves the Heaviside step function (Section 7.7), and therefore $f''(x)$ involves the Dirac delta function. Use the properties of the delta function to evaluate $\int_0^l ff''\,dx$, and find the percent error using this triangular function.]

Functions with discontinuities in their first derivatives have been used as trial variation functions in quantum chemistry; see L. C. Snyder and R. G. Parr, *J. Chem. Phys.*, **34**, 1661 (1961); C. A. Coulson and C. S. Sharma, *Proc. Roy. Soc.*, **272A**, 1 (1963). (b) The variation function ϕ of the first example in Section 8.1 has discontinuities in ϕ' at $x = 0$ and $x = l$, so, strictly speaking, we should use one of the procedures of part (a) of this problem to evaluate $\langle \phi|\hat{H}|\phi \rangle$. Do this and show that (8.12) is obtained.

8.12 (a) For the ground state of the hydrogen atom, use the Gaussian trial function

$$\phi = e^{-cr^2/a_0^2} \qquad (8.97)$$

Find the optimum value of c and the percent error in the energy. (Gaussian variational functions are often used in molecular quantum mechanics; see Section 15.4.) (b) Multiply the function (8.97) by the spherical harmonic Y_2^0, and then minimize the variational integral. This yields an upper bound to the energy of which hydrogen-atom state?

8.13 For the system of Problem 8.5, explain why the variation function $\phi = (2/l)^{1/2} \sin(2\pi x/l)$ gives an upper bound to the energy E_2 of the first excited state. (*Hint:* With the origin at the center of the box, V is an even function.) Use this ϕ to evaluate the variation integral for $b = \hbar^2/ml^2$ and compare with the true value $E_2 = 20.23604\hbar^2/ml^2$.

8.14 Prove that the value of a determinant all of whose elements below the principal diagonal are zero is equal to the product of the diagonal elements.

8.15 Evaluate

$$\begin{vmatrix} 2 & 5 & 1 & 3 \\ 8 & 0 & 4 & -1 \\ 6 & 6 & 6 & 1 \\ 5 & -2 & -2 & 2 \end{vmatrix}$$

8.16 (a) Consider some permutation of the integers $1, 2, 3, \ldots, n$. The permutation is an *even permutation* if an even number of interchanges of pairs of integers restores the permutation to the natural order $1, 2, 3, \ldots, n$. An *odd permutation* requires an odd number of interchanges of pairs to reach the natural order. For example, the permutation 3124 is even, since two interchanges restore it to the natural order: $3124 \rightarrow 1324 \rightarrow 1234$. Write down and classify (even or odd) all permutations of 123. (b) Verify that the definition (8.24) of the third-order determinant is equivalent to

$$\begin{vmatrix} a_{11} & a_{12} & a_{13} \\ a_{21} & a_{22} & a_{23} \\ a_{31} & a_{32} & a_{33} \end{vmatrix} = \sum (\pm 1) a_{1i} a_{2j} a_{3k}$$

where ijk is one of the permutations of the integers 123, the sum is over the 3! different permutations of these integers, and the sign of each term is plus or minus, depending on whether the permutation is even or odd. (c) How would we define the nth-order determinant using this type of definition?

8.17 How many terms are there in the expansion of an nth-order determinant?

8.18 Use Gaussian elimination with detached coefficients to solve

$$2x_1 - x_2 + 4x_3 + 2x_4 = 16$$
$$3x_1 \quad\quad - x_3 + 4x_4 = -5$$
$$2x_1 + x_2 + x_3 - 2x_4 = 8$$
$$-4x_1 + 6x_2 + 2x_3 + x_4 = 3$$

8.19 When Gaussian elimination is programmed for a computer (or done on an electronic calculator), one should use exchanges of equations to avoid dividing by a coefficient that is much smaller in absolute value than the other coefficients in the equations. Explain why division by an extremely small coefficient can lead to large errors in the results.

8.20 Gaussian elimination can be used to efficiently evaluate a determinant, as follows. Divide each element of row 1 of the determinant (8.20) by a_{11} and place the factor a_{11} in front of the determinant (Theorem IV of Section 8.3). Then subtract the appropriate multiples of the row 1 elements from row 2, row 3, . . . , row n to make $a_{21}, a_{31}, \ldots, a_{n1}$ zero (Theorem V). Then divide the second-row elements by a_{22} and insert the factor a_{22} in front of the determinant, and so on. Ultimately, we get a determinant all of whose elements below the principal diagonal are zero. From Problem 8.14, this determinant equals the product of its diagonal elements. Use this procedure to evaluate the determinant in Problem 8.15.

8.21 Write a computer program that uses Gaussian elimination to solve a system of n linear, simultaneous, inhomogeneous equations in n unknowns, where $n \leq 10$. Test it on a couple of examples.

8.22 Solve each set of simultaneous equations using Gauss–Jordan elimination with detached coefficients:

$$x + 2y + 3z = 0 \qquad\qquad x + 2y + 3z = 0$$
$$(a) \ 3x + y + 2z = 0 \qquad (b) \quad x - y + z = 0$$
$$2x + 3y + z = 0 \qquad\qquad 7x - y + 11z = 0$$

8.23 Solve the second-order secular equation (8.57) for the special case where $H_{11} = H_{22}$ and $S_{11} = S_{22}$. (*Reminder:* f_1 and f_2 are real functions.) Then solve for c_1/c_2 for each of the two roots W_1 and W_2.

8.24 For the system of Problem 8.5 with $b = \hbar^2/ml^2$ (a particle in a box with a rectangular bump in the middle), consider the linear variation function

$$\phi = c_1 f_1 + c_2 f_2 = c_1 (2/l)^{1/2} \sin(\pi x/l) + c_2 (2/l)^{1/2} \sin(3\pi x/l)$$

for $0 \leq x \leq l$. (a) Explain why this variation function will give upper bounds to the energies E_1 and E_3 in (8.61). (b) Explain why f_1 and f_2 are orthonormal. (c) Note that $\langle f_i | \hat{H} | f_j \rangle = \langle f_i | \hat{T} | f_j \rangle + \langle f_i | V | f_j \rangle$. Explain why $\hat{T} f_j = \varepsilon_j f_j$ (for $0 \leq x \leq l$), and give the $\hat{T}$ eigenvalues ε_1 and ε_2. Show that $\langle f_i | \hat{T} | f_j \rangle = \delta_{ij} \varepsilon_j$. (d) Using the results of parts (b) and (c) to help evaluate the integrals, set up and solve the secular equation. Then find the variation functions ϕ_1 and ϕ_2 that correspond to W_1 and W_2. Compare W_1 and W_2 with the true energies $E_1 = 5.750345\hbar^2/ml^2$ and $E_3 = 44.808373\hbar^2/ml^2$. Compare W_1 with the value $W = 5.753112\hbar^2/ml^2$ found in Problem 8.5 using $\phi = f_1$. (e) If we want to improve on the results of part (d) by using a three-term linear variation function, what would be a logical choice for f_1, f_2, and f_3?

8.25 Apply the linear variation function

$$\phi = c_1 x^2(l - x) + c_2 x(l - x)^2 , \qquad 0 \leqslant x \leqslant l$$

to the particle in a one-dimensional box. Calculate the percent errors for the $n = 1$ and $n = 2$ energies. Sketch $x^2(l - x)$, $x(l - x)^2$, and the two approximate wave functions you find. (To help sketch the functions, find the nodes and the maxima and minima of each function.)

8.26 Show that if the change of variable $x' \equiv x - \frac{1}{2}l$ (corresponding to shifting the origin to the center of the box) is made in the functions (8.63) then f_1 and f_2 are even functions of x' and f_3 and f_4 are odd.

8.27 Verify the values given for H_{12}, H_{22}, S_{12}, and S_{22} in the example in Section 8.5.

8.28 Derive the roots (8.73) of the odd-function secular equation in the Section 8.5 example.

8.29 Derive the functions ϕ_2 and ϕ_3 in Eq. (8.76) of the Section 8.5 example.

8.30 Let the variation function of (8.40) be complex. Then $c_j = a_j + ib_j$, where a_j and b_j are real numbers. There are $2n$ parameters to be varied, namely the a_j's and b_j's. (a) Use the chain rule to show that the minimization conditions $\partial W/\partial a_i = 0$, $\partial W/\partial b_i = 0$ are equivalent to the conditions $\partial W/\partial c_i = 0$, $\partial W/\partial c_i^* = 0$. (b) Show that minimization of W leads to Eq. (8.54) and its complex conjugate, which may be discarded. Hence Eqs. (8.54) and (8.59) are valid for complex variation functions.

8.31 We wish to prove that the approximate wave functions obtained in the linear variation method are orthogonal. Let the approximate function ϕ_α have the value W_α for the variational integral and the coefficients $c_j^{(\alpha)}$ in (8.40). [We add α to distinguish the n different ϕ's.] We rewrite (8.54) as

$$\sum_k [(\langle f_i|\hat{H}|f_k \rangle - \langle f_i|f_k \rangle W_\alpha)c_k^{(\alpha)}] = 0 , \qquad i = 1, \ldots, n \qquad (8.98)$$

(a) Show that $\langle f_i|\hat{H} - W_\alpha|\phi_\alpha \rangle = 0$ by showing this integral to equal the left side of (8.98). (b) Use the result of (a) to show $\langle \phi_\beta|\hat{H} - W_\alpha|\phi_\alpha \rangle = 0$ and $\langle \phi_\alpha|\hat{H} - W_\beta|\phi_\beta \rangle^* = 0$ for all α and β. (c) Equate the two integrals in (b), and use the Hermitian property of $\hat{H}$ to show that $\langle \phi_\beta|\phi_\alpha \rangle(W_\beta - W_\alpha) = 0$. We conclude that, for $W_\alpha \neq W_\beta$, ϕ_α and ϕ_β are orthogonal. (For $W_\alpha = W_\beta$, we can form orthogonal linear combinations of ϕ_α and ϕ_β that will have the same value for the variational integral.)

8.32 Find $\mathbf{A}^*$, $\tilde{\mathbf{A}}$, and $\mathbf{A}^\dagger$, if

$$\mathbf{A} = \begin{pmatrix} 7 & 3 & 0 \\ 2 - i & 2i & i \\ 1 + i & 4 & 2 \end{pmatrix}$$

8.33 Verify the orthonormality equation (8.93) for the column vectors of a unitary matrix.

8.34 If the functions v and w are normalized and orthogonal, and if v and w are expanded in terms of the complete, orthonormal set $\{f_i\}$ as $v = \sum_i v_i f_i$ and $w = \sum_i w_i f_i$ (where the expansion coefficients v_i and w_i are constants), show that the column vectors $\mathbf{v}$ and $\mathbf{w}$ that consist of the expansion coefficients v_1, v_2, $\ldots$ and w_1, w_2, $\ldots$, respectively, are normalized and orthogonal, as defined by (8.83) and (8.84).

8.35 (a) Find the eigenvalues and normalized eigenvectors of

$$\mathbf{A} = \begin{pmatrix} 2 & 2 \\ 2 & -1 \end{pmatrix}$$

(b) Is $\mathbf{A}$ real and symmetric? Is $\mathbf{A}$ Hermitian? (c) Is the eigenvector matrix $\mathbf{C}$ orthogonal? Is the eigenvector matrix $\mathbf{C}$ unitary? (d) Write down $\mathbf{C}^{-1}$ without doing any calculations. (e) Verify that $\mathbf{C}^{-1}\mathbf{A}\mathbf{C}$ equals the diagonal matrix of eigenvalues.

8.36 For

$$\mathbf{A} = \begin{pmatrix} 2 & -2i \\ 2i & 2 \end{pmatrix}$$

find the eigenvalues and normalized eigenvectors and answer the questions of Problem 8.35(b)–(e).

8.37 Find the eigenvector $\mathbf{c}^{(2)}$ for the eigenvalue λ_2 in the example in Section 8.6.

8.38 Find the eigenvalues and normalized eigenvectors of the matrix

$$\mathbf{A} = \begin{pmatrix} -1 & 0 & -2 \\ 0 & 5 & 0 \\ -2 & 4 & 2 \end{pmatrix}$$

8.39 An efficient way to calculate the inverse of a square matrix $\mathbf{A}$ of order n is: (a) Place the nth-order unit matrix $\mathbf{I}$ at the right of the matrix $\mathbf{A}$ to form an n-row, $2n$-column array, which we denote by $(\mathbf{A}\,\vdots\,\mathbf{I})$. (b) Perform Gauss–Jordan elimination on the rows of $(\mathbf{A}\,\vdots\,\mathbf{I})$ so as to reduce the $\mathbf{A}$ portion of $(\mathbf{A}\,\vdots\,\mathbf{I})$ to the unit matrix. At the end of this process, the array will have the form $(\mathbf{I}\,\vdots\,\mathbf{B})$. The matrix $\mathbf{B}$ is $\mathbf{A}^{-1}$. (If $\mathbf{A}^{-1}$ does not exist, it will be impossible to reduce the $\mathbf{A}$ portion of the array to $\mathbf{I}$.) Use this procedure to find the inverse of the matrix in Problem 8.38.

8.40 Suppose $\hat{A}g_n = a_n g_n$. Let the eigenfunctions g_n be expanded in terms of the complete orthonormal set $\{f_i\}$ according to $g_n = \sum_k c_k^{(n)} f_k$. Substitute this expansion into $\hat{A}g_n = a_n g_n$, multiply by f_i^*, integrate over all space, and show that the set of equations $\sum_k (A_{ik} - a_n \delta_{ik}) c_k^{(n)} = 0$ for $i = 1, 2, 3, \ldots$ is obtained, where $A_{ik} \equiv \langle f_i | \hat{A} | f_k \rangle$. This set of equations has the same form as the set (8.54) with $S_{ik} = \delta_{ik}$. Thus, just as the W values and the coefficients c_k in (8.54) with $S_{ik} = \delta_{ik}$ can be found by finding the eigenvalues and eigenvectors of the $\mathbf{H}$ matrix, the eigenvalues a_n of $\hat{A}$ and the expansion coefficients $c_k^{(n)}$ of the eigenfunctions g_n of $\hat{A}$ can be found by finding the eigenvalues a_n and eigenvectors $\mathbf{c}^{(n)}$ of the matrix $\mathbf{A}$, whose elements are A_{ik}.

9

Perturbation Theory

9.1 INTRODUCTION

We now discuss the second major quantum-mechanical approximation method, perturbation theory.

Suppose we have a system with a time-independent Hamiltonian $\hat{H}$ and we are unable to solve the Schrödinger equation

$$\hat{H}\psi_n = E_n\psi_n \tag{9.1}$$

for the eigenfunctions and eigenvalues of the bound stationary states. Suppose also that the Hamiltonian $\hat{H}$ is only slightly different from the Hamiltonian $\hat{H}^0$ of a system whose Schrödinger equation

$$\hat{H}^0\psi_n^{(0)} = E_n^{(0)}\psi_n^{(0)} \tag{9.2}$$

we are able to solve. An example is the one-dimensional anharmonic oscillator with

$$\hat{H} = -\frac{\hbar^2}{2m}\frac{d^2}{dx^2} + \tfrac{1}{2}kx^2 + cx^3 + dx^4 \tag{9.3}$$

The Hamiltonian (9.3) is closely related to the Hamiltonian

$$\hat{H}^0 = -\frac{\hbar^2}{2m}\frac{d^2}{dx^2} + \tfrac{1}{2}kx^2 \tag{9.4}$$

of the harmonic oscillator. If the constants c and d in (9.3) are small, we expect the eigenfunctions and eigenvalues of the anharmonic oscillator to be closely related to those of the harmonic oscillator.

We shall call the system with Hamiltonian $\hat{H}^0$ the **unperturbed system**; the system with Hamiltonian $\hat{H}$ is the **perturbed system**. The difference between the two Hamiltonians is the **perturbation**, $\hat{H}'$:

$$\hat{H}' \equiv \hat{H} - \hat{H}^0$$

$$\hat{H} = \hat{H}^0 + \hat{H}' \tag{9.5}*$$

(Of course the prime does not refer to differentiation.) For the anharmonic oscillator with Hamiltonian (9.3), the perturbation on the related harmonic oscillator is

$$\hat{H}' = cx^3 + dx^4 \tag{9.6}$$

Our task is to relate the unknown eigenvalues and eigenfunctions of the perturbed system to the known eigenvalues and eigenfunctions of the unperturbed system. To aid in doing so, we shall imagine that the perturbation is applied gradually, giving a continuous change from the unperturbed to the perturbed system. Mathematically, this corresponds to introducing a parameter λ into the Hamiltonian, so that

$$\hat{H} = \hat{H}^0 + \lambda \hat{H}' \tag{9.7}$$

When λ is zero, we have the unperturbed system. As λ increases, the perturbation grows larger, and at $\lambda = 1$ the perturbation is fully "turned on." We have introduced λ as a convenience in relating the perturbed and unperturbed eigenfunctions, and ultimately we shall set $\lambda = 1$, thereby eliminating it.

Sections 9.2 to 9.8 deal with time-independent Hamiltonians and stationary states. Section 9.9 deals with time-dependent perturbations.

9.2 NONDEGENERATE PERTURBATION THEORY

The perturbation treatments of degenerate and nondegenerate energy levels differ. This section examines the effect of a perturbation on a nondegenerate level. If some of the energy levels of the unperturbed system are degenerate while others are nondegenerate, the treatment in this section will apply to the nondegenerate levels only.

Nondegenerate Perturbation Theory. Let $\psi_n^{(0)}$ be the wave function of some particular unperturbed nondegenerate level with energy $E_n^{(0)}$. Let ψ_n be the perturbed wave function into which $\psi_n^{(0)}$ is converted when the perturbation is applied. From (9.1) and (9.7), the Schrödinger equation for the perturbed state is

$$\hat{H}\psi_n = (\hat{H}^0 + \lambda \hat{H}')\psi_n = E_n \psi_n \tag{9.8}$$

Since the Hamiltonian in (9.8) depends on the parameter λ, both the eigenfunction ψ_n and the eigenvalue E_n depend on λ:

$$\psi_n = \psi_n(\lambda, q) \quad \text{and} \quad E_n = E_n(\lambda)$$

where q indicates the system's coordinates. We now expand ψ_n and E_n as Taylor series (Problem 4.1) in powers of λ:

$$\psi_n = \psi_n|_{\lambda=0} + \left.\frac{\partial \psi_n}{\partial \lambda}\right|_{\lambda=0} \lambda + \left.\frac{\partial^2 \psi_n}{\partial \lambda^2}\right|_{\lambda=0} \frac{\lambda^2}{2!} + \cdots \tag{9.9}$$

$$E_n = E_n|_{\lambda=0} + \left.\frac{dE_n}{d\lambda}\right|_{\lambda=0} \lambda + \left.\frac{d^2 E_n}{d\lambda^2}\right|_{\lambda=0} \frac{\lambda^2}{2!} + \cdots \tag{9.10}$$

By hypothesis, when λ goes to zero, ψ_n and E_n go to $\psi_n^{(0)}$ and $E_n^{(0)}$:

$$\psi_n|_{\lambda=0} = \psi_n^{(0)} \quad \text{and} \quad E_n|_{\lambda=0} = E_n^{(0)} \tag{9.11}$$

We introduce the following abbreviations:

$$\psi_n^{(k)} = \frac{1}{k!} \frac{\partial^k \psi_n}{\partial \lambda^k}\bigg|_{\lambda=0}, \qquad E_n^{(k)} = \frac{1}{k!} \frac{d^k E_n}{d\lambda^k}\bigg|_{\lambda=0}, \qquad k = 1, 2, \ldots \quad (9.12)$$

Equations (9.9) and (9.10) become

$$\psi_n = \psi_n^{(0)} + \lambda \psi_n^{(1)} + \lambda^2 \psi_n^{(2)} + \cdots + \lambda^k \psi_n^{(k)} + \cdots \qquad (9.13)$$

$$E_n = E_n^{(0)} + \lambda E_n^{(1)} + \lambda^2 E_n^{(2)} + \cdots + \lambda^k E_n^{(k)} + \cdots \qquad (9.14)$$

We call $\psi_n^{(k)}$ and $E_n^{(k)}$ the **kth-order corrections** to the wave function and energy. We shall assume that the series (9.13) and (9.14) converge for $\lambda = 1$, and we hope that for a small perturbation, taking just the first few terms of the series will provide a good approximation to the true energy and wave function.

We shall take $\psi_n^{(0)}$ to be normalized: $\langle \psi_n^{(0)} | \psi_n^{(0)} \rangle = 1$. Instead of taking ψ_n as normalized, we shall require that ψ_n satisfy

$$\langle \psi_n^{(0)} | \psi_n \rangle = 1$$

If ψ_n does not satisfy this equation, then multiplication of ψ_n by the constant $1/\langle \psi_n^{(0)} | \psi_n \rangle$ gives a perturbed wave function with the desired property. The condition $\langle \psi_n^{(0)} | \psi_n \rangle = 1$, called *intermediate normalization*, simplifies the derivation. Note that multiplication of ψ_n by a constant does not change the energy in the Schrödinger equation $\hat{H}\psi_n = E_n\psi_n$, so use of intermediate normalization does not affect the results for the energy corrections. If desired, at the end of the calculation, the intermediate-normalized ψ_n can be multiplied by a constant to normalize it in the usual sense.

Substitution of (9.13) into the intermediate-normalization equation $1 = \langle \psi_n^{(0)} | \psi_n \rangle$ gives

$$1 = \langle \psi_n^{(0)} | \psi_n^{(0)} \rangle + \lambda \langle \psi_n^{(0)} | \psi_n^{(1)} \rangle + \lambda^2 \langle \psi_n^{(0)} | \psi_n^{(2)} \rangle + \cdots$$

Since this equation is true for all values of λ in the range 0 to 1, the coefficients of like powers of λ on each side of the equation must be equal, as proved after Eq. (4.13). Equating the λ^0 coefficients, we have $1 = \langle \psi_n^{(0)} | \psi_n^{(0)} \rangle$, which is satisfied since $\psi_n^{(0)}$ is normalized. Equating the coefficients of λ^1, of λ^2, and so on, we have

$$\langle \psi_n^{(0)} | \psi_n^{(1)} \rangle = 0, \qquad \langle \psi_n^{(0)} | \psi_n^{(2)} \rangle = 0, \qquad \text{etc.} \qquad (9.15)$$

The corrections to the wave function are orthogonal to $\psi_n^{(0)}$ when intermediate normalization is used.

Substituting (9.13) and (9.14) into (9.8), we have

$$(\hat{H}^0 + \lambda \hat{H}')(\psi_n^{(0)} + \lambda \psi_n^{(1)} + \lambda^2 \psi_n^{(2)} + \cdots)$$

$$= (E_n^{(0)} + \lambda E_n^{(1)} + \lambda^2 E_n^{(2)} + \cdots)(\psi_n^{(0)} + \lambda \psi_n^{(1)} + \lambda^2 \psi_n^{(2)} + \cdots)$$

Collecting like powers of λ, we have

$$\hat{H}^0 \psi_n^{(0)} + \lambda(\hat{H}' \psi_n^{(0)} + \hat{H}^0 \psi_n^{(1)}) + \lambda^2(\hat{H}^0 \psi_n^{(2)} + \hat{H}' \psi_n^{(1)}) + \cdots$$

$$= E_n^{(0)} \psi_n^{(0)} + \lambda(E_n^{(1)} \psi_n^{(0)} + E_n^{(0)} \psi_n^{(1)}) + \lambda^2(E_n^{(2)} \psi_n^{(0)} + E_n^{(1)} \psi_n^{(1)} + E_n^{(0)} \psi_n^{(2)}) + \cdots$$

$$(9.16)$$

Now (assuming suitable convergence) for the two series on each side of (9.16) to be equal to each other for all values of λ, the coefficients of like powers of λ in the two series must be equal.

Equating the coefficients of the λ^0 terms, we have

$$\hat{H}^0 \psi_n^{(0)} = E_n^{(0)} \psi_n^{(0)} \tag{9.17}$$

which is the Schrödinger equation for the unperturbed problem, Eq. (9.2), and gives us no new information.

Equating the coefficients of the λ^1 terms, we have

$$\hat{H}' \psi_n^{(0)} + \hat{H}^0 \psi_n^{(1)} = E_n^{(1)} \psi_n^{(0)} + E_n^{(0)} \psi_n^{(1)}$$

$$\hat{H}^0 \psi_n^{(1)} - E_n^{(0)} \psi_n^{(1)} = E_n^{(1)} \psi_n^{(0)} - \hat{H}' \psi_n^{(0)} \tag{9.18}$$

The First-Order Energy Correction.   To find $E_n^{(1)}$, we multiply (9.18) by $\psi_m^{(0)*}$ and integrate over all space, which gives

$$\langle \psi_m^{(0)} | \hat{H}^0 | \psi_n^{(1)} \rangle - E_n^{(0)} \langle \psi_m^{(0)} | \psi_n^{(1)} \rangle = E_n^{(1)} \langle \psi_m^{(0)} | \psi_n^{(0)} \rangle - \langle \psi_m^{(0)} | \hat{H}' | \psi_n^{(0)} \rangle \tag{9.19}$$

where the bracket notation of Eqs. (7.1) and (7.3) is used. $\hat{H}^0$ is Hermitian, and use of the Hermitian property (7.12) gives for the first term on the left side of (9.19)

$$\langle \psi_m^{(0)} | \hat{H}^0 | \psi_n^{(1)} \rangle = \langle \psi_n^{(1)} | \hat{H}^0 | \psi_m^{(0)} \rangle^* = \langle \psi_n^{(1)} | \hat{H}^0 \psi_m^{(0)} \rangle^*$$

$$= \langle \psi_n^{(1)} | E_m^{(0)} \psi_m^{(0)} \rangle^* = E_m^{(0)*} \langle \psi_n^{(1)} | \psi_m^{(0)} \rangle^* = E_m^{(0)} \langle \psi_m^{(0)} | \psi_n^{(1)} \rangle \tag{9.20}$$

where we used the unperturbed Schrödinger equation $\hat{H}^0 \psi_m^{(0)} = E_m^{(0)} \psi_m^{(0)}$, the fact that $E_m^{(0)}$ is real, and Eq. (7.4). Substitution of (9.20) into (9.19) and use of the orthonormality equation $\langle \psi_m^{(0)} | \psi_n^{(0)} \rangle = \delta_{mn}$ for the unperturbed eigenfunctions gives

$$(E_m^{(0)} - E_n^{(0)}) \langle \psi_m^{(0)} | \psi_n^{(1)} \rangle = E_n^{(1)} \delta_{mn} - \langle \psi_m^{(0)} | \hat{H}' | \psi_n^{(0)} \rangle \tag{9.21}$$

If $m = n$, the left side of (9.21) equals zero, and (9.21) becomes

$$E_n^{(1)} = \langle \psi_n^{(0)} | \hat{H}' | \psi_n^{(0)} \rangle = \int \psi_n^{(0)*} \hat{H}' \psi_n^{(0)} \, d\tau \tag{9.22}^*$$

The first-order correction to the energy is found by averaging the perturbation $\hat{H}'$ over the appropriate unperturbed wave function.

Setting $\lambda = 1$ in (9.14), we have

$$E_n \approx E_n^{(0)} + E_n^{(1)} = E_n^{(0)} + \int \psi_n^{(0)*} \hat{H}' \psi_n^{(0)} \, d\tau \tag{9.23}$$

EXAMPLE   For the anharmonic oscillator with Hamiltonian (9.3), evaluate $E^{(1)}$ for the ground state if the unperturbed system is taken as the harmonic oscillator.

The perturbation is given by Eqs. (9.3) to (9.5) as

$$\hat{H}' = \hat{H} - \hat{H}^0 = cx^3 + dx^4$$

and the first-order energy correction for the state with quantum number v is given by (9.22) as $E_v^{(1)} = \langle \psi_v^{(0)} | cx^3 + dx^4 | \psi_v^{(0)} \rangle$, where $\psi_v^{(0)}$ is the harmonic-oscillator wave function for state v. For the $v = 0$ ground state, use of $\psi_0^{(0)} = (\alpha/\pi)^{1/4} e^{-\alpha x^2/2}$ [Eq. (4.59)] gives

$$E_0^{(1)} = \langle \psi_0^{(0)} | cx^3 + dx^4 | \psi_0^{(0)} \rangle = \left(\frac{\alpha}{\pi}\right)^{1/2} \int_{-\infty}^{\infty} e^{-\alpha x^2} (cx^3 + dx^4)\, dx$$

The integral from $-\infty$ to ∞ of the odd function $cx^3 e^{-\alpha x^2}$ is zero; use of the Appendix integral (A.9) with $n = 2$ and (4.33) for α gives

$$E_0^{(1)} = 2d\left(\frac{\alpha}{\pi}\right)^{1/2} \int_0^{\infty} e^{-\alpha x^2} x^4\, dx = \frac{3d}{4\alpha^2} = \frac{3dh^2}{64\pi^4 v^2 m^2}$$

The unperturbed ground-state energy is $E_0^{(0)} = \frac{1}{2}hv$.

The First-Order Wave-Function Correction.

For $m \neq n$, Eq. (9.21) becomes

$$(E_m^{(0)} - E_n^{(0)})\langle \psi_m^{(0)} | \psi_n^{(1)} \rangle = -\langle \psi_m^{(0)} | \hat{H}' | \psi_n^{(0)} \rangle, \qquad m \neq n \qquad (9.24)$$

To find $\psi_n^{(1)}$, we expand it in terms of the complete, orthonormal set of unperturbed eigenfunctions $\psi_m^{(0)}$ of the Hermitian operator $\hat{H}^0$:

$$\psi_n^{(1)} = \sum_m a_m \psi_m^{(0)}, \qquad \text{where } a_m = \langle \psi_m^{(0)} | \psi_n^{(1)} \rangle \qquad (9.25)$$

where Eq. (7.41) was used for the expansion coefficients a_m. Use of $a_m = \langle \psi_m^{(0)} | \psi_n^{(1)} \rangle$ in (9.24) gives

$$(E_m^{(0)} - E_n^{(0)}) a_m = -\langle \psi_m^{(0)} | \hat{H}' | \psi_n^{(0)} \rangle, \qquad m \neq n$$

By hypothesis, the level $E_n^{(0)}$ is nondegenerate; hence $E_m^{(0)} \neq E_n^{(0)}$ for $m \neq n$, and we may divide by $(E_m^{(0)} - E_n^{(0)})$ to get

$$a_m = \frac{\langle \psi_m^{(0)} | \hat{H}' | \psi_n^{(0)} \rangle}{E_n^{(0)} - E_m^{(0)}}, \qquad m \neq n \qquad (9.26)$$

The coefficients a_m in the expansion (9.25) of $\psi_n^{(1)}$ are given by (9.26) except for a_n, the coefficient of $\psi_n^{(0)}$. From the second equation in (9.25), $a_n = \langle \psi_n^{(0)} | \psi_n^{(1)} \rangle$. Recall that the choice of intermediate normalization for ψ_n makes $\langle \psi_n^{(0)} | \psi_n^{(1)} \rangle = 0$ [Eq. (9.15)]. Hence $a_n = \langle \psi_n^{(0)} | \psi_n^{(1)} \rangle = 0$, and Eqs. (9.25) and (9.26) give the first-order correction to the wave function as

$$\psi_n^{(1)} = \sum_{m \neq n} \frac{\langle \psi_m^{(0)} | \hat{H}' | \psi_n^{(0)} \rangle}{E_n^{(0)} - E_m^{(0)}} \psi_m^{(0)} \qquad (9.27)$$

The symbol $\sum_{m \neq n}$ means we sum over all the unperturbed states except state n.

Setting $\lambda = 1$ in (9.13) and using just the first-order wave-function correction, we have as the approximation to the perturbed wave function

$$\psi_n \approx \psi_n^{(0)} + \sum_{m \neq n} \frac{\langle \psi_m^{(0)} | \hat{H}' | \psi_n^{(0)} \rangle}{E_n^{(0)} - E_m^{(0)}} \psi_m^{(0)} \qquad (9.28)$$

[For $\psi_n^{(2)}$ and the normalization of ψ, see *Kemble*, Chapter XI.]

The Second-Order Energy Correction. Equating the coefficients of the λ^2 terms in (9.16), we get

$$\hat{H}^0\psi_n^{(2)} - E_n^{(0)}\psi_n^{(2)} = E_n^{(2)}\psi_n^{(0)} + E_n^{(1)}\psi_n^{(1)} - \hat{H}'\psi_n^{(1)} \tag{9.29}$$

Multiplication by $\psi_m^{(0)*}$ followed by integration over all space gives

$$\langle\psi_m^{(0)}|\hat{H}^0|\psi_n^{(2)}\rangle - E_n^{(0)}\langle\psi_m^{(0)}|\psi_n^{(2)}\rangle$$
$$= E_n^{(2)}\langle\psi_m^{(0)}|\psi_n^{(0)}\rangle + E_n^{(1)}\langle\psi_m^{(0)}|\psi_n^{(1)}\rangle - \langle\psi_m^{(0)}|\hat{H}'|\psi_n^{(1)}\rangle \tag{9.30}$$

The integral $\langle\psi_m^{(0)}|\hat{H}^0|\psi_n^{(2)}\rangle$ in this equation is the same as the integral in (9.20), except that $\psi_n^{(1)}$ is replaced by $\psi_n^{(2)}$. Replacement of $\psi_n^{(1)}$ by $\psi_n^{(2)}$ in (9.20) gives

$$\langle\psi_m^{(0)}|\hat{H}^0|\psi_n^{(2)}\rangle = E_m^{(0)}\langle\psi_m^{(0)}|\psi_n^{(2)}\rangle \tag{9.31}$$

Use of (9.31) and orthonormality of the unperturbed functions in (9.30) gives

$$(E_m^{(0)} - E_n^{(0)})\langle\psi_m^{(0)}|\psi_n^{(2)}\rangle = E_n^{(2)}\delta_{mn} + E_n^{(1)}\langle\psi_m^{(0)}|\psi_n^{(1)}\rangle - \langle\psi_m^{(0)}|\hat{H}'|\psi_n^{(1)}\rangle \tag{9.32}$$

For $m = n$, the left side of (9.32) is zero and we get

$$E_n^{(2)} = -E_n^{(1)}\langle\psi_n^{(0)}|\psi_n^{(1)}\rangle + \langle\psi_n^{(0)}|\hat{H}'|\psi_n^{(1)}\rangle$$
$$E_n^{(2)} = \langle\psi_n^{(0)}|\hat{H}'|\psi_n^{(1)}\rangle \tag{9.33}$$

since $\langle\psi_n^{(0)}|\psi_n^{(1)}\rangle = 0$ [Eq. (9.15)]. Note from (9.33) that to determine the *second*-order correction to the energy, we have to know only the *first*-order correction to the wave function. In fact, it can be shown that knowledge of $\psi_n^{(1)}$ suffices to determine $E_n^{(3)}$ also; in general, it can be shown that if we know the corrections to the wave function through the kth order then we can compute the corrections to the energy through order $2k + 1$ (see *Bates*, Vol. I, p. 184).

Substitution of (9.27) for $\psi_n^{(1)}$ into (9.33) gives

$$E_n^{(2)} = \sum_{m\neq n} \frac{\langle\psi_m^{(0)}|\hat{H}'|\psi_n^{(0)}\rangle}{E_n^{(0)} - E_m^{(0)}} \langle\psi_n^{(0)}|\hat{H}'|\psi_m^{(0)}\rangle \tag{9.34}$$

since the expansion coefficients a_m [Eq. (9.26)] are constants that can be taken outside the integral. Since $\hat{H}'$ is Hermitian, we have

$$\langle\psi_m^{(0)}|\hat{H}'|\psi_n^{(0)}\rangle\langle\psi_n^{(0)}|\hat{H}'|\psi_m^{(0)}\rangle = \langle\psi_m^{(0)}|\hat{H}'|\psi_n^{(0)}\rangle\langle\psi_m^{(0)}|\hat{H}'|\psi_n^{(0)}\rangle^*$$
$$= |\langle\psi_m^{(0)}|\hat{H}'|\psi_n^{(0)}\rangle|^2$$

and

$$E_n^{(2)} = \sum_{m\neq n} \frac{|\langle\psi_m^{(0)}|\hat{H}'|\psi_n^{(0)}\rangle|^2}{E_n^{(0)} - E_m^{(0)}} \tag{9.35}$$

which is the desired expression for $E_n^{(2)}$ in terms of the unperturbed wave functions and energies.

Inclusion of $E_n^{(2)}$ in (9.14) with $\lambda = 1$ gives as the approximate energy of the perturbed state

$$E_n \approx E_n^{(0)} + H_{nn}' + \sum_{m \neq n} \frac{|H_{mn}'|^2}{E_n^{(0)} - E_m^{(0)}} \tag{9.36}$$

where the integrals are over the unperturbed normalized wave functions.

For formulas for higher-order energy corrections, see *Bates*, Volume I, pages 181–185. (The form of perturbation theory developed in this section is called *Rayleigh–Schrödinger perturbation theory*; other approaches exist.)

Discussion. Equation (9.28) shows that the effect of the perturbation on the wave function $\psi_n^{(0)}$ is to "mix in" contributions from other states $\psi_m^{(0)}$, $m \neq n$. Because of the factor $1/(E_n^{(0)} - E_m^{(0)})$, the most important contributions (aside from $\psi_n^{(0)}$) to the perturbed wave function come from states nearest in energy to state n.

To evaluate the first-order correction to the energy, we must evaluate only the single integral H_{nn}', whereas to evaluate the second-order energy correction, we must evaluate the matrix elements of $\hat{H}'$ between the nth state and all other states m, and then perform the infinite sum in (9.35). In many cases it is impossible to evaluate exactly the second-order energy correction. Third-order and higher-order energy corrections are even more difficult to deal with.

The sums in (9.28) and (9.36) are sums over different states rather than sums over different energy values. If some of the energy levels (other than the nth) are degenerate, we must include a term in the sums for each linearly independent wave function corresponding to the degenerate levels.

The reason we have a sum over states in (9.28) and (9.36) is that we require a complete set of functions for the expansion (9.25), and therefore we must include all linearly independent wave functions in the sum. If the unperturbed problem has continuum wave functions (for example, the hydrogen atom), we must also include an integration over the continuum functions, if we are to have a complete set. If $\psi_E^{(0)}$ denotes an unperturbed continuum wave function of energy $E^{(0)}$, then (9.27) and (9.35) become

$$\psi_n^{(1)} = \sum_{m \neq n} \frac{H_{mn}'}{E_n^{(0)} - E_m^{(0)}} \psi_m^{(0)} + \int \frac{H_{E,n}'}{E_n^{(0)} - E^{(0)}} \psi_E^{(0)} \, dE^{(0)}$$

$$E_n^{(2)} = \sum_{m \neq n} \frac{|H_{mn}'|^2}{E_n^{(0)} - E_m^{(0)}} + \int \frac{|H_{E,n}'|^2}{E_n^{(0)} - E^{(0)}} \, dE^{(0)}$$

where $H_{E,n}' \equiv \langle \psi_E^{(0)} | \hat{H}' | \psi_n^{(0)} \rangle$. The integrals in these equations are over the range of continuum-state energies (for example, from zero to infinity for the hydrogen atom). The existence of continuum states in the unperturbed problem makes evaluation of $E_n^{(2)}$ even more difficult.

The Variation–Perturbation Method. The variation–perturbation method allows one to accurately estimate $E^{(2)}$ and higher-order perturbation-theory energy corrections for the ground state of a system. The method is based on the inequality

$$\langle u|\hat{H}^0 - E_g^{(0)}|u\rangle + \langle u|\hat{H}' - E_g^{(1)}|\psi_g^{(0)}\rangle + \langle \psi_g^{(0)}|\hat{H}' - E_g^{(1)}|u\rangle \geq E_g^{(2)} \quad (9.37)$$

where u is any well-behaved function that satisfies the boundary conditions, and where the subscript g refers to the ground-state. For the proof of (9.37), see *Hameka*, Section 7-9. By taking u to be a trial function with parameters that we vary to minimize the left side of (9.37), we can estimate $E_g^{(2)}$. The function u turns out to be an approximation to $\psi_g^{(1)}$, the first-order correction to the ground-state wave function, and u can then be used to estimate $E_g^{(3)}$ also. Similar variational integrals can be used to find higher-order corrections to the ground-state energy and wave function.

9.3 PERTURBATION TREATMENT OF THE HELIUM-ATOM GROUND STATE

The helium atom has two electrons and a nucleus of charge $+2e$. We shall consider the nucleus to be at rest (Section 6.6) and place the origin of the coordinate system at the nucleus. The coordinates of electrons 1 and 2 are (x_1, y_1, z_1) and (x_2, y_2, z_2); see Fig. 9.1.

If we take the nuclear charge to be $+Ze$ instead of $+2e$, we can treat heliumlike ions such as H^-, Li^+, Be^{2+}. The Hamiltonian operator is

$$\hat{H} = -\frac{\hbar^2}{2m_e}\nabla_1^2 - \frac{\hbar^2}{2m_e}\nabla_2^2 - \frac{Ze'^2}{r_1} - \frac{Ze'^2}{r_2} + \frac{e'^2}{r_{12}} \quad (9.38)$$

where m_e is the mass of the electron, r_1 and r_2 are the distances of electrons 1 and 2 from the nucleus, and r_{12} is the distance from electron 1 to 2. The first two terms are the operators for the electrons' kinetic energy; the third and fourth terms are the potential energies of attraction between the electrons and the nucleus; the final term is the potential energy of interelectronic repulsion [Eq. (6.58)]. Note that the potential energy of a system of interacting particles cannot be written as the sum of potential energies of the individual particles; the potential energy is a property of the system as a whole.

The Schrödinger equation involves six independent variables, three coordinates for each electron. In spherical polar coordinates,

$$\psi = \psi(r_1, \theta_1, \phi_1, r_2, \theta_2, \phi_2) \quad (9.39)$$

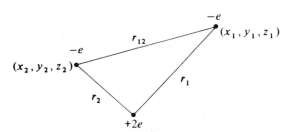

Figure 9.1 Interparticle distances in the helium atom.

The operator ∇_1^2 is given by Eq. (6.6) with r_1, θ_1, ϕ_1 replacing r, θ, ϕ. The variable r_{12} is $r_{12} = [(x_1 - x_2)^2 + (y_1 - y_2)^2 + (z_1 - z_2)^2]^{1/2}$, and by using the relations between Cartesian and spherical polar coordinates, we can express r_{12} in terms of the coordinates of (9.39).

Because of the $1/r_{12}$ term, the Schrödinger equation for helium cannot be separated in any coordinate system, and we must use approximation methods. To use the perturbation method, we must separate the Hamiltonian (9.38) into two parts, $\hat{H}^0$ and $\hat{H}'$, where $\hat{H}^0$ is the Hamiltonian of an exactly solvable problem. If we choose

$$\hat{H}^0 = -\frac{\hbar^2}{2m_e} \nabla_1^2 - \frac{Ze'^2}{r_1} - \frac{\hbar^2}{2m_e} \nabla_2^2 - \frac{Ze'^2}{r_2} \qquad (9.40)$$

$$\hat{H}' = \frac{e'^2}{r_{12}} \qquad (9.41)$$

then $\hat{H}^0$ is the sum of two hydrogenlike Hamiltonians, one for each electron:

$$\hat{H}^0 = \hat{H}_1^0 + \hat{H}_2^0 \qquad (9.42)$$

$$\hat{H}_1^0 \equiv -\frac{\hbar^2}{2m_e} \nabla_1^2 - \frac{Ze'^2}{r_1} , \qquad \hat{H}_2^0 \equiv -\frac{\hbar^2}{2m_e} \nabla_2^2 - \frac{Ze'^2}{r_2} \qquad (9.43)$$

The unperturbed system is a helium atom in which the two electrons exert no forces on each other. Although there is no physical reality to such a system, this does not prevent us from applying perturbation theory to this system.

Since the unperturbed Hamiltonian (9.42) is the sum of the Hamiltonians for two independent particles, we can use the results of Eqs. (6.18) to (6.24) to conclude that the unperturbed wave functions have the form

$$\psi^{(0)}(r_1, \theta_1, \phi_1, r_2, \theta_2, \phi_2) = F_1(r_1, \theta_1, \phi_1) F_2(r_2, \theta_2, \phi_2) \qquad (9.44)$$

and the unperturbed energies are

$$E^{(0)} = E_1 + E_2 \qquad (9.45)$$

$$\hat{H}_1^0 F_1 = E_1 F_1 , \qquad \hat{H}_2^0 F_2 = E_2 F_2 \qquad (9.46)$$

Since $\hat{H}_1^0$ and $\hat{H}_2^0$ are hydrogenlike Hamiltonians, the solutions of (9.46) are the hydrogenlike eigenfunctions and eigenvalues. From Eq. (6.94), we have

$$E_1 = -\frac{Z^2}{n_1^2} \frac{e'^2}{2a_0} , \qquad E_2 = -\frac{Z^2}{n_2^2} \frac{e'^2}{2a_0} \qquad (9.47)$$

$$E^{(0)} = -Z^2 \left(\frac{1}{n_1^2} + \frac{1}{n_2^2} \right) \frac{e'^2}{2a_0} , \qquad \begin{array}{l} n_1 = 1, 2, 3, \ldots \\ n_2 = 1, 2, 3, \ldots \end{array} \qquad (9.48)$$

where a_0 is the Bohr radius. Equation (9.48) gives the zeroth-order energies of states with both electrons bound to the nucleus; we also have continuum states.

The lowest level has $n_1 = 1$, $n_2 = 1$, and its zeroth-order wave function is [Eq. (6.104)]

$$\psi_{1s^2}^{(0)} = \frac{1}{\pi^{1/2}} \left(\frac{Z}{a_0} \right)^{3/2} e^{-Zr_1/a_0} \cdot \frac{1}{\pi^{1/2}} \left(\frac{Z}{a_0} \right)^{3/2} e^{-Zr_2/a_0} \qquad (9.49)$$

where the subscript indicates that both electrons are in hydrogenlike $1s$ orbitals. (Note that the procedure of assigning electrons to orbitals and writing the atomic wave function as the product of one-electron orbital functions is an *approximation*.) The energy of this unperturbed ground state is

$$E_{1s^2}^{(0)} = -Z^2(2)\frac{e'^2}{2a_0} \tag{9.50}$$

The quantity $-\frac{1}{2}e'^2/a_0$ is the ground-state energy of the hydrogen atom (taking the nucleus to be infinitely heavy) and equals -13.606 eV (Section 6.6). For helium, $Z = 2$, and

$$E_{1s^2}^{(0)} = -108.8 \text{ eV} \tag{9.51}$$

How does this zeroth-order energy compare with the true ground-state energy of helium? The experimental first ionization energy of He is 24.6 eV. The second ionization energy of He is easily calculated theoretically, since it is the ionization energy of the hydrogenlike ion He$^+$ and is equal to $2^2(13.606 \text{ eV}) = 54.4 \text{ eV}$. If we choose the zero of energy as the completely ionized atom [this choice is implicit in (9.38)], then the ground-state energy of the helium atom is $-(24.6 + 54.4) \text{ eV} = -79.0 \text{ eV}$. The zeroth-order energy is in error by 38%. We should have expected such a large error, since the perturbation term e'^2/r_{12} is not particularly small.

The next step is to evaluate the first-order perturbation correction to the energy. The unperturbed ground state is nondegenerate, and

$$E^{(1)} = \langle \psi^{(0)} | \hat{H}' | \psi^{(0)} \rangle$$

$$E^{(1)} = \frac{Z^6 e'^2}{\pi^2 a_0^6} \int_0^{2\pi}\int_0^{2\pi}\int_0^\pi\int_0^\pi\int_0^\infty\int_0^\infty e^{-2Zr_1/a_0}e^{-2Zr_2/a_0}\frac{1}{r_{12}}r_1^2 \sin\theta_1$$

$$\times r_2^2 \sin\theta_2 \, dr_1 \, dr_2 \, d\theta_1 \, d\theta_2 \, d\phi_1 \, d\phi_2 \tag{9.52}$$

The volume element for this two-electron problem contains the coordinates of both electrons; $d\tau = d\tau_1 \, d\tau_2$. The evaluation of the integral may be skimmed, if desired; begin reading again at Eq. (9.55).

[Handwritten margin notes: "Volume element" with arrow; "$\int r_{12} \, d\tau \Rightarrow$"; "$r_{12} = r_1^2 r_2^2 \sin\theta_1 \sin\theta_2$"; "$d\tau = r_1^2 r_2^2 \sin\theta_1 \, d\theta_1 \, d\phi_1 \, d\phi_2$"; "$dr_1 \, dr_2 \, d\theta_1 \, d\theta_2 \, d\phi_1 \, d\phi_2$"; "six-fold integral"]

To evaluate the integral in (9.52), we use an expansion of $1/r_{12}$ in terms of spherical harmonics. One can show that (see *Eyring, Walter, and Kimball*, p. 369)

$$\frac{1}{r_{12}} = \sum_{l=0}^\infty \sum_{m=-l}^l \frac{4\pi}{2l+1}\frac{r_<^l}{r_>^{l+1}}[Y_l^m(\theta_1, \phi_1)]^* Y_l^m(\theta_2, \phi_2) \tag{9.53}$$

where $r_<$ means the smaller of r_1 and r_2, and $r_>$ is the larger of r_1 and r_2. Substituting (9.53) into (9.52) and multiplying and dividing by $Y_0^0(Y_0^0)^* = 1/4\pi$, we get

$$E^{(1)} = \frac{16Z^6 e'^2}{a_0^6}\sum_{l=0}^\infty\sum_{m=-l}^l\frac{1}{2l+1}\int_0^\infty\int_0^\infty e^{-2Zr_1/a_0}e^{-2Zr_2/a_0}\frac{r_<^l}{r_>^{l+1}}r_1^2 r_2^2 \, dr_1 \, dr_2$$

$$\times \int_0^{2\pi}\int_0^\pi [Y_l^m(\theta_1, \phi_1)]^* Y_0^0(\theta_1, \phi_1) \sin\theta_1 \, d\theta_1 \, d\phi_1$$

$$\times \int_0^{2\pi}\int_0^\pi [Y_0^0(\theta_2, \phi_2)]^* Y_l^m(\theta_2, \phi_2) \sin\theta_2 \, d\theta_2 \, d\phi_2$$

Recalling the orthonormality of the spherical harmonics [Eq. (7.27)], we have

$$E^{(1)} = \frac{16Z^6 e'^2}{a_0^6} \sum_{l=0}^{\infty} \sum_{m=-l}^{l} \frac{1}{2l+1} \int_0^{\infty} \int_0^{\infty} (\cdots) \, dr_1 \, dr_2 \, \delta_{l,0}\delta_{m,0}\delta_{l,0}\delta_{m,0}$$

The Kronecker deltas make all terms vanish except the single term with $m = 0 = l$, and

$$E^{(1)} = \frac{16Z^6 e'^2}{a_0^6} \int_0^{\infty} \int_0^{\infty} e^{-2Zr_1/a_0} e^{-2Zr_2/a_0} \frac{1}{r_>} r_1^2 r_2^2 \, dr_1 \, dr_2$$

If we integrate first over r_1, then in the range $0 \leqslant r_1 \leqslant r_2$, we have $r_> = r_2$; in the range $r_2 \leqslant r_1 \leqslant \infty$, we have $r_> = r_1$. Thus

$$E^{(1)} = \frac{16Z^6 e'^2}{a_0^6} \int_0^{\infty} e^{-2Zr_2/a_0} r_2^2 \left[\int_0^{r_2} e^{-2Zr_1/a_0} \frac{r_1^2}{r_2} \, dr_1 + \int_{r_2}^{\infty} e^{-2Zr_1/a_0} \frac{r_1^2}{r_1} \, dr_1 \right] dr_2$$

$$E^{(1)} = \frac{16Z^6 e'^2}{a_0^6} \int_0^{\infty} e^{-2Zr_2/a_0} r_2 \left(\int_0^{r_2} e^{-2Zr_1/a_0} r_1^2 \, dr_1 \right) dr_2$$

$$+ \frac{16Z^6 e'^2}{a_0^6} \int_0^{\infty} e^{-2Zr_2/a_0} r_2^2 \left(\int_{r_2}^{\infty} e^{-2Zr_1/a_0} r_1 \, dr_1 \right) dr_2 \tag{9.54}$$

Using the indefinite integrals (A.5) and (A.6) in the Appendix, we can do the r_1 integrals to obtain r_2 integrals, which are evaluated using (A.7). The result is

$$E^{(1)} = \frac{5Z}{8} \left(\frac{e'^2}{a_0} \right) \tag{9.55}$$

Recalling that $\frac{1}{2}e'^2/a_0 = 13.606 \text{ eV}$ and putting $Z = 2$, we find for the first-order perturbation energy correction for the helium ground state:

$$E^{(1)} = \tfrac{10}{4}(13.606 \text{ eV}) = 34.0 \text{ eV}$$

Our approximation to the energy is now

$$E^{(0)} + E^{(1)} = -108.8 \text{ eV} + 34.0 \text{ eV} = -74.8 \text{ eV} \tag{9.56}$$

which, compared with the experimental value of -79.0 eV, is in error by 5.3%.

To evaluate the first-order correction to the wave function and higher-order corrections to the energy requires evaluating the matrix elements of $1/r_{12}$ between the ground unperturbed state and all excited states (including the continuum) and performing the appropriate summations and integrations. No one has yet figured out how to evaluate directly all the contributions to $E^{(2)}$. Note that the effect of $\psi^{(1)}$ is to mix into the wave function contributions from other configurations besides $1s^2$; we call this *configuration interaction*. The largest contribution to the true ground-state wave function of helium will naturally come from the $1s^2$ configuration, which is the unperturbed (zeroth-order) wave function.

$E^{(2)}$ for the helium ground state has been evaluated using the variation–perturbation method, Eq. (9.37). Scherr and Knight used 100-term trial functions to get extremely accurate approximations to the wave-function corrections through sixth order and thus to the energy corrections through thirteenth order [C. W. Scherr and R. E. Knight, *Rev. Mod. Phys.*, **35**, 436 (1963); for calculations of the energy corrections through twenty-first order, see J. Midtal, *Phys. Rev.*, **138**, A1010 (1965)]. The second-order correction $E^{(2)}$ turns out to be

−4.3 eV, and $E^{(3)}$ is +0.1 eV. Through third order, we have for the ground-state energy

$$E \approx -108.8 \text{ eV} + 34.0 \text{ eV} - 4.3 \text{ eV} + 0.1 \text{ eV} = -79.0 \text{ eV}$$

which agrees (within 0.1 eV) with the experimental value −79.0 eV. Including corrections through thirteenth order, Scherr and Knight obtained a ground-state energy for helium of $-2.90372433(e'^2/a_0)$, which is nearly as good as the value $-2.90372438(e'^2/a_0)$ obtained from the purely variational calculations described in the next section.

9.4 VARIATION TREATMENTS OF THE GROUND STATE OF HELIUM

In the last section, we wrote the helium-atom Hamiltonian as $\hat{H} = \hat{H}^0 + \hat{H}'$, where the ground-state eigenfunction $\psi_g^{(0)}$ of $\hat{H}^0$ is (9.49). What happens if we use the zeroth-order, perturbation-theory ground-state wave function $\psi_g^{(0)}$ as the variation function ϕ in the variational integral? The variational integral $\langle \phi | \hat{H} | \phi \rangle = \langle \phi | \hat{H} \phi \rangle$ then becomes

$$\langle \phi | \hat{H} | \phi \rangle = \langle \psi_g^{(0)} | (\hat{H}^0 + \hat{H}') \psi_g^{(0)} \rangle = \langle \psi_g^{(0)} | \hat{H}^0 \psi_g^{(0)} + \hat{H}' \psi_g^{(0)} \rangle$$

$$= \langle \psi_g^{(0)} | E_g^{(0)} \psi_g^{(0)} \rangle + \langle \psi_g^{(0)} | \hat{H}' \psi_g^{(0)} \rangle = E_g^{(0)} + E_g^{(1)} \qquad (9.57)$$

since $\hat{H}^0 \psi_g^{(0)} = E_g^{(0)} \psi_g^{(0)}$, $\langle \psi_g^{(0)} | \psi_g^{(0)} \rangle = 1$, and $E_g^{(1)} = \langle \psi_g^{(0)} | \hat{H}' | \psi_g^{(0)} \rangle$ [Eq. (9.23)]. Use of $\psi_g^{(0)}$ as the variation function gives the same energy result as in first-order perturbation theory.

Now consider variation functions for the helium-atom ground state. If we used $\psi_g^{(0)}$ [Eq. (9.49)] as the trial function, we would get the first-order perturbation result, −74.8 eV. To improve on this result, we introduce a variational parameter into (9.49). We try the normalized function

$$\phi = \frac{1}{\pi} \left(\frac{\zeta}{a_0} \right)^3 e^{-\zeta r_1/a_0} e^{-\zeta r_2/a_0} \qquad (9.58)$$

which is obtained from (9.49) by replacing the true atomic number Z by a variational parameter ζ (zeta). ζ has a simple physical interpretation. Since one electron tends to screen the other from the nucleus, each electron is subject to an effective nuclear charge somewhat less than the full nuclear charge Z. If one electron fully shielded the other from the nucleus, we would have an effective nuclear charge of $Z - 1$; since both electrons are in the same orbital, they will be only partly effective in shielding each other. We thus expect ζ to lie between $Z - 1$ and Z.

We now evaluate the variational integral. To expedite things, we rewrite the helium Hamiltonian (9.38) as

$$\hat{H} = \left[-\frac{\hbar^2}{2m_e} \nabla_1^2 - \frac{\zeta e'^2}{r_1} - \frac{\hbar^2}{2m_e} \nabla_2^2 - \frac{\zeta e'^2}{r_2} \right] + (\zeta - Z) \frac{e'^2}{r_1} + (\zeta - Z) \frac{e'^2}{r_2} + \frac{e'^2}{r_{12}}$$

$$(9.59)$$

where we have added and subtracted the terms involving zeta. The terms in brackets in (9.59) are the sum of two hydrogenlike Hamiltonians for nuclear charge ζ; moreover, the trial function (9.58) is the product of two hydrogenlike $1s$ functions for nuclear charge ζ. Therefore, when these terms operate on ϕ, we have an eigenvalue equation, the eigenvalue being the sum of two hydrogenlike $1s$ energies for nuclear charge ζ:

$$\left[-\frac{\hbar^2}{2m_e}\nabla_1^2 - \frac{\zeta e'^2}{r_1} - \frac{\hbar^2}{2m_e}\nabla_2^2 - \frac{\zeta e'^2}{r_2}\right]\phi = -\zeta^2(2)\frac{e'^2}{2a_0}\phi \qquad (9.60)$$

Using (9.59) and (9.60), we have

$$\int \phi^* \hat{H} \phi \, d\tau = -\zeta^2 \frac{e'^2}{a_0}\int \phi^*\phi \, d\tau + (\zeta - Z)e'^2 \int \frac{\phi^*\phi}{r_1} \, d\tau$$
$$+ (\zeta - Z)e'^2 \int \frac{\phi^*\phi}{r_2} \, d\tau + e'^2 \int \frac{\phi^*\phi}{r_{12}} \, d\tau \qquad (9.61)$$

Let f_1 be a normalized $1s$ hydrogenlike orbital for nuclear charge ζ, occupied by electron 1; let f_2 be the same function for electron 2:

$$f_1 = \frac{1}{\pi^{1/2}}\left(\frac{\zeta}{a_0}\right)^{3/2} e^{-\zeta r_1/a_0}, \qquad f_2 = \frac{1}{\pi^{1/2}}\left(\frac{\zeta}{a_0}\right)^{3/2} e^{-\zeta r_2/a_0} \qquad (9.62)$$

Noting that $\phi = f_1 f_2$, we now evaluate the integrals in (9.61):

$$\int \phi^*\phi \, d\tau = \int\int f_1^* f_2^* f_1 f_2 \, d\tau_1 \, d\tau_2 = \int f_1^* f_1 \, d\tau_1 \int f_2^* f_2 \, d\tau_2 = 1$$

$$\int \frac{\phi^*\phi}{r_1} \, d\tau = \int \frac{f_1^* f_1}{r_1} \, d\tau_1 \int f_2^* f_2 \, d\tau_2 = \int \frac{f_1^* f_1}{r_1} \, d\tau_1$$

$$\int \frac{\phi^*\phi}{r_1} \, d\tau = \frac{1}{\pi}\frac{\zeta^3}{a_0^3}\int_0^\infty e^{-2\zeta r_1/a_0}\frac{r_1^2}{r_1} \, dr_1 \int_0^\pi \sin\theta_1 \, d\theta_1 \int_0^{2\pi} d\phi_1 = \frac{\zeta}{a_0}$$

where the Appendix integral (A.7) was used. Also

$$\int \frac{\phi^*\phi}{r_2} \, d\tau = \int \frac{f_2^* f_2}{r_2} \, d\tau_2 = \int \frac{f_1^* f_1}{r_1} \, d\tau_1 = \frac{\zeta}{a_0}$$

since it doesn't matter whether the label 1 or 2 is used on the dummy variables in the definite integral. Finally, we must evaluate $e'^2 \int (\phi^*\phi/r_{12}) \, d\tau$. This is the same as the integral (9.52) that occurred in the perturbation treatment, except that Z is replaced by ζ. Hence, from (9.55)

$$e'^2 \int \frac{\phi^*\phi}{r_{12}} \, d\tau = \frac{5\zeta e'^2}{8a_0} \qquad (9.63)$$

The variational integral (9.61) thus has the value

$$\int \phi^* \hat{H} \phi \, d\tau = (\zeta^2 - 2Z\zeta + \tfrac{5}{8}\zeta)\frac{e'^2}{a_0} \qquad (9.64)$$

As a check, if we set $\zeta = Z$ in (9.64), we get the first-order perturbation-theory result, (9.50) plus (9.55).

We now vary ζ to minimize the variational integral:

$$\frac{\partial}{\partial \zeta} \int \phi^* \hat{H} \phi \, d\tau = (2\zeta - 2Z + \tfrac{5}{8}) \frac{e'^2}{a_0} = 0$$

$$\zeta = Z - \tfrac{5}{16} \tag{9.65}$$

As anticipated, the effective nuclear charge lies between Z and $Z - 1$. Using (9.65) and (9.64), we get

$$\int \phi^* \hat{H} \phi \, d\tau = (-Z^2 + \tfrac{5}{8}Z - \tfrac{25}{256}) \frac{e'^2}{a_0} = -(Z - \tfrac{5}{16})^2 \frac{e'^2}{a_0} \tag{9.66}$$

Putting $Z = 2$, we get as our approximation to the helium ground-state energy $-(27/16)^2 e'^2/a_0 = -(729/256)27.21 \text{ eV} = -77.5 \text{ eV}$, as compared with the true value of -79.0 eV. Use of ζ instead of Z has reduced the error from 5.3% to 1.9%. In accord with the variation theorem, the true ground-state energy is less than the variational integral.

How can we improve our variational result? We might try a function that had the general form of (9.58), that is, a product of two functions, one for each electron:

$$\phi = u(1)u(2) \tag{9.67}$$

However, we could try a variety of functions u in (9.67), instead of the single exponential used in (9.58). A systematic procedure for finding the function u that gives the lowest value of the variational integral will be discussed in Section 11.1. This procedure shows that for the best possible choice of u in (9.67) the variational integral equals -77.9 eV, which is still in error by 1.4%. We might ask why (9.67) does not cause the variational integral to converge to the true ground-state energy, no matter what form we try for u. The answer is that, when we write the trial function as the product of separate functions for each electron, we are making an approximation. Because of the e'^2/r_{12} term in the Hamiltonian, the Schrödinger equation for helium is not separable, and the true ground-state wave function cannot be written as the product of separate functions for each electron. To reach the true ground-state energy, we must go beyond a function of the form (9.67).

The Bohr model gave the correct energies for the hydrogen atom but failed when applied to helium. Hence, in the early days of quantum mechanics, it was important to show that the new theory could give an accurate treatment of helium. The pioneering work on the helium ground state was done by Hylleraas in the years 1928–1930. He used variational functions that contained the interelectronic distance r_{12} explicitly. This provides an effective way of taking into account the effects of one electron on the motion of the other. One function Hylleraas used is

$$\phi = N[e^{-\zeta r_1/a_0} e^{-\zeta r_2/a_0}(1 + br_{12})] \tag{9.68}$$

where N is the normalization constant and ζ and b are variational parameters. Since

$$r_{12} = [(x_2 - x_1)^2 + (y_2 - y_1)^2 + (z_2 - z_1)^2]^{1/2} \tag{9.69}$$

the function (9.68) goes beyond the simple product form (9.67). Minimization of the variational integral with respect to the parameters gives $\zeta = 1.849$, $b = 0.364/a_0$, and a ground-state energy of -78.7 eV, in error by 0.3 eV. The $1 + br_{12}$ term makes the wave function larger for large values of r_{12}; this is as it should be, because the repulsion between the electrons makes it energetically more favorable for the electrons to avoid each other. Using a more complicated six-term trial function containing r_{12}, Hylleraas obtained an energy only 0.01 eV above the true ground-state energy.

Hylleraas's work has been extended by others. Using a 1078-term variational function, Pekeris found a ground-state energy of $-2.903724375(e'^2/a_0)$ [C. L. Pekeris, *Phys. Rev.*, **115**, 1216 (1959)]. With relativistic and nuclear-motion corrections added, this gave for E_i, the ionization energy of helium, $E_i/hc = 198310.69$ cm^{-1}, compared with the experimental value 198310.82 ± 0.15 cm^{-1}. Using an improved variational function, Schwartz bettered Pekeris's result by obtaining $-2.903724376(e'^2/a_0)$, a result estimated to be within $10^{-9}e'^2/a_0$ of the true nonrelativistic ground-state energy [C. Schwartz, *Phys. Rev.*, **128**, 1146 (1962); see also K. Frankowski and C. L. Pekeris, *Phys. Rev.*, **146**, 46 (1966)].

A variational calculation on the Li ground state using a 60-term function that contained r_{12}, r_{13}, and r_{23} gave an energy of $-7.47802(e'^2/a_0)$, compared with the experimental value $-7.47807(e'^2/a_0)$ [S. Larsson, *Phys. Rev.*, **169**, 49 (1968)]. Variational calculations with functions containing r_{ij} become very difficult for many-electron atoms because of the many terms and difficult integrals that occur.

9.5 PERTURBATION THEORY FOR A DEGENERATE ENERGY LEVEL

We now consider the perturbation treatment of an energy level whose degree of degeneracy is d. We have d linearly independent unperturbed wave functions corresponding to the degenerate level. We shall use the labels $1, 2, \ldots, d$ for the states of the degenerate level, without implying that these are necessarily the lowest-lying states. The unperturbed Schrödinger equation is

$$\hat{H}^0 \psi_n^{(0)} = E_n^{(0)} \psi_n^{(0)} \tag{9.70}$$

with

$$E_1^{(0)} = E_2^{(0)} = \cdots = E_d^{(0)} \tag{9.71}$$

The perturbed problem is

$$\hat{H}\psi_n = E_n \psi_n \tag{9.72}$$

$$\hat{H} = \hat{H}^0 + \lambda \hat{H}' \tag{9.73}$$

As λ goes to zero, the eigenvalues of (9.72) go to the eigenvalues of (9.70); we have $\lim_{\lambda \to 0} E_n = E_n^{(0)}$. Figure 9.2 shows this for a hypothetical system with six states and a threefold-degenerate unperturbed level. Note that the perturbation splits the degenerate energy level. In some cases the perturbation may have no effect on the degeneracy or may only partly remove the degeneracy.

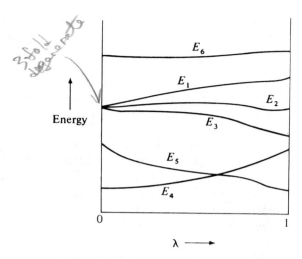

Figure 9.2 Effect of a perturbation on energy levels.

As λ goes to zero, the eigenfunctions satisfying (9.72) approach eigenfunctions satisfying (9.70). Does this mean that $\lim_{\lambda \to 0} \psi_n = \psi_n^{(0)}$? Not necessarily. If $E_n^{(0)}$ is nondegenerate, there is a unique normalized eigenfunction $\psi_n^{(0)}$ of $\hat{H}^0$ with eigenvalue $E_n^{(0)}$, and we can be sure that $\lim_{\lambda \to 0} \psi_n = \psi_n^{(0)}$. However, if $E_n^{(0)}$ is the eigenvalue of the d-fold degenerate level, then (Section 3.6) any linear combination of the form

$$c_1 \psi_1^{(0)} + c_2 \psi_2^{(0)} + \cdots + c_d \psi_d^{(0)} \qquad (9.74)$$

will be a solution of (9.70) with eigenvalue (9.71). The set of linearly independent normalized functions $\psi_1^{(0)}, \psi_2^{(0)}, \ldots, \psi_d^{(0)}$, which we use as eigenfunctions corresponding to the states of the degenerate level, is not unique. Using (9.74), we can construct an infinite number of sets of d linearly independent normalized eigenfunctions for the degenerate level. As far as the unperturbed problem is concerned, one such set is as good as another. For example, for the three degenerate $2p$ states of the hydrogen atom, we can use the $2p_1$, $2p_0$, and $2p_{-1}$ functions, the $2p_x$, $2p_y$, and $2p_z$ functions, or some other set of three linearly independent functions constructed as linear combinations of the members of one of the preceding sets. For the perturbed eigenfunctions that correspond to the d-fold degenerate unperturbed level, all we can say is that as λ approaches zero they each approach a linear combination of unperturbed eigenfunctions:

$$\lim_{\lambda \to 0} \psi_n = \sum_{i=1}^{d} c_i \psi_i^{(0)}, \qquad 1 \leq n \leq d \qquad (9.75)$$

Our first task is thus to determine the *correct* zeroth-order wave functions (9.75) for the perturbation $\hat{H}'$. Calling these correct zeroth-order functions $\phi_n^{(0)}$, we have

$$\phi_n^{(0)} = \sum_{i=1}^{d} c_i \psi_i^{(0)}, \qquad 1 \leq n \leq d \qquad (9.76)$$

Each different function $\phi_n^{(0)}$ has a different set of coefficients in (9.76). The correct set of zeroth-order functions depends on what the perturbation $\hat{H}'$ is.

The treatment of the d-fold degenerate level proceeds like the nondegenerate treatment of Section 9.2, except that instead of $\psi_n^{(0)}$ we use $\phi_n^{(0)}$. Instead of Eqs. (9.13) and (9.14), we have

$$\psi_n = \phi_n^{(0)} + \lambda\psi_n^{(1)} + \lambda^2\psi_n^{(2)} + \cdots, \qquad n = 1, 2, \ldots, d \qquad (9.77)$$

$$E_n = E_d^{(0)} + \lambda E_n^{(1)} + \lambda^2 E_n^{(2)} + \cdots, \qquad n = 1, 2, \ldots, d \qquad (9.78)$$

where (9.71) was used. Substitution into the Schrödinger equation $\hat{H}\psi_n = E_n\psi_n$ gives

$$(\hat{H}^0 + \lambda\hat{H}')(\phi_n^{(0)} + \lambda\psi_n^{(1)} + \lambda^2\psi_n^{(2)} + \cdots)$$
$$= (E_d^{(0)} + \lambda E_n^{(1)} + \lambda^2 E_n^{(2)} + \cdots)(\phi_n^{(0)} + \lambda\psi_n^{(1)} + \lambda^2\psi_n^{(2)} + \cdots)$$

Equating the coefficients of λ^0 in this equation, we get $\hat{H}^0\phi_n^{(0)} = E_d^{(0)}\phi_n^{(0)}$. By the theorem of Section 3.6, each linear combination $\phi_n^{(0)}$ ($n = 1, 2, \ldots, d$) is an eigenfunction of $\hat{H}^0$ with eigenvalue $E_d^{(0)}$, and this equation gives no new information.

Equating the coefficients of the λ^1 terms, we get

$$\hat{H}^0\psi_n^{(1)} + \hat{H}'\phi_n^{(0)} = E_d^{(0)}\psi_n^{(1)} + E_n^{(1)}\phi_n^{(0)}$$

$$\hat{H}^0\psi_n^{(1)} - E_d^{(0)}\psi_n^{(1)} = E_n^{(1)}\phi_n^{(0)} - \hat{H}'\phi_n^{(0)}, \qquad n = 1, 2, \ldots, d \qquad (9.79)$$

We now multiply (9.79) by $\psi_m^{(0)*}$ and integrate over all space, where m is one of the states corresponding to the d-fold degenerate unperturbed level under consideration; that is, $1 \leqslant m \leqslant d$. We get

$$\langle\psi_m^{(0)}|\hat{H}^0|\psi_n^{(1)}\rangle - E_d^{(0)}\langle\psi_m^{(0)}|\psi_n^{(1)}\rangle$$
$$= E_n^{(1)}\langle\psi_m^{(0)}|\phi_n^{(0)}\rangle - \langle\psi_m^{(0)}|\hat{H}'|\phi_n^{(0)}\rangle \qquad 1 \leqslant m \leqslant d \qquad (9.80)$$

From Eq. (9.20), we have $\langle\psi_m^{(0)}|\hat{H}^0|\psi_n^{(1)}\rangle = E_m^{(0)}\langle\psi_m^{(0)}|\psi_n^{(1)}\rangle$. From (9.71), $E_m^{(0)} = E_d^{(0)}$ for $1 \leqslant m \leqslant d$, so $\langle\psi_m^{(0)}|\hat{H}^0|\psi_n^{(1)}\rangle = E_d^{(0)}\langle\psi_m^{(0)}|\psi_n^{(1)}\rangle$, and the left side of (9.80) equals zero. Equation (9.80) becomes

$$\langle\psi_m^{(0)}|\hat{H}'|\phi_n^{(0)}\rangle - E_n^{(1)}\langle\psi_m^{(0)}|\phi_n^{(0)}\rangle = 0 \qquad m = 1, 2, \ldots, d$$

Substitution of the linear combination (9.76) for $\phi_n^{(0)}$ gives

$$\sum_{i=1}^{d} c_i\langle\psi_m^{(0)}|\hat{H}'|\psi_i^{(0)}\rangle - E_n^{(1)}\sum_{i=1}^{d} c_i\langle\psi_m^{(0)}|\psi_i^{(0)}\rangle = 0 \qquad (9.81)$$

The zeroth-order wave functions $\psi_i^{(0)}$ ($i = 1, 2, \ldots, d$) of the degenerate level can always be chosen to be orthonormal, and we shall assume this has been done:

$$\langle\psi_m^{(0)}|\psi_i^{(0)}\rangle = \delta_{mi} \qquad (9.82)$$

for m and i in the range 1 to d. Equation (9.81) becomes

$$\sum_{i=1}^{d} [\langle\psi_m^{(0)}|\hat{H}'|\psi_i^{(0)}\rangle - E_n^{(1)}\delta_{mi}]c_i = 0, \qquad m = 1, 2, \ldots, d \qquad (9.83)$$

This is a set of d linear, homogeneous equations in the d unknowns $c_1, c_2, \ldots, c_d$ [which are the coefficients in the correct zeroth-order wave function $\phi_n^{(0)}$ in (9.76)]. Writing out (9.83), we have

$$(H_{11}' - E_n^{(1)})c_1 + H_{12}'c_2 + \cdots + H_{1d}'c_d = 0$$

$$H_{21}'c_1 + (H_{22}' - E_n^{(1)})c_2 + \cdots + H_{2d}'c_d = 0$$

$$\cdots\cdots\cdots\cdots\cdots\cdots\cdots\cdots\cdots\cdots\cdots \quad (9.84)$$

$$H_{d1}'c_1 + H_{d2}'c_2 + \cdots + (H_{dd}' - E_n^{(1)})c_d = 0$$

$$\text{where} \quad \boxed{H_{mi}' \equiv \langle \psi_m^{(0)} | \hat{H}' | \psi_i^{(0)} \rangle}$$

For this set of linear homogeneous equations to have a nontrivial solution, the determinant of the coefficients must vanish (Section 8.4):

$$\boxed{\det \left(\langle \psi_m^{(0)} | \hat{H}' | \psi_i^{(0)} \rangle - E_n^{(1)} \delta_{mi} \right) = 0} \quad (9.85)$$

the roots,
$E_n^{(1)}$, of the
determinant are
the 1st order energy
corrections if we
get d E_n ω's then
the degeneracy has been
fully removed

$$\begin{vmatrix} H_{11}' - E_n^{(1)} & H_{12}' & \cdots & H_{1d}' \\ H_{21}' & H_{22}' - E_n^{(1)} & \cdots & H_{2d}' \\ \cdots\cdots & \cdots\cdots\cdots & \cdots & \cdots\cdots \\ H_{d1}' & H_{d2}' & \cdots & H_{dd}' - E_n^{(1)} \end{vmatrix} = 0 \quad (9.86)$$

The *secular equation* (9.86) is an algebraic equation of degree d in $E_n^{(1)}$. It has d roots, $E_1^{(1)}, E_2^{(1)}, \ldots, E_d^{(1)}$, which are the first-order corrections to the energy of the d-fold degenerate unperturbed level. If the roots are all different, then the first-order perturbation correction has split the d-fold degenerate unperturbed level into d different perturbed levels of energies (correct through first order):

$$E_d^{(0)} + E_1^{(1)}, \quad E_d^{(0)} + E_2^{(1)}, \quad \ldots, \quad E_d^{(0)} + E_d^{(1)}$$

If two or more roots of the secular equation are equal, the degeneracy is not completely removed in first order. In the rest of this section, we shall assume that all the roots of (9.86) are different.

Having found the d first-order energy corrections, we go back to the set of equations (9.84) to find the unknowns c_i, which determine the correct zeroth-order wave functions. To determine the correct zeroth-order function

$$\phi_n^{(0)} = c_1\psi_1^{(0)} + c_2\psi_2^{(0)} + \cdots + c_d\psi_d^{(0)} \quad (9.87)$$

corresponding to the root $E_n^{(1)}$, we solve (9.84) for $c_2, c_3, \ldots, c_d$ in terms of c_1 and then find c_1 by normalization. Use of (9.82) in $\langle \phi_n^{(0)} | \phi_n^{(0)} \rangle = 1$ gives (Problem 9.13)

$$\sum_{k=1}^{d} |c_k|^2 = 1 \quad (9.88)$$

For each root $E_n^{(1)}$, $n = 1, 2, \ldots, d$, we have a different set of coefficients c_1, $c_2, \ldots, c_d$, giving a different correct zeroth-order wave function.

In the next section, we shall show that

$$E_n^{(1)} = \langle \phi_n^{(0)} | \hat{H}' | \phi_n^{(0)} \rangle \,, \qquad n = 1, 2, \ldots, d \tag{9.89}$$

which is similar to the nondegenerate-case formula (9.22), except that we use the correct zeroth-order functions.

Using procedures similar to those for the degenerate case, one can now find the first-order corrections to the correct zeroth-order wave functions and the second-order energy corrections. For the results, see *Bates*, Volume I, pages 197–198; *Hameka*, pages 230–231.

As an example, consider the effect of a perturbation $\hat{H}'$ on the lowest degenerate energy level of a particle in a cubic box. We have three states corresponding to this level: $\psi_{211}^{(0)}$, $\psi_{121}^{(0)}$, and $\psi_{112}^{(0)}$. These unperturbed wave functions are orthonormal, and the secular equation (9.86) is

$$\begin{vmatrix} \langle 211 | \hat{H}' | 211 \rangle - E_n^{(1)} & \langle 211 | \hat{H} | 121 \rangle & \langle 211 | \hat{H}' | 112 \rangle \\ \langle 121 | \hat{H} | 211 \rangle & \langle 121 | \hat{H}' | 121 \rangle - E_n^{(1)} & \langle 121 | \hat{H}' | 112 \rangle \\ \langle 112 | \hat{H}' | 211 \rangle & \langle 112 | \hat{H}' | 121 \rangle & \langle 112 | \hat{H}' | 112 \rangle - E_n^{(1)} \end{vmatrix} = 0$$

Solving this equation, we find the first-order energy corrections:

$$E_1^{(1)} \,, \qquad E_2^{(1)} \,, \qquad E_3^{(1)} \tag{9.90}$$

The triply degenerate unperturbed level is split into three levels of energies (through first order):

$$(6h^2/8ma^2) + E_1^{(1)} \,, \qquad (6h^2/8ma^2) + E_2^{(1)} \,, \qquad (6h^2/8ma^2) + E_3^{(1)}$$

Using each of the roots (9.90), we get a different set of simultaneous equations (9.84). Solving each set, we find three sets of coefficients, which determine the three correct zeroth-order wave functions.

[If you are familiar with matrix algebra, note that solving (9.86) and (9.84) amounts to finding the eigenvalues and eigenvectors of the matrix whose elements are $\langle \psi_m^{(0)} | \hat{H}' | \psi_i^{(0)} \rangle$.]

9.6 SIMPLIFICATION OF THE SECULAR EQUATION

The secular equation (9.86) is easier to solve if some of the off-diagonal elements of the secular determinant are zero. In the most favorable case, all the off-diagonal elements are zero, and

$$\begin{vmatrix} H_{11}' - E_n^{(1)} & 0 & \cdots & 0 \\ 0 & H_{22}' - E_n^{(1)} & \cdots & 0 \\ \cdots\cdots\cdots\cdots\cdots\cdots\cdots\cdots\cdots\cdots\cdots\cdots \\ 0 & 0 & \cdots & H_{dd}' - E_n^{(1)} \end{vmatrix} = 0 \tag{9.91}$$

$$(H_{11}' - E_n^{(1)})(H_{22}' - E_n^{(1)}) \cdots (H_{dd}' - E_n^{(1)}) = 0$$

$$E_1^{(1)} = H_{11}' \,, \qquad E_2^{(1)} = H_{22}' \,, \qquad \ldots, \qquad E_d^{(1)} = H_{dd}' \tag{9.92}$$

Now we want to determine the correct zeroth-order wave functions. We shall assume that the roots (9.92) are all different. For the root $E_n^{(1)} = H'_{11}$, the system of equations (9.84) is

$$0 = 0$$

$$(H'_{22} - H'_{11})c_2 = 0$$

$$\ldots \ldots \ldots \ldots$$

$$(H'_{dd} - H'_{11})c_d = 0$$

Since we are assuming unequal roots, the quantities $H'_{22} - H'_{11}, \ldots, H'_{dd} - H'_{11}$ are all nonzero. Therefore,

$$c_2 = 0, \quad c_3 = 0, \quad \ldots, \quad c_d = 0$$

The normalization condition (9.88) gives $c_1 = 1$. The correct zeroth-order wave function corresponding to the first-order perturbation energy correction H'_{11} is then [Eq. (9.76)]

$$\phi_1^{(0)} = \psi_1^{(0)}$$

For the root H'_{22}, the same reasoning gives

$$\phi_2^{(0)} = \psi_2^{(0)}$$

Using each of the remaining roots, we find similarly

$$\phi_3^{(0)} = \psi_3^{(0)}, \quad \ldots, \quad \phi_d^{(0)} = \psi_d^{(0)}$$

When the secular determinant is in diagonal form, the initially assumed wave functions $\psi_1^{(0)}, \psi_2^{(0)}, \ldots, \psi_n^{(0)}$ are the correct zeroth-order wave functions for the perturbation $\hat{H}'$.

The converse is also true. If the initially assumed functions are the correct zeroth-order functions, then the secular determinant is in diagonal form. This is seen as follows. From $\phi_1^{(0)} = \psi_1^{(0)}$, we know that the coefficients in the expansion $\phi_1^{(0)} = \sum_{i=1}^{d} c_i \psi_i^{(0)}$ are $c_1 = 1, c_2 = c_3 = \cdots = 0$, so for $n = 1$ the set of simultaneous equations (9.84) becomes

$$H'_{11} - E_1^{(1)} = 0, \quad H'_{21} = 0, \quad \ldots, \quad H'_{d1} = 0$$

Applying the same reasoning to the remaining functions $\phi_n^{(0)}$, we conclude that $H'_{mi} = 0$ for $i \neq m$. Hence, use of the correct zeroth-order functions makes the secular determinant diagonal. Note also that the first-order corrections to the energy can be found by averaging the perturbation over the correct zeroth-order wave functions:

$$E_n^{(1)} = H'_{nn} = \langle \phi_n^{(0)} | \hat{H}' | \phi_n^{(0)} \rangle \tag{9.93}$$

a result mentioned in Eq. (9.89).

Often, instead of being in diagonal form, the secular determinant is in block-diagonal form. For example, we might have

$$\begin{vmatrix} H'_{11} - E_n^{(1)} & H'_{12} & 0 & 0 \\ H'_{21} & H'_{22} - E_n^{(1)} & 0 & 0 \\ 0 & 0 & H'_{33} - E_n^{(1)} & H'_{34} \\ 0 & 0 & H'_{43} & H'_{44} - E_n^{(1)} \end{vmatrix} = 0 \qquad (9.94)$$

The secular determinant in (9.94) has the same form as the secular determinant in the linear-variation secular equation (8.66) with $S_{ij} = \delta_{ij}$. By the same reasoning used to show that two of the variation functions are linear combinations of f_1 and f_2 and two are linear combinations of f_3 and f_4 [Eq. (8.70)], it follows that two of the correct zeroth-order wave functions are linear combinations of $\psi_1^{(0)}$ and $\psi_2^{(0)}$ and two are linear combinations of $\psi_3^{(0)}$ and $\psi_4^{(0)}$:

$$\phi_1^{(0)} = c_1 \psi_1^{(0)} + c_2 \psi_2^{(0)}, \qquad \phi_2^{(0)} = c_1' \psi_1^{(0)} + c_2' \psi_2^{(0)}$$
$$\phi_3^{(0)} = c_3 \psi_3^{(0)} + c_4 \psi_4^{(0)}, \qquad \phi_4^{(0)} = c_3' \psi_3^{(0)} + c_4' \psi_4^{(0)}$$

where primes were used to distinguish different coefficients.

When the secular determinant of degenerate perturbation theory is in block-diagonal form, the secular equation breaks up into two or more smaller secular equations, and the set of simultaneous equations (9.84) for the coefficients c_i breaks up into two or more smaller sets of simultaneous equations.

Conversely, if we have, say, a fourfold-degenerate unperturbed level, and we happen to know that $\phi_1^{(0)}$ and $\phi_2^{(0)}$ are each linear combinations of $\psi_1^{(0)}$ and $\psi_2^{(0)}$ only, while $\phi_3^{(0)}$ and $\phi_4^{(0)}$ are each linear combinations of $\psi_3^{(0)}$ and $\psi_4^{(0)}$ only, we deal with two second-order secular determinants, rather than a fourth-order secular determinant.

How can we choose the right zeroth-order wave functions in advance and thereby simplify the secular equation? Suppose there is an operator $\hat{A}$ that commutes with both $\hat{H}^0$ and $\hat{H}'$. Then we can choose the unperturbed functions to be eigenfunctions of $\hat{A}$. Because $\hat{A}$ commutes with $\hat{H}'$, this choice of unperturbed functions will make the integrals H'_{ij} vanish if $\psi_i^{(0)}$ and $\psi_j^{(0)}$ belong to different eigenvalues of $\hat{A}$ [see Eq. (7.50)]. Thus, if the eigenvalues of $\hat{A}$ for $\psi_1^{(0)}$, $\psi_2^{(0)}, \ldots, \psi_d^{(0)}$ are all different, the secular determinant will be in diagonal form, and we will have the right zeroth-order wave functions. If some of the eigenvalues of $\hat{A}$ are the same, we get block-diagonal rather than diagonal form. In general, the correct zeroth-order functions will be linear combinations of those unperturbed functions that have the same eigenvalue of $\hat{A}$. (This is to be expected since $\hat{A}$ commutes with $\hat{H} = \hat{H}^0 + \hat{H}'$, so the perturbed eigenfunctions of $\hat{H}$ can be chosen to be eigenfunctions of $\hat{A}$.) For an application of these ideas, see Problem 9.15.

9.7   PERTURBATION TREATMENT OF THE FIRST EXCITED STATES OF HELIUM

Section 9.3 applied perturbation theory to the ground state of the helium atom. We now treat the lowest excited states of helium. The unperturbed energies are given by (9.48). The lowest unperturbed excited states have $n_1 = 1$, $n_2 = 2$ or

$n_1 = 2$, $n_2 = 1$, and substitution in (9.48) gives

$$E^{(0)} = -\frac{5Z^2}{8} \left(\frac{e'^2}{a_0}\right) = -\frac{20}{8} 2\left(\frac{e'^2}{2a_0}\right) = -5(13.606 \text{ eV}) = -68.03 \text{ eV} \qquad (9.95)$$

Recall that the $n = 2$ level of a hydrogenlike atom is fourfold degenerate, the $2s$ and three $2p$ states all having the same energy. The first excited unperturbed energy level is eightfold degenerate; the eight unperturbed wave functions are [Eq. (9.44)]

$$\begin{array}{ll}
\psi_1^{(0)} = 1s(1)2s(2) & \psi_5^{(0)} = 1s(1)2p_y(2) \\[6pt]
\psi_2^{(0)} = 2s(1)1s(2) & \psi_6^{(0)} = 2p_y(1)1s(2) \\[6pt]
\psi_3^{(0)} = 1s(1)2p_x(2) & \psi_7^{(0)} = 1s(1)2p_z(2) \\[6pt]
\psi_4^{(0)} = 2p_x(1)1s(2) & \psi_8^{(0)} = 2p_z(1)1s(2)
\end{array} \qquad (9.96)$$

where $1s(1)2s(2)$ signifies the product of a hydrogenlike $1s$ function for electron one and a hydrogenlike $2s$ function for electron two. The explicit form of $\psi_8^{(0)}$, for example, is (Table 6.2)

$$\psi_8^{(0)} = \frac{1}{4(2\pi)^{1/2}} \left(\frac{Z}{a_0}\right)^{5/2} r_1 e^{-Zr_1/2a_0} \cos\theta_1 \cdot \frac{1}{\pi^{1/2}} \left(\frac{Z}{a_0}\right)^{3/2} e^{-Zr_2/a_0}$$

We chose to use the real $2p$ hydrogenlike orbitals, rather than the complex ones.

Since the unperturbed level is degenerate, we must solve a secular equation. The secular equation (9.86) assumes that the functions $\psi_1^{(0)}$, $\psi_2^{(0)}$, ..., $\psi_8^{(0)}$ are orthonormal. This condition is met. For example,

$$\int \psi_1^{(0)*} \psi_1^{(0)} \, d\tau = \int \int 1s(1)^* 2s(2)^* 1s(1)2s(2) \, d\tau_1 \, d\tau_2$$

$$= \int |1s(1)|^2 \, d\tau_1 \int |2s(2)|^2 \, d\tau_2 = 1 \cdot 1 = 1$$

$$\int \psi_3^{(0)*} \psi_5^{(0)} \, d\tau = \int |1s(1)|^2 \, d\tau_1 \int 2p_x(2)^* 2p_y(2) \, d\tau_2 = 1 \cdot 0 = 0$$

where the orthonormality of the hydrogenlike orbitals has been used.

The secular determinant contains $8^2 = 64$ elements. The operator $\hat{H}'$ is Hermitian, and $H'_{ij} = (H'_{ji})^*$. Also, since $\hat{H}'$ and $\psi_1^{(0)}, \ldots, \psi_8^{(0)}$ are all real, we have $(H'_{ji})^* = H'_{ji}$, so $H'_{ij} = H'_{ji}$. The secular determinant is symmetric about the principal diagonal. This cuts the labor of evaluating integrals almost in half.

By using parity considerations, we can show that most of the integrals H'_{ij} are zero. First consider H'_{13}:

$$H'_{13} = \int_{-\infty}^{\infty} \int_{-\infty}^{\infty} \int_{-\infty}^{\infty} \int_{-\infty}^{\infty} \int_{-\infty}^{\infty} \int_{-\infty}^{\infty} 1s(1)2s(2) \frac{e'^2}{r_{12}} 1s(1)2p_x(2) \, dx_1 \, dy_1 \, dz_1 \, dx_2 \, dy_2 \, dz_2$$

An s hydrogenlike function depends only on $r = (x^2 + y^2 + z^2)^{1/2}$ and is therefore an even function. The $2p_x(2)$ function is an odd function of x_2 [Eq. (6.119)]. r_{12} is given by (9.69), and if we invert all six coordinates, r_{12} is unchanged:

$$r_{12} \to [(-x_1 + x_2)^2 + (-y_1 + y_2)^2 + (-z_1 + z_2)^2]^{1/2} = r_{12}$$

Hence, on inverting all six coordinates, the integrand of H'_{13} goes into minus itself. We conclude [Eq. (7.60)] that $H'_{13} = 0$. The same reasoning yields $H'_{14} = H'_{15} = H'_{16} = H'_{17} = H'_{18} = 0$ and $H'_{23} = H'_{24} = H'_{25} = H'_{26} = H'_{27} = H'_{28} = 0$. Now consider H'_{35}:

$$H'_{35} = \int_{-\infty}^{\infty} \cdots \int_{-\infty}^{\infty} 1s(1)2p_x(2) \frac{e'^2}{r_{12}} 1s(1)2p_y(2) \, dx_1 \cdots dz_2$$

Consider the effect of inverting the x coordinates: $x_1 \to -x_1$ and $x_2 \to -x_2$. This transformation will leave r_{12} unchanged. The functions $1s(1)$ and $2p_y(2)$ will be unaffected. However, $2p_x(2)$ will go over to minus itself, so the net effect will be to change the integrand of H'_{35} into minus itself. Therefore [Eq. (7.62)], $H'_{35} = 0$. Likewise, $H'_{36} = H'_{37} = H'_{38} = 0$ and $H'_{45} = H'_{46} = H'_{47} = H'_{48} = 0$. By considering the transformation $y_1 \to -y_1$, $y_2 \to -y_2$, we see that $H'_{57} = H'_{58} = H'_{67} = H'_{68} = 0$. The secular equation is therefore

$$\begin{vmatrix} b_{11} & H'_{12} & 0 & 0 & 0 & 0 & 0 & 0 \\ H'_{12} & b_{22} & 0 & 0 & 0 & 0 & 0 & 0 \\ 0 & 0 & b_{33} & H'_{34} & 0 & 0 & 0 & 0 \\ 0 & 0 & H'_{34} & b_{44} & 0 & 0 & 0 & 0 \\ 0 & 0 & 0 & 0 & b_{55} & H'_{56} & 0 & 0 \\ 0 & 0 & 0 & 0 & H'_{56} & b_{66} & 0 & 0 \\ 0 & 0 & 0 & 0 & 0 & 0 & b_{77} & H'_{78} \\ 0 & 0 & 0 & 0 & 0 & 0 & H'_{78} & b_{88} \end{vmatrix} = 0$$

$$b_{ii} \equiv H'_{ii} - E^{(1)}, \qquad i = 1, 2, \ldots, 8$$

The secular determinant is in block-diagonal form and factors into four determinants, each of second order. We conclude that the correct zeroth-order functions have the form

$$\phi_1^{(0)} = c_1 \psi_1^{(0)} + c_2 \psi_2^{(0)}, \qquad \phi_2^{(0)} = \bar{c}_1 \psi_1^{(0)} + \bar{c}_2 \psi_2^{(0)}$$

$$\phi_3^{(0)} = c_3 \psi_3^{(0)} + c_4 \psi_4^{(0)}, \qquad \phi_4^{(0)} = \bar{c}_3 \psi_3^{(0)} + \bar{c}_4 \psi_4^{(0)}$$

$$\phi_5^{(0)} = c_5 \psi_5^{(0)} + c_6 \psi_6^{(0)}, \qquad \phi_6^{(0)} = \bar{c}_5 \psi_5^{(0)} + \bar{c}_6 \psi_6^{(0)} \qquad (9.97)$$

$$\phi_7^{(0)} = c_7 \psi_7^{(0)} + c_8 \psi_8^{(0)}, \qquad \phi_8^{(0)} = \bar{c}_7 \psi_7^{(0)} + \bar{c}_8 \psi_8^{(0)}$$

where the unbarred coefficients correspond to one root of each second-order determinant and the barred coefficients correspond to the second root.

The first determinant is

$$\begin{vmatrix} H'_{11} - E^{(1)} & H'_{12} \\ H'_{12} & H'_{22} - E^{(1)} \end{vmatrix} = 0 \qquad (9.98)$$

We have

$$H'_{11} = \int_{-\infty}^{\infty} \cdots \int_{-\infty}^{\infty} 1s(1)2s(2) \frac{e'^2}{r_{12}} 1s(1)2s(2) \, dx_1 \cdots dz_2$$

$$H'_{11} = \int \int [1s(1)]^2 [2s(2)]^2 \frac{e'^2}{r_{12}} \, d\tau_1 \, d\tau_2$$

$$H'_{22} = \int \int [1s(2)]^2 [2s(1)]^2 \frac{e'^2}{r_{12}} \, d\tau_1 \, d\tau_2$$

The integration variables are dummy variables and may be given any symbols whatever. Let us relabel the integration variables in H'_{22} as follows: We interchange x_1 and x_2, interchange y_1 and y_2, and interchange z_1 and z_2. This relabeling leaves r_{12} [Eq. (9.69)] unchanged, so

$$H'_{22} = \int \int [1s(1)]^2 [2s(2)]^2 \frac{e'^2}{r_{12}} \, d\tau_2 \, d\tau_1 = H'_{11} \qquad (9.99)$$

The same argument shows that $H'_{33} = H'_{44}$, $H'_{55} = H'_{66}$, and $H'_{77} = H'_{88}$.

We denote H'_{11} by the symbol J_{1s2s}:

$$H'_{11} = J_{1s2s} = \int \int [1s(1)]^2 [2s(2)]^2 \frac{e'^2}{r_{12}} \, d\tau_1 \, d\tau_2 \qquad (9.100)$$

This is an example of a **Coulomb integral**, the name arising from the fact that J_{1s2s} is equal to the electrostatic energy of repulsion between an electron with probability density function $[1s]^2$ and an electron with probability density function $[2s]^2$. The integral H'_{12} is denoted by K_{1s2s}:

$$H'_{12} = K_{1s2s} = \int \int 1s(1)2s(2) \frac{e'^2}{r_{12}} 2s(1)1s(2) \, d\tau_1 \, d\tau_2 \qquad (9.101)$$

This is an **exchange integral**: the functions on the left and right of e'^2/r_{12} differ from each other by an exchange of electrons one and two. The general definitions of the Coulomb integral J_{ij} and the exchange integral K_{ij} are

$$J_{ij} \equiv \langle f_i(1)f_j(2) | e'^2/r_{12} | f_i(1)f_j(2) \rangle, \qquad K_{ij} \equiv \langle f_i(1)f_j(2) | e'^2/r_{12} | f_j(1)f_i(2) \rangle$$

$f_{i/j} \Rightarrow$ spatial (1s, 2p, etc.) orbitals.   (9.102)

where the integrals go over the full range of the spatial coordinates of electrons 1 and 2 and f_i and f_j are spatial orbitals.

Substitution of (9.99) to (9.101) into (9.98) gives

$$\begin{vmatrix} J_{1s2s} - E^{(1)} & K_{1s2s} \\ K_{1s2s} & J_{1s2s} - E^{(1)} \end{vmatrix} = 0 \qquad (9.103)$$

$$(J_{1s2s} - E^{(1)})^2 = (K_{1s2s})^2$$

$$E_1^{(1)} = J_{1s2s} - K_{1s2s}, \qquad E_2^{(1)} = J_{1s2s} + K_{1s2s} \qquad (9.104)$$

We now find the coefficients of the correct zeroth-order wave functions that correspond to these two roots. Use of $E_1^{(1)}$ in (9.84) gives

$$K_{1s2s}c_1 + K_{1s2s}c_2 = 0$$
$$K_{1s2s}c_1 + K_{1s2s}c_2 = 0$$

Hence $c_2 = -c_1$. Normalization gives

$$\langle \phi_1^{(0)} | \phi_1^{(0)} \rangle = \langle c_1\psi_1^{(0)} - c_1\psi_2^{(0)} | c_1\psi_1^{(0)} - c_1\psi_2^{(0)} \rangle = |c_1|^2 + |c_1|^2 = 1$$

$$c_1 = 2^{-1/2}$$

where the orthonormality of $\psi_1^{(0)}$ and $\psi_2^{(0)}$ was used. The zeroth-order wave function corresponding to $E_1^{(1)}$ is then

$$\phi_1^{(0)} = 2^{-1/2}(\psi_1^{(0)} - \psi_2^{(0)}) = 2^{-1/2}[1s(1)2s(2) - 2s(1)1s(2)] \qquad (9.105)$$

HOMO-type

Similarly, one finds the function corresponding to $E_2^{(1)}$ to be

$$\phi_2^{(0)} = 2^{-1/2}(\psi_1^{(0)} + \psi_2^{(0)}) = 2^{-1/2}[1s(1)2s(2) + 2s(1)1s(2)] \qquad (9.106)$$

LUMO-type

We have three more second-order determinants to deal with:

$$\begin{vmatrix} H'_{33} - E^{(1)} & H'_{34} \\ H'_{34} & H'_{33} - E^{(1)} \end{vmatrix} = 0 \qquad (9.107)$$

$$\begin{vmatrix} H'_{55} - E^{(1)} & H'_{56} \\ H'_{56} & H'_{55} - E^{(1)} \end{vmatrix} = 0 \qquad (9.108)$$

$$\begin{vmatrix} H'_{77} - E^{(1)} & H'_{78} \\ H'_{78} & H'_{77} - E^{(1)} \end{vmatrix} = 0 \qquad (9.109)$$

Consider H'_{33} and H'_{55}:

$$H'_{33} = \int_{-\infty}^{\infty} \cdots \int_{-\infty}^{\infty} 1s(1)2p_x(2) \frac{e'^2}{r_{12}} 1s(1)2p_x(2) \, dx_1 \cdots dz_2$$

$$H'_{55} = \int_{-\infty}^{\infty} \cdots \int_{-\infty}^{\infty} 1s(1)2p_y(2) \frac{e'^2}{r_{12}} 1s(1)2p_y(2) \, dx_1 \cdots dz_2$$

These two integrals are equal—the only difference between them involves replacement of $2p_x(2)$ by $2p_y(2)$—and these two orbitals differ only in their orientation in space. More formally, if we relabel the dummy integration variables in H'_{33} according to the scheme $x_2 \to y_2$, $y_2 \to x_2$, $x_1 \to y_1$, $y_1 \to x_1$, then r_{12} is unaffected and H'_{33} is transformed to H'_{55}. Similar reasoning shows H'_{77} is equal to these two integrals. Introducing the symbol J_{1s2p} for these Coulomb integrals, we have   *Coulomb Integrals involving 2p are equal.*

$$H'_{33} = H'_{55} = H'_{77} = J_{1s2p} = \int\int 1s(1)2p_z(2) \frac{e'^2}{r_{12}} 1s(1)2p_z(2) \, d\tau_1 \, d\tau_2$$

Also, the exchange integrals involving the $2p$ orbitals are equal:

$$H'_{34} = H'_{56} = H'_{78} = K_{1s2p} = \int\int 1s(1)2p_z(2) \frac{e'^2}{r_{12}} 2p_z(1)1s(2) \, d\tau_1 \, d\tau_2$$

The three determinants (9.107) to (9.109) are thus identical and have the form

$$
\begin{vmatrix}
J_{1s2p} - E^{(1)} & K_{1s2p} \\
K_{1s2p} & J_{1s2p} - E^{(1)}
\end{vmatrix} = 0
$$

The determinant is similar to (9.103), and by analogy with (9.104)–(9.106), we have

$$E_3^{(1)} = E_5^{(1)} = E_7^{(1)} = J_{1s2p} - K_{1s2p} \tag{9.110}$$

$$E_4^{(1)} = E_6^{(1)} = E_8^{(1)} = J_{1s2p} + K_{1s2p} \tag{9.111}$$

$$
\begin{aligned}
\phi_3^{(0)} &= 2^{-1/2}[1s(1)2p_x(2) - 1s(2)2p_x(1)] \\
\phi_4^{(0)} &= 2^{-1/2}[1s(1)2p_x(2) + 1s(2)2p_x(1)] \\
\phi_5^{(0)} &= 2^{-1/2}[1s(1)2p_y(2) - 1s(2)2p_y(1)] \\
\phi_6^{(0)} &= 2^{-1/2}[1s(1)2p_y(2) + 1s(2)2p_y(1)] \\
\phi_7^{(0)} &= 2^{-1/2}[1s(1)2p_z(2) - 1s(2)2p_z(1)] \\
\phi_8^{(0)} &= 2^{-1/2}[1s(1)2p_z(2) + 1s(2)2p_z(1)]
\end{aligned}
\tag{9.112}
$$

The electrostatic repulsion e'^2/r_{12} between the electrons has partly removed the degeneracy; the hypothetical eightfold-degenerate unperturbed level has been split into two nondegenerate levels associated with the configuration $1s2s$ and two triply degenerate levels associated with the configuration $1s2p$. It might be thought that higher-order energy corrections would further resolve the degeneracy; actually, application of an external magnetic field is required to completely remove the degeneracy. Because the e'^2/r_{12} perturbation has not completely removed the degeneracy, any normalized linear combinations of $\phi_3^{(0)}$, $\phi_5^{(0)}$, and $\phi_7^{(0)}$ and of $\phi_4^{(0)}$, $\phi_6^{(0)}$, and $\phi_8^{(0)}$ can serve as correct zeroth-order wave functions.

To evaluate the Coulomb and exchange integrals in $E^{(1)}$ in (9.104) and (9.110), one uses the $1/r_{12}$ expansion (9.53). The results are (Problem 9.17)

$$J_{1s2s} = \frac{17}{81}\frac{Ze'^2}{a_0} = 11.42 \text{ eV} , \qquad J_{1s2p} = \frac{59}{243}\frac{Ze'^2}{a_0} = 13.21 \text{ eV}$$

$$\tag{9.113}$$

$$K_{1s2s} = \frac{16}{729}\frac{Ze'^2}{a_0} = 1.19 \text{ eV} , \qquad K_{1s2p} = \frac{112}{6561}\frac{Ze'^2}{a_0} = 0.93 \text{ eV}$$

where we used $Z = 2$ and $e'^2/2a_0 = 13.606$ eV. Recalling that $E^{(0)} = -68.03$ eV [Eq. (9.95)], we get (Fig. 9.3)

$$E^{(0)} + E_1^{(1)} = E^{(0)} + J_{1s2s} - K_{1s2s} = -57.8 \text{ eV}$$

$$E^{(0)} + E_2^{(1)} = E^{(0)} + J_{1s2s} + K_{1s2s} = -55.4 \text{ eV}$$

$$E^{(0)} + E_3^{(1)} = E^{(0)} + J_{1s2p} - K_{1s2p} = -55.7_5 \text{ eV}$$

$$E^{(0)} + E_4^{(1)} = E^{(0)} + J_{1s2p} + K_{1s2p} = -53.9 \text{ eV}$$

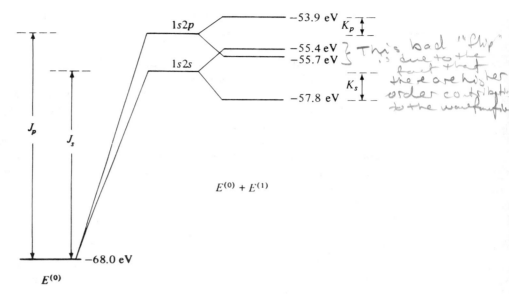

Figure 9.3 The first excited levels of the helium atom.

The first-order energy corrections seem to indicate that the lower of the two levels of the 1s2p configuration lies below the higher of the two levels of the 1s2s configuration. Study of the helium spectrum reveals that this is not so. The error is due to neglect of the higher-order perturbation-energy corrections.

Using the variation–perturbation method (Section 9.2), Knight and Scherr calculated the second- and third-order corrections $E^{(2)}$ and $E^{(3)}$ for these four excited levels. [R. E. Knight and C. W. Scherr, *Rev. Mod. Phys.*, **35**, 431 (1963); for energy corrections through 17th order, see F. C. Sanders and C. W. Scherr, *Phys. Rev.*, **181**, 84 (1969).] Figure 9.4 shows their results (which are within 0.1 eV of the experimental energies). Figure 9.4 shows that Fig. 9.3 is quite inaccurate. Since the perturbation e'^2/r_{12} is not really very small, a perturbation treatment that includes only the $E^{(1)}$ correction does not give accurate results.

The first-order correction to the wave function, $\psi^{(1)}$, will include contributions from other configurations (configuration interaction). When we say that a

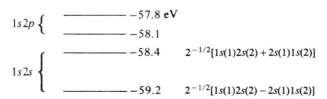

Figure 9.4 $E^{(0)} + E^{(1)} + E^{(2)} + E^{(3)}$ for the first excited levels of helium. Also shown are the correct zeroth-order wave functions for the 1s2s levels.

level belongs to the configuration $1s2s$, we are indicating the configuration that makes the largest contribution to the true wave function.

We started with the eight degenerate zeroth-order functions (9.96). These functions have three kinds of degeneracy. There is the degeneracy between hydrogenlike functions with the same n, but different l; the $2s$ and the $2p$ functions have the same energy. There is the degeneracy between hydrogenlike functions with the same n and l, but different m; the $2p_1$, $2p_0$, and $2p_{-1}$ functions have the same energy. (For convenience we used the real functions $2p_{x,y,z}$, but we could have started with the functions $2p_{1,0,-1}$.) Finally, there is the degeneracy between functions that differ only in the interchange of the two electrons between the orbitals; the functions $\psi_1^{(0)} = 1s(1)2s(2)$ and $\psi_2^{(0)} = 1s(2)2s(1)$ have the same energy. This last kind of degeneracy is called *exchange degeneracy*. When the interelectronic repulsion e'^2/r_{12} was introduced as a perturbation, the exchange degeneracy and the degeneracy associated with the quantum number l were removed. The degeneracy associated with m remained, however; each $1s2p$ helium level is triply degenerate, and we could just as well have used the $2p_1$, $2p_0$, and $2p_{-1}$ orbitals instead of the real orbitals in constructing the correct zeroth-order wave functions. Let us consider the reasons for the removal of the l degeneracy and the exchange degeneracy.

The interelectronic repulsion in helium makes the $2s$ orbital energy less than the $2p$ energy. Figures 6.9 and 6.8 show that a $2s$ electron has a greater probability than a $2p$ electron of being closer to the nucleus than the $1s$ electron(s). A $2s$ electron will not be as effectively shielded from the nucleus by the $1s$ electrons and will therefore have a lower energy than a $2p$ electron. [According to Eq. (6.94), the greater the nuclear charge, the lower the energy.] Mathematically, the difference between the $1s2s$ and the $1s2p$ energies results from the Coulomb integral J_{1s2s} being smaller than J_{1s2p}. These Coulomb integrals represent the electrostatic repulsion between the appropriate charge distributions. When the $2s$ electron penetrates the charge distribution of the $1s$ electron, it feels a repulsion from only the unpenetrated part of the $1s$ charge distribution. Hence the $1s$–$2s$ electrostatic repulsion is less than the $1s$–$2p$ repulsion, and the $1s2s$ levels lie below the $1s2p$ levels. The interelectronic repulsion in many-electron atoms lifts the l degeneracy, and the orbital energies for the same value of n increase with increasing l.

Now consider the removal of the exchange degeneracy. The functions (9.96) with which we began the perturbation treatment have each electron assigned to a definite orbital. For example, the function $\psi_1^{(0)} = 1s(1)2s(2)$ has electron 1 in the $1s$ orbital and electron 2 in the $2s$ orbital; for $\psi_2^{(0)}$ the opposite is true. The secular determinant was not diagonal, so the initial functions were not the correct zeroth-order wave functions. The correct zeroth-order functions do not assign each electron to a definite orbital. Thus the first two correct zeroth-order functions are

$$\phi_1^{(0)} = 2^{-1/2}[1s(1)2s(2) - 1s(2)2s(1)]$$

$$\phi_2^{(0)} = 2^{-1/2}[1s(1)2s(2) + 1s(2)2s(1)]$$

We cannot say which orbital electron 1 is in for either $\phi_1^{(0)}$ or $\phi_2^{(0)}$. This property of the wave functions of systems containing more than one electron results from the indistinguishability of identical particles in quantum mechanics and will be discussed further in Chapter 10. Since the functions $\phi_1^{(0)}$ and $\phi_2^{(0)}$ have different energies, the exchange degeneracy is removed when the correct zeroth-order functions are used. Note that $\phi_1^{(0)}$ becomes zero when the two electrons are at the same point in space, since if $r_1 = r_2$, then $1s(r_1)2s(r_2) = 1s(r_2)2s(r_1)$. Therefore, the probability of having the two electrons very close together is smaller for $\phi_1^{(0)}$ than for $\phi_2^{(0)}$. Hence we expect the average interelectronic repulsion to be less for $\phi_1^{(0)}$, making $\phi_1^{(0)}$ lie lower in energy. (See, however, the discussion of Hund's rule in Section 11.5.)

9.8 COMPARISON OF THE VARIATION AND PERTURBATION METHODS

The perturbation method applies to all the states of an atom or molecule, whereas the variation method is restricted to the ground state (which is the state of greatest chemical interest) and possibly the first few excited states. However, it is usually very difficult to evaluate the infinite sums over discrete states and integrals over continuum states that are necessary to evaluate the second-order and higher-order perturbation energy corrections.

In the perturbation method, one can calculate the energy much more accurately (to order $2k + 1$) than the wave function (to order k). The same situation holds in the variation method, where one can get a rather good energy with a rather inaccurate wave function; if one attempts to calculate properties other than the energy, the results will generally not be as reliable as the calculated energy.

Although most calculations of molecular wave functions have been done using the variation method, there has been a revival of interest in applying the perturbation method to such problems. See Section 15.12.

9.9 TIME-DEPENDENT PERTURBATION THEORY

In spectroscopy, we start with a system in some stationary state, expose it to electromagnetic radiation (light), and then observe whether the system has made a transition to another stationary state. The radiation produces a time-dependent potential-energy term in the Hamiltonian, so we must use the time-dependent Schrödinger equation. The most convenient approach here is an approximate one called time-dependent perturbation theory.

Let the system (atom or molecule) have the time-independent Hamiltonian $\hat{H}^0$ in the absence of the radiation (or other time-dependent perturbation), and let $\hat{H}'(t)$ be the time-dependent perturbation. The time-independent Schrödinger equation for the unperturbed problem is

$$\hat{H}^0 \psi_k^0 = E_k^0 \psi_k^0 \tag{9.114}$$

where E_k^0 and ψ_k^0 are the stationary-state energies and wave functions. The time-dependent Schrödinger equation (7.96) in the presence of the radiation is

$$-\frac{\hbar}{i}\frac{\partial\Psi}{\partial t} = (\hat{H}^0 + \hat{H}')\Psi \tag{9.115}$$

where the state function Ψ depends on the spatial and spin coordinates (symbolized by q) and on the time: $\Psi = \Psi(q, t)$. (See Chapter 10 for a discussion of spin coordinates.)

First suppose that $H'(t)$ is absent. The unperturbed time-dependent Schrödinger equation is

$$-(\hbar/i)\partial\Psi^0/\partial t = \hat{H}^0\Psi^0 \tag{9.116}$$

The system's possible stationary-state functions are given by (7.98) as $\Psi_k^0 = \exp(-iE_k^0 t/\hbar)\psi_k^0$, where the ψ_k^0 functions are the eigenfunctions of $\hat{H}^0$ [Eq. (9.114)]. Each Ψ_k^0 is a solution of (9.116). Moreover, the linear combination

$$\Psi^0 = \sum_k c_k \Psi_k^0 = \sum_k c_k \exp(-iE_k^0 t/\hbar)\psi_k^0 \tag{9.117}$$

with the c_k's being arbitrary time-independent constants, is a solution of the time-dependent Schrödinger equation (9.116), as proved in the discussion leading to Eq. (7.99). The functions Ψ_k^0 form a complete set (since they are the eigenfunctions of the Hermitian operator $\hat{H}^0$), so any solution of (9.116) can be expressed in the form (9.117). Hence (9.117) is the general solution of the time-dependent Schrödinger equation (9.116), where $\hat{H}^0$ is independent of time.

Now suppose that $\hat{H}'(t)$ is present. The function (9.117) is no longer a solution of the time-dependent Schrödinger equation. However, because the unperturbed functions Ψ_k^0 form a complete set, the true state function Ψ can at any instant of time be expanded as a linear combination of the Ψ_k^0 functions according to $\Psi = \sum_k b_k \Psi_k^0$. Because $\hat{H}$ is time dependent, Ψ will change with time and the expansion coefficients b_k will change with time. Therefore,

$$\Psi = \sum_k b_k(t) \exp(-iE_k^0 t/\hbar)\psi_k^0 \tag{9.118}$$

In the limit $\hat{H}'(t) \to 0$, the expansion (9.118) reduces to (9.117).

Substitution of (9.118) into the time-dependent Schrödinger equation (9.115) and use of (9.114) gives

$$-\frac{\hbar}{i}\sum_k \frac{db_k}{dt}\exp(-iE_k^0 t/\hbar)\psi_k^0 + \sum_k E_k^0 b_k \exp(-iE_k^0 t/\hbar)\psi_k^0$$

$$= \sum_k b_k \exp(-iE_k^0 t/\hbar)E_k^0\psi_k^0 + \sum_k b_k \exp(-iE_k^0 t/\hbar)\hat{H}'\psi_k^0$$

$$-\frac{\hbar}{i}\sum_k \frac{db_k}{dt}\exp(-iE_k^0 t/\hbar)\psi_k^0 = \sum_k b_k \exp(-iE_k^0 t/\hbar)\hat{H}'\psi_k^0$$

We now multiply by ψ_m^{0*} and integrate over the spatial and spin coordinates. Using the orthonormality of the unperturbed wave functions, $\langle\psi_m^0|\psi_k^0\rangle = \delta_{mk}$, we get

$$-\frac{\hbar}{i}\sum_k \frac{db_k}{dt}\exp(-iE_k^0 t/\hbar)\delta_{mk} = \sum_k b_k \exp(-iE_k^0 t/\hbar)\langle\psi_m^0|\hat{H}'|\psi_k^0\rangle$$

Because of the δ_{mk} factor, all terms but one in the sum on the left are zero, and the left side equals $-(\hbar/i)(db_m/dt)\exp(-iE_m^0 t/\hbar)$. We get

$$\frac{db_m}{dt} = -\frac{i}{\hbar}\sum_k b_k \exp[i(E_m^0 - E_k^0)t/\hbar]\langle\psi_m^0|\hat{H}'|\psi_k^0\rangle \qquad (9.119)$$

Let us suppose that the perturbation $\hat{H}'(t)$ was applied at time $t = 0$ and that before the perturbation was applied the system was in stationary state n with energy $E_n^{(0)}$. The state function at $t = 0$ is therefore $\Psi = \exp(-iE_n^0 t/\hbar)\psi_n^0$ [Eq. (7.98)], and the $t = 0$ values of the expansion coefficients in (9.118) are thus $b_n(0) = 1$ and $b_k(0) = 0$ for $k \neq n$:

$$b_k(0) = \delta_{kn} \qquad (9.120)$$

To facilitate the solution of (9.119), we shall assume that the perturbation $\hat{H}'$ is small and acts for only a short time. Under these conditions, the change in the expansion coefficients b_k from their initial values at the time the perturbation is applied will be small. To a good approximation, we can replace the expansion coefficients on the right side of (9.119) by their initial values (9.120). This gives

$$\frac{db_m}{dt} \approx -\frac{i}{\hbar}\exp[i(E_m^0 - E_n^0)t/\hbar]\langle\psi_m^0|\hat{H}'|\psi_n^0\rangle$$

Let the perturbation $\hat{H}'$ act from $t = 0$ to $t = t'$. Integrating from $t = 0$ to t' and using (9.120), we get

$$b_m(t') \approx \delta_{mn} - \frac{i}{\hbar}\int_0^{t'}\exp[i(E_m^0 - E_n^0)t/\hbar]\langle\psi_m^0|\hat{H}'|\psi_n^0\rangle\,dt \qquad (9.121)$$

Use of the approximate result (9.121) for the expansion coefficients in (9.118) gives the desired approximation to the state function at time t' for the case that the time-dependent perturbation $\hat{H}'$ is applied at $t = 0$ to a system in stationary state n. [As with time-independent perturbation theory, one can go to higher-order approximations (see *Fong*, pp. 234–244).]

For times after t', the perturbation has ceased to act, and $\hat{H}' = 0$. Equation (9.119) gives $db_m/dt = 0$ for $t > t'$, so $b_m = b_m(t')$ for $t \geq t'$. Therefore, for times after exposure to the perturbation, the state function Ψ is [Eq. (9.118)]

$$\Psi = \sum_m b_m(t')\exp(-iE_m^0 t/\hbar)\psi_m^0, \qquad \text{for } t \geq t' \qquad (9.122)$$

where $b_m(t')$ is given by (9.121). In (9.122), Ψ is a superposition of the eigenfunctions ψ_m^0 of the energy operator $\hat{H}^0$, the expansion coefficients being $b_m\exp(-iE_m^0 t/\hbar)$. [Compare Eqs. (9.122) and (7.65).] The work of Section 7.6 tells us that a measurement of the system's energy at a time after t' will give one of the eigenvalues E_m^0 of the energy operator $\hat{H}^0$, and the probability of getting E_m^0 equals the square of the absolute value of the expansion coefficient that multiplies ψ_m^0; that is, it equals $|b_m(t')\exp(-iE_m^0 t/\hbar)|^2 = |b_m(t')|^2$.

The time-dependent perturbation changes the system's state function from $\exp(-iE_n^0 t/\hbar)\psi_n^0$ to the superposition (9.122). Measurement of the energy then changes Ψ to one of the energy eigenfunctions $\exp(-iE_m^0 t/\hbar)\psi_m^0$ (reduction of the wave function, Section 7.9). The net result is a transition from stationary state n to stationary state m, the probability of such a transition being given by $|b_m(t')|^2$.

9.10 INTERACTION OF RADIATION AND MATTER

We now consider the interaction of an atom (or molecule) with electromagnetic radiation. A proper quantum-mechanical approach would treat both the atom and the radiation quantum mechanically, but we shall simplify things by using the classical picture of the light as an electromagnetic wave of oscillating electric and magnetic fields.

A detailed investigation, which we omit (see *Levine, Molecular Spectroscopy*, Section 3.2), shows that usually the interaction between the radiation's magnetic field and the atom's charges is much weaker than the interaction between the radiation's electric field and the charges, so we shall consider only the latter interaction. (In NMR spectroscopy the important interaction is between the magnetic dipole moments of the nuclei and the radiation's magnetic field. We shall not consider this case.)

Let the electric field $\mathscr{E}$ of the electromagnetic wave point in the x direction only. (This is plane-polarized radiation.) The electric field is defined as the force per unit charge, so the force on charge Q_i is $F = Q_i \mathscr{E}_x = -dV/dx$, where (4.26) was used. Integration gives the potential energy of interaction between the radiation's electric field and the charge as $V = -Q_i \mathscr{E}_x x$, where the arbitrary integration constant was taken as zero. For a system of several charges, $V = -\Sigma_i Q_i x_i \mathscr{E}_x$. This is the time-dependent perturbation $\hat{H}'(t)$. The space and time dependence of the electric field of an electromagnetic wave traveling in the z direction with wavelength λ and frequency ν is given by (*Halliday and Resnick*, Section 41-8) $\mathscr{E}_x = \mathscr{E}_0 \sin(2\pi\nu t - 2\pi z/\lambda)$, where $\mathscr{E}_0$ is the maximum value of $\mathscr{E}_x$ (the amplitude). Therefore,

$$\hat{H}'(t) = -\mathscr{E}_0 \sum_i Q_i x_i \sin(2\pi\nu t - 2\pi z_i/\lambda)$$

where the sum goes over all the electrons and nuclei of the atom or molecule.

Defining ω and ω_{mn} as

$$\omega \equiv 2\pi\nu, \qquad \omega_{mn} \equiv (E_m^0 - E_n^0)/\hbar \qquad (9.123)$$

and substituting $\hat{H}'(t)$ into (9.121), we get the coefficients in the expansion (9.118) of the state function Ψ as

$$b_m \approx \delta_{mn} + \frac{i\mathscr{E}_0}{\hbar} \int_0^{t'} \exp(i\omega_{mn}t) \langle \psi_m^0 | \sum_i Q_i x_i \sin(\omega t - 2\pi z_i/\lambda) | \psi_n^0 \rangle \, dt$$

The integral $\langle \psi_m^0 | \Sigma_i \cdots | \psi_n^0 \rangle$ in this equation is over all space, but significant contributions to its magnitude come only from regions where ψ_m^0 and ψ_n^0 are of significant magnitude. In regions well outside the atom or molecule, ψ_m^0 and ψ_n^0 are vanishingly small, and such regions can be ignored. Let the coordinate origin be chosen within the atom or molecule. Since regions well outside the atom can be ignored, the coordinate z_i can be considered to have a maximum magnitude of the order of several angstroms. For ultraviolet light, the wavelength λ is on the order of 10^3 Å; for visible, infrared, microwave, and radio-frequency radiation, λ is even larger. Hence $2\pi z_i/\lambda$ is very small and can be neglected, and this leaves $\Sigma_i Q_i x_i \sin \omega t$ in the integral.

Use of the identity (Problem 1.22) $\sin \omega t = (e^{i\omega t} - e^{-i\omega t})/2i$ gives

$$b_m(t') \approx \delta_{mn} + \frac{\mathcal{E}_0}{2\hbar} \langle \psi_m^0 | \sum_i Q_i x_i | \psi_n^0 \rangle \int_0^{t'} [e^{i(\omega_{mn}+\omega)t} - e^{i(\omega_{mn}-\omega)t}] \, dt$$

Using $\int_0^{t'} e^{at} \, dt = a^{-1}(e^{at'} - 1)$, we get

$$b_m(t') \approx \delta_{mn} + \frac{\mathcal{E}_0}{2\hbar i} \langle \psi_m^0 | \sum_i Q_i x_i | \psi_n^0 \rangle \left[\frac{e^{i(\omega_{mn}+\omega)t'} - 1}{\omega_{mn} + \omega} - \frac{e^{i(\omega_{mn}-\omega)t'} - 1}{\omega_{mn} - \omega} \right] \quad (9.124)$$

For $m \neq n$, the δ_{mn} term equals zero.

As noted at the end of Section 9.9, $|b_m(t')|^2$ gives the probability of a transition to state m from state n. There are two cases where this probability becomes of significant magnitude. If $\omega_{mn} = \omega$, the denominator of the second fraction in brackets is zero and this fraction's absolute value is large (but not infinite; see Problem 9.20). If $\omega_{mn} = -\omega$, the first fraction has a zero denominator and a large absolute value.

For $\omega_{mn} = \omega$, Eq. (9.123) gives $E_m^0 - E_n^0 = h\nu$. Exposure of the atom to radiation of frequency ν has produced a transition from stationary state n to stationary state m, where (since ν is positive) $E_m^0 > E_n^0$. We might suppose that the energy for this transition came from the absorption by the system of a photon of energy $h\nu$. This supposition is confirmed by a fully quantum mechanical treatment (called *quantum field theory*) in which the radiation is treated quantum mechanically rather than classically. We have ***absorption*** of radiation with a consequent increase in the system's energy.

For $\omega_{mn} = -\omega$, we get $E_n^0 - E_m^0 = h\nu$. Exposure to radiation of frequency ν has induced a transition from stationary state n to stationary state m, where (since ν is positive) $E_n^0 > E_m^0$. The system has gone to a lower energy level, and a quantum-field-theory treatment shows that a photon of energy $h\nu$ is emitted in this process. This is ***stimulated emission*** of radiation. Stimulated emission is the basis of the laser.

A defect of our treatment is that it does not predict ***spontaneous emission***, the emission of a photon by a system not exposed to radiation, the system falling to a lower energy level in the process. Quantum field theory does predict spontaneous emission.

Note from (9.124) that the probability of absorption is proportional to $|\langle \psi_m^0 | \sum_i Q_i x_i | \psi_n^0 \rangle|^2$. The quantity $\sum_i Q_i x_i$ is the x component of the system's dipole-moment operator $\hat{\mathbf{d}}$ (see Section 13.15 for details), which is [Eqs. (13.136) and (13.137)] $\hat{\mathbf{d}} = \mathbf{i} \sum_i Q_i x_i + \mathbf{j} \sum_i Q_i y_i + \mathbf{k} \sum_i Q_i z_i = \mathbf{i}\hat{d}_x + \mathbf{j}\hat{d}_y + \mathbf{k}\hat{d}_z$, where $\mathbf{i}$, $\mathbf{j}$, $\mathbf{k}$ are unit vectors along the axes and $\hat{d}_x$, $\hat{d}_y$, $\hat{d}_z$ are the components of $\hat{\mathbf{d}}$. We assumed polarized radiation with an electric field in the x direction only. If the radiation has electric-field components in the y and z directions also, then the probability of absorption will be proportional to

$$|\langle \psi_m^0 | \hat{d}_x | \psi_n^0 \rangle|^2 + |\langle \psi_m^0 | \hat{d}_y | \psi_n^0 \rangle|^2 + |\langle \psi_m^0 | \hat{d}_z | \psi_n^0 \rangle|^2 = |\langle \psi_m^0 | \hat{\mathbf{d}} | \psi_n^0 \rangle|^2$$

where Eq. (5.25) was used. The integral $\langle \psi_m^0 | \hat{\mathbf{d}} | \psi_n^0 \rangle = \mathbf{d}_{mn}$ is the ***transition (dipole) moment***.

When $\mathbf{d}_{mn} = 0$, the transition between states m and n with absorption or emission of radiation is said to be **forbidden**. **Allowed** transitions have $\mathbf{d}_{mn} \neq 0$. Because of approximations made in the derivation of (9.124), forbidden transitions may have some small probability of occurring.

Consider, for example, the particle in a one-dimensional box (Section 2.2). The transition dipole moment is $\langle \psi_m^0 | Qx | \psi_n^0 \rangle$, where Q is the particle's charge and x is its coordinate and where $\psi_m^0 = (2/l)^{1/2} \sin(m\pi x/l)$ and $\psi_n^0 = (2/l)^{1/2} \sin(n\pi x/l)$. Evaluation of this integral (Problem 9.21) shows it is nonzero only when $m - n = \pm 1, \pm 3, \pm 5, \ldots$ and is zero when $m - n = 0, \pm 2, \ldots$. The **selection rule** for a particle in a one-dimensional box is that the quantum number must change by an odd integer when radiation is absorbed or emitted.

Evaluation of the transition moment for the harmonic oscillator and for the two-particle rigid rotor gives the selection rules stated in Sections 4.3 and 6.4.

The quantity $|b_m|^2$ in (9.124) is sharply peaked at $\omega = \omega_{mn}$ and $\omega = -\omega_{mn}$, but there is a nonzero probability that a transition will occur when ω is not precisely equal to $|\omega_{mn}|$, that is, when $h\nu$ is not precisely equal to $|E_m^0 - E_n^0|$. This fact is related to the energy–time uncertainty relation (5.14). States with a finite lifetime have an uncertainty in their energy.

Radiation is not the only time-dependent perturbation that produces transitions between states. When an atom or molecule comes close to another atom or molecule, it suffers a time-dependent perturbation that can change its state. Selection rules derived for radiative transitions need not apply to collision processes, since $\hat{H}'(t)$ differs for the two processes.

9.11   SUMMARY

For a system whose time-independent Schrödinger equation is $(\hat{H}^0 + \hat{H}')\psi_n = E_n\psi_n$, perturbation theory expresses the energies and wave functions of the nondegenerate levels as $E_n = E_n^{(0)} + E_n^{(1)} + E_n^{(2)} + \cdots$ and $\psi_n = \psi_n^{(0)} + \psi_n^{(1)} + \psi_n^{(2)} + \cdots$, where the unperturbed wave functions $\psi_n^{(0)}$ and energies $E_n^{(0)}$ satisfy $\hat{H}^0\psi_n^{(0)} = E_n^{(0)}\psi_n^{(0)}$. The first-order energy correction is $E_n^{(1)} = \int \psi_n^{(0)*} \hat{H}' \psi_n^{(0)} \, d\tau$. The second-order energy correction is given by (9.35). The first-order correction to the wave function is $\psi_n^{(1)} = \sum_{m \neq n} a_m \psi_m^{(0)}$, where the expansion coefficients a_m are given by (9.26). For a degenerate level with degree of degeneracy d, we have $\psi_n = \phi_n^{(0)} + \psi_n^{(1)} + \cdots$, where the correct zeroth-order wave functions are $\phi_n^{(0)} = \sum_{i=1}^{d} c_i \psi_i^{(0)}$ for $n = 1, \ldots, d$. The first-order energy corrections for the degenerate level are found from the secular equation $\det(\langle \psi_m^{(0)} | \hat{H}' | \psi_i^{(0)} \rangle - E_n^{(1)} \delta_{mi}) = 0$, and then the coefficients c_i are found by solving (9.84).

Perturbation theory was applied to the helium atom, with $\hat{H}'$ taken as e'^2/r_{12}. The unperturbed ground-state wave function is $\psi^{(0)} = 1s(1)1s(2)$. The first-order energy correction was calculated for the ground state. Degenerate perturbation theory was applied to the first group of helium excited states. We found that the $1s2s$ configuration gives rise to two nondegenerate energy levels with correct zeroth-order wave functions $[1s(1)2s(2) \pm 1s(2)2s(1)]/\sqrt{2}$; the $1s2p$ configuration gives rise to two triply degenerate levels with correct zeroth-order wave functions $[1s(1)2p(2) \pm 1s(2)2p(1)]/\sqrt{2}$, where $2p$ can be $2p_x$, $2p_y$, or $2p_z$.

(These conclusions will be modified when electron spin is taken into account in Chapter 10.) The $1s2s$ levels lie below the $1s2p$ levels.

Time-dependent perturbation theory shows that, when an atom or molecule in a stationary state is exposed to electromagnetic radiation of frequency ν, the molecule may make a transition between two stationary states m and n whose energy difference is $h\nu$, provided the transition dipole moment $\langle \psi_m^0 | \hat{\mathbf{d}} | \psi_n^0 \rangle$ is nonzero for states m and n.

PROBLEMS

9.1 For the anharmonic oscillator with Hamiltonian (9.3), evaluate $E^{(1)}$ for the first excited state, taking the unperturbed system as the harmonic oscillator.

9.2 Consider the one-particle, one-dimensional system with potential-energy function

$$V = b \quad \text{for } \tfrac{1}{4}l < x < \tfrac{3}{4}l, \qquad V = 0 \quad \text{for } 0 \leqslant x \leqslant \tfrac{1}{4}l \text{ and } \tfrac{3}{4}l \leqslant x \leqslant l$$

and $V = \infty$ elsewhere, where $b = \hbar^2/ml^2$. Treat the system as a perturbed particle in a box. (a) Find the first-order energy correction for the general stationary state with quantum number n. (b) For the ground state and for the first excited state, compare $E^{(0)} + E^{(1)}$ with the true energies $5.750345\hbar^2/ml^2$ and $20.23604\hbar^2/ml^2$. Explain why $E^{(0)} + E^{(1)}$ for each of these two states is the same as obtained by the variational treatment of Problems 8.5a and 8.13.

9.3 For the perturbed particle in a box of Problem 9.2, find the first-order correction to the wave function of the stationary state with quantum number n.

9.4 Consider the perturbed particle in a box of Problem 9.2. (a) Explain why $\langle \psi_m^{(0)} | \hat{H}' | \psi_n^{(0)} \rangle = 0$ when $n = 1$ and m is an even integer. (b) Use a computer to evaluate $E^{(2)}$ for the ground state by summing over odd values of m in (9.35). Keep adding terms until the last added term is negligibly small in magnitude. Compare $E^{(0)} + E^{(1)} + E^{(2)}$ with the true ground-state energy $5.750345\hbar^2/ml^2$.

9.5 For the perturbed particle in a box of Problem 9.2, explain (without doing any calculations) why we expect $E^{(1)}$ to be greatest for the $n = 1$ state.

9.6 Assume that the charge of the proton is distributed uniformly throughout the volume of a sphere of radius 10^{-13} cm. Use perturbation theory to find the shift in the ground-state hydrogen-atom energy due to the finite proton size. The potential energy experienced by the electron when it has penetrated the nucleus and is at distance r from the nuclear center is $-eQ/4\pi\varepsilon_0 r$, where Q is the amount of proton charge within the sphere of radius r (*Halliday and Resnick*, Section 28-8). The evaluation of the integral is simplified by noting that the exponential factor in ψ is essentially equal to 1 within the nucleus.

9.7 For the anharmonic oscillator with $\hat{H} = -(\hbar^2/2m)d^2/dx^2 + \tfrac{1}{2}kx^2 + cx^3$, take $\hat{H}'$ as cx^3. (a) Find $E^{(1)}$ for the state with quantum number v. (b) Find $E^{(2)}$ for the state with quantum number v. You will need the following integral (*Levine, Molecular Spectroscopy*, p. 154):

$$\langle \psi_{v'}^{(0)} | x^3 | \psi_v^{(0)} \rangle = [(v+1)(v+2)(v+3)/8\alpha^3]^{1/2}\delta_{v',v+3}$$
$$+ 3[(v+1)/2\alpha]^{3/2}\delta_{v',v+1} + 3(v/2\alpha)^{3/2}\delta_{v',v-1}$$
$$+ [v(v-1)(v-2)/8\alpha^3]^{1/2}\delta_{v',v-3}$$

where the $\psi^{(0)}$'s are harmonic-oscillator wave functions and α is defined by (4.33). (c) Which unperturbed states contribute to $\psi_v^{(1)}$?

9.8 When Hylleraas began his calculations on helium, it was not known whether the isolated hydride ion H^- was a stable entity. Calculate the ground-state energy of H^- predicted by the trial function (9.58). Compare the result with the ground-state energy of the hydrogen atom, $-13.60\,eV$, and show that this simple variation function (erroneously) indicates H^- is unstable with respect to ionization into a hydrogen atom and an electron. (More complicated variational functions give a ground-state energy of $-14.35\,eV$.)

9.9 There is more than one way to divide a Hamiltonian $\hat{H}$ into an unperturbed part $\hat{H}^0$ and a perturbation $\hat{H}'$. Instead of the division (9.40) and (9.41), consider the following way of dividing up the helium-atom Hamiltonian:

$$\hat{H}^0 = -\frac{\hbar^2}{2m_e}\nabla_1^2 - \frac{\hbar^2}{2m_e}\nabla_2^2 - \left(Z - \frac{5}{16}\right)\frac{e'^2}{r_1} - \left(Z - \frac{5}{16}\right)\frac{e'^2}{r_2}$$

$$\hat{H}' = -\frac{5}{16}\frac{e'^2}{r_1} - \frac{5}{16}\frac{e'^2}{r_2} + \frac{e'^2}{r_{12}}$$

What are the unperturbed wave functions? Calculate $E^{(0)}$ and $E^{(1)}$ for the ground state. (See Section 9.4.)

9.10 Most (but not all) of the effect of nuclear motion in helium can be corrected for by replacing the electron's mass m_e by the reduced mass (6.59) in the expression for the energy. The energy of helium is proportional to what power of m_e? [See Eq. (9.66).] Use of μ instead of m_e multiplies the energies calculated on the basis of infinite nuclear mass by what factor?

9.11 Calculate $\langle r_1 \rangle$ for the helium trial function (9.58). To save time, use the result of Problem 6.16.

9.12 Show that the secular equation (9.85) can be written as

$$\det\left[\langle \psi_m^{(0)}|\hat{H}|\psi_i^{(0)}\rangle - \delta_{mi}(E_n^{(0)} + E_n^{(1)})\right] = 0$$

9.13 Verify (9.88) for the coefficients in a correct zeroth-order wave function.

9.14 (a) For a particle in a square box of length l with origin at $x = 0$, $y = 0$, write down the wave functions and energy levels. (b) If the system of (a) is perturbed by

$$\hat{H}' = b \qquad \text{for } \tfrac{1}{4}l \le x \le \tfrac{3}{4}l \text{ and } \tfrac{1}{4}l \le y \le \tfrac{3}{4}l$$

where b is a constant, and $\hat{H}' = 0$ elsewhere, find $E^{(1)}$ for the ground state. For the first excited energy level, find the $E^{(1)}$ values and the correct zeroth-order wave functions.

9.15 For a hydrogen atom perturbed by a uniform applied electric field in the z direction, the perturbation Hamiltonian is

$$\hat{H}' = e\mathscr{E}z = e\mathscr{E}r\cos\theta$$

where $\mathscr{E}$ is the magnitude of the electric field. Consider the effect of $\hat{H}'$ on the $n = 2$ energy level, which is fourfold degenerate. Since $\hat{H}'$ commutes with the angular-momentum operator $\hat{L}_z$, the ideas of Section 9.6 lead us to set up the secular determinant using the complex hydrogen-atom orbitals $2s$, $2p_1$, $2p_0$, and $2p_{-1}$, which are eigenfunctions of $\hat{L}_z$. Set up the secular determinant using the fact that matrix elements of $\hat{H}'$ between states with different values of the quantum number m will vanish; also, use parity considerations to show that certain other integrals are zero (see Problem 7.21). Evaluate the nonzero integrals, and find the first-order energy corrections and the correct zeroth-order wave functions. *Hint:* Choose the order of the orbitals so as to make the secular determinant block diagonal.

9.16 Consider the perturbation treatment of the helium configurations $1s3s$, $1s3p$, and $1s3d$. Without setting up the secular equation, but simply by analogy to the results of Section 9.7, write down the 18 correct zeroth-order wave functions. How many energy levels correspond to each of these three configurations, and what is the degeneracy of each energy level? The levels of which configuration lie lowest? highest?

9.17 Verify the result (9.113) for the integral J_{1s2p}.

9.18 We have considered helium configurations in which only one electron is excited. Get a rough estimate of the energy of the $2s^2$ configuration from Eq. (9.48). Compare this with the ground-state energy of the He^+ ion to show that the $2s^2$ helium configuration is unstable with respect to ionization to He^+ and an electron. If we had obtained a more accurate estimate of the $2s^2$ energy by including the first-order energy correction, would this increase or decrease our estimate of the $2s^2$ energy?

9.19 For helium the first-order perturbation energy correction is e'^2/r_{12} averaged over the correct unperturbed wave function. Show that if we evaluate $\langle e'^2/r_{12} \rangle$ using the incorrect zeroth-order functions $1s(1)2s(2)$ or $1s(2)2s(1)$ we get J_{1s2s} in each case. Now show that when we use the correct functions (9.105) and (9.106) to evaluate $\langle e'^2/r_{12} \rangle$ we get $J_{1s2s} \pm K_{1s2s}$ (as found from the secular equation). The exchange-integral contribution to the energy thus arises from the indistinguishability of electrons in the same atom.

9.20 Evaluate $\lim_{s \to 0} (e^{as} - 1)/s$. [$s$ corresponds to $\omega_{mn} \pm \omega$ in (9.124).]

9.21 Evaluate $\langle \psi_m^0 | Qx | \psi_n^0 \rangle$ for the particle in a one-dimensional box.

9.22 Find the selection rules for a charged particle in a three-dimensional box exposed to unpolarized radiation.

9.23 (a) Set f in (7.41) equal to $\hat{B}S$ and operate on each side of the resulting equation with $\hat{A}$. Then multiply by R^* and integrate over all space to obtain the sum rule

$$\sum_i \langle R|\hat{A}|g_i \rangle \langle g_i|\hat{B}|S \rangle = \langle R|\hat{A}\hat{B}|S \rangle$$

where the functions g_i form a complete, orthonormal set, the functions R and S are any two well-behaved functions, the operators $\hat{A}$ and $\hat{B}$ are linear, and the sum is over all members of the complete set. [See also Eq. (7.107). For other sum rules, see A. Dalgarno, *Rev. Mod. Phys.*, **35**, 522 (1963).] (b) An approximate way to evaluate $E_n^{(2)}$ in (9.35) is to replace $E_n^{(0)} - E_m^{(0)}$ by ΔE, where ΔE is some sort of average excitation energy for the problem, whose value can be roughly estimated from the spacings of the unperturbed levels. Use the sum rule of (a) to show that this replacement gives

$$E_n^{(2)} \approx \frac{1}{\Delta E} [\langle n|(\hat{H}')^2|n \rangle - (\langle n|\hat{H}'|n \rangle)^2]$$

where n stands for $\psi_n^{(0)}$.

9.24 True or false: (a) Every linear combination of solutions of the time-dependent Schrödinger equation is a solution of this equation. (b) Every linear combination of solutions of the time-independent Schrödinger equation is a solution of this equation.

10

Electron Spin and the Pauli Principle

10.1 ELECTRON SPIN

All chemists are familiar with the yellow color imparted to a flame by sodium atoms. Examining the spectrum of sodium, we find that the strongest yellow line (the D line) is actually two closely spaced lines. The sodium D line arises from a transition from the excited configuration $1s^2 2s^2 2p^6 3p$ to the ground state. The doublet nature of this and other lines in the Na spectrum indicates a doubling of the expected number of states available to the valence electron.

To explain this *fine structure* of atomic spectra, Uhlenbeck and Goudsmit proposed in 1925 that the electron has an *intrinsic* (built-in) angular momentum in addition to the orbital angular momentum due to its motion about the nucleus. If we picture the electron as a sphere of charge spinning about one of its diameters, we can see how such an intrinsic angular momentum can arise. Hence we have the term *spin angular momentum* or, more simply, *spin*. However, electron "spin" is not a classical effect, and the picture of an electron rotating about one of its axes should not be considered to represent physical reality. The intrinsic angular momentum is real, but there is no easily visualizable model that can explain its origin properly. We cannot hope to obtain a proper understanding of microscopic particles based on models taken from our experience in the macroscopic world. Other elementary particles besides the electron have spin angular momentum.

In 1928, Paul Dirac developed the relativistic quantum mechanics of an electron, and in his treatment electron spin arises naturally. Dirac's theory also indicated the existence of positively charged electrons, positrons, although Dirac did not fully realize this in 1928. Positrons were discovered in 1932. The positron is the antiparticle of the electron.

In the nonrelativistic quantum mechanics to which we are confining ourselves, electron spin must be introduced as an additional hypothesis. We have learned that each physical property has its corresponding linear Hermitian operator in quantum mechanics. For such properties as orbital angular momentum, we can construct the quantum-mechanical operator from the classical expression by replacing p_x, p_y, p_z by the appropriate operators. The inherent spin angular momentum of a microscopic particle has no analog in classical mechanics, so we cannot use this method to construct operators for spin. For our purposes, we shall simply use symbols for the spin operators, without giving an explicit form for them.

Analogous to the orbital angular-momentum operators $\hat{L}^2$, $\hat{L}_x$, $\hat{L}_y$, $\hat{L}_z$, we have the spin angular-momentum operators $\hat{S}^2$, $\hat{S}_x$, $\hat{S}_y$, $\hat{S}_z$, which are postulated to be linear and Hermitian. $\hat{S}^2$ is the operator for the square of the magnitude of the total spin angular momentum of a particle. $\hat{S}_z$ is the operator for the z component of the particle's spin angular momentum. We have

$$\hat{S}^2 = \hat{S}_x^2 + \hat{S}_y^2 + \hat{S}_z^2 \tag{10.1}$$

We postulate that the spin angular-momentum operators obey the same commutation relations as the orbital angular-momentum operators. Analogous to $[\hat{L}_x, \hat{L}_y] = i\hbar \hat{L}_z$, $[\hat{L}_y, \hat{L}_z] = i\hbar \hat{L}_x$, $[\hat{L}_z, \hat{L}_x] = i\hbar \hat{L}_y$ [Eqs. (5.46) and (5.48)], we have

$$[\hat{S}_x, \hat{S}_y] = i\hbar \hat{S}_z \ , \qquad [\hat{S}_y, \hat{S}_z] = i\hbar \hat{S}_x \ , \qquad [\hat{S}_z, \hat{S}_x] = i\hbar \hat{S}_y \tag{10.2}$$

From (10.1) and (10.2), it follows, by the same operator algebra used to obtain (5.49) and (5.50), that

$$[\hat{S}^2, \hat{S}_x] = [\hat{S}^2, \hat{S}_y] = [\hat{S}^2, \hat{S}_z] = 0 \tag{10.3}$$

Since Eqs. (10.1) and (10.2) are of the form of Eqs. (5.114) and (5.115), it follows from the work of Section 5.4 (which depended only on the commutation relations and not the specific forms of the operators) that the eigenvalues of $\hat{S}^2$ are [Eq. (5.149)]

$$s(s + 1)\hbar^2 \ , \qquad s = 0, \tfrac{1}{2}, 1, \tfrac{3}{2}, \ldots \tag{10.4}*$$

and the eigenvalues of $\hat{S}_z$ are [Eq. (5.148)]

$$m_s \hbar \ , \qquad m_s = -s, -s + 1, \ldots, s - 1, s \tag{10.5}*$$

The quantum number s is called the **spin** of the particle. Although nothing in Section 5.4 restricts electrons to a single value for s, experiment shows that all electrons do have a single value for s, namely, $s = \tfrac{1}{2}$. Protons and neutrons also have $s = \tfrac{1}{2}$. Pions have $s = 0$. Photons have $s = 1$. However, Eq. (10.5) does not hold for photons. Photons travel at speed c in vacuum; because of their relativistic nature, it turns out that photons can have either $m_s = +1$ or $m_s = -1$, but not $m_s = 0$ (see *Merzbacher*, Chapter 22). These two m_s values correspond to left circularly polarized and right circularly polarized light.

With $s = \tfrac{1}{2}$, the magnitude of the total spin angular momentum of an electron is given by the square root of (10.4) as

$$[\tfrac{1}{2}(\tfrac{3}{2})\hbar^2]^{1/2} = \tfrac{1}{2}\sqrt{3}\hbar \tag{10.6}$$

For $s = \tfrac{1}{2}$, Eq. (10.5) gives the possible eigenvalues of $\hat{S}_z$ of an electron as $+\tfrac{1}{2}\hbar$ and $-\tfrac{1}{2}\hbar$. The electron spin eigenfunctions that correspond to these $\hat{S}_z$ eigenvalues are denoted by α and β:

$$\hat{S}_z \alpha = +\tfrac{1}{2}\hbar\alpha \qquad (10.7)*$$

$$\hat{S}_z \beta = -\tfrac{1}{2}\hbar\beta \qquad (10.8)*$$

Since $\hat{S}_z$ commutes with $\hat{S}^2$, we can take the eigenfunctions of $\hat{S}_z$ to be eigenfunctions of $\hat{S}^2$ also, with the eigenvalue given by (10.4) with $s = \tfrac{1}{2}$:

$$\hat{S}^2 \alpha = \tfrac{3}{4}\hbar^2\alpha , \qquad \hat{S}^2 \beta = \tfrac{3}{4}\hbar^2\beta \qquad (10.9)$$

$\hat{S}_z$ does not commute with $\hat{S}_x$ or $\hat{S}_y$, so α and β are not eigenfunctions of these operators. The terms *spin up* and *spin down* refer to $m_s = +\tfrac{1}{2}$ and $m_s = -\tfrac{1}{2}$, respectively. See Fig. 10.1. We shall later show that the two possibilities for the quantum number m_s give the doubling of lines in the spectra of the alkali metals.

The wave functions we have dealt with previously are functions of the spatial coordinates of the particle: $\psi = \psi(x, y, z)$. We might ask: What is the variable for the spin eigenfunctions α and β? Sometimes one talks of a spin coordinate ω, without really specifying what this coordinate is. Most often, one takes the spin quantum number m_s as being the variable on which the spin eigenfunctions depend. This procedure is quite unusual as compared with the spatial wave functions; but because we have only two possible electronic spin eigenfunctions and eigenvalues, this is a convenient choice. We have

$$\alpha = \alpha(m_s) , \qquad \beta = \beta(m_s) \qquad (10.10)$$

As usual, we want the eigenfunctions to be normalized. The three variables of a one-particle space wave function range continuously from $-\infty$ to $+\infty$, so normalization means

$$\int_{-\infty}^{\infty} \int_{-\infty}^{\infty} \int_{-\infty}^{\infty} |\psi(x, y, z)|^2 \, dx \, dy \, dz = 1$$

The variable m_s of the electronic spin eigenfunctions takes on only the two discrete values $+\tfrac{1}{2}$ and $-\tfrac{1}{2}$, so normalization of the one-particle spin eigenfunctions means

$$\sum_{m_s=-1/2}^{1/2} |\alpha(m_s)|^2 = 1 , \qquad \sum_{m_s=-1/2}^{1/2} |\beta(m_s)|^2 = 1 \qquad (10.11)$$

Figure 10.1 Possible orientations of the electron spin vector with respect to the z axis. In each case, **S** lies on the surface of a cone whose axis is the z axis.

Since the eigenfunctions α and β correspond to different eigenvalues of the Hermitian operator $\hat{S}_z$, they are orthogonal:

$$\sum_{m_s=-1/2}^{1/2} \alpha^*(m_s)\beta(m_s) = 0 \tag{10.12}$$

When we consider the complete wave function for an electron including both space and spin variables, we shall normalize it according to

$$\sum_{m_s=-1/2}^{1/2} \int_{-\infty}^{\infty} \int_{-\infty}^{\infty} \int_{-\infty}^{\infty} |\psi(x, y, z, m_s)|^2 \, dx \, dy \, dz = 1 \tag{10.13*}$$

The notation

$$\int |\psi(x, y, z, m_s)|^2 \, d\tau$$

will denote summation over the spin variable and integration over the full range of the space variables, as in (10.13). The symbol $\int dv$ will denote integration over the full range of the system's spatial variables.

10.2   SPIN AND THE HYDROGEN ATOM

The wave function specifying the state of an electron depends not only on the coordinates x, y, and z but also on the spin state of the electron. What effect does this have on the wave functions and energy levels of the hydrogen atom?

To a very good approximation, the Hamiltonian for a system of electrons does not involve the spin variables but is a function only of spatial coordinates and derivatives with respect to spatial coordinates. As a result, we can separate the stationary-state wave function of a single electron into a product of space and spin parts:

$$\psi(x, y, z)g(m_s)$$

where $g(m_s)$ is either one of the functions α or β, depending on whether $m_s = \frac{1}{2}$ or $-\frac{1}{2}$. [More generally, $g(m_s)$ might be a linear combination of α and β; $g(m_s) = c_1\alpha + c_2\beta$.] Since the Hamiltonian operator has no effect on the spin function, we have

$$\hat{H}[\psi(x, y, z)g(m_s)] = g(m_s)\hat{H}\psi(x, y, z) = E[\psi(x, y, z)g(m_s)]$$

and we get the same energies as previously found without taking spin into account. The only difference spin makes is to double the possible number of states; instead of the state $\psi(x, y, z)$, we have the two possible states $\psi(x, y, z)\alpha$ and $\psi(x, y, z)\beta$. When we take spin into account, the degeneracy of the hydrogen-atom energy levels is $2n^2$ rather than n^2.

10.3   THE PAULI PRINCIPLE

Suppose we have a system of several identical particles. In classical mechanics the identity of the particles leads to no special consequences. For example, consider identical billiard balls rolling on a billiard table. It is perfectly possible to follow the motion of any individual ball, say by taking a motion picture of the system. We can say that ball number one is moving along a certain path, ball two is on another definite path, and so on, the paths being determined by Newton's laws of motion. Thus, although the balls are identical, we can distinguish among them by specifying the path each takes. The identity of the balls has no special effect on their motions.

In quantum mechanics the uncertainty principle tells us that we cannot follow the exact path taken by a microscopic "particle." If the microscopic particles of the system all have different masses or charges or spins, we can use one of these properties to distinguish the particles from one another. But if they are all identical, then the one way we had in classical mechanics of distinguishing them, namely by specifying their paths, is lost in quantum mechanics because of the uncertainty principle. Therefore, the wave function of a system of interacting identical particles must not distinguish among the particles. For example, in the perturbation treatment of the helium-atom excited states in Chapter 9, we saw that the function $1s(1)2s(2)$, which says that electron 1 is in the $1s$ orbital and electron 2 is in the $2s$ orbital, was not a correct zeroth-order wave function. Rather we had to use the functions $2^{-1/2}[1s(1)2s(2) \pm 1s(2)2s(1)]$, which do not specify which electron is in which orbital. (If the identical particles are well separated from one another so that their wave functions do not overlap, they may be regarded as distinguishable.)

We now derive the restrictions on the wave function due to the requirement of indistinguishability of identical particles in quantum mechanics. The wave function of a system of n identical microscopic particles depends on the space and spin variables of the particles. For particle 1, these variables are x_1, y_1, z_1, m_{s1}. Let q_1 stand for all four of these variables. Thus $\psi = \psi(q_1, q_2, \ldots, q_n)$.

We define the **permutation operator** $\hat{P}_{12}$ as the operator that interchanges all the coordinates of particles 1 and 2:

$$\hat{P}_{12} f(q_1, q_2, q_3, \ldots, q_n) = f(q_2, q_1, q_3, \ldots, q_n) \qquad (10.14)*$$

For example, the effect of $\hat{P}_{12}$ on the function that has electron 1 in a $1s$ orbital with spin up and electron 2 in a $3s$ orbital with spin down is

$$\hat{P}_{12}[1s(1)\alpha(1)3s(2)\beta(2)] = 1s(2)\alpha(2)3s(1)\beta(1)$$

What are the eigenvalues of $\hat{P}_{12}$? Applying $\hat{P}_{12}$ twice has no net effect:

$$\hat{P}_{12}\hat{P}_{12} f(q_1, q_2, \ldots, q_n) = \hat{P}_{12} f(q_2, q_1, \ldots, q_n) = f(q_1, q_2, \ldots, q_n)$$

Therefore, $\hat{P}_{12}^2 = \hat{1}$. Let w_i and c_i denote the eigenfunctions and eigenvalues of $\hat{P}_{12}$. We have $\hat{P}_{12} w_i = c_i w_i$. Application of $\hat{P}_{12}$ to this equation gives $\hat{P}_{12}^2 w_i = c_i \hat{P}_{12} w_i$. Substitution of $\hat{P}_{12}^2 = \hat{1}$ and $\hat{P}_{12} w_i = c_i w_i$ in $\hat{P}_{12}^2 w_i = c_i \hat{P}_{12} w_i$ gives $w_i = c_i^2 w_i$. Since zero is not allowed as an eigenfunction, we can divide by w_i to get $1 = c_i^2$ and $c_i = \pm 1$. The eigenvalues of $\hat{P}_{12}$ (and of any linear operator whose square is the unit operator) are $+1$ and -1.

If w_+ is an eigenfunction of $\hat{P}_{12}$ with eigenvalue $+1$, then

$$\hat{P}_{12}w_+(q_1, q_2, \ldots, q_n) = (+1)w_+(q_1, q_2, \ldots, q_n)$$

$$w_+(q_2, q_1, \ldots, q_n) = w_+(q_1, q_2, \ldots, q_n) \qquad (10.15)^*$$

A function such as w_+ that has the property (10.15) of being unchanged when particles 1 and 2 are interchanged is said to be **symmetric** with respect to interchange of particles 1 and 2. For eigenvalue -1, we have

$$w_-(q_2, q_1, \ldots, q_n) = -w_-(q_1, q_2, \ldots, q_n) \qquad (10.16)^*$$

The function w_- in (10.16) is **antisymmetric** with respect to interchange of particles 1 and 2. There is no necessity for an arbitrary function $f(q_1, q_2, \ldots, q_n)$ to be either symmetric or antisymmetric with respect to interchange of 1 and 2.

Do not confuse the property of being symmetric or antisymmetric with respect to particle interchange with the property of being even or odd with respect to inversion in space. The function $x_1 + x_2$ is symmetric with respect to 1–2 interchange and is an odd function of x_1 and x_2. The function $x_1^2 + x_2^2$ is symmetric with respect to 1–2 interchange and is an even function of x_1 and x_2.

The operator $\hat{P}_{ij}$ is defined by

$$\hat{P}_{ij}f(q_1, \ldots, q_i, \ldots, q_j, \ldots, q_n) = f(q_1, \ldots, q_j, \ldots, q_i, \ldots, q_n) \qquad (10.17)$$

The eigenvalues of $\hat{P}_{ij}$ are, like those of $\hat{P}_{12}$, $+1$ and -1.

We now consider the wave function of a system of n identical microscopic particles. Since the particles are indistinguishable, the way we label them cannot affect the state of the system. Thus the two wave functions

$$\psi(q_1, \ldots, q_i, \ldots, q_j, \ldots, q_n) \quad \text{and} \quad \psi(q_1, \ldots, q_j, \ldots, q_i, \ldots, q_n)$$

must correspond to the same state of the system. Two wave functions that correspond to the same state can differ at most by a multiplicative constant. Hence

$$\psi(q_1, \ldots, q_j, \ldots, q_i, \ldots, q_n) = c\psi(q_1, \ldots, q_i, \ldots, q_j, \ldots, q_n)$$

$$\hat{P}_{ij}\psi(q_1, \ldots, q_i, \ldots, q_j, \ldots, q_n) = c\psi(q_1, \ldots, q_i, \ldots, q_j, \ldots, q_n) \tag{10.18}$$

Equation (10.18) states that ψ is an eigenfunction of $\hat{P}_{ij}$. But we know that the only possible eigenvalues of $\hat{P}_{ij}$ are 1 and -1. We conclude that the wave function for a system of n identical particles must be symmetric or antisymmetric with respect to interchange of any two of the identical particles, i and j. Since the n particles are all identical, we could not have the wave function symmetric with respect to some interchanges and antisymmetric with respect to other interchanges. Thus the wave function of n identical particles must be either symmetric with respect to every possible interchange or antisymmetric with respect to every possible interchange of two particles.

We have shown that there are two possible cases for the wave function of a system of identical particles, the symmetric and the antisymmetric cases. Experimental evidence (such as the periodic table of the elements to be discussed later) shows that for electrons only the antisymmetric case occurs. Thus we have an additional postulate of quantum mechanics, which states that *the wave function of a system of electrons must be antisymmetric with respect to interchange of any two electrons*. This important postulate is called the **Pauli principle**, after the physicist Wolfgang Pauli.

Pauli showed that relativistic quantum field theory indicates that particles with half-integral spin ($s = \frac{1}{2}, \frac{3}{2}$, and so on) require antisymmetric wave functions, while particles of integral spin ($s = 0, 1$, and so on) require symmetric wave functions. Experimental evidence leads to the same conclusion. Particles requiring antisymmetric wave functions, such as electrons, are called **fermions** (after E. Fermi), while particles requiring symmetric wave functions, such as pions, are called **bosons** (after S. N. Bose).

The Pauli principle has an interesting consequence for a system of identical fermions. The antisymmetry requirement means that

$$\psi(q_1, q_2, q_3, \ldots, q_n) = -\psi(q_2, q_1, q_3, \ldots, q_n) \qquad (10.19)$$

Consider the value of ψ when electrons 1 and 2 have the same coordinates, that is, when $x_1 = x_2$, $y_1 = y_2$, $z_1 = z_2$, and $m_{s1} = m_{s2}$. Putting $q_2 = q_1$ in (10.19), we have

$$\psi(q_1, q_1, q_3, \ldots, q_n) = -\psi(q_1, q_1, q_3, \ldots, q_n)$$
$$2\psi = 0$$
$$\psi(q_1, q_1, q_3, \ldots, q_n) = 0 \qquad (10.20)$$

Thus two electrons with the same spin have zero probability of being found at the same point in three-dimensional space. (By "the same spin," we mean the same value of m_s.) Since ψ is a continuous function, Eq. (10.20) means that the probability of finding two electrons with the same spin close to each other in space is quite small. Thus the Pauli antisymmetry principle forces electrons of like spin to keep apart from one another; to describe this, one often speaks of a **Pauli repulsion** between such electrons. This "repulsion" is not a real physical force, but a reflection of the fact that the electronic wave function must be antisymmetric with respect to exchange.

10.4 THE HELIUM ATOM

We now reconsider the helium atom from the standpoint of electron spin and the Pauli principle. In the perturbation treatment of helium in Section 9.3, we found the zeroth-order wave function for the ground state to be $1s(1)1s(2)$. To take spin into account, we must multiply this spatial function by a spin eigenfunction. We therefore consider the possible spin eigenfunctions for two electrons. We shall use the notation $\alpha(1)\alpha(2)$ to indicate a state where electron 1 has spin up and electron 2 has spin up; $\alpha(1)$ stands for $\alpha(m_{s1})$. Since each electron has two possible spin states, we have at first sight the four possible spin functions:

$$\alpha(1)\alpha(2), \qquad \beta(1)\beta(2), \qquad \alpha(1)\beta(2), \qquad \alpha(2)\beta(1)$$

There is nothing wrong with the first two functions, but the third and fourth functions violate the principle of indistinguishability of identical particles. For example, the third function says that electron 1 has spin up and electron 2 has spin down, which *does* distinguish between electrons 1 and 2. More formally, if we apply the permutation operator $\hat{P}_{12}$ to these functions, we find that the first two functions are symmetric with respect to interchange of the two electrons, but the third and fourth functions are neither symmetric nor antisymmetric and so are unacceptable.

What now? Recall that we ran into essentially the same situation in treating the helium excited states (Section 9.7), where we started with the functions $1s(1)2s(2)$ and $2s(1)1s(2)$. We found that these two functions, which distinguished between electrons 1 and 2, were not the correct zeroth-order functions and that the correct zeroth-order functions were $2^{-1/2}[1s(1)2s(2) \pm 2s(1)1s(2)]$. This result suggests pretty strongly that instead of $\alpha(1)\beta(2)$ and $\beta(1)\alpha(2)$, we use the spin functions

$$2^{-1/2}[\alpha(1)\beta(2) \pm \beta(1)\alpha(2)] \tag{10.21}$$

These two functions are the normalized linear combinations of $\alpha(1)\beta(2)$ and $\beta(1)\alpha(2)$ that are eigenfunctions of $\hat{P}_{12}$, that is, are symmetric or antisymmetric. When electrons 1 and 2 are interchanged, $2^{-1/2}[\alpha(1)\beta(2) + \beta(1)\alpha(2)]$ becomes $2^{-1/2}[\alpha(2)\beta(1) + \beta(2)\alpha(1)]$, which is the same as the original function; in contrast, $2^{-1/2}[\alpha(1)\beta(2) - \beta(1)\alpha(2)]$ becomes $2^{-1/2}[\alpha(2)\beta(1) - \beta(2)\alpha(1)]$, which is -1 times the original function. To show that the functions (10.21) are normalized, we have

$$\sum_{m_{s1}} \sum_{m_{s2}} \frac{1}{\sqrt{2}} [\alpha(1)\beta(2) \pm \beta(1)\alpha(2)]^* \frac{1}{\sqrt{2}} [\alpha(1)\beta(2) \pm \beta(1)\alpha(2)]$$

$$= \frac{1}{2} \sum_{m_{s1}} |\alpha(1)|^2 \sum_{m_{s2}} |\beta(2)|^2 \pm \frac{1}{2} \sum_{m_{s1}} \alpha^*(1)\beta(1) \sum_{m_{s2}} \beta^*(2)\alpha(2)$$

$$\pm \frac{1}{2} \sum_{m_{s1}} \beta^*(1)\alpha(1) \sum_{m_{s2}} \alpha^*(2)\beta(2) + \frac{1}{2} \sum_{m_{s1}} |\beta(1)|^2 \sum_{m_{s2}} |\alpha(2)|^2 = 1$$

where we used the orthonormality relations (10.11) and (10.12).

Therefore, the four normalized two-electron spin eigenfunctions with the correct exchange properties are

$$\text{symmetric:} \quad \begin{cases} \alpha(1)\alpha(2) & \textbf{(10.22)}* \\ \beta(1)\beta(2) & \textbf{(10.23)}* \\ [\alpha(1)\beta(2) + \beta(1)\alpha(2)]/\sqrt{2} & \textbf{(10.24)}* \end{cases}$$

$$\text{antisymmetric:} \quad [\alpha(1)\beta(2) - \beta(1)\alpha(2)]/\sqrt{2} \qquad \textbf{(10.25)}*$$

We now include spin in the zeroth-order ground-state wave function. The function $1s(1)1s(2)$ is symmetric with respect to exchange. According to the Pauli principle, the overall wave function including spin must be antisymmetric with respect to interchange of the two electrons. Hence we must multiply the symmetric space function $1s(1)1s(2)$ by an antisymmetric spin function. There is only one antisymmetric two-electron spin function, so the ground-state zeroth-order wave function for the helium atom including spin is

$$\psi^{(0)} = 1s(1)1s(2) \cdot 2^{-1/2}[\alpha(1)\beta(2) - \beta(1)\alpha(2)] \qquad (10.26)$$

The function $\psi^{(0)}$ is an eigenfunction of $\hat{P}_{12}$ with eigenvalue -1, as the Pauli principle requires.

To a very good approximation, the Hamiltonian does not contain spin terms, so the energy is unaffected by inclusion of the spin factor in the ground-state wave function. Also, the ground state of helium is still nondegenerate when spin is considered.

To further demonstrate that the spin factor has no effect on the value of the energy, we shall assume we are doing a variational calculation for the helium ground state using the trial function

$$\phi = f(r_1, r_2, r_{12})2^{-1/2}[\alpha(1)\beta(2) - \beta(1)\alpha(2)]$$

where f is a normalized function symmetric in the coordinates of the two electrons. The variational integral is

$$\int \phi^* \hat{H} \phi \, d\tau = \sum_{m_{s1}} \sum_{m_{s2}} \int \int f^*(r_1, r_2, r_{12}) \frac{1}{\sqrt{2}} [\alpha(1)\beta(2) - \beta(1)\alpha(2)]^*$$

$$\times \hat{H} f(r_1, r_2, r_{12}) \frac{1}{\sqrt{2}} [\alpha(1)\beta(2) - \beta(1)\alpha(2)] \, dv_1 \, dv_2$$

Since $\hat{H}$ has no effect on the spin functions, the variational integral becomes

$$\int \int f^* \hat{H} f \, dv_1 \, dv_2 \sum_{m_{s1}} \sum_{m_{s2}} \tfrac{1}{2} |\alpha(1)\beta(2) - \beta(1)\alpha(2)|^2$$

Since the spin function (10.25) is normalized, the variational integral reduces to

$$\int \phi^* \hat{H} \phi \, d\tau = \int \int f^* \hat{H} f \, dv_1 \, dv_2$$

which is the expression we used before we introduced spin.

Now consider the excited states of helium. We found the lowest excited state to have the zeroth-order space wave function

$$2^{-1/2}[1s(1)2s(2) - 2s(1)1s(2)]$$

Since this space function is antisymmetric, we must multiply it by a symmetric spin function. We can use any one of the three symmetric two-electron spin functions, so instead of the nondegenerate level previously found, we have a triply degenerate level with the three zeroth-order wave functions

$$2^{-1/2}[1s(1)2s(2) - 2s(1)1s(2)]\alpha(1)\alpha(2) \qquad (10.27)$$

$$2^{-1/2}[1s(1)2s(2) - 2s(1)1s(2)]\beta(1)\beta(2) \qquad (10.28)$$

$$2^{-1/2}[1s(1)2s(2) - 2s(1)1s(2)]2^{-1/2}[\alpha(1)\beta(2) + \beta(1)\alpha(2)] \qquad (10.29)$$

For the next excited state, the requirement of antisymmetry of the overall wave function leads to the zeroth-order wave function

$$2^{-1/2}[1s(1)2s(2) + 2s(1)1s(2)]2^{-1/2}[\alpha(1)\beta(2) - \beta(1)\alpha(2)] \qquad (10.30)$$

The same considerations apply for the $1s2p$ states.

10.5 THE PAULI EXCLUSION PRINCIPLE

So far, we have not seen any very spectacular consequences of electron spin and the Pauli principle. In the hydrogen and helium atoms, the spin factors in the wave functions and the antisymmetry requirement simply affect the degeneracy of the levels but do not (except for very small effects to be considered later) affect the previously obtained energies. For lithium, the story is quite different.

The natural perturbation approach to the lithium atom is to take the interelectronic repulsions as a perturbation on the remaining terms in the Hamiltonian. By the same steps used in the treatment of helium, the unperturbed wave functions are products of three hydrogenlike functions. For the ground state,

$$\psi^{(0)} = 1s(1)1s(2)1s(3) \qquad (10.31)$$

and the zeroth-order (unperturbed) energy is [Eq. (9.48)]

$$E^{(0)} = -\left(\frac{1}{1^2} + \frac{1}{1^2} + \frac{1}{1^2}\right)\left(\frac{Z^2 e'^2}{2a_0}\right) = -27\left(\frac{e'^2}{2a_0}\right) = -27(13.606) \text{ eV} = -367.4 \text{ eV}$$

The first-order energy correction is $E^{(1)} = \langle \psi^{(0)}|\hat{H}'|\psi^{(0)}\rangle$. The perturbation $\hat{H}'$ consists of the interelectronic repulsions, and

$$E^{(1)} = \int |1s(1)|^2|1s(2)|^2|1s(3)|^2 \frac{e'^2}{r_{12}} \, dv + \int |1s(1)|^2|1s(2)|^2|1s(3)|^2 \frac{e'^2}{r_{23}} \, dv$$
$$+ \int |1s(1)|^2|1s(2)|^2|1s(3)|^2 \frac{e'^2}{r_{13}} \, dv$$

The way we label the dummy integration variables in these definite integrals cannot affect their value. If we interchange the labels 1 and 3 on the variables in the second integral, it is converted to the first integral. Hence these two integrals are equal. Interchange of the labels 2 and 3 in the third integral shows it to be equal to the first integral also. Hence

$$E^{(1)} = 3 \int \int |1s(1)|^2|1s(2)|^2 \frac{e'^2}{r_{12}} \, dv_1 \, dv_2 \int |1s(3)|^2 \, dv_3$$

The integral over electron 3 gives 1 (normalization). The integral over electrons 1 and 2 was evaluated in the perturbation treatment of helium, and [Eqs. (9.52) and (9.55)]

$$E^{(1)} = 3\left(\frac{5Z}{4}\right)\left(\frac{e'^2}{2a_0}\right) = 153.1 \text{ eV}$$

$$E^{(0)} + E^{(1)} = -214.3 \text{ eV}$$

Since we can use the zeroth-order perturbation wave function as a trial variation function (recall the discussion at the beginning of Section 9.4), $E^{(0)} + E^{(1)}$ must be, according to the variation principle, equal to or greater than the true ground-state energy. The experimental value of the lithium ground-state energy is found by adding up the first, second, and third ionization energies, which gives [C. E. Moore, "Ionization Potentials and Ionization Limits," publication NSRDS-NBS 34 of the National Bureau of Standards (1970)]

$$-(5.39 + 75.64 + 122.45) \, eV = -203.5 \, eV$$

We thus have $E^{(0)} + E^{(1)}$ as less than the true ground-state energy, which is a violation of the variation principle. Moreover, the supposed configuration $(1s)^3$ for the Li ground state is in complete disagreement with the low value of the first ionization potential and with all chemical evidence. If we continued on in this manner, we would have a $(1s)^Z$ ground-state configuration for the element of atomic number Z. We would not get the well-known periodic behavior of the elements.

Of course, our error is failure to consider spin and the Pauli principle. The hypothetical zeroth-order wave function $1s(1)1s(2)1s(3)$ is symmetric with respect to interchange of any two electrons. If we are to satisfy the Pauli principle, we must multiply this symmetric space function by an antisymmetric spin function. It is easy to construct completely symmetric spin functions for three electrons, such as $\alpha(1)\alpha(2)\alpha(3)$. However, try as we may, it is impossible to construct a completely antisymmetric spin function for three electrons.

Let us consider how we can systematically construct an antisymmetric function for three electrons. We shall use f, g, and h to stand for three functions of electronic coordinates, without specifying whether we are considering space coordinates or spin coordinates or both. We start with the function

$$f(1)g(2)h(3) \tag{10.32}$$

which is certainly not antisymmetric. The antisymmetric function we desire must be converted into its negative by each of the permutation operators $\hat{P}_{12}$, $\hat{P}_{13}$, and $\hat{P}_{23}$. Applying each of these operators in turn to $f(1)g(2)h(3)$, we get the functions

$$f(2)g(1)h(3), \qquad f(3)g(2)h(1), \qquad f(1)g(3)h(2) \tag{10.33}$$

We might try to construct the antisymmetric functions as a linear combination of the four functions (10.32) and (10.33), but this attempt would fail. Application of $\hat{P}_{12}$ to the last two functions in (10.33) gives

$$f(3)g(1)h(2) \quad \text{and} \quad f(2)g(3)h(1) \tag{10.34}$$

which are not included in (10.32) or (10.33). We must therefore include all six functions (10.32) to (10.34) in the desired antisymmetric linear combination. These six functions are the six $(3 \cdot 2 \cdot 1)$ possible permutations of the three electrons among the three functions f, g, and h. If $f(1)g(2)h(3)$ is a solution of the Schrödinger equation with eigenvalue E, then, because of the identity of the particles, each of the functions (10.32) to (10.34) is also a solution with the same eigenvalue E (exchange degeneracy), and any linear combination of these functions is an eigenfunction with eigenvalue E.

The antisymmetric linear combination will have the form

$$c_1 f(1)g(2)h(3) + c_2 f(2)g(1)h(3) + c_3 f(3)g(2)h(1) + c_4 f(1)g(3)h(2)$$
$$+ c_5 f(3)g(1)h(2) + c_6 f(2)g(3)h(1) \tag{10.35}$$

Since $f(2)g(1)h(3) = \hat{P}_{12}f(1)g(2)h(3)$, in order to have (10.35) be an eigenfunction of $\hat{P}_{12}$ with eigenvalue -1, we must have $c_2 = -c_1$. Likewise, $f(3)g(2)h(1) = \hat{P}_{13}f(1)g(2)h(3)$ and $f(1)g(3)h(2) = \hat{P}_{23}f(1)g(2)h(3)$, so $c_3 = -c_1$ and $c_4 = -c_1$. Since $f(3)g(1)h(2) = \hat{P}_{12}f(3)g(2)h(1)$, we must have $c_5 = -c_3 = c_1$. Similarly, we find $c_6 = c_1$. We thus arrive at the linear combination

$$c_1[f(1)g(2)h(3) - f(2)g(1)h(3) - f(3)g(2)h(1) - f(1)g(3)h(2)$$
$$+ f(3)g(1)h(2) + f(2)g(3)h(1)] \tag{10.36}$$

which is easily verified to be antisymmetric with respect to 1–2, 1–3, and 2–3 interchange. [Taking all signs as plus in (10.36), we would get a completely symmetric function.]

Let us assume f, g, and h to be orthonormal and choose c_1 so that (10.36) is normalized. Multiplying (10.36) by its complex conjugate, we get many terms, but because of the assumed orthogonality the integrals of all products involving two different terms of (10.36) vanish. For example,

$$\int [f(1)g(2)h(3)]^* f(2)g(1)h(3) \, d\tau$$

$$= \int f^*(1)g(1) \, d\tau_1 \int g^*(2)f(2) \, d\tau_2 \int h^*(3)h(3) \, d\tau_3 = 0 \cdot 0 \cdot 1 = 0$$

Integrals involving the product of a term of (10.36) with its own complex conjugate are equal to 1, because f, g, and h are normalized. Therefore,

$$1 = \int |(10.36)|^2 \, d\tau = |c_1|^2(1 + 1 + 1 + 1 + 1 + 1)$$

$$c_1 = 1/\sqrt{6}$$

We could work with (10.36) as it stands, but its properties are most easily found if we recognize it as simply the expansion [Eq. (8.24)] of the following third-order determinant:

$$\frac{1}{\sqrt{6}} \begin{vmatrix} f(1) & g(1) & h(1) \\ f(2) & g(2) & h(2) \\ f(3) & g(3) & h(3) \end{vmatrix} \tag{10.37*}$$

(For the reader familiar with Problem 8.16, this should come as no surprise.) The antisymmetry property holds for (10.37) because interchange of two electrons amounts to interchanging two rows of the determinant, which multiplies it by -1.

We stated that it is impossible to construct an antisymmetric spin function for three electrons. We now use (10.37) to prove this statement. The functions f, g, and h may each be either α or β. If we take $f = \alpha$, $g = \beta$, $h = \alpha$, then (10.37) becomes

$$\frac{1}{\sqrt{6}} \begin{vmatrix} \alpha(1) & \beta(1) & \alpha(1) \\ \alpha(2) & \beta(2) & \alpha(2) \\ \alpha(3) & \beta(3) & \alpha(3) \end{vmatrix} \qquad (10.38)$$

Although (10.38) is antisymmetric, we must reject it because it is equal to zero. The first and third columns of the determinant are identical, so (Section 8.3) the determinant vanishes. No matter how we choose f, g, and h, at least two columns of the determinant must be equal, so we cannot construct a nonzero antisymmetric three-electron spin function.

We now use (10.37) to construct the zeroth-order ground-state wave function for lithium, including both space and spin variables. The functions f, g, and h will now involve both space and spin variables. We choose

$$f(1) = 1s(1)\alpha(1) \qquad (10.39)$$

We call a function like (10.39) a ***spin-orbital***. A spin-orbital is the product of a one-electron space orbital and a one-electron spin function. If we were to take $g(1) = 1s(1)\alpha(1)$, this would make the first and second columns of (10.37) identical, and the wave function would vanish. This is a particular case of the ***Pauli exclusion principle***: *No two electrons can occupy the same spin-orbital.* Another way of stating this is to say that no two electrons in an atom can have the same values for all their quantum numbers. The Pauli exclusion principle is a consequence of the more general Pauli-principle antisymmetry requirement and is less satisfying than the antisymmetry statement, since the exclusion principle is based on approximate (zeroth-order) wave functions. We therefore take $g(1) = 1s(1)\beta(1)$, which puts two electrons with opposite spin in the $1s$ orbital. For the spin-orbital h, we cannot use either $1s(1)\alpha(1)$ or $1s(1)\beta(1)$, since these choices make the determinant vanish. We take $h(1) = 2s(1)\alpha(1)$, which gives the familiar Li ground-state configuration $1s^2 2s$ and the zeroth-order wave function

$$\psi^{(0)} = \frac{1}{\sqrt{6}} \begin{vmatrix} 1s(1)\alpha(1) & 1s(1)\beta(1) & 2s(1)\alpha(1) \\ 1s(2)\alpha(2) & 1s(2)\beta(2) & 2s(2)\alpha(2) \\ 1s(3)\alpha(3) & 1s(3)\beta(3) & 2s(3)\alpha(3) \end{vmatrix} \qquad (10.40)$$

Note especially that (10.40) is *not* simply a product of space and spin parts (as we found for H and He), but is a linear combination of terms, each of which is a product of space and spin parts.

Since we could just as well have taken $h(1) = 2s(1)\beta(1)$, the ground state of lithium is, like hydrogen, doubly degenerate, corresponding to the two possible orientations of the spin of the $2s$ electron. We might use the usual orbital diagrams

     1s 2s   1s 2s

     ↑↓ ↑   and   ↑↓ ↓

to indicate this. Each space orbital such as $1s$ or $2p_0$ can hold two electrons of opposite spin. A spin-orbital such as $2s\alpha$ can hold one electron.

Although the $1s^2 2p$ configuration will have the same unperturbed energy $E^{(0)}$ as the $1s^2 2s$ configuration, when we take electron repulsion into account by calculating $E^{(1)}$ and higher corrections, we find that the $1s^2 2s$ configuration lies lower for the same reason as in helium.

We close this section by discussing some points about the Pauli exclusion principle, which we restate as follows: *In a system of identical fermions, no two particles can occupy the same state.* If we have a system of n interacting particles (for example, an atom), there is a single wave function (involving $4n$ variables) for the entire system. Because of the interactions between the particles, the wave function cannot be written as the product of wave functions of the individual particles; hence, strictly speaking, we cannot talk of the states of individual particles, only the state of the whole system. If, however, the interactions between the particles are not too large, then as an initial approximation we can neglect them and write the zeroth-order wave function of the system as a product of wave functions of the individual particles. In this zeroth-order wave function, no two fermions can have the same wave function (state).

Since bosons require a wave function symmetric with respect to interchange, there is no restriction on the number of bosons in a given state.

10.6 SLATER DETERMINANTS

Slater pointed out in 1929 that a determinant of the form (10.40) satisfies the antisymmetry requirement for a many-electron atom. A determinant like (10.40) is called a **Slater determinant**. All the elements in a given column of a Slater determinant involve the same spin-orbital, while elements in the same row all involve the same electron. (Since interchanging rows and columns does not affect the value of a determinant, we could write the Slater determinant in another, equivalent form.)

Consider how the zeroth-order helium wave functions that we found previously can be written as Slater determinants. For the ground-state configuration $(1s)^2$, we have the spin-orbitals $1s\alpha$ and $1s\beta$, which give the Slater determinant

$$\frac{1}{\sqrt{2}} \begin{vmatrix} 1s(1)\alpha(1) & 1s(1)\beta(1) \\ 1s(2)\alpha(2) & 1s(2)\beta(2) \end{vmatrix} = 1s(1)1s(2) \frac{1}{\sqrt{2}} [\alpha(1)\beta(2) - \beta(1)\alpha(2)]$$

$$(10.41)$$

which agrees with (10.26). For the states corresponding to the excited configuration $1s2s$, we have the possible spin-orbitals $1s\alpha$, $1s\beta$, $2s\alpha$, $2s\beta$, which give the four Slater determinants

$$D_1 = \frac{1}{\sqrt{2}} \begin{vmatrix} 1s(1)\alpha(1) & 2s(1)\alpha(1) \\ 1s(2)\alpha(2) & 2s(2)\alpha(2) \end{vmatrix} \qquad D_2 = \frac{1}{\sqrt{2}} \begin{vmatrix} 1s(1)\alpha(1) & 2s(1)\beta(1) \\ 1s(2)\alpha(2) & 2s(2)\beta(2) \end{vmatrix}$$

$$D_3 = \frac{1}{\sqrt{2}} \begin{vmatrix} 1s(1)\beta(1) & 2s(1)\alpha(1) \\ 1s(2)\beta(2) & 2s(2)\alpha(2) \end{vmatrix} \qquad D_4 = \frac{1}{\sqrt{2}} \begin{vmatrix} 1s(1)\beta(1) & 2s(1)\beta(1) \\ 1s(2)\beta(2) & 2s(2)\beta(2) \end{vmatrix}$$

Comparison with (10.27) to (10.30) shows that the $1s2s$ zeroth-order wave functions are related to these four Slater determinants as follows:

$$2^{-1/2}[1s(1)2s(2) - 2s(1)1s(2)]\alpha(1)\alpha(2) = D_1 \qquad (10.42)$$

$$2^{-1/2}[1s(1)2s(2) - 2s(1)1s(2)]\beta(1)\beta(2) = D_4 \qquad (10.43)$$

$$2^{-1/2}[1s(1)2s(2) - 2s(1)1s(2)]2^{-1/2}[\alpha(1)\beta(2) + \beta(1)\alpha(2)] = 2^{-1/2}(D_2 + D_3) \qquad (10.44)$$

$$2^{-1/2}[1s(1)2s(2) + 2s(1)1s(2)]2^{-1/2}[\alpha(1)\beta(2) - \beta(1)\alpha(2)] = 2^{-1/2}(D_2 - D_3) \qquad (10.45)$$

(To get a zeroth-order function that is an eigenfunction of the spin and orbital angular-momentum operators, we sometimes have to take a linear combination of the Slater determinants of a configuration; see Chapter 11.)

Next consider some notations used for Slater determinants. Instead of writing α and β for spin functions, one often puts a bar over the space function to indicate the spin function β, while a space function without a bar implies the spin factor α. With this notation, (10.40) is written as

$$\psi^{(0)} = \frac{1}{\sqrt{6}} \begin{vmatrix} 1s(1) & \overline{1s}(1) & 2s(1) \\ 1s(2) & \overline{1s}(2) & 2s(2) \\ 1s(3) & \overline{1s}(3) & 2s(3) \end{vmatrix} \qquad (10.46)$$

Given the spin-orbitals occupied by the electrons, we can readily construct the Slater determinant. Thus it is redundant to write out the full determinant. Instead, a shorthand notation that simply specifies the spin-orbitals is often used. In this notation, (10.46) is written as

$$\psi^{(0)} = |1s\overline{1s}2s| \qquad (10.47)^*$$

where the vertical lines indicate formation of the determinant and multiplication by $1/\sqrt{6}$.

We showed that the factor $1/\sqrt{6}$ normalizes a third-order Slater determinant constructed of orthonormal functions. The expansion of an nth-order determinant has $n!$ terms (Problem 8.17). For an nth-order Slater determinant of orthonormal spin-orbitals, the same reasoning used in the third-order case shows that the normalization constant is $1/\sqrt{n!}$. We always include a factor $1/\sqrt{n!}$ in defining a Slater determinant of order n.

10.7 PERTURBATION TREATMENT OF THE LITHIUM GROUND STATE

Let us carry out a perturbation treatment of the ground state of the lithium atom.

We take

$$\hat{H}^0 = -\frac{\hbar^2}{2m_e}\nabla_1^2 - \frac{\hbar^2}{2m_e}\nabla_2^2 - \frac{\hbar^2}{2m_e}\nabla_3^2 - \frac{Ze'^2}{r_1} - \frac{Ze'^2}{r_2} - \frac{Ze'^2}{r_3}$$

$$\hat{H}' = \frac{e'^2}{r_{12}} + \frac{e'^2}{r_{23}} + \frac{e'^2}{r_{13}}$$

We found in Section 10.5 that to satisfy the Pauli principle the ground-state configuration must be $1s^2 2s$; the correct zeroth-order wave function is (10.40):

$$\psi^{(0)} = 6^{-1/2}[1s(1)1s(2)2s(3)\alpha(1)\beta(2)\alpha(3) - 1s(1)2s(2)1s(3)\alpha(1)\alpha(2)\beta(3)$$

$$- 1s(1)1s(2)2s(3)\beta(1)\alpha(2)\alpha(3) + 1s(1)2s(2)1s(3)\beta(1)\alpha(2)\alpha(3)$$

$$+ 2s(1)1s(2)1s(3)\alpha(1)\alpha(2)\beta(3) - 2s(1)1s(2)1s(3)\alpha(1)\beta(2)\alpha(3)]$$

What is $E^{(0)}$? Each term in $\psi^{(0)}$ contains the product of two $1s$ hydrogenlike functions and one $2s$ hydrogenlike function, multiplied by a spin factor. $\hat{H}^0$ is the sum of three hydrogenlike Hamiltonians, one for each electron, and does not involve spin. Thus $\psi^{(0)}$ is a linear combination of terms, each of which is an eigenfunction of $\hat{H}^0$ with eigenvalue $E_{1s}^{(0)} + E_{1s}^{(0)} + E_{2s}^{(0)}$, where these are hydrogenlike energies; hence $\psi^{(0)}$ is an eigenfunction of $\hat{H}^0$ with eigenvalue $E_{1s}^{(0)} + E_{1s}^{(0)} + E_{2s}^{(0)}$. Therefore [Eq. (6.94)],

$$E^{(0)} = -\left(\frac{1}{1^2} + \frac{1}{1^2} + \frac{1}{2^2}\right)\left(\frac{Z^2 e'^2}{2a_0}\right) = -\frac{81}{4}(13.606\text{ eV}) = -275.5\text{ eV} \quad (10.48)$$

To find $E^{(1)}$, we must evaluate $\langle \psi^{(0)}|\hat{H}'|\psi^{(0)}\rangle$. We begin by grouping together terms in $\psi^{(0)}$ that have the same spin factor:

$$\psi^{(0)} = 6^{-1/2}[1s(1)2s(2)1s(3) - 1s(1)1s(2)2s(3)]\beta(1)\alpha(2)\alpha(3)$$

$$+ 6^{-1/2}[1s(1)1s(2)2s(3) - 2s(1)1s(2)1s(3)]\alpha(1)\beta(2)\alpha(3)$$

$$+ 6^{-1/2}[2s(1)1s(2)1s(3) - 1s(1)2s(2)1s(3)]\alpha(1)\alpha(2)\beta(3) \quad (10.49)$$

$$\psi^{(0)} = a\beta(1)\alpha(2)\alpha(3) + b\alpha(1)\beta(2)\alpha(3) + c\alpha(1)\alpha(2)\beta(3) = A + B + C \quad (10.50)$$

where the space function multiplying the spin function $\beta(1)\alpha(2)\alpha(3)$ in (10.49) is called a and where $A = a\beta(1)\alpha(2)\alpha(3)$, with similar definitions for b, c, B, and C. We have

$$E^{(1)} = \int |\psi^{(0)}|^2 H' \, d\tau$$

$$E^{(1)} = \int |A|^2 H' \, d\tau + \int |B|^2 H' \, d\tau + \int |C|^2 H' \, d\tau + \int A^* B H' \, d\tau$$

$$+ \int B^* C H' \, d\tau + \int A^* C H' \, d\tau + \int AB^* H' \, d\tau + \int BC^* H' \, d\tau + \int AC^* H' \, d\tau$$

$$(10.51)$$

Because of the orthogonality of the different spin functions in A, B, and C, the last six integrals in (10.51) are zero. For example, the integral $\int A^* B H' \, d\tau$ involves summations over spins, as follows:

$$\sum_{m_{s1}} \sum_{m_{s2}} \sum_{m_{s3}} [\beta(1)\alpha(2)\alpha(3)]^*[\alpha(1)\beta(2)\alpha(3)]$$

$$= \sum_{m_{s1}} \beta^*(1)\alpha(1) \sum_{m_{s2}} \alpha^*(2)\beta(2) \sum_{m_{s3}} \alpha^*(3)\alpha(3) = 0 \cdot 0 \cdot 1 = 0$$

Since the spin functions are normalized, summation over spins in the first three integrals in (10.51) gives unity. Therefore,

$$E^{(1)} = \int\int\int a^2 H' \, dv_1 \, dv_2 \, dv_3 + \int\int\int b^2 H' \, dv_1 \, dv_2 \, dv_3 + \int\int\int c^2 H' \, dv_1 \, dv_2 \, dv_3$$

$$(10.52)$$

where spin is no longer involved. By relabeling dummy integration variables, we can prove that the three integrals in (10.52) are equal to one another. Using orthonormality of the $1s$ and $2s$ orbitals and relabeling of integration variables, we can then show that (Problem 10.12)

$$E^{(1)} = 2 \int \int 1s^2(1)2s^2(2) \frac{e'^2}{r_{12}} dv_1\, dv_2 + \int \int 1s^2(1)1s^2(2) \frac{e'^2}{r_{12}} dv_1\, dv_2$$

$$- \int \int 1s(1)2s(2)1s(2)2s(1) \frac{e'^2}{r_{12}} dv_1\, dv_2$$

These integrals are Coulomb and exchange integrals:

$$E^{(1)} = 2J_{1s2s} + J_{1s1s} - K_{1s2s} \tag{10.53}$$

We have [Eqs. (9.52), (9.55), and (9.113)]

$$J_{1s1s} = \frac{5}{8}\frac{Ze'^2}{a_0}, \qquad J_{1s2s} = \frac{17}{81}\frac{Ze'^2}{a_0}, \qquad K_{1s2s} = \frac{16}{729}\frac{Ze'^2}{a_0}$$

$$E^{(1)} = \frac{5965}{972}\left(\frac{e'^2}{2a_0}\right) = 83.5 \text{ eV}$$

The energy through first order is -192.0 eV, as compared with the true ground-state energy of lithium, -203.5 eV. To improve on this result, we must calculate higher-order wave-function and energy corrections. This will mix into the wave function contributions from Slater determinants involving configurations besides $1s^2 2s$ (configuration interaction).

10.8 VARIATION TREATMENTS OF THE LITHIUM GROUND STATE

The zeroth-order perturbation wave function (10.40) uses the full nuclear charge ($Z = 3$) for both the $1s$ and $2s$ orbitals of lithium. We expect that the $2s$ electron, which is partially shielded from the nucleus by the two $1s$ electrons, will see an effective nuclear charge that is much less than 3. Even the $1s$ electrons partially shield each other (recall the treatment of the helium ground state). This reasoning suggests the introduction of two variational parameters b_1 and b_2 into (10.40).

Instead of using the $Z = 3$ $1s$ function in Table 6.2, we take

$$f \equiv \frac{1}{\pi^{1/2}}\left(\frac{b_1}{a_0}\right)^{3/2} e^{-b_1 r/a_0} \tag{10.54}$$

where b_1 is a variational parameter representing an effective nuclear charge for the $1s$ electrons. Instead of the $Z = 3$ $2s$ function in Table 6.2, we use

$$g = \frac{1}{4(2\pi)^{1/2}}\left(\frac{b_2}{a_0}\right)^{3/2}\left(2 - \frac{b_2 r}{a_0}\right)e^{-b_2 r/2a_0} \tag{10.55}$$

Our trial variation function is then

$$\phi = \frac{1}{\sqrt{6}}\begin{vmatrix} f(1)\alpha(1) & f(1)\beta(1) & g(1)\alpha(1) \\ f(2)\alpha(2) & f(2)\beta(2) & g(2)\alpha(2) \\ f(3)\alpha(3) & f(3)\beta(3) & g(3)\alpha(3) \end{vmatrix} \tag{10.56}$$

The use of different charges b_1 and b_2 for the $1s$ and $2s$ orbitals destroys their orthogonality, so (10.56) is not normalized. The best values of the variational parameters are found by setting $\partial W/\partial b_1 = 0$ and $\partial W/\partial b_2 = 0$, where the variational integral W is given by the left side of Eq. (8.9). The results are [E. B. Wilson, Jr., *J. Chem. Phys.*, **1**, 210 (1933)] $b_1 = 2.686$, $b_2 = 1.776$, and $W = -201.2$ eV. W is much closer to the true value -203.5 than the result -192.0 eV found in the last section. The value of b_2 indicates substantial, but not complete, screening of the $2s$ electron by the $1s$ electrons.

We might try other forms for the orbitals besides (10.54) and (10.55) to improve the trial function. However, no matter what orbital functions we try, if we restrict ourselves to a trial function of the form of (10.56), we can never reach the true ground-state energy. To do this, we can introduce r_{12}, r_{23}, and r_{13} into the trial function or use a linear combination of several Slater determinants corresponding to various configurations (configuration interaction).

10.9   SPIN MAGNETIC MOMENT

Recall that the orbital angular momentum **L** of an electron has a magnetic moment $-(e/2m_e)\mathbf{L}$ associated with it [Eq. (6.128)]. It is natural to suppose that there is also a magnetic moment $\boldsymbol{\mu}_S$ associated with the electronic spin angular momentum **S**. We might guess that $\boldsymbol{\mu}_S$ would be $-e/2m_e$ times **S**. Spin is a relativistic phenomenon, however, and we cannot expect $\boldsymbol{\mu}_S$ to be related to **S** in exactly the same way that $\boldsymbol{\mu}_L$ is related to **L**. In fact, Dirac's relativistic treatment of the electron gave the result that (in SI units)

$$\boldsymbol{\mu}_S = -g_e \frac{e}{2m_e} \mathbf{S} = -\frac{e}{m_e} \mathbf{S} \tag{10.57}$$

where Dirac found an electron g factor equal to 2. The magnitude of the spin magnetic moment of an electron is (in SI units)

$$|\boldsymbol{\mu}_S| = g_e \frac{e}{2m_e} |\mathbf{S}| = \sqrt{3}\, \frac{e\hbar}{2m_e} \tag{10.58}$$

Theoretical and experimental work subsequent to Dirac's treatment has shown that g_e is slightly greater than 2 [see P. Kusch, *Physics Today*, Feb. 1966, p. 23]:

$$g_e = 2(1 + \alpha/2\pi + \cdots) = 2.0023$$

where the dots indicate terms involving higher powers of α and where the *fine-structure constant* α is defined as

$$\alpha \equiv \frac{e^2}{4\pi\varepsilon_0\hbar c} \equiv \frac{e'^2}{\hbar c} = 0.007297 \tag{10.59}$$

The ferromagnetism of iron is due to the electron's magnetic dipole moment.

The two possible orientations of an electron's spin and its associated spin magnetic moment with respect to an axis produce two energy levels in an externally applied magnetic field. In electron-spin-resonance (ESR) spectroscopy, one observes transitions between these two levels. ESR spectroscopy is applicable to species such as free radicals and transition-metal ions that have one or more unpaired electron spins and hence have a nonzero total electron spin and spin magnetic moment.

Certain nuclei have nonzero spins and spin magnetic moments. In nuclear-magnetic-resonance (NMR) spectroscopy, one observes transitions between nuclear-spin energy levels for a sample in an applied magnetic field. The proton (spin $\frac{1}{2}$) is the nucleus most commonly studied.

10.10 LADDER OPERATORS FOR ELECTRON SPIN

The spin angular-momentum operators obey the general angular-momentum commutation relations of Section 5.4, and it is often helpful to use spin-angular-momentum ladder operators.

From (5.117) and (5.118), the raising and lowering operators for spin angular momentum are

$$\hat{S}_+ = \hat{S}_x + i\hat{S}_y \quad \text{and} \quad \hat{S}_- = \hat{S}_x - i\hat{S}_y \tag{10.60}$$

and we have [Eqs. (5.119) and (5.120)]

$$\hat{S}_+\hat{S}_- = \hat{S}^2 - \hat{S}_z^2 + \hbar\hat{S}_z \tag{10.61}$$

$$\hat{S}_-\hat{S}_+ = \hat{S}^2 - \hat{S}_z^2 - \hbar\hat{S}_z \tag{10.62}$$

The spin functions α and β are eigenfunctions of $\hat{S}_z$ with eigenvalues $+\frac{1}{2}\hbar$ and $-\frac{1}{2}\hbar$, respectively. Since $\hat{S}_+$ is the raising operator, the function $\hat{S}_+\beta$ is an eigenfunction of $\hat{S}_z$ with eigenvalue $+\frac{1}{2}\hbar$. The most general eigenfunction of $\hat{S}_z$ with this eigenvalue is an arbitrary constant times α. Hence

$$\hat{S}_+\beta = c\alpha \tag{10.63}$$

where c is some constant. To find c, we use Eq. (10.11):

$$1 = \sum_{m_s} [\alpha(m_s)]^*\alpha(m_s) = \sum (\hat{S}_+\beta/c)^*(\hat{S}_+\beta/c)$$

$$|c|^2 = \sum (\hat{S}_+\beta)^*\hat{S}_+\beta = \sum (\hat{S}_+\beta)^*(\hat{S}_x + i\hat{S}_y)\beta$$

$$|c|^2 = \sum (\hat{S}_+\beta)^*\hat{S}_x\beta + i\sum (\hat{S}_+\beta)^*\hat{S}_y\beta \tag{10.64}$$

We now use the Hermitian property of $\hat{S}_x$ and $\hat{S}_y$. For an operator $\hat{A}$ that acts on functions of the continuous variable x, the Hermitian property is

$$\int_{-\infty}^{\infty} f^*(x)\hat{A}g(x)\,dx = \int_{-\infty}^{\infty} g(x)[\hat{A}f(x)]^*\,dx$$

For an operator such as $\hat{S}_x$ that acts on functions of the variable m_s, which takes on discrete values, the Hermitian property is

$$\sum_{m_s} f^*(m_s)\hat{S}_x g(m_s) = \sum_{m_s} g(m_s)[\hat{S}_x f(m_s)]^* \tag{10.65}$$

Taking $f = \hat{S}_+\beta$ and $g = \beta$, we can write (10.64) as

$$c^*c = \sum \beta[\hat{S}_x\hat{S}_+\beta]^* + i\sum \beta[\hat{S}_y\hat{S}_+\beta]^*$$

Taking the complex conjugate of this equation, we have

$$cc^* = \sum \beta^* \hat{S}_x \hat{S}_+ \beta - i \sum \beta^* \hat{S}_y \hat{S}_+ \beta$$

$$|c|^2 = \sum \beta^* (\hat{S}_x - i\hat{S}_y) \hat{S}_+ \beta = \sum \beta^* \hat{S}_- \hat{S}_+ \beta$$

$$|c|^2 = \sum \beta^* (\hat{S}^2 - \hat{S}_z^2 - \hbar \hat{S}_z) \beta$$

$$|c|^2 = \sum \beta^* (\tfrac{3}{4}\hbar^2 - \tfrac{1}{4}\hbar^2 + \tfrac{1}{2}\hbar^2) \beta = \hbar^2 \sum \beta^* \beta = \hbar^2$$

$$|c| = \hbar$$

Choosing the phase of c as zero, we have $c = \hbar$, and (10.63) reads

$$\hat{S}_+ \beta = \hbar \alpha \tag{10.66}$$

A similar calculation gives

$$\hat{S}_- \alpha = \hbar \beta \tag{10.67}$$

Since α is the eigenfunction with the highest possible value of m_s, the operator $\hat{S}_+$ acting on α must annihilate it [Eq. (5.142)]:

$$\hat{S}_+ \alpha = 0$$

Likewise,

$$\hat{S}_- \beta = 0$$

From these last four equations, we get

$$(\hat{S}_+ + \hat{S}_-)\beta = \hbar \alpha \tag{10.68}$$

$$(\hat{S}_+ - \hat{S}_-)\beta = \hbar \alpha \tag{10.69}$$

Use of (10.60) in Eqs. (10.68) and (10.69) gives

$$\hat{S}_x \beta = \tfrac{1}{2}\hbar \alpha , \qquad \hat{S}_y \beta = -\tfrac{1}{2}i\hbar \alpha \tag{10.70}$$

Similarly, we find

$$\hat{S}_x \alpha = \tfrac{1}{2}\hbar \beta , \qquad \hat{S}_y \alpha = \tfrac{1}{2}i\hbar \beta \tag{10.71}$$

Matrix representatives of the spin operators are considered in Problem 10.20.

10.11   SUMMARY

An elementary particle possesses a spin angular momentum of magnitude $[s(s+1)\hbar^2]^{1/2}$ and z component $m_s\hbar$, where $m_s = -s, -s+1, \ldots, s-1, s$. For an electron, $s = \tfrac{1}{2}$. The spin angular-momentum operators $\hat{S}_x$, $\hat{S}_y$, $\hat{S}_z$, and $\hat{S}^2$ obey relations analogous to those obeyed by the orbital angular-momentum operators. The electron spin eigenfunctions corresponding to the states $m_s = \tfrac{1}{2}$ and $m_s = -\tfrac{1}{2}$ are denoted by α and β. We have $\hat{S}_z \alpha = \tfrac{1}{2}\hbar \alpha$, $\hat{S}_z \beta = -\tfrac{1}{2}\hbar \beta$, $\hat{S}^2\alpha = \tfrac{1}{2}(\tfrac{3}{2})\hbar^2\alpha$, and $\hat{S}^2\beta = \tfrac{1}{2}(\tfrac{3}{2})\hbar^2\beta$. The spin functions α and β are orthonormal [Eqs. (10.11) and

(10.12)]. For a one-electron system, the complete stationary-state wave function is the product of a spatial function $\psi(x, y, z)$ and a spin function (α or β or a linear combination of α and β).

According to the Pauli principle, the complete wave function (including both space and spin coordinates) of a system of identical fermions (particles with half-integral spin) must be antisymmetric with respect to interchange of any two fermions. The complete wave function of a system of identical bosons (particles with integral spin) must be symmetric with respect to interchange of any two bosons.

The two-electron spin eigenfunctions consist of the symmetric functions $\alpha(1)\alpha(2)$, $\beta(1)\beta(2)$, and $[\alpha(1)\beta(2) + \beta(1)\alpha(2)]/\sqrt{2}$ and the antisymmetric function $[\alpha(1)\beta(2) - \beta(1)\alpha(2)]/\sqrt{2}$. For the helium atom, each stationary state wave function is the product of a symmetric spatial function and an antisymmetric spin function or an antisymmetric spatial function and a symmetric spin function. Some approximate helium-atom wave functions are Eqs. (10.26) to (10.30).

A spin-orbital is the product of a one-electron spatial wave function and a one-electron spin function. An approximate wave function for a system of electrons can be written as a Slater determinant of spin-orbitals; interchange of two electrons interchanges two rows in the Slater determinant, which multiplies the wave function by -1, ensuring antisymmetry. In such an approximate wave function, no two electrons can be assigned to the same spin-orbital; this is the Pauli exclusion principle and is a consequence of the Pauli-principle antisymmetry requirement.

An electron has a spin magnetic moment $\boldsymbol{\mu}_S$ that is proportional to its spin angular momentum $\mathbf{S}$.

By using ladder operators, we found the effects of $\hat{S}_x$ and $\hat{S}_y$ on α and β.

PROBLEMS

10.1 Calculate the angle that the spin vector $\mathbf{S}$ makes with the z axis for an electron with spin function α.

10.2 Verify that taking the spin functions α and β as $\alpha(m_s) = \delta_{m_s, 1/2}$ and $\beta(m_s) = \delta_{m_s, -1/2}$ gives functions that satisfy the orthonormality conditions (10.11) and (10.12).

10.3 (a) Show that $\hat{P}_{12}$ commutes with the Hamiltonian for the lithium atom. (b) Show that $\hat{P}_{12}$ and $\hat{P}_{23}$ do not commute with each other. (c) Show that $\hat{P}_{12}$ and $\hat{P}_{23}$ commute when they are applied to antisymmetric functions.

10.4 Show that $\hat{P}_{12}$ is Hermitian.

10.5 Which of the following functions are (a) symmetric? (b) antisymmetric? (1) $f(1)g(2)\alpha(1)\alpha(2)$; (2) $f(1)f(2)[\alpha(1)\beta(2) - \beta(1)\alpha(2)]$; (3) $f(1)f(2)f(3)\beta(1)\beta(2)\beta(3)$; (4) $e^{-a(r_1-r_2)}$; (5) $[f(1)g(2) - g(1)f(2)][\alpha(1)\beta(2) - \alpha(2)\beta(1)]$; (6) $r_{12}^2 e^{-a(r_1+r_2)}$.

10.6 Explain why the function $Ne^{-ar_1}e^{-ar_2}(r_1 - r_2)$ should not be used as a trial variation function for the helium-atom ground state.

10.7 If electrons had a spin of zero, what would be the zeroth-order (interelectronic repulsions neglected) wave functions for the ground state and first excited state of lithium?

10.8 The antisymmetrization operator $\hat{A}$ is defined as the operator that antisymmetrizes a product of n one-electron functions and multiplies them by $(n!)^{-1/2}$. For $n = 2$, we have

$$\hat{A}f(1)g(2) = \frac{1}{\sqrt{2}} \begin{vmatrix} f(1) & g(1) \\ f(2) & g(2) \end{vmatrix}$$

(a) For $n = 2$, express $\hat{A}$ in terms of $\hat{P}_{12}$. (b) For $n = 3$, express $\hat{A}$ in terms of $\hat{P}_{12}$, $\hat{P}_{13}$, and $\hat{P}_{23}$.

10.9 A *permanent* is defined by the same expansion as a determinant except that all terms are given a plus sign. Thus the second-order permanent is

$$\begin{vmatrix} \overset{+}{a} & \overset{+}{b} \\ c & d \end{vmatrix} = ad + bc$$

Can you think of a use for permanents in quantum mechanics?

10.10 Use theorems about determinants to show that taking the lithium spin-orbitals in a Slater determinant as $1s\alpha$, $1s\beta$, and $1s(c_1\alpha + c_2\beta)$, where c_1 and c_2 are constants, gives a wave function that equals zero.

10.11 A muon has the same charge and spin as an electron, but a heavier mass. What would be the ground-state configuration of a lithium atom with two electrons and one muon?

10.12 Derive Eq. (10.53) for $E^{(1)}$ of lithium from (10.52).

10.13 If we had incorrectly used as the zeroth-order lithium ground-state wave function the nonantisymmetric function $1s(1)1s(2)2s(3)$, what would $E^{(1)}$ be calculated to be?

10.14 Calculate the magnitude of the spin magnetic moment of an electron.

10.15 (a) Use Eq. (6.131) to find the expression for the energy levels of the electron spin magnetic moment $\boldsymbol{\mu}_S$ in an applied magnetic field **B**. (b) Calculate the ESR absorption frequency of an electron in a magnetic field of 1.00 T.

10.16 Verify Eqs. (10.67) and (10.71).

10.17 (a) If the spin component S_x of an electron is measured, what possible values can result? (b) The functions α and β form a complete set, so any one-electron spin function can be written as a linear combination of them. Use Eqs. (10.70) and (10.71) to construct the two normalized eigenfunctions of $\hat{S}_x$ with eigenvalues $+\frac{1}{2}\hbar$ and $-\frac{1}{2}\hbar$. (c) Suppose a measurement of S_z for an electron gives the value $+\frac{1}{2}\hbar$; if a measurement of S_x is then carried out, give the probabilities for each possible outcome. (d) Do the same as in (b) for $\hat{S}_y$ instead of $\hat{S}_x$. (In the Stern–Gerlach experiment, a beam of particles is sent through an inhomogeneous magnetic field, which splits the beam into several beams each having particles with a different component of magnetic dipole moment in the field direction. For example, a beam of ground-state sodium atoms is split into two beams, corresponding to the two possible orientations of the valence electron's spin. Problem 10.17c corresponds to setting up a Stern–Gerlach apparatus with the field in the z direction and then allowing the $+\frac{1}{2}\hbar$ beam from this apparatus to enter a Stern–Gerlach apparatus that has the field in the x direction.)

10.18 Show that α and β are each eigenfunctions of $\hat{S}_x^2$ (but not of $\hat{S}_x$).

10.19 Let Y_{jm} be the *normalized* eigenfunction of the generalized angular-momentum operators (Section 5.4) $\hat{M}^2$ and $\hat{M}_z$:

$$\hat{M}^2 Y_{jm} = j(j+1)\hbar^2 Y_{jm}, \qquad \hat{M}_z Y_{jm} = m\hbar Y_{jm}$$

From Section 5.4, the effect of $\hat{M}_+$ on Y_{jm} is to increase the $\hat{M}_z$ eigenvalue by $\hbar$:

$$\hat{M}_+ Y_{jm} = A Y_{j,m+1}$$

where A is a constant. Use the same procedure that led to Eqs. (10.66) and (10.67) to show that

$$\hat{M}_+ Y_{jm} = [j(j+1) - m(m+1)]^{1/2} \hbar Y_{j,m+1} \tag{10.72}$$

$$\hat{M}_- Y_{jm} = [j(j+1) - m(m-1)]^{1/2} \hbar Y_{j,m-1} \tag{10.73}$$

(b) Show that (10.72) and (10.73) are consistent with (10.66) and (10.67). (c) With $\mathbf{M} = \mathbf{L}$, the function Y_{jm} is the spherical harmonic $Y_l^m(\theta, \phi)$. Verify (10.72) directly for $l = 2$, $m = -1$. [Actually, for consistency with the phase choice of Eqs. (10.72) and (10.73), we must add the factor $(-i)^{m+|m|}$ to the definition (5.110) of the spherical harmonics; this introduces a minus sign for odd positive values of m.]

10.20 The eigenfunctions α and β of the Hermitian operator $\hat{S}_z$ form a complete, orthonormal set, and any one-electron spin function can be written as $c_1\alpha + c_2\beta$. We saw in Section 7.10 that functions can be represented by column vectors and operators by square matrices. For the representation that uses α and β as the basis functions, (a) write down the column vectors that correspond to the functions α, β, and $c_1\alpha + c_2\beta$; (b) use the results of Section 10.10 to show that the matrices that correspond to $\hat{S}_x$, $\hat{S}_y$, $\hat{S}_z$, and $\hat{S}^2$ are

$$\mathbf{S}_x = \tfrac{1}{2}\hbar \begin{pmatrix} 0 & 1 \\ 1 & 0 \end{pmatrix}, \qquad \mathbf{S}_y = \tfrac{1}{2}\hbar \begin{pmatrix} 0 & -i \\ i & 0 \end{pmatrix}, \qquad \mathbf{S}_z = \tfrac{1}{2}\hbar \begin{pmatrix} 1 & 0 \\ 0 & -1 \end{pmatrix}$$

$$\mathbf{S}^2 = \tfrac{1}{4}\hbar^2 \begin{pmatrix} 3 & 0 \\ 0 & 3 \end{pmatrix}$$

(c) Verify that the matrices in (b) obey $\mathbf{S}_x\mathbf{S}_y - \mathbf{S}_y\mathbf{S}_x = i\hbar\mathbf{S}_z$ [Eq. (10.2)]. (d) Find the eigenvalues and eigenvectors of the $\mathbf{S}_x$ matrix. Compare the results with those of Problem 10.17.

11

Many-Electron Atoms

11.1 THE HARTREE–FOCK SELF-CONSISTENT-FIELD METHOD

For hydrogen, the exact wave function is known. For helium and lithium, very accurate wave functions have been calculated by including interelectronic distances in the variation functions. For atoms of higher atomic number, the best approach to finding a good wave function lies in first calculating an approximate wave function using the Hartree–Fock procedure, which we shall outline in this section. The Hartree–Fock method is the basis for the use of atomic and molecular orbitals in many-electron systems.

The Hamiltonian operator for an n-electron atom is

$$\hat{H} = -\frac{\hbar^2}{2m_e} \sum_{i=1}^{n} \nabla_i^2 - \sum_{i=1}^{n} \frac{Ze'^2}{r_i} + \sum_{i=1}^{n-1} \sum_{j=i+1}^{n} \frac{e'^2}{r_{ij}} \tag{11.1}$$

where an infinitely heavy point nucleus was assumed (Section 6.6). The first sum in (11.1) contains the kinetic-energy operators for the n electrons. The second sum is the potential energy for the attractions between the electrons and the nucleus of charge Ze'; for a neutral atom, $Z = n$. The last sum is the potential energy of the interelectronic repulsions; the restriction $j > i$ avoids counting the same interelectronic repulsion twice and avoids terms like e'^2/r_{ii}. [The Hamiltonian (11.1) is not complete, since it omits spin–orbit and other interactions. The omitted terms are generally small and will be considered in Sections 11.6 and 11.7.]

The Hartree SCF Method. Because of the interelectronic repulsion terms e'^2/r_{ij}, the Schrödinger equation for an atom is not separable. Recalling the perturbation treatment of helium (Section 9.3), we can obtain a zeroth-order wave function by neglecting these repulsions. The Schrödinger equation would then separate into n one-electron hydrogenlike equations. The zeroth-order wave function would be a product of n hydrogenlike (one-electron) orbitals:

$$\psi^{(0)} = f_1(r_1, \theta_1, \phi_1) f_2(r_2, \theta_2, \phi_2) \cdots f_n(r_n, \theta_n, \phi_n) \tag{11.2}$$

where the hydrogenlike orbitals are

$$f = R_{nl}(r) Y_l^m(\theta, \phi) \tag{11.3}$$

For the ground state of the atom, we would feed two electrons with opposite spin into each of the lowest orbitals, in accord with the Pauli exclusion principle, giving rise to the ground-state configuration. Although the approximate wave function (11.2) is qualitatively useful, it is gravely lacking in quantitative accuracy. For one thing, all the orbitals use the full nuclear charge Z. Recalling our variational treatments of helium and lithium, we know we can get a better approximation by using different effective atomic numbers for the different orbitals to account for screening of electrons. The use of effective atomic numbers gives considerable improvement, but we are still far from having an accurate wave function. The next step is to use a variation function that has the same form as (11.2), but is not restricted to hydrogenlike or any other particular form of orbitals. Thus we take

$$\phi = g_1(r_1, \theta_1, \phi_1) g_2(r_2, \theta_2, \phi_2) \cdots g_n(r_n, \theta_n, \phi_n) \qquad (11.4)$$

and we seek to determine the functions $g_1, g_2, \ldots, g_n$ that minimize the variational integral $\int \phi^* \hat{H} \phi \, dv / \int \phi^* \phi \, dv$. Our task is considerably more difficult than in previous variational calculations where we guessed a trial function that included some parameters and then varied the *parameters*. In (11.4) we must vary the *functions* g_i. [After we have found the best possible functions g_i, Eq. (11.4) will still be only an approximate wave function. The many-electron Schrödinger equation is not separable, so the true wave function cannot be written as the product of n one-electron functions.]

Finding the best possible approximate wave function of the form (11.4) is a formidable computational task for a many-electron atom. To simplify matters somewhat, we approximate the best possible orbitals with orbitals that are the product of a radial factor and a spherical harmonic:

$$g_i = h_i(r_i) Y_{l_i}^{m_i}(\theta_i, \phi_i) \qquad (11.5)$$

This approximation is generally made in atomic calculations.

The procedure for calculating the g_i's was introduced by Douglas Hartree in 1928 and is called the **Hartree self-consistent-field (SCF) method**. Hartree arrived at the SCF procedure by intuitive physical arguments. The proof that Hartree's procedure gives the best possible variation function of the form (11.4) was given by Slater and by Fock in 1930. [For the proof and a review of the SCF method, see S. M. Blinder, *Am. J. Phys.*, **33**, 431 (1965).]

Hartree's procedure is as follows. We first guess a product wave function

$$\phi_0 = s_1(r_1, \theta_1, \phi_1) s_2(r_2, \theta_2, \phi_2) \cdots s_n(r_n, \theta_n, \phi_n) \qquad (11.6)$$

where each s_i is a normalized function of r multiplied by a spherical harmonic. A reasonable guess for ϕ_0 would be a product of hydrogenlike orbitals with effective atomic numbers. For the function (11.6), the probability density of electron i is $|s_i|^2$. We now focus attention on electron 1 and regard electrons $2, 3, \ldots, n$ as being smeared out to form a static distribution of electric charge through which electron 1 moves. We are thus averaging out the instantaneous interactions between electron 1 and the other electrons. The potential energy of interaction between point charges Q_1 and Q_2 is given by (6.58) and (1.36) as $V_{12} = Q_1' Q_2' / r_{12} = Q_1 Q_2 / 4\pi\varepsilon_0 r_{12}$. We now take Q_2 and smear it out into a continuous charge distribution such that ρ_2 is the charge density, the charge per unit volume. The infinitesimal charge in the infinitesimal volume dv_2 is $\rho_2\,dv_2$, and summing up the interactions between Q_1 and the infinitesimal elements of charge, we have

$$V_{12} = \frac{Q_1}{4\pi\varepsilon_0} \int \frac{\rho_2}{r_{12}}\,dv_2$$

For electron 2 (with charge $-e$), the charge density of the hypothetical charge cloud is $\rho_2 = -e|s_2|^2$, and for electron 1, $Q_1 = -e$. Hence

$$V_{12} = e'^2 \int \frac{|s_2|^2}{r_{12}}\,dv_2$$

where $e'^2 = e^2/4\pi\varepsilon_0$. Adding in the interactions with the other electrons, we have

$$V_{12} + V_{13} + \cdots + V_{1n} = \sum_{j=2}^{n} e'^2 \int \frac{|s_j|^2}{r_{1j}}\,dv_j$$

The potential energy of interaction between electron 1 and the other electrons and the nucleus is then

$$V_1(r_1, \theta_1, \phi_1) = \sum_{j=2}^{n} e'^2 \int \frac{|s_j|^2}{r_{1j}}\,dv_j - \frac{Ze'^2}{r_1} \tag{11.7}$$

We now make a further approximation beyond assuming the wave function to be a product of one-electron orbitals. We assume that the effective potential acting on an electron in an atom can be adequately approximated by a function of r only. This *central-field approximation* can be shown to be generally accurate. We therefore average $V_1(r_1, \theta_1, \phi_1)$ over the angles to arrive at a potential energy that depends only on r_1:

$$V_1(r_1) = \frac{\int_0^{2\pi}\int_0^{\pi} V_1(r_1, \theta_1, \phi_1)\sin\theta_1\,d\theta_1\,d\phi_1}{\int_0^{2\pi}\int_0^{\pi}\sin\theta\,d\theta\,d\phi} \tag{11.8}$$

We now use $V_1(r_1)$ as the potential energy in a one-electron Schrödinger equation,

$$\left[-\frac{\hbar^2}{2m_e}\nabla_1^2 + V_1(r_1) \right] t_1(1) = \varepsilon_1 t_1(1) \tag{11.9}$$

and solve for $t_1(1)$, which will be an improved orbital for electron 1. In (11.9), ε_1 is the energy of the orbital of electron 1 at this stage of the approximation. Since the potential energy in (11.9) is spherically symmetric, the angular factor in $t_1(1)$ is a spherical harmonic involving quantum numbers l_1 and m_1 (Section 6.1). The radial factor $R(r_1)$ in t_1 is the solution of a one-dimensional Schrödinger equation of the form (6.17). We get a set of solutions $R(r_1)$, where the number of nodes k interior to the boundary points ($r = 0$ and ∞) starts at zero for the lowest energy and increases by 1 for each higher energy (Section 4.2). We now *define* the quantum number n as $n \equiv l + 1 + k$, where $k = 0, 1, 2, \ldots$. We thus have $1s$, $2s$, $2p$, and so on, orbitals (with orbital energy ε increasing with n) just as in hydrogenlike atoms, and the number of interior radial nodes $(n - l - 1)$ is the same as in hydrogenlike atoms (Section 6.6). However, since $V(r_1)$ is not a simple Coulomb potential, the radial factor $R(r_1)$ is not a hydrogenlike function. Of the set of solutions $R(r_1)$, we take the one that corresponds to the orbital we are improving. For example, if electron 1 is a $1s$ electron in the beryllium $1s^2 2s^2$ configuration, then $V_1(r_1)$ is calculated from the guessed orbitals of one $1s$ electron and two $2s$ electrons, and we use the radial solution of (11.9) with $k = 0$ to find an improved $1s$ orbital.

We now go to electron 2 and regard it as moving in a charge cloud of density

$$-e[|t_1(1)|^2 + |s_3(3)|^2 + |s_4(4)|^2 + \cdots + |s_n(n)|^2]$$

due to the other electrons. We calculate an effective potential energy $V_2(r_2)$ and solve a one-electron Schrödinger equation for electron 2 to obtain an improved orbital $t_2(2)$. We continue this process until we have a set of improved orbitals for all n electrons. Then we go back to electron 1 and repeat the process. We continue to calculate improved orbitals until there is no further change from one iteration to the next. The final set of orbitals gives the Hartree self-consistent-field wave function.

How do we get the energy of the atom in the SCF approximation? It seems natural to simply take the sum of the orbital energies of the electrons, $\varepsilon_1 + \varepsilon_2 + \cdots + \varepsilon_n$, but this is wrong. In calculating the orbital energy ε_1, we iteratively solved the one-electron Schrödinger equation (11.9). The potential energy in (11.9) includes, in an average way, the energy of the repulsions between electrons 1 and 2, 1 and 3, ..., 1 and n. When we solve for ε_2, we solve a one-electron Schrödinger equation whose potential energy includes repulsions between electrons 2 and 1, 2 and 3, ..., 2 and n. If we take $\Sigma_i \, \varepsilon_i$, we will count each interelectronic repulsion twice. To correctly obtain the total energy E of the atom, we must take

$$E = \sum_{i=1}^{n} \varepsilon_i - \sum_{i=1}^{n-1} \sum_{j=i+1}^{n} \int \int \frac{e'^2 |g_i(i)|^2 g_j(j)|^2}{r_{ij}} \, dv_i \, dv_j$$

$$E = \sum_{i} \varepsilon_i - \sum_{i} \sum_{j>i} J_{ij} \tag{11.10}$$

where the average repulsions of the electrons in the Hartree orbitals of (11.4) were subtracted from the sum of the orbital energies, and where the notation J_{ij} was used for Coulomb integrals [Eq. (9.100)].

The set of orbitals belonging to a given principal quantum number n constitutes a **shell**. The $n = 1, 2, 3, \ldots$ shells are the $K, L, M, \ldots$ shells, respectively. The orbitals belonging to a given n and a given l constitute a **subshell**. Consider the sum of the Hartree probability densities for the electrons in a filled subshell. Using (11.5), we have

$$2 \sum_{m=-l}^{l} |h_{n,l}(r)|^2 |Y_l^m(\theta, \phi)|^2 = 2|h_{n,l}(r)|^2 \sum_{m=-l}^{l} |Y_l^m(\theta, \phi)|^2 \tag{11.11}$$

where the factor 2 comes from the pair of electrons in each orbital. The spherical-harmonic addition theorem (*Merzbacher*, Section 9.7) shows that the sum on the right side of (11.11) equals $(2l + 1)/4\pi$. Hence the sum of the probability densities is $[(2l + 1)/2\pi]|h_{n,l}(r)|^2$, which is independent of the angles. A closed subshell gives a spherically symmetric probability density, a result called

Unsöld's theorem. For a half-filled subshell, the factor 2 is omitted from (11.11), and here also we get a spherically symmetric probability density.

The Hartree–Fock SCF Method.   The alert reader may have realized that there is something fundamentally wrong with the Hartree product wave function (11.4). Although we have paid some attention to spin and the Pauli principle by putting no more than two electrons in each spatial orbital, any approximation to the true wave function should include spin explicitly and should be antisymmetric to interchange of electrons (Chapter 10). Hence, instead of the spatial orbitals, we must use spin-orbitals and must take an antisymmetric linear combination of products of spin-orbitals. This was pointed out by Fock (and by Slater) in 1930, and an SCF calculation that uses antisymmetrized spin-orbitals is called a ***Hartree–Fock calculation***. We have seen that a Slater determinant of spin-orbitals provides the proper antisymmetry. For example, to carry out a Hartree–Fock calculation for the lithium ground state, we start with the function (10.56), where f and g are guesses for the $1s$ and $2s$ orbitals. We then carry out the SCF iterative process until we get no further improvement in f and g; this gives the lithium ground-state Hartree–Fock wave function.

The differential equations for finding the Hartree–Fock orbitals have the same general form as (11.9):

$$\hat{F}u_i = \varepsilon_i u_i, \qquad i = 1, 2, \ldots, n \tag{11.12}$$

where u_i is the ith spin-orbital, the operator $\hat{F}$, called the ***Fock*** (or ***Hartree–Fock***) ***operator***, is the effective Hartree–Fock Hamiltonian, and the eigenvalue ε_i is the energy of spin-orbital i. However, the Hartree–Fock operator $\hat{F}$ has extra terms as compared with the effective Hartree Hamiltonian given by the bracketed terms in (11.9). The Hartree–Fock expression for the total energy of the atom involves exchange integrals K_{ij} in addition to the Coulomb integrals that occur in the Hartree expression (11.10). See Section 13.16.

The orbital energy ε_i in the Hartree–Fock equations (11.12) can be shown to be a good approximation to the negative of the energy needed to ionize a closed-subshell atom by removing an electron from spin-orbital i (Koopmans' theorem; Section 15.6).

Originally, Hartree–Fock atomic calculations were done by using numerical integration to solve the Hartree–Fock differential equations (11.12), and the resulting orbitals were given as tables of the radial functions for various values of r. In 1951, Roothaan proposed representing the Hartree–Fock orbitals as linear combinations of a complete set of known functions, called ***basis functions***. Thus for lithium we would write the Hartree–Fock $1s$ and $2s$ spatial orbitals as

$$f = \sum_i b_i \chi_i, \qquad g = \sum_i c_i \chi_i \tag{11.13}$$

where the χ_i's are some complete set of functions, and where the b_i's and c_i's are expansion coefficients that are found by the SCF iterative procedure. Since the χ_i (chi i) functions form a complete set, these expansions are valid. The Roothaan expansion procedure allows one to find the Hartree–Fock wave function using matrix algebra (see Section 13.16 for details). The Roothaan procedure is readily implemented on a computer and is the method used nowadays to find Hartree–Fock wave functions.

A commonly used set of basis functions for atomic Hartree–Fock calculations is the set of **Slater-type orbitals** (STOs) whose normalized form is

$$\frac{[2\zeta/a_0]^{n+1/2}}{[(2n)!]^{1/2}} \, r^{n-1} e^{-\zeta r/a_0} Y_l^m(\theta, \phi) \tag{11.14}$$

The set of all such functions with n, l, and m being integers but with ζ having all possible positive values forms a complete set. The parameter ζ is called the **orbital exponent**. To get a truly accurate representation of the Hartree–Fock orbitals, we would have to include an infinite number of Slater orbitals in the expansions; in practice, one can get very accurate results by using only a few judiciously chosen Slater orbitals. (Another possibility is to use Gaussian-type basis functions; see Section 15.4.)

Clementi and Roetti did Hartree–Fock calculations for the ground state and some excited states of the first 54 elements of the periodic table [E. Clementi and C. Roetti, *At. Data Nucl. Data Tables*, **14**, 177 (1974)]. For example, consider the Hartree–Fock ground-state wave function of helium, which has the form [cf. Eq. (10.41)]

$$f(1)f(2) \cdot 2^{-1/2}[\alpha(1)\beta(2) - \alpha(2)\beta(1)]$$

Clementi and Roetti expressed the $1s$ orbital function f as the following combination of five $1s$ Slater-type orbitals:

$$f = \pi^{-1/2} \sum_{i=1}^{5} c_i \left(\frac{\zeta_i}{a_0}\right)^{3/2} e^{-\zeta_i r/a_0}$$

where the expansion coefficients c_i are $c_1 = 0.76838$, $c_2 = 0.22346$, $c_3 = 0.04082$, $c_4 = -0.00994$, $c_5 = 0.00230$ and where the orbital exponents ζ_i are $\zeta_1 = 1.41714$, $\zeta_2 = 2.37682$, $\zeta_3 = 4.39628$, $\zeta_4 = 6.52699$, $\zeta_5 = 7.94252$. [Note that the largest term in the expansion has an orbital exponent that is similar to the orbital exponent (9.65) for the simple trial function (9.58).] The Hartree–Fock energy is -77.9 eV, as compared with the true energy, -79.0 eV. The $1s$ orbital energy corresponding to f was found to be -25.0 eV, as compared with the experimental helium ionization energy of 24.6 eV.

For the lithium ground state, Clementi and Roetti used a basis set consisting of two $1s$ STOs (with different orbital exponents) and four $2s$ STOs (with different orbital exponents). The lithium $1s$ and $2s$ Hartree–Fock orbitals were each expressed as a linear combination of all six of these basis functions. The Hartree–Fock energy is -202.3 eV, as compared with the true energy, -203.5 eV.

Electron densities calculated from Hartree–Fock wave functions are quite accurate. Figure 11.1 compares the radial distribution function of argon (found by integrating the electron density over the angles θ and ϕ and multiplying the result by r^2) calculated by the Hartree–Fock method with the radial distribution function found by electron diffraction. (Recall from Section 6.6 that the radial distribution function is proportional to the probability of finding an electron in a thin spherical shell at a distance r from the nucleus.) Note the electronic shell

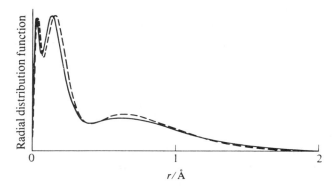

Figure 11.1 Radial distribution function in Ar as a function of r. The broken line is the result of a Hartree–Fock calculation. The solid line is the result of electron-diffraction data. [From L. S. Bartell and L. O. Brockway, *Phys. Rev.*, **90**, 833 (1953). Used by permission.]

structure in Fig. 11.1. Because of the high nuclear charge in $_{18}$Ar, the average distance of the $1s$ electrons from the nucleus is far less than in H or He. Thus there is only a moderate increase in atomic size as we go down a given group in the periodic table. Calculations show that the radius of a sphere containing 98% of the Hartree–Fock electron probability density gives an atomic radius in good agreement with the empirically determined van der Waals radius. [See C. W. Kammeyer and D. R. Whitman, *J. Chem. Phys.*, **56**, 4419 (1972).]

Although the radial distribution function of an atom shows the shell structure, the electron probability density integrated over the angles and plotted versus r does not oscillate. Rather, for ground-state atoms this probability density is a maximum at the nucleus (because of the s electrons) and continually decreases as r increases. Similarly, in molecules the maxima in electron probability density usually occur at the nuclei; see, for example, Fig. 13.6. [For further discussion, see H. Weinstein, P. Politzer, and S. Srebnik, *Theor. Chim. Acta*, **38**, 159 (1975).]

Accurate representation of a many-electron atomic orbital (AO) requires a linear combination of several Slater-type orbitals. For rough calculations, it is convenient to have simple approximations for AO's. We might use hydrogenlike orbitals with effective nuclear charges, but Slater suggested an even simpler method: to approximate an AO by a single function of the form (11.14) with the orbital exponent ζ taken as

$$\zeta = (Z - s)/n \tag{11.15}$$

where Z is the atomic number, n is the orbital's principal quantum number, and s is a screening constant calculated by a set of rules (see Problem 15.47). A Slater orbital replaces the polynomial in r in a hydrogenlike orbital with a single power of r. Hence a single Slater orbital does not have the proper number of radial nodes and does not represent well the inner part of an orbital.

A great deal of computation is required to perform a Hartree–Fock SCF calculation for a many-electron atom. Hartree did several SCF calculations in the 1930s, when electronic computers were not in existence. Fortunately, Hartree's father, a retired engineer, enjoyed numerical computation as a hobby and helped his son. Nowadays computers have replaced Hartree's father.

11.2   ORBITALS AND THE PERIODIC TABLE

The orbital concept and the Pauli exclusion principle allow us to understand the periodic table of the elements. An orbital is a one-electron wave function. We have used orbitals to obtain approximate wave functions for many-electron atoms, writing the wave function as a Slater determinant of one-electron spin-orbitals. In the crudest approximation, we neglect all interelectronic repulsions and obtain hydrogenlike orbitals. The best possible orbitals are the Hartree–Fock SCF functions. We build up the periodic table by feeding electrons into these orbitals, each of which can hold a pair of electrons with opposite spin.

Figure 11.2 shows orbital energies for neutral, ground-state atoms, as calculated by the Thomas–Fermi–Dirac method, which uses ideas of statistical mechanics to find approximate atomic wave functions (*Bethe and Jackiw*, Chapter 5). The AO energies of Fig. 11.2 generally agree well with both Hartree–Fock and experimentally found orbital energies (see J. C. Slater, *Quantum Theory of Matter*, 2nd ed., McGraw-Hill, New York, 1968, pp. 146, 147, 325, 326).

Orbital energies change with changing atomic number Z. As Z increases, the orbital energies decrease because of the increased attraction between the nucleus and the electrons. This decrease is most rapid for the inner orbitals, which are less well shielded from the nucleus.

For $Z > 1$, orbitals with the same value of n but different l have different energies. Thus, for the $n = 3$ orbital energies, we have

$$E_{3s} < E_{3p} < E_{3d} , \qquad Z > 1$$

The splitting of these levels, which are degenerate in the hydrogen atom, arises from the interelectronic repulsions. (Recall the perturbation treatment of helium in Section 9.7.) In the limit $Z \to \infty$, orbitals with the same value of n are again degenerate, because the interelectronic repulsions become insignificant in comparison with the electron–nucleus attractions.

The relative positions of certain orbitals change with changing Z. Thus in hydrogen the $3d$ orbital lies below the $4s$ orbital, but for Z in the range from 7 through 20 the $4s$ is below the $3d$. For large values of Z, the $3d$ is again lower. At $Z = 19$, the $4s$ is lower; hence $_{19}$K has the ground-state configuration $1s^2 2s^2 2p^6 3s^2 3p^6 4s$. Recall that s orbitals are more penetrating than p or d orbitals; this allows the $4s$ orbital to lie below the $3d$ orbital for some values of Z. Note the sudden drop in the $3d$ energy, which starts at $Z = 21$, when filling of the $3d$ orbital begins. The electrons of the $3d$ orbital do not shield each other very well; hence the sudden drop in $3d$ energy. Similar drops occur for other orbitals.

Since $3d$ lies below $4s$ for Z above 20, one might wonder why the ground-state configuration of, say, $_{23}$V is $\ldots 3d^3 4s^2$, rather than $\ldots 3d^5$. It is true that the $3d^5$ configuration has a lower sum of orbital energies than the $3d^3 4s^2$.

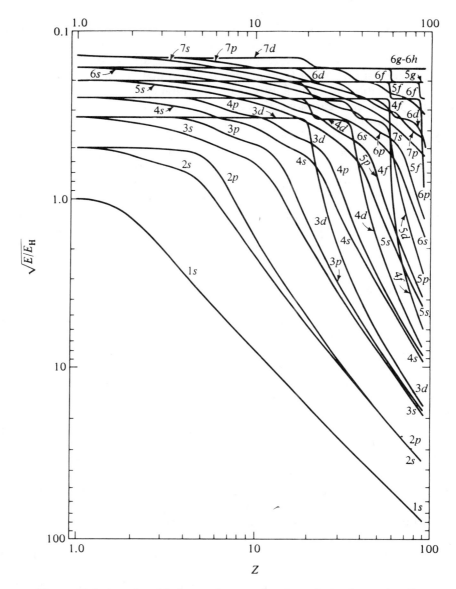

Figure 11.2 Atomic-orbital energies as a function of atomic number for neutral atoms, as calculated by Latter. [Figure redrawn by M. Kasha from R. Latter, *Phys. Rev.*, **99**, 510 (1955). Used by permission.] Note the logarithmic scales. E_H is the ground-state hydrogen-atom energy, -13.6 eV.

However, as we saw in Eq. (11.10) for the Hartree method and will see in Section 13.16 for the Hartree–Fock method, the energy of an atom is not given by the sum of the orbital energies of the electrons. Despite the lower sum of orbital energies for $3d^5$, the energy of the lowest level of the $3d^5$ configuration is higher than the energy of the lowest level of the $3d^3 4s^2$ configuration. (For the ion V^{2+}, the $3d^3$ configuration has both a lower sum of orbital energies and a lower energy than either of the configurations $3d4s^2$ or $3d^2 4s$, so the ground state of V^{2+} has the $3d^3$ configuration.)

Figure 11.2 shows that the separation between ns and np orbitals is much less than that between np and nd orbitals, giving the familiar $ns^2 np^6$ stable octet.

The orbital concept is the basis for most qualitative discussions of the chemistry of atoms and molecules. The use of orbitals, however, is an approximation; to reach the true wave function, we must go beyond a Slater determinant of spin-orbitals. Two review articles on atomic orbitals are R. S. Berry, *J. Chem. Educ.*, **43**, 283 (1966) and I. Cohen and T. Bustard, *J. Chem. Educ.*, **43**, 187 (1966).

11.3 ELECTRON CORRELATION

Energies calculated by the Hartree–Fock method are typically in error by $\frac{1}{2}\%$ to 1%. On an absolute basis this is not much, but for the chemist it is too large. For example, the total energy of the carbon atom is about 1000 eV, and $\frac{1}{2}\%$ of this is 5 eV. Chemical single-bond energies run about 5 eV/molecule. Attempting to calculate a bond energy by taking the difference between Hartree–Fock molecular and atomic energies, which are in error by several electron volts for light atoms, is an unreliable procedure. We must seek a way of improving on Hartree–Fock wave functions and energies. (Our discussion will be applicable to molecules as well as atoms.)

A Hartree–Fock SCF wave function takes into account the interactions between electrons only in an average way. Actually, we must consider the instantaneous interactions between electrons. Since electrons repel each other, they tend to keep out of each other's way. For example, in helium, if one electron is close to the nucleus at a given instant, it is energetically more favorable for the other electron to be far from the nucleus at that instant. One sometimes speaks of a *Coulomb hole* surrounding each electron in an atom; this is a region in which the probability of finding another electron is small. The motions of electrons are correlated with each other, and we speak of ***electron correlation***. We must find a way to introduce the instantaneous electron correlation into the wave function.

Actually, a Hartree–Fock wave function does have some instantaneous electron correlation built into it. A Hartree–Fock function satisfies the antisymmetry requirement of the Pauli principle; therefore [Eq. (10.20)], it vanishes when two electrons with the same spin have the same spatial coordinates. For a Hartree–Fock function, there is little probability of finding electrons of the same spin in the same region of space, so a Hartree–Fock function has some correlation of the motions of electrons with the same spin. This makes the Hartree–Fock energy lower than the Hartree energy. One sometimes refers to a *Fermi hole* around each electron in a Hartree–Fock wave function, thereby indicating a region in which the probability of finding another electron with the same spin is small.

The **correlation energy** is the difference between the exact nonrelativistic energy and the Hartree–Fock energy. For helium, the correlation energy is 1.1 eV (Section 11.1).

We have already indicated two ways in which we may provide for instantaneous electron correlation. The first method is to introduce the interelectronic distances r_{ij} into the wave function (Section 9.4). This method is only practicable for systems with a few electrons.

The second method is configuration interaction. We found (Sections 9.3 and 10.4) the zeroth-order wave function for the helium $1s^2$ ground state to be $1s(1)1s(2)[\alpha(1)\beta(2) - \beta(1)\alpha(2)]/\sqrt{2}$. We remarked that first- and higher-order corrections to the wave function will mix in contributions from excited configurations, giving rise to **configuration interaction** (CI), also called **configuration mixing** (CM).

The most common way to do a configuration-interaction calculation on an atom or molecule uses the variation method. One starts by choosing a basis set of one-electron functions χ_i. In principle, this basis set should be complete. In practice, one is limited to a basis set of finite size; one hopes that a good choice of basis functions will give a good approximation to a complete set. For atomic calculations, STOs [Eq. (11.14)] are often chosen.

The SCF atomic (or molecular) orbitals ϕ_i are written as linear combinations of the basis-set members, and the Hartree–Fock equations (11.12) are solved to give the coefficients in these linear combinations. The number of atomic (or molecular) orbitals obtained equals the number of basis functions used. The lowest-energy orbitals are the occupied orbitals for the ground state. The remaining unoccupied orbitals are called **virtual orbitals**.

Using the set of occupied and virtual spin-orbitals, one can form antisymmetric many-electron functions that have different orbital occupancies. For example, for helium, one can form functions that correspond to the electron configurations $1s^2$, $1s2s$, $1s2p$, $2s^2$, $2s2p$, $2p^2$, $1s3s$, and so on. Moreover, more than one function can correspond to a given electron configuration; recall the functions (10.27) to (10.30) corresponding to the helium $1s2s$ configuration. Each such many-electron function Φ_i is a Slater determinant or a linear combination of a few Slater determinants; use of more than one Slater determinant is required for certain open-shell functions such as (10.44) and (10.45). Each Φ_i is called a **configuration state function** or a **configuration function** or simply a "configuration." (This last name is unfortunate, since it leads to confusion between an electron configuration such as $1s^2$ and a configuration function such as $|1s\overline{1s}|$.)

As we saw in perturbation theory, the true atomic (or molecular) wave function ψ contains contributions from configurations other than the one that makes the main contribution to ψ, so we express ψ as a linear combination of the configuration functions Φ_i:

$$\psi = \sum_i c_i \Phi_i \tag{11.16}$$

We then regard (11.16) as a linear variation function (Section 8.5). Variation of the coefficients c_i to minimize the variational integral leads to the determinantal equation

$$\det (H_{ij} - ES_{ij}) = 0 \qquad (11.17)$$

[Only configuration functions whose angular-momentum eigenvalues are the same as those of the state ψ will contribute to the expansion (11.16); see Section 11.5.]

Because the many-electron configuration functions Φ_i are ultimately based on a one-electron basis set that is a complete set, the set of all possible configuration functions is a complete set for the many-electron problem: any antisymmetric many-electron function (including the exact wave function) can be expressed as a linear combination of the Φ_i functions. [For a proof of this, see P.-O. Löwdin, *Adv. Chem. Phys.*, **2**, 207 (1959); the proof is on pages 260–261.] Therefore, if one starts with a complete one-electron basis set and includes all possible configuration functions, a CI calculation will give the exact atomic (or molecular) wave function ψ for the state under consideration. In practice, one is limited to a finite, incomplete basis set, rather than an infinite, complete basis set. Moreover, even with a modest-size basis set, the number of possible configuration functions is extremely large, and one usually does not include all possible configuration functions. Part of the "art" of the CI method is choosing those configurations that will make the greatest contributions.

Because it generally takes very many configuration functions to give a truly accurate wave function, configuration-interaction calculations for systems with more than a few electrons are difficult.

In summary, to do a CI calculation, we choose a one-electron basis set χ_i, iteratively solve the Hartree–Fock equations (11.12) to determine one-electron atomic (or molecular) orbitals ϕ_i as linear combinations of the basis set, form many-electron configuration functions Φ_i using the orbitals ϕ_i, express the wave function ψ as a linear combination of these configuration functions, solve (11.17) for the energy, and solve the associated simultaneous linear equations for the coefficients c_i in (11.16).

As an example, consider the ground state of beryllium. The Hartree–Fock SCF method would find the best forms for the $1s$ and $2s$ orbitals in the Slater determinant $|1s\overline{1s}2s\overline{2s}|$ and use this for the ground-state wave function. [We are using the notation of Eq. (10.47).] Going beyond the Hartree–Fock method, we would include contributions from excited configuration functions (for example, $|1s\overline{1s}3s\overline{3s}|$) in a linear variation function for the ground state. Bunge did a CI calculation for the beryllium ground state using a linear combination of 650 configuration functions [C. F. Bunge, *Phys. Rev. A*, **14**, 1965 (1976)]. The Hartree–Fock energy is $-14.5730(e'^2/a_0)$, Bunge's CI result is $-14.6669(e'^2/a_0)$, and the exact nonrelativistic energy is $-14.6674(e'^2/a_0)$. Bunge was able to obtain 99.5% of the correlation energy.

11.4 ADDITION OF ANGULAR MOMENTA

For a many-electron atom, the operators for individual angular momenta of the electrons do not commute with the Hamiltonian operator, but their sum does. Hence we want to learn how to add angular momenta.

   Suppose we have a system with two angular-momentum vectors $\mathbf{M}_1$ and $\mathbf{M}_2$. They might be the orbital angular-momentum vectors of two electrons in an atom, or they might be the spin angular-momentum vectors of two electrons, or one might be the spin and the other the orbital angular momentum of a single electron. The eigenvalues of $\hat{M}_1^2$, $\hat{M}_2^2$, $\hat{M}_{1z}$, and $\hat{M}_{2z}$ are $j_1(j_1+1)\hbar^2$, $j_2(j_2+1)\hbar^2$, $m_1\hbar$, and $m_2\hbar$, where the quantum numbers obey the usual restrictions. The components of $\hat{\mathbf{M}}_1$ and $\hat{\mathbf{M}}_2$ obey the angular-momentum commutation relations [Eqs. (5.46), (5.48), and (5.114)]

$$[\hat{M}_{1x}, \hat{M}_{1y}] = i\hbar \hat{M}_{1z}, \text{ etc.} \tag{11.18}$$

$$[\hat{M}_{2x}, \hat{M}_{2y}] = i\hbar \hat{M}_{2z}, \text{ etc.} \tag{11.19}$$

We define the **total angular momentum** $\mathbf{M}$ of the system as the vector sum

$$\mathbf{M} = \mathbf{M}_1 + \mathbf{M}_2 \tag{11.20}$$

$\mathbf{M}$ is a vector with three components:

$$\mathbf{M} = M_x \mathbf{i} + M_y \mathbf{j} + M_z \mathbf{k} \tag{11.21}$$

The vector equation (11.20) gives the three scalar equations

$$M_x = M_{1x} + M_{2x}, \qquad M_y = M_{1y} + M_{2y}, \qquad M_z = M_{1z} + M_{2z} \tag{11.22}$$

For the operator $\hat{M}^2$, we have

$$\hat{M}^2 = \hat{\mathbf{M}} \cdot \hat{\mathbf{M}} = \hat{M}_x^2 + \hat{M}_y^2 + \hat{M}_z^2 \tag{11.23}$$

$$\hat{M}^2 = (\hat{\mathbf{M}}_1 + \hat{\mathbf{M}}_2) \cdot (\hat{\mathbf{M}}_1 + \hat{\mathbf{M}}_2)$$

$$\hat{M}^2 = \hat{M}_1^2 + \hat{M}_2^2 + \hat{\mathbf{M}}_1 \cdot \hat{\mathbf{M}}_2 + \hat{\mathbf{M}}_2 \cdot \hat{\mathbf{M}}_1 \tag{11.24}$$

If $\hat{\mathbf{M}}_1$ and $\hat{\mathbf{M}}_2$ refer to different electrons, they will commute with each other, since each will affect only functions of the coordinates of one electron and not the other. Even if $\hat{\mathbf{M}}_1$ and $\hat{\mathbf{M}}_2$ are the orbital and spin angular momenta of the same electron, they will commute, as one will affect only functions of the spatial coordinates while the other will affect functions of the spin coordinates. Thus Eq. (11.24) becomes

$$\hat{M}^2 = \hat{M}_1^2 + \hat{M}_2^2 + 2\hat{\mathbf{M}}_1 \cdot \hat{\mathbf{M}}_2 \tag{11.25}$$

$$\hat{M}^2 = \hat{M}_1^2 + \hat{M}_2^2 + 2(\hat{M}_{1x}\hat{M}_{2x} + \hat{M}_{1y}\hat{M}_{2y} + \hat{M}_{1z}\hat{M}_{2z}) \tag{11.26}$$

   We now show that the components of the total angular momentum obey the usual angular-momentum commutation relations. We have [Eq. (5.4)]

$$[\hat{M}_x, \hat{M}_y] = [\hat{M}_{1x} + \hat{M}_{2x}, \hat{M}_{1y} + \hat{M}_{2y}]$$

$$= [\hat{M}_{1x}, \hat{M}_{1y} + \hat{M}_{2y}] + [\hat{M}_{2x}, \hat{M}_{1y} + \hat{M}_{2y}]$$

$$= [\hat{M}_{1x}, \hat{M}_{1y}] + [\hat{M}_{1x}, \hat{M}_{2y}] + [\hat{M}_{2x}, \hat{M}_{1y}] + [\hat{M}_{2x}, \hat{M}_{2y}]$$

Since all components of $\hat{\mathbf{M}}_1$ commute with all components of $\hat{\mathbf{M}}_2$, we have

$$[\hat{M}_x, \hat{M}_y] = [\hat{M}_{1x}, \hat{M}_{1y}] + [\hat{M}_{2x}, \hat{M}_{2y}] = i\hbar\hat{M}_{1z} + i\hbar\hat{M}_{2z}$$

$$[\hat{M}_x, \hat{M}_y] = i\hbar\hat{M}_z \qquad (11.27)$$

Cyclic permutation of x, y, and z gives

$$[\hat{M}_y, \hat{M}_z] = i\hbar\hat{M}_x, \qquad [\hat{M}_z, \hat{M}_x] = i\hbar\hat{M}_y \qquad (11.28)$$

The same commutator algebra used to derive (5.116) gives

$$[\hat{M}^2, \hat{M}_x] = [\hat{M}^2, \hat{M}_y] = [\hat{M}^2, \hat{M}_z] = 0 \qquad (11.29)$$

Thus we can simultaneously quantize M^2 and one of its components, say M_z. Since the components of the total angular momentum obey the angular-momentum commutation relations, the work of Section 5.4 shows that the eigenvalues of $\hat{M}^2$ are

$$J(J + 1)\hbar^2, \qquad J = 0, \tfrac{1}{2}, 1, \tfrac{3}{2}, 2, \ldots \qquad \mathbf{(11.30)}^*$$

and the eigenvalues of $\hat{M}_z$ are

$$M_J\hbar, \qquad M_J = -J, -J + 1, \ldots, J - 1, J \qquad \mathbf{(11.31)}^*$$

We want to find out how the total-angular-momentum quantum numbers J and M_J are related to the quantum numbers j_1, j_2, m_1, m_2 of the two angular momenta we are adding in (11.20). We also want the eigenfunctions of $\hat{M}^2$ and $\hat{M}_z$. These eigenfunctions are characterized by the quantum numbers J and M_J, and, using ket notation (Section 7.3) we write them as $|JM_J\rangle$. Similarly, let $|j_1m_1\rangle$ denote the eigenfunctions of $\hat{M}_1^2$ and $\hat{M}_{1z}$ and $|j_2m_2\rangle$ denote the eigenfunctions of $\hat{M}_2^2$ and $\hat{M}_{2z}$. Now it is readily shown (Problem 11.6) that

$$[\hat{M}_x, \hat{M}_1^2] = [\hat{M}_y, \hat{M}_1^2] = [\hat{M}_z, \hat{M}_1^2] = [\hat{M}^2, \hat{M}_1^2] = 0 \qquad (11.32)$$

with similar equations with $\hat{M}_2^2$ replacing $\hat{M}_1^2$. Hence we can have simultaneous eigenfunctions of all four operators $\hat{M}_1^2$, $\hat{M}_2^2$, $\hat{M}^2$, $\hat{M}_z$, and the eigenfunctions $|JM_J\rangle$ can be more fully written as $|j_1j_2JM_J\rangle$. However, one finds that $\hat{M}^2$ does not commute with $\hat{M}_{1z}$ or $\hat{M}_{2z}$ (Problem 11.8), so the eigenfunctions $|j_1j_2JM_J\rangle$ are not necessarily eigenfunctions of $\hat{M}_{1z}$ or $\hat{M}_{2z}$.

If we take the complete set of functions $|j_1m_1\rangle$ for particle 1 and the complete set $|j_2m_2\rangle$ for particle 2 and form all possible products of the form $|j_1m_1\rangle|j_2m_2\rangle$, we will have a complete set of functions for the two particles. Each unknown eigenfunction $|j_1j_2JM_J\rangle$ can then be expanded using this complete set:

$$|j_1j_2JM_J\rangle = \sum c(j_1j_2JM_J; m_1m_2)|j_1m_1\rangle|j_2m_2\rangle \qquad (11.33)$$

where the expansion coefficients are the $c(j_1 \cdots m_2)$'s. The functions $|j_1j_2JM_J\rangle$ are eigenfunctions of the commuting operators $\hat{M}_1^2$, $\hat{M}_2^2$, $\hat{M}^2$, and $\hat{M}_z$ with the following eigenvalues:

$\hat{M}_1^2$	$\hat{M}_2^2$	$\hat{M}^2$	$\hat{M}_z$
$j_1(j_1 + 1)\hbar^2$	$j_2(j_2 + 1)\hbar^2$	$J(J + 1)\hbar^2$	$M_J\hbar$

The functions $|j_1 m_1\rangle |j_2 m_2\rangle$ are eigenfunctions of the commuting operators $\hat{M}_1^2$, $\hat{M}_{1z}$, $\hat{M}_2^2$, $\hat{M}_{2z}$ with the following eigenvalues:

$\hat{M}_1^2$	$\hat{M}_{1z}$	$\hat{M}_2^2$	$\hat{M}_{2z}$
$j_1(j_1+1)\hbar^2$	$m_1\hbar$	$j_2(j_2+1)\hbar^2$	$m_2\hbar$

Since the function $|j_1 j_2 J M_J\rangle$ being expanded in (11.33) is an eigenfunction of $\hat{M}_1^2$ with eigenvalue $j_1(j_1+1)\hbar^2$, we include in the sum only terms that have the same j_1 value as in the function $|j_1 j_2 J M_J\rangle$. (See Theorem 3 at the end of Section 7.3.) Likewise, only terms with the same j_2 value as in $|j_1 j_2 J M_J\rangle$ are included in the sum. Hence the sum goes over only the m_1 and m_2 values. Furthermore, using $\hat{M}_z = \hat{M}_{1z} + \hat{M}_{2z}$, we can prove (Problem 11.7) that the coefficient c vanishes unless

$$m_1 + m_2 = M_J \qquad (11.34)^*$$

To explicitly obtain the total-angular-momentum eigenfunctions, one must evaluate the coefficients in (11.33). These are called *Clebsch–Gordan* or *Wigner* or *vector addition* coefficients. For their evaluation, see *Merzbacher*, Section 16.6. For tables of them, see *Anderson, Introduction to Quantum Chemistry*, pages 332–345.

Thus each total-angular-momentum eigenfunction $|j_1 j_2 J M_J\rangle$ is a linear combination of those product functions $|j_1 m_1\rangle |j_2 m_2\rangle$ whose m values satisfy $m_1 + m_2 = M_J$.

We now find the possible values of the total-angular-momentum quantum number J that arise from the addition of angular momenta with individual quantum numbers j_1 and j_2.

Before discussing the general case, we consider the case with $j_1 = 1$, $j_2 = 2$. The possible values of m_1 are -1, 0, $+1$, and the possible values of m_2 are -2, -1, 0, $+1$, $+2$. If we describe the system by the quantum numbers j_1, j_2, m_1, m_2, then the total number of possible states is fifteen, corresponding to three possibilities for m_1 and five for m_2. Instead, we can describe the system using the quantum numbers j_1, j_2, J, M_J, and we must have the same number of states in this description. Let us tabulate the fifteen possible values of M_J using (11.34):

$m_1 = -1$	0	$+1$	
-3	-2	-1	$-2 = m_2$
-2	-1	0	-1
-1	0	$+1$	0
0	$+1$	$+2$	$+1$
$+1$	$+2$	$+3$	$+2$

where each M_J value in the table is the sum of the m_1 and m_2 values at the top and side. The number of times each value of M_J occurs is

value of M_J	3	2	1	0	-1	-2	-3
number of occurrences	1	2	3	3	3	2	1

The highest value of M_J is $+3$. Since M_J ranges from $-J$ to $+J$, the highest value of J must be 3. Corresponding to $J = 3$, there are seven values of M_J ranging from

−3 to +3. Eliminating these seven values, we are left with

value of M_J	2	1	0	−1	−2
number of occurrences	1	2	2	2	1

The highest remaining value, $M_J = 2$, must correspond to $J = 2$; for $J = 2$, we have five values of M_J, which when eliminated leave

value of M_J	1	0	−1
number of occurrences	1	1	1

These remaining values of M_J clearly correspond to $J = 1$. Thus, for the individual angular-momentum quantum numbers $j_1 = 1$, $j_2 = 2$, the possible values of the total-angular-momentum quantum number J are 3, 2, and 1.

Now consider the general case. There are $2j_1 + 1$ values of m_1 (ranging from $-j_1$ to $+j_1$) and $2j_2 + 1$ values of m_2. Hence there are $(2j_1 + 1)(2j_2 + 1)$ possible states $|j_1 m_1\rangle |j_2 m_2\rangle$ with fixed j_1 and j_2 values. The highest possible values of m_1 and m_2 are j_1 and j_2, respectively. Therefore, the maximum possible value of $M_J = m_1 + m_2$ is $j_1 + j_2$ [Eq. (11.34)]. Since M_J ranges from $-J$ to $+J$, the maximum possible value of J must also be $j_1 + j_2$:

$$J_{max} = j_1 + j_2 \tag{11.35}$$

The second-highest value of M_J is $j_1 + j_2 - 1$, which arises in two ways: $m_1 = j_1 - 1$, $m_2 = j_2$ and $m_1 = j_1$, $m_2 = j_2 - 1$. Linear combinations of these two states must give one state with $J = j_1 + j_2$, $M_J = j_1 + j_2 - 1$ and one state with $J = j_1 + j_2 - 1$, $M_J = j_1 + j_2 - 1$. Continuing in this manner, we find that the possible values of J are

$$j_1 + j_2, \quad j_1 + j_2 - 1, \quad j_1 + j_2 - 2, \quad \ldots, \quad J_{min}$$

where J_{min} is the lowest possible value of J.

We determine J_{min} by the requirement that the total number of states be $(2j_1 + 1)(2j_2 + 1)$. For a particular value of J, there are $2J + 1$ M_J values, and so $2J + 1$ states correspond to each value of J. The total number of states $|j_1 j_2 J M_J\rangle$ for fixed j_1 and j_2 is found by summing the number of states $2J + 1$ for each J from J_{min} to J_{max}:

$$\text{number of states} = \sum_{J=J_{min}}^{J_{max}} (2J + 1) \tag{11.36}$$

This sum goes from J_{min} to J_{max}. Let us now take the lower limit of the sum to be $J = 0$ instead of J_{min}; this change adds to the sum terms with J values of $0, 1, 2, \ldots, J_{min} - 1$. To compensate, we must subtract the corresponding sum that goes from $J = 0$ to $J = J_{min} - 1$. Therefore, (11.36) becomes

$$\text{number of states} = \sum_{J=0}^{J_{max}} (2J + 1) - \sum_{J=0}^{J_{min}-1} (2J + 1)$$

Equation (6.99) with $b = n - 1$ gives $\sum_{J=0}^{b} (2J + 1) = (b + 1)^2$. Therefore,

$$\text{number of states} = (J_{max} + 1)^2 - J_{min}^2 = J_{max}^2 + 2J_{max} + 1 - J_{min}^2$$

Replacing J_{max} by $j_1 + j_2$ [Eq. (11.35)] and equating the number of states to $(2j_1 + 1)(2j_2 + 1) = 4j_1j_2 + 2j_1 + 2j_2 + 1$, we have

$$(j_1 + j_2)^2 + 2(j_1 + j_2) + 1 - J_{min}^2 = 4j_1j_2 + 2j_1 + 2j_2 + 1$$

$$J_{min}^2 = j_1^2 - 2j_1j_2 + j_2^2 = (j_1 - j_2)^2$$

$$J_{min} = \pm(j_1 - j_2) \tag{11.37}$$

If $j_1 = j_2$, then $J_{min} = 0$. If $j_1 \neq j_2$, then one of the values in (11.37) is negative and must be rejected [Eq. (11.30)]. Thus

$$J_{min} = |j_2 - j_1| \tag{11.38}$$

To summarize, we have shown that *the addition of two angular momenta characterized by quantum numbers j_1 and j_2 results in a total angular momentum whose quantum number J has the possible values*

$$J = j_1 + j_2, \quad j_1 + j_2 - 1, \quad \ldots, \quad |j_1 - j_2| \tag{11.39}*$$

EXAMPLE   Find the possible values of the total-angular-momentum quantum number resulting from the addition of angular momenta with quantum numbers: (a) $j_1 = 2$, $j_2 = 3$; (b) $j_1 = 3$, $j_2 = \frac{3}{2}$.

(a) The maximum and minimum J values are given by (11.39) as $j_1 + j_2 = 2 + 3 = 5$ and $|j_1 - j_2| = |2 - 3| = 1$. The possible J values (11.39) are therefore $J = 5$, 4, 3, 2, 1. (b) We have $j_1 + j_2 = 3 + \frac{3}{2} = \frac{9}{2}$ and $|j_1 - j_2| = |3 - \frac{3}{2}| = \frac{3}{2}$. Therefore, $J = \frac{9}{2}, \frac{7}{2}, \frac{5}{2}, \frac{3}{2}$.

EXAMPLE   Find the possible J values when angular momenta with quantum numbers $j_1 = 1$, $j_2 = 2$, and $j_3 = 3$ are added.

To add more than two angular momenta, we apply (11.39) repeatedly. Addition of $j_1 = 1$ and $j_2 = 2$ gives the possible quantum numbers 3, 2, and 1. Addition of j_3 to each of these values gives the following possibilities for the total-angular-momentum quantum number:

$$6, 5, 4, 3, 2, 1, 0; \quad 5, 4, 3, 2, 1; \quad 4, 3, 2 \tag{11.40}$$

We have one set of states with total-angular-momentum quantum number 6, two sets of states with $J = 5$, three sets with $J = 4$, and so on.

11.5 ANGULAR MOMENTUM IN MANY-ELECTRON ATOMS

Total Electronic Orbital and Spin Angular Momenta.   The *total electronic orbital angular momentum* of an n-electron atom is defined as the vector sum of the orbital angular momenta of the individual electrons:

$$\mathbf{L} = \sum_{i=1}^{n} \mathbf{L}_i \tag{11.41}$$

Although the individual orbital-angular-momentum operators $\hat{\mathbf{L}}_i$ do not commute with the atomic Hamiltonian (11.1), one can show (*Bethe and Jackiw*, pp.

102–103) that $\hat{\mathbf{L}}$ does commute with the atomic Hamiltonian [provided spin–orbit interaction (Section 11.6) is neglected]. We can therefore characterize an atomic state by a quantum number L, where $L(L+1)\hbar^2$ is the square of the magnitude of the total electronic orbital angular momentum; the electronic wave function ψ of an atom satisfies $\hat{L}^2\psi = L(L+1)\hbar^2\psi$. The total-electronic-orbital-angular-momentum quantum number L of an atom is indicated by a code letter, as follows:

L	0	1	2	3	4	5	6	7	8	
letter	S	P	D	F	G	H	I	K	L	$(11.42)^*$

The total orbital angular momentum is designated by a capital letter, while lowercase letters are used for orbital angular momenta of individual electrons.

EXAMPLE   Find the possible values of the quantum number L for states of the carbon atom that arise from the electron configuration $1s^2 2s^2 2p3d$.

The s electrons have zero orbital angular momentum and contribute nothing to the total orbital angular momentum. The $2p$ electron has $l=1$ and the $3d$ electron has $l=2$. From the angular-momentum addition rule (11.39), the total-orbital-angular-momentum quantum number ranges from $1+2=3$ to $|1-2|=1$; the possible values of L are $L=3,2,1$. The configuration $1s^2 2s^2 2p3d$ gives rise to P, D, and F states. [The Hartree–Fock central-field approximation has each electron moving in a central-field potential, $V=V(r)$; hence, within this approximation, the individual electronic orbital angular momenta are constant, giving rise to a wave function composed of a single configuration that specifies the individual orbital angular momenta. When we go beyond the SCF central-field approximation, we mix in other configurations so we no longer specify precisely the individual orbital angular momenta; even so, we can still use the rule (11.39) for finding the possible values of the total orbital angular momentum.]

The **total electronic spin angular momentum** $\mathbf{S}$ of an atom is defined as the vector sum of the spins of the individual electrons:

$$\mathbf{S} = \sum_{i=1}^{n} \mathbf{S}_i \qquad (11.43)$$

The atomic Hamiltonian $\hat{H}$ of (11.1) (which omits spin–orbit interaction) does not involve spin and therefore commutes with the total-spin operators $\hat{S}^2$ and $\hat{S}_z$. The fact that $\hat{S}^2$ commutes with $\hat{H}$ is not enough to show that the atomic wave functions ψ are eigenfunctions of $\hat{S}^2$. The Pauli antisymmetry principle requires that each ψ must be an eigenfunction of the permutation operator $\hat{P}_{ij}$ with eigenvalue -1 (Section 10.3). Hence $\hat{S}^2$ must also commute with $\hat{P}_{ij}$ if we are to have simultaneous eigenfunctions of $\hat{H}$, $\hat{S}^2$, and $\hat{P}_{ij}$. Problem 11.16 shows that $[\hat{S}^2, \hat{P}_{ij}] = 0$, so the atomic wave functions are eigenfunctions of $\hat{S}^2$. We have $\hat{S}^2\psi = S(S+1)\hbar^2\psi$, and each atomic state can be characterized by a total-electronic-spin quantum number S.

EXAMPLE   Find the possible values of the quantum number S for states that arise from the electron configuration $1s^2 2s^2 2p3d$.

Consider first the two $1s$ electrons. To satisfy the exclusion principle, one of these electrons must have $m_s = +\frac{1}{2}$ while the other has $m_s = -\frac{1}{2}$. If M_S is the quantum number that specifies the z component of the total spin of the $1s$ electrons, then the only possible value of M_S is $\frac{1}{2} - \frac{1}{2} = 0$ [Eq. (11.34)]. This single value of M_S clearly means that the total spin of the two $1s$ electrons is zero. Thus, although in general when we add the spins $s_1 = \frac{1}{2}$ and $s_2 = \frac{1}{2}$ of two electrons according to the rule (11.39), we get the two possibilities $S = 0$ and $S = 1$, the restriction imposed by the Pauli principle leaves $S = 0$ as the only possibility in this case. Likewise, the spins of the $2s$ electrons add up to zero. The exclusion principle does not restrict the m_s values of the $2p$ and $3d$ electrons. Application of rule (11.39) to the spins $s_1 = \frac{1}{2}$ and $s_2 = \frac{1}{2}$ of the $2p$ and $3d$ electrons gives $S = 0$ and $S = 1$; these are the possible values of the total spin quantum number, since the $1s$ and $2s$ electrons do not contribute to S.

Atomic Terms.   A given electron configuration gives rise in general to several different atomic states, some having the same energy and some having different energies, depending on whether the interelectronic repulsions are the same or different for the states. For example, the $1s2s$ configuration of helium gives rise to four states: the three states with zeroth-order wave functions (10.27) to (10.29) all have the same energy; the single state (10.30) has a different energy. The $1s2p$ electron configuration gives rise to twelve states: the nine states obtained by replacing $2s$ in (10.27) to (10.29) by $2p_x$, $2p_y$, or $2p_z$ have the same energy; the three states obtained by replacing $2s$ in (10.30) by $2p_x$, $2p_y$, or $2p_z$ have the same energy, which differs from the energy of the other nine states.

Thus the atomic states that arise from a given electron configuration can be grouped into sets of states that have the same energy. One can show that states that arise from the same electron configuration and that have the same energy (with spin–orbit interaction neglected) will have the same value of L and the same value of S (see *Kemble*, Section 63a). A set of equal-energy atomic states that arise from the same electron configuration and that have the same L value and the same S value constitutes an atomic **term**. For a fixed L value, the quantum number M_L (where $M_L \hbar$ is the z component of the total electronic orbital angular momentum) takes on $2L + 1$ values ranging from $-L$ to $+L$; for a fixed S value, M_S takes on $2S + 1$ values. The atomic energy does not depend on M_L or M_S, and each term consists of $(2L + 1)(2S + 1)$ atomic states of equal energy. The degeneracy of an atomic term is $(2L + 1)(2S + 1)$ (spin–orbit interaction neglected).

Each term of an atom is designated by a **term symbol** formed by writing the numerical value of the quantity $2S + 1$ as a left superscript on the code letter (11.42) that gives the L value. For example, a term that has $L = 2$ and $S = 1$ has the term symbol ^3D, since $2S + 1 = 3$.

EXAMPLE   Find the terms arising from each of the following electron configurations: (a) $1s2p$; (b) $1s^2 2s^2 2p3d$. Give the degeneracy of each term.

(a) The $1s$ electron has quantum number $l = 0$ and the $2p$ electron has $l = 1$. The addition rule (11.39) gives $L = 1$ as the only possibility. The code letter for $L = 1$ is P. Each electron has $s = \frac{1}{2}$, and (11.39) gives $S = 1, 0$ as the possible S values. The possible values of $2S + 1$ are 3 and 1. The possible terms are thus ^3P

and ^1P. The ^3P term has $L = 1$ and $S = 1$, and its degeneracy is $(2L + 1) \times (2S + 1) = 3(3) = 9$. The ^1P term has $L = 1$ and $S = 0$, and its degeneracy is $(2L + 1)(2S + 1) = 3(1) = 3$. [The nine states of the ^3P term are obtained by replacing $2s$ in (10.27) to (10.29) by $2p_x$, $2p_y$, or $2p_z$. The three states of the ^1P term are obtained by replacing $2s$ in (10.30) by $2p$ functions.]

(b) In the two previous examples in this section, we found that the configuration $1s^2 2s^2 2p 3d$ has the possible L values $L = 3, 2, 1$ and has $S = 1, 0$. The code letters for these L values are F, D, P, and the terms are

$$^1P, \quad ^3P, \quad ^1D, \quad ^3D, \quad ^1F, \quad ^3F \tag{11.44}$$

The degeneracies are found as in part (a) and are 3, 9, 5, 15, 7, and 21, respectively.

Derivation of Atomic Terms. We now examine how to systematically derive the terms that arise from a given electron configuration.

First consider configurations that contain only completely filled subshells. In such configurations, for each electron with $m_s = +\frac{1}{2}$ there is an electron with $m_s = -\frac{1}{2}$. Let the quantum number specifying the z component of the total electronic spin angular momentum be M_S. The only possible value for M_S is zero ($M_S = \sum_i m_{si} = 0$); hence S must be zero. For each electron in a closed subshell with magnetic quantum number m, there is an electron with magnetic quantum number $-m$; for example, for a $2p^6$ configuration we have two electrons with $m = +1$, two with $m = -1$, and two with $m = 0$. Denoting the quantum number specifying the z component of the total electronic orbital angular momentum by M_L, we have $M_L = \sum_i m_i = 0$; we conclude that L must be zero. In summary, a configuration of closed subshells gives rise to only one term: ^1S. For configurations consisting of closed subshells and open subshells, the closed subshells make no contribution to L or S and may be ignored in finding the terms.

We now consider two electrons in different subshells; such electrons are called **nonequivalent**. Nonequivalent electrons have different values of n or l or both, and we need not worry about any restrictions imposed by the exclusion principle when we derive the terms. We simply find the possible values of L from l_1 and l_2 according to (11.39); combining s_1 and s_2 gives $S = 0, 1$. We previously worked out the pd case, which gives the terms in (11.44). If we have more than two nonequivalent electrons, we combine the individual l's to find the values of L, and we combine the individual s's to find the values of S. For example, consider a pdf configuration. The possible values of L are given by (11.40). Combining three spin angular momenta, each of which is $\frac{1}{2}$, gives $S = \frac{3}{2}, \frac{1}{2}, \frac{1}{2}$. Each of the three possibilities in (11.40) with $L = 3$ may be combined with each of the two possibilities for $S = \frac{1}{2}$, giving six ^2F terms. Continuing in this manner, we find that the following terms arise from a pdf configuration: $^2S(2)$, $^2P(4)$, $^2D(6)$, $^2F(6)$, $^2G(6)$, $^2H(4)$, $^2I(2)$, ^4S, $^4P(2)$, $^4D(3)$, $^4F(3)$, $^4G(3)$, $^4H(2)$, ^4I, where the number of times each type of term occurs is in parentheses.

Now consider two electrons in the same subshell (**equivalent** electrons). Equivalent electrons have the same value of n and the same value of l, and the situation is complicated by the necessity to avoid giving two electrons the same four quantum numbers; hence not all the terms derived for nonequivalent

TABLE 11.1   Quantum Numbers for Two Equivalent p Electrons

m_1	m_{s1}	m_2	m_{s2}	$M_L = m_1 + m_2$	$M_S = m_{s1} + m_{s2}$
1	$\frac{1}{2}$	1	$-\frac{1}{2}$	2	0
1	$\frac{1}{2}$	0	$\frac{1}{2}$	1	1
1	$\frac{1}{2}$	0	$-\frac{1}{2}$	1	0
1	$-\frac{1}{2}$	0	$\frac{1}{2}$	1	0
1	$-\frac{1}{2}$	0	$-\frac{1}{2}$	1	-1
1	$\frac{1}{2}$	-1	$\frac{1}{2}$	0	1
1	$\frac{1}{2}$	-1	$-\frac{1}{2}$	0	0
1	$-\frac{1}{2}$	-1	$\frac{1}{2}$	0	0
1	$-\frac{1}{2}$	-1	$-\frac{1}{2}$	0	-1
0	$\frac{1}{2}$	0	$-\frac{1}{2}$	0	0
0	$\frac{1}{2}$	-1	$\frac{1}{2}$	-1	1
0	$\frac{1}{2}$	-1	$-\frac{1}{2}$	-1	0
0	$-\frac{1}{2}$	-1	$\frac{1}{2}$	-1	0
0	$-\frac{1}{2}$	-1	$-\frac{1}{2}$	-1	-1
-1	$\frac{1}{2}$	-1	$-\frac{1}{2}$	-2	0

electrons are possible. As an example, consider the terms arising from two equivalent p electrons, an np^2 configuration. (The carbon ground-state configuration is $1s^2 2s^2 2p^2$.) The possible values of m and m_s for the two electrons are listed in Table 11.1, which also gives M_L and M_S.

Note that certain combinations are missing from this table. For example, $m_1 = 1$, $m_{s1} = \frac{1}{2}$, $m_2 = 1$, $m_{s2} = \frac{1}{2}$ is missing, since it violates the exclusion principle. Another missing combination is $m_1 = 1$, $m_{s1} = -\frac{1}{2}$, $m_2 = 1$, $m_{s2} = \frac{1}{2}$. This combination differs from $m_1 = 1$, $m_{s1} = \frac{1}{2}$, $m_2 = 1$, $m_{s2} = -\frac{1}{2}$ (row 1) solely by interchange of electrons 1 and 2. Each row in Table 11.1 stands for a Slater determinant, which when expanded contains terms for all possible electron interchanges among the spin-orbitals. Two rows that differ from each other solely by interchange of two electrons correspond to the same Slater determinant, and we include only one of them in the table.

The highest value of M_L in Table 11.1 is 2, which must correspond to a term with $L = 2$, a D term. The $M_L = 2$ value occurs in conjunction with $M_S = 0$, indicating that $S = 0$ for the D term; thus we have a ^1D term corresponding to the five states

$$
\begin{aligned}
M_L &= 2 \quad 1 \quad 0 \quad -1 \quad -2 \\
M_S &= 0 \quad 0 \quad 0 \quad \;\;0 \quad \;\;0
\end{aligned}
\tag{11.45}
$$

The highest value of M_S in Table 11.1 is 1, indicating a term with $S = 1$. $M_S = 1$ occurs in conjunction with $M_L = 1, 0, -1$, which indicates a P term. Hence we have a ^3P term corresponding to the nine states

$$
\begin{aligned}
M_L &= 1 \quad 1 \quad \;\;1 \quad 0 \quad 0 \quad \;\;0 \quad -1 \quad -1 \quad -1 \\
M_S &= 1 \quad 0 \quad -1 \quad 1 \quad 0 \quad -1 \quad \;\;1 \quad \;\;0 \quad -1
\end{aligned}
\tag{11.46}
$$

Elimination of the states of (11.45) and (11.46) from Table 11.1 leaves only a

TABLE 11.2 Terms Arising from Various Electron
Configurations

Configuration	Terms
(a) *Equivalent electrons*	
s^2; p^6; d^{10}	^1S
p; p^5	^2P
p^2; p^4	^3P, ^1D, ^1S
p^3	^4S, ^2D, ^2P
d; d^9	^2D
d^2; d^8	^3F, ^3P, ^1G, ^1D, ^1S
d^3; d^7	^4F, ^4P, ^2H, ^2G, ^2F, $^2D(2)$, ^2P
d^4; d^6	$\begin{cases} ^5D,\ ^3H,\ ^3G,\ ^3F(2),\ ^3D,\ ^3P(2) \\ ^1I,\ ^1G(2),\ ^1F,\ ^1D(2),\ ^1S(2) \end{cases}$
d^5	$\begin{cases} ^6S,\ ^4G,\ ^4F,\ ^4D,\ ^4P,\ ^2I,\ ^2H,\ ^2G(2) \\ ^2F(2),\ ^2D(3),\ ^2P,\ ^2S \end{cases}$
(b) *Nonequivalent electrons*	
ss	^1S, ^3S
sp	^1P, ^3P
sd	^1D, ^3D
pp	^3D, ^1D, ^3P, ^1P, ^3S, ^1S

single state, which has $M_L = 0$, $M_S = 0$, corresponding to a ^1S term. Thus a p^2 configuration gives rise to the terms ^1S, ^3P, ^1D. (In contrast, two nonequivalent p electrons give rise to six terms: ^1S, ^3S, ^1P, ^3P, ^1D, ^3D.)

Table 11.2a lists the terms arising from various configurations of equivalent electrons. These results may be derived in the same way that we found the p^2 terms, but this procedure can become quite involved; to derive the terms of the f^7 configuration would require a table with 3432 rows. More efficient methods exist [R. F. Curl and J. E. Kilpatrick, *Am. J. Phys.*, **28**, 357 (1960); K. E. Hyde, *J. Chem. Educ.*, **52**, 87 (1975)]. Note from Table 11.2a that the terms arising from a subshell containing N electrons are the same as the terms for a subshell that is N electrons short of being full; for example, the terms for p^2 and p^4 are the same. We can divide the electrons of a closed subshell into two groups and find the terms for each group; because a closed subshell gives only a ^1S term, the terms for each of these two groups must be the same. Table 11.2b gives the terms arising from some nonequivalent electron configurations.

To deal with a configuration containing both equivalent and nonequivalent electrons, we first find separately the terms from the nonequivalent electrons and the terms from the equivalent electrons. We then take all possible combinations of the L and S values of these two sets of terms. For example, consider an sp^3 configuration. From the s electron, we get a ^2S term. From the three equivalent p electrons, we get the terms ^2P, ^2D, and ^4S. Combining the L and S values of these terms, we have as the terms of an sp^3 configuration

$$^3P,\ ^1P,\ ^3D,\ ^1D,\ ^5S,\ ^3S \qquad (11.47)$$

Hund's Rule. To decide which one of the terms arising from a given electron configuration is lowest in energy, we use the empirical *Hund's rule*: *For terms arising from the same electron configuration, the term with the largest value of S lies lowest; if there is more than one term with the largest S, then the term with the largest S and the largest L lies lowest.*

EXAMPLE Predict the lowest term of (*a*) the carbon ground-state configuration $1s^2 2s^2 2p^2$; (*b*) the configuration $1s^2 2s 2p^3$; (*c*) the configuration $1s^2 2s^2 2p^6 3s^2 3p^6 3d^2 4s^2$.

(*a*) Table 11.2a gives the terms arising from a p^2 configuration as ^3P, ^1D, and ^1S. The term with the largest S will have the largest value of the left superscript $2S + 1$. Hund's rule predicts ^3P as the lowest term. (*b*) The terms of an sp^3 configuration are given by (11.47), and we see that ^5S has the largest spin and is predicted to be lowest. (*c*) Table 11.2a gives the d^2 terms as ^3F, ^3P, ^1G, ^1D, ^1S. Of these terms, ^3F and ^3P have the highest S. ^3F has $L = 3$; ^3P has $L = 1$. Therefore, ^3F is predicted to be lowest.

Hund's rule works very well for the ground-state configuration, but occasionally fails for an excited configuration.

Hund's rule gives only the lowest term of a configuration, and should not be used to decide the order of the remaining terms. For example, for the $1s^2 2s 2p^3$ configuration of carbon, the observed order of the terms is

$$^5S < {}^3D < {}^3P < {}^1D < {}^3S < {}^1P$$

The ^3S term lies above the ^1D term, even though ^3S has the higher spin S. For the $\ldots 3d^2 4s^2$ titanium configuration, the ^1D term lies below ^1G, even though ^1G has the larger L value.

It is not necessary to consult Table 11.2a to find the lowest term of a partly filled subshell configuration. We simply put the electrons in the orbitals so as to give the greatest number of parallel spins. Thus, for a d^3 configuration, we have

$$m: \quad \frac{\uparrow}{+2} \ \frac{\uparrow}{+1} \ \frac{\uparrow}{0} \ \frac{\ }{-1} \ \frac{\ }{-2} \qquad (11.48)$$

The lowest term thus has three parallel spins, so $S = \frac{3}{2}$, giving $2S + 1 = 4$. The maximum value of M_L is 3, corresponding to $L = 3$, an F term. Hund's rule thus predicts ^4F as the lowest term of a d^3 configuration.

The traditional explanation of Hund's rule is as follows: Electrons with the same spin tend to keep out of each other's way (recall the idea of Fermi holes), thereby minimizing the Coulombic repulsion between them. The term that has the greatest number of parallel spins (that is, the greatest value of S) will therefore be lowest in energy. For example, the ^3S term of the helium $1s2s$ configuration has an antisymmetric space function that vanishes when the spatial coordinates of the two electrons are equal; hence the ^3S term is lower than the ^1S term.

This traditional explanation turns out to be wrong in most cases. It is true that the probability that the two electrons are very close together is smaller for the helium ^3S $1s2s$ term than for the ^1S $1s2s$ term. However, calculations with accurate wave functions show that the probability that the two electrons are very far apart is also less for the ^3S term. The net result is that the average distance between the two electrons is slightly less for the ^3S term than for the ^1S term, and the interelectronic repulsion is slightly greater for the ^3S term. The reason the ^3S term lies below the ^1S term is because of a substantially greater electron–nucleus attraction in the ^3S term as compared with the ^1S term. Similar results are found for terms of the atoms beryllium and carbon. [For more details see J. Katriel and R. Pauncz, *Adv. Quantum Chem.*, **10**, 143 (1977).] The following explanation of these results has been proposed [I. Shim and J. P. Dahl, *Theor. Chim. Acta*, **48**, 165 (1978)]. The Pauli-principle "repulsion" between electrons of like spin makes the average angle between the radius vectors of the two electrons larger for the ^3S term than for the ^1S term; this reduces the screening of the nucleus and allows the electrons to get closer to the nucleus in the ^3S term, making the electron–nucleus attraction greater for the ^3S term. [See also R. E. Boyd, *Nature*, **310**, 480 (1984).]

Eigenvalues of Two-Electron Spin Functions.   The helium $1s2s$ configuration gives rise to the term ^3S with degeneracy $(2L + 1)(2S + 1) = 1(3) = 3$ and to the term ^1S with degeneracy $1(1) = 1$. The three helium zeroth-order wave functions (10.27) to (10.29) must correspond to the triply degenerate ^3S term, and the single function (10.30) must correspond to the ^1S term. Since $S = 1$ and $M_S = 1, 0, -1$ for the ^3S term, the three spin functions in (10.27) to (10.29) should be eigenfunctions of $\hat{S}^2$ with eigenvalue $S(S + 1)\hbar^2 = 2\hbar^2$ and eigenfunctions of $\hat{S}_z$ with eigenvalues $M_S\hbar = \hbar$, 0, and $-\hbar$. The spin function in (10.30) should be an eigenfunction of $\hat{S}^2$ and $\hat{S}_z$ with eigenvalue zero in each case, since $S = 0$ and $M_S = 0$ here. We now verify these assertions.

From Eq. (11.43), the total-electron-spin operator is the sum of the spin operators for each electron:

$$\hat{\mathbf{S}} = \hat{\mathbf{S}}_1 + \hat{\mathbf{S}}_2 \tag{11.49}$$

Taking the z components of (11.49), we have

$$\hat{S}_z = \hat{S}_{1z} + \hat{S}_{2z} \tag{11.50}$$

$$
\begin{aligned}
\hat{S}_z \alpha(1)\alpha(2) &= \hat{S}_{1z}\alpha(1)\alpha(2) + \hat{S}_{2z}\alpha(1)\alpha(2) \\
&= \alpha(2)\hat{S}_{1z}\alpha(1) + \alpha(1)\hat{S}_{2z}\alpha(2) \\
&= \tfrac{1}{2}\hbar\alpha(1)\alpha(2) + \tfrac{1}{2}\hbar\alpha(1)\alpha(2) \\
&= \hbar\alpha(1)\alpha(2)
\end{aligned}
\tag{11.51}
$$

where Eq. (10.7) has been used. Similarly, we find

$$\hat{S}_z\beta(1)\beta(2) = -\hbar\beta(1)\beta(2) \tag{11.52}$$

$$\hat{S}_z[\alpha(1)\beta(2) + \beta(1)\alpha(2)] = 0 \tag{11.53}$$

$$\hat{S}_z[\alpha(1)\beta(2) - \beta(1)\alpha(2)] = 0 \tag{11.54}$$

Consider now $\hat{S}^2$. We have [Eq. (11.26)]

$$\hat{S}^2 = (\hat{\mathbf{S}}_1 + \hat{\mathbf{S}}_2) \cdot (\hat{\mathbf{S}}_1 + \hat{\mathbf{S}}_2)$$
$$= \hat{S}_1^2 + \hat{S}_2^2 + 2(\hat{S}_{1x}\hat{S}_{2x} + \hat{S}_{1y}\hat{S}_{2y} + \hat{S}_{1z}\hat{S}_{2z}) \tag{11.55}$$

$$\hat{S}^2\alpha(1)\alpha(2) = \alpha(2)\hat{S}_1^2\alpha(1) + \alpha(1)\hat{S}_2^2\alpha(2) + 2\hat{S}_{1x}\alpha(1)\hat{S}_{2x}\alpha(2)$$
$$+ 2\hat{S}_{1y}\alpha(1)\hat{S}_{2y}\alpha(2) + 2\hat{S}_{1z}\alpha(1)\hat{S}_{2z}\alpha(2)$$

Using Eqs. (10.7) to (10.9) and (10.70) and (10.71), we find

$$\hat{S}^2\alpha(1)\alpha(2) = 2\hbar^2\alpha(1)\alpha(2) \tag{11.56}$$

showing that $\alpha(1)\alpha(2)$ is an eigenfunction of $\hat{S}^2$ corresponding to $S = 1$. Similarly, we find

$$\hat{S}^2\beta(1)\beta(2) = 2\hbar^2\beta(1)\beta(2)$$
$$\hat{S}^2[\alpha(1)\beta(2) + \beta(1)\alpha(2)] = 2\hbar^2[\alpha(1)\beta(2) + \beta(1)\alpha(2)]$$
$$\hat{S}^2[\alpha(1)\beta(2) - \beta(1)\alpha(2)] = 0$$

Thus the spin eigenfunctions in (10.27) to (10.30) correspond to the following values for the total spin quantum numbers:

		S	M_S	
	$\alpha(1)\alpha(2)$	1	1	(11.57)
triplet	$2^{-1/2}[\alpha(1)\beta(2) + \beta(1)\alpha(2)]$	1	0	(11.58)
	$\beta(1)\beta(2)$	1	−1	(11.59)
singlet	$\{2^{-1/2}[\alpha(1)\beta(2) - \beta(1)\alpha(2)]$	0	0	(11.60)

[In the notation of Section 11.4, we are dealing with the addition of two angular momenta with quantum numbers $j_1 = \frac{1}{2}$ and $j_2 = \frac{1}{2}$ to give eigenfunctions with total angular-momentum quantum numbers $J = 1$ and $J = 0$; the coefficients in (11.57) to (11.60) correspond to the coefficients c in (11.33) and are examples of Clebsch–Gordan coefficients.]

Figure 11.3 shows the vector addition of $\mathbf{S}_1$ and $\mathbf{S}_2$ to form $\mathbf{S}$. It might seem surprising that the spin function (11.58), which has the z components of the spins of the two electrons pointing in opposite directions, could have total spin quantum number $S = 1$. Figure 11.3 shows how this is possible.

Atomic Wave Functions.   In Section 10.6, we showed that two of the four zeroth-order wave functions of the $1s2s$ helium configuration could be written as single Slater determinants, but the other two functions had to be expressed as linear combinations of two Slater determinants. Since $\hat{L}^2$ and $\hat{S}^2$ commute with the Hamiltonian (11.1) and with the permutation operator $\hat{P}_{ij}$, the zeroth-order functions should be eigenfunctions of $\hat{L}^2$ and $\hat{S}^2$. The Slater determinants D_2 and D_3 of Section 10.6 are not eigenfunctions of these operators and so are not suitable zeroth-order functions. We have just shown that the linear combinations (10.44) and (10.45) are eigenfunctions of $\hat{S}^2$, and they can also be shown to be eigenfunctions of $\hat{L}^2$.

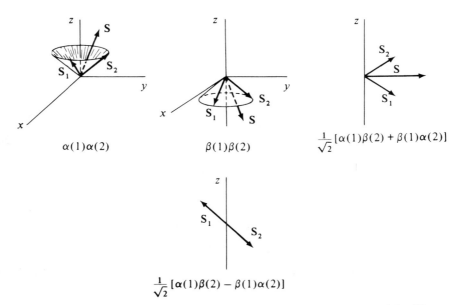

Figure 11.3 Vector addition of the spins of two electrons. For $\alpha(1)\alpha(2)$ and $\beta(1)\beta(2)$, the projections of $\mathbf{S}_1$ and $\mathbf{S}_2$ in the xy plane make an angle of 90° with each other (Problem 11.15c).

For a configuration of closed subshells (for example, the helium ground state), we can write only a single Slater determinant; this determinant is an eigenfunction of $\hat{L}^2$ and $\hat{S}^2$ and is the correct zeroth-order function for the nondegenerate ^1S term. A configuration with one electron outside closed subshells (for example, the boron ground configuration) gives rise to only one term; the Slater determinants for such a configuration differ from one another only in the m and m_s values of this electron and are the correct zeroth-order functions for the states of the term. When all the electrons in singly occupied orbitals have the same spin (either all α or all β), the correct zeroth-order function is a single Slater determinant [for example, see (10.42)]; when this is not true, one has to take a linear combination of Slater determinants to obtain the correct zeroth-order functions, which are eigenfunctions of $\hat{L}^2$ and $\hat{S}^2$. The correct linear combinations can be found by solving the secular equation of degenerate perturbation theory or by operator techniques. Tabulations of the correct combinations for various configurations are available (*Slater, Atomic Structure*, Vol. II). Hartree–Fock calculations of atomic term energies use these linear combinations and find the best possible orbital functions for the Slater determinants.

Each wave function of an atomic term is an eigenfunction of $\hat{L}^2$, $\hat{S}^2$, $\hat{L}_z$, and $\hat{S}_z$. Therefore, when one does a configuration-interaction calculation, only configuration functions that have the same $\hat{L}^2$, $\hat{S}^2$, $\hat{L}_z$, and $\hat{S}_z$ eigenvalues as the state under consideration are included in the expansion (11.16). For example, the helium $1s^2$ ground term is ^1S, which has $L = 0$ and $S = 0$; the electron configuration $1s2p$ produces ^1P and ^3P terms only, and so gives rise to states with $L = 1$

only; no configuration functions arising from the $1s2p$ configuration can occur in the CI wave function (11.16).

Parity of Atomic States.   Consider the atomic Hamiltonian (11.1). We showed in Section 7.5 that the parity operator $\hat{\Pi}$ commutes with the kinetic-energy operator. The quantity $1/r_i$ in (11.1) is

$$r_i^{-1} = (x_i^2 + y_i^2 + z_i^2)^{-1/2}$$

Replacement of each coordinate by its negative leaves $1/r_i$ unchanged. Also

$$r_{ij}^{-1} = [(x_i - x_j)^2 + (y_i - y_j)^2 + (z_i - z_j)^2]^{-1/2}$$

and inversion has no effect on $1/r_{ij}$. Thus $\hat{\Pi}$ commutes with the atomic Hamiltonian, and we can choose atomic wave functions to have definite parity.

For a one-electron atom, the spatial wave function is

$$\psi = R(r) Y_l^m(\theta, \phi)$$

The radial function is unchanged on inversion, and the parity is determined by the angular factor. In Problem 7.21 we showed that Y_l^m is an even function when l is even and is an odd function when l is odd. Thus the states of one-electron atoms have even or odd parity according to whether l is even or odd.

Now consider an n-electron atom. In the Hartree–Fock central-field approximation, we write the wave function as a Slater determinant (or linear combination of Slater determinants) of spin-orbitals. The wave function is the sum of terms, the spatial factor in each term having the form

$$R_1(r_1) \cdots R_n(r_n) Y_{l_1}^{m_1}(\theta_1, \phi_1) \cdots Y_{l_n}^{m_n}(\theta_n, \phi_n)$$

The parity of this product is determined by the spherical-harmonic factors; we see that the product is an even or odd function according to whether $l_1 + l_2 + \cdots + l_n$ is an even or odd number. Therefore, the parity of an atomic state is found by adding the l values of the electrons in the electron configuration that gives rise to the state: if $\sum_i l_i$ is an even number, then ψ is an even function; if $\sum_i l_i$ is odd, ψ is odd. For example, the configuration $1s^2 2s 2p^3$ has $\sum_i l_i = 0 + 0 + 0 + 1 + 1 + 1 = 3$, and all states arising from this electron configuration have odd parity. (Our argument was based on the SCF approximation to ψ, but the conclusions are valid for the true ψ.)

Total Electronic Angular Momentum and Atomic Levels.   The total electronic angular momentum $\mathbf{J}$ of an atom is the vector sum of the total electronic orbital and spin angular momenta:

$$\mathbf{J} = \mathbf{L} + \mathbf{S} \qquad (11.61)^*$$

The operator $\hat{\mathbf{J}}$ for the total electronic angular momentum commutes with the atomic Hamiltonian, and we may characterize an atomic state by a quantum number J, which has the possible values [Eq. (11.39)]

$$L + S, \quad L + S - 1, \quad \ldots, \quad |L - S| \qquad (11.62)$$

We have $\hat{J}^2 \psi = J(J + 1)\hbar^2 \psi$.

For the atomic Hamiltonian (11.1), all states that belong to the same term have the same energy. However, when spin–orbit interaction (Section 11.6) is included in $\hat{H}$, one finds that the value of J affects the energy slightly. Hence states that belong to the same term but that have different values of J will have slightly different energies. The set of states that belong to the same term and that have the same value of J constitute an atomic **level**. The energies of different levels belonging to the same term are slightly different. (See Fig. 11.6.) To denote the level, one adds the J value as a right subscript on the term symbol. Each level is $(2J + 1)$-fold degenerate, corresponding to the $2J + 1$ values of M_J, where $M_J\hbar$ is the z component of the total electronic angular momentum **J**. Each level consists of $2J + 1$ **states** of equal energy.

EXAMPLE   Find the levels of a ^3P term and give the degeneracy of each level.

For a ^3P term, $2S + 1 = 3$ and $S = 1$; also, from (11.42), $L = 1$. With $L = 1$ and $S = 1$, (11.62) gives $J = 2, 1, 0$. The levels are $^3P_2$, $^3P_1$, and $^3P_0$.

The $^3P_2$ level has $J = 2$ and has $2J + 1 = 5$ values of M_J, namely, -2, -1, 0, 1, and 2. The $^3P_2$ level is 5-fold degenerate. The $^3P_1$ level is $2(1) + 1 = 3$-fold degenerate. The $^3P_0$ level has $2J + 1 = 1$ state and is nondegenerate.

The total number of states for the three levels $^3P_2$, $^3P_1$, and $^3P_0$ is $5 + 3 + 1$. When spin–orbit interaction was neglected, we had a ^3P term that consisted of $(2L + 1)(2S + 1) = 3(3) = 9$ equal-energy states. With spin–orbit interaction, the 9-fold-degenerate term splits into three closely spaced levels: $^3P_2$ with 5 states, $^3P_1$ with 3 states, and $^3P_0$ with one state. The total number of states is the same for the term ^3P as for the three levels arising from this term.

The quantity $2S + 1$ is called the **multiplicity** of the term. If $L \geqslant S$, the possible values of J in (11.62) range from $L + S$ to $L - S$ and are $2S + 1$ in number; if $L \geqslant S$, the multiplicity gives the number of levels that arise from a given term. For $L < S$, the values of J range from $S + L$ to $S - L$ and are $2L + 1$ in number; in this case, the multiplicity is greater than the number of levels. For example, if $L = 0$ and $S = 1$ (a ^3S term), the multiplicity is 3, but there is only one possible value for J, namely, $J = 1$. For $2S + 1 = 1, 2, 3, 4, 5, 6, \ldots$, the words **singlet, doublet, triplet, quartet, quintet, sextet,** $\ldots$, are used to designate the multiplicity. The level symbol $^3P_1$ is read as "triplet P one."

For light atoms, the spin–orbit interaction is very small and the separation between levels of a term is very small. Note from Fig. 11.6 in Section 11.7 that the separations between the $^3P_0$, $^3P_1$, and $^3P_2$ levels of the helium $1s2p$ ^3P term are far, far less than the separation between the ^1P and ^3P terms of the $1s2p$ configuration.

Terms and Levels of Hydrogen and Helium.   The hydrogen atom has one electron. Hence $L = l$ and $S = s = \frac{1}{2}$. The possible values of J are $L + \frac{1}{2}$ and $L - \frac{1}{2}$, except for $L = 0$, where $J = \frac{1}{2}$ is the only possibility. Each electron configuration gives rise to only one term, which is composed of one level if $L = 0$ and two levels if $L \neq 0$. The ground-state configuration $1s$ gives the term ^2S, which is composed of the single level $^2S_{1/2}$; the level is twofold degenerate ($M_J = -\frac{1}{2}, \frac{1}{2}$). The $2s$ configuration also gives a $^2S_{1/2}$ level. The $2p$ configuration gives rise to the levels $^2P_{3/2}$ and $^2P_{1/2}$; the $^2P_{3/2}$ level is fourfold degenerate, and the $^2P_{1/2}$ level is twofold degenerate. There are $2 + 4 + 2 = 8$ states with $n = 2$, in agreement with our previous work.

The helium ground-state configuration $1s^2$ is a closed subshell and gives rise to the single level $^1S_0$, which is nondegenerate ($M_J = 0$). The $1s2s$ excited configuration gives rise to the two terms ^1S and ^3S, each of which has one level. The $^1S_0$ level is nondegenerate; the $^3S_1$ level is threefold degenerate. The $1s2p$ configuration gives rise to the terms ^1P and ^3P. The levels of ^3P are $^3P_2$, $^3P_1$, and $^3P_0$; ^1P has the single level $^1P_1$.

Tables of Atomic Energy Levels.   The spectroscopically determined energy levels for atoms with atomic number less than 90 are given in the tables of C. E. Moore and others: C. E. Moore, *Atomic Energy Levels*, National Bureau of Standards Circular 467, vols. I, II, and III, 1949, 1952, and 1958, Washington D.C.; these have been reprinted as Natl. Bur. Stand. Publ. NSRDS-NBS 35, 1971; W. C. Martin et al., *Atomic Energy Levels—The Rare-Earth Elements*, Natl. Bur. Stand. Publ. NSRDS-NBS 60, Washington, D.C., 1978. Revisions of Moore's tables are found in C. E. Moore, Natl. Bur. Stand. Publ. NSRDS-NBS 3, Sections 1–11, 1965–1985 and in papers in the *Journal of Physical and Chemical Reference Data*.

These tables also list the levels of many atomic ions. Spectroscopists use the symbol I to indicate a neutral atom, the symbol II to indicate a singly ionized atom, and so on. Thus C III refers to the C^{2+} ion.

The tables take the zero level of energy at the lowest energy level of the atom and list the level energies E_i as E_i/hc in cm^{-1}, where h and c are Planck's constant and the speed of light. The difference in E/hc values for two levels gives the wave number [Eq. (4.72)] of the spectral transition between the levels (provided the transition is allowed). An energy E of 1 eV corresponds to $E/hc = 8065.5$ cm^{-1} (Problem 11.27).

Figure 11.4 shows some of the term energies of the carbon atom. The separations between levels of each term are too small to be visible in this figure.

11.6 SPIN–ORBIT INTERACTION

The atomic Hamiltonian (11.1) does not involve electron spin. In reality, the existence of spin adds an additional term, usually small, to the Hamiltonian. This term, called the ***spin–orbit interaction***, splits an atomic term into levels. Spin–orbit interaction is a relativistic effect and is properly derived using Dirac's relativistic treatment of the electron. This section gives a qualitative discussion of the origin of spin–orbit interaction.

If we imagine ourselves riding on an electron in an atom, from our viewpoint, the nucleus is moving around the electron (as the sun appears to move around the earth). This apparent motion of the nucleus produces a magnetic field that interacts with the intrinsic (spin) magnetic moment of the electron, giving the spin–orbit interaction term in the Hamiltonian. The interaction energy of a magnetic moment $\mathbf{\mu}$ with a magnetic field $\mathbf{B}$ is given by (6.131) as $-\mathbf{\mu} \cdot \mathbf{B}$. The

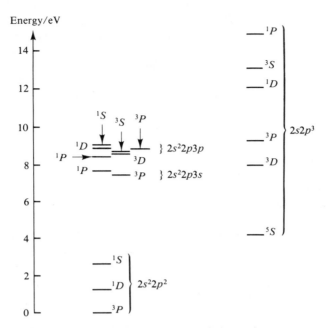

Figure 11.4 Some term energies of the carbon atom.

electron's spin magnetic moment $\boldsymbol{\mu}_S$ is proportional to its spin $\mathbf{S}$ [Eq. (10.57)], and the magnetic field arising from the apparent nuclear motion is proportional to the electron's orbital angular momentum $\mathbf{L}$. Therefore, the spin–orbit interaction is proportional to $\mathbf{L} \cdot \mathbf{S}$. The dot product of $\mathbf{L}$ and $\mathbf{S}$ depends on the relative orientation of these two vectors. The total electronic angular momentum $\mathbf{J} = \mathbf{L} + \mathbf{S}$ also depends on the relative orientation of $\mathbf{L}$ and $\mathbf{S}$, and so the spin–orbit interaction energy depends on J [Eq. (11.67)].

When a proper relativistic derivation of the spin–orbit-interaction term $\hat{H}_{\text{S.O.}}$ in the atomic Hamiltonian is carried out, one finds that for a one-electron atom (see *Bethe and Jackiw*, Chapters 8 and 23)

$$\hat{H}_{\text{S.O.}} = \frac{1}{2m_ec^2} \frac{1}{r} \frac{dV}{dr} \hat{\mathbf{L}} \cdot \hat{\mathbf{S}} \tag{11.63}$$

where V is the potential energy experienced by the electron in the atom. One way to get $\hat{H}_{\text{S.O.}}$ for a many-electron atom is to first neglect $\hat{H}_{\text{S.O.}}$ and carry out an SCF calculation (Section 11.1) using the central-field approximation to get an effective potential energy $V_i(r_i)$ for each electron i in the field of the nucleus and the other electrons viewed as charge clouds [Eqs. (11.7) and (11.8)]. One then sums (11.63) over the electrons to get

$$\hat{H}_{\text{S.O.}} \approx \frac{1}{2m_ec^2} \sum_i \frac{1}{r_i} \frac{dV_i(r_i)}{dr_i} \hat{\mathbf{L}}_i \cdot \hat{\mathbf{S}}_i = \sum_i \xi_i(r_i) \hat{\mathbf{L}}_i \cdot \hat{\mathbf{S}}_i \tag{11.64}$$

where the definition of $\xi_i(r_i)$ is obvious and $\hat{\mathbf{L}}_i$ and $\hat{\mathbf{S}}_i$ are the operators for orbital and spin angular momenta of electron i.

Calculating the spin–orbit interaction energy $E_{s.o.}$ by finding the eigenfunctions and eigenvalues of the operator $\hat{H}_{(11.1)} + \hat{H}_{s.o.}$, where $\hat{H}_{(11.1)}$ is the Hamiltonian of Eq. (11.1), is difficult. One therefore usually estimates $E_{s.o.}$ by using perturbation theory. Except for heavy atoms, the effect of $\hat{H}_{s.o.}$ is small compared with the effect of $\hat{H}_{(11.1)}$, and first-order perturbation theory can be used to estimate $E_{s.o.}$.

Equation (9.22) gives $E_{s.o.} \approx \langle \psi | \hat{H}_{s.o.} | \psi \rangle$, where ψ is an eigenfunction of $\hat{H}_{(11.1)}$. For a one-electron atom,

$$E_{s.o.} \approx \langle \psi | \xi(r)\hat{\mathbf{L}} \cdot \hat{\mathbf{S}} | \psi \rangle \tag{11.65}$$

We have

$$\mathbf{J} \cdot \mathbf{J} = (\mathbf{L} + \mathbf{S}) \cdot (\mathbf{L} + \mathbf{S}) = L^2 + S^2 + 2\mathbf{L} \cdot \mathbf{S}$$

$$\mathbf{L} \cdot \mathbf{S} = \tfrac{1}{2}(J^2 - L^2 - S^2)$$

$$\hat{\mathbf{L}} \cdot \hat{\mathbf{S}} \psi = \tfrac{1}{2}(\hat{J}^2 - \hat{L}^2 - \hat{S}^2)\psi = \tfrac{1}{2}[J(J+1) - L(L+1) - S(S+1)]\hbar^2 \psi$$

since the unperturbed ψ is an eigenfunction of $\hat{L}^2$, $\hat{S}^2$, and $\hat{J}^2$. Therefore,

$$E_{s.o.} \approx \tfrac{1}{2}\langle \xi \rangle \hbar^2 [J(J+1) - L(L+1) - S(S+1)] \tag{11.66}$$

For a many-electron atom, it can be shown (*Bethe and Jackiw*, p. 164) that the spin–orbit interaction energy is

$$E_{s.o.} \approx \tfrac{1}{2}A\hbar^2 [J(J+1) - L(L+1) - S(S+1)] \tag{11.67}$$

where A is a constant for a given term; that is, A depends on L and S but not on J. Equation (11.67) shows that when we include the spin–orbit interaction, the energy of an atomic state depends on its total electronic angular momentum J. Thus each atomic term is split into levels, each level having a different value of J. For example, the $1s^2 2s^2 2p^6 3p$ configuration of sodium has the single term ^2P, which is composed of the two levels $^2P_{3/2}$ and $^2P_{1/2}$. The splitting of these levels gives the observed *fine structure* of the sodium D line (Fig. 11.5). The levels of a given term are said to form its *multiplet structure*.

What about the order of levels within a given term? Since L and S are the

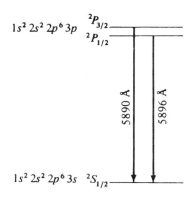

Figure 11.5 Fine structure of the sodium D line.

same for such levels, their relative energies are determined, according to Eq. (11.67), by $AJ(J + 1)$. If A is positive, the level with the lowest value of J lies lowest, and the multiplet is said to be **regular**; if A is negative, the level with the highest value of J lies lowest, and the multiplet is said to be **inverted**. The following rule usually applies to a configuration with only one partly filled subshell: *If this subshell is less than half filled, the multiplet is regular; if this subshell is more than half filled, the multiplet is inverted.* (A few exceptions exist.) For the half-filled case, see Problem 11.26.

EXAMPLE   Find the ground level of the oxygen atom.

The ground electron configuration is $1s^2 2s^2 2p^4$. Table 11.2 gives ^1S, ^1D, and ^3P as the terms of this configuration. By Hund's rule, ^3P is the lowest term. [Alternatively, a diagram like (11.48) could be used to conclude that ^3P is the lowest term.] The ^3P term has $L = 1$ and $S = 1$, so the possible J values are 2, 1, and 0. The levels of ^3P are $^3P_2$, $^3P_1$, and $^3P_0$. The $2p$ subshell is more than half filled, so the rule just given predicts the multiplet is inverted, and the $^3P_2$ level lies lowest. This is the ground level of O.

11.7   THE ATOMIC HAMILTONIAN

The Hamiltonian of an atom can be divided into three parts:

$$\hat{H} = \hat{H}^0 + \hat{H}_{\text{rep}} + \hat{H}_{\text{S.O.}} \tag{11.68}$$

where $\hat{H}^0$ is the sum of hydrogenlike Hamiltonians,

$$\hat{H}^0 = \sum_{i=1}^{n} \left(-\frac{\hbar^2}{2m_e} \nabla_i^2 - \frac{Ze'^2}{r_i} \right) \tag{11.69}$$

$\hat{H}_{\text{rep}}$ consists of the interelectronic repulsions,

$$\hat{H}_{\text{rep}} = \sum_i \sum_{j>i} \frac{e'^2}{r_{ij}} \tag{11.70}$$

and $\hat{H}_{\text{S.O.}}$ is the spin–orbit interaction,

$$\hat{H}_{\text{S.O.}} = \sum_{i=1}^{n} \xi_i \hat{\mathbf{L}}_i \cdot \hat{\mathbf{S}}_i \tag{11.71}$$

If we consider just $\hat{H}^0$, all atomic states corresponding to the same electronic *configuration* are degenerate. Adding in $\hat{H}_{\text{rep}}$, we lift the degeneracy between states with different L or S or both, thus splitting each configuration into *terms*. Next, we add in $\hat{H}_{\text{S.O.}}$, which splits each term into *levels*. Each level is composed of *states* with the same value of J and is $(2J + 1)$-fold degenerate, corresponding to the possible values of M_J.

We can remove the degeneracy of each level by applying an external magnetic field (the *Zeeman effect*). If **B** is the applied field, we have the additional term in the Hamiltonian [Eq. (6.131)]

$$\hat{H}_B = -\hat{\boldsymbol{\mu}} \cdot \mathbf{B} = -(\hat{\boldsymbol{\mu}}_L + \hat{\boldsymbol{\mu}}_S) \cdot \mathbf{B} \tag{11.72}$$

where both the orbital and spin magnetic moments have been included. Using $\boldsymbol{\mu}_L = -(e/2m_e)\mathbf{L}$, $\boldsymbol{\mu}_S = -(e/m_e)\mathbf{S}$, and $\beta_e \equiv e\hbar/2m_e$ [Eqs. (6.128), (10.57), and (6.130)], we have

$$\hat{H}_B = \beta_e \hbar^{-1}(\hat{\mathbf{L}} + 2\hat{\mathbf{S}}) \cdot \mathbf{B} = \beta_e \hbar^{-1}(\hat{\mathbf{J}} + \hat{\mathbf{S}}) \cdot \mathbf{B} = \beta_e B \hbar^{-1}(\hat{J}_z + \hat{S}_z) \quad (11.73)$$

where β_e is the Bohr magneton and the z axis is taken along the direction of the field. If the external field is reasonably weak, its effect will be less than that of the spin–orbit interaction, and the effect of the field can be calculated by use of first-order perturbation theory. Since $\hat{J}_z \psi = M_J \hbar \psi$, the energy of interaction with the applied field is

$$E_B = \langle \psi | \hat{H}_B | \psi \rangle = \beta_e B M_J + \beta_e B \hbar^{-1} \langle S_z \rangle$$

Evaluation of $\langle S_z \rangle$ (*Bethe and Jackiw*, p. 169) gives as the final result:

$$E_B = \beta_e g B M_J \quad (11.74)$$

where g (the *Landé g factor*) is given by

$$g = 1 + \frac{[J(J+1) - L(L+1) + S(S+1)]}{2J(J+1)} \quad (11.75)$$

Thus the external field splits each level into $2J + 1$ states, each state having a different value of M_J.

Figure 11.6 illustrates what happens when we consider successive interactions in an atom, using the $1s2p$ configuration of helium as the example.

We have based the discussion on a scheme in which we first added the individual electronic orbital angular momenta to form a total-orbital-angular-momentum vector and did the same for the spins: $\mathbf{L} = \sum_i \mathbf{L}_i$ and $\mathbf{S} = \sum_i \mathbf{S}_i$. We then combined $\mathbf{L}$ and $\mathbf{S}$ to get $\mathbf{J}$. This scheme is called **Russell–Saunders coupling** (or *L–S coupling*) and is appropriate where the spin–orbit interaction energy is small compared with the interelectronic repulsion energy. The operators $\hat{\mathbf{L}}$ and $\hat{\mathbf{S}}$ commute with $\hat{H}^0 + \hat{H}_{\text{rep}}$, but when $\hat{H}_{\text{S.O.}}$ is included in the Hamiltonian, $\hat{\mathbf{L}}$ and $\hat{\mathbf{S}}$ no longer commute with $\hat{H}$. ($\hat{\mathbf{J}}$ does commute with $\hat{H}^0 + \hat{H}_{\text{rep}} + \hat{H}_{\text{S.O.}}$.) If the spin–orbit interaction is small, then $\hat{\mathbf{L}}$ and $\hat{\mathbf{S}}$ "almost" commute with $\hat{H}$, and L–S coupling is valid.

As the atomic number increases, the average speed v of the electrons increases; as v/c increases, relativistic effects such as the spin–orbit interaction increase. For atoms with very high atomic number, the spin–orbit interaction exceeds the interelectronic repulsion energy, and we can no longer consider $\hat{\mathbf{L}}$ and $\hat{\mathbf{S}}$ to commute with $\hat{H}$; the operator $\hat{\mathbf{J}}$, however, still commutes with $\hat{H}$. In this case we first add in $\hat{H}_{\text{S.O.}}$ to $\hat{H}^0$ and then consider $\hat{H}_{\text{rep}}$. This corresponds to first combining the spin and orbital angular momenta of each electron to give a total angular momentum $\mathbf{j}_i$ for each electron: $\mathbf{j}_i = \mathbf{L}_i + \mathbf{S}_i$. We then add the $\mathbf{j}_i$ to get the total electronic angular momentum: $\mathbf{J} = \sum_i \mathbf{j}_i$. This scheme is called *j–j coupling*. For most heavy atoms, the situation is intermediate between *j–j* and *L–S* coupling, and computations are difficult.

Several other effects should be included in the atomic Hamiltonian. The finite size of the nucleus and the effect of nuclear motion slightly change the

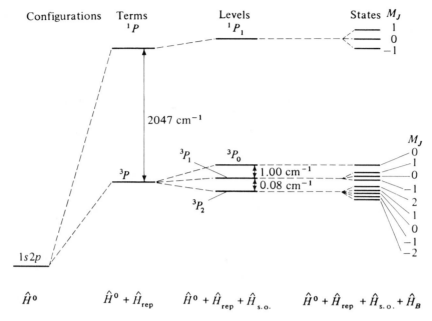

Configurations   Terms   Levels   States M_J

$\hat{H}^0$ $\hat{H}^0 + \hat{H}_{\text{rep}}$ $\hat{H}^0 + \hat{H}_{\text{rep}} + \hat{H}_{\text{s.o.}}$ $\hat{H}^0 + \hat{H}_{\text{rep}} + \hat{H}_{\text{s.o.}} + \hat{H}_B$

Figure 11.6 Effect of inclusion of successive terms in the atomic Hamiltonian for the $1s2p$ helium configuration. $\hat{H}_B$ is not part of the atomic Hamiltonian but is due to an applied magnetic field.

energy (*Bethe and Salpeter*, pp. 102, 166, 351). There is a small relativistic term due to the interaction between the spin magnetic moments of the electrons (*spin–spin interaction*). We should also take into account the relativistic change of electronic mass with velocity; this effect is of appreciable magnitude for inner-shell electrons of heavy atoms, where average electronic speeds are not negligible in comparison with the speed of light.

   If the nucleus has a nonzero spin, the nuclear spin magnetic moment interacts with the electronic spin and orbital magnetic moments, giving rise to atomic *hyperfine structure*. The *nuclear spin angular momentum* **I** adds vectorially to the total electronic angular momentum **J** to give the *total angular momentum* **F** of the atom: $\mathbf{F} = \mathbf{I} + \mathbf{J}$. For example, consider the ground state of the hydrogen atom. The spin of a proton is $\frac{1}{2}$, so $I = \frac{1}{2}$; also, $J = \frac{1}{2}$. Hence the quantum number F can be 0 or 1, corresponding to the proton and electron spins being antiparallel or parallel. The transition $F = 1 \rightarrow 0$ gives a line at 1420 MHz, the famous 21-cm line emitted by hydrogen atoms in outer space. In 1951, Ewen and Purcell (who is also codiscoverer of nuclear magnetic resonance) stuck a horn-shaped antenna out the window of a Harvard physics laboratory and detected this line. The frequency of the hyperfine splitting in the ground state of hydrogen is one of the most accurately measured physical constants: $1420.405751767 \pm 0.000000001$ MHz [L. Essen et al., *Nature*, **229**, 110 (1971)].

11.8   THE CONDON–SLATER RULES

In the Hartree–Fock approximation, the wave function of an atom (or molecule) is a Slater determinant or a linear combination of a few Slater determinants [for example, Eq. (10.44)]. A configuration-interaction wave function such as (11.16) is a linear combination of many Slater determinants. To evaluate the energy and other properties of atoms and molecules using Hartree–Fock or configuration-interaction wave functions, we must be able to evaluate integrals of the form $\langle D'|\hat{A}|D\rangle$, where D and D' are Slater determinants of orthonormal spin-orbitals and $\hat{A}$ is an operator.

Each spin-orbital u_i is a product of a spatial orbital θ_i and a spin function σ_i, where σ_i is either α or β. We have $u_i = \theta_i \sigma_i$ and $\langle u_i(1)|u_j(1)\rangle = \delta_{ij}$, where $\langle u_i(1) u_j(1)\rangle$ involves a sum over the spin coordinate of electron 1 and an integration over its spatial coordinates. If u_i and u_j have different spin functions, then (10.12) ensures the orthogonality of u_i and u_j; if u_i and u_j have the same spin function, their orthogonality is due to the orthogonality of the spatial orbitals θ_i and θ_j.

For an n-electron system, D is

$$D = \frac{1}{\sqrt{n!}} \begin{vmatrix} u_1(1) & u_2(1) & \cdots & u_n(1) \\ u_1(2) & u_2(2) & \cdots & u_n(2) \\ \cdots & \cdots & \cdots & \cdots \\ \cdots & \cdots & \cdots & \cdots \\ \cdots & \cdots & \cdots & \cdots \\ u_1(n) & u_2(n) & \cdots & u_n(n) \end{vmatrix} \tag{11.76}$$

An example with $n = 3$ is Eq. (10.40). D' has the same form as D except that $u_1, u_2, \ldots, u_n$ are replaced by $u_1', u_2', \ldots, u_n'$.

We shall assume that the columns of D and D' are arranged so as to have as many as possible of their left-hand columns match. For example, if we were working with the Slater determinants $|1s\overline{1s}2s3p_0|$ and $|1s\overline{1s}3p_04s|$, we would interchange the third and fourth columns of the first determinant (thereby multiplying it by -1) and let $D = |1s\overline{1s}3p_02s|$ and $D' = |1s\overline{1s}3p_04s|$.

The operator $\hat{A}$ typically has the form

$$\hat{A} = \sum_{i=1}^{n} \hat{f}_i + \sum_{i=1}^{n-1} \sum_{j>i} \hat{g}_{ij} \tag{11.77}$$

where the *one-electron operator* $\hat{f}_i$ involves only coordinate and momentum operators of electron i and the *two-electron operator* $\hat{g}_{ij}$ involves coordinate and momentum operators of electrons i and j. For example, if $\hat{A}$ is the atomic Hamiltonian operator (11.1), then $\hat{f}_i = -(\hbar^2/2m_e)\nabla_i^2 - Ze'^2/r_i$ and $\hat{g}_{ij} = e'^2/r_{ij}$.

Condon and Slater showed that the n-electron integral $\langle D'|\hat{A}|D\rangle$ can be reduced to sums of certain one- and two-electron integrals. The derivation of these Condon–Slater formulas uses the determinant expression of Problem 8.16 together with the orthonormality of the spin-orbitals. (See *Parr*, pp. 23–27 for the derivation.) Table 11.3 gives the Condon–Slater formulas.

In Table 11.3, each matrix element of $\hat{g}_{12}$ involves summation over the spin

[handwritten: Summoves spin & space]

TABLE 11.3 The Condon–Slater Rules

D and D' differ by	$\left\langle D' \left\| \sum_{i=1}^{n} \hat{f}_i \right\| D \right\rangle$	$\left\langle D' \left\| \sum_{i=1}^{n-1} \sum_{j>i} \hat{g}_{ij} \right\| D \right\rangle$
no spin-orbitals	$\sum_{i=1}^{n} \langle u_i(1)|\hat{f}_1|u_i(1)\rangle$	$\sum_{i=1}^{n-1} \sum_{j>i} [\langle u_i(1)u_j(2)|\hat{g}_{12}|u_i(1)u_j(2)\rangle$ $- \langle u_i(1)u_j(2)|\hat{g}_{12}|u_j(1)u_i(2)\rangle]$
one spin-orbital $u'_n \neq u_n$	$\langle u'_n(1)|\hat{f}_1|u_n(1)\rangle$	$\sum_{j=1}^{n-1} [\langle u'_n(1)u_j(2)|\hat{g}_{12}|u_n(1)u_j(2)\rangle$ $- \langle u'_n(1)u_j(2)|\hat{g}_{12}|u_j(1)u_n(2)\rangle]$
two spin-orbitals $u'_n \neq u_n,$ $u'_{n-1} \neq u_{n-1}$	0	$\langle u'_n(1)u'_{n-1}(2)|\hat{g}_{12}|u_n(1)u_{n-1}(2)\rangle$ $- \langle u'_n(1)u'_{n-1}(2)|\hat{g}_{12}|u_{n-1}(1)u_n(2)\rangle$
three or more spin-orbitals	0	0

coordinates of electrons 1 and 2 and integration over the full range of the spatial coordinates of electrons 1 and 2; each matrix element of $\hat{f}_1$ involves summation over the spin coordinate of electron 1 and integration over its spatial coordinates. The variables in the sums and definite integrals are dummy variables.

If the operators $\hat{f}_i$ and $\hat{g}_{ij}$ do not involve spin, the expressions in Table 11.3 can be further simplified. We have $u_i = \theta_i \sigma_i$ and

$$\langle u_i(1)|\hat{f}_1|u_i(1)\rangle = \int \theta_i^*(1)\hat{f}_1\theta_i(1)\, dv_1 \sum_{m_{s1}} \sigma_i^*(1)\sigma_i(1)$$

$$= \int \theta_i^*(1)\hat{f}_1\theta_i(1)\, dv_1 = \langle \theta_i(1)|\hat{f}_1|\theta_i(1)\rangle$$

since σ_i is normalized. Using this result and the orthonormality of σ_i and σ_j, we get for the case $D = D'$ (Problem 11.34)

$$\left\langle D \left| \sum_{i=1}^{n} \hat{f}_i \right| D \right\rangle = \sum_{i=1}^{n} \langle \theta_i(1)|\hat{f}_1|\theta_i(1)\rangle \tag{11.78}$$

$$\left\langle D \left| \sum_{i=1}^{n-1} \sum_{j>i} \hat{g}_{ij} \right| D \right\rangle = \sum_{i=1}^{n-1} \sum_{j>i} [\langle \theta_i(1)\theta_j(2)|\hat{g}_{12}|\theta_i(1)\theta_j(2)\rangle$$

$$- \delta_{m_{s,i}m_{s,j}} \langle \theta_i(1)\theta_j(2)|\hat{g}_{12}|\theta_j(1)\theta_i(2)\rangle] \tag{11.79}$$

where $\delta_{m_{s,i}m_{s,j}}$ is 0 or 1 according to whether $m_{s,i} \neq m_{s,j}$ or $m_{s,i} = m_{s,j}$. Similar equations hold for the other integrals.

Let us apply these equations to evaluate $\langle D|\hat{H}|D\rangle$, where $\hat{H}$ is the Hamiltonian of an n-electron atom with spin–orbit interaction neglected and where D is a Slater determinant of n spin-orbitals. We have $\hat{H} = \sum_i \hat{f}_i + \sum_i \sum_{j>i} \hat{g}_{ij}$, where $\hat{f}_i = -(\hbar^2/2m_e)\nabla_i^2 - Ze'^2/r_i$ and $\hat{g}_{ij} = e'^2/r_{ij}$. Introducing the

Coulomb and exchange integrals of Eq. (9.102) and using (11.78) and (11.79), we have

$$\langle D|\hat{H}|D\rangle = \sum_{i=1}^{n} \langle \theta_i(1)|\hat{f}_1|\theta_i(1)\rangle + \sum_{i=1}^{n-1}\sum_{j>i} (J_{ij} - \delta_{m_{s,i}m_{s,j}}K_{ij}) \qquad (11.80)$$

$$J_{ij} = \langle \theta_i(1)\theta_j(2)|e'^2/r_{12}|\theta_i(1)\theta_j(2)\rangle$$
$$K_{ij} = \langle \theta_i(1)\theta_j(2)|e'^2/r_{12}|\theta_j(1)\theta_i(2)\rangle \qquad (11.81)$$

$$\hat{f}_1 = -(\hbar^2/2m_e)\nabla_1^2 - Ze'^2/r_1 \qquad (11.82)$$

The Kronecker delta in (11.80) results from the orthonormality of the one-electron spin functions.

As an example, consider Li. The SCF approximation to the ground-state ψ is the Slater determinant $D = |1s\overline{1s}2s|$. The spin-orbitals are $u_1 = 1s\alpha$, $u_2 = 1s\beta$, and $u_3 = 2s\alpha$. The spatial orbitals are $\theta_1 = 1s$, $\theta_2 = 1s$, and $\theta_3 = 2s$. We have $J_{12} = J_{1s1s}$ and $J_{13} = J_{23} = J_{1s2s}$. Since $m_{s1} \neq m_{s2}$ and $m_{s2} \neq m_{s3}$, the only exchange integral that appears in the energy expression is $K_{13} = K_{1s2s}$. We get exchange integrals only between spin-orbitals with the same spin. Equation (11.80) gives the SCF energy as

$$E = \langle D|\hat{H}|D\rangle = 2\langle 1s(1)|\hat{f}_1|1s(1)\rangle + \langle 2s(1)|\hat{f}_1|2s(1)\rangle + J_{1s1s} + 2J_{1s2s} - K_{1s2s}$$

The terms involving $\hat{f}_1$ are hydrogenlike energies, and their sum equals $E^{(0)}$ in Eq. (10.48). The remaining terms equal $E^{(1)}$ in Eq. (10.53). As noted at the beginning of Section 9.4, $E^{(0)} + E^{(1)}$ equals the variational integral $\langle D|\hat{H}|D\rangle$, so the Condon–Slater rules have been checked in this case.

For an atom with closed subshells only (for example, ground-state Be with a $1s^2 2s^2$ configuration), the n electrons reside in $n/2$ different spatial orbitals, so that $\theta_1 = \theta_2$, $\theta_3 = \theta_4$, and so on. Let $\phi_1 \equiv \theta_1 = \theta_2$, $\phi_2 \equiv \theta_3 = \theta_4$, ..., $\phi_{n/2} \equiv \theta_{n-1} = \theta_n$. If one rewrites Eq. (11.80) using the ϕ's instead of the θ's, one finds (Problem 11.35) for the SCF energy of the ^1S term produced by a closed-subshell configuration

$$E = \langle D|\hat{H}|D\rangle = 2\sum_{i=1}^{n/2} \langle \phi_i(1)|\hat{f}_1|\phi_i(1)\rangle + \sum_{j=1}^{n/2}\sum_{i=1}^{n/2} (2J_{ij} - K_{ij}) \qquad (11.83)$$

where $\hat{f}_1$ is given by (11.82) and where J_{ij} and K_{ij} have the forms in (11.81) but with θ_i and θ_j replaced by ϕ_i and ϕ_j. Each sum in (11.83) goes over all the $n/2$ different spatial orbitals.

For example, consider the $1s^2 2s^2$ configuration. We have $n = 4$ and the two different spatial orbitals are $1s$ and $2s$. The double sum in (11.83) equals $2J_{1s1s} - K_{1s1s} + 2J_{1s2s} - K_{1s2s} + 2J_{2s1s} - K_{2s1s} + 2J_{2s2s} - K_{2s2s}$. From the definition (11.81), it follows that $J_{ii} = K_{ii}$. The labels 1 and 2 in (11.81) are dummy variables, and interchanging them can have no effect on the integrals. Interchanging 1 and 2 in J_{ij} converts it to J_{ji}; therefore, $J_{ij} = J_{ji}$. The same reasoning gives $K_{ij} = K_{ji}$. Thus

$$J_{ii} = K_{ii}, \qquad J_{ij} = J_{ji}, \qquad K_{ij} = K_{ji} \qquad (11.84)$$

Use of (11.84) gives the Coulomb- and exchange-integrals expression for the

$1s^2 2s^2$ configuration as $J_{1s1s} + J_{2s2s} + 4J_{1s2s} - 2K_{1s2s}$. Between the two electrons in the $1s$ orbital, there is only one Coulombic interaction, and we get the term J_{1s1s}. Each $1s$ electron interacts with two $2s$ electrons, for a total of four $1s$–$2s$ interactions, and we get the term $4J_{1s2s}$. As noted earlier, exchange integrals occur only between spin-orbitals of the same spin. There is an exchange integral between the $1s\alpha$ and $2s\alpha$ spin-orbitals and an exchange integral between the $1s\beta$ and $2s\beta$ spin-orbitals, which gives the $-2K_{1s2s}$ term.

The magnitude of the exchange integrals is generally much less than the magnitude of the Coulomb integrals [for example, see Eq. (9.113)].

11.9 SUMMARY

The Hartree SCF method approximates the atomic wave function as a product of one-electron spatial orbitals [Eq. (11.2)] and finds the best possible forms for the orbitals by an iterative calculation in which each electron is assumed to move in the field produced by the nucleus and a hypothetical charge cloud due to the other electrons.

The more accurate Hartree–Fock method approximates the wave function as an antisymmetrized product (Slater determinant or determinants) of one-electron spin-orbitals and finds the best possible forms for the spatial orbitals in the spin-orbitals. Hartree–Fock calculations are usually done by expanding each orbital as a linear combination of basis functions and iteratively solving the Hartree–Fock equations (11.12). The Slater-type orbitals (11.14) are often used as the basis functions in atomic calculations. The difference between the exact nonrelativistic energy and the Hartree–Fock energy is the correlation energy of the atom (or molecule).

To go beyond the Hartree–Fock approximation and approach the true wave function and energy, we can use configuration interaction (CI), expressing ψ as a linear combination of configuration functions corresponding to various electron configurations [Eq. (11.16)].

Let $\mathbf{M}_1$ and $\mathbf{M}_2$ be two angular momenta with quantum numbers j_1, m_1 and j_2, m_2, and let $\mathbf{M}$ be their sum: $\mathbf{M} = \mathbf{M}_1 + \mathbf{M}_2$. We showed that, for the angular-momentum sum, $\hat{M}^2$ has eigenvalues $J(J+1)\hbar^2$ and $\hat{M}_z$ has eigenvalues $M_J\hbar$, where the possible values of J and M_J are $J = j_1 + j_2, j_1 + j_2 - 1, \ldots, |j_1 - j_2|$ and $M_J = J, J-1, \ldots, -J$.

For a many-electron atom where spin–orbit interaction is small, the individual electronic orbital angular momenta add to give a total electronic orbital angular momentum $(\mathbf{L} = \Sigma_i \mathbf{L}_i)$, and similarly for the spin angular momenta $(\mathbf{S} = \Sigma_i \mathbf{S}_i)$. The eigenvalues of the operators $\hat{L}^2$ and $\hat{S}^2$ for the squares of the magnitudes of the total electronic orbital and spin angular momenta are $L(L+1)\hbar^2$ and $S(S+1)\hbar^2$, respectively. For nonequivalent electrons (electrons in different subshells), the possible L values arising from a given electron configuration are readily found by using the angular-momentum addition rule (11.39), ignoring any closed subshells. For equivalent electrons, the possible L values can be found by consulting Table 11.2a. The possible S values are found using (11.39), ignoring all electrons that are paired.

With spin–orbit interaction neglected, atomic states that have the same energy have the same L value and the same S value. A set of equal-energy states with the same L and the same S constitutes a term. The term is denoted by the term symbol $^{2S+1}(L)$, where $2S+1$ is called the multiplicity and (L) is a code letter that gives the L value [see (11.42)]. The degeneracy of an atomic term is $(2L+1)(2S+1)$.

When spin–orbit interaction is included, each term is split into a number of levels, each level having a different value of the total-electronic-angular-momentum quantum number J. The total electronic angular momentum is $\mathbf{J} = \mathbf{L} + \mathbf{S}$, and the eigenvalues of $\hat{J}^2$ are $J(J+1)\hbar^2$, where J ranges from $L+S$ to $|L-S|$ in integral steps. The symbol for a level is $^{2S+1}(L)_J$. Each level is $(2J+1)$-fold degenerate, corresponding to the $2J+1$ values of M_J, which range from J to $-J$. The degeneracy of an atomic level can be removed by application of an external magnetic field, which splits each level into $2J+1$ states of slightly different energies.

In summary,

$$\text{Configurations} \xrightarrow[\text{repulsions}]{\text{interelectronic}} \text{Terms} \xrightarrow[\text{interaction}]{\text{spin–orbit}} \text{Levels} \xrightarrow[\text{field}]{\text{external}} \text{States}$$

The Condon–Slater rules give the values of one- and two-electron integrals involving Slater determinants.

PROBLEMS

11.1 How many electrons can be put in each of the following: (a) a shell with principal quantum number n; (b) a subshell with quantum numbers n and l; (c) an orbital; (d) a spin-orbital?

11.2 If $R(r_1)$ is the radial factor in the function t_1 in the Hartree differential equation (11.9), write the differential equation satisfied by R.

11.3 Which STOs have the same form as hydrogenlike AOs?

11.4 Estimate the nonrelativistic $1s$ orbital energy in Ar; check with Fig. 11.2.

11.5 At what atomic number does the second crossing of the $3d$ and $4s$ orbital energies occur in Fig. 11.2? Take account of the logarithmic scale. (Atomic spectral data show that the crossing actually occurs between 20 and 21.)

11.6 Verify the angular-momentum commutation relations (11.32).

11.7 Prove that $m_1 + m_2 = M_J$ [Eq. (11.34)], as follows: Apply $\hat{M}_z = \hat{M}_{1z} + \hat{M}_{2z}$ to (11.33), substitute (11.33) in the resulting equation, combine terms, and use the linear independence of the functions involved to deduce (11.34).

11.8 Show that $[\hat{M}^2, \hat{M}_{1z}] = 2i\hbar(\hat{M}_{1x}\hat{M}_{2y} - \hat{M}_{1y}\hat{M}_{2x})$, where $\mathbf{M} = \mathbf{M}_1 + \mathbf{M}_2$.

11.9 Give the possible values of the total-angular-momentum quantum number J that result from the addition of angular momenta with quantum numbers (a) $\frac{3}{2}$ and 4; (b) 2, 3, and $\frac{1}{2}$.

11.10 Verify the terms in Table 11.2b.

11.11 Find the terms that arise from each of the following electron configurations: (a) $1s^2 2s^2 2p^6 3s^2 3p 5g$; (b) $1s^2 2s^2 2p 3p 3d$; (c) $1s^2 2s^2 2p^4 4d$. You may use Table 11.2a for part (c).

11.12 Which of the following electron configurations will contribute configuration functions to a CI calculation of the ground state of He? (a) $1s2s$; (b) $1s2p$; (c) $2s^2$; (d) $2s2p$; (e) $2p^2$; (f) $3d^2$.

11.13 Verify the spin-eigenfunction equations (11.52) to (11.54), (11.56), and the three equations following (11.56).

11.14 Find the $\hat{S}^2$ and $\hat{S}_z$ eigenvalues for the spin function

$$3^{-1/2}[\alpha(1)\alpha(2)\beta(3) + \alpha(1)\beta(2)\alpha(3) + \beta(1)\alpha(2)\alpha(3)]$$

11.15 (a) Calculate the angle in Fig. 11.3 between the z axis and $\mathbf{S}$ for the spin function $\alpha(1)\alpha(2)$. (b) Calculate the angle between $\mathbf{S}_1$ and $\mathbf{S}_2$ for each of the functions (11.57) to (11.60). [*Hint*: One approach is to use the law of cosines. A second approach is to use $\mathbf{S} \cdot \mathbf{S} = (\mathbf{S}_1 + \mathbf{S}_2) \cdot (\mathbf{S}_1 + \mathbf{S}_2)$.] (c) If a vector $\mathbf{A}$ has components (A_x, A_y, A_z), what are the components of the projection of $\mathbf{A}$ in the xy plane? Use the answer to this question to find the angle between the projections of $\mathbf{S}_1$ and $\mathbf{S}_2$ in the xy plane for the function $\alpha(1)\alpha(2)$.

11.16 (a) If $\hat{S}^2 = (\hat{\mathbf{S}}_1 + \hat{\mathbf{S}}_2 + \cdots) \cdot (\hat{\mathbf{S}}_1 + \hat{\mathbf{S}}_2 + \cdots)$, show that $[\hat{S}^2, \hat{P}_{ij}] = 0$. (b) Show that $[\hat{L}^2, \hat{P}_{ij}] = 0$, where $\mathbf{L}$ is the total electronic orbital angular momentum.

11.17 Of the atoms with $Z \leqslant 10$, which have ground states of odd parity?

11.18 Give the number of states that belong to each of the following terms: (a) ^4F; (b) ^1S; (c) ^3P; (d) ^2D.

11.19 How many states belong to each of the following carbon configurations? (a) $1s^22s^22p^2$; (b) $1s^22s^22p3p$.

11.20 True or false? (a) The multiplicity of every term of an atom with an odd number of electrons must be an even number. (b) The multiplicity of every term of an atom with an even number of electrons must be an odd number.

11.21 Give the possible multiplicities of the terms that arise from each of the following electron configurations: (a) f; (b) f^2; (c) f^3; (d) f^7; (e) f^{12}; (f) f^{13}.

11.22 Give the levels that arise from each of the following terms, and give the degeneracy of each level: (a) ^1S; (b) ^2S; (c) ^3F; (d) ^4D.

11.23 For a state belonging to a $^3D_3$ level, give the magnitude of (a) the total electronic orbital angular momentum; (b) the total electronic spin angular momentum; (c) the total electronic angular momentum.

11.24 Give the symbol for the ground level of each of the atoms with $Z \leqslant 10$.

11.25 Give the symbol for the ground level of each of the atoms with $21 \leqslant Z \leqslant 30$. Which one of these atoms has the most degenerate ground level?

11.26 (a) Use a diagram like (11.48) to show that the lowest term of a single half-filled-subshell configuration has $L = 0$. (b) How many levels arise from a term with $L = 0$? Explain why no rule is needed to find the lowest level of the lowest term of a half-filled-subshell configuration.

11.27 Verify that if $E = 1$ eV then $E/hc = 8065.5$ cm^{-1}.

11.28 Consult Moore's table of atomic energy levels (Section 11.5) to find at least three electron configurations of the neutral carbon atom for which Hund's rule does not correctly predict the lowest term.

11.29 Selection rules for spectral transitions of atoms where Russell–Saunders coupling is valid are (*Bethe and Jackiw*, Chapter 11) $\Delta L = 0, \pm 1$; $\Delta S = 0$; $\Delta J = 0, \pm 1$ (but $J = 0$ to $J = 0$ is forbidden); $\Delta(\sum_i l_i) = \pm 1$, meaning that the change in configuration must change the sum of the l values of the electrons by ± 1. For most atomic spectral lines, only one electron changes its subshell; here the $\Delta(\Sigma_i l_i) = \pm 1$ rule becomes $\Delta l = \pm 1$ for the electron making the transition.

For the carbon atom, the energy levels that arise from the $1s^2 2s^2 2p^2$ configuration are

level	$^3P_0$	$^3P_1$	$^3P_2$	$^1D_2$	$^1S_0$
$(E/hc)/\text{cm}^{-1}$	0	16.4	43.4	10192.6	21648.0

and the energy levels of the $1s^2 2s 2p^3$ configuration are

$^5S_2$	$^3D_3$	$^3D_1$	$^3D_2$	$^3P_1$	$^3P_2$	$^3P_0$
33735.2	64086.9	64089.8	64090.9	75254.0	75255.3	75256.1

$^1D_2$	$^3S_1$	$^1P_1$
97878	105798.7	119878

Use the above selection rules to find the wave numbers of all the transitions that are allowed between pairs of these 15 levels.

11.30 Does Fig. 11.6 contain a violation of the rule given in Section 11.6 for determining whether a multiplet is regular or inverted?

11.31 Draw a diagram similar to Fig. 11.6 for the carbon $1s^2 2s^2 2p^2$ configuration. (The ^1S term is the highest.)

11.32 Use Eq. (11.66) to calculate the separation between the $^2P_{3/2}$ and $^2P_{1/2}$ levels of the hydrogen-atom $2p$ configuration. (Because of other relativistic effects, the result will not agree accurately with experiment.)

11.33 Use Eq. (11.74) to calculate the energy separation between the $M_J = \frac{1}{2}$ and $M_J = -\frac{1}{2}$ states of the $2p\ ^2P_{1/2}$ hydrogen-atom level, if a magnetic field of 0.200 T is applied.

11.34 Derive Eqs. (11.78) and (11.79).

11.35 For a closed-subshell configuration, (a) show that the double sum in (11.80) equals

$$\sum_{j>i}^{n/2} \sum_{i=1}^{n/2} (4J_{ij} - 2K_{ij}) + \sum_{i=1}^{n/2} J_{ii}$$

where the Coulomb and exchange integrals are defined in terms of the $n/2$ different spatial orbitals ϕ_i; (b) use (11.84) and the result of part (a) to derive (11.83).

11.36 Use the Condon–Slater rules to prove the orthonormality of two n-electron Slater determinants of orthonormal spin-orbitals.

11.37 Explain why it would be incorrect to calculate the experimental ground-state energy of lithium by taking $E_{2s} + 2E_{1s}$, where E_{2s} is the experimental energy needed to remove the $2s$ electron from lithium and E_{1s} is the experimental energy needed to remove a $1s$ electron from lithium.

12

Molecular Symmetry

12.1 SYMMETRY ELEMENTS AND OPERATIONS

Qualitative information about molecular wave functions and properties can often be obtained from the symmetry of the molecule. By the symmetry of a molecule, we mean the symmetry of the framework formed by the nuclei held fixed in their equilibrium positions. (Our starting point for molecular quantum mechanics will be the Born–Oppenheimer approximation, which regards the nuclei as fixed when solving for the electronic wave function; see Section 13.1.) The symmetry of a molecule can be different in different electronic states. For example, HCN is linear in its ground electronic state, but nonlinear in certain excited states. Unless otherwise specified, we shall be considering the symmetry of the ground electronic state.

Symmetry Elements and Operations. A *symmetry operation* is a transformation of a body such that the final position is physically indistinguishable from the initial position, and the distances between all pairs of points in the body are preserved. For example, consider the trigonal-planar molecule BF_3 (Fig. 12.1a), where for convenience we have numbered the fluorine nuclei. If we rotate the molecule counterclockwise by 120° about an axis through the boron nucleus and perpendicular to the plane of the molecule, the new position will be as in Fig. 12.1b. Since in reality the fluorine nuclei are physically indistinguishable from one another, we have carried out a symmetry operation. The axis about which we rotated is an example of a symmetry element. Symmetry elements and symmetry operations are related but different things, which are often confused. A *symmetry element* is a geometrical entity (point, line, or plane) with respect to which a symmetry operation is carried out.

We say that a body has an *n-fold axis of symmetry* (also called an *n-fold proper axis* or an *n-fold rotation axis*) if rotation about this axis by $360/n$ degrees (where n is an integer) gives a configuration physically indistinguishable from the original position; n is called the *order* of the axis. For example, BF_3 has a threefold axis of symmetry perpendicular to the molecular plane. The symbol for an n-fold rotation axis is C_n. The threefold axis in BF_3 is a C_3 axis. To denote the operation of counterclockwise rotation by $(360/n)°$, we use the symbol $\hat{C}_n$. The "hat" distinguishes symmetry operations from symmetry elements. BF_3 has three more rotation axes; each B—F bond is a twofold symmetry axis (Fig. 12.2).

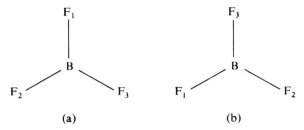

Figure 12.1 (a) The BF_3 molecule. (b) BF_3 after a 120° rotation about the axis through B and perpendicular to the molecular plane.

A second kind of symmetry element is a plane of symmetry. A molecule has a *plane of symmetry* if reflection of all the nuclei through that plane gives a configuration physically indistinguishable from the original one. The symbol for a symmetry plane is σ. (*Spiegel* is the German word for mirror.) The symbol for the operation of reflection is $\hat{\sigma}$. BF_3 has four symmetry planes. The plane of the molecule is a symmetry plane, since nuclei lying in a reflection plane undergo no change in position when a reflection is carried out. The plane passing through the B and F_1 nuclei and perpendicular to the plane of the molecule is a symmetry plane, since reflection in this plane merely interchanges F_2 and F_3. It might be thought that this reflection is the same symmetry operation as rotation by 180° about the C_2 axis passing through B and F_1, which also interchanges F_2 and F_3. This is not so; the reflection carries points lying above the plane of the molecule into points that also lie above the molecular plane, whereas the $\hat{C}_2$ rotation carries points lying above the plane of the molecule into points below the molecular plane. Two symmetry operations are equal only when they represent the same transformation of three-dimensional space. The remaining two symmetry planes in BF_3 pass through B—F_2 and B—F_3 and are perpendicular to the molecular plane.

The third kind of symmetry element is a *center of symmetry*, symbolized by i (no connection with $\sqrt{-1}$). A molecule has a center of symmetry if the operation

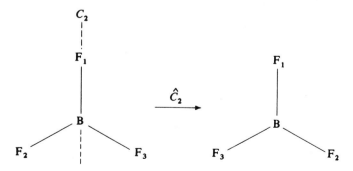

Figure 12.2 A C_2 axis in BF_3.

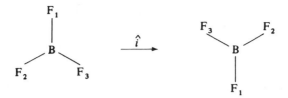

Figure 12.3 Effect of inversion in BF_3.

of inverting all the nuclei through the center gives a configuration indistinguishable from the original one. If we set up a Cartesian coordinate system, the operation of inversion through the origin (symbolized by $\hat{i}$) carries a nucleus originally at (x, y, z) to $(-x, -y, -z)$. Does BF_3 have a center of symmetry? With the origin at the boron nucleus, inversion gives the result shown in Fig. 12.3. Since we get a configuration that is physically distinguishable from the original one, BF_3 does not have a center of symmetry. For SF_6, inversion through the sulfur nucleus is shown in Fig. 12.4, and it is clear that SF_6 has a center of symmetry. (An operation such as $\hat{i}$, $\hat{C}_n$, and so on, may or may not be a symmetry operation; thus $\hat{i}$ is a symmetry operation in SF_6 but not in BF_3.)

The fourth and final kind of symmetry element is an ***n-fold rotation– reflection axis of symmetry*** (also called an *improper axis* or an *alternating axis*), symbolized by S_n. A body has an S_n axis if rotation by $(360/n)°$ (n integral) about the axis, followed by reflection in a plane perpendicular to the axis, carries the body into a position physically indistinguishable from the original one. Clearly, if a body has a C_n axis and also has a plane of symmetry perpendicular to this axis, then the C_n axis is also an S_n axis. Thus the C_3 axis in BF_3 is also an S_3 axis. It is possible to have an S_n axis that is not a C_n axis. An example is CH_4. In Fig. 12.5 we have first carried out a 90° proper rotation ($\hat{C}_4$) about what we assert is an S_4 axis. As can be seen, this operation does not result in an equivalent configuration. When we follow the $\hat{C}_4$ operation by reflection in the plane perpendicular to the axis and passing through the carbon atom, we do get a configuration equivalent to the one existing before we performed the rotation and reflection; hence we have an S_4 axis. The S_4 axis is not a C_4 axis, although it is a C_2 axis. There are two other S_4 axes in methane, each perpendicular to a pair of faces of the cube in which the tetrahedral molecule is inscribed.

The operation of rotation by $(360/n)°$ about an axis, followed by reflection in a plane perpendicular to the axis, is denoted by $\hat{S}_n$. An $\hat{S}_1$ operation is a 360°

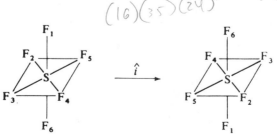

Figure 12.4 Effect of inversion in SF_6.

$(1234) \to (3$
(1423)

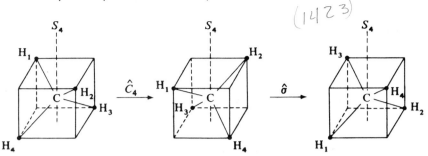

Figure 12.5 An S_4 axis in CH_4.

rotation about an axis, followed by a reflection in a plane perpendicular to the axis. Since a 360° rotation restores the body to its original position, an $\hat{S}_1$ operation is the same as reflection in a plane, $\hat{S}_1 = \hat{\sigma}$; any plane of symmetry has an S_1 axis perpendicular to it.

Consider now the $\hat{S}_2$ operation. We choose the coordinate system so that the S_2 axis is the z axis (Fig. 12.6). Rotation by 180° about the S_2 axis changes the x and y coordinates of a point to $-x$ and $-y$, respectively, and leaves the z coordinate unaffected. Reflection in the xy plane then converts the z coordinate to $-z$. The net effect of the $\hat{S}_2$ operation is to bring a point originally at (x, y, z) to $(-x, -y, -z)$, which amounts to an inversion through the origin: $\hat{S}_2 = \hat{\imath}$. Any axis passing through a center of symmetry is an S_2 axis. Reflection in a plane and inversion are special cases of the $\hat{S}_n$ operation.

The $\hat{S}_n$ operation may seem an arbitrary kind of operation, but it must be included as one of the kinds of symmetry operations. For example, the transformation from the first to the third CH_4 configuration in Fig. 12.5 certainly meets the definition of a symmetry operation, but it is neither a proper rotation nor a reflection nor an inversion.

Performing a symmetry operation on a molecule gives a nuclear configuration physically indistinguishable from the original one. Hence the center of mass must have the same position in space before and after a symmetry operation. For the operation $\hat{C}_n$, the only points that do not move are those on the C_n axis; therefore, a C_n symmetry axis must pass through the molecular center of mass. Similarly, a center of symmetry must coincide with the center of mass; a plane of symmetry and an S_n axis of symmetry must pass through the center of mass. The

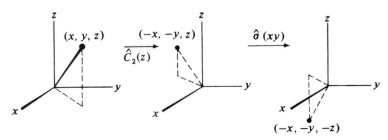

Figure 12.6 The $\hat{S}_2$ operation.

center of mass is the common intersection of all the symmetry elements of the molecule.

In discussing the symmetry of a molecule, we often place it in a Cartesian coordinate system with the molecular center of mass at the origin. The rotational axis of highest order is made the z axis. A plane of symmetry containing this axis is designated σ_v (for *vertical*); a plane of symmetry perpendicular to this axis is designated σ_h (for *horizontal*).

Products of Symmetry Operations. Symmetry operations are operators that transform three-dimensional space, and (as with any operators) we define the **product** of two such operators as meaning successive application of the operators, the operator on the right of the product being applied first. Clearly, the product of any two symmetry operations of a molecule must also be a symmetry operation.

As an example, consider BF_3. The product of the $\hat{C}_3$ operator with itself, $\hat{C}_3\hat{C}_3 = \hat{C}_3^2$, rotates the molecule 240° counterclockwise. If we take $\hat{C}_3\hat{C}_3\hat{C}_3 = \hat{C}_3^3$, we have a 360° rotation, which restores the molecule to its original position. We define the **identity operation** $\hat{E}$ as the operation that does nothing to a body. We have $\hat{C}_3^3 = \hat{E}$. (The symbol comes from the German word *Einheit*, meaning unity.)

Now consider a molecule with a sixfold axis of symmetry, for example, C_6H_6. The operation $\hat{C}_6$ is a 60° rotation, and $\hat{C}_6^2$ is a 120° rotation; hence $\hat{C}_6^2 = \hat{C}_3$. Also $\hat{C}_6^3 = \hat{C}_2$. We conclude that a C_6 symmetry axis is also a C_3 and a C_2 axis. In general, a C_n axis is also a C_m axis if n/m is an integer.

Since two successive reflections in the same plane bring all nuclei back to their original positions, we have $\hat{\sigma}^2 = \hat{E}$. Also $\hat{\imath}^2 = \hat{E}$. More generally, $\hat{\sigma}^n = \hat{E}$, $\hat{\imath}^n = \hat{E}$ for even n, while $\hat{\sigma}^n = \hat{\sigma}$, $\hat{\imath}^n = \hat{\imath}$ for odd n.

Do symmetry operators always commute? Consider SF_6. We examine the products of a $\hat{C}_4$ rotation about the z axis and a $\hat{C}_2$ rotation about the x axis. Figure 12.7 shows that $\hat{C}_4(z)\hat{C}_2(x) \neq \hat{C}_2(x)\hat{C}_4(z)$. Thus symmetry operations do not always commute. Note that we describe symmetry operations with respect to a fixed coordinate system; our convention is that the symmetry elements do not move with the molecule when we perform a symmetry operation but remain fixed in space. For example, the $C_2(x)$ axis does not move when we perform the $\hat{C}_4(z)$ operation.

Symmetry and Dipole Moments. As an application of symmetry, we consider molecular dipole moments. Since a symmetry operation produces a configuration physically indistinguishable from the original one, the direction of the dipole-moment vector must remain unchanged after a symmetry operation. (This is a nonrigorous, unsophisticated argument.) Hence, if we have a proper axis of symmetry, the dipole moment must lie along this axis. If we have two or more noncoincident symmetry axes, the molecule cannot have a dipole moment, since the dipole moment cannot lie on two different axes. CH_4, which has four noncoincident C_3 axes, has no dipole moment. If there is a plane of symmetry, the dipole moment must lie in this plane. If there are several symmetry planes, the dipole moment must lie along the line of intersection of these planes. In H_2O the dipole moment lies on the C_2 axis, which is also the intersection of the two

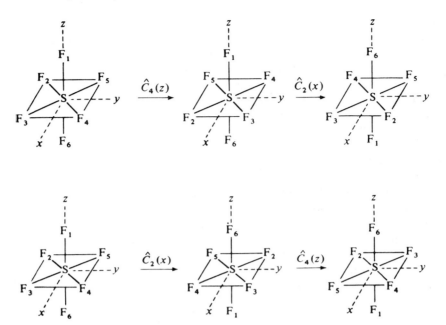

Figure 12.7 Products of two symmetry operations in SF_6. Top: $\hat{C}_2(x)\hat{C}_4(z)$. Bottom: $\hat{C}_4(z)\hat{C}_2(x)$.

symmetry planes (Fig. 12.8). <u>A molecule with a center of symmetry cannot have a dipole moment</u>, since inversion reverses the direction of a vector. A monatomic molecule has a center of symmetry; hence atoms do not have dipole moments. (There is one exception to this statement; see Problem 13.34.) Thus we can use symmetry to discover whether a molecule has a dipole moment; in many cases symmetry considerations also tell us along what line the dipole moment lies.

Symmetry and Optical Activity.   Certain molecules rotate the plane of polarization of plane-polarized light that is passed through them. Experimental evidence and a quantum-mechanical treatment (*Kauzmann*, pp. 703–713) show that the optical rotary powers of two molecules that are mirror images of each

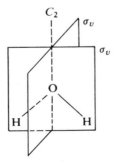

Figure 12.8 The symmetry elements of H_2O.

other are equal in magnitude but opposite in sign. A simple way to see this is from the fact that physical processes occur in the same manner for a system and for its mirror image. Reflection of an optical-rotation experiment in a mirror converts the molecules to their mirror images and converts a clockwise optical rotation to a counterclockwise rotation of equal magnitude. (Interestingly, in elementary-particle decay, nature *does* distinguish between a system and its mirror image. See M. Gardner, *The Ambidextrous Universe*, Mentor—New American Library, New York, 1969.) Hence, if a molecule is its own mirror image, it is optically inactive: $\alpha = -\alpha$, $2\alpha = 0$, $\alpha = 0$, where α is the optical rotary power. If a molecule is not superimposable on its mirror image, it may be optically active. If the conformation of the mirror image differs from that of the original molecule only by rotation about a bond with a low rotational barrier, then we will not have optical activity.

What is the connection between symmetry and optical activity? Consider the $\hat{S}_n$ operation. It consists of a rotation ($\hat{C}_n$) and a reflection ($\hat{\sigma}$). The reflection part of the $\hat{S}_n$ operation converts the molecule to its mirror image, and if the $\hat{S}_n$ operation is a symmetry operation for the molecule, then the $\hat{C}_n$ rotation will superimpose the molecule and its mirror image:

$$\text{molecule} \xrightarrow{\hat{C}_n} \text{rotated molecule} \xrightarrow{\hat{\sigma}} \text{rotated mirror image}$$

We conclude that a molecule with an S_n axis is optically inactive; if the molecule has no S_n axis, it may be optically active.

Since $\hat{S}_1 = \hat{\sigma}$ and $\hat{S}_2 = \hat{i}$, a molecule with either a plane or a center of symmetry is optically inactive. However, an S_n axis of any order rules out optical activity. As an example consider the molecule of Fig. 12.9, which has two mutually perpendicular rings with one atom in common. The molecule has neither a plane nor a center of symmetry but is optically inactive because of the presence of an S_4 axis. As a check, it is easily verified that the mirror image generated by reflection in a plane perpendicular to the S_4 axis is superimposable on the original molecule after a $\hat{C}_4$ rotation.

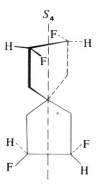

Figure 12.9 A molecule that is optically inactive because of the presence of an S_4 axis.

A molecule can have a symmetry element and still be optically active; if a C_n axis is present and there is no S_n axis, we can have optical activity.

We have made no reference to asymmetric carbon atoms. There are optically active organic compounds that have no asymmetric carbons; there are optically inactive molecules that have two or more asymmetric carbons.

Symmetry Operations and Quantum Mechanics.   What is the relation between the symmetry operations of a molecule and quantum mechanics? To classify the states of a quantum-mechanical system, we consider those operators that commute with the Hamiltonian and with each other. For example, we classified the states of many-electron atoms using the quantum numbers L, S, J, and M_J, which correspond to the operators $\hat{L}^2$, $\hat{S}^2$, $\hat{J}^2$, and $\hat{J}_z$, all of which commute with each other and with the Hamiltonian (omitting spin–orbit interaction). The symmetry operations discussed in this chapter act on *points* in three-dimensional space, transforming each point to a corresponding point. All the quantum-mechanical operators we have discussed act on *functions*, transforming each function to a corresponding function. Corresponding to each symmetry operation $\hat{R}$, we define an operator $\hat{O}_R$ that acts on functions in the following manner. Let $\hat{R}$ bring a point originally at (x, y, z) to the location (x', y', z'):

$$\hat{R}(x, y, z) \rightarrow (x', y', z') \qquad (12.1)$$

The operator $\hat{O}_R$ is defined so that the function $\hat{O}_R f$ has the same value at (x', y', z') that the function f has at (x, y, z):

$$\hat{O}_R f(x', y', z') = f(x, y, z) \qquad (12.2)$$

We shall illustrate this definition with an example. Let $\hat{R}$ be a counterclockwise 90° proper rotation about the z axis: $\hat{R} = \hat{C}_4(z)$. Let f be a $2p_x$ hydrogen orbital:

$$f = 2p_x = Nxe^{-k(x^2+y^2+z^2)^{1/2}}$$

where N and k are constants. A contour showing where f has its maximum positive value is a distorted ellipsoid of revolution about the positive x axis (Section 6.7); likewise, f has its maximum negative value on an ellipsoid lying on the negative x axis. Let us say that these ellipsoids are "centered" about the points $(a, 0, 0)$ and $(-a, 0, 0)$. The operator $\hat{C}_4(z)$ has the following effect (Fig. 12.10):

$$\hat{C}_4(z)(x, y, z) \rightarrow (-y, x, z) \qquad (12.3)$$

For example, the point originally at $(a, 0, 0)$ is moved to $(0, a, 0)$, while the point at $(-a, 0, 0)$ is moved to $(0, -a, 0)$. From (12.2), the function $\hat{O}_{C_4(z)} 2p_x$ must have its contours of maximum positive and negative values centered about $(0, a, 0)$ and $(0, -a, 0)$, respectively. We conclude that

$$\hat{O}_{C_4(z)} 2p_x = 2p_y \qquad (12.4)$$

See Fig. 12.11.

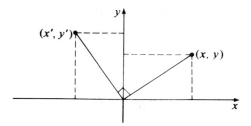

Figure 12.10 Effect of a $\hat{C}_4(z)$ rotation. Use of trigonometry shows that $x' = -y$ and $y' = x$.

For the inversion operation, we have

$$\hat{i}(x, y, z) \rightarrow (-x, -y, -z) \tag{12.5}$$

and (12.2) reads

$$\hat{O}_i f(-x, -y, -z) = f(x, y, z)$$

We now carry out the following renaming of the variables:

$$\bar{x} = -x, \qquad \bar{y} = -y, \qquad \bar{z} = -z$$

Hence

$$\hat{O}_i f(\bar{x}, \bar{y}, \bar{z}) = f(-\bar{x}, -\bar{y}, -\bar{z})$$

The point $(\bar{x}, \bar{y}, \bar{z})$ is a general point in space, and we can drop the bars to get

$$\hat{O}_i f(x, y, z) = f(-x, -y, -z)$$

We conclude that $\hat{O}_i$ is the parity operator: $\hat{O}_i = \hat{\Pi}$.

The wave function of an *n*-particle system is a function of $4n$ variables, and we extend the definition (12.2) of $\hat{O}_R$ to read

$$\hat{O}_R f(x_1', y_1', z_1', m_{s1}, \ldots, x_n', y_n', z_n', m_{sn})$$

$$= f(x_1, y_1, z_1, m_{s1}, \ldots, x_n, y_n, z_n, m_{sn})$$

Note that $\hat{O}_R$ does not affect the spin coordinates. Thus, in looking at the parity of atomic states in Section 11.5, we looked at the spatial factors in each

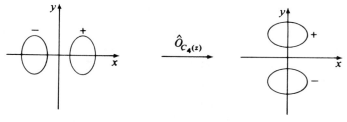

Figure 12.11 Effect of $\hat{O}_{C_4(z)}$ on a p_x orbital.

term of the expansion of the Slater determinant and omitted consideration of the spin factors, since these are unaffected by $\hat{\Pi}$.

When a system is characterized by the symmetry operations $\hat{R}_1, \hat{R}_2, \ldots$, then the corresponding operators $\hat{O}_{R_1}, \hat{O}_{R_2}, \ldots$ commute with the Hamiltonian. (For a detailed proof, see *Schonland*, Sections 7.1–7.3.) For example, if the nuclear framework of a molecule has a center of symmetry, then the parity operator $\hat{\Pi}$ commutes with the Hamiltonian for the electronic motion. We can then choose the electronic states (wave functions) as even or odd, according to the eigenvalue of $\hat{\Pi}$. Of course, all the symmetry operations may not commute among themselves (Fig. 12.7); hence the wave functions cannot in general be chosen as eigenfunctions of all the symmetry operators $\hat{O}_R$. (Further discussion on the relation between symmetry operators and molecular wave functions is given in Section 15.2.)

There is a close connection between symmetry and the *constants of the motion* (these are properties whose operators commute with the Hamiltonian $\hat{H}$). For a system whose Hamiltonian is invariant (that is, doesn't change) under any translation of spatial coordinates, the linear-momentum operator $\hat{p}$ will commute with $\hat{H}$, and p can be assigned a definite value in a stationary state. An example is the free particle. For a system with $\hat{H}$ invariant under any rotation of coordinates, the operators for the angular-momentum components commute with $\hat{H}$, and the total angular momentum and one of its components are specifiable. An example is an atom. A linear molecule has axial symmetry, rather than the spherical symmetry of an atom; here only the axial component of angular momentum can be specified (Chapter 13).

Matrices and Symmetry Operations.   The symmetry operation $\hat{R}$ moves the point originally at x, y, z to the new location x', y', z', where each of x', y', z' is a linear combination of x, y, z (for proof of this see *Schonland*, pp. 52–53):

$$\begin{aligned} x' &= r_{11}x + r_{12}y + r_{13}z \\ y' &= r_{21}x + r_{22}y + r_{23}z \\ z' &= r_{31}x + r_{32}y + r_{33}z \end{aligned} \quad \text{or} \quad \begin{pmatrix} x' \\ y' \\ z' \end{pmatrix} = \begin{pmatrix} r_{11} & r_{12} & r_{13} \\ r_{21} & r_{22} & r_{23} \\ r_{31} & r_{32} & r_{33} \end{pmatrix} \begin{pmatrix} x \\ y \\ z \end{pmatrix}$$

where $r_{11}, r_{12}, \ldots, r_{33}$ are constants whose values depend on the nature of $\hat{R}$. One says that the symmetry operation $\hat{R}$ is ***represented*** by the matrix $\mathbf{R}$ whose elements are $r_{11}, r_{12}, \ldots, r_{33}$. The set of functions x, y, z, whose transformations are described by $\mathbf{R}$, is said to be the ***basis*** for this representation.

For example, from (12.3) and (12.5), for the $\hat{C}_4(z)$ operation, we have $x' = -y$, $y' = x$, $z' = z$; for $\hat{\imath}$, we have $x' = -x$, $y' = -y$, $z' = -z$. The matrices representing $\hat{C}_4(z)$ and $\hat{\imath}$ in the x, y, z basis are

$$\mathbf{C}_4(z) = \begin{pmatrix} 0 & -1 & 0 \\ 1 & 0 & 0 \\ 0 & 0 & 1 \end{pmatrix}, \qquad \mathbf{i} = \begin{pmatrix} -1 & 0 & 0 \\ 0 & -1 & 0 \\ 0 & 0 & -1 \end{pmatrix}$$

If the product $\hat{R}\hat{S}$ of two symmetry operations is $\hat{T}$, then the matrices representing these operations in the x, y, z basis multiply in the same way; that is, if $\hat{R}\hat{S} = \hat{T}$, then $\mathbf{RS} = \mathbf{T}$. (For proof, see *Schonland*, pp. 56–57.)

12.2 SYMMETRY POINT GROUPS

We now consider the possible combinations of symmetry elements. We cannot have arbitrary combinations of symmetry elements in a molecule. For example, suppose a molecule has one and only one C_3 axis. Any symmetry operation must send this axis into itself. The molecule cannot, therefore, have a plane of symmetry at an arbitrary angle to the C_3 axis; any plane of symmetry must either contain this axis or be perpendicular to it. (In BF_3 there are three σ_v planes and one σ_h plane.) The only possibility for a C_n axis noncoincident with the C_3 axis is a C_2 axis perpendicular to the C_3 axis; the corresponding $\hat{C}_2$ operation will send the C_3 axis into itself. Since $\hat{C}_3$ and $\hat{C}_3^2$ are symmetry operations, if we have one C_2 axis perpendicular to the C_3 axis, we must have a total of three such axes (as in BF_3).

The set of all the symmetry operations of a molecule forms a mathematical group. A **group** is a set of entities and a rule for forming the product of these entities, such that certain requirements are met. (The rule for forming the product of symmetry operations is successive performance of them.) We shall not make explicit use of the mathematics of group theory. Note that it is the symmetry *operations*, not the symmetry elements, that constitute the group. For any symmetry operation of a molecule, the point that is the center of mass remains fixed. Hence the symmetry groups of isolated molecules are called **point groups**. For a crystal of infinite extent, we can have symmetry operations (for example, translations) that leave no point fixed, giving rise to *space groups*. We omit consideration of space groups.

Any molecule can be classified as belonging to one of the symmetry point groups that we now list. For convenience we have divided the point groups into four divisions. Script letters denote point groups.

I. *Groups with no C_n axis:* $\mathscr{C}_1$, $\mathscr{C}_s$, $\mathscr{C}_i$

$\mathscr{C}_1$: If a molecule has no symmetry elements at all, it belongs to this group. The only symmetry operation is $\hat{E}$ (which is a $\hat{C}_1$ rotation). CHFClBr belongs to point group $\mathscr{C}_1$.

$\mathscr{C}_s$: A molecule whose only symmetry element is a plane of symmetry belongs to this group. The symmetry operations are $\hat{E}$ and $\hat{\sigma}$. An example is HOCl (Fig. 12.12).

$\mathscr{C}_i$: a molecule whose only symmetry element is a center of symmetry belongs to this group. The symmetry operations are $\hat{i}$ and $\hat{E}$.

II. *Groups with a single C_n axis:* $\mathscr{C}_n$, $\mathscr{C}_{nh}$, $\mathscr{C}_{nv}$, $\mathscr{S}_{2n}$

$\mathscr{C}_n$, $n = 2, 3, 4, \ldots$: A molecule whose only symmetry element is a C_n axis belongs to this group. The symmetry operations are $\hat{C}_n$, $\hat{C}_n^2$, ..., $\hat{C}_n^{n-1}$, $\hat{E}$. A molecule belonging to $\mathscr{C}_2$ is shown in Fig. 12.13.

$\mathscr{C}_{nh}$, $n = 2, 3, 4, \ldots$: If we add a plane of symmetry perpendicular to the C_n axis,

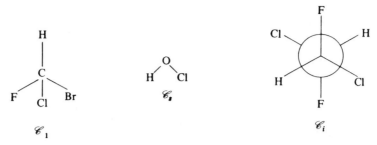

Figure 12.12 Molecules with no C_n axis.

we have a molecule belonging to this group. Since $\hat{\sigma}_h \hat{C}_n = \hat{S}_n$, the C_n axis is also an S_n axis. If n is even, the C_n axis is also a C_2 axis, and we have the symmetry operation

$$\hat{\sigma}_h \hat{C}_2 = \hat{S}_2 = \hat{\imath}$$

Thus, for n even, a molecule belonging to $\mathscr{C}_{nh}$ has a center of symmetry. (The group $\mathscr{C}_{1h}$ is the group $\mathscr{C}_s$ discussed previously.) Examples of molecules belonging to groups $\mathscr{C}_{2h}$ and $\mathscr{C}_{3h}$ are shown in Fig. 12.13.

$\mathscr{C}_{nv}$, $n = 2, 3, 4, \ldots$: A molecule in this group has a C_n axis and n vertical symmetry planes (passing through the C_n axis). (Group $\mathscr{C}_{1v}$ is the group $\mathscr{C}_s$.) Water, with a C_2 axis and two vertical symmetry planes, belongs to $\mathscr{C}_{2v}$. Ammonia belongs to $\mathscr{C}_{3v}$. (See Fig. 12.13.)

$\mathscr{S}_n$, $n = 4, 6, 8, \ldots$: $\mathscr{S}_n$ is the group of symmetry operations associated with an S_n axis. First consider the case of odd n. We have $\hat{S}_n = \hat{\sigma}_h \hat{C}_n$. The operation $\hat{C}_n$ affects the x and y coordinates only, while the $\hat{\sigma}_h$ operation affects the z

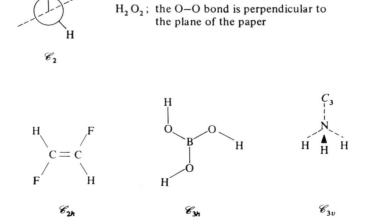

H_2O_2; the O–O bond is perpendicular to the plane of the paper

Figure 12.13 Molecules with a single C_n axis.

coordinate only; hence these operations commute, and we have

$$\hat{S}_n^n = (\hat{\sigma}_h \hat{C}_n)^n = \hat{\sigma}_h \hat{C}_n \hat{\sigma}_h \hat{C}_n \cdots \hat{\sigma}_h \hat{C}_n = \hat{\sigma}_h^n \hat{C}_n^n$$

Now $\hat{C}_n^n = \hat{E}$, and, for odd n, $\hat{\sigma}_h^n = \hat{\sigma}_h$. Hence the symmetry operation $\hat{S}_n^n$ equals $\hat{\sigma}_h$ for odd n, and the group $\mathscr{S}_n$ has a horizontal symmetry plane if n is odd. Also

$$\hat{S}_n^{n+1} = \hat{S}_n^n \hat{S}_n = \hat{\sigma}_h \hat{S}_n = \hat{\sigma}_h \hat{\sigma}_h \hat{C}_n = \hat{C}_n, \qquad n \text{ odd}$$

so that we have a C_n axis if n is odd. We conclude that the group $\mathscr{S}_n$ is identical to the group $\mathscr{C}_{nh}$ if n is odd. Now consider even values of n. Since $\hat{S}_2 = \hat{\imath}$, the group $\mathscr{S}_2$ is identical to $\mathscr{C}_i$. Thus it is only for $n = 4, 6, 8, \ldots$ that we get new groups. The S_{2n} axis is also a C_n axis:

$$\hat{S}_{2n}^2 = \hat{\sigma}_h^2 \hat{C}_{2n}^2 = \hat{E} \hat{C}_n = \hat{C}_n$$

The spiran of Fig. 12.9 belongs to the group $\mathscr{S}_4$.

III. *Groups with one C_n axis and n C_2 axes: $\mathscr{D}_n$, $\mathscr{D}_{nh}$, $\mathscr{D}_{nd}$*

$\mathscr{D}_n$, $n = 2, 3, 4, \ldots$: A molecule with a C_n axis and n C_2 axes perpendicular to the C_n axis (and no symmetry planes) belongs to group $\mathscr{D}_n$. The angle between adjacent C_2 axes is π/n radians. For the group $\mathscr{D}_2$, we have three mutually perpendicular C_2 axes, and the symmetry operations are $\hat{E}$, $\hat{C}_2(x)$, $\hat{C}_2(y)$, $\hat{C}_2(z)$.

$\mathscr{D}_{nh}$, $n = 2, 3, 4, \ldots$: This is the group of a molecule with a C_n axis, n C_2 axes, and a σ_h symmetry plane perpendicular to the C_n axis. As in $\mathscr{C}_{nh}$, the C_n axis is also an S_n axis. If n is even, the C_n axis is a C_2 and an S_2 axis, and we have a center of symmetry. Molecules in $\mathscr{D}_{nh}$ also have n vertical planes of symmetry, each such plane passing through the C_n axis and a C_2 axis. We now prove this assertion. We set up a coordinate system with the C_n axis as the z axis and let one of the C_2 axes be the x axis (Fig. 12.14). This makes the xy plane the σ_h symmetry plane. Looking at the effect of the product $\hat{\sigma}(xy)\hat{C}_2(x)$ on a point originally at (x, y, z), we have

$$(x, y, z) \xrightarrow{\hat{C}_2(x)} (x, -y, -z) \xrightarrow{\hat{\sigma}(xy)} (x, -y, z)$$

We also have

$$(x, y, z) \xrightarrow{\hat{\sigma}(xz)} (x, -y, z)$$

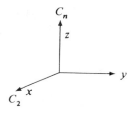

Figure 12.14 Two of the symmetry axes in a $\mathscr{D}_{nh}$ molecule.

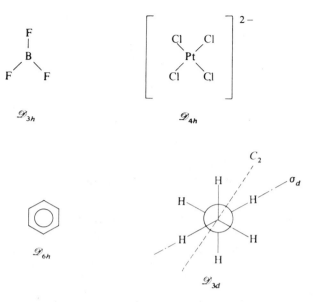

Figure 12.15 Molecules with a C_n axis and n C_2 axes.

Since $\hat{\sigma}(xy)\hat{C}_2(x)$ and $\hat{\sigma}(xz)$ both bring a point originally at (x, y, z) to the final position $(x, -y, z)$, they are equal:

$$\hat{\sigma}(xy)\hat{C}_2(x) = \hat{\sigma}(xz)$$

$\hat{C}_2(x)$ and $\hat{\sigma}(xy)$ are symmetry operations, and their product must be a symmetry operation; hence the xz plane is a symmetry plane. The same argument holds for any C_2 axis, so we have n σ_v planes. BF_3 belongs to $\mathscr{D}_{3h}$; $PtCl_4^{2-}$ belongs to $\mathscr{D}_{4h}$; benzene belongs to $\mathscr{D}_{6h}$ (Fig. 12.15).

$\mathscr{D}_{nd}$, $n = 2, 3, 4, \ldots$: A molecule with a C_n axis, n C_2 axes, and n vertical planes of symmetry, which pass through the C_n axis and bisect the angles between adjacent C_2 axes, belongs to this group. The n vertical planes are called *diagonal* planes and are symbolized by σ_d. The C_n axis can be shown to be an S_{2n} axis. The staggered conformation of ethane is an example of group $\mathscr{D}_{3d}$ (Fig. 12.15). [The symmetry of molecules with internal rotation (for example, ethane) actually requires special consideration; see H. C. Longuet-Higgins, *Mol. Phys.*, **6**, 445 (1963).]

IV. *Groups with more than one C_n axis, $n > 2$: $\mathscr{T}_d$, $\mathscr{T}$, $\mathscr{T}_h$, $\mathscr{O}_h$, $\mathscr{O}$, $\mathscr{I}_h$, $\mathscr{I}$, $\mathscr{K}_h$*

These groups are related to the symmetries of the Platonic solids, solids bounded by congruent regular polygons and having congruent polyhedral angles. There are five such solids: the tetrahedron has four triangular faces, the cube has six square faces, the octahedron has eight triangular faces, the pentagonal dodecahedron has twelve pentagonal faces, and the icosahedron has twenty triangular faces.

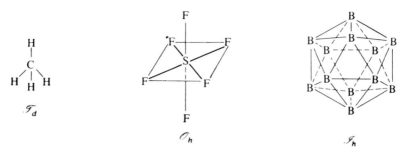

Figure 12.16 Molecules with more than one C_n axis, $n > 2$. (For $B_{12}H_{12}^{2-}$, the hydrogen atoms have been omitted for clarity.)

$\mathscr{T}_d$: The symmetry operations of a regular tetrahedron constitute this group. The prime example is CH_4. The symmetry elements of methane are four C_3 axes (each C—H bond), three S_4 axes, which are also C_2 axes (Fig. 12.5), and six symmetry planes, each such plane containing two C—H bonds. (The number of combinations of 4 things taken 2 at a time is $4!/2!2! = 6$.)

$\mathscr{O}_h$: The symmetry operations of a cube or a regular octahedron constitute this group. The cube and octahedron are said to be *dual* to each other; if we connect the midpoints of adjacent faces of a cube, we get an octahedron, and vice versa. Hence the cube and octahedron have the same symmetry elements and operations. A cube has six faces, eight vertices, and twelve edges. It has these symmetry elements: a center of symmetry, three C_4 axes passing through the centers of opposite faces of the cube (these are also S_4 and C_2 axes), four C_3 axes passing through opposite corners of the cube (these are also S_6 axes), six C_2 axes connecting the midpoints of pairs of opposite edges, three planes of symmetry parallel to pairs of opposite faces, and six planes of symmetry passing through pairs of opposite edges. Octahedral molecules such as SF_6 belong to $\mathscr{O}_h$.

$\mathscr{I}_h$: The symmetry operations of a regular pentagonal dodecahedron or icosahedron (which are dual to each other) constitute this group. The $B_{12}H_{12}^{2-}$ ion belongs to group $\mathscr{I}_h$; the twelve boron atoms lie at the vertices of a regular icosahedron (Fig. 12.16).

$\mathscr{K}_h$: This is the group of symmetry operations of a sphere. (*Kugel* is the German word for sphere.) An atom belongs to this group.

For completeness, we mention the remaining groups related to the Platonic solids; these groups are chemically unimportant. The groups $\mathscr{T}$, $\mathscr{O}$, and $\mathscr{I}$ are the groups of symmetry proper *rotations* of a tetrahedron, cube, and icosahedron, respectively; these groups do not have the symmetry reflections and improper rotations of these solids or the inversion operation of the cube and icosahedron. The group $\mathscr{T}_h$ contains the symmetry rotations of a tetrahedron, the inversion operation, and certain reflections and improper rotations.

What groups do linear molecules belong to? A rotation by any angle about the internuclear axis of a linear molecule is a symmetry operation. A regular polygon of n sides has a C_n axis, and taking the limit as $n \to \infty$ we get a circle, which has a C_∞ axis. The internuclear axis of a linear molecule is a C_∞ axis. Any plane containing this axis is a symmetry plane. If the linear molecule does not have a center of symmetry (for example, CO, HCN), it belongs to the group $\mathscr{C}_{\infty v}$. If the linear molecule has a center of symmetry (for example, H_2, C_2H_2), then it also has a σ_h symmetry plane and an infinite number of C_2 axes perpendicular to the molecular axis; hence it belongs to $\mathscr{D}_{\infty h}$.

How do we find what point group a molecule belongs to? One way is to find all the symmetry elements and then compare with the above list of groups. A more systematic procedure is given in Fig. 12.17 [J. B. Calvert, *Am. J. Phys.*, **31**, 659 (1963)]. This procedure is based on the four divisions of point groups.

We begin by checking whether or not the molecule is linear; linear molecules are classified in $\mathscr{D}_{\infty h}$ or $\mathscr{C}_{\infty v}$ according to whether or not there is a center of symmetry. If the molecule is nonlinear, we look for two or more rotational axes of threefold or higher order; if these are present, the molecule is classified in one of the groups related to the symmetry of the regular polyhedra (division IV). If these axes are not present, we look for any C_n axis at all. If there is no C_n axis, the molecule belongs to one of the groups $\mathscr{C}_s$, $\mathscr{C}_i$, $\mathscr{C}_1$ (division I). If there is at least one C_n axis, we pick the C_n axis of highest order as the main symmetry axis before proceeding to the next step. (Sometimes there will be three mutually perpendicular C_2 axes; in these cases we may pick any one of these axes as the main axis.) We next check for n C_2 axes at right angles to the main C_n axis. If these are present, we have one of the division III groups; if these are absent, we have one of the division II groups. If we find the n C_2 axes, we look for a symmetry plane perpendicular to the main C_n axis; if it is present, the group is $\mathscr{D}_{nh}$. If it is absent, we check for n planes of symmetry containing the main C_n axis (if we have three mutually perpendicular C_2 axes, we must try each axis as the main axis in looking for the two σ_v planes; the three C_2 axes are equivalent in the groups $\mathscr{D}_{nh}$ and $\mathscr{D}_n$, but not in $\mathscr{D}_{nd}$). If we find n σ_v planes, the group is $\mathscr{D}_{nd}$; otherwise it is $\mathscr{D}_n$. If we do not have n C_2 axes perpendicular to the main C_n axis, we classify the molecule in one of the groups $\mathscr{C}_{nh}$, $\mathscr{C}_{nv}$, $\mathscr{S}_{2n}$, or $\mathscr{C}_n$, by looking first for a σ_h plane, then for n σ_v planes, and finally, if these are absent, checking whether or not the C_n axis is an S_{2n} axis. The procedure of Fig. 12.17 does not locate all symmetry elements; after classifying a molecule, check that all the required symmetry elements are indeed present. Although the above procedure might seem involved, it is really quite simple and is easily memorized.

The most common error students make in classifying a molecule is to miss the n C_2 axes perpendicular to the C_n axis of a molecule belonging to $\mathscr{D}_{nd}$. For example, it is easy to see that the C=C=C axis of allene is a C_2 axis, but the other two C_2 axes (Fig. 12.18) are often overlooked. Molecules with two equal halves "staggered" with respect to each other generally belong to $\mathscr{D}_{nd}$. Models may be helpful for those with visualization difficulties.

12.3 SUMMARY

A symmetry operation transforms an object into a position that is physically indistinguishable from the original position and preserves the distances between

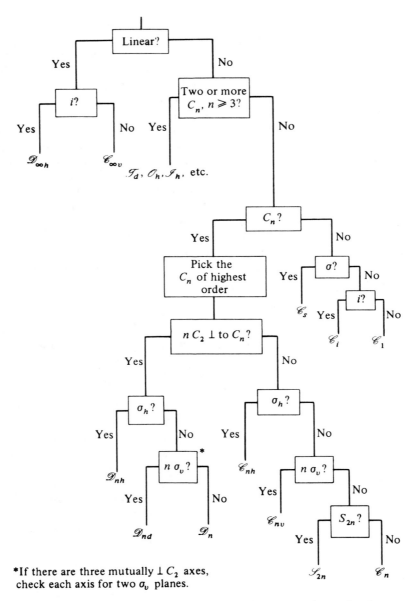

Figure 12.17 How to determine the point group of a molecule.

all pairs of points in the object. A symmetry element is a geometrical entity with respect to which a symmetry operation is performed. For molecules, the four kinds of symmetry elements are an n-fold axis of symmetry (C_n), a plane of symmetry (σ), a center of symmetry (i), and an n-fold rotation–reflection axis of symmetry (S_n). The product of symmetry operations means successive performance of them. We have $\hat{C}_n^n = \hat{E}$, where $\hat{E}$ is the identity operation; also,

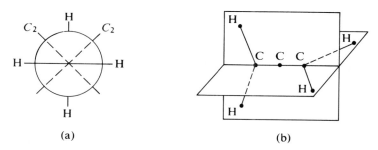

Figure 12.18 Two views of allene. The C=C=C axis is perpendicular to the plane of the paper in (a).

$\hat{S}_1 = \hat{\sigma}$, and $\hat{S}_2 = \hat{\imath}$, where the inversion operation moves a point at x, y, z to $-x, -y, -z$. Two symmetry operations may or may not commute.

For the symmetry operation $\hat{R}$ that brings a point at x, y, z to x', y', z', the operator $\hat{O}_R$ is defined by the equation $\hat{O}_R f(x', y', z') = f(x, y, z)$. If a molecule has the symmetry operations $\hat{R}_1, \hat{R}_2, \ldots$, then the operators $\hat{O}_{R_1}, \hat{O}_{R_2}, \ldots$, commute with the molecular Hamiltonian $\hat{H}$. If $\hat{R}_1, \hat{R}_2, \ldots$ all commute with one another, then the molecular wave functions can be taken as eigenfunctions of $\hat{O}_{R_1}, \hat{O}_{R_2}, \ldots$.

The set of all symmetry operations of a molecule constitutes a mathematical point group. The possible point groups of molecules are as follows. I. Groups with no C_n axis: $\mathscr{C}_1, \mathscr{C}_s, \mathscr{C}_i$. II. Groups with a single C_n axis: $\mathscr{C}_n, \mathscr{C}_{nh}, \mathscr{C}_{nv}, \mathscr{S}_{2n}$. III. Groups with one C_n axis and n C_2 axes: $\mathscr{D}_n, \mathscr{D}_{nh}, \mathscr{D}_{nd}$. IV. Groups with more than one C_n axis, $n > 2$: $\mathscr{T}_d, \mathscr{O}_h, \mathscr{I}_h$, and others.

PROBLEMS

12.1 Two people play the following game. Each in turn places a penny on the surface of a large chessboard. The pennies can be put anywhere on the board, as long as they do not overlap previously placed pennies. A penny may overlap more than one square. Once placed, a penny cannot be moved. When one of the players finds there is no room to place another penny on the board, she loses. With best play, will the person placing the first penny or her opponent win? Give the winning strategy.

12.2 Give all the symmetry elements of each of the following molecules: (a) H_2S; (b) NH_3; (c) CHF_3; (d) $HOCl$; (e) 1,3,5-trichlorobenzene; (f) CH_2F_2; (g) CHFClBr.

12.3 List all the symmetry operations of each of the molecules in Problem 12.2.

12.4 What symmetry operation is each of the following products of operations equal to? (a) $\hat{\sigma}^4$; (b) $\hat{\sigma}^7$; (c) $\hat{C}_4^2$; (d) $\hat{C}_4^6$; (e) $\hat{S}_4^2$; (f) $\hat{S}_6^3$; (g) $\hat{C}_{12}^3$; (h) $\hat{\imath}^3$.

12.5 Use Fig. 12.7 to state what symmetry operation in SF_6 each of the following products of symmetry operations is equal to. (a) $\hat{C}_2(x)\hat{C}_4(z)$; (b) $\hat{C}_4(z)\hat{C}_2(x)$.

12.6 For SF_6, which of the following pairs of operations commute? (a) $\hat{C}_4(z)$, $\hat{\sigma}_h$; (b) $\hat{C}_4(z)$, $\hat{\sigma}(yz)$; (c) $\hat{C}_2(z)$, $\hat{C}_2(x)$; (d) $\hat{\sigma}_h$, $\hat{\sigma}(yz)$; (e) $\hat{\imath}$, $\hat{\sigma}_h$.

12.7 What information does symmetry give about the dipole moment of each of the molecules in Problem 12.2?

12.8 Verify directly that the molecule in Fig. 12.9 is superimposable on its mirror image.

12.9 (a) Does H_2O_2 (Fig. 12.13) have an S_n axis? (b) Is it optically active? Explain.

12.10 For each of the following symmetry operations, find the matrix representative in the x, y, z basis. (a) $\hat{E}$; (b) $\hat{\sigma}(xy)$; (c) $\hat{\sigma}(yz)$; (d) $\hat{C}_2(x)$; (e) $\hat{S}_4(z)$; (f) $\hat{C}_3(z)$.

12.11 (a) Use SF_6 (Fig. 12.7) to verify that $\hat{C}_2(x)\hat{\sigma}(xy) = \hat{\sigma}(xz)$. (b) Write down the matrix representatives in the x, y, z basis of the three operations in part (a). Verify that these matrices multiply the same way the symmetry operations multiply.

12.12 (a) What are the eigenvalues of $\hat{O}_{C_4}$? (b) Is this operator Hermitian?

12.13 Do the same as in Problem 12.12 for $\hat{O}_{C_2}$.

12.14 To what function is a $2p_z$ hydrogenlike orbital converted by (a) $\hat{O}_{C_4(z)}$; (b) $\hat{O}_{C_4(y)}$?

12.15 Consider the following transformation:

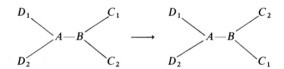

where C_1 and C_2 are identical and D_1 and D_2 are identical. Does it meet the definition of a symmetry operation (Section 12.1)? If so, express it in terms of some combination of the four kinds of symmetry operations discussed.

12.16 It is common to use rotation–inversion axes (rather than rotation–reflection axes) to classify the symmetry of crystals. Any S_n axis is equivalent to a rotation–inversion axis (symbolized by $\bar{p}$) whose order p may differ from n. A rotation–inversion operation consists of rotation by $2\pi/p$ radians followed by inversion. Show that

$$\hat{S}_n(z) = \hat{\imath}[\hat{C}_n(z)\hat{C}_2(z)]$$

Thus we have the following correspondence:

$$\frac{S_n}{\bar{p}} \left| \frac{1}{2} \right| \frac{2}{1} \left| \frac{3}{6} \right| \frac{4}{4} \left|\right.$$

Give the next three pairs of entries in this table.

12.17 (a) What Platonic solid is dual to the regular tetrahedron? (b) How many vertices does a pentagonal dodecahedron have?

12.18 For which point groups can a molecule have a dipole moment?

12.19 For which point groups can a molecule be optically active?

12.20 (a) For what values of n does the presence of an S_n axis imply the presence of a plane of symmetry? (b) For what values of n does the presence of an S_n axis imply the presence of a center of symmetry? (c) The group $\mathscr{D}_{nd}$ has an S_{2n} axis; for what values of n does it have a center of symmetry?

12.21 Give the point group of each of the following molecules. (a) CH_4; (b) CH_3F; (c) CH_2F_2; (d) CHF_3; (e) SF_6; (f) SF_5Br; (g) *trans*-SF_4Br_2; (h) CDH_3.

12.22 Give the point group of (a) $CH_2{=}CH_2$; (b) $CH_2{=}CHF$; (c) $CH_2{=}CF_2$; (d) *cis*-CHF=CHF; (e) *trans*-CHF=CHF.

12.23 Give the point group of (a) benzene; (b) fluorobenzene; (c) *o*-difluorobenzene; (d) *m*-difluorobenzene; (e) *p*-difluorobenzene; (f) 1,3,5-trifluorobenzene; (g) 1,4-difluoro-2,5-dibromobenzene; (h) naphthalene; (i) 2-chloronaphthalene.

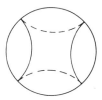

Figure 12.19 A baseball. The dashed and solid parts of the seam are in different hemispheres.

12.24 Give the point group of (a) HCN; (b) H_2S; (c) CO_2; (d) CO; (e) C_2H_2; (f) CH_3OH; (g) ND_3; (h) OCS; (i) P_4; (j) PCl_3; (k) PCl_5; (l) $B_{12}Cl_{12}^{2-}$; (m) UF_6; (n) Ar.

12.25 Given the point group of (a) FeF_6^{3-}; (b) IF_5; (c) $CH_2{=}C{=}CH_2$; (d) C_8H_8, cubane; (e) $C_6H_6Cr(CO)_3$; (f) B_2H_6; (g) XeF_4; (h) F_2O; (i) spiropentane; (j) $B_{10}H_{10}^{2-}$ (for the structure, see E. L. Muetterties and W. H. Knoth, *Chem. Eng. News*, May 9, 1966, p. 88).

12.26 The structure of ferrocene, $C_5H_5FeC_5H_5$, is an iron atom sandwiched midway between two parallel regular pentagons. For the eclipsed conformation, the vertices of the two pentagons are aligned; for the staggered conformation, one pentagon is rotated $2\pi/10$ radians with respect to the other. Electron diffraction results show that the gas-phase equilibrium conformation is the eclipsed one, with a quite low barrier to internal rotation of the rings. [A. Haaland and J. E. Nilsson, *Acta Chem. Scand.*, **22**, 2653 (1968).] What is the point group of (a) eclipsed ferrocene; (b) staggered ferrocene?

12.27 What is the point group of the tris(ethylenediamine)cobalt(III) complex ion? (Each $NH_2CH_2CH_2NH_2$ group occupies two adjacent positions of the octahedral coordination sphere.)

12.28 Give the point group of (a) a square-based pyramid; (b) a right circular cone; (c) a square lamina; (d) a square lamina with the top and bottom sides painted different colors; (e) a right circular cylinder; (f) a right circular cylinder with the two ends painted different colors; (g) a right circular cylinder with a stripe painted parallel to the axis; (h) a snowflake; (i) a doughnut; (j) a baseball (Fig. 12.19); (k) a $2p_z$ orbital; (1) a human being (ignore internal organs and slight external left–right asymmetries).

13

Electronic Structure of Diatomic Molecules

13.1 THE BORN–OPPENHEIMER APPROXIMATION

We now begin the study of molecular quantum mechanics. If we assume the nuclei and electrons to be point masses and neglect spin–orbit and other relativistic interactions (Sections 11.6 and 11.7), then the molecular Hamiltonian is

$$\hat{H} = -\frac{\hbar^2}{2} \sum_\alpha \frac{1}{m_\alpha} \nabla_\alpha^2 - \frac{\hbar^2}{2m_e} \sum_i \nabla_i^2 + \sum_\alpha \sum_{\beta > \alpha} \frac{Z_\alpha Z_\beta e'^2}{r_{\alpha\beta}} - \sum_\alpha \sum_i \frac{Z_\alpha e'^2}{r_{i\alpha}} + \sum_j \sum_{i>j} \frac{e'^2}{r_{ij}}$$

(13.1)

where α and β refer to nuclei and i and j refer to electrons. The first term in (13.1) is the operator for the kinetic energy of the nuclei. The second term is the operator for the kinetic energy of the electrons. The third term is the potential energy of the repulsions between the nuclei, $r_{\alpha\beta}$ being the distance between nuclei α and β with atomic numbers Z_α and Z_β. The fourth term is the potential energy of the attractions between the electrons and the nuclei, $r_{i\alpha}$ being the distance between electron i and nucleus α. The last term is the potential energy of the repulsions between the electrons, r_{ij} being the distance between electrons i and j.

As an example, consider H_2. Let α and β be the two protons, let 1 and 2 be the two electrons, and let m_p be the proton mass; the H_2 molecular Hamiltonian is

$$\hat{H} = -\frac{\hbar^2}{2m_p} \nabla_\alpha^2 - \frac{\hbar^2}{2m_p} \nabla_\beta^2 - \frac{\hbar^2}{2m_e} \nabla_1^2 - \frac{\hbar^2}{2m_e} \nabla_2^2$$
$$+ \frac{e'^2}{r_{\alpha\beta}} - \frac{e'^2}{r_{1\alpha}} - \frac{e'^2}{r_{1\beta}} - \frac{e'^2}{r_{2\alpha}} - \frac{e'^2}{r_{2\beta}} + \frac{e'^2}{r_{12}}$$

(13.2)

The wave functions and energies of a molecule are found from the Schrödinger equation:

$$\hat{H}\psi(q_i, q_\alpha) = E\psi(q_i, q_\alpha)$$

(13.3)

where q_i and q_α symbolize the electronic and nuclear coordinates, respectively. The molecular Hamiltonian (13.1) is formidable enough to strike terror in the heart of any quantum chemist. Fortunately, there exists a highly accurate, simplifying approximation. The key lies in the fact that nuclei are much heavier than electrons: $m_\alpha \gg m_e$. Hence the electrons move much faster than the nuclei, and to a good approximation as far as the electrons are concerned, we can regard the nuclei as fixed while the electrons carry out their motions. Speaking classically, during the time of a cycle of electronic motion, the change in nuclear configuration is negligible. Thus, considering the nuclei as fixed, we omit the nuclear kinetic-energy terms from (13.1) to obtain the Schrödinger equation for electronic motion:

$$(\hat{H}_{el} + V_{NN})\psi_{el} = U\psi_{el} \qquad (13.4)^*$$

where the *purely electronic Hamiltonian* $\hat{H}_{el}$ is

$$\hat{H}_{el} = -\frac{\hbar^2}{2m_e}\sum_i \nabla_i^2 - \sum_\alpha \sum_i \frac{Z_\alpha e'^2}{r_{i\alpha}} + \sum_j \sum_{i>j} \frac{e'^2}{r_{ij}} \qquad (13.5)$$

The electronic Hamiltonian including nuclear repulsion is $\hat{H}_{el} + V_{NN}$. The nuclear-repulsion term V_{NN} is given by

$$V_{NN} = \sum_\alpha \sum_{\beta>\alpha} \frac{Z_\alpha Z_\beta e'^2}{r_{\alpha\beta}} \qquad (13.6)$$

The energy U in (13.4) is the *electronic energy including internuclear repulsion*. The internuclear distances $r_{\alpha\beta}$ in (13.4) are not variables, but are each fixed at some constant value. Of course, there are an infinite number of possible nuclear configurations, and for each of these we may solve the electronic Schrödinger equation (13.4) to get a set of electronic wave functions (and corresponding electronic energies); each member of the set corresponds to a different molecular electronic state. The electronic wave functions and energies thus depend parametrically on the nuclear configuration:

$$\psi_{el} = \psi_{el,n}(q_i; q_\alpha) \quad \text{and} \quad U = U_n(q_\alpha)$$

where n symbolizes the electronic quantum numbers.

The variables in the electronic Schrödinger equation (13.4) are the electronic coordinates. The quantity V_{NN} is independent of these coordinates and is a constant for a given nuclear configuration. Now it is easily proved (Problem 4.32) that the omission of a constant term C from the Hamiltonian does not affect the wave functions and simply decreases each energy eigenvalue by C. Hence, if V_{NN} is omitted from (13.4), we get

$$\hat{H}_{el}\psi_{el} = E_{el}\psi_{el} \qquad (13.7)$$

where the *purely electronic energy* E_{el} is related to the electronic energy including internuclear repulsion by

$$U = E_{el} + V_{NN} \qquad (13.8)$$

We can thus omit the internuclear repulsion from the electronic Schrödinger equation and simply add it to E_{el} after solving (13.7).

For the hydrogen molecule, with the two protons at a fixed distance $r_{\alpha\beta} = R$, the purely electronic Hamiltonian is given by (13.2) with the first, second, and fifth terms omitted. The nuclear repulsion V_{NN} equals e'^2/R. The purely electronic Hamiltonian involves the six electronic coordinates x_1, y_1, z_1, x_2, y_2, z_2 and involves the nuclear coordinates as parameters.

The electronic Schrödinger equation (13.4) can be dealt with by approximate methods to be discussed later. If we plot the electronic energy including nuclear repulsion for a bound state of a diatomic molecule against the internuclear distance R, we find a curve like Fig. 13.1. At $R = 0$, the internuclear repulsion causes U to go to infinity. The internuclear separation at the minimum in this curve is called the ***equilibrium internuclear distance*** R_e. The difference between the limiting value of U at infinite internuclear separation and its value at R_e is called the ***equilibrium dissociation energy*** D_e:

$$D_e \equiv U(\infty) - U(R_e) \qquad (13.9)$$

When nuclear motion is considered (Section 13.2), one finds that the equilibrium dissociation energy D_e differs from the molecular ground-vibrational-state dissociation energy D_0. The lowest state of nuclear motion has zero rotational energy [as shown by Eq. (6.47)] but has a nonzero vibrational energy—the zero-point energy. If we use the harmonic-oscillator approximation for the vibration of a diatomic molecule (Section 4.3), then this zero-point energy is $\frac{1}{2}h\nu$. This zero-point energy raises the energy for the ground state of nuclear motion $\frac{1}{2}h\nu$ above the minimum in the $U(R)$ curve, so D_0 is less than D_e and $D_0 \approx D_e - \frac{1}{2}h\nu$. Different electronic states of the same molecule have different values of R_e, D_e, D_0, and ν.

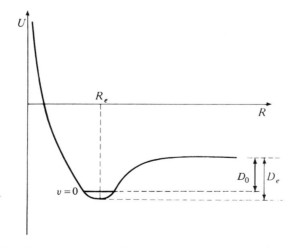

Figure 13.1 Electronic energy including internuclear repulsion as a function of the internuclear distance R for a diatomic-molecule bound electronic state.

Consider a gas composed of diatomic molecules AB. In the limit of absolute zero temperature, all the AB molecules are in their ground states of electronic and nuclear motion, so $D_0 N_A$ (where N_A is the Avogadro constant and D_0 is for the ground electronic state of AB) is the change in the thermodynamic internal energy U and enthalpy H for dissociation of 1 mole of ideal-gas diatomic molecules: $N_A D_0 = \Delta U_0^\circ = \Delta H_0^\circ$ for $AB(g) \rightarrow A(g) + B(g)$.

For some diatomic-molecule electronic states, solution of the electronic Schrödinger equation gives a $U(R)$ curve with no minimum. Such states are not bound and the molecule will dissociate. Examples are some of the states in Fig. 13.4.

Assuming that we have solved the electronic Schrödinger equation (which is quite an assumption), we next consider nuclear motions. According to our picture, the electrons move much faster than the nuclei. When the nuclei change their configuration slightly, say from q'_α to q''_α, the electrons immediately adjust to the change, with the electronic wave function changing from $\psi_{el}(q_i; q'_\alpha)$ to $\psi_{el}(q_i; q''_\alpha)$ and the electronic energy changing from $U(q'_\alpha)$ to $U(q''_\alpha)$. Thus, as the nuclei move, the electronic energy varies smoothly as a function of the parameters defining the nuclear configuration, and $U(q_\alpha)$ becomes, in effect, the potential energy for the nuclear motion. The electrons act like a spring connecting the nuclei; as the internuclear distance changes, the energy stored in the spring changes. Hence the Schrödinger equation for nuclear motion is

$$\hat{H}_N \psi_N = E \psi_N \qquad (13.10)^*$$

$$\hat{H}_N = -\frac{\hbar^2}{2} \sum_\alpha \frac{1}{m_\alpha} \nabla_\alpha^2 + U(q_\alpha) \qquad (13.11)$$

The variables in the nuclear Schrödinger equation are the nuclear coordinates, symbolized by q_α. The energy eigenvalue E in (13.10) is the *total* energy of the molecule, since the Hamiltonian (13.11) includes operators for both nuclear energy and electronic energy. E is simply a number and does not depend on any coordinates. Note that for each electronic state of a molecule we must solve a different nuclear Schrödinger equation, since U differs from state to state. In this chapter we shall concentrate on the electronic Schrödinger equation (13.4).

The approximation of separating electronic and nuclear motions is called the **Born–Oppenheimer approximation** and is basic to quantum chemistry. [The American physicist J. Robert Oppenheimer (1904–1967) was a graduate student of Born in 1927. During World War II, Oppenheimer directed the Los Alamos laboratory that developed the atomic bomb.] Born and Oppenheimer's mathematical treatment showed that the true molecular wave function is adequately approximated as

$$\psi(q_i, q_\alpha) = \psi_{el}(q_i; q_\alpha) \psi_N(q_\alpha) \qquad (13.12)$$

if $(m_e/m_\alpha)^{1/4} \ll 1$. The Born–Oppenheimer approximation introduces little error for the ground electronic states of diatomic molecules. Corrections for excited electronic states are larger than for the ground state, but still are usually small as compared with the errors introduced by the approximations used to solve the electronic Schrödinger equation of a many-electron molecule. Hence we shall not worry about corrections to the Born–Oppenheimer approximation. For further discussion of the Born–Oppenheimer approximation, see J. Goodisman, *Diatomic Interaction Potential Theory*, Academic Press, New York, 1973, Volume 1, Chapter 1.

13.2 NUCLEAR MOTION IN DIATOMIC MOLECULES

Most of our attention in this chapter will be on the electronic Schrödinger equation for diatomic molecules, but this section examines nuclear motion in a bound electronic state of a diatomic molecule. From (13.10) and (13.11), the Schrödinger equation for nuclear motion in a diatomic-molecule bound electronic state is

$$\left[-\frac{\hbar^2}{2m_\alpha} \nabla_\alpha^2 - \frac{\hbar^2}{2m_\beta} \nabla_\beta^2 + U(R) \right] \psi_N = E\psi_N \tag{13.13}$$

where α and β are the nuclei, and the nuclear-motion wave function ψ_N is a function of the nuclear coordinates x_α, y_α, z_α, x_β, y_β, and z_β.

The potential energy $U(R)$ is a function of only the relative coordinates of the two nuclei, and the work of Section 6.3 shows that the two-particle Schrödinger equation (13.13) can be reduced to two separate one-particle Schrödinger equations, one for translational energy of the entire molecule, and one for internal motion of the nuclei relative to each other. We have

$$\psi_N = \psi_{N,\text{tr}}\psi_{N,\text{int}} \quad \text{and} \quad E = E_{\text{tr}} + E_{\text{int}} \tag{13.14}$$

The translational energy levels can be taken as the energy levels of a particle in a three-dimensional box whose dimensions are those of the container holding the gas of diatomic molecules.

The Schrödinger equation for $\psi_{N,\text{int}}$ is [Eq. (6.43)]

$$\left[-\frac{\hbar^2}{2\mu} \nabla^2 + U(R) \right] \psi_{N,\text{int}} = E_{\text{int}}\psi_{N,\text{int}} , \qquad \mu \equiv m_\alpha m_\beta/(m_\alpha + m_\beta) \tag{13.15}$$

where $\psi_{N,\text{int}}$ is a function of the coordinates of one nucleus relative to the other. The best coordinates to use are the spherical polar coordinates of one nucleus relative to the other (Fig. 6.5 with m_N and m_e replaced by m_α and m_β). The radius r in relative spherical polar coordinates is the internuclear distance R, and we shall denote the relative angular coordinates by θ_N and ϕ_N. Since the potential energy in (13.15) depends on R only, this is a central-force problem, and the work of Section 6.1 shows that

$$\psi_{N,\text{int}} = P(R)Y_J^M(\theta_N, \phi_N) , \qquad J = 0, 1, 2, \ldots , \qquad M = -J, \ldots, J \tag{13.16}$$

where the Y_J^M functions are the spherical harmonic functions with quantum numbers J and M.

From (6.17), the radial function $P(R)$ is found by solving

$$-\frac{\hbar^2}{2\mu} \left[P''(R) + \frac{2}{R} P'(R) \right] + \frac{J(J+1)\hbar^2}{2\mu R^2} P(R) + U(R)P(R) = E_{\text{int}}P(R) \tag{13.17}$$

This differential equation is simplified by defining $F(R)$ as

$$F(R) \equiv RP(R) \tag{13.18}$$

Substitution of $P = F/R$ into (13.17) gives (Problem 13.5)

$$-\frac{\hbar^2}{2\mu} F''(R) + \left[U(R) + \frac{J(J+1)\hbar^2}{2\mu R^2} \right] F(R) = E_{\text{int}} F(R) \tag{13.19}$$

which looks like a one-dimensional Schrödinger equation with the effective potential energy $U(R) + J(J+1)\hbar^2/2\mu R^2$.

The most fundamental way to solve (13.19) is as follows: (a) solve the electronic Schrödinger equation (13.7) at several values of R to obtain E_{el} of the particular molecular electronic state one is interested in; (b) add $Z_\alpha Z_\beta e'^2/R$ to each E_{el} value to obtain U at these R values; (c) devise a mathematical function $U(R)$ whose parameters are adjusted to give a good fit to the calculated U values; (d) insert the function $U(R)$ found in (c) into the nuclear-motion radial Schrödinger equation (13.19) and solve (13.19) by numerical methods.

A commonly used fitting procedure for step (c) is the method of *cubic splines*, for which computer programs exist (see *Press et al.*, Chapter 3; *Shoup*, Chapter 6). To fit the known data points (x_1, y_1), (x_2, y_2), . . . , (x_m, y_m) by the cubic-spline method, one finds a set of $m - 1$ cubic polynomials such that the first polynomial goes from (x_1, y_1) to (x_2, y_2), the second goes from (x_2, y_2) to (x_3, y_3), and so on, and such that the first and second derivatives of one polynomial equal the first and second derivatives, respectively, of the next polynomial at the point where the two polynomials join.

As to step (d), numerical solution of the one-dimensional Schrödinger equation (13.19) is a bit tricky, since both the eigenfunctions F and the eigenvalues E_{int} are unknown. Two accurate computer-based numerical methods to solve the one-dimensional Schrödinger equation are the *Cooley–Numerov method* [see J. Tellinghuisen, *J. Chem. Educ.*, **66**, 51 (1989)] and the *finite-element method* [see D. J. Searles and E. I. von Nagy-Felsobuki, *Am. J. Phys.*, **56**, 444 (1988)].

The solutions $F(R)$ of the radial equation (13.19) for a given J are characterized by a quantum number v, where v is the number of nodes in $F(R)$; $v = 0, 1, 2, \ldots$. The energy levels E_{int} found from the condition that $P(R) = F(R)/R$ be quadratically integrable depend on the quantum number J, which occurs in (13.19), and depend on v, which characterizes $F(R)$; $E_{\text{int}} = E_{v,J}$. The angular factor $Y_J^M(\theta_N, \phi_N)$ in (13.16) is a function of the angular coordinates; changes in θ_N and ϕ_N with R held fixed correspond to changes in the spatial orientation of the diatomic molecule, which is rotational motion. The quantum numbers J and M are rotational quantum numbers. Note that Y_J^M is the wave function of a rigid two-particle rotor [Eq. (6.46)]. A change in the R coordinate is a change in the internuclear distance, which is a vibrational motion, and the quantum number v, which characterizes $F(R)$, is a vibrational quantum number.

Since accurate solution of the electronic Schrödinger equation [step (a)] is difficult, one usually uses simpler, less-accurate procedures than that of steps (a) to (d). The simplest approach is to expand $U(R)$ in a Taylor series about R_e

(Problem 4.1):

$$U(R) = U(R_e) + U'(R_e)(R - R_e) + \tfrac{1}{2}U''(R_e)(R - R_e)^2$$
$$+ \tfrac{1}{6}U'''(R_e)(R - R_e)^3 + \cdots \tag{13.20}$$

At the equilibrium internuclear distance R_e, the slope of the $U(R)$ curve is zero (Figure 13.1), so $U'(R_e) = 0$. We can anticipate that the molecule will vibrate about the equilibrium distance R_e. For R close to R_e, $(R - R_e)^3$ and higher powers will be small, and we shall neglect these terms. Defining the *equilibrium force constant* k_e as $k_e \equiv U''(R_e)$, we have

$$U(R) \approx U(R_e) + \tfrac{1}{2}k_e(R - R_e)^2 = U(R_e) + \tfrac{1}{2}k_e q^2 \tag{13.21}$$

$$k_e \equiv U''(R_e) \quad \text{and} \quad q \equiv R - R_e$$

We have approximated $U(R)$ by a parabola [Fig. 4.5 with $V \equiv U(R) - U(R_e)$ and $x \equiv R - R_e$].

With the change of independent variable $q \equiv R - R_e$, (13.19) becomes

$$-\frac{\hbar^2}{2\mu} S''(q) + \left[U(R_e) + \tfrac{1}{2}k_e q^2 + \frac{J(J + 1)\hbar^2}{2\mu(q + R_e)^2} \right] S(q) \approx E_{\text{int}} S(q) \tag{13.22}$$

$$\text{where} \quad S(q) \equiv F(R) \tag{13.23}$$

Expanding $1/(q + R_e)^2$ in a Taylor series, we have (Problem 13.6)

$$\frac{1}{(q + R_e)^2} = \frac{1}{R_e^2(1 + q/R_e)^2} = \frac{1}{R_e^2}\left(1 - 2\frac{q}{R_e} + 3\frac{q^2}{R_e^2} - \cdots\right) \approx \frac{1}{R_e^2} \tag{13.24}$$

We are assuming that $R - R_e = q$ is small, so all terms after the 1 have been neglected in (13.24). Substitution of (13.24) into (13.22) and rearrangement gives

$$-\frac{\hbar^2}{2\mu} S''(q) + \tfrac{1}{2}k_e q^2 S(q) \approx \left[E_{\text{int}} - U(R_e) - \frac{J(J + 1)\hbar^2}{2\mu R_e^2} \right] S(q) \tag{13.25}$$

Equation (13.25) is the same as the Schrödinger equation for a one-dimensional harmonic oscillator with coordinate q, mass μ, potential energy $\tfrac{1}{2}k_e q^2$, and energy eigenvalues $E_{\text{int}} - U(R_e) - J(J + 1)\hbar^2/2\mu R_e^2$. [The boundary conditions for (13.25) and (4.34) are not the same, but this difference is unimportant and can be ignored (*Levine, Molecular Spectroscopy*, p. 147).] We can therefore set the terms in brackets in (13.25) equal to the harmonic-oscillator eigenvalues, and we have

$$E_{\text{int}} - U(R_e) - J(J + 1)\hbar^2/2\mu R_e^2 \approx (v + \tfrac{1}{2})h\nu_e$$

$$E_{\text{int}} \approx U(R_e) + (v + \tfrac{1}{2})h\nu_e + J(J + 1)\hbar^2/2\mu R_e^2 \tag{13.26}$$

$$\nu_e = (k_e/\mu)^{1/2}/2\pi, \qquad v = 0, 1, 2, \ldots \tag{13.27}$$

where (4.25) was used for ν_e, the *equilibrium* (or *harmonic*) *vibrational frequency*. The molecular internal energy E_{int} is the sum of the electronic energy $U(R_e)$ (which differs for different electronic states of the same molecule), the vibrational energy $(v + \tfrac{1}{2})h\nu_e$, and the rotational energy $J(J + 1)\hbar^2/2\mu R_e^2$. The approximations (13.21) and (13.24) correspond to a harmonic-oscillator, rigid-rotor approximation. From (13.26) and (13.14), the molecular energy $E = E_{tr} + E_{int}$ is the sum of translational, rotational, vibrational, and electronic energies.

From (13.14), (13.16), (13.18), and (13.23), the nuclear-motion wave function is

$$\psi_N \approx \psi_{N,tr} S_v(R - R_e) R^{-1} Y_J^M(\theta_N, \phi_N) \qquad (13.28)$$

where $S_v(R - R_e)$ is a harmonic-oscillator eigenfunction with quantum number v.

Comparison with the observed vibration–rotation energy levels of diatomic molecules shows that (13.26) gives rather poor agreement with experiment. The accuracy can be improved by the addition of the first- and second-order perturbation-theory energy corrections due to the terms neglected in (13.21) and (13.24). When this is done (see *Levine, Molecular Spectroscopy*, Section 4.2), the energy contains additional terms corresponding to vibrational anharmonicity [Eq. (4.68)], vibration–rotation interaction, and rotational centrifugal distortion of the molecule.

13.3 ATOMIC UNITS

Most quantum chemists report the results of their calculations using a system of units called *atomic units*.

First consider the cgs Gaussian system of units. The hydrogenlike-atom Hamiltonian in these units is (assuming infinite nuclear mass)

$$-(\hbar^2/2m_e)\nabla^2 - Ze'^2/r$$

The system of atomic units that is based on Gaussian units is defined as follows. The unit of mass is the electron's mass m_e, rather than the gram; the unit of charge is the proton's charge e', rather than the statcoulomb; the unit of angular momentum is $\hbar$, rather than the $g\,cm^2/s$. (The atomic unit of mass used in quantum chemistry should not be confused with the quantity 1 amu, which is one-twelfth the mass of a ^{12}C atom.) When we switch to atomic units, $\hbar$, m_e, and e' each have a numerical value of 1. Hence, to change a formula from cgs Gaussian units to atomic units, we simply set these quantities equal to 1. Thus in atomic units the hydrogenlike Hamiltonian is $-\frac{1}{2}\nabla^2 - Z/r$, where r is now measured in atomic units of length, rather than in centimeters. The ground-state energy of the hydrogenlike atom is given by (6.94) as $-\frac{1}{2}Z^2(e'^2/a_0)$. Since [Eq. (6.106)] $a_0 = \hbar^2/m_e e'^2$, the numerical value of a_0 in atomic units is 1, and the ground-state energy of the hydrogenlike atom has the numerical value (neglecting nuclear motion) $-\frac{1}{2}Z^2$ in atomic units.

The atomic unit of energy, e'^2/a_0, is called the *hartree* (symbol E_h):

$$1 \text{ hartree} \equiv E_h \equiv e'^2/a_0 = 27.2114 \text{ eV} \qquad (13.29)$$

The ground-state energy of the hydrogen atom is $-\frac{1}{2}$ hartree if nuclear motion is neglected.

The atomic unit of length is called the *bohr*:

$$1 \text{ bohr} \equiv a_0 = 0.529177 \text{ Å} \qquad (13.30)$$

Now suppose we start with SI units. The hydrogenlike Hamiltonian in SI units is $-(\hbar^2/2m_e)\nabla^2 - Ze^2/4\pi\varepsilon_0 r$. The system of atomic units based on SI units is defined as follows. The units of mass, charge, and angular momentum are defined as the electron's mass m_e, the proton's charge e, and $\hbar$, respectively (rather than the kilogram, the coulomb, and the $\mathrm{kg\,m^2/s}$); the unit of permittivity is $4\pi\varepsilon_0$, rather than the $\mathrm{C^2\,N^{-1}\,m^{-2}}$. (In Gaussian units, charge is expressible in terms of mass, length, and time, whereas in SI units this is not so. So in SI atomic units, we define four quantities as compared with three for Gaussian atomic units.) When we switch to atomic units, $\hbar$, m_e, e, and $4\pi\varepsilon_0$ each have a numerical value of 1. In SI atomic units, the hydrogenlike Hamiltonian is $-\frac{1}{2}\nabla^2 - Z/r$; the Bohr radius $a_0 = 4\pi\varepsilon_0\hbar^2/m_e e^2$ has the numerical value 1; the hydrogenlike ground-state energy is $-\frac{1}{2}Z^2$.

Although the hydrogenlike Schrödinger equation in atomic units is generally written as $-\frac{1}{2}\nabla^2\psi - (Z/r)\psi = E\psi$, a more accurate way to write this equation would be (Problem 13.8) $-\frac{1}{2}\nabla^{\dagger 2}\psi^\dagger - (Z/r^\dagger)\psi^\dagger = E^\dagger\psi^\dagger$, where $r^\dagger$ and $E^\dagger$ are dimensionless quantities defined as $r^\dagger \equiv r/a_0$ and $E^\dagger \equiv E/E_\mathrm{h}$, where $\nabla^{\dagger 2}$ is given by Eq. (6.6) with r replaced by $r^\dagger$, and $\psi^\dagger$ is found by replacing r with $r^\dagger$ in ψ. However, no one bothers to include the daggers.

Use of atomic units saves time by eliminating m_e, e', and $\hbar$ from equations. Also, if the results of a quantum-mechanical calculation are reported in electron volts, ergs, or joules, the value given depends on the currently accepted values of the physical constants; reporting results in atomic units avoids this.

Atomic units will be used in Chapters 13, 15, and 16.

13.4   THE HYDROGEN MOLECULE ION

We now begin the study of the electronic energies of molecules. We shall use the Born–Oppenheimer approximation, keeping the nuclei fixed while we solve, as best we can, the Schrödinger equation for the motion of the electrons. We shall usually be considering an isolated molecule, ignoring intermolecular interactions. Our results will be most applicable to molecules in the gas phase at low pressure.

We start with diatomic molecules, the simplest of which is $\mathrm{H_2^+}$, the hydrogen molecule ion, consisting of two protons and one electron. Just as the one-electron hydrogen atom serves as a starting point in the discussion of many-electron atoms, the one-electron hydrogen molecular ion furnishes many ideas useful for discussing many-electron diatomic molecules. The electronic Schrödinger equation for $\mathrm{H_2^+}$ is separable, and we can obtain exact solutions for the eigenfunctions and eigenvalues.

Figure 13.2 shows $\mathrm{H_2^+}$. The nuclei are at a and b; R is the internuclear distance; r_a and r_b are the distances from the electron to nuclei a and b. Since the nuclei are fixed, we have a one-particle problem whose Hamiltonian is [Eq. (13.5)]

$$-\frac{\hbar^2}{2m_e}\nabla^2 - \frac{e'^2}{r_a} - \frac{e'^2}{r_b}$$

The first term is the electronic kinetic-energy operator; the second and third terms are the attractions between the electron and the nuclei. In atomic units the purely

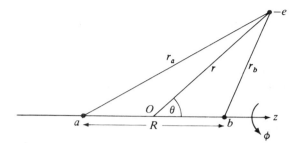

Figure 13.2 Interparticle distances in H_2^+.

electronic Hamiltonian for H_2^+ is

$$\hat{H}_{el} = -\tfrac{1}{2}\nabla^2 - \frac{1}{r_a} - \frac{1}{r_b} \tag{13.31}$$

In Fig. 13.2 the coordinate origin is on the internuclear axis, midway between the nuclei, with the z axis lying along the internuclear axis. We might attempt to solve the electronic Schrödinger equation using the spherical polar coordinates (r, θ, ϕ) of the electron, expressing r_a and r_b in these coordinates (Problem 13.11). We would find that the Schrödinger equation is not separable in spherical polar coordinates. However, separation of variables is possible using *confocal elliptic coordinates* ξ, η, and ϕ. The coordinate ϕ is the angle of rotation of the electron about the internuclear (z) axis, the same as in spherical polar coordinates. The coordinates ξ (xi) and η (eta) are defined by

$$\xi \equiv \frac{r_a + r_b}{R} , \qquad \eta \equiv \frac{r_a - r_b}{R} \tag{13.32}$$

The ranges of these coordinates are (Problem 13.11)

$$0 \leqslant \phi \leqslant 2\pi , \qquad 1 \leqslant \xi \leqslant \infty , \qquad -1 \leqslant \eta \leqslant 1 \tag{13.33}$$

We must put the Hamiltonian (13.31) into these coordinates. We have

$$r_a = \tfrac{1}{2}R(\xi + \eta) , \qquad r_b = \tfrac{1}{2}R(\xi - \eta) \tag{13.34}$$

We also need the expression for the Laplacian in confocal elliptic coordinates. One way to find this is to express ξ, η, and ϕ in terms of x, y, and z, the Cartesian coordinates of the electron, and then use the chain rule to find $\partial/\partial x$, $\partial/\partial y$, and $\partial/\partial z$ in terms of $\partial/\partial\xi$, $\partial/\partial\eta$, and $\partial/\partial\phi$. We then form $\nabla^2 \equiv \partial^2/\partial x^2 + \partial^2/\partial y^2 + \partial^2/\partial z^2$. The derivation of ∇^2 is omitted. (For a discussion, see *Margenau and Murphy*, Chapter 5.) Substitution of ∇^2 and (13.34) into (13.31) gives $\hat{H}_{el}$ of H_2^+ in confocal elliptic coordinates. The result is (in atomic units)

$$\hat{H}_{el} = -\frac{2}{R^2(\xi^2 - \eta^2)} \left[(\xi^2 - 1)\frac{\partial^2}{\partial\xi^2} + 2\xi\frac{\partial}{\partial\xi} + (1 - \eta^2)\frac{\partial^2}{\partial\eta^2} - 2\eta\frac{\partial}{\partial\eta} \right.$$
$$\left. + \left(\frac{1}{\xi^2 - 1} + \frac{1}{1 - \eta^2}\right)\frac{\partial^2}{\partial\phi^2} \right] - \frac{2}{R(\xi + \eta)} - \frac{2}{R(\xi - \eta)} \tag{13.35}$$

For the hydrogen atom, whose Hamiltonian has spherical symmetry, $\hat{L}^2$ and $\hat{L}_z$ both commute with $\hat{H}$. The H_2^+ ion does not have spherical symmetry, and one finds that $[\hat{L}^2, \hat{H}_{el}] \neq 0$ for H_2^+. However, H_2^+ does have axial symmetry, and we shall show that $\hat{L}_z$ commutes with $\hat{H}_{el}$ of H_2^+. The operator $\hat{L}_z = -i\hbar(\partial/\partial\phi)$ [Eq. (5.67)] involves only ϕ; hence, in calculating the commutator, only the part of $\hat{H}_{el}$ that involves ϕ need be considered. We have in atomic units

$$[\hat{L}_z, \hat{H}_{el}] = \left[-i\frac{\partial}{\partial\phi}, -\frac{2}{R^2(\xi^2 - \eta^2)}\left(\frac{1}{\xi^2 - 1} + \frac{1}{1 - \eta^2}\right)\frac{\partial^2}{\partial\phi^2} \right] = 0$$

Therefore, the electronic wave functions can be chosen to be eigenfunctions of $\hat{L}_z$; the eigenfunctions of $\hat{L}_z$ are [Eq. (5.81)]

$$\text{constant} \cdot (2\pi)^{-1/2}e^{im\phi}, \qquad \text{where} \quad m = 0, \pm1, \pm2, \pm3, \ldots \qquad (13.36)$$

The z component of electronic orbital angular momentum in H_2^+ is $m\hbar$ or m in atomic units. The total electronic orbital angular momentum is not a constant for H_2^+.

The "constant" in (13.36) is a constant only as far as $\partial/\partial\phi$ is concerned, so the H_2^+ wave functions have the form $\psi_{el} = F(\xi, \eta)(2\pi)^{-1/2}e^{im\phi}$. We now try a separation of variables:

$$\psi_{el} = L(\xi)M(\eta)(2\pi)^{-1/2}e^{im\phi} \qquad (13.37)$$

Substitution of (13.37) into $\hat{H}_{el}\psi_{el} = E_{el}\psi_{el}$ gives an equation in which the variables are separable. We are led to two ordinary differential equations, one for $L(\xi)$ and one for $M(\eta)$ (Problem 13.9). Solving these equations, one finds that the condition that ψ_{el} be well behaved requires that, for each fixed value of R, only certain values of E_{el} are allowed; this gives a set of different electronic states. There is no algebraic formula for E_{el}; it must be calculated numerically for each desired value of R for each state. In addition to the quantum number m, the H_2^+ electronic wave functions are characterized by the quantum numbers n_ξ and n_η, which give the number of nodes in the $L(\xi)$ and $M(\eta)$ factors in ψ_{el}.

The process of solving for $L(\xi)$ and $M(\eta)$ is very complicated and is omitted. For the procedure used and for tables of H_2^+ electronic wave functions and energies, see D. R. Bates, K. Ledsham, and A. L. Stewart, *Phil. Trans. Roy. Soc.*, **A246**, 215 (1953). The wave functions in Bates et al. are not normalized. For normalization constants, see E. M. Roberts, M. R. Foster, and F. F. Selig, *J. Chem. Phys.*, **37**, 485 (1962). For accurate ground-state energies, see H. Wind, *J. Chem. Phys.*, **42**, 2371 (1965).

For the ground electronic state, the quantum number m is zero. At $R = \infty$, the H_2^+ ground state is dissociated into a proton and a ground-state hydrogen atom; hence $E_{el}(\infty) = -\frac{1}{2}$ hartree. At $R = 0$, the two protons have come together to form the He^+ ion with ground-state energy: $-\frac{1}{2}(2)^2$ hartrees $= -2$ hartrees. Addition of the internuclear repulsion $1/R$ (in atomic units) to $E_{el}(R)$ gives the $U(R)$ potential-energy curve for nuclear motion. Plots of the ground-state $E_{el}(R)$ and $U(R)$, as found from solution of the electronic Schrödinger equation, are shown in Fig. 13.3. At $R = \infty$, the internuclear repulsion is 0, and U is $-\frac{1}{2}$ hartree.

The $U(R)$ curve is found to have a minimum at [see L. J. Schaad and W. V.

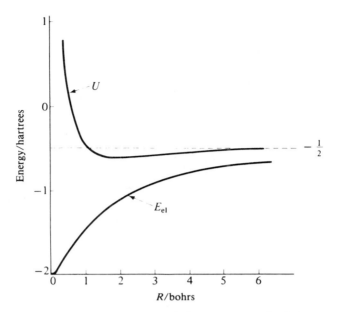

Figure 13.3 Electronic energy with (U) and without (E_{el}) internuclear repulsion for the H_2^+ ground electronic state.

Hicks, *J. Chem. Phys.*, **53**, 851 (1970)]

$$R_e = 1.9972 \text{ bohrs} \approx 2.00 \text{ bohrs} = 1.06 \text{ Å}$$

indicating that the H_2^+ ground electronic state is a stable bound state. The calculated value of E_{el} at 1.9972 bohrs is -1.1033 hartrees; addition of the internuclear repulsion $1/R$ gives $U(R_e) = -0.6026$ hartree, compared with -0.5000 hartree at $R = \infty$. The ground-state binding energy is thus $D_e = 0.1026$ hartree $= 2.79$ eV. This corresponds to 64.4 kcal/mol $= 269$ kJ/mol. The binding energy is only 17% of the total energy at the equilibrium internuclear distance. Thus a reasonably small error in the total energy can correspond to a large error in the binding energy. For heavier molecules the situation is even worse, since chemical binding energies are of the same order of magnitude for most diatomic molecules, but the total electronic energy increases markedly for heavier molecules.

Note that the single electron in H_2^+ is sufficient to give a stable bound state.

Figure 13.4 shows the $U(R)$ curves for the first several electronic energy levels of H_2^+, as found by solving the electronic Schrödinger equation.

The angle ϕ occurs in $\hat{H}_{el}$ of H_2^+ only as $\partial^2/\partial\phi^2$ [Eq. (13.35)]. When ψ_{el} of (13.37) is substituted into $\hat{H}_{el}\psi_{el} = E_{el}\psi_{el}$, the $e^{im\phi}$ factor cancels, and we are led to differential equations for $L(\xi)$ and $M(\eta)$ in which the m quantum number occurs only as m^2. Since E_{el} is found from the $L(\xi)$ and $M(\eta)$ differential equations, E_{el} depends on m^2, and each electronic level with $m \neq 0$ is doubly degenerate, corresponding to states with quantum numbers $+|m|$ and $-|m|$. In the standard notation for diatomic molecules [F. A. Jenkins, *J. Opt. Soc. Am.*,

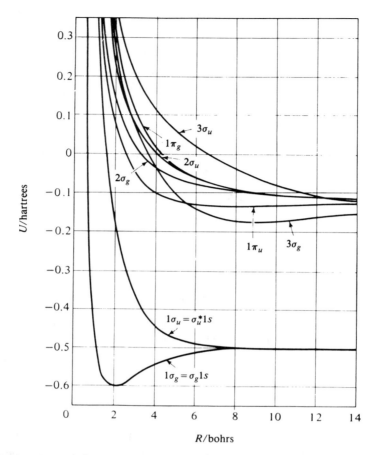

Figure 13.4 $U(R)$ curves for several H_2^+ electronic states. [Curves taken from J. C. Slater, *Quantum Theory of Molecules and Solids*, vol. 1, McGraw-Hill, New York, 1963. Used by permission.]

43, 425 (1953)], the absolute value of m is called λ:

$$\lambda \equiv |m|$$

(Some texts define λ as identical to m.) Similar to the s, p, d, f notation for hydrogen-atom states, a letter code is used to specify λ, the absolute value (in atomic units) of the component along the molecular axis of the electron's orbital angular momentum:

λ	0	1	2	3	4
letter	σ	π	δ	ϕ	γ

$(13.38)^*$

Thus the lowest H_2^+ electronic state is a σ state.

Besides classifying the states of H_2^+ according to λ, we can also classify them according to their parity (Section 7.5). From Fig. 13.11, inversion of the electron's coordinates through the origin O changes ϕ to $\phi + \pi$, r_a to r_b, and r_b to

r_a; this leaves the potential-energy part of the electronic Hamiltonian (13.31) unchanged. We previously showed the kinetic-energy operator to be invariant under inversion. Hence the parity operator commutes with the Hamiltonian (13.31), and the H_2^+ electronic wave functions can be classified as either even or odd. For even electronic wave functions, we use the subscript g (from the German word *gerade*, meaning even); for odd wave functions, we use u (from *ungerade*).

The lowest σ_g energy level in Fig. 13.4 is labeled $1\sigma_g$, the next-lowest σ_g level at small R is labeled $2\sigma_g$, and so on. The lowest σ_u level is labeled $1\sigma_u$, and so on. The alternative notation $\sigma_g 1s$ indicates that this level dissociates to a $1s$ hydrogen atom. The meaning of the star in $\sigma_u^* 1s$ will be explained later.

For completeness we must take spin into account by multiplying each spatial H_2^+ electronic wave function by α or β, depending on whether the component of electron spin along the internuclear axis is $+\frac{1}{2}$ or $-\frac{1}{2}$ (in atomic units). Inclusion of spin doubles the degeneracy of all levels.

Because R_e of the $1\sigma_g$ H_2^+ ground electronic level differs greatly from the R_e values of the $2\sigma_g$ and $1\pi_u$ excited levels (Fig. 13.4), there is very little overlap between the vibrational wave function of $1\sigma_g$ and that of either excited level. Therefore, the transition dipole moment integral $\langle \psi_i | \hat{\mathbf{d}} | \psi_j \rangle$ (Section 9.10) is very small when i is the ground electronic state, and electronic absorption spectra from the ground electronic state of H_2^+ are too weak to be observed. [By measuring the translational energy of H_2^+ ions scattered by helium atoms, a transition between $1\sigma_g$ and $1\pi_u$ has been detected; see N. J. Kirchner et al., *Phys. Rev. Lett.*, **52**, 26 (1984).] H_2^+ is readily detected in a mass spectrometer.

The H_2^+ ground electronic state has $R_e = 2.00$ bohrs $= 2.00\hbar^2/m_e e'^2$. The negative muon (symbol μ^-) is a short-lived (half-life 2×10^{-6} s) elementary particle whose charge is the same as that of an electron but whose mass m_μ is 207 times m_e. When a beam of negative muons (produced when ions accelerated to high speed collide with ordinary matter) enters H_2 gas, a series of processes leads to the formation of muomolecular ions that consist of two protons and one muon. This species, symbolized by $(p\mu p)^+$, is an H_2^+ ion in which the electron has been replaced by a muon. Its R_e is $2.00\hbar^2/m_\mu e'^2 = 2.00\hbar^2/207m_e e'^2 = (2.00/207)$ bohr $= 0.0051$ Å. The two nuclei in this muoion are 207 times closer than in H_2^+. The magnitude of the vibrational-wave-function factor $S_v(R - R_e)$ in (13.28) is small but not entirely negligible for $R - R_e = -0.0051$ Å; so there is some probability for the nuclei in $(p\mu p)^+$ to come in contact, and we have the possibility that nuclear fusion might occur. The isotopic nuclei ^{2}H (deuterium, D), and ^{3}H (tritium, T) undergo fusion much more readily than protons, so instead of H_2 gas, one uses a mixture of D_2 and T_2 gases. After fusion occurs, the muon is released and can then be recaptured to catalyze another fusion. Under the right conditions, one muon can catalyze 150 fusions on average before it decays. Unfortunately, at present, more energy is needed to produce the muon beam than is released by the fusion. (See J. Rafelski and S. E. Jones, *Scientific American*, July 1987, page 84.)

In the rest of this chapter, the subscript el will be dropped from the electronic wave function, Hamiltonian, and energy. It will be understood in Chapters 13, 15, and 16 that ψ means ψ_{el}.

13.5 APPROXIMATE TREATMENTS OF THE
H₂⁺ GROUND ELECTRONIC STATE

For a many-electron atom, the self-consistent-field (SCF) method is used to construct an approximate wave function as a Slater determinant of (one-electron) spin-orbitals. The one-electron spatial part of a spin-orbital is an atomic orbital (AO). We took each AO as a product of a spherical harmonic and a radial factor; as an initial approximation to the radial factors, we can use hydrogenlike radial functions with effective nuclear charges.

For many-electron molecules, which (unlike H_2^+) cannot be solved exactly, we want to use many of the ideas of the SCF treatment of atoms. We shall write an approximate molecular electronic wave function as a Slater determinant of (one-electron) spin-orbitals. The one-electron spatial part of a molecular spin-orbital will be called a *molecular orbital* (MO). Because of the Pauli principle, each MO can hold no more than two electrons, just as for AOs. What kind of functions do we use for the MOs? Ideally, the analytic form of each MO is found by an SCF calculation (Sections 13.16 to 13.18). In this section, we seek a simple approximation for the MOs that will enable us to gain some qualitative understanding of chemical bonding. Just as we took the angular part of each AO to be the same kind of function (a spherical harmonic) as in the one-electron hydrogenlike atom, we shall take the angular part of each diatomic MO to be $(2\pi)^{-1/2}e^{im\phi}$, as in H_2^+. However, the ξ and η factors in the H_2^+ wave functions are complicated functions and do not readily lend themselves to use in MO calculations. We therefore seek simpler functions that will provide reasonably accurate approximations to the H_2^+ wave functions and that can be used to construct molecular orbitals for many-electron diatomic molecules. With this discussion as motivation for looking at approximate solutions in a case where the Schrödinger equation is exactly solvable, we consider approximate treatments of H_2^+.

We shall use the variation method, writing down some function containing several parameters and varying them to minimize the variational integral. This will give an approximation to the ground-state wave function and an upper bound to the ground-state energy. By use of the factor $e^{im\phi}$ in the trial function, we can get an upper bound to the energy of the lowest H_2^+ level for any given value of m (see Section 8.2). By using linear variation functions, we can get approximations for excited states.

The H_2^+ ground state has $m = 0$, and the wave function depends only on ξ and η. We could try any well-behaved function of these coordinates as a trial variation function. We shall, however, use a more systematic approach based on the idea of a molecule as being formed from the interaction of atoms.

Consider what the H_2^+ wave function would look like for large values of the internuclear separation R. When the electron is near nucleus a, nucleus b is so far away that we essentially have a hydrogen atom with origin at a; thus, when r_a is small, the ground-state H_2^+ electronic wave function should resemble the ground-state hydrogen-atom wave function of Eq. (6.104). We have $Z = 1$, and the Bohr radius a_0 has the numerical value 1 in atomic units; hence (6.104) becomes

$$\pi^{-1/2}e^{-r_a} \tag{13.39}$$

Similarly, we conclude that when the electron is near nucleus b the H_2^+ ground-state wave function will be approximated by

$$\pi^{-1/2}e^{-r_b} \tag{13.40}$$

These considerations suggest that we try as a variation function

$$c_1\pi^{-1/2}e^{-r_a} + c_2\pi^{-1/2}e^{-r_b} \tag{13.41}$$

where c_1 and c_2 are variational parameters. When the electron is near nucleus a, the variable r_a is small and r_b is large, and the first term in (13.41) predominates, giving a function resembling (13.39). The function (13.41) is a linear variation function, and we are led to solve a secular equation, which has the form (8.57), where the subscripts 1 and 2 refer to the functions (13.39) and (13.40).

We can also approach the problem using perturbation theory. We take the unperturbed system as the H_2^+ molecule with $R = \infty$. For $R = \infty$, the electron can be bound to nucleus a with wave function (13.39), or it can be bound to nucleus b with wave function (13.40). In either case the energy is $-\frac{1}{2}$ hartree, and we have a doubly degenerate unperturbed energy level. Bringing the nuclei in from infinity gives rise to a perturbation that splits the doubly degenerate unperturbed level into two levels; this is illustrated by the $U(R)$ curves for the two lowest H_2^+ electronic states, which both dissociate to a ground-state hydrogen atom (see Fig. 13.4). The correct zeroth-order wave functions for the perturbed levels are linear combinations of the form (13.41), and we are led to a secular equation of the form (8.57), with W replaced by $E^{(0)} + E^{(1)}$ (see Problem 9.12).

Before proceeding with the solution of (8.57), let us improve the trial function (13.41). Consider the limiting behavior of the H_2^+ ground-state electronic wave function as R goes to zero. In this limit we get the He^+ ion, which has the ground-state wave function [put $Z = 2$ in (6.104)]

$$2^{3/2}\pi^{-1/2}e^{-2r} \tag{13.42}$$

From Fig. 13.2 we see that as R goes to zero both r_a and r_b go to r; hence as R goes to zero the trial function (13.41) goes to $(c_1 + c_2)\pi^{-1/2}e^{-r}$. Comparing with (13.42), we see that our trial function has the wrong limiting behavior at $R = 0$; it should go to e^{-2r}, not e^{-r}. We can fix things by multiplying r_a and r_b in the exponentials by a variational parameter k, which will be some function of R; $k = k(R)$. For the correct limiting behavior at $R = 0$ and at $R = \infty$, we have $k(0) = 2$ and $k(\infty) = 1$ for the H_2^+ ground electronic state. Physically, k is some sort of effective nuclear charge, which increases as the nuclei come together. We thus take as the trial function

$$\phi = c_a 1s_a + c_b 1s_b \tag{13.43}$$

where the c's are variational parameters and

$$1s_a = k^{3/2}\pi^{-1/2}e^{-kr_a}, \qquad 1s_b = k^{3/2}\pi^{-1/2}e^{-kr_b} \tag{13.44}$$

The factor $k^{3/2}$ normalizes $1s_a$ and $1s_b$ [see Eq. (6.104)]. The molecular-orbital function (13.43) is a *linear combination of atomic orbitals*, an LCAO-MO. The trial function (13.43) was first used by Finkelstein and Horowitz in 1928.

For the function (13.43), the secular equation (8.57) is

$$\begin{vmatrix} H_{aa} - WS_{aa} & H_{ab} - WS_{ab} \\ H_{ba} - WS_{ba} & H_{bb} - WS_{bb} \end{vmatrix} = 0 \qquad (13.45)$$

The integrals H_{aa} and H_{bb} are

$$H_{aa} = \int 1s_a^* \hat{H} 1s_a \, dv \, , \qquad H_{bb} = \int 1s_b^* \hat{H} 1s_b \, dv \qquad (13.46)$$

where the H_2^+ electronic Hamiltonian operator $\hat{H}$ is given by (13.31). These two integrals are called *Coulomb integrals*. [Their form differs considerably from that of the Coulomb integral (9.100); the designation of (13.46) as Coulomb integrals is not really appropriate.] We can relabel the variables in a definite integral without affecting its value; changing a to b and b to a changes $1s_a$ to $1s_b$ but leaves $\hat{H}$ unaffected (this would not be true for a heteronuclear diatomic molecule). Hence $H_{aa} = H_{bb}$. We have

$$H_{ab} = \int 1s_a^* \hat{H} 1s_b \, dv \, , \qquad H_{ba} = \int 1s_b^* \hat{H} 1s_a \, dv \qquad (13.47)$$

Since $\hat{H}$ is Hermitian and the functions in these integrals are real, we conclude that $H_{ab} = H_{ba}$. The integral H_{ab} is called a *resonance* (or *bond*) *integral*. Since $1s_a$ and $1s_b$ are normalized and real, we have

$$S_{aa} = \int 1s_a^* 1s_a \, dv = 1 = S_{bb}$$

$$S_{ab} = \int 1s_a^* 1s_b \, dv = S_{ba} \qquad (13.48)^*$$

S_{ab} is an **overlap integral**.

The secular equation (13.45) becomes

$$\begin{vmatrix} H_{aa} - W & H_{ab} - S_{ab}W \\ H_{ab} - S_{ab}W & H_{aa} - W \end{vmatrix} = 0 \qquad (13.49)$$

$$H_{aa} - W = \pm(H_{ab} - S_{ab}W) \qquad (13.50)$$

$$W_1 = \frac{H_{aa} + H_{ab}}{1 + S_{ab}} \, , \qquad W_2 = \frac{H_{aa} - H_{ab}}{1 - S_{ab}} \qquad (13.51)$$

These two roots are upper bounds to the energies of the ground and first excited electronic states of H_2^+. We shall see that H_{ab} is negative, so W_1 is the lower-energy root.

We now find the coefficients in (13.43) for each of the roots of the secular equation. From Eq. (8.55) we have

$$(H_{aa} - W)c_a + (H_{ab} - S_{ab}W)c_b = 0$$

Substituting in W_1 from (13.51) [or using (13.50)], we get

$$c_a/c_b = 1 \qquad (13.52)$$

$$\phi_1 = c_a(1s_a + 1s_b) \qquad (13.53)$$

We fix c_a by normalization:

$$|c_a|^2 \int (1s_a^2 + 1s_b^2 + 2 \cdot 1s_a 1s_b)\, dv = 1$$

$$|c_a| = \frac{1}{(2 + 2S_{ab})^{1/2}}$$

The normalized trial function corresponding to the energy W_1 is thus

$$\phi_1 = \frac{1s_a + 1s_b}{\sqrt{2}(1 + S_{ab})^{1/2}} \qquad (13.54)$$

For the root W_2, we find $c_b = -c_a$ and

$$\phi_2 = \frac{1s_a - 1s_b}{\sqrt{2}(1 - S_{ab})^{1/2}} \qquad (13.55)$$

Equations (13.54) and (13.55) come as no surprise. Since the nuclei are identical, we expect $|\phi|^2$ to remain unchanged on interchanging a and b; in other words, we expect no polarity in the bond.

Let us now evaluate H_{aa}, H_{ab}, and S_{ab}. [This can be skimmed, if desired; begin reading again after Eq. (13.62).] We begin with S_{ab}. From (13.44) and (13.32), we have

$$1s_a 1s_b = k^3 \pi^{-1} e^{-k(r_a + r_b)} = k^3 \pi^{-1} e^{-kR\xi}$$

We need the expression for the volume element dv in elliptic coordinates. This turns out to be (*Eyring, Walter, and Kimball,* Appendix III)

$$dv = \tfrac{1}{8} R^3 (\xi^2 - \eta^2)\, d\xi\, d\eta\, d\phi \qquad (13.56)$$

We have

$$S_{ab} = \frac{k^3 R^3}{8\pi} \int_0^{2\pi} d\phi \int_{-1}^{1} \left[\int_1^\infty e^{-kR\xi}(\xi^2 - \eta^2)\, d\xi \right] d\eta \qquad (13.57)$$

Using the Appendix integral (A.10), we find

$$\int_1^\infty e^{-kR\xi}\xi^2\, d\xi - \int_1^\infty e^{-kR\xi}\eta^2\, d\xi = \frac{2e^{-kR}}{k^3 R^3}\left(1 + kR + \frac{k^2 R^2}{2}\right) - \eta^2 \frac{e^{-kR}}{kR}$$

Substituting in (13.57) and doing the eta integration, we get

$$S_{ab} = e^{-kR}(1 + kR + \tfrac{1}{3}k^2 R^2) \qquad (13.58)$$

Now for H_{aa}. Adding and subtracting k/r_a in (13.31), we get

$$\hat{H} = \left(-\tfrac{1}{2}\nabla^2 - \frac{k}{r_a}\right) + \frac{k-1}{r_a} - \frac{1}{r_b} = \hat{H}_a + \frac{k-1}{r_a} - \frac{1}{r_b} \qquad (13.59)$$

where $\hat{H}_a$ is the Hamiltonian operator for a hydrogenlike atom of nuclear charge k located at a. Since $1s_a$ is a $1s$ hydrogenlike function for nuclear charge k, we have (Section 13.3)

$$\hat{H}_a 1s_a = -\tfrac{1}{2}k^2 1s_a$$

Hence

$$H_{aa} = \int 1s_a^* \left(\hat{H}_a + \frac{k-1}{r_a} - \frac{1}{r_b} \right) 1s_a \, dv$$

$$H_{aa} = -\tfrac{1}{2}k^2 + (k-1) \int \frac{1s_a^* 1s_a}{r_a} \, dv - \int \frac{1s_a^* 1s_a}{r_b} \, dv \qquad (13.60)$$

The first integral in (13.60) can be done in elliptical coordinates, but it is faster to use spherical polar coordinates with origin at a:

$$\int \frac{1s_a^2}{r_a} \, dv = \frac{k^3}{\pi} \int_0^\infty \frac{e^{-2kr_a}}{r_a} r_a^2 \, dr_a \int_0^{2\pi} d\phi \int_0^\pi \sin\theta \, d\theta = k$$

where Eq. (A.7) has been used. Equations (13.56), (13.44), and (13.34) give for the second integral in (13.60)

$$\int \frac{1s_a^2}{r_b} \, dv = \frac{k^3 R^2}{4\pi} \int_0^{2\pi} d\phi \int_{-1}^1 e^{-kR\eta} \left[\int_1^\infty e^{-kR\xi} (\xi + \eta) \, d\xi \right] d\eta = \frac{1}{R} - e^{-2kR} \left(\frac{1}{R} + k \right)$$

Therefore,

$$H_{aa} = \tfrac{1}{2}k^2 - k - \frac{1}{R} + e^{-2kR} \left(k + \frac{1}{R} \right) \qquad (13.61)$$

The evaluation of H_{ab} is left as an exercise. We find

$$H_{ab} = -\tfrac{1}{2}k^2 S_{ab} - k(2-k)(1+kR)e^{-kR} \qquad (13.62)$$

Substituting the values for the integrals into (13.51), we get

$$W_{1,2} = -\tfrac{1}{2}k^2 + \frac{k^2 - k - R^{-1} + R^{-1}(1+kR)e^{-2kR} \pm k(k-2)(1+kR)e^{-kR}}{1 \pm e^{-kR}(1+kR+k^2R^2/3)}$$

$$\qquad (13.63)$$

where the upper signs are for W_1. Since $\hat{H}$ in (13.59) omits the internuclear repulsion $1/R$, W_1 and W_2 are approximations to the purely electronic energy E_{el}, and $1/R$ must be added to $W_{1,2}$ to get $U(R)$ [Eq. (13.8)].

The final task is to vary the parameter k at many fixed values of R so as to minimize first $W_1(R)$ and then $W_2(R)$. This can be done numerically using a computer (Problem 13.14) or analytically (Problem 13.13). The results are that, for the $1s_a + 1s_b$ function (13.54), k increases monotonically from 1 to 2 as R decreases from ∞ to 0; for the $1s_a - 1s_b$ function (13.55), k decreases almost monotonically from 1 to 0.4 as R decreases from ∞ to 0 [C. A. Coulson, *Trans. Faraday Soc.*, **33**, 1479 (1937)]. Since $0 < k \leq 2$ and $S_{ab} > 0$, Eq. (13.62) shows that the resonance integral H_{ab} is always negative. Therefore, W_1 in (13.51) corresponds to the ground electronic state $\sigma_g 1s$ of H_2^+. For the ground state, one finds $k(R_e) = 1.24$.

We might ask why the variational parameter k for the $\sigma_u^* 1s$ state goes to 0.4, rather than to 2, as R goes to zero. The answer is that this state of H_2^+ does not go to the ground state $(1s)$ of He^+ as R goes to zero. The $\sigma_u^* 1s$ state has odd parity and must correlate with an odd state of He^+. The lowest odd states of He^+ are the $2p$ states (Section 11.5); since the $\sigma_u^* 1s$ state has zero electronic orbital angular momentum along the internuclear (z) axis, this state must go to an atomic $2p$ state with $m = 0$, that is, to the $2p_0 = 2p_z$ state.

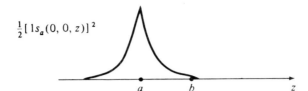

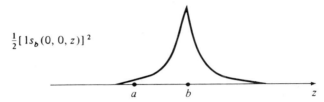

Figure 13.5 Atomic probability densities for H_2^+. Note the cusps at the nuclei.

Having found $k(R)$ for each root, one calculates W_1 and W_2 from (13.63) and adds $1/R$ to get the $U(R)$ curves. The calculated ground-state $U(R)$ curve has a minimum at 2.00 bohrs (Problem 13.15), in agreement with the true R_e value 2.00 bohrs, and has $U(R_e) = -15.96$ eV, giving a predicted D_e of 2.36 eV, as compared with the true value 2.79 eV. (If we omit varying k but simply set it equal to 1, we get $R_e = 2.49$ bohrs and $D_e = 1.76$ eV.)

Now consider the appearance of the trial functions for the $\sigma_g 1s$ and $\sigma_u^* 1s$ states at intermediate values of R. Figure 13.5 shows the values of the functions $(1s_a)^2$ and $(1s_b)^2$ at points on the internuclear axis (see also Fig. 6.7). For the $\sigma_g 1s$ function $1s_a + 1s_b$, we get a buildup of electronic probability density between the nuclei, as shown in Fig. 13.6. It is especially significant that the buildup of charge between the nuclei is greater than that obtained by simply taking the sum of the separate atomic charge densities. The probability density for an electron in a $1s_a$ atomic orbital is $(1s_a)^2$. If we add the probability density for half an electron in a $1s_a$ AO and half an electron in a $1s_b$ AO, we get

$$\tfrac{1}{2}(1s_a^2 + 1s_b^2) \tag{13.64}$$

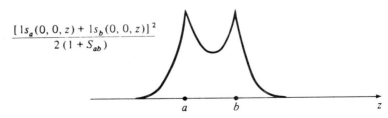

Figure 13.6 Probability density along the internuclear axis for the LCAO-MO function $N(1s_a + 1s_b)$.

However, in quantum mechanics, we do not add the separate atomic probability densities. Instead, we add the wave functions, as in (13.54). The H_2^+ ground-state probability density is then

$$\phi_1^2 = \frac{1}{2(1 + S_{ab})} [1s_a^2 + 1s_b^2 + 2(1s_a 1s_b)] \tag{13.65}$$

The difference between (13.65) and (13.64) is

$$\frac{1}{2(1 + S_{ab})} [2(1s_a 1s_b) - S_{ab}(1s_a^2 + 1s_b^2)] \tag{13.66}$$

Putting $R = 2.00$ and $k = 1.24$ in Eq. (13.58), we find that the overlap integral S_{ab} equals 0.46 at R_e. (It might be thought that because of the orthogonality of different AOs, the integral S_{ab} should be zero. However, the AOs $1s_a$ and $1s_b$ are eigenfunctions of *different* Hamiltonian operators, one for a hydrogen atom at a and one for a hydrogen atom at b; hence the orthogonality theorem does not apply.) Consider now the relative magnitudes of the two terms in brackets in (13.66) for points on the molecular axis. To the left of nucleus a, the function $1s_b$ is very small; to the right of nucleus b, the function $1s_a$ is very small. Hence in the regions outside the nuclei, the product $1s_a 1s_b$ is small, and the second term in (13.66) is dominant. This gives a subtraction of electronic charge density in the regions outside the nuclei, as compared with the sum of the densities of the individual atoms. Now consider the region between the nuclei. At the midpoint of the internuclear axis (and anywhere on the plane perpendicular to the axis and bisecting it), we have $1s_a = 1s_b$, and the bracketed terms in (13.66) become $2(1s_a)^2 - 0.92(1s_a)^2 \approx 1s_a^2$, which is positive. We thus get a buildup of charge probability density between the nuclei in the molecule, as compared with the sum of the densities of the individual atoms. This buildup of electronic charge between the nuclei allows the electron to feel the attractions of both nuclei at the same time, which lowers its potential energy. The greater the overlap in the internuclear region between the atomic orbitals forming the bond, the greater the charge buildup in this region, giving the familiar *principle of maximum overlap*. [Our argument has been based on the approximate wave function (13.54). If we plot Fig. 13.6 using the true wave function, we get a very similar graph, except that the true wave function is somewhat larger between the nuclei and somewhat smaller outside the internuclear region than (13.54).]

The preceding discussion seems to attribute the bonding in H_2^+ mainly to the lowering in the average electronic potential energy that results from having the shared electron interact with two nuclei instead of one. This, however, is an incomplete picture. Calculations on H_2^+ by Feinberg and Ruedenberg show that the decrease in electronic potential energy due to the sharing is of the same order of magnitude as the nuclear repulsion energy $1/R$ and hence is insufficient by itself to give binding. Two other effects also contribute to the bonding. The increase in atomic orbital exponent ($k = 1.24$ at R_e versus 1.0 at ∞) causes charge to accumulate near the nuclei (as well as in the internuclear region), and this further lowers the electronic potential energy. Moreover, the buildup of charge in the internuclear region makes $\partial \psi / \partial z$ zero at the midpoint of the molecular axis and small in the region close to this point; hence the z component of the average electronic kinetic energy [which can be expressed as $\frac{1}{2} \int |\partial \psi / \partial z|^2 \, d\tau$; Problem 7.4b] is lowered as compared with the atomic $\langle T_z \rangle$. (However, the *total* average electronic kinetic energy is raised; see Section 14.2.) For further details, see M. J. Feinberg and K. Ruedenberg, *J. Chem. Phys.*, **54**, 1495 (1971). Wilson and Goddard have also emphasized the important role of kinetic energy in bonding. [C. W. Wilson and W. A. Goddard, *Chem. Phys. Letters*, **5**, 45 (1970); *Theor. Chim. Acta*, **26**, 195 (1972); W. A. Goddard and C. W. Wilson, *Theor. Chim. Acta*, **26**, 211 (1972).]

Bader, however, has strongly criticized the views of Feinberg and Ruedenberg. Bader states (among other points) that H_2^+ and H_2 are atypical and that, in contrast to the increase of charge density in the immediate vicinity of the nuclei in H_2 and H_2^+, molecule formation that involves atoms other than H is usually accompanied by a substantial reduction in charge density in the immediate vicinity of the nuclei. See R. F. W. Bader in *The Force Concept in Chemistry*, B. M. Deb, ed., Van Nostrand Reinhold, New York, 1981, pp. 65–67, 71, 95–100, 113–115. Further study is needed before the origin of the covalent bond can be considered a settled question.

The $\sigma_u^* 1s$ trial function $1s_a - 1s_b$ is proportional to $e^{-r_a} - e^{-r_b}$. On a plane perpendicular to the internuclear axis and midway between the nuclei, we have $r_a = r_b$, so this plane is a nodal plane for the $\sigma_u^* 1s$ function. We do not get a buildup of charge between the nuclei for this state, and the $U(R)$ curve has no minimum. We say that the $\sigma_g 1s$ orbital is **bonding** and the $\sigma_u^* 1s$ orbital is **antibonding**. (See Fig. 13.7.)

Reflection of the electron's coordinates in the σ_h symmetry plane perpendicular to the molecular axis and midway between the nuclei converts r_a to r_b and r_b to r_a and leaves ϕ unchanged [Eq. (13.79)]. The operator $\hat{O}_{\sigma_h}$ (Section 12.1) commutes with the electronic Hamiltonian (13.31) and with the parity (inversion) operator. Hence we can choose the H_2^+ wave functions to be eigenfunctions of this reflection operator as well as of the parity operator. Since the square of this reflection operator is the unit operator, its eigenvalues must be $+1$ and -1 (Section 7.5). States of H_2^+ for which the wave function changes sign upon reflection in this plane (eigenvalue -1) are indicated by a star as a superscript to the letter that specifies λ; states whose wave functions are unchanged on reflection in this plane are left unstarred. Since orbitals with eigenvalue -1 for this reflection have a nodal plane between the nuclei, starred orbitals are antibonding.

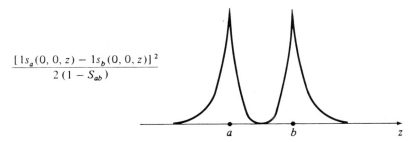

$$\frac{[1s_a(0, 0, z) - 1s_b(0, 0, z)]^2}{2(1 - S_{ab})}$$

Figure 13.7 Probability density along the internuclear axis for the LCAO-MO function $N'(1s_a - 1s_b)$.

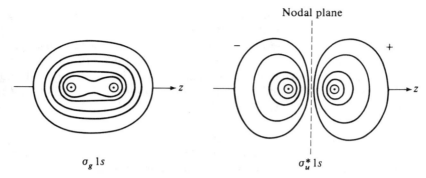

Figure 13.8 Contours of constant $|\psi|$ for the $\sigma_g 1s$ and $\sigma_u^* 1s$ MOs. The three-dimensional contour surfaces are generated by rotating these figures about the z axis. Note the resemblance of the antibonding-MO contours to those of a $2p_z$ AO.

Instead of using graphs, we can make contour diagrams of the orbitals (Section 6.7); see Fig. 13.8.

We might compare the preceding treatment of H_2^+ with the perturbation treatment of the helium $1s2s$ levels (Section 9.7). There we started with the degenerate functions $1s(1)2s(2)$ and $1s(2)2s(1)$. Because of the symmetry of the Hamiltonian with respect to interchange of the identical electrons, we found the correct zeroth-order functions to be $[1s(1)2s(2) \pm 1s(2)2s(1)]/\sqrt{2}$. For H_2^+, we started with the degenerate functions $1s_a$ and $1s_b$. Because of the symmetry of the electronic Hamiltonian with respect to the identical nuclei, we found the correct zeroth-order functions to be $(1s_a \pm 1s_b)/\sqrt{2}(1 \pm S_{ab})^{1/2}$.

Sometimes the binding in H_2^+ is attributed to the resonance integral H_{ab}, since in the approximate treatment we have given, it provides most of the binding energy. This viewpoint is misleading; in the exact treatment of Section 13.4, there arose no such resonance integral. The resonance integral simply arises out of the nature of the LCAO approximation we used.

In summary, we have formed the two H_2^+ MOs (13.54) and (13.55), one bonding and one antibonding, from the AOs $1s_a$ and $1s_b$. The MO energies are given by Eq. (13.51) as

$$W_{1,2} = H_{aa} \pm \frac{H_{ab} - H_{aa}S_{ab}}{1 \pm S_{ab}} \qquad (13.67)$$

where $H_{aa} = \langle 1s_a|\hat{H}|1s_a\rangle$, with $\hat{H}$ being the purely electronic Hamiltonian of H_2^+. The integral H_{aa} would be the molecule's purely electronic energy if the electron's wave function in the molecule were $1s_a$. In a sense, H_{aa} is the energy of the $1s_a$ orbital in the molecule; in the limit $R = \infty$, H_{aa} becomes the $1s$ AO energy in the H atom. In the molecule, H_{aa} is substantially lower than the electronic energy of an H atom because the electron is attracted to both nuclei. A diagram of MO formation from AOs is given in Fig. 13.21. To get $U(R)$, the electronic energy including nuclear repulsion, we must add $1/R$ to (13.67).

We have described the H_2^+ states according to the state of the hydrogen atom obtained on dissociation; this is a **separated-atoms description**. Alternatively, we can use the state of the atom formed as the internuclear distance goes to zero; this is a **united-atom description**. We have seen that for the two lowest electronic states of H_2^+ the united-atom states are the $1s$ and $2p_0$ states of He^+. The united-atom designation is put on the left of the symbol for λ. The $\sigma_g 1s$ state thus has the united-atom designation $1s\sigma_g$. The $\sigma_u^* 1s$ state has the united-atom designation $2p\sigma_u^*$. It is not necessary to write this state as $2p_0\sigma_u^*$, because the fact that it is a σ state tells us that it correlates with the united-atom $2p_0$ state. For the united-atom description, the subscripts g and u are redundant, in that molecular states correlating with $s, d, g, \ldots$ atomic states must be g, while states correlating with $p, f, h, \ldots$ atomic states must be u. From the separated-atoms states, we cannot tell whether the molecular wave function is g or u. Thus from the $1s$ separated-atoms state we formed both a g and a u function for H_2^+.

Before constructing approximate molecular orbitals for other H_2^+ states, we consider how the trial function (13.54) can be improved. From the viewpoint of perturbation theory, (13.54) is the correct zeroth-order wave function. We know that the perturbation of molecule formation will mix in other hydrogen-atom states besides $1s$. Dickinson in 1933 used a trial function with some $2p_0$ character mixed in (since the ground state of H_2^+ is a σ state, it would be wrong to mix in $2p_{\pm 1}$ functions); he took

$$\phi = [1s_a + c(2p_0)_a] + [1s_b + c(2p_0)_b] \tag{13.68}$$

where c is a variational parameter and where (Table 6.2)

$$1s_a = k^{3/2}\pi^{-1/2}e^{-kr_a}, \qquad (2p_0)_a = (2p_z)_a = \frac{\zeta^{5/2}}{4(2\pi)^{1/2}} r_a e^{-\zeta r_a/2} \cos\theta_a$$

with k and ζ being two other variational parameters. We have similar expressions for $1s_b$ and $(2p_0)_b$. The angles θ_a and θ_b refer to two sets of spherical polar coordinates, one set at each nucleus; see Fig. 13.9. The definitions of θ_a and θ_b correspond to using a right-handed coordinate system on atom a and a left-handed system on atom b. The coefficient c goes to zero as R goes to either zero or infinity.

The mixing together of two or more AOs on the same atom is called

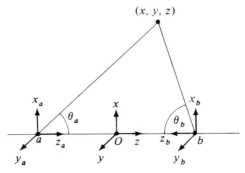

Figure 13.9 Coordinate systems for a homonuclear diatomic molecule.

hybridization. The function $1s + c2p_0$ is a hybridized atomic orbital. Since the $2p_0$ function is positive in one lobe and negative in the other, the inclusion of $2p_0$ produces additional charge buildup between the nuclei, giving a greater binding energy. The hybridization allows for the polarization of the $1s_a$ and $1s_b$ atomic orbitals that occurs on molecule formation. The function (13.68) gives a $U(R)$ curve with a minimum at 2.01 bohrs. At this distance, the parameters have the values $k = 1.246$, $\zeta = 2.965$, and $c = 0.138$ [F. Weinhold, *J. Chem. Phys.*, **54**, 530 (1971)]. The calculated D_e is 2.73 eV, close to the true value 2.79 eV.

Many other variational functions have been used for the H_2^+ ground state. For a partial listing, see M. Geller et al., *J. Chem. Phys.*, **36**, 2693 (1962); F. Weinhold and A. B. Chinen, *J. Chem. Phys.*, **56**, 3798 (1972).

One final point. The approximate wave functions in this chapter are written in atomic units. When rewriting these functions in ordinary units, we must remember that wave functions are not dimensionless. A one-particle wave function ψ has units of length$^{-3/2}$. This follows from the fact that $|\psi|^2 \, dx \, dy \, dz$ is a probability and probabilities are dimensionless. The AOs $1s_a$ and $1s_b$ that occur in the functions (13.54) and (13.55) are given by (13.44) in atomic units; in ordinary units, $1s_a = (k/a_0)^{3/2} \pi^{-1/2} e^{-kr_a/a_0}$.

13.6 MOLECULAR ORBITALS FOR H_2^+ EXCITED STATES

In the last section, we used the approximate functions (13.54) and (13.55) for the two lowest H_2^+ electronic states. Now we construct approximate functions for further excited states so as to build up a supply of H_2^+-like molecular orbitals. We shall then use these MOs to give a qualitative discussion of many-electron diatomic molecules, just as we used hydrogenlike AOs to discuss many-electron atoms.

To get approximations to higher H_2^+ MOs, we can use the method of the linear variation function. We saw that it was natural to take our variation functions for H_2^+ as linear combinations of hydrogenlike atomic-orbital functions, giving LCAO-MOs. To get approximate MOs for higher states, we add in more AOs to the linear combination. Thus, to get approximate wave functions for the six lowest H_2^+ σ states, we use a linear combination of the three lowest $m = 0$ hydrogenlike functions on each atom:

$$\phi = c_1 1s_a + c_2 2s_a + c_3 (2p_0)_a + c_4 1s_b + c_5 2s_b + c_6 (2p_0)_b$$

As found in the last section for the function (13.43), the symmetry of the homonuclear diatomic molecule makes the coefficients of the atom b orbitals equal to ± 1 times the corresponding atom a orbital coefficients:

$$\phi = [c_1 1s_a + c_2 2s_a + c_3 (2p_0)_a] \pm [c_1 1s_b + c_2 2s_b + c_3 (2p_0)_b] \quad (13.69)$$

where the upper sign goes with the even (g) states.

Consider the relative magnitudes of the coefficients in (13.69). For the two electronic states that dissociate into a $1s$ hydrogen atom, we expect that c_1 will be considerably greater than c_2 or c_3, since c_2 and c_3 vanish in the limit of R going to infinity. Thus the Dickinson function (13.68) has the $2p_0$ coefficient equal to one-seventh the $1s$ coefficient at R_e. (This function does not include a $2s$ term, but if it did, we no doubt would find its coefficient to be small compared with the $1s$ coefficient.) As a first approximation, we therefore set c_2 and c_3 equal to zero, taking

$$\phi = c_1(1s_a \pm 1s_b) \tag{13.70}$$

as an approximation for the wave functions of these two states (as we already have done). From the viewpoint of perturbation theory, if we take the separated atoms as the unperturbed problem, the functions (13.70) are the correct zeroth-order wave functions.

The same argument for the two states that dissociate to a $2s$ hydrogen atom gives as approximate wave functions for them

$$\phi = c_2(2s_a \pm 2s_b) \tag{13.71}$$

since c_1 and c_3 will be small for these states. The functions (13.71) are only an approximation to what we would find if we carried out the linear variation treatment; to find rigorous upper bounds to the energies of these two H_2^+ states, we must use the trial function (13.69) and solve the appropriate sixth-order determinantal equation.

In general we have two H_2^+ states correlating with each separated-atoms state, and rough approximations to the wave functions of these two states will be the LCAO functions $f_a + f_b$ and $f_a - f_b$, where f is a hydrogenlike wave function. The functions (13.70) give the $\sigma_g 1s$ and $\sigma_u^* 1s$ states. Similarly, the functions (13.71) give the $\sigma_g 2s$ and $\sigma_u^* 2s$ molecular orbitals. The outer contour lines for these orbitals are like those for the corresponding MOs made from $1s$ AOs. However, since the $2s$ AO has a nodal sphere while the $1s$ AO does not, each of these MOs has one more nodal surface than the corresponding $\sigma_g 1s$ or $\sigma_u^* 1s$ MO.

Next we have the combinations

$$(2p_0)_a \pm (2p_0)_b = (2p_z)_a \pm (2p_z)_b \tag{13.72}$$

giving the $\sigma_g 2p$ and $\sigma_u^* 2p$ molecular orbitals (Fig. 13.10); these are σ MOs even though they correlate with $2p$ separated AOs, since they have $m = 0$.

The preceding discussion is oversimplified. For the hydrogen atom, the $2s$ and $2p$ AOs are degenerate, and so we can expect the correct zeroth-order functions for the $\sigma_g 2s$, $\sigma_u^* 2s$, $\sigma_g 2p$, and $\sigma_u^* 2p$ MOs of H_2^+ to each be mixtures of $2s$ and $2p$ AOs rather than containing only $2s$ or $2p$ character. [In the $R \to \infty$ limit, H_2^+ consists of an H atom perturbed by the essentially uniform electric field of a far-distant proton. Problem 9.15 showed that the correct zeroth-order functions for the $n = 2$ levels of an H atom in a uniform electric field in the z direction are $2^{-1/2}(2s + 2p_0)$, $2^{-1/2}(2s - 2p_0)$, $2p_1$, and $2p_{-1}$. Thus, for H_2^+, $2s$ and $2p_0$ in Eqs. (13.71) and (13.72) should be replaced by $2s + 2p_0$ and $2s - 2p_0$.] For molecules that dissociate into many-electron atoms, the separated-atoms $2s$ and $2p$ AOs are not degenerate but do lie close together in energy; hence the first-order corrections to the wave functions will mix substantial $2s$ character into the $\sigma 2p$ MOs and substantial $2p$ character into the $\sigma 2s$ MOs. Thus the designation of an MO as $\sigma 2s$ or $\sigma 2p$ should not be taken too literally. For H_2^+ and H_2, the united-atom designations of the MOs are preferable to the separated-atoms designations, but we shall use mostly the latter.

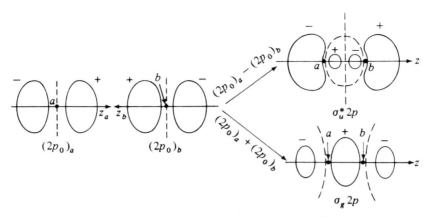

Figure 13.10 Formation of $\sigma_g 2p$ and $\sigma_u^* 2p$ MOs from $2p_z$ AOs. The dashed lines indicate nodal surfaces. The signs on the contours give the sign of the wave function. The contours are symmetric about the z axis. (Because of substantial $2s$–$2p$ hybridization, these contours are not accurate representations of true MO shapes. For accurate contours, see the reference for Fig. 13.19.)

For the other two $2p$ atomic orbitals, we can use either the $2p_{+1}$ and $2p_{-1}$ complex functions or the $2p_x$ and $2p_y$ real functions. If we want MOs that are eigenfunctions of $\hat{L}_z$, we will choose the complex p orbitals. We then have the four MOs

$$(2p_{+1})_a + (2p_{+1})_b \tag{13.73}$$

$$(2p_{+1})_a - (2p_{+1})_b \tag{13.74}$$

$$(2p_{-1})_a + (2p_{-1})_b \tag{13.75}$$

$$(2p_{-1})_a - (2p_{-1})_b \tag{13.76}$$

From Eq. (6.114) we have, since $\phi_a = \phi_b = \phi$,

$$(2p_{+1})_a + (2p_{+1})_b = \tfrac{1}{8}\pi^{-1/2}(r_a e^{-r_a/2}\sin\theta_a + r_b e^{-r_b/2}\sin\theta_b)e^{i\phi} \tag{13.77}$$

Since $\lambda = |m| = 1$, this is a π orbital. The inversion operation amounts to the coordinate transformation (Fig. 13.11)

$$r_a \rightarrow r_b, \qquad r_b \rightarrow r_a, \qquad \phi \rightarrow \phi + \pi \tag{13.78}$$

We have $e^{i(\phi+\pi)} = (\cos\pi + i\sin\pi)e^{i\phi} = -e^{i\phi}$. From Fig. 13.11 we see that inversion converts θ_a to θ_b and vice versa. Thus inversion converts (13.77) to its negative, meaning it is a u orbital. Reflection in the plane perpendicular to the axis and midway between the nuclei causes the following transformations (Problem 13.19):

$$r_a \rightarrow r_b, \qquad r_b \rightarrow r_a, \qquad \phi \rightarrow \phi, \qquad \theta_a \rightarrow \theta_b, \qquad \theta_b \rightarrow \theta_a \tag{13.79}$$

This leaves (13.77) unchanged, so we have an unstarred (bonding) orbital. The designation of (13.77) is then $\pi_u 2p_{+1}$.

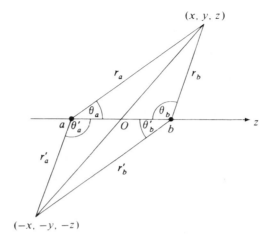

Figure 13.11 The effect of inversion of the electron's coordinates in H_2^+. We have $r_a' = r_b$, $r_b' = r_a$, and $\phi' = \phi + \pi$.

The function (13.77) is complex; taking its absolute value, we can plot the orbital contours of constant probability density (Section 6.7). Since $|e^{i\phi}| = 1$, the probability density is independent of ϕ, giving a density that is symmetric about the z (internuclear) axis. Figure 13.12 shows a cross section of this orbital in a plane containing the nuclei. The three-dimensional shape is found by rotating this figure about the z axis, creating a sort of fat doughnut. (Sometimes this orbital is pictured as having the shape of a right circular cylinder; this is inaccurate, since the probability density is not constant as we move parallel to the z axis.)

The molecular orbital (13.75) differs from (13.77) only in having $e^{i\phi}$ replaced by $e^{-i\phi}$ and is designated $\pi_u 2p_{-1}$. The coordinate ϕ enters the Hamiltonian (13.35) as $\partial^2/\partial\phi^2$. Since $\partial^2 e^{i\phi}/\partial\phi^2 = \partial^2 e^{-i\phi}/\partial\phi^2$, the states (13.73) and (13.75) have the same energy. Recall (Section 13.4) that the $\lambda = 1$ energy levels are doubly degenerate, corresponding to $m = \pm 1$. Since $|e^{i\phi}| = |e^{-i\phi}|$, the $\pi_u 2p_{+1}$ and $\pi_u 2p_{-1}$ MOs have the same shapes, just as the $2p_{+1}$ and $2p_{-1}$ AOs have the same shapes.

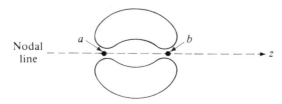

Nodal
line

Figure 13.12 Cross section of the $\pi_u 2p_{+1}$ (or $\pi_u 2p_{-1}$) molecular orbital. To obtain the three-dimensional contour surface, rotate the figure about the z axis. The z axis is a nodal line for this MO (as it is for the $2p_{+1}$ AO).

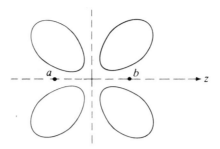

Figure 13.13 Cross section of the $\pi_g^* 2p_{+1}$ (or $\pi_g^* 2p_{-1}$) MO. To obtain the three-dimensional contour surface, rotate the figure about the z axis. The z axis and the xy plane are nodes.

The functions (13.74) and (13.76) give the $\pi_g^* 2p_{+1}$ and $\pi_g^* 2p_{-1}$ MOs. These functions do not give charge buildup between the nuclei; see Fig. 13.13.

Let us now consider the more familiar alternative of using the $2p_x$ and $2p_y$ AOs to make our MOs. The linear combination

$$(2p_x)_a + (2p_x)_b \tag{13.80}$$

gives the $\pi_u 2p_x$ MO (Fig. 13.14). This MO is not symmetric about the internuclear axis but builds up electronic charge probability density in two lobes, one above and one below the yz plane, which is a nodal plane for this function. The

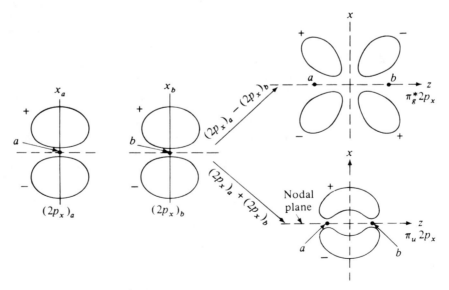

Figure 13.14 Formation of the $\pi_u 2p_x$ and $\pi_g^* 2p_x$ MOs. Since $\phi = 0$ in the xz plane, the cross sections of these MOs in the xz plane are the same as for the corresponding $\pi_u 2p_{+1}$ and $\pi_g^* 2p_{+1}$ MOs. However, the $\pi 2p_x$ MOs are not symmetrical about the z axis. Rather, they consist of blobs of probability density above and below the nodal yz plane.

wave function has opposite signs on each side of this plane. The linear combination

$$(2p_x)_a - (2p_x)_b \tag{13.81}$$

gives the $\pi_g^* 2p_x$ MO (Fig. 13.14). Since the $2p_y$ functions differ from the $2p_x$ functions solely by a rotation of 90° about the internuclear axis, they give molecular orbitals differing from those of Fig. 13.14 by a 90° rotation about the z axis. The linear combinations

$$(2p_y)_a + (2p_y)_b \tag{13.82}$$

$$(2p_y)_a - (2p_y)_b \tag{13.83}$$

give the $\pi_u 2p_y$ and $\pi_g^* 2p_y$ molecular orbitals. The MOs (13.80) and (13.82) have the same energy; the MOs (13.81) and (13.83) have the same energy. (Note that the g $\pi 2p$ MOs are antibonding, while the u $\pi 2p$ MOs are bonding.)

Just as the $2p_x$ and $2p_y$ AOs are linear combinations of the $2p_{+1}$ and $2p_{-1}$ AOs [Eqs. (6.118) and (6.120)], the $\pi_u 2p_x$ and $\pi_u 2p_y$ MOs are linear combinations of the $\pi_u 2p_{+1}$ and $\pi_u 2p_{-1}$ MOs. We can use any linear combination of the eigenfunctions of a degenerate energy level and still have an energy eigenfunction. Just as the $2p_{+1}$ and $2p_{-1}$ AOs are eigenfunctions of $\hat{L}_z$ and the $2p_x$ and $2p_y$ AOs are not, the $\pi_u 2p_{+1}$ and $\pi_u 2p_{-1}$ MOs are eigenfunctions of $\hat{L}_z$ and the $\pi_u 2p_x$ and $\pi_u 2p_y$ MOs are not. Although the molecular orbitals $\pi_u 2p_{+1}$, $\pi_u 2p_{-1}$, $\pi_u 2p_x$, $\pi_u 2p_y$ all have the same energy, they represent different states of the system, since they have different wave functions. The question arises: Given an H_2^+ molecule in the $\pi_u 2p$ energy level, what is its wave function (state)? The answer is that the state depends on the past history of the system. Thus, if we have just measured L_z for a $\pi_u 2p$ H_2^+ molecule and found the result $+\hbar$, then we know that the molecule is in the state $\pi_u 2p_{+1}$. This assertion holds just after the measurement, but not just before. If the state just before the measurement were either $\pi_u 2p_x$ or $\pi_u 2p_y$, then (since these states are superpositions of the $\pi_u 2p_{+1}$ and $\pi_u 2p_{-1}$ states with coefficients equal in magnitude) we would have had a 50% chance of finding each of the results $+\hbar$ and $-\hbar$; the measurement of L_z changes the state to either $\pi_u 2p_{+1}$ or $\pi_u 2p_{-1}$, depending on what the outcome of the measurement is (reduction of the wave function; Section 7.9). For the H_2^+ $\pi_u 2p$ energy level, we can use the pair of real MOs (13.80) and (13.82), or the pair of complex MOs (13.73) and (13.75), or any two linearly independent linear combinations of these functions.

We have shown the correlation of the H_2^+ MOs with the separated-atoms AOs. We can also show how they correlate with the united-atom AOs. As R goes to zero, the $\sigma_u^* 1s$ MO (Fig. 13.8) resembles more and more the $2p_z$ AO, with which it correlates. Similarly, the $\pi_u 2p$ MOs correlate with p united-atom states, while the $\pi_g^* 2p$ MOs correlate with d united-atom states.

We can treat a molecule by applying perturbation theory to the corresponding united atom, taking the perturbation $\hat{H}'$ as the difference between the molecular and united-atom Hamiltonians: $\hat{H}' = \hat{H}_{mol} - \hat{H}_{UA}$. Using perturbation theory, one finds that the purely electronic energy of a diatomic molecule has the following form at small values of the internuclear distance R [W. A. Bingel, *J.*

Chem. Phys., **30**, 1250 (1959); I. N. Levine, *J. Chem. Phys.*, **40**, 3444 (1964); **41**, 2044 (1964); W. Byers Brown and E. Steiner, *J. Chem. Phys.*, **44**, 3934 (1966)]:

$$E_{el} = E_{UA} + aR^2 + bR^3 + cR^4 + dR^5 + eR^5 \ln R + \cdots \qquad (13.84)$$

where E_{UA} is the united-atom energy and $a, b, c, d, e, \ldots$ are constants.

13.7 MOLECULAR-ORBITAL CONFIGURATIONS OF HOMONUCLEAR DIATOMIC MOLECULES

We now use the H_2^+ MOs developed in the last section to discuss many-electron homonuclear diatomic molecules. If we ignore the interelectronic repulsions, the zeroth-order wave function is a Slater determinant of H_2^+-like one-electron spin-orbitals. We approximate the spatial part of the H_2^+ spin-orbitals by the LCAO-MOs of the last section. Treatments that go beyond this crude first approximation will be discussed later.

The sizes and energies of the MOs vary with varying internuclear distance for each molecule and vary as we go from one molecule to another. Thus we saw how the orbital exponent k in the H_2^+ trial function (13.54) varied with R. As we go to molecules with higher nuclear charge, the parameter k for the $\sigma_g 1s$ MO will increase, giving a more compact MO. We want to consider the order of the MO energies. Because of the variation of these energies with R and variations from molecule to molecule, numerous crossings occur, just as for atomic-orbital energies (Fig. 11.2). Hence we cannot give a definitive order. However, the following is the order in which the MOs fill as we go across the periodic table:

$$\sigma_g 1s < \sigma_u^* 1s < \sigma_g 2s < \sigma_u^* 2s < \pi_u 2p_{+1} = \pi_u 2p_{-1} < \sigma_g 2p < \pi_g^* 2p_{+1}$$

$$= \pi_g^* 2p_{-1} < \sigma_u^* 2p$$

Each bonding orbital fills before the corresponding antibonding orbital. The $\pi_u 2p$ orbitals are close in energy to the $\sigma_g 2p$ orbital, and it was formerly believed that the $\sigma_g 2p$ MO filled first.

Besides the separated-atoms designation, there are other ways of referring to these MOs; see Table 13.1. The second column of this table gives the

TABLE 13.1 Molecular-Orbital Nomenclature for Homonuclear Diatomic Molecules

Separated-Atoms Description	United-Atom Description	Numbering by Symmetry
$\sigma_g 1s$	$1s\sigma_g$	$1\sigma_g$
$\sigma_u^* 1s$	$2p\sigma_u^*$	$1\sigma_u$
$\sigma_g 2s$	$2s\sigma_g$	$2\sigma_g$
$\sigma_u^* 2s$	$3p\sigma_u^*$	$2\sigma_u$
$\pi_u 2p$	$2p\pi_u$	$1\pi_u$
$\sigma_g 2p$	$3s\sigma_g$	$3\sigma_g$
$\pi_g^* 2p$	$3d\pi_g^*$	$1\pi_g$
$\sigma_u^* 2p$	$4p\sigma_u^*$	$3\sigma_u$

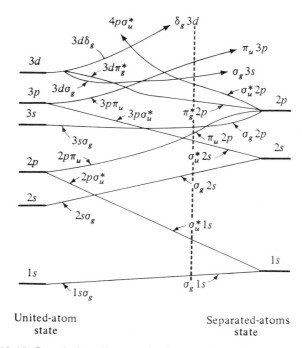

Figure 13.15 Correlation diagram for homonuclear diatomic MOs. (This diagram does not hold for H_2^+.) The dashed vertical line corresponds to the order in which the MOs fill.

united-atom designations. The nomenclature of the third column uses $1\sigma_g$ for the lowest σ_g MO, $2\sigma_g$ for the second lowest σ_g MO, and so on.

Figure 13.15 shows how these MOs correlate with the separated-atoms and united-atom AOs. Because of the variation of MO energies from molecule to molecule, this diagram is not quantitative. (The word correlation is being used here to mean a correspondence; this is a different meaning than in the term electron correlation.)

Recall (Problem 7.21 and Figure 6.13) that $s, d, g, \ldots$ united-atom AOs are even functions and hence correlate with *gerade* (g) MOs, whereas $p, f, h, \ldots$ AOs are odd functions and hence correlate with *ungerade* (u) MOs.

A useful principle in drawing orbital correlation diagrams is the *noncrossing rule*, which states that for MO correlation diagrams of many-electron diatomic molecules, the energies of MOs with the same symmetry cannot cross. For diatomic MOs the word *symmetry* refers to whether the orbital is g or u and whether it is $\sigma, \pi, \delta, \ldots$. For example, two σ_g MOs cannot cross on a correlation diagram. From the noncrossing rule, we conclude that the lowest MO of a given symmetry type must correlate with the lowest united-atom AO of that symmetry, and similarly for higher orbitals. [A similar noncrossing rule holds for potential-energy curves $U(R)$ for different electronic states of a many-electron diatomic molecule.] The proof of the noncrossing rule is a bit subtle; see C. A. Mead, *J. Chem. Phys.*, **70**, 2276 (1979) for a thorough discussion.

Just as we discussed atoms by filling in the AOs, giving rise to atomic configurations such as $1s^2 2s^2$, we shall discuss homonuclear diatomic molecules by filling in the MOs, giving rise to molecular electronic configurations such as $(\sigma_g 1s)^2 (\sigma_u^* 1s)^2$. (Recall that with a single atomic configuration there is associated a hierarchy of terms, levels, and states; the same is true for a molecular configuration; see Section 13.8.)

Figure 13.16 shows the homonuclear diatomic MOs formed from the $1s$, $2s$, and $2p$ AOs.

For H_2^+ we have the ground-state configuration $\sigma_g 1s$, which gives a one-electron bond. For excited states the electron is in one of the higher MOs.

For H_2 we put the two electrons in the $\sigma_g 1s$ MO with opposite spins, giving the ground-state configuration $(\sigma_g 1s)^2$. The two bonding electrons give a single bond. The ground-state dissociation energy D_e is 4.75 eV. In the crudest approximation, the H_2 dissociation energy would be twice the H_2^+ dissociation energy, or 5.6 eV. Because of interelectronic repulsion, the true binding energy is somewhat less than this.

Now consider He_2. Two electrons go in the $\sigma_g 1s$ MO, thereby filling it. The other two go in the next highest MO, $\sigma_u^* 1s$. The ground-state configuration is $(\sigma_g 1s)^2 (\sigma_u^* 1s)^2$. With two bonding and two antibonding electrons, we expect no net bonding, in agreement with the well-known fact that the ground electronic state of He_2 is unstable, showing no significant minimum in the potential-energy curve. However, if an electron is excited from the antibonding $\sigma_u^* 1s$ MO to a higher MO that is bonding, the molecule will have three bonding electrons and

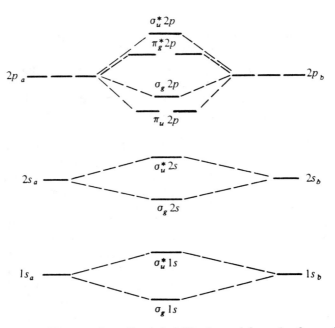

Figure 13.16 Homonuclear diatomic MOs formed from $1s$, $2s$, and $2p$ AOs.

only one antibonding electron. We therefore expect that He_2 has bound excited states, with a significant minimum in the $U(R)$ curve of each such state. Indeed, about two dozen such bound excited states of He_2 have been spectroscopically observed in gas discharge tubes. Of course, such excited states decay to the ground electronic state, and then the molecule dissociates.

The repulsion of two $1s^2$ helium atoms can be ascribed mainly to the Pauli repulsion between electrons with parallel spins (Section 10.3). Each helium atom has a pair of electrons with opposite spin, and each pair tends to exclude the other pair from occupying the same region of space.

Removal of an antibonding electron from He_2 gives the He_2^+ ion, with ground-state configuration $(\sigma_g 1s)^2 (\sigma_u^* 1s)$ and one net bonding electron. Ground-state properties of this molecule are quite close to those for H_2^+; see Table 13.2.

Li_2 has the ground-state configuration $(\sigma_g 1s)^2 (\sigma_u^* 1s)^2 (\sigma_g 2s)^2$ with two net bonding electrons, leading to the description of the molecule as containing an Li—Li single bond. Experimentally, Li_2 is found to be a stable species. In Li_2 the orbital exponent of the $1s$ AOs is considerably greater than in H_2^+ or H_2, because of the increase in the nuclear charges from 1 to 3. This shrinks the $1s_a$ and $1s_b$ AOs in closer to the corresponding nuclei; there is thus only very slight overlap between these two AOs, and the integrals S_{ab} and H_{ab} are very small for these AOs. As a result the energies of the $\sigma_g 1s$ and $\sigma_u^* 1s$ MOs in Li_2 are nearly equal to each other and to the energy of a $1s$ Li AO. (For very small R, the $1s_a$ and $1s_b$ AOs do overlap appreciably and their energies then differ considerably.) The Li_2 ground-state configuration is frequently written as $KK(\sigma_g 2s)^2$ to indicate the negligible change in inner-shell orbital energies on molecule formation, which is in accord with the chemist's usual idea of bonding involving only the valence electrons. The orbital exponent of the $2s$ AOs in Li_2 is not much greater than 1, because these electrons are screened from the nucleus by the $1s$ electrons.

Be_2 has the ground-state configuration $KK(\sigma_g 2s)^2 (\sigma_u^* 2s)^2$ with no net bonding electrons.

B_2 has the ground-state configuration $KK(\sigma_g 2s)^2 (\sigma_u^* 2s)^2 (\pi_u 2p)^2$ with two net bonding electrons, indicating a stable ground state, as is found experimentally. The bonding electrons are π electrons, which is at variance with the notion that single bonds are always σ bonds. We have two degenerate $\pi_u 2p$ MOs. Recall that when we had an atomic configuration such as $1s^2 2s^2 2p^2$ we obtained several terms, which because of interelectronic repulsions had different energies. We saw that the term with the highest total spin was generally the lowest (Hund's rule). With the molecular configuration of B_2 given above, we also have a number of terms. Since the lower (σ) MOs are all filled, their electrons must be paired and contribute nothing to the total spin. If the two $\pi_u 2p$ electrons are both in the same MO (for example, both in $\pi_u 2p_{+1}$), their spins must be paired (antiparallel), giving a total molecular electronic spin of zero. If, however, we have one electron in the $\pi_u 2p_{+1}$ MO and the other in the $\pi_u 2p_{-1}$ MO, their spins can be parallel, giving a net spin of 1; by Hund's rule, this term will be lowest, and the ground term of B_2 will have multiplicity $2S + 1 = 3$.

Boron is high melting and reactive, and it is very hard to produce substantial amounts of boron vapor to investigate the B_2 molecule. Knight and co-workers focused laser light on solid boron over which Ne gas flowed. This produced atomic B vapor (with a small amount of B_2). The mixture of boron and neon gas flowed over a copper surface at 5 K. As the boron and neon condensed on the copper, substantial amounts of B_2 formed. Investigation of the electron-spin-resonance spectrum of the B_2 trapped in the solid neon showed that the B_2 ground term is a triplet with $S = 1$ [L. B. Knight et al., *J. Am. Chem. Soc.*, **109**, 3521 (1987)]. That the triplet term of the B_2 ... $(\pi_u 2p)^2$ configuration is the ground term had previously been predicted by a highly accurate CI calculation [M. Dupuis and B. Liu, *J. Chem. Phys.*, **68**, 2902 (1978)].

C_2 has the ground-state configuration $KK(\sigma_g 2s)^2(\sigma_u^* 2s)^2(\pi_u 2p)^4$ with four net bonding electrons, indicating a stable ground state with a double bond. As mentioned, the $\pi_u 2p$ and $\sigma_g 2p$ MOs have nearly the same energy in many molecules. It was formerly believed that the ground-state configuration of C_2 was $KK(\sigma_g 2s)^2(\sigma_u^* 2s)^2(\pi_u 2p)^3(\sigma_g 2p)$. This configuration would allow two of the electrons to have their spins parallel, giving a triplet term; because of the minimization of the repulsion between these two electrons, the triplet term of this configuration was thought to be lower than the $(\pi_u 2p)^4$ configuration, which can give only a singlet term. (Recall that the ground-state configuration of Cr is $3d^5 4s$, rather than $3d^4 4s^2$.) However, the electronic emission spectrum of C_2 observed using a carbon furnace or carbon arc shows the $(\pi_u 2p)^4$ singlet term to be the ground term by a small margin (0.09 eV) [E. A. Ballik and D. A. Ramsay, *J. Chem. Phys.*, **31**, 1128 (1959); *Astrophys. J.*, **137**, 61, 84 (1963)].

The N_2 ground-state electronic configuration is

$$KK(\sigma_g 2s)^2(\sigma_u^* 2s)^2(\pi_u 2p)^4(\sigma_g 2p)^2$$

The six net bonding electrons indicate a triple bond, in accord with the Lewis structure :N≡N:.

The O_2 ground-state configuration is

$$KK(\sigma_g 2s)^2(\sigma_u^* 2s)^2(\sigma_g 2p)^2(\pi_u 2p)^4(\pi_g^* 2p)^2$$

Spectroscopic evidence indicates that in O_2 (and in F_2) the $\sigma_g 2p$ MO is lower in energy than the $\pi_u 2p$ MO. The four net bonding electrons give a double bond. The $\pi_g^* 2p_{+1}$ and $\pi_g^* 2p_{-1}$ MOs have the same energy, and by putting one electron in each with parallel spins we get a triplet term. By Hund's rule this is the ground term. This explanation of the paramagnetism of O_2 was one of the early triumphs of molecular-orbital theory.

For F_2 the ground-state configuration is

$$KK(\sigma_g 2s)^2(\sigma_u^* 2s)^2(\sigma_g 2p)^2(\pi_u 2p)^4(\pi_g^* 2p)^4$$

The two net bonding electrons give a single bond.

For Ne_2 we have

$$KK(\sigma_g 2s)^2(\sigma_u^* 2s)^2(\sigma_g 2p)^2(\pi_u 2p)^4(\pi_g^* 2p)^4(\sigma_u^* 2p)^2$$

There are no net bonding electrons, and thus there is no chemical bonding, in agreement with experiment.

We can go on to describe homonuclear diatomic molecules formed from atoms of the next period. Thus the lowest electron configuration of Na_2 is $KKLL(\sigma_g 3s)^2$. However, there are some differences as compared with the corresponding molecules of the preceding period. For Al_2, the ground term is the triplet term of the $\ldots (\sigma_g 3p)(\pi_u 3p)$ configuration, which lies a mere 0.02 eV below the triplet term of the $\ldots (\pi_u 3p)^2$ configuration [C. W. Bauschlicher et al., *J. Chem. Phys.*, **86**, 7007 (1987)]. For Si_2, the ground term is the triplet term of the $\ldots (\sigma_g 3p)^2(\pi_u 3p)^2$ configuration, which lies 0.05 eV below the triplet term of the $\ldots (\sigma_g 3p)(\pi_u 3p)^3$ configuration [C. W. Bauschlicher and S. R. Langhoff, *J. Chem. Phys.*, **87**, 2919 (1987); M. R. Nimlos et al., *J. Chem. Phys.*, **87**, 5116 (1987)]. These results were established with massive CI calculations done on supercomputers.

Table 13.2 lists D_e and R_e for the ground states of some homonuclear diatomic molecules; it also lists the *bond order*, which is one-half the difference between the number of bonding and antibonding electrons. The dissociation energy increases and the bond length decreases as the bond order increases. The spectroscopic designations in this table will be explained in the next section.

Bonding MOs produce charge buildup between the nuclei, whereas antibonding MOs produce charge depletion between the nuclei. Hence removal of an electron from a bonding MO usually decreases D_e, whereas removal of an electron from an antibonding MO increases D_e. (Note that as R decreases in Fig. 13.15, the energies of bonding MOs decrease, while the energies of antibonding MOs increase.) For example, the highest filled MO in N_2 is bonding, and Table

TABLE 13.2 Properties of Homonuclear Diatomic Molecules in Their Ground Electronic States

Molecule	Ground Term	Bond Order	D_e/eV	$R_e/\text{Å}$
H_2^+	$^2\Sigma_g^+$	$\frac{1}{2}$	2.79	1.06
H_2	$^1\Sigma_g^+$	1	4.75	0.741
He_2^+	$^2\Sigma_u^+$	$\frac{1}{2}$	2.5	1.08
He_2	$^1\Sigma_g^+$	0	0.0009	3.0
Li_2	$^1\Sigma_g^+$	1	1.07	2.67
Be_2	$^1\Sigma_g^+$	0	0.10	2.45
B_2	$^3\Sigma_g^-$	1	3.1	1.59
C_2	$^1\Sigma_g^+$	2	6.3	1.24
N_2^+	$^2\Sigma_g^+$	$2\frac{1}{2}$	8.85	1.12
N_2	$^1\Sigma_g^+$	3	9.91	1.10
O_2^+	$^2\Pi_g$	$2\frac{1}{2}$	6.78	1.12
O_2	$^3\Sigma_g^-$	2	5.21	1.21
F_2	$^1\Sigma_g^+$	1	1.66	1.41
Ne_2	$^1\Sigma_g^+$	0	0.0036	3.1

Data from K. P. Huber and G. Herzberg, *Constants of Diatomic Molecules* (vol. IV of *Molecular Spectra and Molecular Structure*), Van Nostrand Reinhold, New York, 1979; and (for Be_2) V. E. Bondybey, *Chem. Phys. Lett.*, **109**, 463 (1984).

13.2 shows that in going from the ground state of N_2 to that of N_2^+ the dissociation energy decreases (and the bond length increases). In contrast, the highest filled MO of O_2 is antibonding, and in going from O_2 to O_2^+ the dissociation energy increases (and R_e decreases). The designation of bonding or antibonding is not relevant to the effect of the electrons on the total energy of the molecule. Energy is always required to ionize a stable molecule, no matter which electron is removed. Hence both bonding and antibonding electrons in a stable molecule decrease the total molecular energy.

If the interaction between two ground-state He atoms were strictly repulsive (as predicted by MO theory), the atoms in He gas would not attract one another at all and the gas would never liquefy. Of course, helium gas can be liquefied. Configuration-interaction calculations and direct experimental evidence from scattering experiments show that as two He atoms approach each other there is an initial weak attraction, with the potential energy reaching a minimum at 3.0 Å of 0.0009 eV below the separated-atoms energy; at distances less than 3.0 Å, the force becomes increasingly repulsive because of overlap of the electron probability densities. The initial attraction (called a *London* or *dispersion force*) results from instantaneous correlation between the motions of the electrons in one atom and the motions of the electrons in the second atom. Therefore, a calculation that includes electron correlation is needed to deal with dispersion attractions.

The general term for all kinds of intermolecular forces is *van der Waals forces*. Except for highly polar molecules, the dispersion force is the largest contributor to intermolecular attractions. The dispersion force increases as the number of electrons increases, so boiling points tend to increase as the molecular weight increases.

The slight minimum in the potential-energy curve at relatively large intermolecular separation produced by the dispersion force can be sufficiently deep to allow the existence at low temperatures of molecules bound by the dispersion interaction. Such species are called *van der Waals molecules*. For example argon gas at 100 K has a small concentration of Ar_2 van der Waals molecules for which $D_e = 0.012$ eV and $R_e = 3.76$ Å; Ar_2 has seven bound vibrational levels ($v = 0, \ldots, 6$). For He_2, calculations show that the zero-point vibrational energy is approximately equal to the depth of the potential well, and whether a bound He_2 van der Waals molecule exists is uncertain; calculations based on the experimentally determined potential-energy curve of He_2 strongly suggest that He_2 does have a single extremely weakly bound state [Y. H. Uang and W. C. Stwalley, *J. Chem. Phys.*, **76**, 5069 (1982); R. A. Aziz et al., *Mol. Phys.*, **61**, 1487 (1987)].

Examples of diatomic van der Waals molecules and their R_e and D_e values include Ne_2, 3.1 Å, 0.0036 eV; HeNe, 3.2 Å, 0.0012 eV; Ca_2, 4.28 Å, 0.13 eV; Mg_2, 3.89 Å, 0.053 eV. Observed polyatomic van der Waals molecules include $(O_2)_2$, H_2-N_2, Ar–HCl, and $(Cl_2)_2$. Note that for van der Waals bonding, R_e is significantly greater and D_e very substantially less than the corresponding values for chemically bound molecules. The Be_2 bond length of 2.45 Å is much shorter than is typical for van der Waals molecules, and the nature of the binding in Be_2 is not fully understood. For more on van der Waals molecules, see *Chem. Rev.*, **88**, 813–988 (1988).

13.8 MOLECULAR ELECTRONIC TERMS

We now consider the terms arising from a given molecular electronic configuration.

For atoms, each set of degenerate atomic orbitals constitutes an atomic subshell. For example, the $2p_{+1}$, $2p_0$, and $2p_{-1}$ AOs constitute the $2p$ subshell. An atomic electronic configuration is defined by giving the number of electrons in each subshell; for example, $1s^2 2s^2 2p^4$. For molecules, each set of degenerate molecular orbitals constitutes a molecular subshell. For example, the $\pi_u 2p_{+1}$ and $\pi_u 2p_{-1}$ MOs constitute the $\pi_u 2p$ subshell. Each σ subshell consists of one MO, while each $\pi, \delta, \phi, \ldots$ subshell consists of two MOs; σ subshells are filled with two electrons, while non-σ subshells hold up to four electrons. We define a molecular electronic *configuration* by giving the number of electrons in each subshell, for example,

$$(\sigma_g 1s)^2 (\sigma_u^* 1s)^2 (\sigma_g 2s)^2 (\sigma_u^* 2s)^2 (\pi_u 2p)^3$$

For H_2^+ we saw that, although $\hat{L}^2$ does not commute with $\hat{H}$, the operator $\hat{L}_z$ does commute with $\hat{H}$. For a many-electron diatomic molecule, one finds that the operator for the axial component of the total electronic orbital angular momentum commutes with $\hat{H}$. The component of electronic orbital angular momentum along the molecular axis has the possible values $M_L \hbar$, where $M_L = 0, \pm 1, \pm 2, \pm \ldots$. To calculate M_L, we simply add algebraically the m's of the individual electrons. Analogous to the symbol λ for the one-electron molecule, Λ is defined as

$$\Lambda \equiv |M_L| \tag{13.85}$$

(Some people define Λ as equal to M_L.) The following code is used to indicate the value of Λ:

Λ	0	1	2	3	4
letter	Σ	Π	Δ	Φ	Γ

For $\Lambda \neq 0$, there are two possible values of M_L: $+\Lambda$ and $-\Lambda$. As in H_2^+, the electronic energy depends on M_L^2, so there is a double degeneracy associated with the two values of M_L. (Note that lowercase letters refer to individual electrons, while capital letters refer to the whole molecule.)

Just as in atoms, the individual electronic spins add vectorially to give a total electronic spin $\mathbf{S}$, whose magnitude has the possible values $[S(S+1)]^{1/2}\hbar$, with $S = 0, \frac{1}{2}, 1, \frac{3}{2}, \ldots$. The component of $\mathbf{S}$ along an axis has the possible values $M_S \hbar$, where $M_S = S, S-1, \ldots, -S$. As in atoms, the quantity $2S+1$ is called the *multiplicity* and is written as a left superscript to the code letter for Λ. Diatomic electronic states that arise from the same electron configuration and that have the same value for Λ and the same value for S are said to belong to the same electronic *term*. We now consider how the terms belonging to a given electronic configuration are derived. (We are assuming Russell–Saunders coupling, which holds for molecules composed of atoms of not-too-high atomic number.)

A filled molecular subshell consists of one or two filled molecular orbitals.

The Pauli principle requires that, for two electrons in the same molecular orbital, one have $m_s = +\frac{1}{2}$ and the other have $m_s = -\frac{1}{2}$. Hence the quantum number M_S, which is the algebraic sum of the individual m_s values, must be zero for a filled-subshell molecular configuration; therefore, we must have $S = 0$ for a configuration containing only filled molecular subshells. A filled σ subshell has two electrons with $m = 0$, so M_L is zero. A filled π subshell has two electrons with $m = +1$ and two electrons with $m = -1$, so M_L (which is the algebraic sum of the m's) is zero. The same situation holds for filled $\delta, \phi, \ldots$ subshells. Thus a closed-subshell molecular configuration has both S and Λ equal to zero and gives rise to only a $^1\Sigma$ term. An example is the ground electronic configuration of H_2. (Recall that a filled-subshell atomic configuration gives only a ^1S term.) In deriving molecular terms, we need consider only electrons outside filled subshells.

A single σ electron has $s = \frac{1}{2}$, so S must be $\frac{1}{2}$, and we get a $^2\Sigma$ term. An example is the ground electronic configuration of H_2^+. A single π electron gives a $^2\Pi$ term. And so on.

Now consider more than one electron. Electrons that are in different subshells are called *nonequivalent*, just as for atoms. For such electrons we do not have to worry about giving two of them the same set of quantum numbers, and the terms are easily derived. Consider two nonequivalent σ electrons, a $\sigma\sigma$ configuration. Since both m's are zero, we have $M_L = 0$. Each s is $\frac{1}{2}$, so S can be 1 or 0. We thus have the terms $^1\Sigma$ and $^3\Sigma$. Similarly, a $\sigma\pi$ configuration gives $^1\Pi$ and $^3\Pi$ terms.

For a $\pi\delta$ configuration, we have singlet and triplet terms. The π electron can have $m = \pm 1$, and the δ electron can have $m = \pm 2$. The possible values for M_L are thus $+3, -3, +1,$ and -1. This gives $\Lambda = 3$ or 1, and we have the terms $^1\Pi$, $^3\Pi$, $^1\Phi$, $^3\Phi$. (In atoms we add the *vectors* $\mathbf{L}_i$ to get the total $\mathbf{L}$; hence a pd atomic configuration gives P, D, and F terms. In molecules, however, we add the z *components* of the orbital angular momenta; this is an algebraic rather than a vectorial addition, so a $\pi\delta$ molecular configuration gives Π and Φ terms and no Δ terms.)

For a $\pi\pi$ configuration of two nonequivalent electrons, each electron has $m = \pm 1$, and we have the M_L values $2, -2, 0, 0$. The values of Λ are 2, 0, and 0; the terms are $^1\Delta$, $^3\Delta$, $^1\Sigma$, $^3\Sigma$, $^1\Sigma$, and $^3\Sigma$. The values $+2$ and -2 correspond to the two degenerate states of the same Δ term. However, Σ terms are nondegenerate (apart from spin degeneracy), and the two values of M_L that are zero indicate two different Σ terms (which become four Σ terms when we consider spin).

Consider the forms of the wave functions for the terms. We shall call the two π subshells π and π' and shall use a subscript to indicate the m value. For the Δ terms, both electrons have $m = +1$ or both have $m = -1$. For $M_L = +2$, we might write as the spatial factor in the wave function $\pi_{+1}(1)\pi'_{+1}(2)$ or $\pi_{+1}(2)\pi'_{+1}(1)$. However, these functions are neither symmetric nor antisymmetric with respect to exchange of the indistinguishable electrons and are unacceptable. Instead, we must take the two linear combinations (we shall not bother with normalization constants)

$$^1\Delta: \quad \pi_{+1}(1)\pi'_{+1}(2) + \pi_{+1}(2)\pi'_{+1}(1) \tag{13.86}$$

$$^3\Delta: \quad \pi_{+1}(1)\pi'_{+1}(2) - \pi_{+1}(2)\pi'_{+1}(1) \tag{13.87}$$

Similarly, with both electrons having $m = -1$, we have the spatial factors

$$^1\Delta: \quad \pi_{-1}(1)\pi'_{-1}(2) + \pi_{-1}(2)\pi'_{-1}(1) \tag{13.88}$$

$$^3\Delta: \quad \pi_{-1}(1)\pi'_{-1}(2) - \pi_{-1}(2)\pi'_{-1}(1) \tag{13.89}$$

The functions (13.86) and (13.88) are symmetric with respect to exchange. They therefore go with the antisymmetric two-electron spin factor (11.60), which has $S = 0$. Thus (13.86) and (13.88) are the spatial factors in the wave functions for the two states of the doubly degenerate $^1\Delta$ term. The antisymmetric functions (13.87) and (13.89) must go with the symmetric two-electron spin functions (11.57), (11.58), and (11.59), giving the six states of the $^3\Delta$ term. These states all have the same energy (if we neglect spin–orbit interaction).

Now consider the wave functions of the Σ terms. These have one electron with $m = +1$ and one electron with $m = -1$. We start with the four functions

$$\pi_{+1}(1)\pi'_{-1}(2)\,, \qquad \pi_{+1}(2)\pi'_{-1}(1)\,, \qquad \pi_{-1}(1)\pi'_{+1}(2)\,, \qquad \pi_{-1}(2)\pi'_{+1}(1)$$

Combining them to get symmetric and antisymmetric functions, we have

$$^1\Sigma^+: \quad \pi_{+1}(1)\pi'_{-1}(2) + \pi_{+1}(2)\pi'_{-1}(1) + \pi_{-1}(1)\pi'_{+1}(2) + \pi_{-1}(2)\pi'_{+1}(1)$$

$$^1\Sigma^-: \quad \pi_{+1}(1)\pi'_{-1}(2) + \pi_{+1}(2)\pi'_{-1}(1) - \pi_{-1}(1)\pi'_{+1}(2) - \pi_{-1}(2)\pi'_{+1}(1)$$

$$^3\Sigma^+: \quad \pi_{+1}(1)\pi'_{-1}(2) - \pi_{+1}(2)\pi'_{-1}(1) + \pi_{-1}(1)\pi'_{+1}(2) - \pi_{-1}(2)\pi'_{+1}(1)$$

$$^3\Sigma^-: \quad \pi_{+1}(1)\pi'_{-1}(2) - \pi_{+1}(2)\pi'_{-1}(1) - \pi_{-1}(1)\pi'_{+1}(2) + \pi_{-1}(2)\pi'_{+1}(1)$$

$$\tag{13.90}$$

The first two functions in (13.90) are symmetric; they therefore go with the antisymmetric singlet spin function (11.60). Clearly, these two spatial functions have different energies. The last two functions in (13.90) are antisymmetric and hence are the spatial factors in the wave functions of the two $^3\Sigma$ terms. These four functions are found to have eigenvalue $+1$ or -1 with respect to reflection of electronic coordinates in the xz σ_v symmetry plane containing the molecular (z) axis (Problem 13.23); the superscripts $+$ and $-$ refer to this eigenvalue.

Examination of the Δ terms (13.86) to (13.89) shows that they are not eigenfunctions of the symmetry operator $\hat{O}_{\sigma_v}$. Since a twofold degeneracy (apart from spin degeneracy) is associated with these terms, there is no necessity that their wave functions be eigenfunctions of this operator; however, since $\hat{O}_{\sigma_v}$ commutes with the Hamiltonian, we can *choose* the eigenfunctions to be eigenfunctions of $\hat{O}_{\sigma_v}$. Thus we can combine the functions (13.86) and (13.88), which belong to a degenerate energy level, as follows:

$$(13.86) + (13.88) \quad \text{and} \quad (13.86) - (13.88)$$

These two linear combinations are eigenfunctions of $\hat{O}_{\sigma_v}$ with eigenvalues $+1$ and -1, and we could refer to them as $^1\Delta^+$ and $^1\Delta^-$ states. Since they have the same energy, there is no point in using the $+$ and $-$ superscripts. Thus the $+$ and $-$ designations are used only for Σ terms. However, when one considers the interaction between the molecular rotational angular momentum and the electronic orbital angular momentum, there is a very slight splitting (called Λ-*type doubling*) of the two states of a $^1\Delta$ term; it turns out that the correct zeroth-order wave functions for this perturbation are the linear combinations that are eigenfunctions of $\hat{O}_{\sigma_v}$, so in this case there is a point to distinguishing between Δ^+ and Δ^- states. Note that the linear combinations (13.86) $\pm$ (13.88), which are eigenfunctions of $\hat{O}_{\sigma_v}$, are not eigenfunctions of $\hat{L}_z$ but are superpositions of $\hat{L}_z$ eigenfunctions with eigenvalues $+2$ and -2.

We can distinguish $+$ and $-$ terms for one-electron configurations. The wave function of a single σ electron has no phi factor and hence must correspond to a Σ^+ term. For a π electron, the MOs that are eigenfunctions of $\hat{L}_z$ are the π_{+1} and π_{-1} functions (whose probability densities are each symmetric about the z axis; Fig. 13.12). The π_{+1} and π_{-1} functions are not eigenfunctions of $\hat{O}_{\sigma_v}$, but the linear combinations $\pi_{+1} + \pi_{-1} = \pi_x$ and $\pi_{+1} - \pi_{-1} = \pi_y$ are; the π_x and π_y MOs (whose probability densities are not symmetric about the z axis; Fig. 13.14) are the correct zeroth-order functions if the perturbation of the electronic wave functions due to molecular rotation is considered. The π_x and π_y MOs have eigenvalues $+1$ and -1, respectively, for reflection in the xz plane and eigenvalues -1 and $+1$, respectively, for reflection in the yz plane. (The operators $\hat{L}_z$ and $\hat{O}_{\sigma_v}$ do not commute; Problem 13.24. Hence we cannot have all the eigenfunctions of $\hat{H}$ being eigenfunctions of both these operators as well. However, since each of these operators commutes with the electronic Hamiltonian and since there is no element of choice in the wave function of a nondegenerate level, all the σ MOs must be eigenfunctions of both $\hat{L}_z$ and $\hat{O}_{\sigma_v}$.)

Electrons in the same molecular subshell are called *equivalent*. For equivalent electrons, there are fewer terms than for the corresponding nonequivalent electron configuration, because of the Pauli principle. Thus, for a π^2 configuration of two equivalent π electrons, four of the eight functions (13.86) to (13.90) vanish; the remaining functions give a $^1\Delta$ term, a $^1\Sigma^+$ term, and a $^3\Sigma^-$ term. Alternatively, we can make a table similar to Table 11.1 and use it to derive the terms for equivalent electrons.

Table 13.3 lists terms arising from various electron configurations. A filled subshell always gives the single term $^1\Sigma^+$. A π^3 configuration gives the same result as a π configuration.

For homonuclear diatomic molecules, a g or u right subscript is added to the

TABLE 13.3 Electronic Terms of Diatomic Molecules

Configuration	Terms
$\sigma\sigma$	$^1\Sigma^+, {}^3\Sigma^+$
$\sigma\pi; \sigma\pi^3$	$^1\Pi, {}^3\Pi$
$\pi\pi; \pi\pi^3$	$^1\Sigma^+, {}^3\Sigma^+, {}^1\Sigma^-, {}^3\Sigma^-, {}^1\Delta, {}^3\Delta$
$\pi\delta; \pi^3\delta; \pi\delta^3$	$^1\Pi, {}^3\Pi, {}^1\Phi, {}^3\Phi$
σ	$^2\Sigma^+$
$\sigma^2; \pi^4; \delta^4$	$^1\Sigma^+$
$\pi; \pi^3$	$^2\Pi$
π^2	$^1\Sigma^+, {}^3\Sigma^-, {}^1\Delta$
$\delta; \delta^3$	$^2\Delta$
δ^2	$^1\Sigma^+, {}^3\Sigma^-, {}^1\Gamma$

term symbol to show the parity of the electronic states belonging to the term. Terms arising from an electron configuration that has an odd number of electrons in molecular orbitals of odd parity are odd (u); all other terms are even (g). This is the same rule as for atoms.

The term symbols given in Table 13.2 are readily derived from the MO configurations. For example, O_2 has a π^2 configuration, which gives the three terms $^1\Sigma_g^+$, $^3\Sigma_g^-$, and $^1\Delta_g$. Hund's rule tells us that $^3\Sigma_g^-$ is the lowest term, as listed. The $v = 0$ levels of the $^1\Delta_g$ and $^1\Sigma_g^+$ O_2 terms lie 0.98 eV and 1.6 eV, respectively, above the $v = 0$ level of the ground $^3\Sigma_g^-$ term. Singlet O_2 is a reaction intermediate in many organic, biochemical, and inorganic reactions. [See H. H. Wasserman and R. W. Murray, eds., *Singlet Oxygen*, Academic Press, New York, 1979; B. Ranby and J. F. Rabeck, eds., *Singlet Oxygen*, Wiley, New York, 1978.]

Most stable diatomic molecules have a $^1\Sigma^+$ ground term ($^1\Sigma_g^+$ for homonuclear diatomics). Exceptions include B_2, Al_2, Si_2, and O_2, and NO, which has a $^2\Pi$ ground term.

Spectroscopists prefix the ground term of a molecule by the symbol X. Excited terms of the same multiplicity as the ground term are designated as A, B, C, . . . , while excited terms of different multiplicity from the ground term are designated as a, b, c, ... Exceptions are C_2 and N_2, where the ground terms are $^1\Sigma_g^+$ but the letters A, B, C, . . . are used for excited triplet terms.

Just as for atoms, spin–orbit interaction can split a molecular term into closely spaced energy levels, giving a multiplet structure to the term. The projection of the total electronic spin S on the molecular axis is $M_S \hbar$; in molecules the quantum number M_S is called Σ (not to be confused with the symbol meaning $\Lambda = 0$):

$$\Sigma = S, S - 1, \ldots, -S$$

The axial components of electronic orbital and spin angular momenta add, giving as the total axial component of electronic angular momentum $(\Lambda + \Sigma)\hbar$. (Recall that Λ is the absolute value of M_L; we consider Σ to be positive when it has the same direction as Λ, and negative when it has the opposite direction as Λ.) The possible values of $\Lambda + \Sigma$ are

$$\Lambda + S, \Lambda + S - 1, \ldots, \Lambda - S$$

The value of $\Lambda + \Sigma$ is written as a right subscript to the term symbol to distinguish the energy levels of the term. Thus a $^3\Delta$ term has $\Lambda = 2$ and $S = 1$ and gives rise to the levels $^3\Delta_3$, $^3\Delta_2$, and $^3\Delta_1$. In a sense, $\Lambda + \Sigma$ is the analog in molecules of the quantum number J in atoms. However, $\Lambda + \Sigma$ is the quantum number of the z *component* of total electronic angular momentum and therefore can take on negative values. Thus a $^4\Pi$ term has the four levels $^4\Pi_{5/2}$, $^4\Pi_{3/2}$, $^4\Pi_{1/2}$, and $^4\Pi_{-1/2}$. The absolute value of $\Lambda + \Sigma$ is called Ω:

$$\Omega \equiv |\Lambda + \Sigma| \qquad (13.91)$$

It is the value of $\Lambda + \Sigma$, and not the value of Ω, which is written as a subscript on the term symbol.

The spin–orbit interaction energy in diatomic molecules can be shown to be well approximated by $A\Lambda\Sigma$, where A depends on Λ and on the internuclear distance R but not on Σ. The spacing between levels of the multiplet is thus constant. When A is positive, the level with the lowest value of $\Lambda + \Sigma$ lies lowest, and the multiplet is *regular*. When A is negative, the multiplet is *inverted*. Note that for $\Lambda \neq 0$ the multiplicity $2S + 1$ always equals the number of multiplet components; this is not always true for atoms.

Each energy level of a multiplet with $\Lambda \neq 0$ is doubly degenerate, corresponding to the two values for M_L. Thus a $^3\Delta$ term has six different wave functions [Eqs. (13.87), (13.89), (11.57) to (11.59)] and therefore six different molecular electronic states. Spin–orbit interaction splits the $^3\Delta$ term into three levels, each doubly degenerate. The double degeneracy of the levels is removed by the Λ-type doubling mentioned previously.

For Σ terms ($\Lambda = 0$), the spin–orbit interaction is very small (zero in the first approximation), and the quantum numbers Σ and Ω are not defined.

A $^1\Sigma$ term always corresponds to a single nondegenerate energy level.

13.9   THE HYDROGEN MOLECULE

The hydrogen molecule is the simplest molecule containing an electron-pair bond. We shall devote most of our attention to the ground electronic state.

The purely electronic Hamiltonian (13.5) for H_2 is

$$\hat{H} = -\tfrac{1}{2}\nabla_1^2 - \tfrac{1}{2}\nabla_2^2 - \frac{1}{r_{a1}} - \frac{1}{r_{a2}} - \frac{1}{r_{b1}} - \frac{1}{r_{b2}} + \frac{1}{r_{12}} \tag{13.92}$$

where 1 and 2 are the electrons and a and b are the nuclei (Fig. 13.17). Just as in the helium atom, the $1/r_{12}$ interelectronic-repulsion term prevents the Schrödinger equation from being separable. We therefore use approximation methods.

We start with the molecular-orbital approach. The ground-state electronic configuration of H_2 is $(\sigma_g 1s)^2$, and we can write an approximate wave function as the Slater determinant

$$\frac{1}{\sqrt{2}} \begin{vmatrix} \sigma_g 1s(1)\alpha(1) & \sigma_g 1s(1)\beta(1) \\ \sigma_g 1s(2)\alpha(2) & \sigma_g 1s(2)\beta(2) \end{vmatrix} = \sigma_g 1s(1)\sigma_g 1s(2)2^{-1/2}[\alpha(1)\beta(2) - \beta(1)\alpha(2)]$$

$$= f(1)f(2)2^{-1/2}[\alpha(1)\beta(2) - \beta(1)\alpha(2)] \tag{13.93}$$

which is similar to (10.26) for the helium atom. To save time, we write f instead

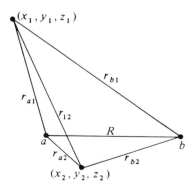

Figure 13.17 Interparticle distances in H_2.

of $\sigma_g 1s$. As we saw in Section 10.4, omission of the spin factor does not affect the variational integral for a two-electron problem. Hence we want to choose f so as to minimize

$$\frac{\int\int f^*(1)f^*(2)\hat{H}f(1)f(2)\,dv_1\,dv_2}{\int\int |f(1)|^2|f(2)|^2\,dv_1\,dv_2}$$

where the integration is over the space coordinates of the two electrons. Ideally, f should be found by an SCF calculation. For simplicity we can use an H_2^+-like MO. (The H_2 Hamiltonian becomes the sum of two H_2^+ Hamiltonians if we omit the $1/r_{12}$ term.) We saw in Section 13.5 that the function [Eq. (13.54)]

$$\frac{k^{3/2}}{(2\pi)^{1/2}(1+S_{ab})^{1/2}}\,(e^{-kr_a}+e^{-kr_b})$$

gave a good approximation to the ground-state H_2^+ wave function. Hence we try as a variation function ϕ for H_2 the product of two such LCAO functions, one for each electron:

$$\phi = \frac{\zeta^3}{2\pi(1+S_{ab})}\,(e^{-\zeta r_{a1}}+e^{-\zeta r_{b1}})(e^{-\zeta r_{a2}}+e^{-\zeta r_{b2}}) \tag{13.94}$$

$$\phi = \frac{1}{2(1+S_{ab})}\,[1s_a(1)+1s_b(1)][1s_a(2)+1s_b(2)] \tag{13.95}$$

where the effective nuclear charge ζ will differ from k for H_2^+. Since

$$\hat{H} = \hat{H}_1^0 + \hat{H}_2^0 + 1/r_{12}$$

where $\hat{H}_1^0$ and $\hat{H}_2^0$ are H_2^+ Hamiltonians for each electron, we have

$$\int\int \phi^*\hat{H}\phi\,dv_1\,dv_2 = 2W_1 + \int\int \frac{\phi^2}{r_{12}}\,dv_1\,dv_2$$

where W_1 is given by (13.63) with k replaced by ζ. The evaluation of the $1/r_{12}$ integral is complicated and is omitted [see *Slater, Quantum Theory of Molecules and Solids*, Volume 1, page 65, and Appendix 6]. Coulson performed the variational calculation in 1937, using (13.94). [For the literature references of the H_2 calculations mentioned in this and later sections, see the bibliography in A. D. Mclean, A. Weiss, and M. Yoshimine, *Rev. Mod. Phys.*, **32**, 211 (1960).] Coulson found $R_e = 0.732$ Å, which is close to the true value 0.741 Å; the minimum in the calculated $U(R)$ curve gave $D_e = 3.49$ eV, as compared with the true value 4.75 eV (Table 13.2). (Of course, the percent error in the total electronic energy is much less than the percent error in D_e, but D_e is the quantity of chemical interest.) The value of ζ at 0.732 Å is 1.197, which is less than k for H_2^+. We attribute this to the screening of the nuclei from each electron by the other electron.

How can we improve on the above simple MO result? We can look for the best possible MO function f in (13.93) to get the Hartree–Fock wave function for H_2. This was done by Kolos and Roothaan [W. Kolos and C. C. J. Roothaan, *Rev. Mod. Phys.*, **32**, 219 (1960)]. They expanded f in elliptic coordinates [Eq. (13.33)]. Since $m = 0$ for the ground state, the $e^{im\phi}$ factor in the SCF MO is equal to 1 and f is a function of ξ and η only. The expansion used is

$$f = e^{-\alpha\xi} \sum_{p,q} a_{pq} \xi^p \eta^q$$

where p and q are integers and α and a_{pq} are variational parameters. The Hartree–Fock results are $R_e = 0.733$ Å and $D_e = 3.64$ eV, which is not much improvement over the value 3.49 eV given by the simple LCAO molecular orbital. The correlation energy for H_2 is thus 1.11 eV, about the same as the value 1.1 eV for the two-electron helium atom (Section 11.1). To get a truly accurate binding energy, we must go beyond the SCF approximation of writing the wave function in the form $f(1)f(2)$. We use the same methods we used for atoms: configuration interaction and introduction of r_{12} into the trial function.

First consider configuration interaction (CI). To reach the exact ground-state wave function, we include contributions from SCF (or other) functions for all the excited states with the same symmetry as the ground state. In the first approximation, only contributions from the lowest-lying excited states are included. The first excited configuration of H_2 is $(\sigma_g 1s)(\sigma_u^* 1s)$, which gives the terms $^1\Sigma_u^+$ and $^3\Sigma_u^+$. (We have one g and one u electron, so the terms are of odd parity.) The ground-state configuration $(\sigma_g 1s)^2$ is a $^1\Sigma_g^+$ state. Hence we do not get any contribution from the $(\sigma_g 1s)(\sigma_u^* 1s)$ states, since they have different parity from the ground state. Next consider the configuration $(\sigma_u^* 1s)^2$. This is a closed-subshell configuration having the single state $^1\Sigma_g^+$. This is of the right symmetry to contribute to the ground-state wave function. As a simple CI trial function, we can take a linear combination of the MO wave functions for the $(\sigma_g 1s)^2$ and $(\sigma_u^* 1s)^2$ configurations. To simplify things, we will use the LCAO-MOs as approximations to the MOs. Thus we take

$$\phi = \sigma_g 1s(1)\sigma_g 1s(2) + c\sigma_u^* 1s(1)\sigma_u^* 1s(2) \qquad (13.96)$$

where $\sigma_g 1s$ and $\sigma_u^* 1s$ are given by (13.54) and (13.55) with a variable-orbital exponent and c is a variational parameter. This calculation was performed by Weinbaum in 1933. The result is a bond length of 0.757 Å and a dissociation energy of 4.03 eV, a considerable improvement over the Hartree–Fock result $D_e = 3.64$ eV. The orbital exponent has the optimum value 1.19. We can improve on this result in two ways: by using a better form for the MOs of each configuration and by including more configuration functions. Hagstrom did a CI calculation in which the MOs were represented by expansions in elliptic coordinates; with 33 configuration functions, he found $D_e = 4.71$ eV, quite close to the true value 4.75 eV. [S. Hagstrom and H. Shull, *Rev. Mod. Phys.*, **35**, 624 (1963).]

Now consider the use of r_{12} in H_2 trial functions. The first really accurate calculation of the hydrogen-molecule ground state was done by James and Coolidge in 1933. They used the trial function

$$\exp\left[-\delta(\xi_1 + \xi_2)\right] \sum c_{mnjkp}[\xi_1^m \xi_2^n \eta_1^j \eta_2^k + \xi_1^n \xi_2^m \eta_1^k \eta_2^j]r_{12}^p$$

where the summation is over integral values of m, n, j, k, and p. The variational parameters are δ and the c_{mnjkp}. The James and Coolidge function is symmetric with respect to interchange of electrons 1 and 2, as it should be, since we have an antisymmetric ground-state spin function. With 13 terms in the sum, James and

Coolidge found $D_e = 4.72$ eV, only 0.03 eV in error. Their work has been extended by Kolos, Wolniewicz, and co-workers, who used as many as 249 terms in the sum. Since it is D_0 that is determined from the observed electronic spectrum, they used the Cooley–Numerov method (Section 13.2) to calculate the vibrational levels from their theoretical $U(R)$ curve and then calculated D_0. Including relativistic corrections and corrections to the Born–Oppenheimer approximation, they found $D_0/hc = 36118.1$ cm^{-1}, in agreement with the spectroscopically determined value 36118.6 ± 0.5 cm^{-1} [W. Kolos et al., *J. Chem. Phys.*, **84**, 3278 (1986)]. Their wave function is essentially indistinguishable from the true wave function, but its great complexity allows no simple physical interpretation.

13.10 THE VALENCE-BOND TREATMENT OF H_2

The first quantum-mechanical treatment of the hydrogen molecule was by Heitler and London in 1927. Their ideas have been extended to give a general theory of chemical bonding, known as the **valence-bond** (VB) theory, or the HLSP theory, after Heitler, London, Slater, and Pauling. The valence-bond method is more closely related to the chemist's idea of molecules as consisting of atoms held together by localized bonds than is the molecular-orbital method. The VB method views molecules as composed of atomic cores (nuclei plus inner-shell electrons) and bonding valence electrons. For H_2, both electrons are valence electrons.

The first step in the Heitler–London treatment of the H_2 ground state is to approximate the molecule as two ground-state hydrogen atoms. The wave function for two such noninteracting atoms is

$$f_1 = 1s_a(1)1s_b(2)$$

where a and b refer to the nuclei and 1 and 2 refer to the electrons. Of course, the function

$$f_2 = 1s_a(2)1s_b(1)$$

is also a valid wave function. This then suggests the trial variation function

$$c_1 f_1 + c_2 f_2 = c_1 1s_a(1)1s_b(2) + c_2 1s_a(2)1s_b(1) \tag{13.97}$$

This linear variation function leads to the determinantal secular equation $\det(H_{ij} - S_{ij}W) = 0$ [Eq. (8.58)], where $H_{11} = \langle f_1|\hat{H}|f_1\rangle$, $S_{11} = \langle f_1|f_1\rangle$, and so on.

We can also consider the problem using perturbation theory (as Heitler and London did). A ground-state hydrogen molecule dissociates to two neutral ground-state hydrogen atoms. We therefore take as our unperturbed problem two ground-state hydrogen atoms at infinite separation. One possible zeroth-order (unperturbed) wave function is $1s_a(1)1s_b(2)$; however, electron 2 could just as well be bound to nucleus a, giving the unperturbed wave function $1s_a(2)1s_b(1)$. These two unperturbed wave functions belong to a doubly degenerate energy level (exchange degeneracy). Under the perturbation of molecule formation, the doubly degenerate level is split into two levels, and the *correct* zeroth-order wave functions are linear combinations of the two unperturbed wave functions:

$$c_1 1s_a(1)1s_b(2) + c_2 1s_a(2)1s_b(1)$$

This leads to a 2×2 secular determinant that is the same as (8.57), except that W is replaced by $E^{(0)} + E^{(1)}$; see Problem 9.12.

We now solve the secular equation. The Hamiltonian is Hermitian, all functions are real, and f_1 and f_2 are normalized; hence

$$H_{12} = H_{21}, \qquad S_{12} = S_{21}, \qquad S_{11} = S_{22} = 1$$

Consider H_{11} and H_{22}:

$$H_{11} = \langle 1s_a(1)1s_b(2)|\hat{H}|1s_a(1)1s_b(2)\rangle$$

$$H_{22} = \langle 1s_a(2)1s_b(1)|\hat{H}|1s_a(2)1s_b(1)\rangle$$

Interchange of the coordinate labels 1 and 2 in H_{22} converts H_{22} to H_{11}, since this relabeling leaves $\hat{H}$ unchanged. Hence $H_{11} = H_{22}$. The secular equation $\det(H_{ij} - S_{ij}W) = 0$ becomes

$$\begin{vmatrix} H_{11} - W & H_{12} - WS_{12} \\ H_{12} - WS_{12} & H_{11} - W \end{vmatrix} = 0 \qquad (13.98)$$

This equation has the same form as Eq. (13.49), and by analogy to Eqs. (13.51), (13.54), and (13.55) the approximate energies and wave functions are

$$W_1 = \frac{H_{11} + H_{12}}{1 + S_{12}}, \qquad W_2 = \frac{H_{11} - H_{12}}{1 - S_{12}} \qquad (13.99)$$

$$\phi_1 = \frac{f_1 + f_2}{\sqrt{2}(1 + S_{12})^{1/2}}, \qquad \phi_2 = \frac{f_1 - f_2}{\sqrt{2}(1 - S_{12})^{1/2}} \qquad (13.100)$$

For S_{12} we have

$$S_{12} = \int f_1^* f_2 \, dv = \int \int 1s_a(1)1s_b(2)1s_a(2)1s_b(1) \, dv_1 \, dv_2$$

$$S_{12} = \langle 1s_a(1)|1s_b(1)\rangle\langle 1s_a(2)|1s_b(2)\rangle = S_{ab}^2$$

where the overlap integral S_{ab} is defined by (13.48).

The numerators of (13.100) are

$$f_1 \pm f_2 = 1s_a(1)1s_b(2) \pm 1s_a(2)1s_b(1)$$

From our previous discussion, we know that the ground state of H_2 is a $^1\Sigma$ state with the antisymmetric spin factor (11.60) and a symmetric spatial factor. Hence ϕ_1 must be the ground state. The Heitler–London ground-state wave function is

$$\frac{1s_a(1)1s_b(2) + 1s_a(2)1s_b(1)}{\sqrt{2}(1 + S_{ab}^2)^{1/2}} \frac{1}{\sqrt{2}}[\alpha(1)\beta(2) - \alpha(2)\beta(1)] \qquad (13.101)$$

The Heitler–London wave functions for the three states of the lowest $^3\Sigma$ term are

$$\frac{1s_a(1)1s_b(2) - 1s_a(2)1s_b(1)}{\sqrt{2}(1 - S_{ab}^2)^{1/2}} \begin{cases} \alpha(1)\alpha(2) \\ 2^{-1/2}[\alpha(1)\beta(2) + \beta(1)\alpha(2)] \\ \beta(1)\beta(2) \end{cases} \qquad (13.102)$$

Now consider the ground-state energy expression. We divide the molecular electronic Hamiltonian into the sum of two hydrogen-atom Hamiltonians plus perturbing terms:

$$\hat{H} = \hat{H}_a(1) + \hat{H}_b(2) + \hat{H}'$$

$$\hat{H}_a(1) = -\tfrac{1}{2}\nabla_1^2 - \frac{1}{r_{a1}}, \qquad \hat{H}_b(2) = -\tfrac{1}{2}\nabla_2^2 - \frac{1}{r_{b2}}, \qquad \hat{H}' = -\frac{1}{r_{b1}} - \frac{1}{r_{a2}} + \frac{1}{r_{12}}$$

We then have

$$H_{11} = \langle 1s_a(1)1s_b(2)|\hat{H}_a(1) + \hat{H}_b(2) + H'|1s_a(1)1s_b(2)\rangle \qquad (13.103)$$

For the integral involving $\hat{H}_a(1)$, we have

$$\langle 1s_a(1)1s_b(2)|\hat{H}_a(1)|1s_a(1)1s_b(2)\rangle = \langle 1s_a(1)|\hat{H}_a(1)|1s_a(1)\rangle \langle 1s_b(2)|1s_b(2)\rangle$$

The Heitler–London calculation does not introduce an effective nuclear charge into the 1s function; hence $1s_a(1)$ is an eigenfunction of $\hat{H}_a(1)$ with eigenvalue $-\tfrac{1}{2}$ hartree, the hydrogen-atom ground-state energy. The 1s function is normalized, and we conclude that the $\hat{H}_a(1)$ integral equals $-\tfrac{1}{2}$ in atomic units. Similarly, the $\hat{H}_b(2)$ integral in (13.103) equals $-\tfrac{1}{2}$. Defining the *Coulomb integral Q* as

$$Q \equiv \langle 1s_a(1)1s_b(2)|\hat{H}'|1s_a(1)1s_b(2)\rangle \qquad (13.104)$$

we have

$$H_{11} = Q - 1$$

For H_{12} we have

$$H_{12} = H_{21} = \langle 1s_a(2)1s_b(1)|\hat{H}_a(1) + \hat{H}_b(2) + \hat{H}'|1s_a(1)1s_b(2)\rangle$$

The $\hat{H}_a(1)$ integral is easily evaluated as

$$\langle 1s_a(2)|1s_b(2)\rangle \langle 1s_b(1)|\hat{H}_a(1)|1s_a(1)\rangle = -\tfrac{1}{2}S_{ab}^2$$

Defining the *exchange integral A* as

$$A \equiv \langle 1s_a(2)1s_b(1)|\hat{H}'|1s_a(1)1s_b(2)\rangle \qquad (13.105)$$

we have

$$H_{12} = A - S_{ab}^2$$

Substitution into (13.99) gives

$$W_1 = -1 + \frac{Q + A}{1 + S_{ab}^2}, \qquad W_2 = -1 + \frac{Q - A}{1 - S_{ab}^2} \qquad (13.106)$$

The quantity -1 hartree in these expressions is the energy of two ground-state hydrogen atoms. To obtain the $U(R)$ potential-energy curves, we add the internuclear repulsion $1/R$ to these expressions.

Many of the integrals needed to evaluate W_1 and W_2 have been evaluated in the treatment of H$_2^+$ in Section 13.5; the only new integrals are those involving $1/r_{12}$. The hardest one is the two-center, two-electron exchange integral:

$$\int\int 1s_a(1)1s_b(2) \frac{1}{r_{12}} 1s_a(2)1s_b(1)\, dv_1\, dv_2$$

(*Two-center* means that the integrand contains functions centered on two different nuclei, *a* and *b*; *two-electron* means that the coordinates of two electrons occur in the integrand.) This must be evaluated using an expansion for $1/r_{12}$ in confocal elliptic coordinates, similar to the expansion (9.53) in spherical polar coordinates. Details of the integral evaluations are given in *Slater, Quantum Theory of Molecules and Solids*, Volume 1, Appendix 6. The results of the Heitler–London treatment are $D_e = 3.15$ eV, $R_e = 0.87$ Å. The agreement with the experimental values $D_e = 4.75$ eV, $R_e = 0.741$ Å is only fair. In this treatment, most of the binding energy is provided by the exchange integral A. The Heitler–London treatment of H_2 bears some resemblance to the Heisenberg treatment of the helium $1s2s$ configuration given in Section 9.7. However, for H_2 the exchange integral is negative, whereas for helium it is positive.

Consider some improvements on the Heitler–London function (13.101). One obvious step is the introduction of an orbital exponent ζ in the $1s$ function. This was done by Wang in 1928. The optimum value of ζ is 1.166 at R_e, and D_e and R_e are improved to 3.78 eV and 0.744 Å. Recall that Dickinson in 1933 improved the Finkelstein–Horowitz H_2^+ trial function by mixing in some $2p_z$ character into the atomic orbitals (hybridization). In 1931 Rosen used this idea to improve the Heitler–London–Wang function. He took the trial function

$$\phi = \phi_a(1)\phi_b(2) + \phi_a(2)\phi_b(1)$$

where the atomic orbital ϕ_a is given by $\phi_a = e^{-\zeta r_a}(1 + cz_a)$, with a similar expression for ϕ_b. This allows for the polarization of the AOs on molecule formation. The result is a binding energy of 4.04 eV. Another improvement, the use of ionic structures, will be considered in the next section. Still another improvement, the generalized valence-bond method, is discussed in Section 15.17.

13.11 COMPARISON OF THE MO AND VB THEORIES

Let us compare the molecular-orbital and valence-bond treatments of the hydrogen-molecule ground state.

If ϕ_a symbolizes an atomic orbital centered on nucleus *a*, the spatial factor of the unnormalized LCAO-MO wave function for the H_2 ground state is

$$[\phi_a(1) + \phi_b(1)][\phi_a(2) + \phi_b(2)] \tag{13.107}$$

In the simplest treatment, ϕ is a $1s$ AO. The function (13.107) equals

$$\phi_a(1)\phi_a(2) + \phi_b(1)\phi_b(2) + \phi_a(1)\phi_b(2) + \phi_b(1)\phi_a(2) \tag{13.108}$$

What is the physical significance of the terms? The last two terms have each electron in an atomic orbital centered on a different nucleus; these are covalent terms, corresponding to equal sharing of the electrons between the atoms. The first two terms have both electrons in AOs centered on the same nucleus; these are ionic terms, corresponding to the chemical structures

$$H^-\ H^+ \quad \text{and} \quad H^+\ H^-$$

The covalent and ionic terms occur with equal weight, so that according to this simple MO function there is a 50–50 chance as to whether the H_2 ground state dissociates to two neutral hydrogen atoms or to a proton and a hydride ion. Actually, the H_2 ground state dissociates to two neutral hydrogen atoms. Thus the simple MO function gives the wrong limiting value of the energy as R goes to infinity.

How can we remedy this? Since H_2 is nonpolar, chemical intuition tells us that ionic terms should make substantially less contribution to the wave function than covalent terms. The simplest procedure is to omit the ionic terms of the MO function (13.108). This gives

$$\phi_a(1)\phi_b(2) + \phi_b(1)\phi_a(2) \tag{13.109}$$

We immediately recognize (13.109) as being the Heitler–London function (13.101).

Although interelectronic repulsion causes the electrons to avoid each other, there is some probability of finding both electrons near the same nucleus, corresponding to an ionic structure. Therefore, instead of simply dropping the ionic terms from (13.108), we might try

$$\phi_{VB,imp} = \phi_a(1)\phi_b(2) + \phi_b(1)\phi_a(2) + \delta[\phi_a(1)\phi_a(2) + \phi_b(1)\phi_b(2)] \tag{13.110}$$

where $\delta(R)$ is a variational parameter and where the subscript indicates an improved VB function. In the language of valence-bond theory, this trial function represents *ionic–covalent resonance*. Of course, the ground-state wave function of H_2 does not undergo a time-dependent change back and forth from a covalent function corresponding to the structure H—H to ionic functions. Rather (in the particular approximation we are considering), the wave function is a time-independent mixture of covalent and ionic functions. Since H_2 dissociates to neutral atoms, we know that $\delta(\infty) = 0$. A variational calculation done by Weinbaum in 1933 using $1s$ AOs with an orbital exponent gave the result that at R_e the parameter δ has the value 0.26; the orbital exponent was found to be 1.19, and the dissociation energy was calculated as 4.03 eV, a modest improvement over the Heitler–London–Wang value of 3.78 eV. With δ equal to zero in (13.110), we get the VB function (13.109); with δ equal to 1, we get the LCAO-MO function (13.108). The optimum value of δ turns out to be closer to zero than to 1, and, in fact, the Heitler–London–Wang VB function gives a better dissociation energy than the LCAO-MO function.

Let us compare the improved valence-bond trial function (13.110) with the simple LCAO-MO function improved by configuration interaction. The LCAO-MO CI trial function (13.96) has the (unnormalized) form

$$\phi_{MO,imp} = [\phi_a(1) + \phi_b(1)][\phi_a(2) + \phi_b(2)] + \gamma[\phi_a(1) - \phi_b(1)][\phi_a(2) - \phi_b(2)]$$

Since we have not yet normalized this function, there is no harm in multiplying it by the constant $1/(1-\gamma)$. Doing so and rearranging terms, we get

$$\phi_{MO,imp} = \phi_a(1)\phi_b(2) + \phi_b(1)\phi_a(2) + \frac{1+\gamma}{1-\gamma}[\phi_a(1)\phi_a(2) + \phi_b(1)\phi_b(2)]$$

There is also no harm done if we define a new constant δ as $\delta = (1+\gamma)/(1-\gamma)$.

We see then that this improved MO function and the improved VB function (13.110) are *identical*. Weinbaum viewed his H_2 calculation as a valence-bond calculation with inclusion of ionic terms. We have shown that we can just as well view the Weinbaum calculation as an MO calculation with configuration interaction. (This was the viewpoint adopted in Section 13.9.)

The MO function (13.108) underestimates electron correlation, in that it says that structures with both electrons on the same atom are just as likely as structures with each electron on a different atom. The VB function (13.109) overestimates electron correlation, in that it has no contribution from structures with both electrons on the same atom. In MO theory, electron correlation is introduced by configuration interaction. In VB theory, electron correlation is reduced by ionic–covalent resonance. The simple VB method is more reliable at large R than the simple MO method, since the latter predicts the wrong dissociation products.

To further fix the differences between the MO and VB approaches, consider how each method divides the H_2 electronic Hamiltonian into unperturbed and perturbation Hamiltonians. For the MO method, we write

$$\hat{H} = \left[\left(-\tfrac{1}{2}\nabla_1^2 - \frac{1}{r_{a1}} - \frac{1}{r_{b1}} \right) + \left(-\tfrac{1}{2}\nabla_2^2 - \frac{1}{r_{a2}} - \frac{1}{r_{b2}} \right) \right] + \frac{1}{r_{12}}$$

where the unperturbed Hamiltonian consists of the bracketed terms. In MO theory the unperturbed Hamiltonian for H_2 is the sum of two H_2^+ Hamiltonians, one for each electron. Accordingly, the zeroth-order MO wave function is a product of two H_2^+-like wave functions, one for each electron. Since the H_2^+ functions are complicated, we approximate the H_2^+-like MOs as LCAOs. The effect of the $1/r_{12}$ perturbation is taken into account in an average way through use of self-consistent-field molecular orbitals; to take instantaneous electron correlation into account, we use configuration interaction, getting a kind of calculation called LCAO-MO SCF CI.

For the valence-bond method, the terms in the Hamiltonian are grouped in either of two ways:

$$\hat{H} = \left[\left(-\tfrac{1}{2}\nabla_1^2 - \frac{1}{r_{a1}} \right) + \left(-\tfrac{1}{2}\nabla_2^2 - \frac{1}{r_{b2}} \right) \right] - \frac{1}{r_{a2}} - \frac{1}{r_{b1}} + \frac{1}{r_{12}}$$

$$\hat{H} = \left[\left(-\tfrac{1}{2}\nabla_1^2 - \frac{1}{r_{b1}} \right) + \left(-\tfrac{1}{2}\nabla_2^2 - \frac{1}{r_{a2}} \right) \right] - \frac{1}{r_{a1}} - \frac{1}{r_{b2}} + \frac{1}{r_{12}}$$

The unperturbed system is two hydrogen atoms. We have two zeroth-order functions consisting of products of hydrogen-atom wave functions, and these belong to a degenerate level. The correct ground-state zeroth-order function is the linear combination (13.101).

Most quantitative molecular calculations are done using the MO method, because it is computationally much simpler than the VB method. The MO method was developed by Hund, Mulliken, and Lennard-Jones in the late 1920s. Originally, it was used largely for qualitative descriptions of molecules, but the electronic digital computer has made possible the calculation of accurate MO functions (Section 13.17).

13.12   MO AND VB WAVE FUNCTIONS FOR HOMONUCLEAR DIATOMIC MOLECULES

The molecular-orbital approximation puts the electrons of a molecule in molecular orbitals, which extend over the whole molecule. As an approximation to the molecular orbitals, we usually use linear combinations of atomic orbitals. The valence-bond method puts the electrons of a molecule in atomic orbitals and constructs the molecular wave function by allowing for "exchange" of the valence electron pairs between the atomic orbitals of the bonding atoms. We have compared the two methods for H_2; we now consider other homonuclear diatomic molecules.

We begin with the ground state of He_2. The separated helium atoms each have the ground-state configuration $1s^2$. This closed-subshell configuration does not have any unpaired electrons to form valence bonds, and the VB wave function is simply the antisymmetrized product of the atomic-orbital functions:

$$\frac{1}{\sqrt{24}} \begin{vmatrix} 1s_a(1) & \overline{1s_a}(1) & 1s_b(1) & \overline{1s_b}(1) \\ 1s_a(2) & \overline{1s_a}(2) & 1s_b(2) & \overline{1s_b}(2) \\ 1s_a(3) & \overline{1s_a}(3) & 1s_b(3) & \overline{1s_b}(3) \\ 1s_a(4) & \overline{1s_a}(4) & 1s_b(4) & \overline{1s_b}(4) \end{vmatrix} \tag{13.111}$$

The $1s$ function in this wave function is a helium-atom $1s$ function, which ideally is an SCF atomic function but can be approximated by a hydrogenlike function with an effective nuclear charge. The subscripts a and b refer to the two atoms, and the bar indicates spin function β. In the shorthand notation of (10.47), the wave function (13.111) is

$$|1s_a\overline{1s_a}1s_b\overline{1s_b}| \tag{13.112}$$

The valence-bond wave function for He_2 has each electron paired with another electron in an orbital on the *same* atom and hence predicts no bonding.

In the MO approach, He_2 has the ground-state configuration $(\sigma_g 1s)^2(\sigma_u^* 1s)^2$. With no net bonding electrons, no bonding is predicted, in agreement with the VB method. The MO approximation to the wave function is

$$|\sigma_g 1s\, \overline{\sigma_g 1s}\, \sigma_u^* 1s\, \overline{\sigma_u^* 1s}| \tag{13.113}$$

The simplest way to approximate the (unnormalized) MOs is to take them as linear combinations of the helium-atom AOs: $\sigma_g 1s = 1s_a + 1s_b$ and $\sigma_u^* 1s = 1s_a - 1s_b$. With this approximation, (13.113) becomes

$$|(1s_a + 1s_b)\overline{(1s_a + 1s_b)}(1s_a - 1s_b)\overline{(1s_a - 1s_b)}| \tag{13.114}$$

We can add or subtract one column of the determinant from another column without changing the determinant's value. If we add column 1 to column 3 and column 2 to column 4, we simplify (13.114) to

$$4|(1s_a + 1s_b)\overline{(1s_a + 1s_b)}1s_a\overline{1s_a}|$$

We now subtract column 3 from column 1 and column 4 from column 2 to get

$$4|1s_b\overline{1s_b}1s_a\overline{1s_a}| \tag{13.115}$$

The interchange of columns 1 and 3 and of columns 2 and 4 multiplies the determinant by $(-1)^2$, so (13.115) is equal to

$$4|1s_a\overline{1s_a}1s_b\overline{1s_b}|$$

which is identical (after normalization) to the valence-bond function (13.112). This result is easily generalized to the statement that the simple VB and simple LCAO-MO methods give the same approximate wave functions for diatomic molecules formed from separated atoms with completely filled atomic subshells. For example, the two methods give the same wave function for the Be_2 ground state. We could now substitute the trial function (13.112) into the variational integral and calculate the repulsive curve for the interaction of two ground-state helium atoms.

Before going on to Li_2, let us express the Heitler–London valence-bond functions for H_2 as Slater determinants. The ground-state Heitler–London function (13.101) can be written as

$$\tfrac{1}{2}(1 + S_{ab}^2)^{-1/2}\left\{\begin{vmatrix} 1s_a(1)\alpha(1) & 1s_b(1)\beta(1) \\ 1s_a(2)\alpha(2) & 1s_b(2)\beta(2) \end{vmatrix} - \begin{vmatrix} 1s_a(1)\beta(1) & 1s_b(1)\alpha(1) \\ 1s_a(2)\beta(2) & 1s_b(2)\alpha(2) \end{vmatrix}\right\}$$

$$= (2 + 2S_{ab}^2)^{-1/2}\{|1s_a\overline{1s_b}| - |\overline{1s_a}1s_b|\} \qquad (13.116)$$

In each Slater determinant, the electron on atom a is paired with an electron of opposite spin on atom b, corresponding to the Lewis structure H—H. The Heitler–London functions (13.102) for the lowest H_2 triplet state can also be written as Slater determinants. Omitting normalization constants, we write the Heitler–London H_2 functions as

$$\text{Singlet:}\quad |1s_a\overline{1s_b}| - |\overline{1s_a}1s_b| \qquad (13.117)$$

$$\text{Triplet:}\quad \begin{cases} |1s_a 1s_b| \\ |1s_a\overline{1s_b}| + |\overline{1s_a}1s_b| \\ |\overline{1s_a}\,\overline{1s_b}| \end{cases} \qquad (13.118)$$

Now consider Li_2. The ground-state configuration of Li is $1s^2 2s$, and the Lewis structure of Li_2 is Li—Li, with the two $2s$ Li electrons paired and the $1s$ electrons remaining in the inner shell of each atom. The part of the valence-bond wave function involving the $1s$ electrons will be like the He_2 function (13.112), while the part of the VB wave function involving the $2s$ electrons (which form the bond) will be like the Heitler–London H_2 function (13.117). Of course, because of the indistinguishability of the electrons, there is complete electronic democracy, and we must allow every electron to be in every orbital; hence we write the ground-state valence-bond function for Li_2 using 6×6 Slater determinants:

$$|1s_a\overline{1s_a}1s_b\overline{1s_b}2s_a\overline{2s_b}| - |1s_a\overline{1s_a}1s_b\overline{1s_b}\,\overline{2s_a}2s_b| \qquad (13.119)$$

We have written down (13.119) simply by analogy to (13.112) and (13.117); for a fuller justification of it, we should show that it is an eigenfunction of the spin operators $\hat{S}^2$ and $\hat{S}_z$ with eigenvalue zero for each operator, which corresponds to a singlet state. This can be shown, but we omit doing so. To save space, (13.119) is sometimes written as

$$|1s_a \overline{1s_a} 1s_b \overline{1s_b} \overbrace{2s_a 2s_b}|$$ (13.120)

where the curved line indicates the pairing (bonding) of the $2s_a$ and $2s_b$ AOs. The MO wave function for the Li_2 ground state is

$$|\sigma_g 1s \, \overline{\sigma_g 1s} \, \sigma_u^* 1s \, \overline{\sigma_u^* 1s} \, \sigma_g 2s \, \overline{\sigma_g 2s}|$$ (13.121)

If we approximate the two lowest MOs by $1s_a \pm 1s_b$ and carry out the same manipulations we did for the He_2 MO function, we can write (13.121) as

$$|1s_a \overline{1s_a} 1s_b \overline{1s_b} \sigma_g 2s \, \overline{\sigma_g 2s}|$$

Recall the notation $KK(\sigma_g 2s)^2$ for the Li_2 ground-state configuration.

Now consider the VB treatment of the N_2 ground state. The lowest configuration of N is $1s^2 2s^2 2p^3$; Hund's rule gives the ground level as $^4S_{3/2}$, with one electron in each of the three $2p$ AOs. We can thus pair the two $2p_x$ electrons, the two $2p_y$ electrons, and the two $2p_z$ electrons to form a triple bond; the Lewis structure is :N≡N:. How is this Lewis structure translated into the VB wave function? In the VB method, opposite spins are given to orbitals bonded together; we have three such pairs of orbitals and two ways to give opposite spins to the electrons of each bonding pair of AOs. Hence there are $2^3 = 8$ possible Slater determinants that we can write. We begin with

$$D_1 = |1s_a \overline{1s_a} 2s_a \overline{2s_a} 1s_b \overline{1s_b} 2s_b \overline{2s_b} 2p_{xa} \overline{2p_{xb}} 2p_{ya} \overline{2p_{yb}} 2p_{za} \overline{2p_{zb}}|$$

In all eight determinants, the first eight columns will remain unchanged, and to save space we write D_1 as

$$D_1 = |\cdots 2p_{xa} \overline{2p_{xb}} 2p_{ya} \overline{2p_{yb}} 2p_{za} \overline{2p_{zb}}|$$ (13.122)

Reversing the spins of the electrons in $2p_{xa}$ and $2p_{xb}$, we get

$$D_2 = |\cdots \overline{2p_{xa}} 2p_{xb} 2p_{ya} \overline{2p_{yb}} 2p_{za} \overline{2p_{zb}}|$$ (13.123)

There are six other determinants formed by interchanges of spins within the three pairs of bonding orbitals, and the VB wave function is a linear combination of eight determinants (Problem 13.29). The following rule (see *Kauzmann*, pages 421–422) gives a VB wave function that is an eigenfunction of $\hat{S}^2$ with eigenvalue 0 (as is desired for the ground state): the coefficient of each determinant is $+1$ or -1 according to whether the number of spin interchanges required to generate the determinant from D_1 is even or odd, respectively. Thus D_2 has coefficient -1. [Compare also (13.117).] Clearly, the single-determinant ground-state N_2 MO function is easier to handle than the eight-determinant VB function.

The VB method places great emphasis on pairing of electrons. In treating O_2, whose ground state is a triplet, the VB method runs into difficulties. It *is* possible to give a VB explanation of why O_2 has a triplet ground state, but the reasoning is involved [see B. J. Moss et al., *J. Chem. Phys.*, **63**, 4632 (1975)] in contrast to the simple MO explanation.

13.13 EXCITED STATES OF H₂

We have concentrated mostly on the ground electronic states of diatomic molecules. In this section we consider some of the excited states of H_2. Figure 13.18
gives the potential-energy curves for some of the H_2 electronic energy levels.

The lowest MO configuration is $(1\sigma_g)^2$, where the notation of the third
column of Table 13.1 is used. This closed-subshell configuration gives only a
nondegenerate $^1\Sigma_g^+$ level, designated $X^1\Sigma_g^+$. The LCAO-MO function is (13.93).

The next-lowest MO configuration is $(1\sigma_g)(1\sigma_u)$, which gives rise to the
terms $^1\Sigma_u^+$ and $^3\Sigma_u^+$ (Table 13.3). Since there is no axial electronic orbital angular
momentum, each of these terms corresponds to one level. Spectroscopists have
named these electronic levels $B^1\Sigma_u^+$ and $b^3\Sigma_u^+$. By Hund's rule, the b level lies
below the B level. The LCAO-MO functions for these levels are [see Eqs.
(10.27)–(10.30)]

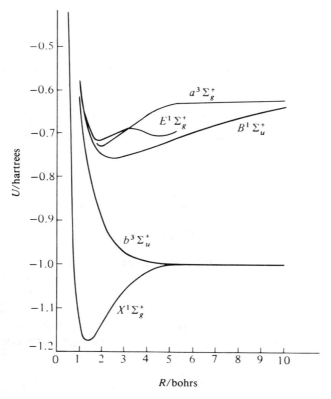

Figure 13.18 $U(R)$ curves for some electronic states of H_2. [See W.
Kolos and L. Wolniewicz, *J. Chem. Phys.*, **43**, 2429 (1965); **45**, 509
(1966); J. Gerhauser and H. S. Taylor, *J. Chem. Phys.*, **42**, 3621
(1965).]

$$b^3\Sigma_u^+: \quad 2^{-1/2}[1\sigma_g(1)1\sigma_u(2) - 1\sigma_g(2)1\sigma_u(1)] \begin{cases} \alpha(1)\alpha(2) \\ 2^{-1/2}[\alpha(1)\beta(2) + \alpha(2)\beta(1)] \\ \beta(1)\beta(2) \end{cases}$$

$$B^1\Sigma_u^+: \quad 2^{-1/2}[1\sigma_g(1)1\sigma_u(2) + 1\sigma_g(2)1\sigma_u(1)]2^{-1/2}[\alpha(1)\beta(2) - \alpha(2)\beta(1)]$$

where $1\sigma_g \approx N(1s_a + 1s_b)$ and $1\sigma_u \approx N'(1s_a - 1s_b)$. The $b^3\Sigma_u^+$ level is triply degenerate. The $B^1\Sigma_u^+$ level is nondegenerate. The Heitler–London wave functions for the b level are given by (13.102). Both these levels have one bonding and one antibonding electron, and we would expect the potential-energy curves for both levels to be repulsive. Actually, the B level has a minimum in its $U(R)$ curve. The stability of this state should caution us against drawing too hasty conclusions from very approximate wave functions.

We expect the next-lowest configuration to be $(1\sigma_g)(2\sigma_g)$, giving rise to $^1\Sigma_g^+$ and $^3\Sigma_g^+$ levels. These levels of H_2 are designated $E^1\Sigma_g^+$ and $a^3\Sigma_g^+$. By Hund's rule, the triplet lies lower. The E state has two substantial minima in its $U(R)$ curve.

Although the $2\sigma_u$ MO fills before the two $1\pi_u$ MOs in going across the periodic table, the $1\pi_u$ MOs lie below the $2\sigma_u$ MO in H_2. The configuration $(1\sigma_g)(1\pi_u)$ gives rise to the terms $^1\Pi_u$ and $^3\Pi_u$, the triplet lying lower. These terms are designated $C^1\Pi_u$ and $c^3\Pi_u$. The c term gives rise to the levels $c^3\Pi_{2u}$, $c^3\Pi_{1u}$, and $c^3\Pi_{0u}$. These levels lie so close together that they are usually not resolved in spectroscopic work. The C level shows a slight hump in its potential-energy curve at large R. Each level is twofold degenerate, which gives a total of eight electronic states arising from the $(1\sigma_g)(1\pi_u)$ configuration.

13.14 ELECTRON PROBABILITY DENSITY

We now consider how the wave function of a system is related to the electron probability density. We want the probability of finding an electron in the rectangular volume element at point (x, y, z) in space with edges dx, dy, dz. The electronic wave function ψ is a function of the space and spin coordinates of the n electrons. (For simplicity the parametric dependence on the nuclear configuration will not be explicitly indicated.) We know that

$$|\psi(x_1, \ldots, z_n, m_{s1}, \ldots, m_{sn})|^2 \, dx_1 \, dy_1 \, dz_1 \cdots dx_n \, dy_n \, dz_n \quad (13.124)$$

is the probability of simultaneously finding electron 1 with spin m_{s1} in the volume element $dx_1 \, dy_1 \, dz_1$ at (x_1, y_1, z_1), electron 2 with spin m_{s2} in the volume element $dx_2 \, dy_2 \, dz_2$ at (x_2, y_2, z_2), and so on. Since we are not interested in what spin the electron we find at (x, y, z) has, we sum the probability (13.124) over all possible spin states of all electrons, to give the probability of simultaneously finding each electron in the appropriate volume element with no regard for spin:

$$\sum_{m_{s1}} \cdots \sum_{m_{sn}} |\psi|^2 \, dx_1 \cdots dz_n \quad (13.125)$$

Suppose we want the probability of finding electron 1 in the volume element $dx\,dy\,dz$ at (x, y, z). For this probability we do not care where electrons 2 through n are. We therefore add the probabilities for all possible locations for these electrons. This amounts to integrating (13.125) over the coordinates of electrons $2, 3, \ldots, n$:

$$\left[\sum_{\text{all } m_s} \int \cdots \int |\psi(x, y, z, x_2, y_2, z_2, \ldots, x_n, y_n, z_n, m_{s1}, \ldots, m_{sn})|^2 \right.$$

$$\left. \times\, dx_2 \cdots dz_n \right] dx\,dy\,dz \qquad (13.126)$$

where there is a $(3n - 3)$-fold integration over x_2 through z_n.

Now suppose we ask for the probability of finding electron 2 in the volume element $dx\,dy\,dz$ at (x, y, z). By analogy to (13.126), this is

$$\left[\sum_{\text{all } m_s} \int \cdots \int |\psi(x_1, y_1, z_1, x, y, z, x_3, \ldots, z_n, m_{s1}, \ldots, m_{sn})|^2 \right.$$

$$\left. \times\, dx_1\,dy_1\,dz_1\,dx_3 \cdots dz_n \right] dx\,dy\,dz \qquad (13.127)$$

Of course, electrons do not come with labels, and because of this indistinguishability (Chapter 10) we know that the probabilities (13.126) and (13.127) must be equal. This equality is readily proved. ψ is antisymmetric with respect to electron exchange, and therefore $|\psi|^2$ is unchanged by such an exchange. Interchanging the space and spin coordinates of electrons 1 and 2 in ψ in (13.127) and doing some relabeling of dummy variables, we see that (13.127) is equal to (13.126). Thus (13.126) gives the probability of finding any one particular electron in the volume element. Since we have n electrons, the probability of finding *an* electron in the volume element is n times (13.126). (In drawing this conclusion, we are assuming that the probability of finding more than one electron in the infinitesimal volume element is negligible compared with the probability of finding one electron; this is certainly valid, since the probability of finding two electrons will involve the product of six infinitesimal quantities as compared with the product of three infinitesimal quantities for the probability of finding one electron.)

Thus the probability density ρ for finding an electron in the neighborhood of point (x, y, z) is

$$\rho(x, y, z) = n \sum_{\text{all } m_s} \int \cdots \int |\psi(x, y, z, x_2, \ldots, z_n, m_{s1}, \ldots, m_{sn})|^2 \, dx_2 \cdots dz_n \qquad (13.128)$$

The atomic units of ρ are electrons/bohr^3.

Experimental determination of ρ of a molecule involves measurement of x-ray diffraction intensities of molecular solids or electron-diffraction intensities of gases. See P. Coppens and M. B. Hall (eds.), *Electron Distributions and the Chemical Bond*, Plenum, New York, 1982; D. A. Kohl and L. S. Bartell, *J. Chem. Phys.*, **51**, 2891, 2896 (1969).

To illustrate (13.128), we calculate the electron density for the simple VB and MO ground-state H_2 functions. The wave function is a product of a spatial factor and the spin function (11.60). (For more than two electrons, the wave function cannot be factored into a simple product of space and spin parts; see Chapter 10.) Summation of (11.60) over m_{s1} and m_{s2} gives unity (Section 10.4). Thus (13.128) becomes for H_2

$$\rho(x, y, z) = 2 \int \int \int |\phi(x, y, z, x_2, y_2, z_2)|^2 \, dx_2 \, dy_2 \, dz_2$$

where ϕ is the spatial factor. The valence-bond function (13.101) gives

$$|\phi(1, 2)|^2 = \frac{[1s_a(1)]^2[1s_b(2)]^2 + [1s_b(1)]^2[1s_a(2)]^2 + 2 \cdot 1s_a(1)1s_a(2)1s_b(1)1s_b(2)}{2(1 + S_{ab}^2)}$$

$$\rho_{VB} = \frac{2}{2(1 + S_{ab}^2)} \left[1s_a^2 \int [1s_b(2)]^2 \, dv_2 + 1s_b^2 \int [1s_a(2)]^2 \, dv_2 \right.$$

$$\left. + 2 \cdot 1s_a 1s_b \int 1s_a(2)1s_b(2) \, dv_2 \right]$$

$$\rho_{VB} = \frac{1s_a^2 + 1s_b^2 + 2S_{ab}1s_a 1s_b}{1 + S_{ab}^2} \tag{13.129}$$

The MO function (13.95) gives (Problem 13.31b)

$$\rho_{MO} = \frac{1s_a^2 + 1s_b^2 + 2 \cdot 1s_a 1s_b}{1 + S_{ab}} \tag{13.130}$$

One finds (Problem 13.32) that $\rho_{MO} > \rho_{VB}$ at the midpoint of the bond, so the MO function (which underestimates electron correlation) piles up more charge between the nuclei than the VB function.

The MO probability density (13.130) is twice the probability density for the H_2^+-like $1s_A + 1s_B$ MO [Eq. (13.65)]. One can prove that, for a many-electron MO wave function, ρ is found by multiplying the probability-density function of each MO by the number of electrons occupying it and summing the results:

$$\rho(x, y, z) = \sum_j n_j |\phi_j|^2 \tag{13.131}$$

where the sum is over the different orthogonal spatial MOs, and n_j (whose possible values are 0, 1, or 2) is the number of electrons in the MO ϕ_j. [We used (13.131) in Eq. (11.11).]

13.15 DIPOLE MOMENTS

We now show how to calculate molecular dipole moments from wave functions.

The classical expression for the electric dipole moment $\mathbf{d}_{cl}$ of a set of discrete charges Q_i is

$$\mathbf{d}_{cl} = \sum_i Q_i \mathbf{r}_i \tag{13.132}$$

where $\mathbf{r}_i$ is the position vector from the origin to the ith charge. The electric dipole moment is a vector; its x component is

$$d_{x,cl} = \sum_i Q_i x_i$$

with similar expressions for the other components. (The symbol $\boldsymbol{\mu}$ is commonly used for the electric dipole moment, but this can be confused with the symbol for magnetic dipole moment.) For a continuous charge distribution with charge density $\rho_Q(x, y, z)$, $\mathbf{d}_{cl}$ is found by summing over the infinitesimal elements of charge $dQ_i = \rho_Q(x, y, z)\, dx\, dy\, dz$:

$$\mathbf{d}_{cl} = \int \rho_Q(x, y, z)\mathbf{r}\, dx\, dy\, dz , \qquad \text{where} \quad \mathbf{r} = x\mathbf{i} + y\mathbf{j} + z\mathbf{k} \qquad (13.133)$$

Now consider the quantum-mechanical definition of the electric dipole moment. Suppose we apply a uniform external electric field $\mathbf{E}$ to an atom or molecule and ask for the effect on the energy of the system. To form the Hamiltonian operator, we first need the classical expression for the energy. The electric field strength $\mathbf{E}$ is defined as $\mathbf{E} \equiv \mathbf{F}/Q$, where $\mathbf{F}$ is the force the field exerts on a charge Q. We take the z direction as the direction of the applied field: $\mathbf{E} = \mathscr{E}_z\mathbf{k}$. The potential energy V is [Eq. (4.26)]

$$dV/dz = -F_z = -Q\mathscr{E}_z \quad \text{and} \quad V = -Q\mathscr{E}_z z$$

This is the potential energy of a single charge in the field. For a system of charges,

$$V = -\mathscr{E}_z \sum_i Q_i z_i \qquad (13.134)$$

where z_i is the z coordinate of charge Q_i. The extension of (13.134) to the case where the electric field points in an arbitrary direction follows from (4.26) and is

$$V = -\mathscr{E}_x \sum_i Q_i x_i - \mathscr{E}_y \sum_i Q_i y_i - \mathscr{E}_z \sum_i Q_i z_i = -\mathbf{E} \cdot \mathbf{d}_{cl} \qquad (13.135)$$

This is the classical-mechanical expression for the energy of an electric dipole in a uniform applied electric field.

To calculate the quantum-mechanical expression, we use perturbation theory. The perturbation operator $\hat{H}'$ corresponding to (13.135) is

$$\hat{H}' = -\mathbf{E} \cdot \hat{\mathbf{d}}$$

where the *electric dipole-moment operator* $\hat{\mathbf{d}}$ is

$$\hat{\mathbf{d}} = \sum_i Q_i \hat{\mathbf{r}}_i = \mathbf{i}\hat{d}_x + \mathbf{j}\hat{d}_y + \mathbf{k}\hat{d}_z \qquad (13.136)$$

$$\hat{d}_x = \sum_i Q_i x_i , \qquad \hat{d}_y = \sum_i Q_i y_i , \qquad \hat{d}_z = \sum_i Q_i z_i \qquad (13.137)$$

The first-order correction to the energy is [Eq. (9.22)]

$$E^{(1)} = -\mathbf{E} \cdot \int \psi^{(0)*}\hat{\mathbf{d}}\psi^{(0)}\, d\tau \qquad (13.138)$$

where $\psi^{(0)}$ is the unperturbed wave function. Comparison of (13.138) and (13.135) shows that the quantum-mechanical quantity that corresponds to $\mathbf{d}_{cl}$ is the integral

$$\mathbf{d} = \int \psi^{(0)*}\hat{\mathbf{d}}\psi^{(0)}\, d\tau \qquad (13.139)$$

The quantity $\mathbf{d}$ in (13.139) is the quantum-mechanical *electric dipole moment* of the system.

An objection to taking (13.139) as the dipole moment is that we considered only the first-order energy correction. If we had included $E^{(2)}$ in (13.138), the comparison with (13.135) would not have given (13.139) as the dipole moment. Actually, (13.139) is the dipole moment of the system in the absence of an applied electric field and is the *permanent* electric dipole moment. Application of the field distorts the wave function from $\psi^{(0)}$, giving rise to an *induced* electric dipole moment in addition to the permanent dipole moment; the induced dipole moment corresponds to the energy correction $E^{(2)}$. (For the details, see *Merzbacher*, Section 17.4.) The induced dipole moment $\mathbf{d}_{ind}$ is related to the applied electric field $\mathbf{E}$ by

$$\mathbf{d}_{ind} = \alpha \mathbf{E} \tag{13.140}$$

where α is the ***polarizability*** of the atom or molecule. The greater the polarizability of molecule B, the greater the London dispersion force (Section 13.7) between two B molecules.

The shift in the energy of a quantum-mechanical system caused by an applied electric field is called the *Stark effect*. The *first-order* (or *linear*) Stark effect is given by (13.138), and from (13.139) it vanishes for a system with no permanent electric dipole moment. The *second-order* (or *quadratic*) Stark effect is given by the energy correction $E^{(2)}$ and is proportional to the square of the applied field.

The electric dipole-moment operator (13.136) is an odd function of the coordinates. If the wave function in (13.139) is either even or odd, then the integrand in (13.139) is an odd function, and the integral over all space vanishes. We conclude that the permanent electric dipole moment $\mathbf{d}$ is zero for states of definite parity.

The permanent electric dipole moment of a molecule whose electronic state is ψ_{el} is

$$\mathbf{d} = \int \psi_{el}^* \hat{\mathbf{d}} \psi_{el} \, d\tau_{el} \tag{13.141}$$

We have seen that the electronic wave functions of homonuclear diatomic molecules can be classified as g or u, according to their parity. Hence a homonuclear diatomic molecule has a zero permanent electric dipole moment, a not too astonishing result. The same holds true for any molecule with a center of symmetry. The electric dipole-moment operator for a molecule includes summation over both the electronic and nuclear charges:

$$\hat{\mathbf{d}} = \sum_i (-e\mathbf{r}_i) + \sum_\alpha Z_\alpha e \mathbf{r}_\alpha \tag{13.142}$$

where $\mathbf{r}_\alpha$ is the vector from the origin to the nucleus of atomic number Z_α and $\mathbf{r}_i$ is the vector to electron i. Since both the dipole-moment operator (13.142) and the electronic wave function depend on the parameters defining the nuclear configuration, the molecular electronic dipole moment $\mathbf{d}$ depends on the nuclear configuration. To indicate this, the quantity (13.141) can be called the dipole-moment *function* of the molecule. In writing (13.141), we ignored the nuclear motion. When the dipole moment of a molecule is experimentally determined, what is measured is the quantity (13.141) averaged over the zero-point vibrations (assuming the temperature is not high enough for there to be appreciable population of higher vibrational levels). We might use $\mathbf{d}_0$ and $\mathbf{d}_e$ to indicate the dipole moment averaged over zero-point vibrations and the dipole moment at the equilibrium nuclear configuration, respectively.

Since the second sum in (13.142) is independent of the electronic coordinates, we have

$$\mathbf{d} = \int \psi_{el}^* \sum_i (-e\mathbf{r}_i) \psi_{el} \, d\tau_{el} + \sum_\alpha Z_\alpha e \mathbf{r}_\alpha \int \psi_{el}^* \psi_{el} \, d\tau_{el} = -e \int |\psi_{el}|^2 \sum_i \mathbf{r}_i \, d\tau_{el} + e \sum_\alpha Z_\alpha \mathbf{r}_\alpha$$

Because of the indistinguishability of the electrons, we can write this equation as

$$\mathbf{d} = -en \int |\psi_{el}|^2 \mathbf{r}_1 \, d\tau_{el} + e \sum_{\alpha} Z_{\alpha} \mathbf{r}_{\alpha} \qquad (13.143)$$

where n is the number of electrons in the molecule and $\mathbf{r}_1$ is the position vector of electron 1. Introducing the electronic probability density (13.128), we write

$$\mathbf{d} = -e \int \int \int \rho(x, y, z) \mathbf{r} \, dx \, dy \, dz + e \sum_{\alpha} Z_{\alpha} \mathbf{r}_{\alpha} \qquad (13.144)$$

Equation (13.144) is what would be obtained if we pretended that the electrons were smeared out into a continuous charge distribution whose charge density is given by (13.128) and we used the classical equation (13.133) to calculate $\mathbf{d}$.

13.16   THE HARTREE–FOCK METHOD FOR MOLECULES

A very important development in quantum chemistry has been the computation of accurate self-consistent-field wave functions for many diatomic and polyatomic molecules. The principles of molecular SCF calculations are essentially the same as for atomic SCF calculations (Section 11.1). We shall restrict ourselves to closed-subshell configurations. For open subshells, the formulas are more complicated.

The molecular Hartree–Fock wave function is written as an antisymmetrized product (Slater determinant) of spin-orbitals, each spin-orbital being a product of a spatial orbital ϕ_i and a spin function (either α or β).

The expression for the Hartree–Fock molecular electronic energy E_{HF} is given by the variation theorem as $E_{HF} = \langle D | \hat{H}_{el} + V_{NN} | D \rangle$, where D is the Slater-determinant Hartree–Fock wave function and $\hat{H}_{el}$ and V_{NN} are given by (13.5) and (13.6). Since V_{NN} does not involve electronic coordinates and D is normalized, we have $\langle D | V_{NN} | D \rangle = V_{NN} \langle D | D \rangle = V_{NN}$. The operator $\hat{H}_{el}$ is the sum of one-electron operators $\hat{f}_i$ and two-electron operators $\hat{g}_{ij}$; we have $\hat{H}_{el} = \sum_i \hat{f}_i + \sum_j \sum_{i>j} \hat{g}_{ij}$, where $\hat{f}_i = -\frac{1}{2} \nabla_i^2 - \sum_{\alpha} Z_{\alpha}/r_{i\alpha}$ and $\hat{g}_{ij} = 1/r_{ij}$. The Hamiltonian $\hat{H}_{el}$ is the same as the Hamiltonian $\hat{H}$ for an atom except that $\sum_{\alpha} Z_{\alpha}/r_{i\alpha}$ replaces Z/r_i in $\hat{f}_i$. Hence Eq. (11.83) can be used to give $\langle D | \hat{H}_{el} | D \rangle$. Therefore, the Hartree–Fock energy of a diatomic or polyatomic molecule with only closed subshells is

$$E_{HF} = 2 \sum_{i=1}^{n/2} H_{ii}^{core} + \sum_{i=1}^{n/2} \sum_{j=1}^{n/2} (2J_{ij} - K_{ij}) + V_{NN} \qquad (13.145)$$

$$H_{ii}^{core} \equiv \langle \phi_i(1) | \hat{H}_{(1)}^{core} | \phi_i(1) \rangle \equiv \langle \phi_i(1) | -\tfrac{1}{2} \nabla_1^2 - \sum_{\alpha} Z_{\alpha}/r_{1\alpha} | \phi_i(1) \rangle \qquad (13.146)$$

$$J_{ij} \equiv \langle \phi_i(1)\phi_j(2) | 1/r_{12} | \phi_i(1)\phi_j(2) \rangle, \qquad K_{ij} \equiv \langle \phi_i(1)\phi_j(2) | 1/r_{12} | \phi_j(1)\phi_i(2) \rangle \qquad (13.147)$$

where the one-electron-operator symbol was changed from $\hat{f}_i$ to $\hat{H}_{(i)}^{core}$. The **one-electron core Hamiltonian** $\hat{H}_{(i)}^{core}$ omits the interactions of electron i with the other electrons. The sums over i and j are over the $n/2$ occupied spatial orbitals ϕ_i

of the n-electron molecule. In the Coulomb integrals J_{ij} and the exchange integrals K_{ij}, the integration goes over the spatial coordinates of electrons 1 and 2.

The Hartree–Fock method looks for those orbitals ϕ_i that minimize the variational integral E_{HF}. Of course, each MO is taken to be normalized: $\langle \phi_i(1)|\phi_i(1)\rangle = 1$. Moreover, for computational convenience one takes the MOs to be orthogonal: $\langle \phi_i(1)|\phi_j(1)\rangle = 0$ for $i \neq j$. It might be thought that a lower energy could be obtained if the orthogonality restriction were omitted, but this is not so. A closed-subshell antisymmetric wave function is a Slater determinant, and one can use the properties of determinants to show that a Slater determinant of nonorthogonal orbitals is equal to a Slater determinant in which the orbitals have been orthogonalized by the Schmidt or some other procedure; see Section 15.8 and F. W. Bobrowicz and W. A. Goddard, Chapter 4, Section 3.1 of *Schaefer, Methods of Electronic Structure Theory*. In effect, the Pauli antisymmetry requirement removes nonorthogonalities from the orbitals.

The derivation of the equation that determines the orthonormal ϕ_i's that minimize E_{HF} is complicated and is omitted. (For the derivation, see *Lowe*, Appendix 7; *Parr*, pages 21–23.) One finds that the closed-subshell orthogonal Hartree–Fock MOs satisfy

$$\hat{F}(1)\phi_i(1) = \varepsilon_i \phi_i(1) \tag{13.148}$$

where ε_i is the orbital energy and where the **(Hartree–) Fock operator** $\hat{F}$ is (in atomic units)

$$\hat{F}(1) = \hat{H}_{(1)}^{core} + \sum_{j=1}^{n/2} [2\hat{J}_j(1) - \hat{K}_j(1)] \tag{13.149}$$

$$\hat{H}_{(1)}^{core} \equiv -\tfrac{1}{2}\nabla_1^2 - \sum_\alpha \frac{Z_\alpha}{r_{1\alpha}} \tag{13.150}$$

where the **Coulomb operator** $\hat{J}_j$ and the **exchange operator** $\hat{K}_j$ are defined by

$$\hat{J}_j(1)f(1) = f(1) \int |\phi_j(2)|^2 \frac{1}{r_{12}} \, dv_2 \tag{13.151}$$

$$\hat{K}_j(1)f(1) = \phi_j(1) \int \frac{\phi_j^*(2)f(2)}{r_{12}} \, dv_2 \tag{13.152}$$

where f is an arbitrary function and the integrals are definite integrals over all space.

The first term on the right of (13.150) is the operator for the kinetic energy of one electron; the second term is the potential-energy operators for the attractions between one electron and the nuclei. The Coulomb operator $\hat{J}_j(1)$ is the potential energy of interaction between electron 1 and a smeared-out electron with electronic density $-|\phi_j(2)|^2$; the factor 2 occurs because there are two electrons in each spatial orbital. The exchange operator has no simple physical interpretation but arises from the requirement that the wave function be antisymmetric with respect to electron exchange. The exchange operators are absent from the Hartree equations (11.9). The Hartree–Fock MOs ϕ_i in (13.148) are eigenfunctions of the same operator $\hat{F}$, the eigenvalues being the orbital energies ε_i.

The orthogonality of the MOs greatly simplifies MO calculations, causing many integrals to vanish. In contrast, the VB method uses atomic orbitals, and AOs centered on different atoms are not orthogonal. MO calculations are mathematically simpler than VB calculations, and most molecular calculations are MO calculations.

The true Hamiltonian operator and wave function involve the coordinates of all n electrons. The Hartree–Fock Hamiltonian operator $\hat{F}$ is a one-electron operator (that is, it involves the coordinates of only one electron), and (13.148) is a one-electron differential equation. This has been indicated in (13.148) by writing $\hat{F}$ and ϕ_i as functions of the coordinates of electron 1; of course, the coordinates of any electron could have been used. The operator $\hat{F}$ is peculiar in that it depends on its own eigenfunctions [see Eqs. (13.149) to (13.152)], which are not known initially. Hence the Hartree–Fock equations must be solved by an iterative process.

To obtain the expression for the orbital energies ε_i, we multiply (13.148) by $\phi_i^*(1)$ and integrate over all space; using the fact that ϕ_i is normalized and using the result of Problem 13.37, we obtain $\varepsilon_i = \int \phi_i^*(1)\hat{F}(1)\phi_i(1)\,dv_1$ and

$$\varepsilon_i = \langle \phi_i(1)|\hat{H}^{\mathrm{core}}_{(1)}|\phi_i(1)\rangle + \sum_j [2\langle \phi_i(1)|\hat{J}_j(1)|\phi_i(1)\rangle - \langle \phi_i(1)|\hat{K}_j(1)|\phi_i(1)\rangle]$$

$$\varepsilon_i = H^{\mathrm{core}}_{ii} + \sum_{j=1}^{n/2}(2J_{ij} - K_{ij}) \tag{13.153}$$

where H^{core}_{ii}, J_{ij}, and K_{ij} are defined by (13.146) and (13.147).

Summation of (13.153) over the $n/2$ occupied orbitals gives

$$\sum_{i=1}^{n/2}\varepsilon_i = \sum_{i=1}^{n/2}H^{\mathrm{core}}_{ii} + \sum_{i=1}^{n/2}\sum_{j=1}^{n/2}(2J_{ij} - K_{ij}) \tag{13.154}$$

Solving this equation for $\sum_i H^{\mathrm{core}}_{ii}$ and substituting the result into (13.145), we obtain the Hartree–Fock energy as

$$E_{\mathrm{HF}} = 2\sum_{i=1}^{n/2}\varepsilon_i - \sum_{i=1}^{n/2}\sum_{j=1}^{n/2}(2J_{ij} - K_{ij}) + V_{NN} \tag{13.155}$$

Since there are two electrons per MO, the quantity $2\sum_i \varepsilon_i$ is the sum of the orbital energies. Subtraction of the double sum in (13.155) avoids counting each interelectronic repulsion twice, as discussed in Section 11.1.

A key development that helped make feasible the calculation of accurate SCF wave functions was Roothaan's 1951 proposal to expand the spatial orbitals ϕ_i as linear combinations of a set of one-electron basis functions χ_s:

$$\phi_i = \sum_{s=1}^{b} c_{si}\chi_s \tag{13.156}$$

To exactly represent the MOs ϕ_i, the basis functions χ_s should form a complete set. This requires an infinite number of basis functions. In practice, one must use a finite number b of basis functions. If b is large enough and the functions χ_s well chosen, one can represent the MOs with negligible error. (To avoid confusion, we shall use the letters r, s, t, and u to label the basis functions χ, and the letters i, j, k, and l to label the MOs ϕ.)

Substitution of the expansion (13.156) into the Hartree–Fock equations (13.148) gives

$$\sum_s c_{si} \hat{F} \chi_s = \varepsilon_i \sum_s c_{si} \chi_s$$

Multiplication by χ_r^* and integration gives

$$\sum_{s=1}^{b} c_{si}(F_{rs} - \varepsilon_i S_{rs}) = 0 , \qquad r = 1, 2, \ldots, b \qquad (13.157)$$

$$F_{rs} \equiv \langle \chi_r | \hat{F} | \chi_s \rangle , \qquad S_{rs} \equiv \langle \chi_r | \chi_s \rangle \qquad (13.158)$$

The equations (13.157) form a set of b simultaneous linear homogeneous equations in the b unknowns c_{si}, $s = 1, 2, \ldots, b$, that describe the MO ϕ_i in (13.156). For a nontrivial solution, we must have

$$\det (F_{rs} - \varepsilon_i S_{rs}) = 0 \qquad (13.159)$$

This is a secular equation whose roots give the orbital energies ε_i. The (*Hartree–Fock–*) *Roothaan equations* (13.157) must be solved by an iterative process, since the F_{rs} integrals depend on the orbitals ϕ_i (through the dependence of $\hat{F}$ on the ϕ_i's), which in turn depend on the unknown coefficients c_{si}.

One starts with guesses for the occupied-MO expressions as linear combinations of the basis functions, as in (13.156). This initial set of MOs is used to compute the Fock operator $\hat{F}$ from (13.149) to (13.152). The matrix elements (13.158) are computed, and the secular equation (13.159) is solved to give an initial set of ε_i's; these ε_i's are used to solve (13.157) for an improved set of coefficients, giving an improved set of MOs, which are then used to compute an improved $\hat{F}$, and so on. One keeps going until no further improvement in MO coefficients and energies occurs from one cycle to the next. The calculations are done using a computer. (The most efficient way to solve the Roothaan equations is to use matrix-algebra methods; see below.)

Now consider the basis functions used. Generally, each MO is written as a linear combination of one-electron functions (orbitals) centered on each atom. For diatomic molecules, one can use Slater functions [Eq. (11.14)] for the AOs. To have a complete set of AO basis functions, an infinite number of Slater orbitals are needed, but the true molecular Hartree–Fock wave function can be closely approximated with a reasonably small number of carefully chosen Slater orbitals. A *minimal basis set* for a molecular SCF calculation consists of a single basis function for each inner-shell AO and each valence-shell AO of each atom. An *extended* basis set is a set that is larger than a minimal set. Minimal-basis-set SCF calculations are easier than extended-basis-set calculations, but the latter are considerably more accurate.

We have used the terms SCF wave function and Hartree–Fock wave function interchangeably. In practice, the term *SCF wave function* is applied to any wave function obtained by iterative solution of the Roothaan equations, whether or not the basis set is large enough to give a really accurate approximation to the Hartree–Fock SCF wave function. There is only one true Hartree–

Fock SCF wave function, which is the best possible wave function that can be written as a Slater determinant of spin-orbitals. Some of the extended-basis-set calculations approach the true Hartree–Fock wave function quite closely; such functions are called "near Hartree–Fock wave functions" or, less cautiously, "Hartree–Fock wave functions."

The Roothaan Matrix Elements. To solve the Roothaan equations (13.157), we first must express the matrix elements (integrals) F_{rs} in terms of the basis functions χ. The Fock operator $\hat{F}$ is given by (13.149), and

$$F_{rs} = \langle \chi_r(1)|\hat{F}(1)|\chi_s(1)\rangle = \langle \chi_r(1)|\hat{H}^{core}_{(1)}|\chi_s(1)\rangle$$

$$+ \sum_{j=1}^{n/2} [2\langle \chi_r(1)|\hat{J}_j(1)\chi_s(1)\rangle - \langle \chi_r(1)|\hat{K}_j(1)\chi_s(1)\rangle] \quad (13.160)$$

Replacement of f by χ_s in (13.151), followed by use of the expansion (13.156), gives

$$\hat{J}_j(1)\chi_s(1) = \chi_s(1) \int \frac{\phi_j^*(2)\phi_j(2)}{r_{12}} \, dv_2 = \chi_s(1) \sum_t \sum_u c_{tj}^* c_{uj} \int \frac{\chi_t^*(2)\chi_u(2)}{r_{12}} \, dv_2$$

Multiplication by $\chi_r^*(1)$ and integration over the coordinates of electron 1 gives

$$\langle \chi_r(1)|\hat{J}_j(1)\chi_s(1)\rangle = \sum_t \sum_u c_{tj}^* c_{uj} \int \int \frac{\chi_r^*(1)\chi_s(1)\chi_t^*(2)\chi_u(2)}{r_{12}} \, dv_1 \, dv_2$$

$$\langle \chi_r(1)|\hat{J}_j(1)\chi_s(1)\rangle = \sum_{t=1}^{b} \sum_{u=1}^{b} c_{tj}^* c_{uj}(rs|tu) \quad (13.161)$$

where the **two-electron repulsion integral** is defined as

$$(rs|tu) \equiv \int \int \frac{\chi_r^*(1)\chi_s(1)\chi_t^*(2)\chi_u(2)}{r_{12}} \, dv_1 \, dv_2 \quad (13.162)^*$$

The widely used notation of (13.162) should not be misinterpreted as an overlap integral. Other notations, some mutually contradictory, are used for electron repulsion integrals, so it is always wise to check an author's definition.

Similarly, replacement of f by χ_s in (13.152) leads to (Problem 13.38)

$$\langle \chi_r(1)|\hat{K}_j(1)\chi_s(1)\rangle = \sum_{t=1}^{b} \sum_{u=1}^{b} c_{tj}^* c_{uj}(ru|ts) \quad (13.163)$$

Substituting (13.163) and (13.161) into (13.160) and changing the order of summation, we get the desired expression for F_{rs} in terms of integrals over the basis functions χ:

$$F_{rs} = H_{rs}^{core} + \sum_{t=1}^{b} \sum_{u=1}^{b} \sum_{j=1}^{n/2} c_{tj}^* c_{uj}[2(rs|tu) - (ru|ts)] \quad (13.164)$$

$$F_{rs} = H_{rs}^{core} + \sum_{t=1}^{b} \sum_{u=1}^{b} P_{tu}[(rs|tu) - \tfrac{1}{2}(ru|ts)] \quad (13.165)$$

$$P_{tu} \equiv 2 \sum_{j=1}^{n/2} c_{tj}^* c_{uj} , \qquad t = 1, 2, \ldots, b , \quad u = 1, 2, \ldots, b \qquad (13.166)$$

The quantities P_{tu} are called *electron-density matrix elements* or *charge-density, bond-order matrix elements*. Substitution of the expansion (13.156) into Eq. (13.131) for the electron probability density ρ gives for a closed-subshell molecule:

$$\rho = 2 \sum_{j=1}^{n/2} \phi_j^* \phi_j = 2 \sum_{r=1}^{b} \sum_{s=1}^{b} \sum_{j=1}^{n/2} c_{rj}^* c_{sj} \chi_r^* \chi_s = \sum_{r=1}^{b} \sum_{s=1}^{b} P_{rs} \chi_r^* \chi_s \qquad (13.167)$$

To express the Hartree–Fock energy in terms of integrals over the basis functions χ, we first solve (13.154) for $\sum_i \sum_j (2 J_{ij} - K_{ij})$ and substitute the result into (13.155) to get

$$E_{HF} = \sum_{i=1}^{n/2} \varepsilon_i + \sum_{i=1}^{n/2} H_{ii}^{core} + V_{NN}$$

We have, using the expansion (13.156),

$$H_{ii}^{core} = \langle \phi_i | \hat{H}^{core} | \phi_i \rangle = \sum_r \sum_s c_{ri}^* c_{si} \langle \chi_r | \hat{H}^{core} | \chi_s \rangle = \sum_r \sum_s c_{ri}^* c_{si} H_{rs}^{core}$$

$$E_{HF} = \sum_{i=1}^{n/2} \varepsilon_i + \sum_r \sum_s \sum_{i=1}^{n/2} c_{ri}^* c_{si} H_{rs}^{core} + V_{NN}$$

$$E_{HF} = \sum_{i=1}^{n/2} \varepsilon_i + \frac{1}{2} \sum_{r=1}^{b} \sum_{s=1}^{b} P_{rs} H_{rs}^{core} + V_{NN} \qquad (13.168)$$

EXAMPLE Do an SCF calculation for the helium-atom ground state using a basis set of two $1s$ STOs with orbital exponents $\zeta_1 = 1.45$ and $\zeta_2 = 2.91$. [By trial and error, these have been found to be the optimum ζ's to use for this basis set; see C. Roetti and E. Clementi, *J. Chem. Phys.*, **60**, 4725 (1974).]

From (11.14), the normalized basis functions are (in atomic units)

$$\chi_1 = 2 \zeta_1^{3/2} e^{-\zeta_1 r} Y_0^0 , \quad \chi_2 = 2 \zeta_2^{3/2} e^{-\zeta_2 r} Y_0^0 , \qquad \zeta_1 = 1.45 , \quad \zeta_2 = 2.91 \qquad (13.169)$$

To solve the Roothaan equations (13.157), we need the integrals F_{rs} and S_{rs}. The overlap integrals S_{rs} are

$$S_{11} = \langle \chi_1 | \chi_1 \rangle = 1 , \qquad S_{22} = \langle \chi_2 | \chi_2 \rangle = 1$$

$$S_{12} = S_{21} = \langle \chi_1 | \chi_2 \rangle = 4 \zeta_1^{3/2} \zeta_2^{3/2} \int_0^\infty e^{-(\zeta_1 + \zeta_2) r} r^2 \, dr = \frac{8 \zeta_1^{3/2} \zeta_2^{3/2}}{(\zeta_1 + \zeta_2)^3} = 0.8366$$

where the Appendix integral (A.7) was used.

The integrals F_{rs} are given by (13.165) and depend on H_{rs}^{core}, P_{tu}, and $(rs|tu)$. From (13.150), $\hat{H}^{core} = -\frac{1}{2} \nabla^2 - 2/r = -\frac{1}{2} \nabla^2 - \zeta/r + (\zeta - 2)/r$. The integrals H_{rs}^{core} are evaluated the same way that similar integrals were evaluated in the variation treatment of He in Section 9.4. We find (Problem 13.39)

$$H_{11}^{core} = \langle \chi_1 | \hat{H}^{core} | \chi_1 \rangle = -\tfrac{1}{2}\zeta_1^2 + (\zeta_1 - 2)\zeta_1 = \tfrac{1}{2}\zeta_1^2 - 2\zeta_1 = -1.8488$$

$$H_{22}^{core} = \tfrac{1}{2}\zeta_2^2 - 2\zeta_2 = -1.5860$$

$$H_{12}^{core} = H_{21}^{core} = \langle \chi_1 | \hat{H}^{core} | \chi_2 \rangle = -\tfrac{1}{2}\zeta_2^2 S_{12} + \frac{4(\zeta_2 - 2)\zeta_1^{3/2}\zeta_2^{3/2}}{(\zeta_1 + \zeta_2)^2}$$

$$= \frac{\zeta_1^{3/2}\zeta_2^{3/2}(4\zeta_1\zeta_2 - 8\zeta_1 - 8\zeta_2)}{(\zeta_1 + \zeta_2)^3} = -1.8826$$

Many of the electron-repulsion integrals $(rs|tu)$ are equal to one another. For real basis functions, one can show that (Problem 13.40)

$$(rs|tu) = (sr|tu) = (rs|ut) = (sr|ut) = (tu|rs) = (ut|rs) = (tu|sr) = (ut|sr)$$
$$(13.170)$$

The electron-repulsion integrals are evaluated using the $1/r_{12}$ expansion (9.53). One finds [see Eq. (9.55) and Problem 13.41]

$$(11|11) = \tfrac{5}{8}\zeta_1 = 0.9062 , \qquad (22|22) = \tfrac{5}{8}\zeta_2 = 1.8188$$

$$(11|22) = (22|11) = (\zeta_1^4\zeta_2 + 4\zeta_1^3\zeta_2^2 + \zeta_1\zeta_2^4 + 4\zeta_1^2\zeta_2^3)/(\zeta_1 + \zeta_2)^4 = 1.1826$$

$$(12|12) = (21|12) = (12|21) = (21|21) = 20\zeta_1^3\zeta_2^3/(\zeta_1 + \zeta_2)^5 = 0.9536$$

$$(11|12) = (11|21) = (12|11) = (21|11)$$

$$= \frac{16\zeta_1^{9/2}\zeta_2^{3/2}}{(3\zeta_1 + \zeta_2)^4}\left[\frac{12\zeta_1 + 8\zeta_2}{(\zeta_1 + \zeta_2)^2} + \frac{9\zeta_1 + \zeta_2}{2\zeta_1^2}\right] = 0.9033$$

$$(12|22) = (22|12) = (21|22) = (22|21)$$

$$= \text{the } (11|12) \text{ expression with 1 and 2 interchanged} = 1.2980$$

To start the calculation, we need an initial guess for the ground-state AO expansion coefficients in (13.156) so that we can get an initial estimate of the density matrix elements P_{tu} in (13.165). We saw in Section 9.4 that the optimum orbital exponent for a helium AO that consists of one 1s STO is $\tfrac{27}{16} = 1.6875$. Since the orbital exponent ζ_1 is much closer to 1.6875 than is ζ_2, we expect that the coefficient of χ_1 in $\phi_1 = c_{11}\chi_1 + c_{21}\chi_2$ will be substantially larger than the coefficient of χ_2. Let us take as an initial guess $c_{11}/c_{21} \approx 2$. The normalization condition $\int |\phi_1|^2 \, d\tau = 1$ gives for real coefficients (Problem 13.43)

$$c_{21} = (1 + k^2 + 2kS_{12})^{-1/2} , \qquad \text{where } k \equiv c_{11}/c_{21} \qquad (13.171)$$

Substitution of $k = 2$ and $S_{12} = 0.8366$ gives $c_{21} \approx 0.3461$ and $c_{11} \approx 2c_{21} = 0.6922$. With $n = 2$ and $b = 2$, Eq. (13.166) gives

$$P_{11} = 2c_{11}^*c_{11} , \quad P_{12} = 2c_{11}^*c_{21} , \quad P_{21} = P_{12}^* , \quad P_{22} = 2c_{21}^*c_{21} \qquad (13.172)$$

The initial guess $c_{11} \approx 0.6922$, $c_{21} \approx 0.3461$ gives as the initial density matrix elements

$$P_{11} \approx 0.9583 , \quad P_{12} = P_{21} \approx 0.4791 , \quad P_{22} \approx 0.2396$$

The Fock matrix elements are found from (13.165) with $b = 2$. Using (13.170) and $P_{12} = P_{21}$ for real functions, we get (Problem 13.42a)

$$F_{11} = H_{11}^{core} + \tfrac{1}{2}P_{11}(11|11) + P_{12}(11|12) + P_{22}[(11|22) - \tfrac{1}{2}(12|21)]$$

$$F_{12} = F_{21} = H_{12}^{core} + \tfrac{1}{2}P_{11}(12|11) + P_{12}[\tfrac{3}{2}(12|12) - \tfrac{1}{2}(11|22)] + \tfrac{1}{2}P_{22}(12|22)$$

$$F_{22} = H_{22}^{core} + P_{11}[(22|11) - \tfrac{1}{2}(21|12)] + P_{12}(22|12) + \tfrac{1}{2}P_{22}(22|22)$$

Substitution of the values of the H_{rs}^{core} and $(rs|tu)$ integrals listed previously gives (Problem 13.42b)

$$F_{11} = -1.8488 + 0.4531P_{11} + 0.9033P_{12} + 0.7058P_{22} \qquad (13.173)$$

$$F_{12} = F_{21} = -1.8826 + 0.4516_5 P_{11} + 0.8391P_{12} + 0.6490P_{22} \qquad (13.174)$$

$$F_{22} = -1.5860 + 0.7058P_{11} + 1.2980P_{12} + 0.9094P_{22} \qquad (13.175)$$

Substitution of the initial guess for the P_{tu}'s into (13.173) to (13.175) gives as the initial estimates of the F_{rs}'s

$$F_{11} \approx -0.813 , \quad F_{12} = F_{21} \approx -0.892 , \quad F_{22} \approx -0.070$$

The initial estimate of the secular equation $\det(F_{rs} - S_{rs}\varepsilon_i) = 0$ is

$$\begin{vmatrix} -0.813 - \varepsilon_i & -0.892 - 0.8366\varepsilon_i \\ -0.892 - 0.8366\varepsilon_i & -0.070 - \varepsilon_i \end{vmatrix} \approx 0$$

$$0.3001\varepsilon_i^2 - 0.609_5\varepsilon_i - 0.739 \approx 0$$

$$\varepsilon_1 \approx -0.854 , \quad \varepsilon_2 \approx 2.885$$

Substitution of the lower root ε_1 into the Roothaan equation (13.157) with $r = 2$ gives

$$c_{11}(F_{21} - \varepsilon_1 S_{21}) + c_{21}(F_{22} - \varepsilon_1 S_{22}) \approx 0$$

$$-0.177_5 c_{11} + 0.784c_{21} \approx 0$$

$$c_{11}/c_{21} \approx 4.42$$

Substitution of $k = 4.42$ and $S_{12} = 0.8366$ in the normalization condition (13.171) gives

$$c_{21} \approx 0.189 , \quad c_{11} = kc_{21} \approx 0.836$$

Substitution of these improved coefficients into (13.172) gives as the improved density matrix elements

$$P_{11} \approx 1.398 , \quad P_{12} = P_{21} \approx 0.316 , \quad P_{22} \approx 0.071$$

Substitution of these improved P_{tu}'s into (13.173) to (13.175) gives as the improved F_{rs} values

$$F_{11} \approx -0.880 , \quad F_{12} = F_{21} \approx -0.940 , \quad F_{22} \approx -0.124_6$$

The improved secular equation is

$$\begin{vmatrix} -0.880 - \varepsilon_i & -0.940 - 0.8366\varepsilon_i \\ -0.940 - 0.8366\varepsilon_i & -0.124_6 - \varepsilon_i \end{vmatrix} \approx 0$$

$$\varepsilon_1 \approx -0.918 , \quad \varepsilon_2 \approx 2.810$$

The improved ε_1 value gives $c_{11}/c_{21} \approx 4.61$ and

$$c_{11} \approx 0.842 , \quad c_{21} \approx 0.183$$

Another cycle of calculation yields (Problem 13.44)

$$P_{11} = 1.418 , \quad P_{12} = P_{21} = 0.308 , \quad P_{22} = 0.067$$

$$F_{11} = -0.881 , \quad F_{12} = F_{21} = -0.940 , \quad F_{22} = -0.124_5 \qquad (13.176)$$

$$\varepsilon_1 = -0.918 , \quad \varepsilon_2 = 2.809 \qquad (13.177)$$

$$c_{11} = 0.842 , \quad c_{21} = 0.183$$

These last c's are the same as those for the previous cycle, so the calculation has converged and we are finished. The He ground-state SCF AO for this basis set is

$$\phi_1 = 0.842\chi_1 + 0.183\chi_2$$

The SCF energy is found from (13.168) with $n = 2$ and $b = 2$ as

$$E_{\mathrm{HF}} = -0.918 + \tfrac{1}{2}[1.418(-1.8488) + 2(0.308)(-1.8826) + 0.067(-1.5860)] + 0$$

$$= -2.862 \text{ hartrees} = -77.9 \text{ eV}$$

A more precise calculation with $\zeta_1 = 1.45363$ and $\zeta_2 = 2.91093$ gives an SCF energy of -2.8616726 hartrees, as compared with the limiting Hartree–Fock energy -2.8616799 hartrees found with five basis functions [C. Roetti and E. Clementi, *J. Chem. Phys.*, **60**, 4725 (1974)].

Matrix Form of the Roothaan Equations. The Roothaan equations are most efficiently solved using matrix methods. The Roothaan equations (13.157) read

$$\sum_{s=1}^{b} F_{rs}c_{si} = \sum_{s=1}^{b} S_{rs}c_{si}\varepsilon_i , \qquad r = 1, 2, \dots , b$$

The coefficients c_{si} relate the MOs ϕ_i to the basis functions χ_s according to $\phi_i = \sum_s c_{si}\chi_s$. Let **C** be the square matrix of order b whose elements are the coefficients c_{si}. Let **F** be the square matrix of order b whose elements are $F_{rs} = \langle \chi_r|\hat{F}|\chi_s \rangle$. Let **S** be the square matrix whose elements are $S_{rs} = \langle \chi_r|\chi_s \rangle$. Let $\boldsymbol{\varepsilon}$ be the diagonal square matrix whose diagonal elements are the orbital energies $\varepsilon_1, \varepsilon_2, \dots, \varepsilon_b$ so that the elements of $\boldsymbol{\varepsilon}$ are $\varepsilon_{mi} = \delta_{mi}\varepsilon_i$, where δ_{mi} is the Kronecker delta.

Use of the matrix multiplication rule (7.103) gives the (s, i)th element of the matrix product $\mathbf{C\varepsilon}$ as $(\mathbf{C\varepsilon})_{si} = \sum_m c_{sm}\varepsilon_{mi} = \sum_m c_{sm}\delta_{mi}\varepsilon_i = c_{si}\varepsilon_i$. Hence the Roothaan equations read

$$\sum_{s=1}^{b} F_{rs}c_{si} = \sum_{s=1}^{b} S_{rs}(\mathbf{C\varepsilon})_{si} \qquad (13.178)$$

From the matrix multiplication rule, the left side of (13.178) is the (r, i)th element of $\mathbf{FC}$, and the right side equals the (r, i)th element of $\mathbf{S(C\varepsilon)}$. Since the general element of $\mathbf{FC}$ equals the general element of $\mathbf{SC\varepsilon}$, these matrices are equal:

$$\mathbf{FC} = \mathbf{SC\varepsilon} \qquad (13.179)$$

This is the matrix form of the Roothaan equations.

The set of basis functions χ_s used to expand the MOs is not an orthogonal set. However, one can use the Schmidt or some other procedure to form orthogonal linear combinations of the basis functions to give a new set of basis functions χ'_s that is an orthonormal set: $\chi'_s = \Sigma_t \, b_{ts}\chi_t$ and $S'_{rs} = \langle \chi'_r | \chi'_s \rangle = \delta_{rs}$. (See *Szabo and Ostlund*, Section 3.4.5, for details of the orthogonalization procedure.) With this orthonormal basis set, the overlap matrix is a unit matrix, and the Roothaan equations (13.179) have the simpler form

$$\mathbf{F'C'} = \mathbf{C'\varepsilon} \qquad (13.180)$$

where $F'_{rs} = \langle \chi'_r | \hat{F} | \chi'_s \rangle$ and $\mathbf{C'}$ is the matrix of the coefficients that relate the MOs ϕ_i to the orthonormal basis functions: $\phi_i = \Sigma_s c'_{si}\chi'_s$.

The matrix equation (13.180) has the same form as Eq. (8.87), which is $\mathbf{HC} = \mathbf{CW}$, where $\mathbf{C}$ and $\mathbf{W}$ [defined by (8.86)] are the eigenvector matrix and eigenvalue matrix, respectively, of $\mathbf{H}$. Thus, the orbital energies ε_i are the eigenvalues of the Fock matrix $\mathbf{F'}$ and each column of $\mathbf{C'}$ is an eigenvector of $\mathbf{F'}$. Because the Fock operator $\hat{F}$ is Hermitian, the Fock matrix $\mathbf{F'}$ is a Hermitian matrix. As noted in the paragraph preceding Eq. (8.94), the eigenvector matrix $\mathbf{C'}$ of the Hermitian matrix $\mathbf{F'}$ can be chosen to be unitary, meaning that its inverse equals its conjugate transpose [Eq. (8.92)] $\mathbf{C'}^{-1} = \mathbf{C'}^\dagger$. (With a unitary coefficient matrix $\mathbf{C'}$, the MOs ϕ_i are orthonormal; see Problem 13.48.) Multiplication of (13.180) on the left by $\mathbf{C'}^{-1} = \mathbf{C'}^\dagger$ gives [see Eqs. (8.88) and (8.94)]

$$\mathbf{C'}^\dagger \mathbf{F'C'} = \varepsilon \qquad (13.181)$$

which has the same form as Eq. (8.94).

To do an SCF calculation, one chooses a basis set χ_s, evaluates the H^{core}_{rs} and $(rs|tu)$ integrals, transforms to orthogonal basis functions $\chi'_s = \Sigma_t \, b_{ts}\chi_t$, makes an initial guess for the coefficients c_{si} in $\phi_i = \Sigma_s c_{si}\chi_s$, uses the transformation $\chi'_s = \Sigma_t \, b_{ts}\chi_t$ to find the initial guess for the coefficients c'_{si} in $\phi_i = \Sigma_s c'_{si}\chi'_s$, uses this estimate of the c'_{si}'s to evaluate the initial estimate of the integrals $F'_{rs} = \langle \chi'_r | \hat{F} | \chi'_s \rangle$, uses a matrix diagonalization method (Section 8.6) to find the eigenvalue and eigenvector matrices ε and $\mathbf{C'}$ of the initial $\mathbf{F'}$ matrix, uses the coefficients from this $\mathbf{C'}$ matrix to calculate an improved $\mathbf{F'}$ matrix, which is again diagonalized to give a further improved $\mathbf{C'}$ matrix, which is used to give a further improved $\mathbf{F'}$ matrix, and so on. When the calculation has converged, the final $\mathbf{C'}$ matrix elements give the MOs in terms of the orthogonal basis functions: $\phi_i = \Sigma_s c'_{si}\chi'_s$; use of $\chi'_s = \Sigma_t \, b_{ts}\chi_t$ then gives the MOs in terms of the original nonorthogonal basis functions.

13.17 SCF WAVE FUNCTIONS FOR DIATOMIC MOLECULES

This section presents some examples of SCF wave functions for diatomic molecules. (For a summary of SCF calculations on diatomic molecules, see *Mulliken and Ermler, Diatomic Molecules*.)

SCF wave functions using a minimal basis set were calculated by Ransil for several light diatomic molecules [B. J. Ransil, *Rev. Mod. Phys.*, **32**, 245 (1960)]. As an example, the SCF MOs for the ground state of Li_2 [MO configuration $(1\sigma_g)^2(1\sigma_u)^2(2\sigma_g)^2$] at $R = R_e$ are

$$1\sigma_g = 0.706(1s_a + 1s_b) + 0.009(2s_{\perp a} + 2s_{\perp b}) + 0.0003(2p\sigma_a + 2p\sigma_b)$$

$$1\sigma_u = 0.709(1s_a - 1s_b) + 0.021(2s_{\perp a} - 2s_{\perp b}) + 0.003(2p\sigma_a - 2p\sigma_b) \qquad (13.182)$$

$$2\sigma_g = -0.059(1s_a + 1s_b) + 0.523(2s_{\perp a} + 2s_{\perp b}) + 0.114(2p\sigma_a + 2p\sigma_b)$$

The AO functions in these equations are STOs, except for $2s_\perp$. A Slater-type $2s$ AO has no radial nodes and is not orthogonal to a $1s$ STO. The Hartree–Fock $2s$ AO has one radial node ($n - l - 1 = 1$) and is orthogonal to the $1s$ AO. We can form an orthogonalized $2s$ orbital with the proper number of nodes by taking the following normalized linear combination of $1s$ and $2s$ STOs of the same atom (Schmidt orthogonalization):

$$2s_\perp = (1 - S^2)^{-1/2}(2s - S \cdot 1s) \qquad (13.183)$$

where S is the overlap integral $\langle 1s|2s \rangle$. Ransil expressed the Li_2 orbitals using the (nonorthogonal) $2s$ STO, but since the orthogonalized $2s_\perp$ function gives a better representation of the $2s$ AO, the orbitals have been rewritten using $2s_\perp$. This changes the $1s$ and $2s$ coefficients, but the actual orbital is, of course, unchanged; see Problem 13.49. The notation $2p\sigma$ for an AO indicates that the p orbital points along the molecular (z) axis; that is, a $2p\sigma$ AO is a $2p_z$ AO. (The $2p_x$ and $2p_y$ AOs are called $2p\pi$ AOs.) The optimum orbital exponents for the orbitals in (13.182) are $\zeta_{1s} = 2.689$, $\zeta_{2s} = 0.634$, $\zeta_{2p\sigma} = 0.761$. Since six AOs were used as basis functions, the Roothaan equations yielded approximations for the six lowest MOs of ground-state Li_2; only three of these MOs are occupied. The expressions for the other three can be found in Ransil's paper.

Our previous simple expressions for these MOs were

$$1\sigma_g = \sigma_g 1s = 2^{-1/2}(1s_a + 1s_b)$$

$$1\sigma_u = \sigma_u^* 1s = 2^{-1/2}(1s_a - 1s_b)$$

$$2\sigma_g = \sigma_g 2s = 2^{-1/2}(2s_a + 2s_b)$$

Comparison of these with (13.182) shows the simple LCAO functions to be reasonable first approximations to the minimal-basis-set SCF MOs. The approximation is best for the $1\sigma_g$ and $1\sigma_u$ MOs, whereas the $2\sigma_g$ MO has a substantial $2p\sigma$ AO contribution in addition to the $2s$ AO contributions. For this reason the notation of the third column of Table 13.1 is preferable to the separated-atoms MO notation. The substantial amount of $2s$–$2p\sigma$ hybridization is to be expected, since the $2s$ and $2p$ AOs are close in energy [see Eq. (9.27)]; the hybridization allows for the polarization of the $2s$ AOs in forming the molecule.

Let us compare the $3\sigma_g$ MO of the F_2 ground state at R_e as calculated by Ransil using a minimal basis set with that calculated by Wahl using an extended basis set [A. C. Wahl, *J. Chem. Phys.*, **41**, 2600 (1964)]:

$$3\sigma_{g,min} = 0.038(1s_a + 1s_b) - 0.184(2s_a + 2s_b) + 0.648(2p\sigma_a + 2p\sigma_b)$$

$$\zeta_{1s} = 8.65 , \qquad \zeta_{2s} = 2.58 , \qquad \zeta_{2p\sigma} = 2.49$$

$$3\sigma_{g,ext} = 0.048(1s_a + 1s_b) + 0.003(1s_a' + 1s_b') - 0.257(2s_a + 2s_b)$$
$$+ 0.582(2p\sigma_a + 2p\sigma_b) + 0.307(2p\sigma_a' + 2p\sigma_b') + 0.085(2p\sigma_a'' + 2p\sigma_b'')$$
$$- 0.056(3s_a + 3s_b) + 0.046(3d\sigma_a + 3d\sigma_b) + 0.014(4f\sigma_a + 4f\sigma_b)$$

$$\zeta_{1s} = 8.27 , \qquad \zeta_{1s'} = 13.17 , \qquad \zeta_{2s} = 2.26$$

$$\zeta_{2p\sigma} = 1.85 , \qquad \zeta_{2p\sigma'} = 3.27 , \qquad \zeta_{2p\sigma''} = 5.86$$

$$\zeta_{3s} = 4.91 , \qquad \zeta_{3d\sigma} = 2.44 , \qquad \zeta_{4f\sigma} = 2.83$$

Just as several STOs are needed to give an accurate representation of Hartree–Fock AOs (Section 11.1), one needs more than one STO of a given n and l in the linear combination of STOs that is to accurately represent the Hartree–Fock MO. The primed and double-primed AOs in the extended-basis-set function are STOs with different orbital exponents. The $3d\sigma$ and $4f\sigma$ AOs are AOs with quantum number $m = 0$, that is, the $3d_0$ and $4f_0$ AOs. The total energies found are -197.877 and -198.768 hartrees for the minimal and extended calculations, respectively. The experimental energy of F_2 at R_e is -199.670 hartrees, so the error for the minimal calculation is twice that of the extended calculation. The extended-basis-set calculation is believed to give a wave function quite close to the true Hartree–Fock wave function; therefore, the correlation energy in F_2 is about 24 eV.

In discussing H_2^+ and H_2, we saw how hybridization (the mixing of different AOs of the same atom) improves molecular wave functions. There is a tendency to think of hybridization as occurring only for certain molecular geometries. The SCF calculations make clear that all MOs are hybridized to some extent. Thus any diatomic-molecule σ MO is a linear combination of $1s$, $2s$, $2p_0$, $3s$, $3p_0$, $3d_0$, . . . AOs of the separated atoms.

To aid in deciding which AOs contribute to a given diatomic MO, we use two rules. First, only σ-type AOs (s, $p\sigma$, $d\sigma$, and so on) can contribute to a σ MO; only π-type AOs ($p\pi$, $d\pi$, and so on) can contribute to a π MO; and so on. Second, only AOs of reasonably similar energy make substantial contributions to a given MO. (For examples, see the minimal- and extended-basis-set MOs quoted above.)

Wahl has plotted the contours of the near Hartree–Fock molecular orbitals of homonuclear diatomic molecules from H_2 through F_2. Figure 13.19 shows these plots for Li_2.

Of course, Hartree–Fock wave functions are only approximations to the true wave functions. It is possible to prove that a Hartree–Fock wave function gives a very good approximation to the electron probability density $\rho(x, y, z)$ for nuclear configurations in the region of the equilibrium configuration. A molecular

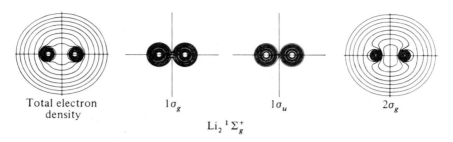

Total electron density $1\sigma_g$ $1\sigma_u$ $2\sigma_g$

$Li_2\ {}^1\Sigma_g^+$

Figure 13.19 Hartree–Fock MO electron-density contours for the ground electronic state of Li_2, as calculated by Wahl. [A. C. Wahl, *Science*, **151**, 961 (1966); *Scientific American*, April 1970, p. 54; *Atomic and Molecular Structure: 4 Wall Charts*, McGraw-Hill, New York, 1970.]

property that involves only one-electron operators can be expressed as an integral involving ρ. Consequently, such properties are accurately calculated using Hartree–Fock wave functions. An example is the molecular dipole moment [Eq. (13.144)], for which the success of the near Hartree–Fock wave functions has been impressive. For example, the LiH dipole moment calculated with a near Hartree–Fock ψ is 6.00 D (debyes) [S. Green, *J. Chem. Phys.*, **54**, 827 (1971)], compared with the experimental value of 5.83 D. (One debye = 10^{-18} statcoul cm.) For NaCl, the calculated and experimental dipole moments are 9.18 D and 9.02 D [R. L. Matcha, *J. Chem. Phys.*, **48**, 335 (1968)]. An error of about 0.2 D is typical in such calculations, but where the dipole moment is small, the percent error can be large. An extreme example is CO, for which the experimental moment is 0.11 D with the polarity C^-O^+, but the near-Hartree–Fock moment is 0.27 D with the wrong polarity C^+O^-. However, a configuration-interaction wave function gives 0.12 D with the correct polarity [S. Green, *J. Chem. Phys.*, **54**, 827 (1971)].

Molecular Hartree–Fock calculations do not give accurate dissociation energies. For example, an extended-basis-set calculation [P. E. Cade et al., *J. Chem. Phys.*, **44**, 1973 (1966)] gives $D_e = 5.3$ eV for N_2, as compared with the true value 9.9 eV. (To calculate the Hartree–Fock D_e, the molecular energy at the minimum in the $U(R)$ Hartree–Fock curve is subtracted from the sum of the Hartree–Fock energies of the separated atoms.) For F_2 the true D_e is 1.66 eV, while the Hartree–Fock D_e is -1.4 eV; in other words, the Hartree–Fock calculation predicts the separated atoms to be more stable than the fluorine molecule. A related defect of Hartree–Fock molecular wave functions is that the energy approaches the wrong limit as $R \rightarrow \infty$. Recall the MO discussion of H_2.

13.18 MO TREATMENT OF HETERONUCLEAR DIATOMIC MOLECULES

The treatment of heteronuclear diatomic molecules is similar to that for homonuclear diatomic molecules. We first consider the MO description.

Suppose the two atoms have atomic numbers that differ only slightly; an example is CO. We could consider CO as being formed from the isoelectronic molecule N_2 by a gradual transfer of nuclear charge from one nucleus to the other. During this hypothetical transfer, the original N_2 MOs would slowly vary to give finally the CO MOs. We therefore expect the CO molecular orbitals to bear some resemblance to those of N_2. For a heteronuclear diatomic molecule such as CO, the symbols used for the MOs are similar to those for homonuclear diatomics. However, for a heteronuclear diatomic, the electronic Hamiltonian (13.5) is not invariant with respect to inversion of the electronic coordinates (that is, $\hat{H}_{el}$ does not commute with $\hat{\Pi}$), and the g, u property of the MOs disappears. The correlation between the N_2 and CO subshell designations is

N_2	$1\sigma_g$	$1\sigma_u$	$2\sigma_g$	$2\sigma_u$	$1\pi_u$	$3\sigma_g$	$1\pi_g$	$3\sigma_u$
CO	1σ	2σ	3σ	4σ	1π	5σ	2π	6σ

MOs of the same symmetry are numbered in order of increasing energy. Because of the absence of the g, u property, the numbers of corresponding homonuclear and heteronuclear MOs differ. Figure 13.20 is a sketch of a contour of the CO $1\pi_{\pm1}$ MOs as determined by an extended-basis-set SCF calculation [W. M. Huo, *J. Chem. Phys.*, **43**, 624 (1965)]. Note its resemblance to the contour of Fig. 13.12, which is for the $1\pi_{u,\pm1}$ MOs of a homonuclear diatomic molecule.

The ground-state configuration of CO is $1\sigma^2 2\sigma^2 3\sigma^2 4\sigma^2 1\pi^4 5\sigma^2$, as compared with the N_2 configuration $(1\sigma_g)^2(1\sigma_u)^2(2\sigma_g)^2(2\sigma_u)^2(1\pi_u)^4(3\sigma_g)^2$.

As in homonuclear diatomics, the heteronuclear diatomic MOs are approximated as linear combinations of atomic orbitals. The coefficients are found by solving the Roothaan equations (13.157). For example, a minimal-basis-set SCF calculation using Slater AOs (with nonoptimized exponents given by Slater's rules) gives for the CO 5σ, 1π, and 2π MOs at $R = R_e$ [B. J. Ransil, *Rev. Mod. Phys.*, **32**, 245 (1960)]:

$$5\sigma = 0.027(1s_C) + 0.011(1s_O) + 0.739(2s_{\perp C}) + 0.036(2s_{\perp O})$$
$$- 0.566(2p\sigma_C) - 0.438(2p\sigma_O)$$
$$1\pi = 0.469(2p\pi_C) + 0.771(2p\pi_O)$$
$$2\pi = 0.922(2p\pi_C) - 0.690(2p\pi_O)$$

The expressions for the π MOs are simpler than those for the σ MOs because s and $p\sigma$ AOs cannot contribute to π MOs. For comparison, the corresponding MOs in N_2 at $R = R_e$ are (Ransil, op. cit.):

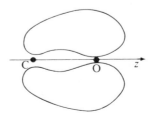

Figure 13.20 Cross section of a contour of the $1\pi_{\pm1}$ MOs in CO.

$$3\sigma_g = 0.030(1s_a + 1s_b) + 0.395(2s_{\perp a} + 2s_{\perp b}) - 0.603(2p\sigma_a + 2p\sigma_b)$$

$$1\pi_u = 0.624(2p\pi_a + 2p\pi_b)$$

$$1\pi_g = 0.835(2p\pi_a - 2p\pi_b)$$

The resemblance of CO and N_2 MOs is apparent. The 1σ MO in CO is found to be nearly the same as a $1s$ oxygen-atom AO; the 2σ MO in CO is essentially a carbon-atom $1s$ AO.

In general, for a heteronuclear diatomic molecule AB where the valence AOs of each atom are of s and p type and where the valence AOs of A do not differ greatly in energy from the valence AOs of B, we can expect the Figure 13.16 pattern of

$$\sigma s < \sigma^* s < \pi p < \sigma p < \pi^* p < \sigma^* p$$

valence-shell MOs formed from s and p valence-shell AOs to hold reasonably well. Figure 13.16 would be modified in that each valence AO of the more electronegative atom would lie below the corresponding valence AO of the other atom.

When the valence-shell AO energies of B lie very substantially below those of A, the s and $p\sigma$ valence AOs of B lie below the s valence-shell AO of A, and this affects which AOs contribute to each MO. Consider the molecule BF, for example. A minimal-basis-set calculation [Ransil, *Rev. Mod. Phys.*, **32**, 245 (1960)] gives the 1σ MO as essentially $1s_F$ and the 2σ MO as essentially $1s_B$. The 3σ MO is predominantly $2s_F$, with small amounts of $2s_B$, $2p\sigma_B$, and $2p\sigma_F$. The 4σ MO is predominantly $2p\sigma_F$, with significant amounts of $2s_B$ and $2s_F$ and a small amount of $2p\sigma_B$; this is quite different from N_2, where the corresponding MO is formed predominantly from the $2s$ AOs on each N. The 1π MO is a bonding combination of $2p\pi_B$ and $2p\pi_F$. The 5σ MO is predominantly $2s_B$, with a substantial contribution from $2p\sigma_B$ and a significant contribution from $2p\sigma_F$; this is unlike the corresponding MO in N_2, where the largest contributions are from $2p\sigma$ MOs on each atom. The 2π MO is an antibonding combination of $2p\pi_B$ and $2p\pi_F$. The 6σ MO has important contributions from $2p\sigma_B$, $2s_B$, $2s_F$, and $2p\sigma_F$.

We see from Fig. 11.2 that the $2p_F$ AO lies well below the $2s_B$ AO. This causes the $2p\sigma_F$ AO to contribute substantially to lower-lying MOs and the $2s_B$ AO to contribute substantially to higher-lying MOs, as compared with what happens in N_2. (This effect occurs in CO, although to a lesser extent. Note the very substantial contribution of $2s_C$ to the 5σ MO; also, the 4σ MO in CO has a very substantial contribution from $2p\sigma_O$.)

For a diatomic molecule AB where each atom has s and p valence-shell AOs (this excludes H and transition elements) and where the A and B valence AOs differ widely in energy, we may expect the pattern of valence MOs to be $\sigma < \sigma < \pi < \sigma < \pi < \sigma$, but it is not so easy to guess which AOs contribute to the various MOs or the bonding or antibonding character of the MOs. By feeding the valence electrons into these MOs, we can make a plausible guess as to the number of unpaired electrons and the ground term of the AB molecule (Problem 13.26).

Diatomic hydrides are a special case, since H has only a $1s$ valence AO. Consider HF as an example. The ground-state configurations of the atoms are $1s$

for H and $1s^2 2s^2 2p^5$ for F. We expect the filled $1s$ and $2s$ F subshells to take little part in the bonding. The four $2p\pi$ fluorine electrons are nonbonding (there are no π valence AOs on H). The hydrogen $1s$ AO and the fluorine $2p\sigma$ AO have the same symmetry (σ) and have rather similar energies (Fig. 11.2), and a linear combination of these two AOs will form a σ MO for the bonding electron pair:

$$\phi = c_1(1s_H) + c_2(2p\sigma_F)$$

where the contributions of $1s_F$ and $2s_F$ to this MO have been neglected. Since fluorine is more electronegative than hydrogen, we expect that $c_2 > c_1$. (In addition, the $1s_H$ and $2p\sigma_F$ AOs form an antibonding MO, which is unoccupied in the ground state.)

The picture of HF just given is only a crude qualitative approximation. A minimal-basis-set SCF calculation using Slater orbitals with optimized exponents gives as the MOs of HF [B. J. Ransil, *Rev. Mod. Phys.*, **32**, 245 (1960)]

$$1\sigma = 1.000(1s_F) + 0.012(2s_{\perp F}) + 0.002(2p\sigma_F) - 0.003(1s_H)$$

$$2\sigma = -0.018(1s_F) + 0.914(2s_{\perp F}) + 0.090(2p\sigma_F) + 0.154(1s_H)$$

$$3\sigma = -0.023(1s_F) - 0.411(2s_{\perp F}) + 0.711(2p\sigma_F) + 0.516(1s_H)$$

$$1\pi_{+1} = (2p\pi_{+1})_F$$

$$1\pi_{-1} = (2p\pi_{-1})_F$$

The ground-state MO configuration of HF is $1\sigma^2 2\sigma^2 3\sigma^2 1\pi^4$. The 1σ MO is virtually identical with the $1s$ fluorine AO. The 2σ MO is pretty close to the $2s$ fluorine AO. The 1π MOs are required by symmetry to be the same as the corresponding fluorine π AOs. The bonding 3σ MO has its largest contribution from the $2p\sigma$ fluorine and $1s$ hydrogen AOs, as would be expected from the discussion of the preceding paragraph. However, there is a substantial contribution to this MO from the $2s$ fluorine AO. (Since a single $2s$ function is only an approximation to the $2s$ AO of F, we cannot use this calculation to say exactly how much $2s$ AO character the 3σ HF molecular orbital has.)

Accurate MO expressions require the solution of the Roothaan equations for their determination. For qualitative discussion (but not quantitative work), it is useful to have simple approximations for heteronuclear diatomic MOs. In the crudest approximation, we can take each valence MO of a heteronuclear diatomic molecule as a linear combination of two AOs ϕ_a and ϕ_b, one on each atom. (As the discussions of CO and BF show, this approximation is often quite inaccurate.) From the two AOs, we can form two MOs:

$$c_1\phi_a + c_2\phi_b \quad \text{and} \quad c_1'\phi_a + c_2'\phi_b$$

The lack of symmetry in the heteronuclear diatomic makes the coefficients unequal in magnitude. The coefficients are determined by solving the secular equation [see Eq. (13.45)]

$$\begin{vmatrix} H_{aa} - W & H_{ab} - WS_{ab} \\ H_{ab} - WS_{ab} & H_{bb} - W \end{vmatrix} = 0$$

$$(H_{aa} - W)(H_{bb} - W) - (H_{ab} - WS_{ab})^2 = 0 \qquad (13.184)$$

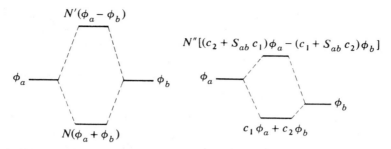

Figure 13.21 Formation of bonding and antibonding MOs from AOs in the homonuclear and heteronuclear cases (see Problem 13.50).

where $\hat{H}$ is some sort of effective one-electron Hamiltonian. Suppose that $H_{aa} > H_{bb}$, and let $f(W)$ be defined as the left side of (13.184). The overlap integral S_{ab} is less than 1 (except at $R = 0$). [A rigorous proof of this follows from the Schwarz inequality, Eq. (3-114) in *Margenau and Murphy*.] The coefficient of W^2 in $f(W)$ is $(1 - S_{ab}^2) > 0$; hence we have $f(\infty) = f(-\infty) = +\infty > 0$. For $W = H_{aa}$ or H_{bb}, the first product in (13.184) vanishes; hence $f(H_{aa}) < 0$ and $f(H_{bb}) < 0$. The roots of (13.184) occur where $f(W)$ equals 0; hence, by continuity, one root must be between $+\infty$ and H_{aa} and the other between H_{bb} and $-\infty$. Therefore, the orbital energy of one MO is less than both H_{aa} and H_{bb} (the energies of the two AOs in the molecule; Section 13.5), and the energy of the other MO is greater than both H_{aa} and H_{bb}. One bonding and one antibonding MO are formed from the two AOs. Figure 13.21 shows the formation of bonding and antibonding MOs from two AOs, for the homonuclear and heteronuclear cases. These figures are gross oversimplifications, since a given MO has contributions from many AOs, not just two.

The coefficients c_1 and c_2 in the bonding heteronuclear MO in Fig. 13.21 are both positive, so as to build up charge between the nuclei. For the antibonding heteronuclear MO, the coefficients of ϕ_a and ϕ_b have opposite signs, causing charge depletion between the nuclei.

13.19 VB TREATMENT OF HETERONUCLEAR DIATOMIC MOLECULES

We shall illustrate the valence-bond approach to the ground states of heteronuclear diatomic molecules by considering HF. We expect a single bond to be formed by the pairing of the hydrogen $1s$ electron and the unpaired fluorine $2p\sigma$ electron. The Heitler–London function corresponding to this pairing is [Eq. (13.119)]

$$\phi_{\text{cov}} = |\cdots 1s_H \overline{2p\sigma_F}| - |\cdots \overline{1s_H} 2p\sigma_F| \tag{13.185}$$

where the dots stand for $1s_F \overline{1s_F} 2s_F \overline{2s_F} 2p\pi_{xF} \overline{2p\pi_{xF}} 2p\pi_{yF} \overline{2p\pi_{yF}}$. This function is essentially covalent, the electrons being shared by the two atoms. However, the high electronegativity of fluorine leads us to include a contribution from an ionic structure as well. An ionic valence-bond function has the form $\phi_a(1)\phi_a(2)$ [Eq.

(13.110)]. Introduction of the required antisymmetric spin factor gives as the valence-bond function for an ionic structure in HF:

$$\phi_{ion} = |\cdots 2p\sigma_F \overline{2p\sigma_F}|$$

The VB wave function is then written as

$$\phi = c_1 \phi_{cov} + c_2 \phi_{ion} \tag{13.186}$$

The optimum values of c_1 and c_2 are found by the variation method; this leads to the usual secular equation. We have ionic–covalent "resonance," involving the structures H—F and H^+F^-; the true molecular structure is intermediate between the covalent and ionic structures. A term $c_3|1s_H \overline{1s_H}|$ corresponding to the ionic structure H^-F^+ could also be included in the wave function, but this should make only a slight contribution for HF. For molecules that are less ionic, both ionic structures might well be included.

For a molecule like NaCl, which is highly ionic, we expect the valence-bond function to have $c_2 \gg c_1$. It might be thought that NaCl would dissociate to Na^+ and Cl^- ions, but this is not accurate. The ionization potential of Na is 5.1 eV, while the electron affinity of Cl is only 3.6 eV. Hence, in the gas phase the neutral separated ground-state atoms $Na + Cl$ are more stable than the ground-state separated ions $Na^+ + Cl^-$. (In aqueous solution the ions are more stable because of the hydration energy, which makes the separated ions more stable than even the diatomic NaCl molecule.) If the nuclei are slowly pulled apart, a gas-phase NaCl molecule will dissociate to neutral atoms; this means that as R increases from R_e, the ratio c_2/c_1 in (13.186) must decrease, becoming zero at $R = \infty$. For intermediate values of R, the Coulombic attraction between the ions is greater than the 1.5-eV difference between the ionization potential and electron affinity, and the molecule is largely ionic; for very large R, the Coulombic attraction between the ions is less than 1.5 eV, and the molecule is largely covalent. However, if the nuclei in NaCl are pulled apart very rapidly, then the electrons will not have a chance to adjust their wave function from the ionic to the covalent wave function, and both bonding electrons will go with the chlorine nucleus, giving dissociation into ions.

Cesium has the lowest ionization potential, 3.9 eV. Chlorine has the highest electron affinity, 3.6 eV. (The electron affinity of fluorine is 3.45 eV.) Thus, even for CsCl and CsF, the separated ground-state neutral atoms are more stable than the separated ground-state ions. There are, however, cases of excited states of diatomic molecules whose potential-energy curves correspond to dissociation to ions.

13.20 THE VALENCE-ELECTRON APPROXIMATION

Suppose we want to treat K_2, which has 38 electrons. In the MO method, we would start by writing down a 38×38 Slater determinant of molecular orbitals. We would then approximate the MOs by functions containing variational parameters and go on to minimize the variational integral. Clearly, the large number of electrons makes this a formidable task. One way to simplify the problem is to divide the electrons into two groups: the 36 *core* electrons and the two 4*s* *valence* electrons, which provide bonding; we then attempt to treat the valence electrons separately from the core, taking the molecular energy as the sum of core- and valence-electron energies. This approach, introduced in the 1930s, is called the *valence-electron approximation*.

The simplest approach is to regard the core electrons as point charges coinciding with the nucleus. For K_2 this would give a Hamiltonian for the two valence electrons that is identical with the electronic Hamiltonian for H_2. If we then go ahead and minimize the variational integral for the valence electrons in K_2, with no restrictions on the valence-electron trial functions, we will clearly be in trouble; such a procedure will cause the valence-electrons' MO to "collapse" to the $\sigma_g 1s$ MO, since the core electrons are considered absent. To avoid this collapse, one can impose the constraint that the variational functions used for the valence electrons be orthogonal to the orbitals of the core electrons. Of course, the task of keeping the valence orbitals orthogonal to the core orbitals means more work. A somewhat different approach is to drop the approximation of treating the core electrons as coinciding with the nucleus, and to treat them as a charge distribution that provides some sort of effective repulsive potential for the motion of the valence electrons. This leads to an effective Hamiltonian for the valence electrons, which is then used in the variational integral. The valence-electron approximation is widely used in approximate treatments of polyatomic molecules (Chapter 16).

13.21 CI WAVE FUNCTIONS

To overcome the deficiencies of the Hartree–Fock wave function (for example, improper behavior as $R \to \infty$ and incorrect D_e values), one can introduce configuration interaction (CI), thus going beyond the Hartree–Fock approximation. Recall (Section 11.3) that in a molecular CI calculation one begins with a set of basis functions χ_i, does an SCF calculation to find SCF occupied and virtual (unoccupied) MOs, uses these MOs to form configuration (state) functions Φ_i, writes the molecular wave function ψ as a linear combination $\sum_i a_i \Phi_i$ of the configuration functions, and uses the variation method to find the a_i's. In calculations on diatomic molecules, the basis functions can be Slater-type AOs, some centered on one atom, the remainder on the second atom.

The number of MOs produced equals the number of basis functions used. The type of MOs produced depends on the type of basis functions used. For example, if we include only s AOs in the basis set, we get only σ MOs, and no π, $\delta, \ldots$ MOs.

The configuration functions in a CI calculation are classified as *singly excited*, *doubly excited*, *triply excited*, . . . , according to whether 1, 2, 3, . . . electrons are excited from occupied to unoccupied (virtual) orbitals. For example, the H_2 configuration function $2^{-1/2} |\sigma_u^* 1s \, \overline{\sigma_u^* 1s}|$ used in Eq. (13.96) is doubly excited (another term sometimes used is *doubly substituted*).

In the CI expansion $\psi = \sum_i a_i \Phi_i$, one includes only configuration functions that have the same symmetry properties (symmetry eigenvalues) as the state that is being approximated by the expansion. [This follows from Theorem 3 proved in Section 7.3.] For example, the ground state of H_2 is a $^1\Sigma_g^+$ state, and a CI calculation of the H_2 ground state would include only configuration functions that correspond to $^1\Sigma_g^+$ terms. A $\sigma_g\sigma_u$ electron configuration would have states of odd parity (u) and would not be included in ψ; a $\sigma_g\pi_g$ configuration would produce only Π terms (Table 13.3) and would not be included; for a π_g^2 or π_u^2 configuration, only the configuration function corresponding to the $^1\Sigma_g^+$ term (Table 13.3) would contribute to ψ.

The number of possible configuration functions with the proper symmetry increases very rapidly as the number of electrons and the number of basis functions increase; for n electrons and b basis functions, the number of configuration functions turns out to be roughly proportional to b^n. A CI calculation that includes all possible configuration functions with proper symmetry is called a *full* CI calculation. Because of the huge number of configuration functions, full CI calculations are out of the question except for small molecules (small n) and small basis sets (small b).

One must therefore decide which types of configuration functions are likely to make the largest contributions to ψ, and one includes only these. We expect the unexcited configuration function (the SCF wave function) to make the largest contribution. The question then is, which types of excited configurations make significant contributions to ψ? To answer this, we consider the effects of instantaneous electron correlation as a perturbation on the Hartree–Fock wave function. One finds that the first-order correction to the unperturbed (Hartree–Fock) wave function of a closed-shell state contains only doubly excited configuration functions (for justification, see Section 15.12). Thus we expect the most important correction to the Hartree–Fock wave function to come from doubly excited configuration functions. Although singly excited configuration functions are less important than double excitations in affecting the wave function, it turns out (see I. Shavitt, in *Schaefer, Methods of Electronic Structure Theory*, page 255) that single excitations have a significant effect on one-electron properties. [A *one-electron property* is one calculated as $\langle\psi|\hat{B}|\psi\rangle$, where the operator $\hat{B}$ is a sum of terms, each of which involves only a single particle; an example is the dipole moment; Eqs. (13.141) and (13.142).] Therefore, one usually includes single excitations in a CI calculation, and the most common type of CI calculation (designated SDCI or CISD or CI-SD) includes the singly and doubly excited configuration functions. (It turns out that the second-order correction to the Hartree–Fock function includes single, double, triple, and quadruple excitations.)

In the majority of calculations (for example, molecular dissociation, excitation of valence-shell electrons to produce excited electronic states), one is looking at energy changes in processes affecting primarily the valence-shell electrons, so one expects the correlation energies involving the inner-shell electrons to change only slightly. Hence one often makes the further approximation of considering only configuration functions involving excitation of valence-shell electrons.

EXAMPLE   The He SCF calculation in Section 13.16 used a basis set of two STOs χ_1 and χ_2. For the helium ground state treated with this basis set, (*a*) write down the configuration state functions (CSFs) that are present in the wave function in a full CI treatment, and (*b*) carry out a CI calculation that includes only doubly excited CSFs.

(*a*) Since we used two basis functions, the SCF calculation yielded two SCF orbitals ϕ_1 and ϕ_2. In the ground state of this two-electron atom, only ϕ_1 is occupied, and ϕ_2 is an unoccupied (virtual) orbital. We found $\phi_1 = 0.842\chi_1 + 0.183\chi_2$, but didn't bother to find ϕ_2. Use of the root $\varepsilon_2 = 2.809$ [Eq. (13.177)] and the final F_{rs}'s of (13.176) in the Roothaan equation (13.157) with $r = 1$ gives $-3.690c_{12} - 3.290c_{22} = 0$, and $c_{12}/c_{22} = -0.892$. The normalization condition (13.171) rewritten for c_{22} gives $c_{22} = 1.816$; hence $c_{12} = -1.620$. The SCF orbitals are

$$\phi_1 = c_{11}\chi_1 + c_{21}\chi_2 = 0.842\chi_1 + 0.183\chi_2$$

$$\phi_2 = c_{12}\chi_1 + c_{22}\chi_2 = -1.620\chi_1 + 1.816\chi_2$$

(Note that ϕ_2 has a node; ϕ_2 is an approximation to the $2s$ AO of helium.) The terms that arise from placing two electrons into these two s orbitals are a ^1S term with both electrons in ϕ_1, a ^1S term with both electrons in ϕ_2, a ^1S term with one electron in ϕ_1 and one in ϕ_2, and a ^3S term with one electron in ϕ_1 and one in ϕ_2. Since the ground state is ^1S, we include only the ^1S CSFs, which are [see Eqs. (10.41), (10.45), and (11.60)]:

$$\Phi_1 = |\phi_1\overline{\phi_1}|, \qquad \Phi_2 = |\phi_2\overline{\phi_2}|$$
$$\Phi_3 = \tfrac{1}{2}[\phi_1(1)\phi_2(2) + \phi_2(1)\phi_1(2)][\alpha(1)\beta(2) - \beta(1)\alpha(2)] \tag{13.187}$$

(*b*) The CSF Φ_2 is doubly substituted, and Φ_3 is singly substituted, so we include only Φ_2 in addition to Φ_1. The variational function is $\psi = a_1\Phi_1 + a_2\Phi_2$, and the secular equation (11.17) is

$$\begin{vmatrix} \langle\Phi_1|\hat{H}|\Phi_1\rangle - E & \langle\Phi_1|\hat{H}|\Phi_2\rangle \\ \langle\Phi_2|\hat{H}|\Phi_1\rangle & \langle\Phi_2|\hat{H}|\Phi_2\rangle - E \end{vmatrix} = 0 \tag{13.188}$$

The orthogonality of the orbitals ϕ_1 and ϕ_2 (Section 13.16) ensures that $\langle\Phi_1|\Phi_2\rangle = 0$.

The function Φ_1 is the SCF wave function, so $\langle\Phi_1|\hat{H}|\Phi_1\rangle$ is equal to the SCF energy -2.862 hartrees calculated in the previous example: $\langle\Phi_1|\hat{H}|\Phi_1\rangle = -2.862$.

In evaluating the integrals in the secular determinant, we can omit the spin factors in the wave functions, since summation over these gives 1.

The He Hamiltonian (9.38) is $\hat{H} = \hat{H}_{(1)}^{\text{core}} + \hat{H}_{(2)}^{\text{core}} + 1/r_{12}$, and

$$\langle\Phi_2|\hat{H}|\Phi_2\rangle = \langle\phi_2(1)\phi_2(2)|\hat{H}_{(1)}^{\text{core}} + \hat{H}_{(2)}^{\text{core}} + 1/r_{12}|\phi_2(1)\phi_2(2)\rangle$$

$$= \langle\phi_2(1)|\hat{H}_{(1)}^{\text{core}}|\phi_2(1)\rangle\langle\phi_2(2)|\phi_2(2)\rangle$$

$$+ \langle\phi_2(2)|\hat{H}_{(2)}^{\text{core}}|\phi_2(2)\rangle\langle\phi_1(1)|\phi_1(1)\rangle$$

$$+ \langle\phi_2(1)\phi_2(2)|1/r_{12}|\phi_2(1)\phi_2(2)\rangle$$

where $\langle\phi_2(2)|\phi_2(2)\rangle = 1 = \langle\phi_1(1)|\phi_1(1)\rangle$. To evaluate the integrals over the orbitals ϕ_i in terms of integrals over the basis functions χ_r, we substitute the expansion $\phi_i = \Sigma_r \, c_{ri}\chi_r$ [Eq. (13.156)] into the integrals. We get

$$\langle \phi_i(1)|\hat{H}^{core}_{(1)}|\phi_j(1)\rangle = \sum_r \sum_s c^*_{ri}c_{sj}\langle \chi_r(1)|\hat{H}^{core}_{(1)}|\chi_s(1)\rangle$$

$$= \sum_{r=1}^{b} \sum_{s=1}^{b} c^*_{ri}c_{sj}H^{core}_{rs} \qquad (13.189)$$

$$\langle \phi_i(1)\phi_j(2)|1/r_{12}|\phi_k(1)\phi_l(2)\rangle$$

$$= \sum_r \sum_s \sum_t \sum_u c^*_{ri}c^*_{sj}c_{tk}c_{ul}\langle \chi_r(1)\chi_s(2)|1/r_{12}|\chi_t(1)\chi_u(2)\rangle$$

$$= \sum_{r=1}^{b} \sum_{t=1}^{b} \sum_{s=1}^{b} \sum_{u=1}^{b} c^*_{ri}c_{tk}c^*_{sj}c_{ul}(rt|su) \qquad (13.190)$$

Using these equations, we find (Problem 13.55)

$$\langle \Phi_2|\hat{H}|\Phi_2\rangle = 2(c^2_{12}H^{core}_{11} + 2c_{12}c_{22}H^{core}_{12} + c^2_{22}H^{core}_{22}) + c^4_{12}(11|11)$$

$$+ 4c^3_{12}c_{22}(11|12) + 2c^2_{12}c^2_{22}(11|22) + 4c^2_{12}c^2_{22}(12|12)$$

$$+ 4c_{12}c^3_{22}(12|22) + c^4_{22}(22|22) = 3.22_6$$

Similarly, we find (Problem 13.55)

$$\langle \Phi_2|\hat{H}|\Phi_1\rangle = \langle \Phi_1|\hat{H}|\Phi_2\rangle = c^2_{11}c^2_{12}(11|11) + 2(c^2_{11}c_{12}c_{22} + c_{11}c^2_{12}c_{21})(11|12)$$

$$+ 2c_{11}c_{12}c_{21}c_{22}(11|22)$$

$$+ (c^2_{11}c^2_{22} + 2c_{11}c_{21}c_{12}c_{22} + c^2_{12}c^2_{21})(12|12)$$

$$+ 2(c_{11}c_{21}c^2_{22} + c_{12}c_{22}c^2_{21})(12|22) + c^2_{21}c^2_{22}(22|22)$$

$$= 0.289_5$$

where the integral values were taken from the previous example. The $\hat{H}^{core}$ integrals vanish for $\langle \Phi_2|\hat{H}|\Phi_1\rangle$ because of the orthogonality of ϕ_1 and ϕ_2.

The secular equation (13.188) is

$$\begin{vmatrix} -2.862 - E & 0.289_5 \\ 0.289_5 & 3.22_6 - E \end{vmatrix} = 0$$

$$E = -2.876, \; 3.24$$

The lower root gives

$$E = -2.876 \text{ hartrees} = -78.25 \text{ eV}$$

as compared with the SCF energy of -77.87 eV and the true energy of -79.00 eV. This CI calculation has recovered 34% of the correlation energy. The CI ground-state ψ is found to be (Problem 13.54) $\psi = 0.9989\Phi_1 - 0.0474\Phi_2$.

One finds that inclusion of the singly excited CSF Φ_3 gives only a very slight further improvement (see *Jørgensen and Oddershede*, page 38). Significant further improvement requires redoing the SCF calculation with a larger basis set, which will generate more virtual orbitals so that many more CSFs can be included in the CI calculation.

The CI procedure just discussed calculates SCF occupied and virtual orbitals from the basis functions and uses these SCF MOs to form configuration state functions. The convergence rate of this procedure is very slow, and very large numbers of CSFs must be included for accurate results. Accurate CI calculations use from 10^3 to 10^7 CSFs, depending on the size of the molecule. For large molecules, even 10^7 CSFs will not yield accurate results, but the use of more than 10^7 (or 10^8) CSFs is impractical. A major reason for the very slow convergence is that the excited (virtual) SCF orbitals have much of their probability density at large distances from the nuclei, whereas the ground-state wave function has most of its probability density reasonably near the nuclei.

Actually, there is no necessity to use SCF MOs in a CI calculation. Any set of MOs calculated from the basis set will produce the same final wave function provided a full CI calculation is carried out. Moreover, if the non-SCF MOs are well chosen, they can produce much faster convergence to the true wave function than is obtained with SCF MOs, thereby allowing substantially fewer CSFs to be included in ψ. Two approaches that use this idea are the multiconfiguration SCF method and the method of natural orbitals.

In the **multiconfiguration SCF** (MCSCF) **method**, one writes the molecular wave function as a linear combination of CSFs Φ_i and varies not only the expansion coefficients a_i in $\psi = \sum_i a_i \Phi_i$, but also the forms of the molecular orbitals in the CSFs. The MOs are varied by varying the expansion coefficients c_{ri} that relate the MOs ϕ_i to the basis functions χ_r. For example, if we were to do an MCSCF calculation for the He ground state using the basis functions χ_1 and χ_2 of (13.169) and including only the CSFs Φ_1 and Φ_2 of (13.187), we would write as the MCSCF wave function

$$\psi = a_1 \Phi_1 + a_2 \Phi_2 = a_1 |\phi_1 \overline{\phi_1}| + a_2 |\phi_2 \overline{\phi_2}|$$
$$= a_1 |(c_{11}\chi_1 + c_{21}\chi_2)(\overline{c_{11}\chi_1 + c_{21}\chi_2})| + a_2 |(c_{12}\chi_1 + c_{22}\chi_2)(\overline{c_{12}\chi_1 + c_{22}\chi_2})|$$

and we would simultaneously vary the coefficients a_1, a_2, c_{11}, c_{21}, c_{12}, and c_{22} (subject to the conditions of orthonormality of ϕ_1 and ϕ_2 and normalization of ψ) to minimize the variational integral. The MCSCF values found for these six coefficients will differ from the corresponding values found in the SCF plus CI calculations we did, and the energy will be lower, since the c_{ri}'s will be optimal for the MCSCF ψ, rather than being optimal for the SCF ψ.

The optimum MCSCF orbitals can be found by an iterative process somewhat similar to the iterative process used to find SCF wave functions; see A. C. Wahl and G. Das, Chapter 3 in *Schaefer, Methods of Electronic Structure Theory*. By optimizing the orbitals, one can get good results with inclusion of relatively few CSFs. Because the orbitals are varied, the amount of calculation required in the MCSCF procedure is great, but advances in methods of computing MCSCF wave functions [R. Shepard, *Adv. Chem. Phys.*, **69**, 63 (1987)] have led to wide use of the MCSCF and related methods.

Wahl, Das, and co-workers have developed an MCSCF approach to the calculation of diatomic D_e's, which they call the *optimized valence configuration* (OVC) *method*. Since the atomic cores remain relatively unchanged on molecule formation, the OVC method includes only CSFs with valence electrons excited.

To correctly take into account the changes that occur as the molecule forms, one must add those CSFs needed to give proper dissociation of the molecule to Hartree–Fock atoms in their correct states, together with a few other configuration functions that make important contributions in the molecule near R_e. Although the OVC CI ψ gives only a part (about 25% for F_2) of the molecular correlation energy (which is the error in the Hartree–Fock ψ; Section 11.3), it includes that part of the correlation energy that varies with R and hence gives a good $U(R)$ curve and a good D_e. The OVC method has been applied with success to several diatomics. (See A. C. Wahl and G. Das, Chapter 3 in *Schaefer, Methods of Electronic Structure Theory*.) The OVC method is not applicable to excited electronic states or to polyatomic molecules.

A commonly used kind of MCSCF method is the ***complete active space SCF*** (CASSCF) method [B. O. Roos, *Adv. Chem. Phys.*, **69**, 399 (1987)]. [The *fully optimized reaction space* (FORS) *method* is essentially equivalent to the CASSCF method.] Here, as usual, one writes the orbitals ϕ_i to be used in the CSFs as linear combinations of basis functions: $\phi_i = \sum_{r=1}^{b} c_{ri} \chi_r$. One divides the orbitals in the CSFs into ***inactive*** and ***active*** orbitals. The inactive orbitals are kept doubly occupied in all CSFs. The electrons not in the inactive orbitals are called ***active*** electrons. One writes the wave function as a linear combination of all CSFs Φ_i that can be formed by distributing the active electrons among the active orbitals in all possible ways and that have the same spin and symmetry eigenvalues as the state to be treated; $\psi = \sum_i a_i \Phi_i$. One then does an MCSCF calculation to find the optimum coefficients c_{ri} and a_i. A reasonable choice is to take the active orbitals as those MOs that arise from the valence orbitals of the atoms that form the molecule.

For example, the C_2 ground-state configuration is

$$(1\sigma_g)^2(1\sigma_u)^2(2\sigma_g)^2(2\sigma_u)^2(1\pi_u)^4$$

The $2s$ and $2p$ carbon AOs give rise to the $2\sigma_g$, $2\sigma_u$, $1\pi_{ux}$, $1\pi_{uy}$, $3\sigma_g$, $1\pi_{gx}$, $1\pi_{gy}$, and $3\sigma_u$ MOs, where the last four are unoccupied in the ground state. One might thus take the inactive orbitals as the $1\sigma_g$ and $1\sigma_u$ MOs (giving eight active electrons) and the active orbitals as the $2\sigma_g$, $2\sigma_u$, $1\pi_u$, $3\sigma_g$, $1\pi_g$, and $3\sigma_u$ MOs. A CASSCF calculation on the C_2 ground electronic state used a basis set of 82 functions and took the inactive orbitals as $1\sigma_g$ and $1\sigma_u$ and the active orbitals as $2\sigma_g$, $2\sigma_u$, $1\pi_u$, $3\sigma_g$, $1\pi_g$, $3\sigma_u$, $4\sigma_g$, and $4\sigma_u$ [W. P. Kraemer and B. O. Roos, *Chem. Phys.*, **118**, 345 (1987)]. Distribution of the eight active electrons among the ten active orbitals gave a wave function consisting of 1900 CSFs. The electronic energy was calculated for several points in the neighborhood of R_e, and the calculated $U(R)$ curve was substituted into the Schrödinger equation for nuclear motion, which was solved numerically (Section 13.2) to find vibrational–rotational energy levels, from which spectroscopic constants were found. The results were (experimental values in parentheses): $R_e/\text{Å} = 1.25$ (1.24), $D_e/\text{eV} = 6.06$ (6.3), $\tilde{\nu}_e/\text{cm}^{-1} = 1836$ (1855), and $\tilde{\nu}_e x_e/\text{cm}^{-1} = 14.9$ (13.4), where $\tilde{\nu}_e x_e \equiv \nu_e x_e/c$ [see Eq. (4.68)]. For comparison, the Hartree–Fock ψ gives $D_e = 0.8$ eV, $\tilde{\nu}_e/\text{cm}^{-1} = 1905$, $\tilde{\nu}_e x_e/\text{cm}^{-1} = 12.1$, and $R_e = 1.24$ Å.

With modern computational techniques, very large MCSCF wave functions can be calculated. A CASSCF calculation on FeO included 178916 CSFs and fully optimized the orbitals [P. J. Knowles and H.-J. Werner, *Chem. Phys. Lett.*, **115**, 259 (1985)]. For large molecules, use of all the valence orbitals as active gives rise to too many CSFs to be handled, and so the number of active orbitals must be limited in such cases.

A widely used method that combines the MCSCF and conventional CI methods is the ***multireference CI*** (MRCI) ***method.*** In the conventional CI method, one starts with the SCF wave function Φ_1 (which is called the *reference* function) and moves electrons out of occupied orbitals of Φ_1 into virtual SCF orbitals to produce CSFs $\Phi_2, \Phi_3, \ldots$, and one writes the wave function as $\psi = \sum_i a_i \Phi_i$; one then varies the a_i's to minimize the variational integral. Recall that for H_2, although the Hartree–Fock function Φ_1 gives a reasonably good representation of the wave function in the region of R_e, in the limit as $R \to \infty$ we must take the variational function as a linear combination of two CSFs in order to get the proper dissociation behavior. For N_2, which has a triple bond, one finds that a linear combination of 10 CSFs is needed to give proper behavior at large R. In the MRCI method, one first does an MCSCF calculation to produce a wave function that consists of several CSFs $\Phi_1, \Phi_2, \ldots, \Phi_m$ with optimized orbitals and that has the proper behavior for all nuclear configurations. One then takes this MCSCF function and moves electrons out of occupied orbitals of the CSFs $\Phi_1, \Phi_2, \ldots, \Phi_m$ (called the *reference* CSFs) into virtual orbitals to produce further CSFs $\Phi_{m+1}, \ldots, \Phi_n$. One writes $\psi = \sum_{i=1}^{n} a_i \Phi_i$ and does a CI calculation to find the optimum a_i's. Typically, the reference CSFs will contain singly and doubly excited CSFs and, in the final wave function, one will consider single and double excitations from the reference CSFs. Thus, the final MRCI wave function will include some triple and quadruple excitations.

CASSCF wave functions are often used as the starting point for MRCI calculations. The calculations that determined the ordering of the lowest states of Si_2 and Al_2 (Section 13.7) were CASSCF/MRCI calculations.

Besides the MCSCF approach, another alternative to the use of SCF MOs in CI calculations is provided by natural orbitals. For a CI wave function, which is a linear combination of Slater determinants, the electron probability density ρ can be shown to have the form [P.-O. Löwdin, *Rev. Mod. Phys.*, **32**, 328 (1960)]

$$\rho(x, y, z) = \sum_i \sum_j a_{ij} \phi_i^*(x, y, z) \phi_j(x, y, z)$$

where the ϕ's are all the MOs that appear in the Slater determinants of the CI wave function and the a_{ij}'s are a set of numbers. As we shall see in Section 15.8, we can take a linear combination of MOs to form a new set of MOs without changing the overall wave function. It can be shown that there exists a set of MOs θ_i, called the *natural orbitals*, that have the property that when the wave function is expressed using the θ_i's the probability density has the simple form

$$\rho(x, y, z) = \sum_i b_i |\theta_i(x, y, z)|^2$$

where the *occupation numbers* b_i lie between 0 and 2 and need not be integers. Note the resemblance of this expression to (13.131), which is for an SCF wave function.

It turns out that a CI calculation using natural orbitals converges much faster than one using SCF orbitals, so one needs to include far fewer CSFs to obtain a given level of accuracy. Unfortunately, the natural orbitals are defined in terms of the final CI wave function and cannot be calculated until a CI calculation using SCF orbitals has been completed. Hence several schemes have been devised to calculate approximate natural orbitals and use these in CI calculations.

For example, the *iterative natural-orbital* (INO) *method* starts by calculating a CI wave function using a manageable number of CSFs. From this CI wave function, one calculates approximate natural orbitals and uses these orbitals to get an improved CI wave function; this process is then repeated to get further improvement. As an example, an INO calculation on the LiH ground state using only 45 CSFs obtained 87% of the correlation energy and gave a slightly lower energy than an ordinary CI wave function of 939 CSFs; see C. F. Bender and E. R. Davidson, *J. Phys. Chem.*, **70**, 2675 (1966).

In a CI calculation that uses SCF MOs, there are two major computational tasks. One task is to transform the known integrals over the basis functions χ_r into integrals over the SCF orbitals using Eqs. (13.189) and (13.190). Calculation of the integrals $\langle \phi_i \phi_j | 1/r_{12} | \phi_k \phi_l \rangle$ from the $(rs|tu)$ integrals is especially time consuming. For a basis set of b functions $\chi_1, \ldots, \chi_b$, there are b orbitals ϕ_i and approximately $b^4/8$ different $\langle \phi_i \phi_j | 1/r_{12} | \phi_k \phi_l \rangle$ integrals to be computed [the factor $1/8$ arises from equalities similar to (13.170)], and there are b^4 terms in the sums on the right side of (13.190). Hence the number of computations is apparently about $b^8/8$. For 100 basis functions, $b^8/8$ is about 10^{15}. Fortunately, use of a clever procedure can reduce the number of computations to about b^5 (see *Hehre* et al., Section 3.3.4). The second task is to solve the CI secular equation to find the lowest-energy eigenvalue and the corresponding set of expansion coefficients. In an SCF calculation that uses 100 basis functions, one must solve for the eigenvalues and eigenvectors of a matrix whose order is 100. This is readily done using standard matrix diagonalization methods (Section 8.6). In an accurate CI calculation, one might use 10^5 CSFs to expand the wave function, and one must find the lowest eigenvalue and corresponding eigenvector of a matrix whose order is 10^5. Special techniques have been devised to find the lowest eigenvalue of a very large matrix (see Shavitt in *Schaefer, Methods of Electronic Structure Theory*, pages 228–238).

In a CI calculation with 10^6 CSFs, there are 10^{12} matrix elements H_{ij} between CSFs, which is too many to be stored in the internal memory of the computer. To avoid the problems of dealing with large matrices, Roos developed the ***direct configuration interaction method*** in 1972 (see B. O. Roos and P. E. M. Siegbahn, in *Schaefer, Methods of Electronic Structure Theory*, Chapter 7). The direct CI method avoids explicit calculation of the integrals H_{ij} between CSFs and avoids the need to solve the secular equation (11.17); instead, the CI expansion coefficients in (11.16) and the energy are calculated directly from the one- and two-electron integrals over the basis functions. The direct CI method allows conventional CI calculations with more than 10^7 CSFs and MCSCF calculations with more than 10^5 CSFs [H.-J. Werner, *Adv. Chem. Phys.*, **69**, 1 (1987)].

Another technique used to speed up CI calculations is the *graphical unitary-group approach* (GUGA); for details, see B. R. Brooks and H. F.

Schaefer, *J. Chem. Phys.*, **70**, 5092 (1979); B. R. Brooks et al., *J. Chem. Phys.*, **72**, 3837 (1980). GUGA has been incorporated into the direct CI method [P. E. M. Siegbahn, *J. Chem. Phys.*, **72**, 1647 (1980)].

Although calculation of SCF wave functions for closed-shell, reasonably small molecules is essentially a routine procedure, each CI calculation of a molecule presents its own special problems. To obtain reliable results, one must use sound judgment in choosing the basis set, the molecular orbitals, the configuration functions to be included, and the procedure to be used.

Numerous CI calculations have been performed on diatomic molecules to yield potential-energy curves, dipole moments, and dissociation energies for ground and excited states. For discussions of results, see *Mulliken and Ermler, Diatomic Molecules* and *Schaefer, The Electronic Structure of Atoms and Molecules.*

13.22 SUMMARY

Because electrons are much lighter than nuclei, we can use the Born–Oppenheimer approximation to deal with molecules. This approximation takes the molecular wave function ψ as the product of wave functions for electronic motion and for nuclear motion: $\psi = \psi_{el}(q_i; q_\alpha)\psi_N(q_\alpha)$, where q_i and q_α are the electronic and nuclear coordinates, respectively. We solve the electronic Schrödinger equation for fixed positions of the nuclei; this equation is $(\hat{H}_{el} + V_{NN})\psi_{el} = U\psi_{el}$, where V_{NN} is the internuclear-repulsion term and $\hat{H}_{el}$ is the sum of operators for electronic kinetic energy, electron–nuclear attractions, and electron–electron repulsions. U is the electronic energy including internuclear repulsions. After finding $U(q_\alpha)$, we solve the nuclear-motion Schrödinger equation, which is $(\hat{T}_N + U)\psi_N = E\psi_N$, where $\hat{T}_N$ is the nuclear-kinetic-energy operator and E is the (total) molecular energy. Solution of the nuclear-motion Schrödinger equation shows that the molecular energy is approximately the sum of translational, rotational, vibrational, and electronic energies.

Quantum chemists use atomic units, in which energies are measured in hartrees and lengths in bohrs [Eqs. (13.29) and (13.30)].

The electronic Schrödinger equation for H_2^+ can be solved exactly in confocal elliptic coordinates ξ, η, and ϕ to give wave functions that are eigenfunctions of $\hat{L}_z$, the operator for the component of electronic orbital angular momentum along the internuclear axis. The letters $\sigma, \pi, \delta, \phi, \ldots$ denote $\lambda \equiv |m|$ values of $0, 1, 2, 3, \ldots$, respectively, where $m\hbar$ is the $\hat{L}_z$ eigenvalue.

Approximate wave functions for the two lowest electronic states of H_2^+ are $N(1s_a + 1s_b)$ and $N'(1s_a - 1s_b)$, where $1s_a$ and $1s_b$ are $1s$ AOs centered on nuclei a and b, respectively.

Approximate MOs for homonuclear diatomic molecules were constructed as $N(1s_a \pm 1s_b)$, $N(2s_a \pm 2s_b)$, and so on, and in the separated-atoms notation, these were called $\sigma_g 1s$, $\sigma_u^* 1s$, $\sigma_g 2s$, $\sigma_u^* 2s$, and so on, where σ means $\lambda = 0$, g or u denotes even or odd functions, and the star denotes an antibonding MO. An alternative notation, $1\sigma_g$, $1\sigma_u$, $2\sigma_g$, $2\sigma_u$, $\ldots$, numbers the MOs of each symmetry type in order of increasing energy. Using these approximate MOs, we examined the electron configurations, bond orders, and spin multiplicities of the ground terms of H_2, He_2, Li_2, . . . , Ne_2.

A molecular electronic configuration gives rise to one or more electronic terms. Each diatomic-molecule term designation has the form $^{2S+1}(\Lambda)$, where S is the total electronic spin quantum number and (Λ) is a code letter (Σ, Π, Δ, Φ, . . .) that gives the $|M_L|$ value (0, 1, 2, 3, . . .), where $M_L\hbar$ is the component of electronic orbital angular momentum along the internuclear axis. Σ terms are designated Σ^+ or Σ^-, according to whether the eigenvalue of ψ_{el} for reflection in a plane containing the molecular axis is $+1$ or -1. For homonuclear diatomic molecules, a g or u subscript is added to the term symbol to show whether ψ_{el} is even or odd.

The simple LCAO-MO wave function for the H_2 ground electronic state is

$$\sigma_g 1s(1)\sigma_g 1s(2)2^{-1/2}[\alpha(1)\beta(2) - \beta(1)\alpha(2)]$$

$$\approx N[1s_a(1) + 1s_b(1)][1s_a(2) + 1s_b(2)]2^{-1/2}[\alpha(1)\beta(2) - \beta(1)\alpha(2)]$$

The simple covalent VB wave function for ground-state H_2 is constructed by interchange of the bonding electrons between the bonding atoms to give

$$N[1s_a(1)1s_b(2) + 1s_a(2)1s_b(1)]2^{-1/2}[\alpha(1)\beta(2) - \beta(1)\alpha(2)]$$

The H_2 MO function gives equal weight to covalent and ionic terms and thus underestimates electron correlation, which acts to keep electrons apart and makes the ionic terms less important than the covalent terms. The MO function can be improved by configuration interaction; the most important excited CSF is that from the doubly excited electron configuration $(\sigma_u^* 1s)^2$. The H_2 covalent VB wave function can be improved by the addition of a contribution from ionic terms.

The MO wave function for a $^1\Sigma$ diatomic-molecule state is a single Slater determinant of spin-orbitals. The VB wave function is a linear combination of Slater determinants that involve interchanges of spins within the pairs of bonding AOs.

The electron probability density ρ of an n-electron molecule is found by summation of $|\psi|^2$ over spins, integration over the spatial coordinates of $n-1$ electrons, and multiplication by n [Eq. (13.128)]. The dipole moment of a molecule is given by (13.144).

The best possible forms for the MOs ϕ_i of a molecule are the solutions of the Hartree–Fock equations $\hat{F}\phi_i = \varepsilon_i\phi_i$, where $\hat{F}$ is the Fock operator [Eqs. (13.149) to (13.152)] and ε_i is the orbital energy. To solve the Hartree–Fock equations, we expand ϕ_i using a set of basis AOs: $\phi_i = \Sigma_s c_{si}\chi_s$; this leads to the Roothaan equations (13.157) for the coefficients c_{si} and orbital energies ε_i. Since the Fock operator $\hat{F}$ and its matrix elements F_{rs} [Eq. (13.165)] depend on the occupied MOs, which are unknown, the Roothaan equations are solved by an iterative process that starts with an initial guess for the occupied MOs.

A minimal basis set consists of one basis function for each inner shell and each valence AO. Examples of minimal- and extended basis set SCF wave functions for diatomic molecules were given in Sections 13.17 and 13.18. SCF wave functions give pretty accurate molecular geometries and dipole moments, but very inaccurate dissociation energies, due to improper behavior as $R \to \infty$.

To improve on SCF wave functions, we can use configuration interaction, expressing the wave function as a linear combination of configuration state functions (CSFs) Φ_i. After the ground-state CSF, the most-important contributions come from doubly excited CSFs. The most common kind of CI calculation includes only singly and doubly excited CSFs.

An MCSCF wave function is a linear combination of several CSFs, $\psi = \Sigma_i a_i \Phi_i$, in which the forms of the MOs and the coefficients a_i have been optimized. A commonly used type of MCSCF function is a CASSCF wave function.

A multireference CI (MRCI) wave function is a linear combination of an MCSCF wave function and CSFs formed by excitation of electrons from occupied orbitals of the MCSCF function.

PROBLEMS

13.1 For the H_2 ground electronic state, $D_0 = 4.4781$ eV. Find ΔH_0° for $H_2(g) \rightarrow 2H(g)$ in kJ/mol.

13.2 Use the D_0 value of H_2 (4.478 eV) and the D_0 value of H_2^+ (2.651 eV) to calculate the first ionization energy of H_2 (that is, the energy needed to remove an electron from H_2).

13.3 The infrared absorption spectrum of $^1H^{35}Cl$ has its strongest band at 8.65×10^{13} Hz. For this molecule, $D_0 = 4.43$ eV. (a) Find D_e for $^1H^{35}Cl$. (b) Find D_0 for $^2H^{35}Cl$.

13.4 (a) Verify that if anharmonicity is taken into account by inclusion of the $\nu_e x_e$ term in the vibrational energy then $D_e = D_0 + \frac{1}{2}h\nu_e - \frac{1}{4}h\nu_e x_e$. (b) The $^7Li^1H$ ground electronic state has $D_0 = 2.4287$ eV, $\nu_e/c = 1405.65$ cm^{-1}, and $\nu_e x_e/c = 23.20$ cm^{-1}, where c is the speed of light. (These last two quantities are usually designated ω_e and $\omega_e x_e$ in the literature.) Calculate D_e for $^7Li^1H$.

13.5 Verify that the change of variable (13.18) transforms the radial Schrödinger equation from (13.17) to (13.19).

13.6 Verify the Taylor-series expansion (13.24).

13.7 Give the numerical value in atomic units of each of the following quantities: (a) proton mass; (b) electron charge; (c) Planck's constant; (d) He$^+$ ground-state energy, assuming infinite nuclear mass; (e) one second; (f) c (speed of light); (g) hydrogen-atom ground-state energy, taking internal nuclear motion into account; (h) one debye.

13.8 Show that the substitutions $E^\dagger \equiv E/E_h$, $x^\dagger \equiv x/a_0$, $y^\dagger \equiv y/a_0$, $z^\dagger \equiv z/a_0$, $\psi^\dagger(x^\dagger, y^\dagger, z^\dagger) \equiv \psi(x, y, z)$ in $-(\hbar^2/2m_e)\nabla^2\psi - (Ze'^2/r)\psi = E\psi$ gives $-\frac{1}{2}\nabla^{\dagger 2}\psi^\dagger - (Z/r^\dagger)\psi^\dagger = E^\dagger\psi^\dagger$.

13.9 Show that the substitution (13.37) in the H_2^+ electronic Schrödinger equation leads to two ordinary differential equations.

13.10 In confocal elliptic coordinates, what is the shape of surfaces of constant (a) ξ; (b) η; (c) ϕ?

13.11 (a) From Fig. 13.2 show that

$$R\xi = [r^2 \sin^2 \theta + (z + \tfrac{1}{2}R)^2]^{1/2} + [r^2 \sin^2 \theta + (z - \tfrac{1}{2}R)^2]^{1/2}$$

and that η is given by a similar expression with the plus sign between the brackets replaced by a minus sign. (b) Verify the ranges given for the coordinates in Eq. (13.33).

13.12 Verify Eq. (13.62) for H_{ab}.

13.13 (a) Show that the variational integral W_1 for the ground state of H_2^+ [Eq. (13.63)] can be written as $W_1 = k^2 F(t) + kG(t)$, where $t \equiv kR$ and where F and G are certain functions of t. (b) Show that the minimization condition $\partial W_1 / \partial k = 0$ leads to

$$k = - \frac{G(t) + tG'(t)}{2F(t) + tF'(t)}$$

Using this equation, we can find k for a given value of t; we then use $R = t/k$ to find the value of R corresponding to our value of k.

13.14 Write a computer program that will calculate the optimum value of the orbital exponent k for the H_2^+ trial function (13.54) for a given R value. Have the program calculate $W_1 + 1/R$ [where W_1 is given by (13.63)] for k ranging from 0 to 3 in steps of 0.1 and have the program pick the k value that gives the smallest $W_1 + 1/R$. Let this k value be k'. Rerun the program to calculate $W_1 + 1/R$ for k ranging from $k' - 0.1$ to $k' + 0.1$ in steps of 0.01 and again find the optimum k. Write the program so that the value of R, the initial and final values of k, and the interval between k values are input from the keyboard for each run. Use the program to find k to the nearest 0.01 for R values of 0.5, 1.0, 2.0, 3.0, 4.0, and 6.0. Use the calculated energies to plot the $U(R)$ curve predicted by this trial function. (For more efficient ways to find the minimum of a function of one variable, see *Press et al.*, Chapter 10.)

13.15 Write a computer program that will simultaneously find the values of k and R that minimize $W_1 + 1/R$ in (13.63). On the first run, have R range from 0.1 to 6 in steps of 0.1 and have k range from 0 to 3 in steps of 0.1; compute $W_1 + 1/R$ for all possible pairs of k and R values in these ranges and have the computer find the pair of k and R values that gives the smallest $W_1 + 1/R$. Let these values be k' and R'. Rerun the program with k and R ranging from $k' - 0.1$ to $k' + 0.1$ and $R' - 0.1$ to $R' + 0.1$, respectively, each in steps of 0.01; and so on. Run the program enough times to find the optimum k and R, each accurate to 0.0001. (For much more efficient ways to find the minimum of a function of two variables, see *Press et al.*, Chapter 10.)

13.16 Consider the trial function (13.55) for the first excited state of H_2^+. Find the limit of this function as R goes to zero. Compare the result with the $2p_z$ hydrogenlike function.

13.17 Calculate the energy at $R = 2.00$ bohrs for the H_2^+ trial function $e^{-k\xi}$. Calculate the percent error in (a) the electronic energy including nuclear repulsion; (b) the purely electronic energy; (c) the dissociation energy.

13.18 Write a computer program to calculate (at $R = R_e$) values of the H_2^+ MOs (13.54) and (13.55) at points in a plane containing the nuclei. Use the output to make contour plots of the MOs. Label the contours in units of electrons/bohr^3 (probability density).

13.19 Verify Eq. (13.79) for a σ_h reflection.

13.20 Which species of each pair has the greater D_e? (a) Li_2 or Li_2^+; (b) C_2 or C_2^+; (c) O_2 or O_2^+; (d) F_2 or F_2^+.

13.21 How many independent electronic wave functions correspond to each of the following diatomic-molecule terms: (a) $^1\Sigma^-$; (b) $^3\Sigma^+$; (c) $^3\Pi$; (d) $^1\Phi$; (e) $^6\Delta$?

13.22 Give the levels belonging to each of the terms in Problem 13.21.

13.23 Show that the four functions of (13.90) have the indicated eigenvalues with respect to a $\sigma_v(xz)$ reflection of electronic coordinates. Start by showing that this reflection converts ϕ to $-\phi$ and leaves r_a and r_b unchanged.

13.24 Show that for a diatomic molecule $[\hat{L}_z, \hat{O}_{\sigma_v}] \neq 0$.

13.25 Use simple MO theory and Table 13.3 to predict the bond order, the number of unpaired electrons, and the ground term for each of the following molecules: (a) S_2; (b) S_2^+; (c) S_2^-; (d) N_2^+; (e) N_2^-; (f) F_2^+; (g) F_2^-; (h) Ne_2^+; (i) Na_2^+; (j) Na_2^-; (k) H_2^-; (l) C_2^+; (m) C_2; (n) C_2^-.

13.26 Use simple MO theory to predict the number of unpaired electrons and the ground term of each of the following: (a) BF; (b) BN; (c)BeS; (d) BO; (e) NO; (f) CF; (g) CP; (h) NBr; (i) LiO; (j) ClO; (k) BrCl. Compare your results with the experimentally observed ground terms: (a) $^1\Sigma^+$; (b) $^3\Pi$; (c) $^1\Sigma^+$; (d) $^2\Sigma^+$; (e) $^2\Pi$; (f) $^2\Pi$; (g) $^2\Sigma^+$; (h) $^3\Sigma^-$; (i) $^2\Pi$; (j) $^2\Pi$; (k) $^1\Sigma^+$.

13.27 The ground state of H_2 has $^1\Sigma_g^+$ symmetry. What restriction does this impose on the values of m, n, j, and k in the James and Coolidge wave function (Section 13.9).

13.28 In applying quantum chemistry to chemical kinetics, which would be the more accurate approximation, the simple MO or the simple VB method?

13.29 Write down abbreviated expressions for the remaining six determinants of the N_2 VB function of Section 13.12. Use the rule given in that section to find the coefficient of each determinant in the wave function.

13.30 (a) Show that the simple MO wave function for the $b\,^3\Sigma_u^+$ level of H_2 is the same as the Heitler–London VB function for this level. (b) Show that the simple MO wave function for the $B\,^1\Sigma_u^+$ level of H_2 contains only ionic terms.

13.31 (a) Use (13.128) to find the electron probability density for the simple MO wave function of the $b\,^3\Sigma_u^+$ level of H_2. Verify that (13.131) holds for this case. (b) Verify Eq. (13.130).

13.32 Show that ρ_{MO} of (13.130) is greater than ρ_{VB} of (13.129) at the midpoint of the line joining the nuclei.

13.33 Show that the dipole moment (13.132) of a system of charges is independent of the choice of the coordinate origin, *provided* the system has no net charge.

13.34 (a) Explain why the permanent dipole moment of a many-electron atom is always zero. (b) Explain why the permanent electric dipole moment of H can be nonzero for certain excited states. (c) Show qualitatively that two of the four correct zeroth-order functions of Problem 9.15 give nonzero permanent dipole moments.

13.35 Prove that the one-electron Hartree–Fock operator (13.149) is Hermitian.

13.36 Explain the origin of the extra terms in the molecular Hartree–Fock operator (13.149) as compared to the atomic Hartree operator of (11.9) and (11.7).

13.37 Verify that the Coulomb and exchange integrals J_{ij} and K_{ij} can be written in terms of the Coulomb and exchange operators of Section 13.16 as

$$J_{ij} = \langle \phi_i(1) | \hat{J}_j(1) | \phi_i(1) \rangle, \qquad K_{ij} = \langle \phi_i(1) | \hat{K}_j(1) | \phi_i(1) \rangle$$

13.38 Verify Eq. (13.163) for the $\hat{K}_j$ integral over basis functions.

13.39 Verify the equations for H_{11}^{core} and H_{22}^{core} in the Section 13.16 example.

13.40 Verify the equalities (13.170) for electron-repulsion integrals.

13.41 Use Eq. (9.55) to verify the expressions for the integrals $(11|11)$ and $(22|22)$ in the Section 13.16 example.

13.42 (a) Verify the equations for F_{11}, F_{12}, and F_{22} that immediately precede Eq. (13.173). (b) Verify Eqs. (13.173) to (13.175) for F_{11}, F_{12}, and F_{22}.

13.43 Derive (13.171) from the normalization condition for ϕ_1.

13.44 Verify the numerical results for P_{11}, P_{12}, P_{22}, F_{11}, F_{12}, F_{22}, ε_1, ε_2, c_{11}, and c_{21} obtained on the last cycle of calculation in the Section 13.16 example [Eqs. (13.176), (13.177), and the preceding and following equation].

13.45 Repeat the He SCF calculation of Section 13.16 using the same basis functions but starting with the initial guess $c_{11} = c_{21}$ and c_{21} determined by the normalization condition (13.171).

13.46 (a) Write a computer program that will perform the helium-atom SCF calculation of Section 13.16. Have the input to the program be ζ_1, ζ_2, and the initial guess for c_{11}/c_{21}. Do not use the Section 13.16 values of the integrals, but have the program calculate all integrals from ζ_1 and ζ_2. Have the program print c_{11}, c_{21}, ε_1, and ε_2 for each cycle of calculation. Use the convergence criterion that c_{11} and c_{21} each differ from the c_{11} and c_{21} values of the previous cycle by less than 10^{-4} atomic units. (b) Use your program with $\zeta_1 = 1.45$ and $\zeta_2 = 2.91$ to find the number of iterations needed to reach convergence for each of these initial guesses for c_{11}/c_{21}: 100, 10, 1, 0, -1, -10, -100. (c) Run the program with $\zeta_1 = 1.45363$ and $\zeta_2 = 2.91093$ to verify the SCF energy given for these ζ's at the end of the Section 13.16 example. (d) Run the program with ζ_1 changed by $+0.01$ and by -0.01 from the value in (c), and verify that each E_{HF} obtained is higher than that in (c). Repeat for ζ_2.

13.47 Calculate ρ for the He SCF wave function of the Section 13.16 example at $r = 0$ and at $r = 1$ bohr.

13.48 Given that $\phi_i = \Sigma_s c'_{si} \chi'_s$, where the χ' functions are orthonormal, show that the orbitals ϕ_i form an orthonormal set if the matrix $\mathbf{C}'$ of coefficients c'_{si} is unitary.

13.49 (a) Verify that the $2s_\perp$ AO in (13.183) is orthogonal to the $1s$ AO and is normalized. (b) Let an MO ϕ have the form $\phi = a(1s) + b(2s) + \cdots$ when expressed using a nonorthogonal $2s$ STO and the form $\phi = c(1s) + d(2s_\perp) + \cdots$ when expressed using an orthogonalized $2s$ orbital. Show that $c = a + Sb$ and $d = b(1 - S^2)^{1/2}$, where $S = \langle 1s|2s \rangle$. (c) Let ζ_1 and ζ_2 be the orbital exponents of $1s$ and $2s$ STOs, respectively. Show that $S = 24 \zeta_1^{3/2} \zeta_2^{5/2} / 3^{1/2} (\zeta_1 + \zeta_2)^4$.

13.50 (a) Use orthogonality (Section 8.5 and Problem 8.31) to derive the expression given in Fig. 13.21 for the heteronuclear antibonding MO. Then do the same for the homonuclear antibonding MO. (b) Verify that the functions (13.54) and (13.55) are orthogonal.

13.51 True or false: (a) If $|\psi_1| = |\psi_2|$, then ψ_1 and ψ_2 must represent the same state. (b) The probability density for the wave function (13.186) is the sum of the probability densities for the functions $c_1 \phi_{cov}$ and $c_2 \phi_{ion}$.

13.52 For NaCl, $R_e = 2.36$ Å. The ionization potential of Na is 5.14 eV, and the electron affinity of Cl is 3.61 eV. Use the simple model of NaCl as a pair of spherical ions in contact to estimate D_e and the dipole moment d of NaCl. Compare with the experimental values $D_e = 4.25$ eV and $d = 9.0$ D. [One debye (D) $= 10^{-18}$ statC cm.]

13.53 For a CI calculation of the H_2 ground electronic state, which of the following electron configurations will produce a CSF that will contribute to the wave function? (a) $(1\sigma_g)(2\sigma_g)$; (b) $(1\sigma_g)(2\sigma_u)$; (c) $(1\sigma_u)^2$; (d) $(1\pi_u)(1\pi_g)$; (e) $(1\pi_u)(3\sigma_u)$; (f) $(1\pi_u)^2$; (g) $(1\pi_u)(2\pi_u)$.

13.54 Verify the expression given for the CI ground-state wave function in the Section 13.21 example.

13.55 In the Section 13.21 CI example, verify the expressions given in terms of integrals over the basis functions for (a) $\langle \Phi_2|\hat{H}|\Phi_2 \rangle$; (b) $\langle \Phi_2|\hat{H}|\Phi_1 \rangle$.

13.56 For a CASSCF wave function in which the active orbitals are taken as those that arise from the $2s$ and $2p$ AOs, state the number of active electrons and the maximum number of electrons excited to virtual orbitals in the ground-state wave function for (a) C_2; (b) N_2; (c) O_2; (d) F_2.

14

The Virial Theorem and the Hellmann–Feynman Theorem

14.1 THE VIRIAL THEOREM

This chapter discusses two theorems that aid in understanding chemical bonding. We begin with the virial theorem.

Let $\hat{H}$ be the (time-independent) Hamiltonian of a system in the bound stationary state ψ:

$$\hat{H}\psi = E\psi \tag{14.1}$$

Let $\hat{A}$ be a linear, time-independent operator. Consider the integral

$$\int \psi^*[\hat{H}, \hat{A}]\psi \, d\tau = \langle \psi|\hat{H}\hat{A} - \hat{A}\hat{H}|\psi \rangle = \langle \psi|\hat{H}|\hat{A}\psi \rangle - E\langle \psi|\hat{A}|\psi \rangle \tag{14.2}$$

where (14.1) was used. Since $\hat{H}$ is Hermitian, we have

$$\langle \psi|\hat{H}|\hat{A}\psi \rangle = \langle \hat{A}\psi|\hat{H}|\psi \rangle^* = E^*\langle \hat{A}\psi|\psi \rangle^* = E\langle \psi|\hat{A}\psi \rangle = E\langle \psi|\hat{A}|\psi \rangle$$

and Eq. (14.2) becomes

$$\int \psi^*[\hat{H}, \hat{A}]\psi \, d\tau = 0 \tag{14.3}$$

Equation (14.3) is the *hypervirial theorem*. [For some of its applications, see J. O. Hirschfelder, *J. Chem. Phys.*, **33**, 1462 (1960); J. H. Epstein and S. T. Epstein, *Am. J. Phys.*, **30**, 266 (1962).] In deriving (14.3), we used the Hermitian property of $\hat{H}$. The proof that $\hat{p}_x$ and $\hat{p}_x^2$ are Hermitian, and hence that $\hat{H}$ is Hermitian, requires that ψ vanish at $\pm\infty$ [see Eq. (7.17)]. Hence the hypervirial theorem does not apply to continuum stationary states, for which ψ does not vanish at ∞.

We now derive the virial theorem from (14.3). We choose $\hat{A}$ to be

$$\sum_i \hat{q}_i \hat{p}_i = -i\hbar \sum_i q_i \frac{\partial}{\partial q_i} \tag{14.4}$$

where the sum runs over the $3n$ Cartesian coordinates of the n particles. (Particle 1 has Cartesian coordinates q_1, q_2, q_3 and corresponding linear-momentum components p_1, p_2, p_3. Throughout this chapter the symbol q will indicate a *Cartesian* coordinate.) We want to evaluate $[\hat{H}, \hat{A}]$. Using (5.4), (5.5), (5.8), and (5.9), we get

$$\left[\hat{H}, \sum_i \hat{q}_i \hat{p}_i\right] = \sum_i [\hat{H}, \hat{q}_i \hat{p}_i] = \sum_i \hat{q}_i [\hat{H}, \hat{p}_i] + \sum_i [\hat{H}, \hat{q}_i]\hat{p}_i$$

$$= i\hbar \sum_i q_i \frac{\partial V}{\partial q_i} - i\hbar \sum_i \frac{1}{m_i} \hat{p}_i^2 = i\hbar \sum_i q_i \frac{\partial V}{\partial q_i} - 2i\hbar \hat{T} \quad (14.5)$$

where $\hat{T}$ and $\hat{V}$ are the kinetic- and potential-energy operators for the system. Substitution of (14.5) into (14.3) gives

$$\left\langle \psi \left| \sum_i q_i \frac{\partial V}{\partial q_i} \right| \psi \right\rangle = 2\langle \psi | \hat{T} | \psi \rangle \quad (14.6)$$

Using $\langle B \rangle$ for the quantum-mechanical average of B, we write (14.6) as

$$\left\langle \sum_i q_i \frac{\partial V}{\partial q_i} \right\rangle = 2\langle T \rangle \quad (14.7)$$

Equation (14.7) is the quantum-mechanical **virial theorem**. Note that its validity is restricted to bound stationary states. [The name of the theorem comes from a similar theorem in classical mechanics, which has the same form as (14.7) except that the averages are time averages. The word *vires* is Latin for "forces"; the derivatives of the potential energy give the negatives of the force components.]

For certain systems the form of V causes the virial theorem to take on a particularly simple form. To discuss these systems, we introduce the idea of a homogeneous function. A function $f(x_1, x_2, \ldots, x_j)$ of several variables is said to be **homogeneous of degree** n if it satisfies

$$f(sx_1, sx_2, \ldots, sx_j) = s^n f(x_1, x_2, \ldots, x_j) \quad \textbf{(14.8)*}$$

where s is an arbitrary parameter. For example, the function

$$g = \frac{1}{x^3} + \frac{1}{y^3} + \frac{1}{z^3} + \frac{x}{y^2 z^2}$$

is homogeneous of degree -3, since

$$g(sx, sy, sz) = \frac{1}{s^3 x^3} + \frac{1}{s^3 y^3} + \frac{1}{s^3 z^3} + \frac{sx}{s^2 y^2 s^2 z^2} = s^{-3} g(x, y, z)$$

Euler's theorem on homogeneous functions states that, if $f(x_1, \ldots, x_j)$ is homogeneous of degree n, then

$$\sum_{k=1}^{j} x_k \frac{\partial f}{\partial x_k} = nf \quad (14.9)$$

The theorem is proved as follows. Let

$$u_1 \equiv sx_1, \quad u_2 \equiv sx_2, \quad \ldots, \quad u_j \equiv sx_j$$

Use of the chain rule gives for the partial derivative of the left side of (14.8) with respect to s

$$\frac{\partial f(u_1, \ldots, u_j)}{\partial s} = \frac{\partial f}{\partial u_1}\frac{\partial u_1}{\partial s} + \frac{\partial f}{\partial u_2}\frac{\partial u_2}{\partial s} + \cdots + \frac{\partial f}{\partial u_j}\frac{\partial u_j}{\partial s}$$

$$= x_1 \frac{\partial f}{\partial u_1} + x_2 \frac{\partial f}{\partial u_2} + \cdots + x_j \frac{\partial f}{\partial u_j} = \sum_{k=1}^{j} x_k \frac{\partial f}{\partial u_k}$$

The partial derivative of Eq. (14.8) with respect to s is thus

$$\sum_{k=1}^{j} x_k \frac{\partial f(u_1, \ldots, u_j)}{\partial u_k} = ns^{n-1} f(x_1, \ldots, x_j) \tag{14.10}$$

Now let $s = 1$, so that $u_i = x_i$; Eq. (14.10) then gives (14.9). This completes the proof. As an example of Euler's theorem, let $f = x^2 + y^2 + z^2$. We have

$$\sum_k x_k \frac{\partial f}{\partial x_k} = x \cdot 2x + y \cdot 2y + z \cdot 2z = 2f$$

Now we return to the virial theorem (14.7). If V happens to be a homogeneous function of degree n when expressed in Cartesian coordinates, then Euler's theorem gives

$$\sum_i q_i \frac{\partial V}{\partial q_i} = nV \tag{14.11}$$

and the virial theorem (14.7) simplifies to

$$2\langle T \rangle = n\langle V \rangle \tag{14.12}*$$

for a bound stationary state. Since (Problem 6.29)

$$\langle T \rangle + \langle V \rangle = E \tag{14.13}$$

we can write (14.12) in two other forms:

$$\langle V \rangle = \frac{2E}{n+2} \tag{14.14}$$

$$\langle T \rangle = \frac{nE}{n+2} \tag{14.15}$$

EXAMPLE Apply the virial theorem to (a) the one-dimensional harmonic oscillator; (b) the hydrogen atom; (c) a many-electron atom.

(a) For the one-dimensional harmonic oscillator, $V = \frac{1}{2}kx^2$, which is homogeneous of degree $n = 2$. Equations (14.12) and (14.14) give

$$\langle T \rangle = \langle V \rangle = \frac{1}{2}E = \frac{1}{2}h\nu(v + \frac{1}{2}) \tag{14.16}$$

This was verified for the ground state in Problem 4.11.

(b) For the H atom, $V = -e'^2/(x^2 + y^2 + z^2)^{1/2}$ in Cartesian coordinates. V is a homogeneous function of degree -1. Hence

$$2\langle T \rangle = -\langle V \rangle \tag{14.17}$$

which was verified for the ground state in Problem 6.30. For any hydrogen-atom stationary state,

$$\langle V \rangle = 2E \quad \text{and} \quad \langle T \rangle = -E \qquad (14.18)$$

(c) For a many-electron atom with spin–orbit interaction neglected,

$$V = -Ze'^2 \sum_{i=1}^{n} \frac{1}{(x_i^2 + y_i^2 + z_i^2)^{1/2}} + \sum_i \sum_{j>i} \frac{e'^2}{[(x_i - x_j)^2 + (y_i - y_j)^2 + (z_i - z_j)^2]^{1/2}}$$

Replacing each of the 3n coordinates by s times the coordinate, we find that V is homogeneous of degree -1. Hence Eqs. (14.17) and (14.18) hold for any atom.

Now consider molecules. In the Born–Oppenheimer approximation, the molecular wave function is [Eq. (13.12)]

$$\psi = \psi_{el}(q_i; q_\alpha)\psi_N(q_\alpha)$$

where q_i and q_α symbolize the electronic and nuclear coordinates, respectively. ψ_{el} is found by solving the electronic Schrödinger equation (13.7):

$$\hat{H}_{el}\psi_{el}(q_i; q_\alpha) = E_{el}(q_\alpha)\psi_{el}(q_i; q_\alpha)$$

where E_{el} is the electronic energy and where

$$\hat{H}_{el} = \hat{T}_{el} + \hat{V}_{el} \qquad (14.19)$$

$$\hat{T}_{el} = -\frac{\hbar^2}{2m_e} \sum_i \left(\frac{\partial^2}{\partial x_i^2} + \frac{\partial^2}{\partial y_i^2} + \frac{\partial^2}{\partial z_i^2} \right) \qquad (14.20)$$

$$\hat{V}_{el} = -\sum_\alpha \sum_i \frac{Z_\alpha e'^2}{[(x_i - x_\alpha)^2 + (y_i - y_\alpha)^2 + (z_i - z_\alpha)^2]^{1/2}}$$

$$+ \sum_i \sum_{j>i} \frac{e'^2}{[(x_i - x_j)^2 + (y_i - y_j)^2 + (z_i - z_j)^2]^{1/2}} \qquad (14.21)$$

Let the system be in the electronic stationary state ψ_{el}. If we put the subscript el on $\hat{H}$ and ψ in (14.1) and regard the variables of ψ_{el} to be the electronic coordinates q_i (with the nuclear coordinates q_α being parameters), then the derivation of the virial theorem (14.7) is seen to be valid for the electronic kinetic- and potential-energy operators, and we have

$$2\langle \psi_{el}|\hat{T}_{el}|\psi_{el}\rangle = \left\langle \psi_{el}\left| \sum_i q_i \frac{\partial V_{el}}{\partial q_i}\right|\psi_{el}\right\rangle \qquad (14.22)$$

Viewed as a function of the electronic coordinates, V_{el} is not a homogeneous function, since

$$\frac{1}{[(sx_i - x_\alpha)^2 + (sy_i - y_\alpha)^2 + (sz_i - z_\alpha)^2]^{1/2}}$$

$$\neq \frac{1}{s[(x_i - x_\alpha)^2 + (y_i - y_\alpha)^2 + (z_i - z_\alpha)^2]^{1/2}}$$

Thus the virial theorem for the average electronic kinetic and potential energies of a molecule will not have the simple form (14.17), which holds for atoms. We can, however, view V_{el} as a function of both the electronic and the nuclear Cartesian coordinates. From this viewpoint V_{el} *is* a homogeneous function of degree -1, since

$$\frac{1}{[(sx_i - sx_\alpha)^2 + (sy_i - sy_\alpha)^2 + (sz_i - sz_\alpha)^2]^{1/2}}$$
$$= \frac{1}{s[(x_i - x_\alpha)^2 + (y_i - y_\alpha)^2 + (z_i - z_\alpha)^2]^{1/2}}$$

Therefore, considering V_{el} as a function of both electronic and nuclear coordinates and applying Euler's theorem (14.9), we have

$$\sum_i q_i \frac{\partial V_{el}}{\partial q_i} + \sum_\alpha q_\alpha \frac{\partial V_{el}}{\partial q_\alpha} = -V_{el}$$

Using this equation in (14.22), we get

$$2\langle \psi_{el}|\hat{T}_{el}|\psi_{el}\rangle = -\langle \psi_{el}|\hat{V}_{el}|\psi_{el}\rangle - \left\langle \psi_{el}\left|\sum_\alpha q_\alpha \frac{\partial V_{el}}{\partial q_\alpha}\right|\psi_{el}\right\rangle \qquad (14.23)$$

which contains an additional term as compared to the atomic virial theorem (14.17).

Consider this extra term. We have

$$\left\langle \psi_{el}\left|\sum_\alpha q_\alpha \frac{\partial V_{el}}{\partial q_\alpha}\right|\psi_{el}\right\rangle = \sum_\alpha q_\alpha \int \psi_{el}^* \frac{\partial V_{el}}{\partial q_\alpha} \psi_{el}\, d\tau_{el}$$

where the nuclear coordinate q_α has been taken outside the integral over electronic coordinates. In Section 14.3 we shall show that [see the sentence after Eq. (14.72)]

$$\int \psi_{el}^* \frac{\partial V_{el}}{\partial q_\alpha} \psi_{el}\, d\tau_{el} = \frac{\partial E_{el}}{\partial q_\alpha} \qquad (14.24)$$

[Equation (14.24) is an example of the Hellmann–Feynman theorem.] Using these last two equations in the molecular electronic virial theorem (14.23), we get

$$2\langle \psi_{el}|\hat{T}_{el}|\psi_{el}\rangle = -\langle \psi_{el}|\hat{V}_{el}|\psi_{el}\rangle - \sum_\alpha q_\alpha \frac{\partial E_{el}}{\partial q_\alpha}$$

where the q_α are the nuclear *Cartesian* coordinates. We abbreviate this equation as

$$2\langle T_{el}\rangle = -\langle V_{el}\rangle - \sum_\alpha q_\alpha \frac{\partial E_{el}}{\partial q_\alpha} \qquad (14.25)$$

Using

$$\langle T_{el}\rangle + \langle V_{el}\rangle = E_{el} \qquad (14.26)$$

we can eliminate either $\langle T_{el}\rangle$ or $\langle V_{el}\rangle$ from (14.25).

Now consider a diatomic molecule. The electronic energy is a function of R, the internuclear distance: $E_{el} = E_{el}(R)$. The summation in (14.25) is over the nuclear Cartesian coordinates x_a, y_a, z_a, x_b, y_b, z_b. We have

$$\frac{\partial E_{el}}{\partial x_a} = \frac{dE_{el}}{dR}\frac{\partial R}{\partial x_a}, \qquad \frac{\partial E_{el}}{\partial x_b} = \frac{dE_{el}}{dR}\frac{\partial R}{\partial x_b}$$

$$R = [(x_a - x_b)^2 + (y_a - y_b)^2 + (z_a - z_b)^2]^{1/2}$$

$$\frac{\partial R}{\partial x_a} = \frac{x_a - x_b}{R}, \qquad \frac{\partial R}{\partial x_b} = \frac{x_b - x_a}{R} \qquad (14.27)$$

with similar equations for the y and z coordinates. The sum in (14.25) becomes

$$\sum_\alpha q_\alpha \frac{\partial E_{el}}{\partial q_\alpha} = \frac{1}{R}\frac{dE_{el}}{dR}[x_a(x_a - x_b) + x_b(x_b - x_a) + y_a(y_a - y_b)$$

$$+ y_b(y_b - y_a) + z_a(z_a - z_b) + z_b(z_b - z_a)]$$

$$\sum_\alpha q_\alpha \frac{\partial E_{el}}{\partial q_\alpha} = R\frac{dE_{el}}{dR}$$

The virial theorem (14.25) for a diatomic molecule becomes

$$2\langle T_{el}\rangle = -\langle V_{el}\rangle - R\frac{dE_{el}}{dR} \qquad (14.28)$$

Using (14.26), we have the two alternative forms

$$\langle T_{el}\rangle = -E_{el} - R\frac{dE_{el}}{dR} \qquad (14.29)$$

$$\langle V_{el}\rangle = 2E_{el} + R\frac{dE_{el}}{dR} \qquad (14.30)$$

In deriving the molecular electronic virial theorem (14.25), we omitted the internuclear repulsion

$$V_{NN} = \sum_\beta \sum_{\alpha > \beta} \frac{Z_\alpha Z_\beta e'^2}{[(x_\alpha - x_\beta)^2 + (y_\alpha - y_\beta)^2 + (z_\alpha - z_\beta)^2]^{1/2}} \qquad (14.31)$$

from the electronic Hamiltonian (14.19) to (14.21). Let

$$V = V_{el} + V_{NN}$$

where V_{el} is given by (14.21). We can rewrite the electronic Schrödinger equation $\hat{H}_{el}\psi_{el} = E_{el}\psi_{el}$ as [Eq. (13.4)]

$$(\hat{T}_{el} + \hat{V})\psi_{el} = U(q_\alpha)\psi_{el}$$

where

$$U(q_\alpha) = E_{el}(q_\alpha) + V_{NN}$$

$U(q_\alpha)$ is the potential-energy function for nuclear motion. Now consider what happens to the right side of Eq. (14.25) when we add in V_{NN} to V_{el} and E_{el}. We have

$$-\int \psi_{el}^*(\hat{V}_{el} + V_{NN})\psi_{el}\, d\tau_{el} - \sum_\alpha q_\alpha \frac{\partial U}{\partial q_\alpha}$$

$$= -\langle \psi_{el}|\hat{V}_{el}|\psi_{el}\rangle - V_{NN} - \sum_\alpha q_\alpha \frac{\partial E_{el}}{\partial q_\alpha} - \sum_\alpha q_\alpha \frac{\partial V_{NN}}{\partial q_\alpha} \qquad (14.32)$$

Since V_{NN} is a homogeneous function of the nuclear Cartesian coordinates of degree -1, Euler's theorem gives

$$\sum_\alpha q_\alpha \frac{\partial V_{NN}}{\partial q_\alpha} = -V_{NN}$$

and (14.32) becomes

$$-\langle \psi_{el} | \hat{V}_{el} + V_{NN} | \psi_{el} \rangle - \sum_\alpha q_\alpha \frac{\partial U}{\partial q_\alpha} = -\langle \psi_{el} | \hat{V}_{el} | \psi_{el} \rangle - \sum_\alpha q_\alpha \frac{\partial E_{el}}{\partial q_\alpha} \qquad (14.33)$$

Substitution of (14.33) into (14.25) gives

$$2\langle \psi_{el} | \hat{T}_{el} | \psi_{el} \rangle = -\langle \psi_{el} | \hat{V}_{el} + V_{NN} | \psi_{el} \rangle - \sum_\alpha q_\alpha \frac{\partial U}{\partial q_\alpha}$$

$$2\langle T_{el} \rangle = -\langle V \rangle - \sum_\alpha q_\alpha \frac{\partial U}{\partial q_\alpha} \qquad (14.34)$$

so the molecular electronic virial theorem holds whether or not we include the internuclear repulsion. Corresponding to Eqs. (14.28) to (14.30) for diatomic molecules, we have

$$2\langle T_{el} \rangle = -\langle V \rangle - R(dU/dR) \qquad (14.35)$$

$$\langle T_{el} \rangle = -U - R(dU/dR) \qquad (14.36)$$

$$\langle V \rangle = 2U + R(dU/dR) \qquad (14.37)$$

The potential energy $V = V_{el} + V_{NN}$ takes the zero of energy with all particles (electrons and nuclei) at infinite separation from one another. Therefore, $U(R)$ in (14.35) to (14.37) does not go to zero at $R = \infty$ but goes to the sum of the energies of the separated atoms, which is negative.

The molecular electronic virial theorem was first derived by Slater.

For polyatomic molecules, (14.34) can be written as

$$2\langle T_{el} \rangle = -\langle V \rangle - \sum_\alpha \sum_{\beta > \alpha} R_{\alpha\beta}(\partial U/\partial R_{\alpha\beta})$$

where the sum runs over either all internuclear distances or the bond lengths only; proofs are given in R. G. Parr and J. E. Brown, *J. Chem. Phys.*, **49**, 4849 (1968) and B. Nelander, *J. Chem. Phys.*, **51**, 469 (1969).

The true wave functions for a system with V a homogeneous function of the coordinates must satisfy the form of the virial theorem (14.12). We now ask: What determines whether an *approximate* wave function for such a system satisfies (14.12)? The answer is that, by inserting a variational parameter as a multiplier of each Cartesian coordinate and choosing this parameter to minimize the variational integral, we can make any trial variation function satisfy the virial theorem. (For the proof, see *Kauzmann*, page 229.) This process is called *scaling*, and the variational parameter multiplying each coordinate is called a *scale factor*. For a molecular trial function, the scaling parameter must be inserted in front of the nuclear Cartesian coordinates, as well as in front of the electronic coordinates.

Consider some examples. The zeroth-order perturbation wave function (9.49) for the heliumlike atom has no scale factor and so does not satisfy the virial theorem. If we were to calculate $\langle T \rangle$ and $\langle V \rangle$ for (9.49), we would find $2\langle T \rangle \neq -\langle V \rangle$; see Problem 14.4. The Heitler–London trial function for H_2, Eq. (13.101), has no scale factor and does not satisfy the virial theorem. The Heitler–London–Wang function, which uses a variationally determined orbital exponent, satisfies the virial theorem. Hartree–Fock wave functions satisfy the virial theorem; note the scale factor in the Slater basis functions (11.14).

14.2   THE VIRIAL THEOREM AND CHEMICAL BONDING

We now use the virial theorem to examine the changes in electronic kinetic and potential energy that occur when a covalent chemical bond is formed in a diatomic molecule. For formation of a stable bond, the $U(R)$ curve must have a substantial minimum. At this minimum we have

$$\left. \frac{dU}{dR} \right|_{R_e} = 0 \tag{14.38}$$

and Eqs. (14.35) to (14.37) become

$$2\langle T_{el} \rangle |_{R_e} = -\langle V \rangle |_{R_e} \tag{14.39}$$

$$\langle T_{el} \rangle |_{R_e} = -U(R_e) \tag{14.40}$$

$$\langle V \rangle |_{R_e} = 2U(R_e) \tag{14.41}$$

These equations resemble those for atoms [Eqs. (14.17) and (14.18)]. At $R = \infty$ we have the separated atoms, and the atomic virial theorem gives

$$2\langle T_{el} \rangle |_{\infty} = -\langle V \rangle |_{\infty} \tag{14.42}$$

$$\langle T_{el} \rangle |_{\infty} = -U(\infty) \tag{14.43}$$

$$\langle V \rangle |_{\infty} = 2U(\infty) \tag{14.44}$$

$U(\infty)$ is the sum of the energies of the two separated atoms. We have

$$\langle T_{el} \rangle |_{R_e} - \langle T_{el} \rangle |_{\infty} = U(\infty) - U(R_e) \tag{14.45}$$

$$\langle V \rangle |_{R_e} - \langle V \rangle |_{\infty} = 2[U(R_e) - U(\infty)] \tag{14.46}$$

For bonding, we have $U(R_e) < U(\infty)$. Therefore, Eqs. (14.45) and (14.46) show that the average molecular potential energy at R_e is *less* than the sum of the potential energies of the separated atoms, while the average molecular kinetic energy is *greater* at R_e than at ∞. The decrease in potential energy is twice the increase in kinetic energy, and results from allowing the electrons to feel the attractions of both nuclei and perhaps from an increase in orbital exponents in the molecule (Section 13.5). The equilibrium dissociation energy (13.9) is $D_e = \frac{1}{2}[\langle V \rangle |_{\infty} - \langle V \rangle |_{R_e}]$.

Consider the behavior of the average potential and kinetic energies for large R. The interactions between atoms at large distances are called *van der Waals forces*. For two neutral atoms, at least one of which is in an S state, quantum-mechanical perturbation theory shows that the van der Waals force of attraction is proportional to $1/R^7$, and the potential energy behaves like

$$U(R) \approx U(\infty) - \frac{A}{R^6} , \qquad R \text{ large} \qquad (14.47)$$

where A is a positive constant. (See *Kauzmann*, Chapter 13.) This expression was first derived by London, and van der Waals forces between neutral atoms are called *London forces* or *dispersion forces*. (Recall the discussion near the end of Section 13.7.)

Using (14.47), (14.36), (14.37), (14.43), and (14.44), we have

$$\langle V \rangle \approx \langle V \rangle|_\infty + \frac{4A}{R^6} , \qquad R \text{ large} \qquad (14.48)$$

$$\langle T_{el} \rangle \approx \langle T_{el} \rangle|_\infty - \frac{5A}{R^6} , \qquad R \text{ large} \qquad (14.49)$$

Hence, as R decreases from infinity, the average potential energy at first increases, while the average kinetic energy at first decreases. The combination of these conclusions with our conclusions about $\langle V \rangle|_{R_e}$ and $\langle T_{el} \rangle|_{R_e}$ shows that $\langle V \rangle$ must go through a maximum somewhere between R_e and infinity and $\langle T_{el} \rangle$ must go through a minimum in this region.

For R much less than R_e, we can use (13.84) and (13.8) to write

$$U(R) \approx \frac{Z_a Z_b e'^2}{R} + E_{UA} + aR^2 , \qquad R \text{ small} \qquad (14.50)$$

where E_{UA} is the united-atom energy. The virial theorem then gives

$$\langle T_{el} \rangle \approx -E_{UA} - 3aR^2 , \qquad R \text{ small}$$

$$\langle V \rangle \approx \frac{Z_a Z_b e'^2}{R} + 2E_{UA} + 4aR^2 , \qquad R \text{ small}$$

The virial theorem holds for the united atom:

$$\langle T_{el} \rangle|_0 = -E_{UA} , \qquad \langle V_{el} \rangle|_0 = 2E_{UA}$$

and we have

$$\langle T_{el} \rangle \approx \langle T_{el} \rangle|_0 - 3aR^2 , \qquad R \text{ small} \qquad (14.51)$$

$$\langle V \rangle \approx \frac{Z_a Z_b e'^2}{R} + \langle V_{el} \rangle|_0 + 4aR^2 , \qquad R \text{ small} \qquad (14.52)$$

The average value of $\langle V \rangle$ goes to infinity as R goes to zero, because of the internuclear repulsion.

Having found the general behavior of $\langle V \rangle$ and $\langle T_{el} \rangle$ as functions of R, we now draw Fig. 14.1. This figure is not for any particular molecule but resembles the known curves for H_2 and H_2^+ [W. Kolos and L. Wolniewicz, *J. Chem. Phys.*, **41**, 3663 (1964); *Slater, Quantum Theory of Molecules and Solids*, Volume 1, page 36].

What explanation can we give for the change in average kinetic and

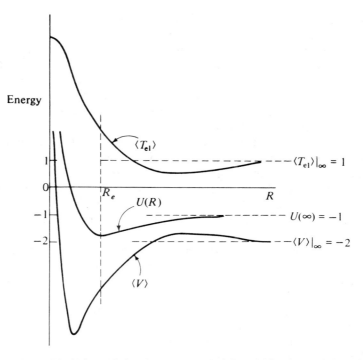

Figure 14.1 Variation of the average potential and kinetic energies of a diatomic molecule. The unit of energy is taken as the electronic kinetic energy of the separated atoms.

potential energy with R? Consider H_2^+. The electronic potential-energy function is

$$V_{el} = -\frac{e'^2}{r_a} - \frac{e'^2}{r_b} \tag{14.53}$$

If we plot the values of V_{el} for points on the molecular axis for a large value of R, we get a curve like Fig. 14.2, which resembles two hydrogen-atom potential-energy curves (Fig. 6.6) placed side by side. We have seen that the overlapping of

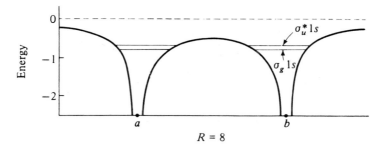

Figure 14.2 Potential energy along the internuclear axis for electronic motion in H_2^+ at a large internuclear separation. Atomic units are used.

the $1s$ AOs occurring in molecule formation leads to a buildup of charge probability density between the nuclei for the ground state. However, Fig. 14.2 shows that the potential energy is relatively *high* in the region midway between the nuclei when R is large. Thus we get an initial increase in $\langle V \rangle$ as R decreases from infinity. Now consider the kinetic energy. The uncertainty principle (5.13) gives $(\Delta x)^2(\Delta p_x)^2 \geq \hbar^2/4$. For a stationary state, $\langle p_x \rangle$ is zero [see Eq. (3.92) and Problem 14.6] and (5.11) gives $(\Delta p_x)^2 = \langle p_x^2 \rangle$. Hence a small value of $(\Delta x)^2$ means a large value of $\langle p_x^2 \rangle$ and a large value of the average kinetic energy (which equals $\langle p^2 \rangle / 2m$). Thus a compact ψ_{el} corresponds to a large electronic kinetic energy. In the separated atoms, the wave function is concentrated in two rather small regions about each nucleus (Fig. 6.7). In the initial stages of molecule formation, the buildup of probability density between the nuclei corresponds to having a wave function that is less compact than it was in the separated atoms. Thus, as R decreases from infinity, the electronic kinetic energy initially decreases. The energies E_{el} of the two lowest H_2^+ states have been indicated in Fig. 14.2. For large R the region between the nuclei is classically forbidden, but it is accessible according to quantum mechanics (tunneling).

Now consider what happens as R decreases further. Plotting (14.53) for an intermediate value of R, we find that now the region between the nuclei is a region of *low* potential energy, since an electron in this region feels substantial attractions from both nuclei. (See Fig. 14.3.) Hence at intermediate values of R the overlap charge buildup between the nuclei lowers the potential energy. For intermediate values of R, the wave function has become more compact as compared with large R, which gives an increase in $\langle T_{el} \rangle$ as R is reduced. In fact, we see from Fig. 14.1 and Eq. (14.45) that $\langle T_{el} \rangle$ is *greater* at R_e in the molecule than in the separated atoms; hence the molecular wave function at R_e is more compact than the separated-atoms wave functions.

For very small R, the average potential energy goes to infinity, because of the internuclear repulsion. *However*, for $R = R_e$, Fig. 14.1 shows that $\langle V \rangle$ is still decreasing sharply with decreasing R, and it is the increase in $\langle T_{el} \rangle$, and not the nuclear repulsion, that causes the $U(R)$ curve to turn up as R becomes less than R_e. The squeezing of the molecular wave function into a smaller region with the associated increase in $\langle T_{el} \rangle$ is more important than the internuclear repulsion in causing the initial repulsion between the atoms.

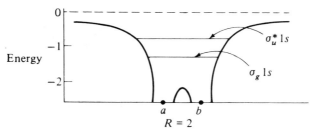

Figure 14.3 Potential energy along the internuclear axis for electronic motion in H_2^+ at an intermediate internuclear distance.

14.3   THE HELLMANN–FEYNMAN THEOREM

Consider a system with a time-independent Hamiltonian $\hat{H}$ that involves parameters. An obvious example is the molecular electronic Hamiltonian (13.5), which depends parametrically on the nuclear coordinates. However, the Hamiltonian of any system contains parameters. For example, the one-dimensional harmonic-oscillator Hamiltonian is

$$-\frac{\hbar^2}{2m}\frac{d^2}{dx^2} + \tfrac{1}{2}kx^2 \tag{14.54}$$

The force constant k is a parameter, as is the mass m. Although $\hbar$ is a constant, we can consider it as a parameter also. The stationary-state energies E_n are functions of the same parameters as $\hat{H}$. For example, for the harmonic oscillator

$$E_n = (v + \tfrac{1}{2})h\nu = (v + \tfrac{1}{2})\hbar(k/m)^{1/2} \tag{14.55}$$

The stationary-state wave functions also depend on the parameters in $\hat{H}$. We now investigate how E_n varies with each of the parameters. More specifically, let λ be one of the parameters; we ask for $\partial E_n/\partial\lambda$, where the partial derivative is taken with all other parameters held constant.

We begin with the Schrödinger equation

$$\hat{H}\psi_n = E_n\psi_n \tag{14.56}$$

where the ψ_n are the normalized stationary-state eigenfunctions. Because of normalization, we have

$$E_n = \int \psi_n^* \hat{H}\psi_n \, d\tau \tag{14.57}$$

$$\frac{\partial E_n}{\partial\lambda} = \frac{\partial}{\partial\lambda} \int \psi_n^* \hat{H}\psi_n \, d\tau \tag{14.58}$$

The integral in (14.57) is a definite integral over all space, and its value depends parametrically on λ, since $\hat{H}$ and ψ_n depend on λ. Provided the integrand is well behaved, we can find the integral's derivative with respect to a parameter by differentiating the integrand with respect to the parameter and then integrating. (Recall Problem 8.9b.) Thus

$$\frac{\partial E_n}{\partial\lambda} = \int \frac{\partial}{\partial\lambda} (\psi_n^* \hat{H}\psi_n) \, d\tau = \int \frac{\partial\psi_n^*}{\partial\lambda} \hat{H}\psi_n \, d\tau + \int \psi_n^* \frac{\partial}{\partial\lambda} (\hat{H}\psi_n) \, d\tau \tag{14.59}$$

We have

$$\frac{\partial}{\partial\lambda} (\hat{H}\psi_n) = \frac{\partial}{\partial\lambda} (\hat{T}\psi_n) + \frac{\partial}{\partial\lambda} (\hat{V}\psi_n) \tag{14.60}$$

The potential-energy operator is just multiplication by V, so

$$\frac{\partial}{\partial\lambda} (\hat{V}\psi_n) = \frac{\partial V}{\partial\lambda} \psi_n + V \frac{\partial\psi_n}{\partial\lambda} \tag{14.61}$$

The parameter λ will occur in the kinetic-energy operator as part of the factor multiplying one or more of the derivatives with respect to the coordinates. For example, taking λ as the mass of the particle, we have for a one-particle problem

$$\hat{T} = -\frac{\hbar^2}{2\lambda}\left(\frac{\partial^2}{\partial x^2} + \frac{\partial^2}{\partial y^2} + \frac{\partial^2}{\partial z^2}\right)$$

$$\frac{\partial}{\partial\lambda}(\hat{T}\psi) = -\frac{\hbar^2}{2}\frac{\partial}{\partial\lambda}\left[\frac{1}{\lambda}\left(\frac{\partial^2\psi}{\partial x^2} + \frac{\partial^2\psi}{\partial y^2} + \frac{\partial^2\psi}{\partial z^2}\right)\right]$$

$$= \frac{\hbar^2}{2\lambda^2}\left(\frac{\partial^2\psi}{\partial x^2} + \frac{\partial^2\psi}{\partial y^2} + \frac{\partial^2\psi}{\partial z^2}\right) - \frac{\hbar^2}{2\lambda}\left(\frac{\partial^2}{\partial x^2} + \frac{\partial^2}{\partial y^2} + \frac{\partial^2}{\partial z^2}\right)\left(\frac{\partial\psi}{\partial\lambda}\right)$$

since we can change the order of the partial differentiations without affecting the result. We can write this last equation as

$$\frac{\partial}{\partial\lambda}(\hat{T}\psi_n) = \left(\frac{\partial\hat{T}}{\partial\lambda}\right)\psi_n + \hat{T}\left(\frac{\partial\psi_n}{\partial\lambda}\right) \tag{14.62}$$

where $(\partial\hat{T}/\partial\lambda)$ is found by differentiating $\hat{T}$ with respect to λ just as if it were a function instead of an operator. Although we got (14.62) by considering a specific $\hat{T}$ and λ, the same arguments show it to be generally valid. Combining (14.62) and (14.61), we write

$$\frac{\partial}{\partial\lambda}(\hat{H}\psi_n) = \left(\frac{\partial\hat{H}}{\partial\lambda}\right)\psi_n + \hat{H}\left(\frac{\partial\psi_n}{\partial\lambda}\right) \tag{14.63}$$

Equation (14.59) becomes

$$\frac{\partial E_n}{\partial\lambda} = \int \frac{\partial\psi_n^*}{\partial\lambda}\hat{H}\psi_n\,d\tau + \int \psi_n^*\frac{\partial\hat{H}}{\partial\lambda}\psi_n\,d\tau + \int \psi_n^*\hat{H}\frac{\partial\psi_n}{\partial\lambda}\,d\tau \tag{14.64}$$

For the first integral in (14.64), we have

$$\int \frac{\partial\psi_n^*}{\partial\lambda}\hat{H}\psi_n\,d\tau = E_n\int \frac{\partial\psi_n^*}{\partial\lambda}\psi_n\,d\tau \tag{14.65}$$

The Hermitian property of $\hat{H}$ and (14.56) give for the last integral in (14.64)

$$\int \psi_n^*\hat{H}\frac{\partial\psi_n}{\partial\lambda}\,d\tau = \int \frac{\partial\psi_n}{\partial\lambda}(\hat{H}\psi_n)^*\,d\tau = E_n\int \psi_n^*\frac{\partial\psi_n}{\partial\lambda}\,d\tau$$

Therefore,

$$\frac{\partial E_n}{\partial\lambda} = \int \psi_n^*\frac{\partial\hat{H}}{\partial\lambda}\psi_n\,d\tau + E_n\int \frac{\partial\psi_n^*}{\partial\lambda}\psi_n\,d\tau + E_n\int \psi_n^*\frac{\partial\psi_n}{\partial\lambda}\,d\tau \tag{14.66}$$

The wave function is normalized; hence

$$\int \psi_n^*\psi_n\,d\tau = 1$$

$$\frac{\partial}{\partial\lambda}\int \psi_n^*\psi_n\,d\tau = 0$$

$$\int \frac{\partial\psi_n^*}{\partial\lambda}\psi_n\,d\tau + \int \psi_n^*\frac{\partial\psi_n}{\partial\lambda}\,d\tau = 0 \tag{14.67}$$

Using (14.67) in (14.66), we obtain

$$\frac{\partial E_n}{\partial \lambda} = \int \psi_n^* \frac{\partial \hat{H}}{\partial \lambda} \psi_n \, d\tau \qquad (14.68)^*$$

Equation (14.68) is called the **generalized Hellmann–Feynman theorem**. [For a discussion of the origin of the Hellmann–Feynman and related theorems, see J. I. Musher, *Am. J. Phys.*, **34**, 267 (1966).]

EXAMPLE   Apply the generalized Hellmann–Feynman theorem to the one-dimensional harmonic oscillator with λ taken as the force constant.

For the harmonic oscillator, $\hat{H} = -(\hbar^2/2m)(d^2/dx^2) + \frac{1}{2}kx^2$ and $\partial \hat{H}/\partial k = \frac{1}{2}x^2$. The energy levels are $E_v = (v + \frac{1}{2})h\nu = (v + \frac{1}{2})h(k/m)^{1/2}/2\pi$. We have $\partial E_v/\partial k = \frac{1}{2}(v + \frac{1}{2})hk^{-1/2}m^{-1/2}/2\pi = \frac{1}{2}(v + \frac{1}{2})h\nu/k$. Substitution in (14.68) gives

$$\int_{-\infty}^{\infty} \psi_v^* x^2 \psi_v \, dx = (v + \frac{1}{2})h\nu/k \qquad (14.69)$$

We have found the average value of x^2 for any harmonic-oscillator stationary state without evaluating any integrals. This result was also obtained from the virial theorem; see Eq. (14.16). For a third derivation, see *Eyring, Walter, and Kimball*, page 79.

Application of the Hellmann–Feynman theorem to the hydrogenlike atom, with Z as the parameter, gives (Problem 14.11a)

$$\int r^{-1} |\psi|^2 \, d\tau = \left\langle \frac{1}{r} \right\rangle = \frac{Z}{n^2} \left(\frac{1}{a_0} \right) \qquad (14.70)$$

This result was also obtained from the virial theorem; see Eq. (14.18).

The first-order perturbation theory result Eq. (9.22) is a special case of the Hellmann–Feynman theorem (Problem 14.13).

The Hellmann–Feynman theorem with E_n being the Hartree–Fock energy is obeyed by Hartree–Fock (as well as exact) wave functions. [See R. E. Stanton, *J. Chem. Phys.*, **36**, 1298 (1962).]

14.4 THE ELECTROSTATIC THEOREM

Hellmann and Feynman independently applied Eq. (14.68) to molecules, taking λ as a nuclear Cartesian coordinate. We now consider their results.

[Hellmann was a German physical chemist. The Nazis had him dismissed from his position, and he then settled in Russia. He was executed as a "German spy" in the Stalinist purges of the late 1930s. His results appear in his textbook *Einführung in die Quantenchemie*, Deuticke, Leipzig, 1937, page 285. The American physicist Richard Feynman is noted for his contributions to quantum field theory. His paper "Forces in Molecules," *Phys. Rev.*, **56**, 340 (1939), consists of work he did as an undergraduate at MIT.]

As usual, we are using the Born–Oppenheimer approximation, solving the electronic Schrödinger equation for a fixed nuclear configuration [Eq. (13.4)]:

$$\hat{H}\psi_{\text{el}} = (\hat{T}_{\text{el}} + \hat{V})\psi_{\text{el}} = U\psi_{\text{el}}$$

The potential energy V is

$$\hat{V} = \hat{V}_{el} + \hat{V}_{NN}$$

$\hat{T}_{el}$, $\hat{V}_{el}$, and $\hat{V}_{NN}$ are given by (14.20), (14.21), and (14.31). The Hamiltonian $\hat{H}$ depends on the nuclear Cartesian coordinates as parameters. If x_δ is the x coordinate of nucleus δ, the generalized Hellmann–Feynman theorem (14.68) gives

$$\frac{\partial U}{\partial x_\delta} = \int \psi_{el}^* \frac{\partial \hat{H}}{\partial x_\delta} \psi_{el} \, d\tau_{el} \tag{14.71}$$

The kinetic-energy part of $\hat{H}$ is independent of the nuclear Cartesian coordinates, as can be seen from (14.20). Hence

$$\frac{\partial U}{\partial x_\delta} = \int \psi_{el}^* \frac{\partial V}{\partial x_\delta} \psi_{el} \, d\tau_{el} \tag{14.72}$$

[If we had omitted V_{NN} from V, we would have obtained Eq. (14.24), which was used in deriving the molecular electronic virial theorem.] We have

$$\frac{\partial V}{\partial x_\delta} = \frac{\partial V_{el}}{\partial x_\delta} + \frac{\partial V_{NN}}{\partial x_\delta} \tag{14.73}$$

From (14.21) we get

$$\frac{\partial V_{el}}{\partial x_\delta} = -\sum_i \frac{Z_\delta(x_i - x_\delta)e'^2}{r_{i\delta}^3} \tag{14.74}$$

where $r_{i\delta}$ is the distance from electron i to nucleus δ. To evaluate $\partial V_{NN}/\partial x_\delta$, we need consider only those internuclear repulsion terms that involve nucleus δ; hence

$$\frac{\partial V_{NN}}{\partial x_\delta} = \frac{\partial}{\partial x_\delta} \sum_{\alpha \neq \delta} \frac{Z_\alpha Z_\delta e'^2}{[(x_\alpha - x_\delta)^2 + (y_\alpha - y_\delta)^2 + (z_\alpha - z_\delta)^2]^{1/2}} = \sum_{\alpha \neq \delta} Z_\alpha Z_\delta e'^2 \frac{x_\alpha - x_\delta}{R_{\alpha\delta}^3}$$

where $R_{\alpha\delta}$ is the distance between nuclei α and δ. Noting that $\partial V_{NN}/\partial x_\delta$ does not involve the electronic coordinates and that ψ_{el} is normalized, we see that (14.72) becomes

$$\frac{\partial U}{\partial x_\delta} = -\sum_i Z_\delta e'^2 \int |\psi_{el}|^2 \frac{x_i - x_\delta}{r_{i\delta}^3} \, d\tau_{el} + \sum_{\alpha \neq \delta} Z_\alpha Z_\delta e'^2 \frac{x_\alpha - x_\delta}{R_{\alpha\delta}^3} \tag{14.75}$$

Consider the integral in (14.75). The factor multiplying $|\psi_{el}|^2$ depends only on the coordinates of electron i, and this integral equals (recall that $\int d\tau$ implies a summation over all spin variables, as well as integration over all space coordinates)

$$\int_{-\infty}^{\infty} \int_{-\infty}^{\infty} \int_{-\infty}^{\infty} \left[\sum_{\text{all } m_s} \int' |\psi_{el}|^2 \, dv_{el}' \right] \frac{x_i - x_\delta}{r_{i\delta}^3} \, dx_i \, dy_i \, dz_i$$

where by $\int' dv_{el}'$ we mean an integration over the $3n - 3$ spatial coordinates of all electrons except electron i. From Eq. (13.128) we see that the quantity in brackets is equal to ρ/n, where ρ is the electron probability density of the

n-electron molecule. Hence (14.75) simplifies to

$$\frac{\partial U}{\partial x_\delta} = -\frac{1}{n} \sum_{i=1}^{n} Z_\delta e'^2 \int \rho(x_i, y_i, z_i) \frac{x_i - x_\delta}{r_{i\delta}^3} dv_i + \sum_{\alpha \neq \delta} Z_\alpha Z_\delta e'^2 \frac{x_\alpha - x_\delta}{R_{\alpha\delta}^3} \qquad (14.76)$$

The integration variables in each of the n definite integrals in (14.76) are dummy variables, and the integrals have the same value no matter what the value of i is. Thus each term in the first sum in (14.76) has the same value, and since there are n terms in this sum, we get

$$\frac{\partial U}{\partial x_\delta} = -Z_\delta e'^2 \int \int \int \rho(x, y, z) \frac{x - x_\delta}{r_\delta^3} dx\, dy\, dz + \sum_{\alpha \neq \delta} Z_\alpha Z_\delta e'^2 \frac{x_\alpha - x_\delta}{R_{\alpha\delta}^3} \qquad (14.77)$$

where the superfluous subscript i was dropped. The variable r_δ is the distance between nucleus δ and point (x, y, z) in space:

$$r_\delta = [(x - x_\delta)^2 + (y - y_\delta)^2 + (z - z_\delta)^2]^{1/2}$$

What is the significance of (14.77)? In the Born–Oppenheimer approximation, the function $U(x_\alpha, y_\alpha, z_\alpha, x_\beta, \ldots)$ is the potential-energy function for nuclear motion, the nuclear Schrödinger equation being

$$\left(-\frac{\hbar^2}{2} \sum_\alpha \frac{1}{m_\alpha} \nabla_\alpha^2 + U\right)\psi_N = E\psi_N$$

The quantity $-\partial U/\partial x_\delta$ can thus be viewed [see Eq. (5.32)] as the x component of the effective force on nucleus δ due to the other nuclei and the electrons. In addition to (14.77), we have two corresponding equations for $\partial U/\partial y_\delta$ and $\partial U/\partial z_\delta$; if $\mathbf{F}_\delta$ is the effective force on nucleus δ, then

$$\mathbf{F}_\delta = -\mathbf{i}\frac{\partial U}{\partial x_\delta} - \mathbf{j}\frac{\partial U}{\partial y_\delta} - \mathbf{k}\frac{\partial U}{\partial z_\delta} \qquad (14.78)$$

$$\mathbf{F}_\delta = -Z_\delta e'^2 \int \int \int \rho(x, y, z) \frac{\mathbf{r}_\delta}{r_\delta^3} dx\, dy\, dz + e'^2 \sum_{\alpha \neq \delta} Z_\alpha Z_\delta \frac{\mathbf{R}_{\alpha\delta}}{R_{\alpha\delta}^3} \qquad (14.79)$$

where $\mathbf{r}_\delta$ is the vector from point (x, y, z) to nucleus δ,

$$\mathbf{r}_\delta = \mathbf{i}(x_\delta - x) + \mathbf{j}(y_\delta - y) + \mathbf{k}(z_\delta - z)$$

and where $\mathbf{R}_{\alpha\delta}$ is the vector from nucleus α to nucleus δ:

$$\mathbf{R}_{\alpha\delta} = \mathbf{i}(x_\delta - x_\alpha) + \mathbf{j}(y_\delta - y_\alpha) + \mathbf{k}(z_\delta - z_\alpha)$$

Equation (14.79) has a simple physical interpretation. Let us imagine the electrons smeared out into a charge distribution whose density is $-e\rho(x, y, z)$. The force on nucleus δ exerted by the infinitesimal element of electronic charge $-e\rho\, dx\, dy\, dz$ is [Eq. (6.56)]

$$-Z_\delta e'^2 \frac{\mathbf{r}_\delta}{r_\delta^3} \rho\, dx\, dy\, dz \qquad (14.80)$$

and integration of (14.80) shows that the total force exerted on δ by this hypothetical electron smear is given by the first term on the right of (14.79). The second term on the right of (14.79) is clearly the Coulomb's law force on nucleus δ due to the electrostatic repulsions of the other nuclei.

Thus *the effective force acting on a nucleus in a molecule can be calculated by simple electrostatics as the sum of the Coulombic forces exerted by the other nuclei and by a hypothetical electron cloud whose charge density* $-e\rho(x, y, z)$ *is found by solving the electronic Schrödinger equation.* This statement is the **Hellmann–Feynman electrostatic theorem**. The electron probability density depends on the parameters defining the nuclear configuration:

$$\rho = \rho(x, y, z; x_\alpha, y_\alpha, z_\alpha, x_\beta, \ldots)$$

It is quite reasonable that the electrostatic theorem follows from the Born–Oppenheimer approximation, since the rapid motion of the electrons allows the electronic wave function and probability density to adjust immediately to changes in nuclear configuration; the rapid motion of the electrons causes the sluggish nuclei to "see" the electrons as a charge cloud, rather than as discrete particles. The fact that the effective forces on the nuclei are electrostatic affirms that there are no "mysterious quantum-mechanical forces" acting in molecules.

Let us consider the implications of the electrostatic theorem for chemical bonding in diatomic molecules. We take the internuclear axis as the z axis (Fig. 14.4). By symmetry the x and y components of the effective forces on the two nuclei are zero. For the z force component on nucleus a, we have [Eq. (14.79)]

$$F_{z,a} = -Z_a e'^2 \int\int\int \rho \frac{z_a - z}{r_a^3} \, dx \, dy \, dz - \frac{Z_a Z_b e'^2}{R^2}$$

where R is the internuclear distance. From Fig. 14.4 we see that $r_a \cos\theta_a = -z_a + z$ (z_a is negative). Hence

$$F_{z,a} = Z_a e'^2 \int\int\int \rho \frac{\cos\theta_a}{r_a^2} \, dx \, dy \, dz - \frac{Z_a Z_b e'^2}{R^2} \qquad (14.81)$$

Similarly, we find

$$F_{z,b} = -Z_b e'^2 \int\int\int \rho \frac{\cos\theta_b}{r_b^2} \, dx \, dy \, dz + \frac{Z_a Z_b e'^2}{R^2} \qquad (14.82)$$

Using Eq. (14.27), we have

$$F_{z,a} = -\frac{dU(R)}{dR} \frac{\partial R}{\partial z_a} = -\frac{dU}{dR} \frac{z_a - z_b}{R} = \frac{dU}{dR} \frac{\partial R}{\partial z_b} = -F_{z,b} \qquad (14.83)$$

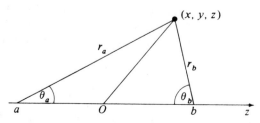

Figure 14.4 Coordinate system for a diatomic molecule. The origin is at O.

The effective forces on nuclei a and b are equal in magnitude and opposite in direction.

From (14.81) the z component of the effective force on nucleus a due to the element of electronic charge in the region about (x, y, z) is

$$e'^2 Z_a \rho \, \frac{\cos \theta_a}{r_a^2} \, dx \, dy \, dz \qquad (14.84)$$

Similarly, the z component of force on nucleus b due to this charge is

$$-e'^2 Z_b \rho \, \frac{\cos \theta_b}{r_b^2} \, dx \, dy \, dz \qquad (14.85)$$

A positive value of (14.84) or (14.85) corresponds to a force in the $+z$ direction, that is, to the right in Fig. 14.4. When the force on nucleus a is algebraically greater than the force on nucleus b, then the element of electronic charge tends to draw a toward b; hence electronic charge that is binding is located in the region where

$$e'^2 Z_a \rho \, \frac{\cos \theta_a}{r_a^2} \, dx \, dy \, dz > -e'^2 Z_b \rho \, \frac{\cos \theta_b}{r_b^2} \, dx \, dy \, dz \qquad (14.86)$$

Since the probability density ρ is nonnegative, division by ρ preserves the direction of the inequality sign, and the binding region of space is where

$$Z_a \, \frac{\cos \theta_a}{r_a^2} + Z_b \, \frac{\cos \theta_b}{r_b^2} > 0 \qquad (14.87)$$

When the force on b is algebraically greater than that on a, the electronic charge element tends to draw b away from a; the antibinding region of space is thus characterized by a negative value for the left side of (14.87). The surfaces for which the left side of (14.87) equals zero divide space into the **binding** and **antibinding regions**. This concept of binding and antibinding regions was proposed by Berlin. [T. Berlin, *J. Chem. Phys.*, **19**, 208 (1951); Berlin's ideas are extended to polyatomic molecules in T. Koga et al., *J. Am. Chem. Soc.*, **100**, 7522 (1978).]

Figures 14.5 and 14.6 show the binding and antibinding regions for a

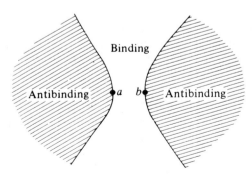

Figure 14.5 Cross section of binding and antibinding regions in a homonuclear diatomic molecule. To obtain the three-dimensional regions, rotate the figure about the internuclear axis.

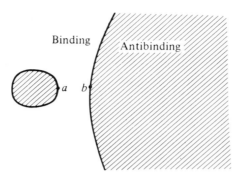

Figure 14.6 Binding and antibinding regions for a heteronuclear diatomic molecule with $Z_b > Z_a$.

homonuclear and a heteronuclear diatomic molecule. As might be expected, the binding region for a homonuclear diatomic molecule lies between the nuclei. Charge in this region tends to draw the nuclei together. We have seen time and again that bonding leads to a transfer of charge probability density into the region between the nuclei, because of the overlap between the bonding AOs. Electronic charge that is "behind" the nuclei (to the left of nucleus a or to the right of nucleus b in Fig. 14.5) exerts a greater attraction on the nucleus that is nearer to it than on the other nucleus and thus tends to pull the nuclei apart.

Bader, Henneker, and Cade have taken the electron probability densities for homonuclear diatomic molecules at R_e as calculated from Hartree–Fock functions and subtracted off the probability densities for the corresponding separated atoms, as calculated from atomic Hartree–Fock wave functions [R. F. W. Bader, W. H. Henneker, and P. E. Cade, *J. Chem. Phys.*, **46**, 3341 (1967)]. They then plotted contours of this *density difference* $\Delta\rho$ for Li_2 through F_2. A contour with positive $\Delta\rho$ corresponds to a buildup in charge over that of the separated atoms. Not surprisingly they found a charge buildup in the central part of the binding region, between the nuclei. However, they also found that molecule formation is accompanied by a charge buildup in most of the volume of the antibinding regions also. (The buildup is at the expense of charge probability density in the regions near the nuclei and containing the boundaries between binding and antibinding regions; see Fig. 3 of the paper.) Of course, the charge buildup between the nuclei is such that its contribution to the attractive force between the nuclei exceeds the Hellmann–Feynman repulsive force due to the charge buildup in the antibinding region. At $R = R_e$, the effective force on each nucleus is zero ($\partial U/\partial R|_{R_e} = 0$), and the hypothetical electron smear must exert a net attractive Hellmann–Feynman force to counterbalance the internuclear repulsion. As mentioned (Section 13.17), Hartree–Fock wave functions provide accurate electron probability-density functions. However, the difference between the molecular and atomic ρ's is sensitive to small errors in the ρ's, and one might therefore question the accuracy of Hartree–Fock $\Delta\rho$ maps. Calculations on several diatomic molecules show that, when configuration-interaction wave functions are used in place of the Hartree–Fock wave functions, the $\Delta\rho$ contour values change by typically a few to 10% [M. E. Stephens and P. J. Becker, *Mol. Phys.*, **49**, 65 (1983); R. Moszynski and K. Szalewicz, *J. Phys. B*, **20**, 4347 (1987)]. This indicates that the Hartree–Fock $\Delta\rho$ maps are pretty accurate at R_e.

In discussing things from the Hellmann–Feynman viewpoint, we seem to be considering chemical bonding solely in terms of potential energy, whereas the virial-theorem discussion talked about both potential and kinetic energy. For the purposes of the Hellmann–Feynman discussion, we are imagining the electrons to be smeared out into a continuous charge distribution; hence we make no reference to electronic kinetic energy. The use of the electrostatic theorem to explain chemical bonding has been criticized by some quantum chemists on the grounds that it hides the role of kinetic energy in bonding. [See the references cited after Eq. (13.66).]

For further applications of the Hellmann–Feynman electrostatic theorem, see B. M. Deb, ed., *The Force Concept in Chemistry*, Van Nostrand Reinhold, New York, 1981.

14.5 SUMMARY

For a bound stationary state, the quantum-mechanical virial theorem states that $2\langle T\rangle = \sum_i \langle q_i(\partial V/\partial q_i)\rangle$, where the sum is over the Cartesian coordinates of all the particles. If V is a homogeneous function of degree n, then $2\langle T\rangle = n\langle V\rangle$. For a diatomic molecule, the virial theorem becomes $\langle T_{\mathrm{el}}\rangle = -U - R(dU/dR)$ and $\langle V\rangle = 2U + R(dU/dR)$, where $U(R)$ is the potential-energy function for nuclear motion. The virial theorem shows that at R_e, $\langle V\rangle$ of a diatomic molecule is less than the total $\langle V\rangle$ of the separated atoms, and $\langle T_{\mathrm{el}}\rangle$ is greater than the total $\langle T_{\mathrm{el}}\rangle$ of the separated atoms.

For a bound stationary state, the generalized Hellmann–Feynman theorem is $\partial E_n/\partial\lambda = \int \psi_n^*(\partial\hat{H}/\partial\lambda)\psi_n\, d\tau$, where λ is a parameter in the Hamiltonian. Taking λ as a nuclear coordinate, we are led to the Hellmann–Feynman electrostatic theorem, which states that the force on a nucleus in a molecule is the sum of the electrostatic forces exerted by the other nuclei and the electron charge density.

PROBLEMS

14.1 Which of the following functions are homogeneous? Give the degree of homogeneity. (a) $x + 3yz$; (b) 179; (c) x^2/yz^3; (d) $(ax^3 + bxy^2)^{1/2}$.

14.2 Let 1 and 2 be two bound stationary states of an atom, with $E_2 > E_1$. For which state is the average electronic kinetic energy larger?

14.3 Show that the hypervirial theorem follows from Eq. (7.109).

14.4 (a) Calculate $\langle T\rangle$ and $\langle V\rangle$ for the helium-atom trial function (9.58). All the needed integrals were evaluated in Chapter 9. (b) Verify that the virial theorem is satisfied for $\zeta = Z - 5/16$ but not for $\zeta = Z$.

14.5 A particle is subject to the potential energy $V = ax^4 + by^4 + cz^4$. If its ground-state energy is 10 eV, calculate $\langle T\rangle$ and $\langle V\rangle$ for the ground state.

14.6 Prove that for a bound stationary state: (a) $\langle p_x \rangle = 0$; (b) $\langle \partial V/\partial x \rangle = 0$. *Hint:* Use certain equations in Section 5.1.

14.7 The $U(R)$ curve for a repulsive state can be roughly approximated by the function $ae^{-bR} - c$, where a, b, and c are positive constants with $a > c$. (This function omits the van der Waals minimum and fails to go to infinity at $R = 0$.) Sketch U, $\langle T_{el} \rangle$, and $\langle V \rangle$ as functions of R for this function.

14.8 Prove that $\partial \langle V \rangle / \partial R$ must be nonnegative at $R = R_e$; that is, $\langle V \rangle$ cannot be increasing with decreasing R as we go through the minimum in the $U(R)$ curve (Fig. 14.1). State and prove the corresponding theorem for $\langle T_{el} \rangle$.

14.9 Let ψ be the complete wave function for a molecule, with the Born–Oppenheimer approximation $\psi = \psi_{el}\psi_N$ not necessarily holding. Is it true that

$$2\langle \psi | \hat{T}_{el} + \hat{T}_N | \psi \rangle = -\langle \psi | \hat{V} | \psi \rangle$$

where $\hat{T}_{el}$ and $\hat{T}_N$ are the kinetic-energy operators for the electrons and nuclei and $\hat{V}$ is the complete potential-energy operator? Justify your answer.

14.10 The Fues potential-energy function for nuclear vibration of a diatomic molecule is $U(R) = U(\infty) + D_e(-2R_e/R + R_e^2/R^2)$. Find the expressions for $\langle T_{el} \rangle$ and $\langle V \rangle$ predicted by this potential and comment on the results.

14.11 (a) Use the generalized Hellmann–Feynman theorem to calculate $\langle 1/r \rangle$ for the hydrogen-atom bound states. (b) We now want an expression for $\langle 1/r^2 \rangle$, and we proceed as follows. Equation (6.62) can be written as $\hat{H}_r R(r) = ER(r)$, where the "radial Hamiltonian" is given by deleting the function R from the left side of (6.62). Since $\hat{H}_r$ does not affect the angles θ and ϕ, we can multiply the above equation by Y_l^m to get $\hat{H}_r \psi = E\psi$. If we view the quantum number l as a parameter in $\hat{H}_r$, then the generalized Hellmann–Feynman theorem gives $\langle \psi | \partial \hat{H}_r / \partial l | \psi \rangle = \partial E/\partial l$. Use this equation and Eq. (6.91) to show that for hydrogen-atom bound states

$$\left\langle \frac{1}{r^2} \right\rangle = \frac{2Z^2}{(2l+1)n^3} \left(\frac{1}{a_0^2} \right)$$

14.12 Use the generalized Hellmann–Feynman theorem to find $\langle p_x^2 \rangle$ for the one-dimensional harmonic-oscillator stationary states. Check that the result obtained agrees with the virial theorem.

14.13 Differentiate Eqs. (9.7) and (9.14) with respect to λ, substitute the results into the generalized Hellmann–Feynman theorem, and let $\lambda = 0$ to derive $E_n^{(1)} = \langle \psi_n^{(0)} | \hat{H}' | \psi_n^{(0)} \rangle$.

14.14 The R_e values for the ground electronic states of HF, HCl, HBr, and HI are 0.9, 1.3, 1.4, and 1.6 Å. The surface enclosing the antibinding region "behind" the proton in these molecules intersects the internuclear axis at two points, one of which is the proton location. Calculate the distance between these two points of intersection for each of the hydrogen halides.

15

Ab Initio Treatments of Polyatomic Molecules

15.1 AB INITIO METHODS AND SEMIEMPIRICAL METHODS

The search for accurate electronic wave functions of polyatomic molecules uses mainly the MO method. The presence of several nuclei causes greater computational difficulties than for diatomic molecules. Moreover, the electronic wave function of a diatomic molecule is a function of only one parameter—the internuclear distance; in contrast, the electronic wave function of a polyatomic molecule depends simultaneously on several parameters—the bond distances, bond angles, and dihedral angles of rotation about single bonds (these angles define the molecular conformation). A full theoretical treatment of a polyatomic molecule involves calculation of the electronic wave function for a range of each of these parameters; the equilibrium bond distances and angles are then found as those values that minimize the electronic energy (including nuclear repulsion).

Molecular quantum-mechanical methods are classified as either ab initio or semiempirical. *Semiempirical* methods use a simpler Hamiltonian than the correct molecular Hamiltonian and use parameters whose values are adjusted to fit experimental data or the results of ab initio calculations. An example of a semiempirical method is the Hückel MO treatment of conjugated hydrocarbons (Section 16.3), which uses a one-electron Hamiltonian and takes the bond integrals as adjustable parameters rather than quantities to be calculated theoretically. In contrast, an *ab initio* calculation uses the correct Hamiltonian and does not use experimental data other than the values of the fundamental physical constants. A Hartree–Fock SCF calculation seeks the antisymmetrized product Φ of one-electron functions that minimizes $\int \Phi^* \hat{H} \Phi \, d\tau$, where $\hat{H}$ is the true Hamiltonian, and is thus an ab initio calculation. (Ab initio is Latin for "from the beginning" and indicates a calculation based on fundamental principles.) The term ab initio should not be interpreted to mean "100% correct." An ab initio SCF calculation uses the approximation of taking ψ as an antisymmetrized product of one-electron spin-orbitals and uses a finite (and hence incomplete) basis set.

15.2 CLASSIFICATION OF ELECTRONIC TERMS OF POLYATOMIC MOLECULES

For polyatomic molecules the operator $\hat{S}^2$ for the square of the total electronic spin angular momentum commutes with the electronic Hamiltonian, and, as for diatomic molecules, the electronic terms of polyatomic molecules are classified as singlets, doublets, triplets, and so on, according to the value of $2S + 1$. (The commutation of $\hat{S}^2$ and $\hat{H}_{el}$ holds provided spin–orbit interaction is omitted from the Hamiltonian; for molecules containing heavy atoms, spin–orbit interaction is considerable, and S is not a good quantum number.)

For linear polyatomic molecules, the operator $\hat{L}_z$ for the axial component of the total electronic orbital angular momentum commutes with the electronic Hamiltonian, and the same term classifications are used as for diatomic molecules; we have such possibilities as $^1\Sigma^+$, $^1\Sigma^-$, $^3\Sigma^+$, $^1\Pi$, and so on. For linear polyatomic molecules with a center of symmetry, the g, u classification is added.

For nonlinear polyatomic molecules, there is no orbital angular-momentum operator that commutes with the electronic Hamiltonian, and the angular-momentum classification of electronic terms cannot be used. Operators that do commute with the electronic Hamiltonian are the symmetry operators $\hat{O}_R$ of the molecule (Section 12.1), and the electronic states of polyatomic molecules are classified according to the behavior of the electronic wave function on application of these operators. Consider H_2O as an example.

In its equilibrium configuration, water belongs to group $\mathcal{C}_{2v}$ with the symmetry operations

$$\hat{E} \qquad \hat{C}_2(z) \qquad \hat{\sigma}_v(xz) \qquad \hat{\sigma}_v(yz) \tag{15.1}$$

The standard convention [R. S. Mulliken, *J. Chem. Phys.*, **23**, 1997 (1955); **24**, 1118 (1956)] takes the molecular plane as the yz plane (Fig. 15.1). We readily find that each of the symmetry operations commutes with the other three. Therefore, the electronic wave functions can be chosen as simultaneous eigenfunctions of all four symmetry operators. Since $\hat{O}_E$ is the unit operator, we have $\hat{O}_E \psi_{el} = \psi_{el}$. Each of the remaining symmetry operators satisfies $\hat{O}_R^2 = \hat{1}$, and so each has as its eigenvalues $+1$ and -1 [Eq. (7.55)]. Therefore, each electronic wave function of

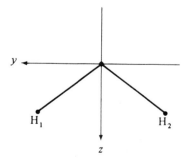

Figure 15.1 Coordinate axes for the H_2O molecule. The x axis is perpendicular to the molecular plane.

H_2O is an eigenfunction of $\hat{O}_E$ with eigenvalue $+1$ and an eigenfunction of each of the other three symmetry operators with the eigenvalues $+1$ or -1.

How many different combinations of these symmetry eigenvalues are there for H_2O? At first sight, we might think there are $(1)(2)(2)(2) = 8$ possible sets. However, certain sets can be ruled out, as we now show. Let the product of the symmetry operations $\hat{R}$ and $\hat{S}$ be the symmetry operation $\hat{T}$; $\hat{R}\hat{S} = \hat{T}$. Let ψ_{el} be an eigenfunction of $\hat{O}_R$, $\hat{O}_S$, and $\hat{O}_T$ with eigenvalues r, s, and t, respectively. Since the symmetry operators multiply the same way the symmetry operations do, we have

$$t\psi_{el} = \hat{O}_T\psi_{el} = \hat{O}_R(\hat{O}_S\psi_{el}) = s\hat{O}_R\psi_{el} = rs\psi_{el}$$

Dividing by ψ_{el}, we have $rs = t$ if $\hat{R}\hat{S} = \hat{T}$. Hence the eigenvalues of the symmetry operators must multiply in the same way the symmetry operations do. Now consider H_2O. Let us examine the set of symmetry eigenvalues

$$1 \qquad -1 \qquad -1 \qquad -1 \tag{15.2}$$

where the eigenvalues are listed in the order corresponding to (15.1). We have $\hat{C}_2(z)\hat{\sigma}_v(xz) = \hat{\sigma}_v(yz)$; the set (15.2), however, has the product of the $\hat{O}_{C_2}$ and $\hat{O}_{\sigma_v(xz)}$ eigenvalues as $(-1)(-1) = 1$, which differs from the $\hat{O}_{\sigma_v(yz)}$ eigenvalue in (15.2). The set (15.2) must be discarded. Of the eight possible symmetry-eigenvalue combinations, only four are found to multiply properly; these four sets are (Problem 15.1)

	$\hat{E}$	$\hat{C}_2(z)$	$\hat{\sigma}_v(xz)$	$\hat{\sigma}_v(yz)$	
A_1	1	1	1	1	
A_2	1	1	-1	-1	(15.3)
B_1	1	-1	1	-1	
B_2	1	-1	-1	1	

In this table the sets have been labeled A_1, A_2, B_1, and B_2. The letter A or B indicates whether the symmetry eigenvalue for the highest-order $\hat{C}_n$ or $\hat{S}_n$ operation of the molecule [$\hat{C}_2(z)$ for water] is $+1$ or -1, respectively. The subscripts 1 and 2 distinguish sets having the same letter label. Each possible set of symmetry eigenvalues in (15.3) is called a **symmetry species** (or *symmetry type*). (The group-theory term is *irreducible representation*.) The symmetry species with all symmetry eigenvalues $+1$ (A_1 for H_2O) is called the **totally symmetric species**.

Each molecular electronic term of H_2O is designated by giving the symmetry species of the electronic wave functions of the term, with the spin multiplicity $2S + 1$ as a left superscript. For example, an electronic state of water with two electrons unpaired and with the electronic wave function unchanged by all four symmetry operators belongs to a $^3A_1$ term. (The subscript 1 is not an angular-momentum eigenvalue, but is part of the symmetry-species label.)

We now consider the **orbital degeneracy** of molecular electronic terms; this is degeneracy connected with the electrons' spatial (orbital) motion, as distinguished from spin degeneracy. Thus $^1\Pi$ and $^3\Pi$ terms of linear molecules are orbitally degenerate, while $^1\Sigma$ and $^3\Sigma$ terms are orbitally nondegenerate. Consider an operator $\hat{F}$ that commutes with the molecular electronic Hamiltonian and that does not involve spin; we have

$$\hat{F}\hat{H}_{el}\psi_{el,i} = \hat{F}E_{el}\psi_{el,i}$$

$$\hat{H}_{el}(\hat{F}\psi_{el,i}) = E_{el}(\hat{F}\psi_{el,i}) \tag{15.4}$$

Thus $\hat{F}\psi_{el,i}$ is an eigenfunction of $\hat{H}_{el}$ with eigenvalue E_{el}. Let the orbital degeneracy of the term to which $\psi_{el,i}$ belongs be n. It follows from (15.4) that $\hat{F}\psi_{el,i}$ must be some linear combination of the n orbitally degenerate wave functions of the term that have the same value of S_z as $\psi_{el,i}$:

$$\hat{F}\psi_{el,i} = \sum_{j=1}^{n} c_{ij,F}\psi_{el,j} \tag{15.5}$$

where the c's are certain constants. (These arguments were previously given in Section 7.4.) For a level that is not orbitally degenerate ($n = 1$), Eq. (15.5) reduces to

$$\hat{F}\psi_{el,i} = c_F\psi_{el,i} \tag{15.6}$$

and here $\psi_{el,i}$ *must* be an eigenfunction of $\hat{F}$. For $n > 1$, we can *choose* n linear combinations of the $\psi_{el,j}$'s that are eigenfunctions of $\hat{F}$, but there is no *necessity* for the eigenfunctions of a degenerate level to be eigenfunctions of $\hat{F}$. If we have two operators $\hat{F}$ and $\hat{G}$ that commute with $\hat{H}_{el}$, but not with each other, we can pick the linear combinations to be eigenfunctions of $\hat{F}$ or of $\hat{G}$, but not in general of both $\hat{F}$ and $\hat{G}$ simultaneously.

Now consider symmetry operators. For water all the symmetry operators commute among themselves, and so each electronic wave function is simultaneously an eigenfunction of all the symmetry operators. If all the electronic wave functions were orbitally nondegenerate, then it would automatically follow from (15.6) that they were simultaneously eigenfunctions of all the symmetry operators. We therefore suspect (but have not proved) that the electronic wave functions of water are all orbitally nondegenerate. This statement is in fact correct. The letters A and B designate symmetry species of orbitally nondegenerate electronic terms. For any molecule all of whose symmetry operators commute with one another, the electronic wave functions are each simultaneous eigenfunctions of all symmetry operators, and the only symmetry species are nondegenerate A and B species.

For some point groups, the symmetry operators do not all commute; an example is $\mathcal{O}_h$ (Fig. 12.7). When the symmetry operators do not all commute, some of the electronic terms are orbitally degenerate. A symmetry operator applied to an electronic wave function of an orbitally degenerate term converts it to a linear combination of the wave functions of the term [Eq. (15.5)]. The effects of the symmetry operator $\hat{O}_R$ on the wave functions of an n-fold orbitally degenerate electronic term are specified by the n^2 numbers

$$c_{ij,R}, \qquad i = 1, \ldots, n, \qquad j = 1, \ldots, n \tag{15.7}$$

The n^2 numbers (15.7) when arranged in an $n \times n$ square array form a matrix (Section 7.10). If there are h symmetry operations in the molecular point group, then the h matrices of coefficients in (15.5) constitute the **symmetry species** of the degenerate-term wave functions in (15.5). The following letter labels are used for the symmetry species, according to the orbital degeneracy n:

$$
\begin{array}{c|c|c|c|c|c}
n & 1 & 2 & 3 & 4 & 5 \\
\hline
Letter & A, B & E & T & G & H
\end{array}
\tag{15.8}
$$

Numerical subscripts distinguish different symmetry species having the same letter designation. For molecules with a center of symmetry, a g or u subscript is added, depending on whether the wave function has the eigenvalue $+1$ or -1 for inversion of all electronic spatial coordinates. The possible symmetry species can be found in a systematic way using group theory (see *Schonland*). As an example, group theory shows the possible symmetry species of a $\mathscr{D}_{6h}$ molecule to be

$$
\begin{array}{cccccc}
A_{1g} & A_{2g} & B_{1g} & B_{2g} & E_{1g} & E_{2g} \\
A_{1u} & A_{2u} & B_{1u} & B_{2u} & E_{1u} & E_{2u}
\end{array}
\tag{15.9}
$$

We shall not give the numbers $c_{ij,R}$ that specify these symmetry species, except to note that A_{1g} is the totally symmetric symmetry species, with all c's equal to 1.

It is an empirical fact that for most molecules in their electronic ground states the wave function belongs to the (nondegenerate) totally symmetric species; also the electronic spins are usually all paired in the ground state, and the ground state is a singlet. For water the ground electronic state is $^1A_1$; for benzene it is $^1A_{1g}$.

We have been using the equilibrium-geometry point groups of H_2O and C_6H_6. For nonequilibrium nuclear configurations, the symmetry is in general less than that of $\mathscr{C}_{2v}$ or $\mathscr{D}_{6h}$; for reasonably small departures from equilibrium, however, the symmetry behavior should be given to a good approximation by the symmetry species of the equilibrium-geometry point group. We also note that excited states sometimes differ in their equilibrium point group from the ground electronic state. For example, the point group of NH_3 is $\mathscr{D}_{3h}$ for several excited electronic states. (For properties of molecular electronic states, see G. Herzberg, *Electronic Spectra of Polyatomic Molecules*, Van Nostrand, New York, 1966, Appendix VI.)

Corresponding to a given molecular electronic term, there are in general several electronic states. The interactions between electronic spin and electronic orbital motion and between electronic and nuclear motions split the energies of these states. These splittings are usually small.

15.3   THE SCF MO TREATMENT OF POLYATOMIC MOLECULES

The purely electronic nonrelativistic Hamiltonian for a polyatomic molecule is (in atomic units)

$$
\hat{H}_{el} = -\frac{1}{2}\sum_i \nabla_i^2 - \sum_i \sum_\alpha \frac{Z_\alpha}{r_{i\alpha}} + \sum_i \sum_{j>i} \frac{1}{r_{ij}}
\tag{15.10}
$$

If interelectronic repulsions are neglected, the zeroth-order wave function is the product of one-electron spatial functions (molecular orbitals). Allowance for electron spin and the Pauli principle gives a zeroth-order wave function that is an antisymmetrized product of molecular spin-orbitals, each spin-orbital being a product of a spatial MO and a spin function. The best possible variation function that has the form of an antisymmetrized product of spin-orbitals is the Hartree–Fock SCF function. [Improvements beyond the Hartree–Fock stage require some method (Sections 15.11 to 15.14) to allow for electron correlation.] The MOs are usually expressed as linear combinations of basis functions, the coefficients being found by solution of the Roothaan equations (Section 13.16). If a sufficiently large basis set is used, the MOs are accurate approximations to the Hartree–Fock MOs. If a minimal basis set is used, the MOs are only rough approximations to the Hartree–Fock MOs but are still referred to as SCF molecular orbitals.

How are polyatomic MOs classified? As might be expected, the MOs of a polyatomic molecule show the same kinds of possible symmetry behavior as the overall electronic wave function does (Section 15.2); for the proof, see C. C. J. Roothaan, *Rev. Mod. Phys.*, **23**, 69 (1951). The MOs are therefore classified according to the symmetry species of the molecular point group. For example, the MOs of H_2O have the possible symmetry species a_1, a_2, b_1, and b_2 of (15.3). Lowercase letters are used for MO symmetry species. To distinguish MOs of the same symmetry species, we number them in order of increasing energy. Thus the lowest three a_1 MOs of water are called $1a_1$, $2a_1$, and $3a_1$. This nomenclature is similar to that of the third column in Table 13.1.

Each MO holds two electrons of opposite spin. MOs having the same energy constitute a *shell*. A shell that consists of a single MO of symmetry species a or b holds two electrons; a shell that consists of two e MOs having the same energy holds four electrons; and so on. For water there are no degenerate symmetry species, and each shell holds two electrons. For benzene there are some doubly degenerate symmetry species [see (15.9)], so some of the benzene MOs occur in pairs having the same energy. Specification of the number of electrons in each shell specifies the molecular *electronic configuration*. Just as in atoms and diatomic molecules, a given electron configuration of a polyatomic molecule gives rise to one or more electronic terms. [For example, an $(e_{1g})^2$ configuration of a $\mathscr{D}_{6h}$ molecule gives the terms $^1A_{1g}$, $^1E_{2g}$, and $^3A_{2g}$.] The systematic method for finding the terms of a configuration uses group theory and will not be discussed here. For tables of the terms arising from various configurations, see G. Herzberg, *Electronic Spectra of Polyatomic Molecules*, Van Nostrand, New York, 1966, pp. 330–334, 570–573. *A closed-shell configuration gives rise to a single nondegenerate term whose multiplicity is 1 and whose symmetry species is the totally symmetric one.* Most polyatomic molecules have a closed-shell ground state; here, the MO wave function is a single Slater determinant. For states arising from open-shell configurations, the MO wave function may require a linear combination of several Slater determinants.

SCF-MO Wave Functions for Open-Shell States.   For SCF MO calculations on closed-shell states of molecules and atoms, electrons paired with each other are almost always given precisely the same spatial orbital function. A Hartree–Fock wave function in which electrons whose spins are paired occupy the same spatial orbital is called a *restricted Hartree–Fock* (RHF) wave function. (The unmodified term "Hartree–Fock wave function" is understood to mean the RHF wave function.)

Although the RHF wave function is generally used for closed-shell states, two different approaches are widely used for open-shell states. In the **restricted open-shell Hartree–Fock** (ROHF) method, electrons that are paired with each other are given the same spatial orbital function. For example, the ROHF wave function of the Li ground state is $|1s\,\overline{1s}\,2s|$, where the two $1s$ electrons occupy the same spatial MO. The $2s$ electron in this ROHF function has been given spin α. Since electrons with the same spin tend to keep away from each other (Pauli repulsion; Section 10.3), the interaction between the $2s\alpha$ and $1s\alpha$ electrons differs from the interaction between the $2s\alpha$ and $1s\beta$ electrons, and it seems reasonable to give the two $1s$ electrons slightly different spatial orbitals, which we shall call $1s$ and $1s'$. This gives the **unrestricted Hartree–Fock** (UHF) wave function for the Li ground state as $|1s\,\overline{1s'}\,2s|$, where $1s \neq 1s'$. In a UHF wave function, the spatial orbitals of spin-α electrons are allowed to differ from those of spin-β electrons.

The UHF wave function gives a slightly lower energy than the ROHF wave function and is much more useful in predicting electron-spin-resonance spectra (see *Szabo and Ostlund*, Section 3.8.6). The main problem with the UHF wave function is that it is not an eigenfunction of the spin operator $\hat{S}^2$ (nor can it be made an eigenfunction of $\hat{S}^2$ by taking a linear combination of a few UHF functions), whereas the true wave function and the ROHF wave function are eigenfunctions of $\hat{S}^2$. When a UHF wave function is found, one calculates $\langle S^2 \rangle$ for the UHF function; if the deviation of $\langle S^2 \rangle$ from $S(S+1)\hbar^2$ is substantial, the UHF wave function should be viewed with suspicion.

A common method of allowing for electron correlation is a perturbation-theory treatment in which the Hartree–Fock wave function is taken as the zeroth-order function (Section 15.12). The UHF wave function readily lends itself to such a perturbation treatment (whereas the ROHF function does not), so UHF functions are often used to deal with radicals.

15.4 BASIS FUNCTIONS

Most molecular quantum-mechanical methods, whether SCF, CI, perturbation theory (Section 15.12), or coupled cluster (Section 15.13), begin the calculation with the choice of a basis set. The use of an adequate basis set is an essential requirement for success of the calculation.

For diatomic molecules, the basis functions are usually taken as atomic orbitals, some centered on one atom, the remainder centered on the other atom; each AO can be represented as a linear combination of one or more Slater-type orbitals (STOs). An STO centered on atom a has the form $N r_a^{n-1} e^{-\zeta r_a} Y_l^m(\theta_a, \phi_a)$ [Eq. (11.14)]. For nonlinear molecules, the real form of the STOs, with Y_l^m replaced by $(Y_l^{m*} \pm Y_l^m)/2^{1/2}$ (Section 6.6), is used. Each MO ϕ_i is expressed as $\phi_i = \sum_r c_{ri} \chi_r$, where the χ_r's are the STO basis functions. We have LC-STO MOs.

For polyatomic molecules, the LC-STO method uses STOs centered on each of the atoms. The presence of more than two atoms causes difficulties in evaluating the needed integrals. For a triatomic molecule, one must deal with three-center, two-center, and one-center integrals. For a molecule with four or more atoms, one also has four-center integrals, but the number of centers in any one integral does not exceed four. Solution of the Roothaan equations requires evaluation of the electron-repulsion integrals $(rs|tu)$ and the H_{rs}^{core} integrals [Eq. (13.165)]. If the basis functions $\chi_r(1)$, $\chi_s(1)$, $\chi_t(2)$, and $\chi_u(2)$ are each centered on a different nucleus, then $(rs|tu)$ is a four-center integral. The H_{rs}^{core} integrals involve either one or two centers.

For the basis set $\chi_1, \chi_2, \ldots, \chi_b$, there are b different possibilities for each basis function in $(rs|tu)$, and use of the identities $(rs|tu) = (sr|tu) = \cdots$ [Eq. (13.170)] shows that there are about $b^4/8$ different electron-repulsion integrals to be evaluated. Accurate SCF molecular calculations on small- to medium-size molecules might use from 20 to 200 basis functions, producing from 20000 to 2×10^8 electron-repulsion integrals. Computer evaluation of three- and four-center integrals over STO basis functions is very time consuming. This has led to two other choices of basis functions: the one-center expansion method and the Gaussian-function method.

The **one-center expansion** (OCE) **method** takes each MO as a linear combination of STOs, all of which are centered at the same point in space. For example, a one-center MO calculation of CH_4 takes each MO as a linear combination of functions, all of which centered on the carbon atom. With the OCE method, all integrals are one-center, but the method has drawbacks. The presence of a nucleus at a point in space introduces a cusp (Fig. 6.7) into the electronic wave function at that point. The representation of an MO as a linear combination of orbitals all centered at a single nucleus makes it difficult to get the proper cusp behavior at the other nuclei. A one-center wave function for CH_4 is inaccurate in the regions near the hydrogen nuclei, unless very many basis functions are used. The OCE method is most applicable to molecules with the formula AH_n; here, most of the electron probability density is located near atom A and the hydrogen atoms introduce only small cusps into the wave function. Because of generally discouraging results and the limited applicability of the method, there is currently not much interest in the OCE method.

To simplify molecular integral evaluation, Boys proposed in 1950 the use of **Gaussian-type functions** (GTFs) instead of STOs for the atomic orbitals in an LCAO wave function. A **Cartesian Gaussian** centered on atom a is defined as

$$g_{ijk} = Nx_a^i y_a^j z_a^k e^{-\alpha r_a^2} \tag{15.11}$$

where N is the normalization constant, i, j, and k are nonnegative integers, and α is a positive orbital exponent. When $i + j + k = 0$ (that is, $i = 0$, $j = 0$, $k = 0$), the GTF is called an s-type Gaussian; when $i + j + k = 1$, we have a p-type Gaussian, which contains the factor x_a, y_a, or z_a. When $i + j + k = 2$, we have a d-type Gaussian. There are six d-type Gaussians, with the factors x_a^2, y_a^2, z_a^2, $x_a y_a$, $x_a z_a$, and $y_a z_a$. If desired, five linear combinations (having the factors $x_a y_a$, $x_a z_a$, $y_a z_a$, $x_a^2 - y_a^2$, and $3z_a^2 - r_a^2$) can be formed to have the same angular behavior as the five real $3d$ AOs; a sixth combination with the factor $x_a^2 + y_a^2 + z_a^2 = r_a^2$ is like a $3s$ function. This sixth combination is sometimes omitted from the basis set. In general, linear combinations of Cartesian Gaussians can be formed to have the form

$$Nr_a^l e^{-\alpha r_a^2}(Y_l^{m*} \pm Y_l^m)/2^{1/2}$$

Note the absence of the principal quantum number n. Any s AO (whether $1s$ or $2s$ or ...) is represented by a linear combination of several Gaussians with different orbital exponents, each Gaussian having the form $\exp(-\alpha r_a^2)$; any atomic p_x orbital is represented by a linear combination of Gaussians, each of the form $x_a \exp(-\alpha r_a^2)$; and so on. The Cartesian Gaussians form a complete set.

An occasionally used alternative to Cartesian Gaussians is *spherical Gaussians*, whose real form is

$$Nr_a^{n-1} e^{-\alpha r_a^2}(Y_l^{m*} \pm Y_l^m)/2^{1/2}$$

The behavior of the Gaussian exponential factor is shown in Fig. 4.3a, where the origin is at nucleus a. A Gaussian function does not have the desired cusp at the nucleus and hence gives a poor representation of an AO for small values of r_a. To get an accurate representation of an AO, we must use a linear combination of several Gaussians. Therefore, an LC-GTF SCF MO calculation involves evaluation of very many more integrals than the corresponding LC-STO SCF MO calculation, since the number of two-electron integrals is proportional to the fourth power of the number of basis functions. However, Gaussian integral evaluation takes much less computer time than Slater integral evaluation. This is because the product of two Gaussian functions centered at two different points is equal to a single Gaussian centered at a third point. Thus all three- and four-center two-electron repulsion integrals are reduced to two-center integrals.

Let us discuss some of the terminology used to describe STO basis sets. A **minimal** (or **minimum**) **basis set** consists of one STO for each inner-shell and valence-shell AO of each atom (Section 13.16). For example, for C_2H_2 a minimal basis set consists of $1s$, $2s$, $2p_x$, $2p_y$, and $2p_z$ AOs on each carbon and a $1s$ STO on each hydrogen; there are five STOs on each C and one on each H, for a total of twelve basis functions. This set contains two s-type STOs and one set of p-type STOs on each carbon and one s-type STO on each hydrogen; such a set is denoted by $(2s1p)$ for the carbon functions and $(1s)$ for the hydrogen functions, a notation which is further abbreviated to $(2s1p/1s)$. The numbers of basis functions in a minimal set for the first part of the periodic table are

H, He	Li–Ne	Na–Ar	K, Ca	Sc–Kr
1	5	9	13	18

A **double-zeta** (DZ) **basis set** is obtained by replacing each STO of a minimal basis set by two STOs that differ in their orbital exponents ζ (zeta). (Recall that a single STO is not an accurate representation of an AO; use of two STOs gives substantial improvement.) For example, for C_2H_2 a double-zeta set consists of two $1s$ STOs on each H, two $1s$ STOs, two $2s$ STOs, two $2p_x$, two $2p_y$, and two $2p_z$ STOs on each carbon, for a total of twenty-four basis functions; this is a $(4s2p/2s)$ basis set. (Recall that we did a double-zeta SCF calculation on He in Section 13.16.) Since each basis function χ_r in $\phi_i = \Sigma_i c_{ri}\chi_r$ has its own independently determined variational coefficient c_{ri}, the number of variational parameters in a double-zeta-basis-set wave function is twice that in a minimal-basis-set wave function.

A *split-valence* (SV) **basis set** uses two STOs for each valence AO but only one STO for each inner-shell AO. This basis set is minimal for the inner-shell AOs and double zeta for the valence AOs.

AOs are distorted (or polarized) upon molecule formation. To allow for this polarization, one must introduce basis-function STOs whose l quantum numbers are greater than the maximum l in the corresponding free atom. A common kind of basis set is a *double-zeta plus polarization set* (DZ + P or DZP), which adds to a double-zeta set a set of five $3d$ functions on each "first-row" and each "second-row" atom and a set of three $2p$ functions $(2p_x, 2p_y, 2p_z)$ on each hydrogen atom. (In the quantum-chemistry literature, Li–Ne are called the "first-row" elements, even though they are actually the second row of the periodic table. This terminology will be used in Chapter 15.) A DZ + P basis set for $C_2H_5OSiH_3$ is designated as $(6s4p1d/4s2p1d/2s1p)$, where the slashes separate the functions for atoms of different rows of the periodic table in decreasing order. (For increased accuracy, higher-l polarization functions can be added.)

Now consider Gaussian-basis-set terminology. Instead of using the individual Gaussian functions (15.11) as basis functions, the current practice is to take each basis function as a linear combination of a small number of Gaussians, according to

$$\chi_r = \sum_u d_{ur} g_u \tag{15.12}$$

where the g_u's are Cartesian Gaussians [Eq. (15.11)] centered on the same atom and having the same i, j, k values as one another, but different α's. The coefficients d_{ur} are constants that are held fixed during the calculation. In (15.12), χ_r is called a **contracted** Gaussian-type function (CGTF) and the g_u's are called **primitive** Gaussians. By using contracted Gaussians instead of primitive Gaussians as the basis set, the number of variational coefficients to be determined is reduced, which gives large savings in computational time with little loss in accuracy if the contraction coefficients d_{ur} are well chosen.

When contracted Gaussians are used, a minimal basis set consists of one contracted Gaussian function for each inner-shell and valence AO, a DZ basis set consists of two contracted Gaussians for each such AO, and a DZ + P set adds contracted Gaussians with higher l values to the DZ set.

Several methods exist to form contracted Gaussian sets. Minimal CGTF sets are usually formed by fitting STOs. One starts with a minimal basis set of one STO per AO, with the STO orbital exponents fixed at values found to work well in calculations on small molecules. Each STO is then approximated as a linear combination of N Gaussian functions, where the coefficients in the linear combination and the Gaussian orbital exponents are chosen to give the best least-squares fit to the STO. Most commonly, $N = 3$, giving a set of CGTFs called STO-3G. Since a linear combination of three Gaussians is only an approximation to an STO, the STO-3G basis set gives results not quite as good as a minimal-basis-set STO calculation. A minimal basis set of STOs for a compound containing only first-row elements and hydrogen is denoted by $(2s1p/1s)$. Since each STO is replaced by a linear combination of three primitive Gaussians (which is one contracted Gaussian), the STO-3G basis set for a compound of first-row atoms and H is denoted by $(6s3p/3s)$ contracted to $[2s1p/1s]$, where parentheses indicate the primitive Gaussians and brackets indicate the contracted Gaussians.

Another way to form contracted Gaussians is to start with atomic GTF SCF calculations. Huzinaga used a $(9s5p)$ basis set of uncontracted Gaussians to do SCF calculations on the atoms Li–Ne. For example, for the ground state of the O atom, the optimized orbital exponents of the nine s-type basis GTFs were found to be [S. Huzinaga, *J. Chem. Phys.*, **42**, 1293 (1965)]

g_1	g_2	g_3	g_4	g_5	g_6	g_7	g_8	g_9
7817	1176	273.2	81.2	27.2	9.53	3.41	0.940	0.285

The expansion coefficients for the $1s$ SCF oxygen AO were found to be

g_1	g_2	g_3	g_4	g_5	g_6	g_7	g_8	g_9
0.0012	0.009	0.043	0.144	0.356	0.461	0.140	−0.0006	0.001

and the $2s$ SCF AO coefficients are

g_1	g_2	g_3	g_4	g_5	g_6	g_7	g_8	g_9
−0.0003	−0.002	−0.010	−0.036	−0.095	−0.196	−0.037	0.596	0.526

Suppose we want to form a split-valence $[3s2p]$ set of contracted GTFs for O. We see that the g_1, g_2, g_3, g_4, g_5, and g_7 coefficients are much larger for the $1s$ AO than for the $2s$ AO, the g_8 and g_9 coefficients are much larger for the $2s$ AO than for the $1s$ AO, and g_6 makes substantial contributions to both $1s$ and $2s$. We might therefore take the contracted $1s$ basis function as

$$1s = N(0.0012g_1 + 0.009g_2 + 0.043g_3 + 0.144g_4 + 0.356g_5 + 0.461g_6 + 0.140g_7)$$

where the normalization constant N is needed because g_8 and g_9 have been omitted. For an SV set, we need two basis functions for the $2s$ AO. These will be formed from g_6, g_8, and g_9, which are the main contributors to the $2s$ AO. Of these three, the function g_9 has the smallest orbital exponent and so falls off most slowly as r increases. (g_9 is called a **diffuse** function.) The outer region of an AO changes the most upon molecule formation, and to allow for this change, we can take the diffuse function g_9 as one of the basis functions, giving as the $2s$ oxygen contracted basis functions

$$2s = N'(-0.196g_6 + 0.596g_8), \qquad 2s' = g_9$$

The $2p$ and $2p'$ CGTFs can be formed similarly.

Dunning's DZ $[4s2p]$ and Dunning and Hay's SV $[3s2p]$ contractions of Huzinaga's $(9s5p)$ first-row-atom AOs are widely used in molecular calculations [T. H. Dunning, *J. Chem. Phys.*, **53**, 2823 (1970); T. H. Dunning and P. J. Hay in *Schaefer, Methods of Electronic Structure Theory*, pp. 1–27].

The 3-21G and 4-31G sets are commonly used split-valence basis sets of CGTFs. In the 3-21G set, each inner-shell AO ($1s$ for Li–Ne; $1s$, $2s$, $2p_x$, $2p_y$, $2p_z$ for Na–Ar; and so on) is represented by a single CGTF that is a linear combination of three primitive Gaussians; for each valence-shell AO ($1s$ for H; $2s$ and the $2p$'s for Li–Ne; . . . ; $4s$ and the $4p$'s for K, Ca, Ga–Kr; $4s$, the $4p$'s, and the five $3d$'s for Sc–Zn), there are two basis functions, one of which is a CGTF that is a linear combination of two Gaussian primitives and one which is a single diffuse Gaussian. The 4-31G set uses four primitives in each inner-shell CGTF and represents each valence-shell AO by one CGTF with three primitives and one Gaussian with one primitive. The orbital exponents and contraction coefficients d_{ur} in these basis sets were determined by using these basis sets to minimize the SCF energies of atoms.

The 6-31G* basis set (defined for the atoms H through Ar) is a split-valence set with some polarization functions added; it uses a linear combination of six primitives in each inner-shell AO and adds a single set of six d-type Cartesian Gaussian polarization functions for each nonhydrogen atom. The 6-31G** basis set adds to the 6-31G* set a set of three p-type Gaussian polarization functions on each hydrogen atom. The orbital exponents of the polarization functions in these two basis sets were determined as the average of the optimum values found in calculations on small molecules. The 6-31G* and 6-31G** sets are sometimes denoted as 6-31G(d) and 6-31G(d,p), respectively. In the 6-31G* basis set, a phosphorus atom has 19 basis functions centered on it ($1s$, $2s$, $2p_x$, $2p_y$, $2p_z$, $3s$, $3s'$, $3p_x$, $3p_y$, $3p_z$, $3p'_x$, $3p'_y$, $3p'_z$, and six d's.)

For second-row atoms, d orbitals contribute significantly to the bonding. To allow for this, the 3-21G$^{(*)}$ basis set is constructed by the addition to the 3-21G set of a set of six d-type Gaussian functions on each second-row atom. For H–Ne, the 3-21G$^{(*)}$ set is the same as the 3-21G set.

Anions, compounds with lone pairs, and hydrogen-bonded dimers have significant electron density at large distances from the nuclei. To improve the accuracy for such compounds, the 3-21+G and 6-31+G* basis sets are formed from the 3-21G and 6-31G* sets by the addition of four highly diffuse functions (s, p_x, p_y, p_z) on each nonhydrogen atom; a highly diffuse function is one with a very small orbital exponent (typically, 0.01 to 0.1). The 3-21++G and 6-31++G* sets also include a highly diffuse s function on each hydrogen atom.

The STO-3G, 3-21G, 3-21G$^{(*)}$, 4-31G, 6-31G*, and 6-31G** basis sets were developed by Pople and co-workers (see *Hehre et al.*, Section 4.3) and are available in the widely used ab initio programs GAUSSIAN 82, GAUSSIAN 86, GAUSSIAN 88, and GAUSSIAN 90. The latest versions of GAUSSIAN are sold by Gaussian, Inc., of Pittsburgh, Pennsylvania. Other ab initio programs for molecular calculations are GAMESS, GRADSCF, CADPAC, and HONDO.

Because of the time savings given by Gaussians in multicenter-integral evaluation, most current molecular ab initio calculations use contracted-Gaussian basis sets. Most semiempirical methods (which neglect large classes of integrals) use STOs. Two review articles on ab initio basis sets are S. Wilson, *Adv. Chem. Phys.*, **67**, 439 (1987); E. R. Davidson and D. Feller, *Chem. Rev.*, **86**, 681 (1986).

15.5 REDUCTION OF THE NUMBER OF INTEGRALS

The calculation of the approximately $b^4/8$ two-electron integrals $(rs|tu)$ over the b basis functions consumes a major part of the time needed in an SCF MO calculation. Several techniques are used to reduce the number of integrals evaluated.

Molecular symmetry is used to identify integrals that are equal so that only one of them need be evaluated. For example, in H_2O, the integrals

($H_1 1s\ O2s|H_2 1s\ H_2 1s$) and ($H_2 1s\ O2s|H_1 1s\ H_1 1s$) are equal, provided the O–H_1 and O–H_2 bond distances are equal. Use of symmetry cuts the number of integrals to be evaluated in H_2O approximately in half.

In a large molecule, any one atom is far from most of the other atoms, and so a large fraction of the two-electron integrals are negligibly small for large molecules; $(rs|tu)$ will be very small if $\chi_r(1)$ and $\chi_s(1)$ or $\chi_t(2)$ and $\chi_u(2)$ are centered on widely separated nuclei. Hence, many programs test each $(rs|tu)$ integral to get its order of magnitude before it is calculated accurately. Integrals smaller than a certain threshold value can be neglected without affecting the accuracy of the overall calculation. Although the number of two-electron integrals increases as b^4, the number of such integrals whose value exceeds a fixed threshold increases only as b^2 for large molecules; calculations on large molecules with integrals of value less than 10^{-9} hartree neglected showed that the time to do an SCF calculation increased only as $b^{2.3}$ [R. Ahlrichs et al., *Chem. Phys. Lett.*, **162**, 165 (1989)].

Many $(rs|tu)$ integrals involve basis functions representing inner-shell orbitals. These orbitals are little changed on molecule formation, and one can eliminate the need to explicitly represent them by using an ***effective core potential*** (ECP) or ***pseudopotential*** (Section 13.20). The ECP is a one-electron operator that replaces those two-electron Coulomb and exchange operators in the valence-electrons' Hartree–Fock equation $\hat{F}\phi_i = \varepsilon_i\phi_i$ that arise from interactions between the core electrons and the valence electrons. ECPs are derived from ab initio calculations on atoms. For compounds of main-group elements, calculations that use properly chosen ECPs give almost the same results as comparable all-electron ab initio calculations. For transition elements, obtaining accurate results with ECPs is harder. ECPs are reviewed in M. Krauss and W. J. Stevens, *Ann. Rev. Phys. Chem.*, **35**, 357 (1984).

Not only must the integrals $(rs|tu)$ be calculated, they must be stored and then recalled from memory as their values are needed in each SCF iteration (recall the SCF example in Section 13.16). Typically, 5 to 50 iterations are needed to achieve SCF convergence. For the large basis sets used in modern ab initio calculations, the number of $(rs|tu)$ values to be stored exceeds the internal (core) memory capacity of most computers, and the $(rs|tu)$ values must be stored on external memory disk drives. (Certain supercomputers with very large internal memories allow all the integrals to be stored in core memory.) Locating and reading in the value of an integral from external memory is a relatively slow process. Moreover, with very large basis sets, the number of integrals to be stored may even exceed the capacity of the available external memory.

To avoid the use of external storage memory, Almlöf developed the ***direct SCF method*** (not to be confused with the direct CI method of Section 13.21), in which no $(rs|tu)$ integrals are stored, but each two-electron integral is recomputed each time its value is needed. The direct SCF method allows ab initio calculations of large molecules and is an option in GAUSSIAN 88 and 90. Using a workstation computer and the direct SCF method, Ahlrichs and co-workers computed an ab initio SCF wave function for the 840-electron molecule $Si_{24}O_{60}H_{24}$ with a 3-21G basis set of 900 CGTFs (contracted from 1620 primitives) in 440 minutes [R. Ahlrichs et al., *Chem. Phys. Lett.*, **162**, 165 (1989)]. Almlöf and Lüthi used a supercomputer and the direct SCF method to find an ab initio SCF wave function for the 930-electron aromatic hydrocarbon $C_{150}H_{30}$ (a graphitelike species with H's attached to peripheral carbons to avoid dangling bonds on the C's); their DZ basis set contained 1560 CGTFs (and 2490 primitives) [A. Almlöf and H. P. Lüthi in K. F. Jensen and D. G. Truhlar (eds.), *Supercomputer Research in Chemistry and Chemical Engineering*, American Chemical Society, Washington, D.C., 1987, pp. 35–48]. The high symmetry of these species (point groups $\mathcal{O}_h$ and $\mathcal{D}_{6h}$) greatly speeds up the calculations; SCF calculations on unsymmetrical molecules this large are not currently possible.

The **pseudospectral** (PS) **method** for solving the Hartree–Fock equations (developed by Friesner and co-workers) uses both a basis-set expansion of each MO and a representation of each MO as a set of numerical values at chosen grid points in three-dimensional space. This allows the SCF equations to be solved without the necessity of explicitly evaluating two-electron integrals. The method was found to be 2 to 9 times as fast as conventional ab initio SCF programs in single-geometry test calculations on six molecules and may be useful for SCF calculations on very large molecules [M. N. Ringnalda, M. Belhadj, and R. A. Friesner, *J. Chem. Phys.*, **93**, 3397 (1990)].

15.6 THE SCF MO TREATMENT OF H₂O

For a minimal-basis-set MO treatment of H_2O (Fig. 15.1), we start with the $O1s$, $O2s$, $O2p_x$, $O2p_y$, and $O2p_z$ inner-shell and valence oxygen AOs and the H_11s and H_21s valence AOs of the hydrogen atoms. Linear combinations of these seven basis AOs give LCAO approximations to the seven lowest MOs of water. As stated in Section 15.3, the MOs can be chosen so that upon application of the molecular symmetry operators each MO transforms according to one of the symmetry species of the molecular point group. To aid in choosing the right linear combinations of AOs, we examine the symmetry behavior of the AOs.

The $1s$ and $2s$ oxygen AOs are spherically symmetric; rotation about the $C_2(z)$ axis and reflection in the xz or yz plane has no effect on them. They thus belong to the totally symmetric symmetry species a_1 of (15.3). The effect of a $\hat{C}_2(z)$ rotation on the oxygen $2p_y$ AO is shown in Fig. 15.2; this rotation sends $O2p_y$ into its negative (see also Fig. 12.11). Reflection in the molecular plane sends $O2p_y$ into itself, while reflection in the xz plane sends it into its negative.

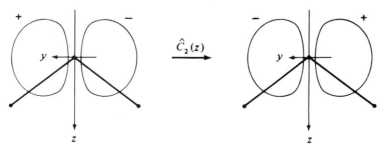

Figure 15.2 The effect of $\hat{C}_2(z)$ on the $2p_y$ oxygen AO in H_2O.

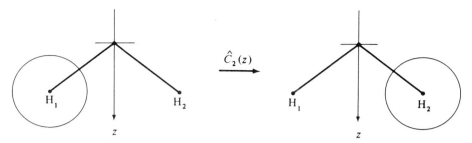

Figure 15.3 The effect of a $\hat{C}_2(z)$ rotation on the $H_1 1s$ AO in H_2O.

The symmetry eigenvalues of $O2p_y$ are 1, -1, -1, and 1; the symmetry species of this AO is b_2. Similarly, the symmetry species of $O2p_x$ and $O2p_z$ are found to be b_1 and a_1, respectively.

Now for the hydrogen $1s$ AOs. Reflection in the yz plane leaves each of them unchanged. However, rotation by 180° about the z axis sends $H_1 1s$ into $H_2 1s$ and vice versa (Fig. 15.3); the result is the same for reflection in the xz plane. The $H_1 1s$ and $H_2 1s$ functions are *not* eigenfunctions of $\hat{O}_{C_2(z)}$ and $\hat{O}_{\sigma_v(xz)}$, and they therefore do not transform according to any of the symmetry species of H_2O.

As a preliminary step in finding the MOs of a molecule, it is helpful (but not essential) to construct linear combinations of the original basis AOs such that each linear combination does transform according to one of the molecular symmetry species. Such linear combinations are called **symmetry orbitals** or **symmetry-adapted basis functions**. The symmetry orbitals are used as the basis functions χ_s in the expansions $\phi_i = \Sigma_s c_{si} \chi_s$ [Eq. (13.156)] of the MOs ϕ_i. *The use of basis functions that transform according to the molecular symmetry species simplifies the calculation by putting the secular determinant in block-diagonal form*; this will be illustrated below.

Each oxygen AO transforms according to one of the symmetry species of water and can serve as a symmetry orbital. However, neither of the two hydrogen $1s$ AOs belongs to a symmetry species of water, and we must construct two symmetry orbitals from these AOs. Consider the following linear combinations:

$$H_1 1s + H_2 1s \tag{15.13}$$

$$H_1 1s - H_2 1s \tag{15.14}$$

We have

$$\hat{O}_{C_2(z)}(H_1 1s + H_2 1s) = H_2 1s + H_1 1s$$

$$\hat{O}_{C_2(z)}(H_1 1s - H_2 1s) = H_2 1s - H_1 1s$$

Thus (15.13) and (15.14) are eigenfunctions of $\hat{O}_{C_2(z)}$ with eigenvalues $+1$ and -1, respectively. Examination of the effects of the other three symmetry operators shows (15.13) and (15.14) to belong to the symmetry species a_1 and b_2, respectively. We shall not bother to normalize the symmetry orbitals (15.13) and

(15.14). The seven basis symmetry functions and their symmetry species are then

χ_1	χ_2	χ_3	χ_4	χ_5	χ_6	χ_7
$H_1 1s + H_2 1s$	$O1s$	$O2s$	$O2p_z$	$H_1 1s - H_2 1s$	$O2p_y$	$O2p_x$
a_1	a_1	a_1	a_1	b_2	b_2	b_1

$$\text{(15.15)}$$

Now consider the SCF secular determinant $\det(F_{rs} - \varepsilon_i S_{rs})$ [Eq. (13.159)]. We assert that

$$F_{rs} \equiv \langle \chi_r | \hat{F} | \chi_s \rangle = 0 \tag{15.16}$$

whenever χ_r and χ_s belong to different symmetry species. This result follows from the theorem [Eq. (7.50)] that $\langle g_j | \hat{B} | g_k \rangle = 0$ if g_j and g_k are eigenfunctions of a Hermitian operator $\hat{A}$ with different eigenvalues, where $\hat{A}$ commutes with $\hat{B}$. The symmetry orbitals of water are eigenfunctions of the symmetry operators, each of which commutes with the electronic Hamiltonian and with the Fock operator $\hat{F}$. Symmetry orbitals χ_r and χ_s that belong to different symmetry species differ in at least one symmetry eigenvalue. Hence (15.16) follows. Moreover, since two eigenfunctions of a Hermitian operator that correspond to different eigenvalues are orthogonal, we have

$$S_{rs} \equiv \langle \chi_r | \chi_s \rangle = 0 \tag{15.17}$$

whenever χ_r and χ_s belong to different symmetry species. From (15.16) and (15.17), it follows that the use of symmetry orbitals puts the secular determinant of water in block-diagonal form, each block corresponding to a different symmetry species; the blocks are 4×4, 2×2, and 1×1. [For molecules with degenerate symmetry species (E, T, and so on), the symmetry orbitals of the degenerate species are not necessarily eigenfunctions of the symmetry operators; nevertheless, the symmetry orbitals still put the secular determinant in block-diagonal form, as can be shown using group theory.]

The set of Roothaan simultaneous equations (13.157) then breaks up into one set of four simultaneous equations, one set of two simultaneous equations, and one set of one "simultaneous" equation (Section 9.6). The first set contains matrix elements involving only the four a_1 symmetry orbitals. Therefore, four of the lowest seven water MOs are linear combinations of the four a_1 symmetry orbitals; these four MOs must have a_1 symmetry. Similarly, we have two MOs of b_2 symmetry and one MO of b_1 symmetry. The symmetry orbitals are *not* (in general) the MOs, but each MO must be a linear combination of those symmetry orbitals having the same symmetry species as the MO. The forms of the lowest MOs of H_2O are then

$$
\begin{aligned}
\phi_1 &= c_{11}\chi_1 + c_{21}\chi_2 + c_{31}\chi_3 + c_{41}\chi_4 & \phi_5 &= c_{55}\chi_5 + c_{65}\chi_6 \\
\phi_2 &= c_{12}\chi_1 + c_{22}\chi_2 + c_{32}\chi_3 + c_{42}\chi_4 & \phi_6 &= c_{56}\chi_5 + c_{66}\chi_6 \\
\phi_3 &= c_{13}\chi_1 + c_{23}\chi_2 + c_{33}\chi_3 + c_{43}\chi_4 & \phi_7 &= \chi_7 \\
\phi_4 &= c_{14}\chi_1 + c_{24}\chi_2 + c_{34}\chi_3 + c_{44}\chi_4
\end{aligned}
\tag{15.18}
$$

The next step in the SCF MO calculation is to choose explicit forms for the seven AOs. The orbital energies and the coefficients of the symmetry orbitals are then found using Roothaan's equations.

Pitzer and Merrifield did an H_2O minimal-basis-set calculation, representing each AO by a single STO [R. M. Pitzer and D. P. Merrifield, *J. Chem. Phys.*, **52**, 4782 (1970); S. Aung, R. M. Pitzer, and S. I. Chan, *J. Chem. Phys.*, **49**, 2071 (1968)]. They optimized the orbital exponents, finding 1.27 for H1s, 7.66 for O1s, 2.25 for O2s, and 2.21 for O2p. (To optimize the exponents, one must repeat the entire SCF iterative calculation for several different sets of orbital exponents to locate the set that gives the minimum energy. Since orbital-exponent optimization is time consuming, it is only feasible for small molecules.) The orbital energies in hartrees are found to be the following: $1a_1$, -20.56; $2a_1$, -1.28; $1b_2$, -0.62; $3a_1$, -0.47; $1b_1$, -0.40. The ground-state electronic configuration of this ten-electron molecule is

$$(1a_1)^2(2a_1)^2(1b_2)^2(3a_1)^2(1b_1)^2 \tag{15.19}$$

The ground state has a closed-subshell configuration and is a $^1A_1$ state.

The five lowest SCF MOs found by Pitzer and Merrifield at the experimental geometry are

$$1a_1 = 1.000(O1s) + 0.015(O2s_\perp) + 0.003(O2p_z) - 0.004(H_1 1s + H_2 1s)$$

$$2a_1 = -0.027(O1s) + 0.820(O2s_\perp) + 0.132(O2p_z) + 0.152(H_1 1s + H_2 1s)$$

$$1b_2 = 0.624(O2p_y) + 0.424(H_1 1s - H_2 1s) \tag{15.20}$$

$$3a_1 = -0.026(O1s) - 0.502(O2s_\perp) + 0.787(O2p_z) + 0.264(H_1 1s + H_2 1s)$$

$$1b_1 = O2p_x$$

The $O2s_\perp$ orbital in (15.20) is an orthogonalized orbital [Eq. (13.183)]:

$$O2s_\perp = 1.028[O2s - 0.2313(O1s)] \tag{15.21}$$

where O2s is the ordinary 2s STO:

$$O2s = 2.25^{5/2}\pi^{-1/2}3^{-1/2}r_O \exp(-2.25r_O)$$

where r_O is the distance to the oxygen nucleus. The MO approximation to the ground-state wave function of H_2O is the 10×10 Slater determinant

$$|1a_1\overline{1a_1}2a_1\overline{2a_1}1b_2\overline{1b_2}3a_1\overline{3a_1}1b_1\overline{1b_1}| \tag{15.22}$$

Consider the nature of the MOs. The lowest MO, $1a_1$, is essentially a pure nonbonding 1s oxygen AO, which is hardly surprising.

The oxygen part of the $2a_1$ MO is mostly O2s with some $O2p_z$ mixed in. This mixing in (hybridization) of $O2p_z$ adds to the value of the O2s AO along the positive z axis (which points toward the hydrogens; Fig. 15.1) and subtracts from O2s along the negative z axis. The combination of the hybridized O2s and $O2p_z$ orbitals with the $H_1 1s$ and $H_2 1s$ orbitals in the $2a_1$ MO then gives electron probability-density buildup in the region enclosed by the three nuclei. Therefore, the $2a_1$ MO contributes to the bonding in water.

Consider the $1b_2$ MO. The $2p_y$ oxygen AO has its positive lobe on the H_1

side of the molecule, and so the positive lobe of $O2p_y$ adds to H_11s in the $1b_2$ MO, giving electron charge buildup between the H_1 and O nuclei. Similarly, the negative lobe of $O2p_y$ adds to $-H_21s$, giving charge buildup between O and H_2 in this MO. Hence $1b_2$ is a bonding MO.

The hybridization of the $2s$ and $2p_z$ oxygen AOs in $3a_1$ builds up electron probability density in the region around the negative z axis, away from the hydrogens, giving this MO substantial lone-pair character. On the positive side of the z axis, the oxygen $2s$ and $2p_z$ AOs tend to cancel each other, and we get little bonding overlap between the oxygen hybrid and the hydrogen AOs. Alternatively, we can look separately at the overlap between $O2s_\perp$ and the hydrogens (which is negative or antibonding) and the overlap between $O2p_z$ and the hydrogens (which is positive or bonding); because of the approximate cancellation of these overlaps (see Section 15.7), the $3a_1$ MO has little bonding character and is best described as mainly a lone-pair MO.

The $1b_1$ MO is a nonbonding lone-pair $2p_x$ oxygen AO.

Figure 15.4 shows the shapes of the bonding MOs $2a_1$ and $1b_2$. Note that their symmetry eigenvalues are as given by (15.3). [For accurately plotted H_2O MO contours, see T. H. Dunning, R. M. Pitzer, and S. Aung, *J. Chem. Phys.*, **57**, 5044 (1972).]

Each of the SCF bonding MOs in (15.20) is delocalized over the entire molecule and does not resemble a chemical bond. Hence the reader may be wondering about the relation of these MOs to the picture presented in most freshman chemistry books, where one bonding MO points along the O—H_1 bond and the other points along the O—H_2 bond. We shall discuss this point in Section 15.8.

The unoccupied $4a_1$ and $2b_2$ MOs of water calculated by Pitzer and Merrifield (for nonoptimized, Slater-rule exponents) are

$$4a_1 = 0.08(O1s) + 0.84(O2s_\perp) + 0.70(O2p_z) - 0.75(H_11s + H_21s)$$
$$2b_2 = 0.99(O2p_y) - 0.89(H_11s - H_21s)$$

$$(15.23)$$

For these two MOs, the opposite signs of the oxygen and hydrogen AO coefficients give charge depletion between the nuclei. These MOs are antibonding.

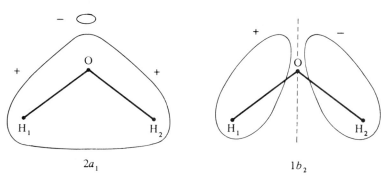

$2a_1$ $1b_2$

Figure 15.4 Sketches of the two main bonding MOs of H_2O.

The unoccupied orbitals (15.23) are called virtual orbitals. Because they were calculated for the electron configuration (15.19), they are not accurate representations of the SCF orbitals actually occupied in H_2O excited electronic states. When the electron configuration changes, the interorbital electronic interactions change, thereby changing the forms of all the SCF orbitals. We can, however, use the virtual orbitals and their calculated energies as rough approximations to the higher SCF orbitals and energies.

The formation of the H_2O MOs from the separated-atoms AOs is illustrated schematically in Fig. 15.5; the dashed lines indicate which AOs contribute significantly to each MO. Note that only AOs of roughly the same energy combine to a significant degree in a given MO. [This fact is explained by the $1/(E_n^{(0)} - E_m^{(0)})$ term in Eq. (9.28).] We can thus get a qualitative idea of the MOs without doing any calculations by using the rules that only symmetry orbitals of the same symmetry species combine and that only AOs of comparable energy (Fig. 11.2) contribute significantly to a given MO.

Table 15.1 lists some of the many SCF calculations on the H_2O ground state. The experimental equilibrium electronic energy of H_2O is -76.480 hartrees. All the calculations listed are nonrelativistic. The only significant relativistic correction will occur in the $1s$ inner-shell electrons of oxygen (see Section 11.7); since this shell remains essentially unchanged on molecule formation, one can use

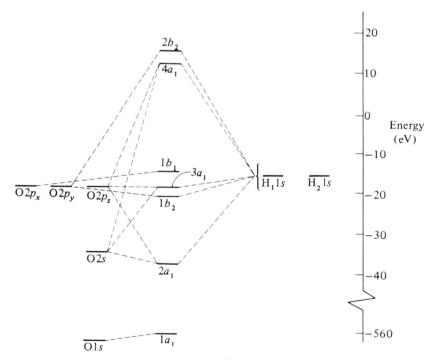

Figure 15.5 Formation of the H_2O MOs from the minimal-basis AOs. The five lowest MOs are filled in the ground state. (Note the break in the scale.)

TABLE 15.1 H_2O SCF MO Calculations[a]

Reference[b]	Basis Set[c]	Energy/E_h	μ/D	θ	R_{OH}/Å
Lathan et al.	STO-3G	−74.97	1.69	100.0°	0.990
Binkley et al.	3-21G	−75.59	2.44	107.6°	0.967
Pitzer, Merrifield	Minimal STO	−75.70	1.92	100.3°	0.990
Lathan et al.	4-31G	−75.91	2.52	111.2°	0.950
Moccia	28 1-center STOs	−75.92	2.08	106.5°	0.960
Dunning	DZ CGTFs	−76.009	2.68		
Harihan, Pople	6-31G*	−76.010	2.19	105.5°	0.947
Amos	6-31G**	−76.024	2.15	106.0°	0.943
Bauschlicher, Taylor	DZP$_1$ CGTFs	−76.041			
Scuseria, Schaefer	DZP$_2$ CGTFs	−76.047	2.18	106.6°	0.944
Dunning et al.	41 CGTFs	−76.062	2.08	106.6°	0.941
Rosenberg et al.	39 STOs	−76.064	2.00	106.1°	0.940
Feller et al.	120 CGTFs	−76.0672	1.98		
Feller et al.	140 CGTFs	−76.0673	1.98		
Estimated Hartree–Fock values[d]		−76.0675		106.3°	0.940
Nonrelativistic fixed-nuclei energy		−76.440			
Experimental values		−76.480	1.85	104.5°	0.958

[a]Energy/E_h is the total electronic energy including nuclear repulsion in hartrees at either the calculated or the experimental equilibrium geometry; μ, θ, and R_{OH} are the calculated electric dipole moment, equilibrium bond angle, and equilibrium bond length.

[b]W. J. Lathan et al., *J. Am. Chem. Soc.*, **93**, 6377 (1971); J. S. Binkley, J. A. Pople, and W. J. Hehre, *J. Am. Chem. Soc.*, **102**, 939 (1980); R. M. Pitzer and D. P. Merrifield, *J. Chem. Phys.*, **52**, 4782 (1970); W. J. Lathan et al., op. cit.; R. Moccia, *J. Chem. Phys.*, **40**, 2186 (1964); T. H. Dunning, *J. Chem. Phys.*, **53**, 2823 (1970); P. C. Harihan and J. A. Pople, *Mol. Phys.*, **27**, 209 (1974); R. D. Amos, *J. Chem. Soc., Faraday Trans. 2*, **83**, 1595 (1987); C. W. Bauschlicher and P. A. Taylor, *J. Chem. Phys.*, **85**, 2279 (1986); G. E. Scuseria and H. F. Schaefer, III, *Chem. Phys. Lett.*, **146**, 23 (1988); T. H. Dunning, R. M. Pitzer, and S. Aung, *J. Chem. Phys.*, **57**, 5044 (1972); B. J. Rosenberg and I. Shavitt, *J. Chem. Phys.*, **63**, 2162 (1975) and B. J. Rosenberg, W. C. Ermler, and I. Shavitt, *J. Chem. Phys.*, **65**, 4072 (1976); D. Feller, C. M. Boyle, and E. R. Davidson, *J. Chem. Phys.*, **86**, 3424 (1987).

[c]The DZP$_1$ and DZP$_2$ sets are DZ + P sets that differ only in the orbital exponents of the polarization functions.

[d]The Hartree–Fock energy is given for the experimental geometry. The Hartree–Fock geometry is from R. D. Amos, *J. Chem. Soc., Faraday Trans. 2*, **83**, 1595 (1987).

the relativistic energy correction calculated for the O atom as the relativistic correction for H_2O. In addition, there is a small correction due to motion of the center of mass of the nuclei relative to the center of mass of the molecule. When these two corrections are subtracted from the experimental energy of −76.480 hartrees, we obtain the nonrelativistic fixed-nuclei H_2O energy as −76.440 hartrees.

All the SCF MO calculations listed are multicenter except for Moccia's one-center calculation. The size and quality of the basis set used determine how closely the Hartree–Fock limit is approached.

Although the SCF MO calculations give good geometries and fairly good dipole moments, they give poor binding energies. For the minimal-basis Pitzer–Merrifield calculation, the difference between the energy of two isolated hydrogen atoms and one oxygen atom (as calculated with the minimal basis set) and the calculated H_2O energy gives a binding energy of only 4.5 eV, compared with the experimental value 10.1 eV. The near-Hartree–Fock calculations of Dunning, Pitzer, and Aung give a binding energy of 6.9 eV. This large error in dissociation energy is typical of Hartree–Fock calculations, as we saw for diatomics. When H_2O is formed from 2H + O, two new electron pairs are formed. Two electrons paired in the same MO move through the same region of space, and hence the correlation energy of such a pair is high. The Hartree–Fock calculation does not take into account this extra correlation in the molecule (as compared with the separated atoms) and hence gives too small a binding energy.

Hartree–Fock orbital energies have experimental as well as theoretical significance. In 1933, Koopmans gave arguments that indicate that the energy required to remove an electron from a closed-shell atom or molecule is reasonably well approximated by minus the orbital energy ε of the AO or MO from which the electron is removed. A partial justification of this result is the fact that, if we neglect the change in the form of the MOs that occurs when the molecule is ionized, then the difference between the Hartree–Fock energies of the ion and the neutral closed-shell molecule can be shown to equal the orbital energy of the MO from which the electron was removed (see Problem 15.13).

The energy needed to remove an electron from an MO of a molecule can be found experimentally using photoelectron spectroscopy; here, one uses photons of known energy to knock electrons out of gas-phase molecules and measures the kinetic energies of the emitted electrons. A comparison of minus the near-Hartree–Fock orbital energies (Rosenberg and Shavitt, Table 15.1) with the experimentally observed ionization energies from the various H_2O MOs follows, where the energies are in electron volts and the experimental values are in parentheses: $1a_1$, 559.5 (539.7); $2a_1$, 36.7 (32.2); $1b_2$, 19.5 (18.5); $3a_1$, 15.9 (14.7); $1b_1$, 13.8 (12.6). Koopmans'-theorem ionization potentials are somewhat inaccurate because of (1) neglect of the change in the forms of the MOs that occurs on ionization and (2) neglect of the change in correlation energy between the neutral molecule and the ion.

15.7 POPULATION ANALYSIS

A widely used (and widely criticized) method to analyze SCF wave functions is population analysis, introduced by Mulliken. He proposed a method that apportions the electrons of an n-electron molecule into net populations n_r in the basis functions χ_r and overlap populations $n_{r\text{-}s}$ for all possible pairs of basis functions.

For the set of basis functions $\chi_1, \chi_2, \ldots, \chi_b$, each MO ϕ_i has the form $\phi_i = \sum_s c_{si}\chi_s = c_{1i}\chi_1 + c_{2i}\chi_2 + \cdots + c_{bi}\chi_b$. For simplicity, we shall assume that the c_{si}'s and χ_s's are real. The probability density associated with one electron in ϕ_i is

$$|\phi_i|^2 = c_{1i}^2\chi_1^2 + c_{2i}^2\chi_2^2 + \cdots + 2c_{1i}c_{2i}\chi_1\chi_2 + 2c_{1i}c_{3i}\chi_1\chi_3 + 2c_{2i}c_{3i}\chi_2\chi_3 + \cdots$$

Integrating this equation over three-dimensional space and using the fact that ϕ_i and the χ_s's are normalized, we get

$$1 = c_{1i}^2 + c_{2i}^2 + \cdots + 2c_{1i}c_{2i}S_{12} + 2c_{1i}c_{3i}S_{13} + 2c_{2i}c_{3i}S_{23} + \cdots \quad (15.24)$$

where the S's are overlap integrals: $S_{12} = \int \chi_1 \chi_2 \, dv_1 \, dv_2$, and so on. Mulliken proposed that the terms in (15.24) be apportioned as follows. One electron in the MO ϕ_i contributes c_{1i}^2 to the net population in χ_1, c_{2i}^2 to the net population in χ_2, and so on, and contributes $2c_{1i}c_{2i}S_{12}$ to the overlap population between χ_1 and χ_2, $2c_{1i}c_{3i}S_{13}$ to the overlap population between χ_1 and χ_3, and so on.

Let there be n_i electrons in the MO ϕ_i ($n_i = 0, 1, 2$) and let $n_{r,i}$ and $n_{r\text{-}s,i}$ symbolize the contributions of electrons in the MO ϕ_i to the net population in χ_r and to the overlap population between χ_r and χ_s, respectively. We have

$$n_{r,i} = n_i c_{ri}^2 , \qquad n_{r\text{-}s,i} = n_i(2c_{ri}c_{si}S_{rs}) \qquad (15.25)$$

By summing over the occupied MOs, we obtain the Mulliken **net population** n_r in χ_r and the **overlap population** $n_{r\text{-}s}$ for the pair χ_r and χ_s as

$$n_r = \sum_i n_{r,i} \quad \text{and} \quad n_{r\text{-}s} = \sum_i n_{r\text{-}s,i}$$

The sum of all the net and overlap populations equals the total number of electrons in the molecule (Problem 15.16): $\sum_r n_r + \sum_{r>s} \sum_s n_{r\text{-}s} = n$.

EXAMPLE   For the H_2O MOs in (15.20), calculate the net and overlap population contributions from the $2a_1$ MO and find n_r for each basis function. Use $H_1 1s$ and $H_2 1s$ as basis functions, rather than the symmetry-adapted basis functions.

To find overlap populations, we need the overlap integrals. Since an orthogonalized $2s$ AO is used, all the basis functions centered on oxygen are mutually orthogonal, and the overlap populations are zero for all pairs of functions centered on oxygen. (For extended basis sets, this is not true. For example, a DZ basis set uses two $1s$-type functions on oxygen and these functions are not orthogonal to each other.) Overlap integrals between basis STOs centered on different atoms can be found by interpolation in the tables of R. S. Mulliken et al., *J. Chem. Phys.*, **17**, 1248 (1949). One finds (Problem 15.17)

$$\langle H_1 1s | O 1s \rangle = \langle H_2 1s | O 1s \rangle = 0.054 , \quad \langle H_1 1s | O 2s_\perp \rangle = \langle H_2 1s | O 2s_\perp \rangle = 0.471$$

$$\langle H_1 1s | O 2p_y \rangle = -\langle H_2 1s | O 2p_y \rangle = 0.319 , \quad \langle H_1 1s | O 2p_z \rangle = \langle H_2 1s | O 2p_z \rangle = 0.247$$

$$\langle H_1 1s | H_2 1s \rangle = 0.238$$

The contributions to the net populations in the basis functions from the two electrons in the $2a_1$ MO are given by (15.25) and (15.20) as $n_{O1s,2a_1} = 2(-0.027)^2 = 0.0015$, $n_{O2s_\perp,2a_1} = 2(0.820)^2 = 1.345$, $n_{O2p_z,2a_1} = 0.035$, $n_{H_1 1s,2a_1} = 2(0.152)^2 = 0.046$, $n_{H_2 1s,2a_1} = 0.046$. The $2a_1$ contributions to the nonzero overlap populations are given by (15.25) and (15.20) as

$$n_{O1s\text{-}H_1 1s,2a_1} = 2(2)(-0.027)(0.152)(0.054) = -0.0009 = n_{O1s\text{-}H_2 1s,2a_1}$$

$$n_{O2s_\perp\text{-}H_1 1s,2a_1} = 0.235 = n_{O2s_\perp\text{-}H_2 1s,2a_1}$$

$$n_{O2p_z\text{-}H_1 1s,2a_1} = 0.020 = n_{O2p_z\text{-}H_2 1s,2a_1} , \quad n_{H_1 1s\text{-}H_2 1s,2a_1} = 0.022$$

The net population of $O1s$ is found from (15.25) and (15.20) as the sum of contributions from each occupied MO:

$$n_{O1s} = 2(1.000)^2 + 2(-0.027)^2 + 2(-0.026)^2 = 2.00$$

The net populations for the other basis functions are (Problem 15.18a)

$$n_{O2s_\perp} = 1.85\,, \quad n_{O2p_x} = 2.00\,, \quad n_{O2p_y} = 0.78\,, \quad n_{O2p_z} = 1.27$$

$$n_{H_1 1s} = 0.54_5\,, \quad n_{H_2 1s} = 0.54_5$$

To decide whether the MO ϕ_i in a covalent molecule is bonding, we examine the sum of those overlap-population contributions $n_{r\text{-}s,i}$ for which the basis functions χ_r and χ_s lie on different atoms. If this interatomic overlap population contribution is substantially positive, the MO is bonding; if it is substantially negative, the MO is antibonding. If it is zero or near zero, the MO is nonbonding.

For example, for the $3a_1$ MO in (15.20), overlap of O1s with $H_1 1s$ contributes $2(2)(-0.026)(0.264)(0.054) = -0.001_5$, overlap of O1s with $H_2 1s$ contributes -0.001_5, overlap of $O2s_\perp$ with $H_1 1s$ contributes $2(2)(-0.502) \times (0.264)(0.471) = -0.250$, overlap of $O2s_\perp$ with $H_2 1s$ contributes -0.250, overlap of $O2p_z$ with $H_1 1s$ contributes 0.205 and with $H_2 1s$ contributes 0.205, and overlap of $H_1 1s$ with $H_2 1s$ contributes $2(2)(0.264)^2(0.238) = 0.066$. Summing, we get an interatomic overlap population of -0.03 for the $3a_1$ MO. This value is near zero, indicating a nonbonding (lone-pair) MO. The interatomic overlap population for $2a_1$ is found to be 0.53 and that for $1b_2$ is 0.50 (Problem 15.18b). These are bonding MOs. For the inner-shell $1a_1$ MO, we get 0.00.

Instead of apportioning the electrons into net populations in basis functions and overlap populations for pairs of basis functions, it is convenient for some purposes to apportion the electrons among the basis functions only, with no overlap populations. Mulliken proposed that this be done by splitting each overlap population $n_{r\text{-}s}$ equally between the basis functions χ_r and χ_s. For each basis function χ_r, this gives a **gross population** N_r in χ_r that equals the net population n_r plus one-half the sum of the overlap populations between χ_r and all other basis functions:

$$N_r = n_r + \frac{1}{2} \sum_{s \neq r} n_{r\text{-}s} \tag{15.26}$$

The sum of all the gross populations equals the number of electrons in the molecule: $\sum_{r=1}^{b} N_r = n$.

For example, the contribution to the gross population of $O2s_\perp$ from the $2a_1$ MO in (15.20) is

$$N_{O2s_\perp, 2a_1} = 2[(0.820)^2 + (0.820)(0.152)(0.471) + (0.820)(0.152)(0.471)] = 1.58$$

One finds these other contributions to the gross population of $O2s_\perp$: 0.00 from $1a_1$, 0.25 from $3a_1$, zero from $1b_2$ and $1b_1$. Addition of these contributions gives a gross population (or occupation number) of 1.83 for $O2s_\perp$. Carrying out the calculation for the other basis functions, one finds (Problem 15.18c) the following gross populations (where the contributions are listed in the order $1a_1$, $2a_1$, $1b_2$, $3a_1$, $1b_1$): $N_{O1s} = 2.00 + 0.00 + 0 + 0.00 + 0 = 2.00$; $N_{O2s_\perp} = 1.83$; $N_{O2p_x} = 0 + 0 + 0 + 0 + 2 = 2$; $N_{O2p_y} = 0 + 0 + 1.12 + 0 + 0 = 1.12$; $N_{O2p_z} = 0 + 0.05_5 + 0 + 1.44_5 + 0 = 1.50$; $N_{H_1 1s} = 0.00 + 0.18_4 + 0.44_2 + 0.150 + 0 = 0.77_6$; $N_{H_2 1s} = 0.77_6$.

Addition of the gross populations for all basis functions centered on atom B gives the **gross atomic population** N_B for atom B:

$$N_B = \sum_{r \in B} N_r$$

where the notation $r \in B$ denotes all basis functions centered on atom B. Provided all basis functions are atom centered (this is usually true), the sum of the gross atomic populations equals the number of electrons in the molecule. The **net charge** q_B on atom B with atomic number Z_B is defined as

$$q_B \equiv Z_B - N_B$$

For example, for the Pitzer–Merrifield H_2O wave function, the gross atomic populations, found by summing the gross populations of the basis functions on each atom, are $N_O = 2.00 + 1.83 + 2 + 1.12 + 1.50 = 8.45$, $N_{H_1} = 0.77_6 = N_{H_2}$; the net charges are $q_O = 8 - 8.45 = -0.45$, $q_{H_1} = 1 - 0.77_6 = 0.22_4 = q_{H_2}$. As expected, the oxygen is negatively charged.

One should not put too much reliance on numbers calculated by population analysis. Mulliken's assignment of half the overlap population to each basis function is arbitrary and sometimes leads to unphysical results (see *Mulliken and Ermler, Diatomic Molecules*, pages 36–38, 88–89). Moreover, a small change in basis set can produce a large change in the calculated net charges. For example, net charges on each H atom in CH_4, NH_3, and H_2O calculated by the STO-3G and 3-21G basis sets are (*Hehre* et al., Section 6.6.2)

	CH_4	NH_3	H_2O
STO-3G	0.06	0.16	0.18
3-21G	0.20	0.28	0.36

Comparison of values calculated with the same basis set correctly shows the increasing charge on each H atom as the electronegativity increases from C to N to O, but comparison of values calculated with different basis sets could erroneously lead one to say that the C—H bond in CH_4 is more polar than the O—H bond in H_2O.

Many other methods have been proposed to assign charges to atoms in molecules. See P. Politzer et al., *Theor. Chim. Acta*, **38**, 101 (1975); J. Cioslowski, *J. Am. Chem. Soc.*, **111**, 8333 (1989); *Hehre* et al., Section 6.6.2.

15.8 LOCALIZED MOs

The idea of a chemical *bond* between a pair of atoms in a molecule is fundamental to chemistry. The experimental evidence supporting this concept is substantial. One can assign bond energies to various kinds of bonds and obtain good estimates for the heats of formation of most molecules by adding the energies of the individual bonds. Other molecular properties (for example, magnetic susceptibility, dipole moment) can also be analyzed as the sum of contributions from individual bonds (and lone pairs). Examination of the infrared spectrum of a compound containing an —O—H group shows a characteristic band near 3600 cm^{-1}; this O—H stretching vibrational band occurs at nearly the same frequency no matter whether it is HOH or HOCl or CH_3OH that is observed. The length of an O—H bond is nearly constant from molecule to molecule, about 0.96 Å.

The MO picture of H_2O presented in Section 15.6 appears to be gravely deficient, in that it is seemingly inconsistent with the existence of individual bonds in the molecule. Each of the bonding MOs in (15.20) is delocalized over the entire molecule; if one were to compare the bonding MOs for HOH and HOCl, one would find them to be quite different. Yet we know that the OH parts of these two molecules are similar. Actually, MO theory *can* explain the observed near invariance of a given kind of chemical bond, as we now show.

The MO approximation to the ground state of water is a Slater determinant of the form

$$|\phi_1\overline{\phi_1}\phi_2\overline{\phi_2}\phi_3\overline{\phi_3}\phi_4\overline{\phi_4}\phi_5\overline{\phi_5}| \tag{15.27}$$

Putting (15.27) into words, we say that two electrons are in each of the orthonormal MOs ϕ_1, ϕ_2, ϕ_3, ϕ_4, and ϕ_5. However, this MO description is not unique. The addition of a multiple of one column of a determinant to another column leaves the determinant unchanged in value. Adding column 7 to column 9 and column 8 to column 10 in (15.27), we get

$$|\phi_1\overline{\phi_1}\phi_2\overline{\phi_2}\phi_3\overline{\phi_3}\phi_4\overline{\phi_4}(\phi_4 + \phi_5)(\overline{\phi_4 + \phi_5})| \tag{15.28}$$

This determinant leads to the description of the molecule as having the MOs ϕ_1, ϕ_2, ϕ_3, ϕ_4, and $(\phi_4 + \phi_5)$ each doubly occupied. Despite the different verbal descriptions, the wave functions (15.27) and (15.28) are identical.

[In discussing SCF calculations, we used an orthogonalized $2s$ AO of the form $a(1s) + b(2s)$, where $2s$ is a (nodeless) $2s$ Slater-type orbital. This procedure is justified by use of the freedom of adding a multiple of one column of a determinant to another; the determinantal wave function for the oxygen atom is the same whether it is the $2s$ or the orthogonalized $2s_\perp$ AO that is used.]

An objection that can be raised to (15.28) is that the MO $(\phi_4 + \phi_5)$ is neither normalized nor orthogonal to ϕ_4. Consider, however, the Slater determinant

$$|\phi_1\overline{\phi_1}\phi_2\overline{\phi_2}\phi_3\overline{\phi_3}(b\phi_4 + c\phi_5)(\overline{b\phi_4 + c\phi_5})(c\phi_4 - b\phi_5)(\overline{c\phi_4 - b\phi_5})| \tag{15.29}$$

where b and c are any two real constants such that

$$b^2 + c^2 = 1 \tag{15.30}$$

We now show (15.29) and (15.27) to be the same wave function. Multiplication of columns 7 and 8 of (15.27) by b and columns 9 and 10 by $-b^{-1}$ gives

$$|\cdots b\phi_4(\overline{b\phi_4})(-b^{-1}\phi_5)(\overline{-b^{-1}\phi_5})|$$

This step multiplies (15.27) by $b^2(-b^{-1})^2 = 1$. Next, $-bc$ times column 9 is added to column 7 and $-bc$ times column 10 is added to column 8 to give

$$|\cdots (b\phi_4 + c\phi_5)(\overline{b\phi_4 + c\phi_5})(-b^{-1}\phi_5)(\overline{-b^{-1}\phi_5})|$$

Finally, c/b times column 7 is added to column 9, c/b times column 8 is added to column 10, and (15.30) is used. This gives

$$|\cdots (b\phi_4 + c\phi_5)(\overline{b\phi_4 + c\phi_5})(c\phi_4 - b\phi_5)(\overline{c\phi_4 - b\phi_5})|$$

This completes the proof. The orthonormality of the MOs $(c\phi_4 - b\phi_5)$ and $(b\phi_4 + c\phi_5)$ readily follows from (15.30) and the orthonormality of ϕ_4 and ϕ_5. Thus we can describe the molecule as having the orthonormal MOs

$$\phi_1, \quad \phi_2, \quad \phi_3, \quad b\phi_4 + c\phi_5, \quad c\phi_4 - b\phi_5 \qquad (15.31)$$

each doubly occupied.

There are an infinite number of other MO descriptions consistent with (15.27): Let d and e be any two real constants such that $d^2 + e^2 = 1$. If the procedure used to go from (15.27) to (15.29) is applied to columns 5, 6, 7, and 8 of (15.29), we end up with

$$|\phi_1\overline{\phi_1}\phi_2\overline{\phi_2}(d\phi_3 + be\phi_4 + ce\phi_5)(\overline{d\phi_3 + be\phi_4 + ce\phi_5})$$

$$\times (e\phi_3 - bd\phi_4 - cd\phi_5)(\overline{e\phi_3 - bd\phi_4 - cd\phi_5})(c\phi_4 - b\phi_5)(\overline{c\phi_4 - b\phi_5})|$$

We can thus describe the electronic configuration as having the orthonormal MOs

$$\phi_1, \phi_2, (d\phi_3 + be\phi_4 + ce\phi_5), (e\phi_3 - bd\phi_4 - cd\phi_5), (c\phi_4 - b\phi_5)$$

each doubly occupied. Continuing on in this manner, we can derive orthonormal MOs, each of which is a linear combination of all the original MOs ϕ_1, ϕ_2, ϕ_3, ϕ_4, ϕ_5. For the general conditions that must be satisfied by the coefficients in these linear combinations, see *Szabo and Ostlund*, Section 3.2.3.

Thus for a given closed-subshell electronic state of a molecule, there are many possible MO descriptions. The delocalized MOs (15.20) of H_2O are uniquely determined as the solutions of the SCF equations [Eq. (13.148)]

$$\hat{F}\phi_i = \varepsilon_i \phi_i, \qquad i = 1, 2, \ldots \qquad (15.32)$$

Hence the reader may be wondering how we can have other sets of MOs that also minimize the variational integral. Actually, Eq. (15.32) is not the most general equation satisfied by the MOs that minimize the variational integral. Instead, one finds in deriving the SCF equations that the Hartree–Fock MOs must satisfy

$$\hat{F}\phi_i = \sum_{j=1}^{N} \lambda_{ji}\phi_j, \qquad i = 1, \ldots, N \qquad (15.33)$$

where the sum runs over the occupied MOs. The λ_{ji}'s are certain constants (Lagrangian multipliers), which can be chosen arbitrarily, subject to the MO orthonormality requirement. Different choices of the λ_{ji}'s give different sets of MOs, but each such set minimizes the variational integral. For a closed-subshell configuration, one possible choice of the λ_{ji}'s can be shown to be

$$\lambda_{ji} = \varepsilon_i \delta_{ij} \qquad (15.34)$$

This choice reduces (15.33) to (15.32). [For the proof of (15.33) and (15.34), see S. M. Blinder, *Am. J. Phys.*, **33**, 431 (1965); *Szabo and Ostlund*, Chapter 3.]

The delocalized MOs [such as (15.20) for H_2O] that satisfy (15.32) are called the **canonical** SCF MOs. The canonical MOs are eigenfunctions of the Fock operator, which commutes with the molecular symmetry operators, and hence the canonical MOs each transform according to one of the possible molecular symmetry species. A set of MOs like (15.31) satisfies (15.33) but not (15.32) and does not necessarily transform according to the molecular symmetry species.

For an open-subshell configuration, one can also find canonical delocalized MOs that satisfy (15.32) and that transform according to the molecular symmetry species; however, the form of $\hat{F}$ is more complicated than for the closed-subshell case. [See C. C. J. Roothaan, *Rev. Mod. Phys.*, **32**, 179 (1960).]

Of the possible MO sets formed as linear combinations of the delocalized canonical MOs of water, a set that would appeal to a chemist is one for which the charge probability density of each bonding MO is localized in the region of one of the O—H bonds. There are many ways of taking linear combinations of the delocalized MOs to get such *localized* MOs. A natural requirement that the two localized bonding MOs in water should satisfy is that they be equivalent to each other; that is, each localized bonding MO of water should be transformed into the other by the $\hat{C}_2(z)$ symmetry operation (Fig. 15.7). Localized MOs that are permuted among one another by a symmetry operation that permutes equivalent chemical bonds are called *equivalent orbitals*. As we shall see, localized MOs reconcile MO theory with the chemist's intuitive picture of chemical bonding.

For water, chemical intuition (H—Ö—H) suggests the localized MOs to be a pair of equivalent bonding orbitals $b(OH_1)$ and $b(OH_2)$, an inner-shell $1s$ oxygen orbital $i(O)$, and two lone-pair equivalent MOs $l_1(O)$ and $l_2(O)$ on oxygen.

The most "localized" MOs are those that are most separated from one another; by this we mean that set of MOs for which the interelectronic repulsion between the different MOs viewed as "charge clouds" is a minimum. The energy of repulsion between the charge clouds of electron 1 in the MO ϕ_i and electron 2 in the MO ϕ_j is a Coulomb integral of the form (9.100). The total interorbital charge-cloud repulsion energy is then

$$4 \sum_i \sum_{j>i} \int \int |\phi_i(1)|^2 |\phi_j(2)|^2 \frac{1}{r_{12}} \, dv_1 \, dv_2 \qquad (15.35)$$

where the sums run over the occupied MOs. (The factor 4 comes from the four interorbital repulsions of the electrons in each MO pair.) We define the localized MOs as those orthonormal MOs that minimize (15.35). [It turns out that minimization of (15.35) implies the maximization of the total *intra*orbital electron repulsion and the minimization of the magnitude of the (interorbital) exchange energy.] This definition was originally suggested by Lennard-Jones and Pople and has been applied to many molecules by Edmiston and Ruedenberg [C. Edmiston and K. Ruedenberg, *Rev. Mod. Phys.*, **35**, 457 (1963); *J. Chem. Phys.*, **43**, S97 (1965); P.-O. Löwdin, ed., *Quantum Theory of Atoms, Molecules, and the Solid State*, Academic Press, New York, 1966, pages 263–280].

The MOs that minimize (15.35) are called the *energy-localized* MOs. For molecules with symmetry, we expect that the energy-localized MOs will also be equivalent orbitals, and this is borne out by the calculated energy-localized MOs. Energy-localized MOs are thus a generalization of equivalent MOs.

Liang and Taylor calculated the energy-localized MOs for H_2O, starting with the minimal-basis canonical MOs (15.20). In terms of the canonical (delocalized) MOs, Liang and Taylor found [J. H. Liang, Ph.D. thesis, Ohio State University, 1970; quoted in F. Franks, ed., *Water*, Vol. 1, Plenum, 1972, p. 42]

$$i(O) = 0.99(1a_1) - 0.12(2a_1) \qquad\qquad + 0.06(3a_1)$$

$$b(OH_1) = 0.05(1a_1) + 0.57(2a_1) + 0.71(1b_2) + 0.42(3a_1)$$

$$b(OH_2) = 0.05(1a_1) + 0.57(2a_1) - 0.71(1b_2) + 0.42(3a_1) \qquad (15.36)$$

$$l_1(O) = 0.08(1a_1) + 0.42(2a_1) \qquad\qquad - 0.57(3a_1) - 0.71(1b_1)$$

$$l_2(O) = 0.08(1a_1) + 0.42(2a_1) \qquad\qquad - 0.57(3a_1) + 0.71(1b_1)$$

The lone-pair $1b_1$ MO ($2p_{xO}$) is equally divided between the equivalent lone-pair localized orbitals ($\sqrt{\tfrac{1}{2}} = 0.71$). The inner-shell $1a_1$ MO contributes substantially to only the $i(O)$ MO. The bonding $1b_2$ MO is equally divided between the two bonding localized MOs. The bonding $2a_1$ MO makes substantial contributions to the two bonding localized MOs and smaller, but still substantial, contributions to the lone-pair localized MOs. The largely lone-pair $3a_1$ MO makes substantial contributions to the lone-pair localized MOs and lesser contributions to the bonding localized MOs.

In terms of the AOs, the energy-localized MOs of water are

$$i(O) = -0.007(H_1 1s) - 0.007(H_2 1s) + 0.99(O1s) - 0.12(O2s_\perp) + 0.03(O2p_z)$$

$$b(OH_1) = 0.50(H_1 1s) - 0.10(H_2 1s) + 0.02(O1s) + 0.25(O2s_\perp)$$
$$+ 0.407(O2p_z) + 0.441(O2p_y)$$

$$b(OH_2) = -0.10(H_1 1s) + 0.50(H_2 1s) + 0.02(O1s) + 0.25(O2s_\perp)$$
$$+ 0.407(O2p_z) - 0.441(O2p_y) \qquad (15.37)$$

$$l_1(O) = -0.09(H_1 1s) - 0.09(H_2 1s) + 0.09(O1s) + 0.63(O2s_\perp)$$
$$- 0.39(O2p_z) - 0.71(O2p_x)$$

$$l_2(O) = -0.09(H_1 1s) - 0.09(H_2 1s) + 0.09(O1s) + 0.63(O2s_\perp)$$
$$- 0.39(O2p_z) + 0.71(O2p_x)$$

The MO wave function

$$|i(O)\overline{i(O)}b(OH_1)\overline{b(OH_1)}b(OH_2)\overline{b(OH_2)}l_1(O)\overline{l_1(O)}l_2(O)\overline{l_2(O)}|$$

is identical to (15.22).

Let us analyze these localized MOs for water. The $i(O)$ MO is nearly a pure $1s$ inner-shell oxygen AO.

To define the angle between the two localized bonding MOs, we draw the line from O to H_1 along which the electron probability density in $b(OH_1)$ is a maximum, and we draw a similar line from O to H_2. The angle between these lines where they intersect at the O nucleus defines the angle between the localized bonding MOs. (If this angle differs significantly from the angle defined by straight lines between the nuclei, the bonds are said to be *bent*.) The angle between the localized bonding MOs is determined mainly by the oxygen $2p_y$ and $2p_z$ AO contributions (with a small influence by the hydrogen $1s$ AOs). For $b(OH_1)$ the $O2p_y O2p_z$ contribution contains the factor $0.407z_O + 0.441y_O$. Let us carry out a rotation of coordinates in the zy plane by an angle $\alpha = \arctan(0.441/0.407) = 47\frac{1}{2}°$, as in Fig. 15.6. The relation between coordinates in the unrotated system and the rotated $z'y'$ system is given by the well-known formulas

$$z' = z \cos \alpha + y \sin \alpha$$
$$y' = -z \sin \alpha + y \cos \alpha \qquad (15.38)$$

From (15.38) we have

$$0.407z_O + 0.441y_O = 0.600z_O'$$

Multiplication by the exponential factor of the $2p$ oxygen AO then gives

$$0.407(O2p_z) + 0.441(O2p_y) = 0.60(O2p_{z'}) \qquad (15.39)$$

In other words, the hybridized $2p_z 2p_y$ AO on the left of (15.39) is the same function as 0.60 times a $2p$ AO inclined at an angle of $47\frac{1}{2}°$ with the z axis. Thus the bonding $2p_z 2p_y$ hybrid AOs of oxygen in $b(OH_1)$ and $b(OH_2)$ point in the general direction of the hydrogen atoms. The contribution from the hydrogen atoms to $b(OH_1)$ is mostly from $H_1 1s$, and the overlap between $H_1 1s$ and the $2p_z 2p_y$ oxygen hybrid then forms the $O—H_1$ chemical bond. The angle between the two hybrid oxygen AOs in $b(OH_1)$ and $b(OH_2)$ is 95°, and this is approximately the angle between the localized bonding MOs.

Energy-localized H_2O bonding MOs calculated from an extended-basis-set SCF MO wave function were found to have an angle of 96° between the oxygen hybrids that contribute to these MOs and an angle of 103° between the localized MOs themselves, which is nearly the same as the $104\frac{1}{2}°$ molecular bond angle, indicating that the bonds in water are not bent to any significant degree. [W. von Niessen, *Theor. Chim. Acta*, **29**, 29 (1973).] In contrast, in the strained molecule cyclopropane, the angle between the carbon hybrids contributing to the carbon–carbon bonding energy-localized MOs deviates outward by 28° from the internuclear lines [M. D. Newton, E. Switkes, and W. N. Lipscomb, *J. Chem. Phys.*, **53**, 2645 (1970)].

The $b(OH_1)$ energy-localized MO is not completely confined to the region of the $O—H_1$ bond. We see from (15.37) that this MO has a small contribution from the $H_2 1s$ AO. (The ratio of the contributions of two AOs to an MO is given essentially by the square of the ratio of their coefficients.) Moreover, consider the contributions of $O2s_\perp$, $O2p_y$, and $O2p_z$ to $b(OH_1)$. The $2p_y 2p_z$ hybrid has a

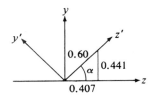

Figure 15.6 A coordinate rotation in the yz plane.

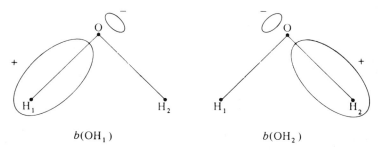

$b(OH_1)$ $b(OH_2)$

Figure 15.7 Rough sketches of the localized bonding MOs in H_2O.

positive lobe in the region of the OH_1 bond, and the contribution of the $O2s_\perp$ AO reinforces this positive lobe; overlap with $H_1 1s$ then gives the O—H_1 bond. The $2p_y 2p_z$ hybrid has a negative lobe on the side of the oxygen opposite the OH_1 bond; this negative lobe is partly, but not completely, canceled by the $O2s_\perp$ contribution. Thus the $b(OH_1)$ MO has a "tail" on the side of the O away from H_1, this tail being somewhat distorted toward H_2 by the $-0.10H_2 1s$ contribution (Fig. 15.7). (For accurately plotted contours, see F. Franks, ed., *Water*, Vol. 1, Plenum, New York, 1972, p. 46.) Despite this tail, this MO is far more localized than any of the bonding canonical MOs in (15.20). Because $b(OH_1)$ is mostly localized in the OH_1 region, we expect only a small change in its form on going from HOH to, say, HOCl. The observed near invariance of the O—H bond from molecule to molecule is explained by MO theory using localized bonding MOs. (Neglect of the contribution of $H_2 1s$ to the $b(OH_1)$ localized MO gives a two-center orbital called a *bond orbital*.)

Finally, consider the two lone-pair MOs $l_1(O)$ and $l_2(O)$. These MOs are mainly localized on the oxygen atom and are equivalent to each other. They are directed away from the hydrogen atoms and project above and below the molecular (yz) plane (Fig. 15.8). The $\hat{\sigma}_v(yz)$ operation interchanges the lone-pair MOs. The angle between them is approximately $2 \arctan(0.71/0.39) = 122°$. The angle between localized lone-pair MOs calculated from a more accurate wave function is $114°$; von Niessen, *Theor. Chim. Acta*, **29**, 29 (1973).

Edmiston and Ruedenberg calculated energy-localized MOs for several diatomic molecules. Consider Li_2. Chemical intuition (Li—Li) suggests the

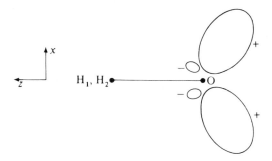

Figure 15.8 Rough sketches of the localized lone-pair MOs of H_2O.

localized MOs to be an inner-shell $1s$ AO on Li_a, an equivalent inner-shell AO on Li_b, and a bonding MO extending over the molecule. In terms of Ransil's minimal-basis-set SCF MOs with optimized orbital exponents (Section 13.17), Edmiston and Ruedenberg found the energy-localized Li_2 bonding MO to be very nearly the same as the $2\sigma_g$ MO and found the inner-shell localized MOs to be very nearly a $1s$ AO on Li_a and a $1s$ AO on Li_b.

For N_2 we expect (:N≡N:) the localized MOs to be an inner-shell $1s$ AO on each atom, a lone-pair $2s$ AO on each atom, and three bonding MOs spread over the two atoms. The canonical-MO picture is that the triple bond consists of one σ bond and two π bonds, as in Section 13.7. The energy-localized bonding MOs were found by Edmiston and Ruedenberg to be three equivalent banana-shaped bond orbitals spaced 120° apart from one another; the AOs that make significant contributions to the localized bonding MOs are the $2s$, $2p\sigma$, $2p\pi_x$, and $2p\pi_y$ orbitals of each atom. The $i(N_a)$ and $i(N_b)$ localized MOs were found to be nearly pure $1s$ nitrogen AOs. Each of the $l(N_a)$ and $l(N_b)$ localized MOs is a hybrid of the $2s$ and $2p\sigma$ AOs of the relevant nitrogen atom, with the $2s$ AO making the larger contribution to the MO. Each lone-pair localized MO is directed away from the other nitrogen atom.

The concept of localized MOs is not as widely applicable as that of delocalized canonical MOs. Delocalized MOs are valid for any molecule. However (as noted in Section 11.5), the Hartree–Fock wave functions of nonclosed-shell electronic states are, in most cases, linear combinations of *several* Slater determinants [for example, see (10.44) and (10.45)], and the above localization procedure is not applicable to the open-shell orbitals in these wave functions. Thus, in a molecule in an excited electronic state with an open-shell configuration, the electrons in the incompletely filled MOs are delocalized over much of the molecule.

The great success of the concept of chemical bonds between pairs of atoms is a reflection of the fact that the electronic ground states of most molecules have closed-shell configurations, for which the localized MO description is just as valid as the delocalized MO description.

For a closed-shell ground-state molecule, those properties that involve only the ground-state wave function can be calculated just as well with either the localized or the delocalized MO description. Such properties include electron probability density, dipole moment, geometry, and heat of formation. Properties of a molecule that involve the wave function of the ground state and also the wave function of an open-shell excited state or the wave function of an open-shell ion cannot be calculated using a localized MO description. Such properties include the ultraviolet absorption spectrum and molecular ionization energies.

Localized MOs are approximately transferable from molecule to molecule, which is not true for canonical MOs. The localized $b(CH)$ MO in CH_4 is quite similar to the localized $b(CH)$ MOs in C_2H_6 and in CH_3OH [S. Rothenberg, *J. Chem. Phys.*, **51**, 3389 (1969); *J. Am. Chem. Soc.*, **93**, 68 (1971)].

Calculation of the Edmiston–Ruedenberg energy-localized MOs is very time consuming. Boys proposed a method to find localized MOs that is computationally much faster than the Edmiston–Ruedenberg method and that gives similar results in most cases; see D. A. Kleier, *J. Chem. Phys.*, **61**, 3905 (1974).

15.9 THE SCF MO TREATMENT OF METHANE, ETHANE, AND ETHYLENE

Methane. The AOs for a minimal-basis-set SCF calculation of CH_4 are the carbon $1s$, $2s$, $2p_x$, $2p_y$, and $2p_z$ AOs and a $1s$ AO on each hydrogen atom. The point group of methane is $\mathscr{T}_d$. Group theory (see *Cotton* or *Schonland*) gives the possible symmetry species as A_1, A_2, E, T_1, and T_2. We shall not worry about specifying the symmetry behavior that corresponds to each symmetry species and shall use the species mainly as labels for the MOs. As usual, we set up the coordinate system with the z axis coinciding with the highest-order C_n or S_n axis; for methane this is an S_4 axis. The coordinates of the hydrogen atoms are (q, q, q), $(q, -q, -q)$, $(-q, q, -q)$, and $(-q, -q, q)$, where $2q$ is the edge of the cube in which the molecule is inscribed (Fig. 15.9). Note the equivalence of the x, y, and z axes.

The carbon atom is at the center of the molecule, and the carbon $1s$ and $2s$ AOs are each sent into themselves by every symmetry operation. These AOs transform according to the totally symmetric species A_1. The carbon $2p_x$, $2p_y$, and $2p_z$ AOs are given by x, y, or z times a radial function; their symmetry behavior is the same as that of the functions x, y, and z, respectively. From the formulas for rotation of coordinates [Eq. (15.50)], we see that any proper rotation sends each of the functions x, y, and z into some linear combination of x, y, and z. Any improper rotation is the product of some proper rotation and an inversion (Problem 12.16); the inversion simply converts each coordinate to its negative. Hence the three carbon $2p$ orbitals are sent into linear combinations of one another by each symmetry operation. They must therefore transform according to one of the triply degenerate symmetry species. Further investigation (which is omitted) shows the symmetry species of the $2p$ AOs to be T_2.

Just as in water, each $1s$ hydrogen AO in methane does not transform according to any of the molecular symmetry species, and it is convenient to form symmetry-adapted basis functions by taking linear combinations of the $1s$ AOs. One obvious symmetry function is

$$\chi_1 = H_1 1s + H_2 1s + H_3 1s + H_4 1s \tag{15.40}$$

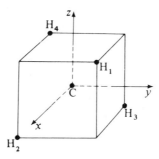

Figure 15.9 Coordinate axes for CH_4. The origin is at the center of the cube.

Since each methane symmetry operator permutes the hydrogen $1s$ orbitals among themselves, (15.40) is sent into itself by each symmetry operation and belongs to the totally symmetric species A_1. We need three more symmetry functions. The construction of these is not obvious without the use of group theory, and we shall simply write down the results. The remaining three orthogonal (unnormalized) symmetry-adapted basis functions can be taken as

$$\chi_2 = H_1 1s + H_2 1s - H_3 1s - H_4 1s \tag{15.41}$$

$$\chi_3 = H_1 1s - H_2 1s + H_3 1s - H_4 1s \tag{15.42}$$

$$\chi_4 = H_1 1s - H_2 1s - H_3 1s + H_4 1s \tag{15.43}$$

Each of these three functions is transformed into some linear combination of the three functions by each symmetry operation. For example, a $\hat{C}_3$ rotation about the CH_1 bond permutes the hydrogen atoms as follows: $1 \rightarrow 1$, $2 \rightarrow 3$, $3 \rightarrow 4$, $4 \rightarrow 2$. The corresponding $\hat{O}_{C_3}$ operator transforms χ_2, χ_3, and χ_4 into χ_3, χ_4, and χ_2, respectively. These three symmetry functions therefore transform according to one of the triply degenerate symmetry species. The function χ_2 has positive signs for the hydrogen AOs with a positive x coordinate and negative signs for the hydrogen AOs with a negative x coordinate; we thus expect χ_2 to have the same symmetry behavior as the function x. Similarly, χ_3 and χ_4 behave as y and z, respectively. (As an example, note that a 120° counterclockwise rotation about the CH_1 bond has the following effect on the three unit vectors: $\mathbf{i} \rightarrow \mathbf{j}$, $\mathbf{j} \rightarrow \mathbf{k}$, $\mathbf{k} \rightarrow \mathbf{i}$; this is the same behavior shown by the functions χ_2, χ_3, and χ_4 for this rotation.) These three symmetry orbitals thus transform according to the same symmetry species as $C2p_x$, $C2p_y$, and $C2p_z$, that is, the species T_2.

The symmetry-adapted basis functions are

Symmetry function	χ_1	χ_2	χ_3	χ_4	$C1s$	$C2s$	$C2p_x$	$C2p_y$	$C2p_z$
Symmetry species	a_1	t_2	t_2	t_2	a_1	a_1	t_2	t_2	t_2

The nine lowest MOs therefore consist of three a_1 and six t_2 MOs. The six t_2 MOs belong to triply degenerate levels and thus fall into two different shells $1t_2$ and $2t_2$; each such shell contains three MOs of equal orbital energy and each shell holds six electrons. SCF MO calculations give the three lowest shells as $1a_1$, $2a_1$, and $1t_2$, with energies -11.20, -0.93, and -0.54 hartrees, respectively. The ground state of methane thus has the closed-shell configuration $(1a_1)^2(2a_1)^2(1t_2)^6$ and is a $^1A_1$ state.

Pitzer did SCF calculations on CH_4 at several CH distances, using a minimal-basis set of STOs [R. M. Pitzer, *J. Chem. Phys.*, **46**, 4871 (1967)]. He found the optimized orbital exponents to be 1.17 for $H1s$, 5.68 for $C1s$, 1.76 for $C2s$, and 1.76 for $C2p_z$, which may be compared with the values 1.0, 5.7, 1.625, and 1.625 given by Slater's rules (Problem 15.47). Pitzer found the minimum in energy to be at a bond distance of 1.089 Å, close to the experimental value 1.085 Å. The MOs at the experimental equilibrium bond length are

$$1a_1 = -0.005(H_1 1s + H_2 1s + H_3 1s + H_4 1s) + 1.001(C1s) + 0.025(C2s_\perp)$$

$$2a_1 = 0.186(H_1 1s + H_2 1s + H_3 1s + H_4 1s) - 0.064(C1s) + 0.584(C2s_\perp)$$

$$1t_{2x} = 0.318(H_1 1s + H_2 1s - H_3 1s - H_4 1s) + 0.554(C2p_x)$$

$$1t_{2y} = 0.318(H_1 1s - H_2 1s + H_3 1s - H_4 1s) + 0.554(C2p_y) \qquad (15.44)$$

$$1t_{2z} = 0.318(H_1 1s - H_2 1s - H_3 1s + H_4 1s) + 0.554(C2p_z)$$

The $1a_1$ MO is essentially the carbon $1s$ AO. The $2a_1$ MO is a bonding combination of the carbon $2s$ AO and the symmetry orbital (15.40); this MO has charge buildup between the carbon atom and each of the four hydrogen atoms. The $1t_{2x}$ MO is a bonding MO; the function (15.41) is positive on the positive half of the x axis and negative on the negative half of the x axis, and its overlap with $C2p_x$ gives charge buildup in the regions about the x axis. Similarly, the bonding $1t_{2y}$ and $1t_{2z}$ MOs have charge buildup in the regions about the y and z axes, respectively.

We now consider the four localized (equivalent) bonding MOs of methane. Because of the tetrahedral symmetry, each of these orbitals *must* point along a CH bond, since otherwise they would not be equivalent to one another. (This is not true for water, where the equivalence requirement is satisfied by any two bonding MOs that make the same angle with the C_2 axis.) Each localized bonding MO is some linear combination of the five canonical occupied MOs:

$$b(CH_1) = a(1a_1) + b(2a_1) + d(1t_{2x}) + e(1t_{2y}) + f(1t_{2z}) \qquad (15.45)$$

with similar expressions for $b(CH_2)$, $b(CH_3)$, and $b(CH_4)$. Since $1a_1$ is a nonbonding low-energy inner-shell orbital, we expect $|a| \ll |b|$. The $1a_1$ and $2a_1$ canonical MOs are directed equally to all four hydrogens, and so varying a or b in (15.45) does not direct $b(CH_1)$ preferentially to any one hydrogen. The $1t_{2x}$, $1t_{2y}$, and $1t_{2z}$ MOs are directed along the x, y, and z axes, respectively, so that by adjusting d, e, and f appropriately, we can get $b(CH_1)$ to be localized mainly in the region of the CH_1 bond. To fix d, e, and f, we use some properties of direction cosines.

If line L passes through the origin and makes the angles α, β, and γ with the x, y, and z axes, respectively, then the quantities

$$l \equiv \cos\alpha, \qquad m \equiv \cos\beta, \qquad n \equiv \cos\gamma \qquad (15.46)$$

are the *direction cosines* of L. If (x_L, y_L, z_L) is a point on L, then clearly

$$x_L = r\cos\alpha, \qquad y_L = r\cos\beta, \qquad z_L = r\cos\gamma \qquad (15.47)$$

where r is the distance from the origin. From $x^2 + y^2 + z^2 = r^2$, it follows that

$$l^2 + m^2 + n^2 = 1 \qquad (15.48)$$

Let the lines L_1 and L_2 go from the origin to (x_1, y_1, z_1) and (x_2, y_2, z_2), respectively. If θ_{12} is the angle between L_1 and L_2, then [Eq. (5.20)]

$$\cos\theta_{12} = \frac{x_1 x_2 + y_1 y_2 + z_1 z_2}{r_1 r_2}$$

$$\cos\theta_{12} = l_1 l_2 + m_1 m_2 + n_1 n_2 \qquad (15.49)$$

Direction cosines are useful in discussing changes in coordinate axes. Let the $x'y'z'$ and the xyz Cartesian coordinate systems have a common origin, and let the $x'y'z'$ axes be obtained from the xyz axes by rotation, reflection, inversion, or some combination of these operations. Let the direction cosines of the x' axis with respect to the xyz system be l_1, m_1, n_1; let l_2, m_2, n_2 and l_3, m_3, n_3 be the direction cosines of the y' and z' axes, respectively. Let the vector $\mathbf{s}$ have coordinates (x, y, z) and (x', y', z') in the unrotated and rotated coordinate systems. We have $x' = \mathbf{s} \cdot \mathbf{i}'$, where $\mathbf{i}'$ is a unit vector along the x' axis; since $\mathbf{i}'$ is of unit length, it follows from (15.47) and (15.46) that the coordinates of $\mathbf{i}'$ in the xyz system are l_1, m_1, and n_1. Hence

$$x' = l_1 x + m_1 y + n_1 z$$
$$y' = l_2 x + m_2 y + n_2 z \qquad (15.50)$$
$$z' = l_3 x + m_3 y + n_3 z$$

where the y' and z' equations follow from $y' = \mathbf{s} \cdot \mathbf{j}'$, $z' = \mathbf{s} \cdot \mathbf{k}'$. [Equation (15.38) is a special case of (15.50).] Since the angle between any pair of the x', y', and z' axes is 90°, it follows from (15.49) that

$$l_1 l_2 + m_1 m_2 + n_1 n_2 = 0$$
$$l_1 l_3 + m_1 m_3 + n_1 n_3 = 0 \qquad (15.51)$$
$$l_2 l_3 + m_2 m_3 + n_2 n_3 = 0$$

Now we return to the determination of d, e, and f. The MOs t_{2x}, t_{2y}, and t_{2z} are directed along the x, y, and z axes, respectively, and the contributions of the carbon $2p_x$, $2p_y$, and $2p_z$ AOs to these MOs are

$$xe^{-\zeta r}, \qquad ye^{-\zeta r}, \qquad ze^{-\zeta r} \qquad (15.52)$$

The contribution of the hydrogen AOs to the t_2 MOs has a more complicated form (the hydrogens are not at the coordinate origin), but we need not explicitly consider the hydrogen part of the MOs; this is because the hydrogen symmetry orbitals (15.41) to (15.43) have the same directional properties as the corresponding carbon $2p$ AOs (15.52) with which each is combined in the $1t_2$ MOs [Eq. (15.44)]. The linear combination (15.45) has as its carbon $2p$ contribution

$$(dx + ey + fz)e^{-\zeta r} \qquad (15.53)$$

Let l_1, m_1, and n_1 be the direction cosines of the CH_1 line. We assert that if d, e, and f are chosen as proportional to these direction cosines, then $b(CH_1)$ will be directed toward H_1. To verify this, we set $d:e:f = l_1:m_1:n_1$ in (15.53) to get

$$c(l_1 x + m_1 y + n_1 z)e^{-\zeta r} = cx'e^{-\zeta r} \qquad (15.54)$$

where c is some constant and the x' axis runs from C to H_1.

Similarly, by picking d, e, and f proportional to the direction cosines of the other CH lines, we form localized orbitals along these bonds. From (15.47) the direction cosines of the CH lines are

$$CH_1: 3^{-1/2}, 3^{-1/2}, 3^{-1/2} \qquad CH_2: 3^{-1/2}, -3^{-1/2}, -3^{-1/2}$$

$$CH_3: -3^{-1/2}, 3^{-1/2}, -3^{-1/2} \qquad CH_4: -3^{-1/2}, -3^{-1/2}, 3^{-1/2} \tag{15.55}$$

To satisfy the equivalence requirement, the values of a and b in (15.45) must be the same for each bonding localized MO. The equivalent localized MOs for methane thus have the forms

$$b(CH_1) = a(1a_1) + b(2a_1) + 3^{-1/2}c(1t_{2x} + 1t_{2y} + 1t_{2z})$$

$$b(CH_2) = a(1a_1) + b(2a_1) + 3^{-1/2}c(1t_{2x} - 1t_{2y} - 1t_{2z})$$

$$b(CH_3) = a(1a_1) + b(2a_1) + 3^{-1/2}c(-1t_{2x} + 1t_{2y} - 1t_{2z}) \tag{15.56}$$

$$b(CH_4) = a(1a_1) + b(2a_1) + 3^{-1/2}c(-1t_{2x} - 1t_{2y} + 1t_{2z})$$

Orthonormality of the bonding localized MOs (15.56) requires that

$$a^2 + b^2 + c^2 = 1 \quad \text{and} \quad a^2 + b^2 - \tfrac{1}{3}c^2 = 0 \tag{15.57}$$

Hence

$$c = \tfrac{1}{2}\sqrt{3}, \qquad (a^2 + b^2)^{1/2} = \tfrac{1}{2} \tag{15.58}$$

The equivalence, direction, and orthonormality requirements have fixed all but one parameter (the ratio a/b) in the localized bonding MOs of methane.

The $1t_{2x}$ MO points equally in the $+x$ and $-x$ directions. Similarly, the $1t_{2y}$ and $1t_{2z}$ MOs point equally on both sides of the carbon atom. This is not true of the bonding localized MOs: The $2a_1$ MO is positive in most of the bonding region between the carbon atom and the hydrogen atoms. [It is negative in the region very near the carbon atom, because of the $-0.06(C1s)$ term and the negative portion of the orthogonalized C2s AO; see Eq. (15.21).] The linear combination $1t_{2x} + 1t_{2y} + 1t_{2z}$ points equally in the $(1,1,1)/\sqrt{3}$ and the $(-1,-1,-1)/\sqrt{3}$ directions. If b in (15.56) is taken as positive (as we have taken c), then the $2a_1$ MO adds to the positive half of this linear combination of the $1t_2$ MOs and cancels much of the negative half of this linear combination. With b and c having the same sign, the $b(CH_1)$ MO points mostly in the $(1,1,1)/\sqrt{3}$ direction, with only a small "tail" in the $(-1,-1,-1)/\sqrt{3}$ direction.

Pitzer's SCF calculation gives the energy-localized bonding and inner-shell methane MOs as

$$b(CH_1) = 0.055(1a_1) + 0.497(2a_1) + \tfrac{1}{2}(1t_{2x} + 1t_{2y} + 1t_{2z})$$

$$b(CH_2) = 0.055(1a_1) + 0.497(2a_1) + \tfrac{1}{2}(1t_{2x} - 1t_{2y} - 1t_{2z})$$

$$b(CH_3) = 0.055(1a_1) + 0.497(2a_1) + \tfrac{1}{2}(-1t_{2x} + 1t_{2y} - 1t_{2z}) \tag{15.59}$$

$$b(CH_4) = 0.055(1a_1) + 0.497(2a_1) + \tfrac{1}{2}(-1t_{2x} - 1t_{2y} + 1t_{2z})$$

$$i(C) = 0.994(1a_1) - 0.111(2a_1)$$

From (15.44) and (15.59), we have

$$i(C) = 1.002(C1s) - 0.040(C2s_\perp) - 0.025(H_1 1s + H_2 1s + H_3 1s + H_4 1s)$$

$$b(CH_1) = 0.024(C1s) + 0.292(C2s_\perp) + 0.569(H_1 1s) - 0.066(H_2 1s + H_3 1s + H_4 1s)$$

$$+ 0.277(C2p_x + C2p_y + C2p_z) \tag{15.60}$$

From (15.52) to (15.54), the linear combination $dp_x + ep_y + fp_z$ is equal to a rotated p orbital pointing along the x' axis and containing the factor $dx + ey + fz = c(l_1 x + m_1 y + n_1 z) = cx'$, where l_1, m_1, n_1 are the direction cosines of the x' axis relative to the x, y, z coordinates. Substitution of $l_1 = d/c$, $m_1 = e/c$, $n_1 = f/c$ into $l_1^2 + m_1^2 + n_1^2 = 1$ [Eq. (15.48)] gives $c = (d^2 + e^2 + f^2)^{1/2}$. Therefore, a localized MO that contains the terms $aC2s_\perp + dC2p_x + eC2p_y + fC2p_z$ can be viewed as containing the terms $aC2s_\perp + (d^2 + e^2 + f^2)^{1/2}C2p'_x$, and we say the hybridization of the carbon AO in this MO is

$$sp^{(d^2+e^2+f^2)/a^2} \tag{15.61}$$

For example, for the localized bonding MO (15.60) of CH_4, $a^2 = 0.0853$, $d^2 + e^2 + f^2 = 0.230$, and this MO is an $sp^{2.7}$ hybrid, which is close to the traditional sp^3 hybridization of VB theory (Section 15.16).

Ethane.   The most fascinating property of C_2H_6 is the barrier to internal rotation about the carbon–carbon single bond. The staggered conformation of a C_2H_6 molecule is more stable than the eclipsed conformation by $0.125 \, eV = 0.00461$ hartree, which is equivalent to 2.89 kcal/mol. In 1936, J. D. Kemp and Kenneth Pitzer discovered this fact in examining thermodynamic data for ethane. The experimental electronic energy (including nuclear repulsion) of ethane is -80 hartrees $= -2200 \, eV$. The correlation energy runs one-half to one percent of the total energy of an atom or molecule. Thus we expect the Hartree–Fock energy of ethane to differ by 10 or 20 eV from the true molecular electronic energy. This difference is far greater than the barrier height. At first sight, it seems hopeless to expect an SCF MO calculation to give a meaningful result for the ethane barrier.

The minimal-basis-set AOs for ethane are the hydrogen $1s$ orbitals and the carbon $1s$, $2s$, and $2p$ orbitals, a total of $6(1) + 2(5) = 16$ basis AOs. To calculate the barrier in ethane, we must calculate the energy of the staggered and the eclipsed conformations, which requires two separate SCF calculations. One first forms appropriate linear combinations of the hydrogen AOs and of the carbon AOs to get symmetry-adapted basis functions. The Roothaan equations are then solved iteratively to give the basis-function coefficients and orbital energies, and the total molecular energy is found. Table 15.2 lists the energy results from several SCF MO calculations. All the calculations listed are many-center ones.

The pioneering ethane calculation is by Russell Pitzer (Kenneth Pitzer's son) and W. N. Lipscomb. They used a minimal basis set of 16 STOs; the orbital exponents were not optimized but were chosen according to Slater's rules; all integrals were evaluated accurately. The Pitzer calculation is similar to the Pitzer–Lipscomb calculation, except that the orbital exponents used are those that minimize the energy of methane.

The first three calculations listed used the experimentally observed equilibrium geometry for staggered ethane and assumed the geometry of the eclipsed

TABLE 15.2 SCF MO Calculations of Ethane

Reference[a]	Basis Functions	Eclipsed Energy (hartrees)	Staggered Energy (hartrees)	Barrier (kcal/mol)
Pitzer, Lipscomb	16 STOs	−78.98593	−78.99115	3.3
Pitzer	16 STOs	−79.09233	−79.09797	3.5
Clementi, Davis	32 CGTFs	−79.10247	−79.10824	3.6
Veillard	68 CGTFs	−79.23410	−79.23899	3.07
Clementi, Popkie	102 CGTFs	−79.25364	−79.25875	3.21
Experimental values			−79.84	2.89

[a]R. M. Pitzer and W. N. Lipscomb, *J. Chem. Phys.*, **39**, 1995 (1963); R. M. Pitzer, *J. Chem. Phys.*, **47**, 965 (1967); E. Clementi and D. R. Davis, *J. Chem. Phys.*, **45**, 2593 (1966); A. Veillard, *Theor. Chim. Acta*, **18**, 21 (1970); E. Clementi and H. Popkie, *J. Chem. Phys.*, **57**, 4870 (1972). Experimental barrier: N. Moazzen-Ahmadi et al., *J. Chem. Phys.*, **88**, 563 (1988).

form to be that given by rigidly rotating one methyl group with respect to the other. Veillard varied the bond angles and the CC bond length for each conformation to minimize the energy. He found an increase of 0.02 Å in the CC length and a decrease of $\frac{1}{2}°$ in the HCH angle on going from the staggered to the eclipsed form; this change affected the barrier by $\frac{1}{2}$ kcal/mol. Clementi and Popkie's near-Hartree–Fock calculation also uses geometry optimization.

All the SCF MO calculations give good values of the ethane rotational barrier. Why should this be so? The answer lies in the correlation energy. Electrons paired in the same localized orbital move through the same region of space. Hence the intraorbital correlation for such a pair should be substantially greater than the interorbital correlation energy for two electrons in different localized MOs. In line with this, it was formerly believed that the total interorbital molecular correlation energy was much less than the total intraorbital correlation energy. However, there are many more interorbital pair correlations than intra-orbital correlations, and it is now recognized that the total interorbital correlation is not negligible and in some cases can be of comparable magnitude to the total intraorbital correlation. [See E. Steiner, *J. Chem. Phys.*, **54**, 1114 (1971).] Hence we must consider both kinds of correlation in ethane. Rotation of a methyl group in ethane does not change any of the bonds, and thus intraorbital correlation should be essentially the same in the staggered and eclipsed forms. Moreover, we expect most of the interorbital correlations to be essentially unchanged in the two forms. In particular, correlations between the C—H bonds within each methyl group should remain virtually the same, and so should correlations between the C—C and C—H bonding pairs. It is only the correlations between C—H pairs of different methyl groups that should change, and these make the smallest contributions to interorbital correlation. Thus we expect the total correlation energy to be only very slightly changed from staggered to eclipsed. Thus the energy error in an SCF calculation is about the same for the two forms, and the Hartree–Fock energy difference should yield a good estimate of the barrier. None of the SCF calculations listed in Table 15.2 have actually reached the Hartree–Fock limit, but they still yield good values for the barrier. Recall that Hartree–Fock calculations give poor values for dissociation energies. This is because the number of electron pairs changes in forming a chemical bond from atoms, thereby changing the correlation energy substantially.

Rotational barriers in other molecules have been studied by the SCF MO method (see P. W. Payne and L. C. Allen in *Schaefer, Applications of Electronic Structure Theory*, Chapter 2). Some results are (values in kcal/mol): CH_3OH, 1.4 calculated versus 1.1 experimental; CH_3NH_2, 2.4 calculated versus 2.0 experimental; CH_3CHO, 1.1 calculated versus 1.2 experimental; CH_3SiH_3, 1.4 calculated versus 1.7 experimental.

Table 15.2 shows that for ethane a reasonably accurate barrier can be obtained with an SCF calculation that has a minimal basis set and does not use geometry optimization, and this is true for many barriers investigated. For H_2O_2, such calculations give quite poor barrier results. However, when a large basis set (including polarization functions) and geometry optimization are used, the H_2O_2 barriers can be calculated by the Hartree–Fock method, as shown by the following results [T. H. Dunning and N. W. Winter, *J. Chem. Phys.*, **63**, 1847 (1975); see also D. Cremer, *J. Chem. Phys.*, **69**, 4440 (1978)]: calculated *cis* barrier (0° dihedral angle) 8.4 kcal/mol versus 7.3 kcal/mol experimental; calculated *trans* barrier (180° dihedral angle) 1.1 kcal/mol versus 1.1 kcal/mol experimental; calculated equilibrium dihedral angle 114° versus 112° experimental.

The physical origin of the ethane rotational barrier is attributed to the Pauli-exclusion-principle repulsion (Section 10.3) between the eclipsing localized C—H bonding electron pairs in eclipsed ethane; see O. J. Sovers, C. W. Kern, R. M. Pitzer, and M. Karplus, *J. Chem. Phys.*, **49**, 2592 (1968); R. M. Stevens and M. Karplus, *J. Am. Chem. Soc.*, **94**, 5140 (1972); R. M. Pitzer, *Acc. Chem. Res.*, **16**, 207 (1983). For a different explanation, see R. F. Bader et al., *J. Am. Chem. Soc.*, **112**, 6530 (1990).

Ethylene. For C_2H_4, the ground-state equilibrium-geometry point group is $\mathscr{D}_{2h}$. The standard choice of coordinate axes is shown in Fig. 15.10. (The standard convention has often been ignored, and one finds calculations with the molecular plane being the *xz* or the *xy* plane; this affects the naming of the MO symmetry species, and so different workers use different symbols for the same MO.)

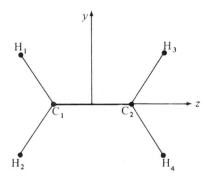

Figure 15.10 Coordinate axes for C_2H_4. The *x* axis is perpendicular to the molecular plane.

There are eight symmetry operations for $\mathscr{D}_{2h}$. Each symmetry operation commutes with every other symmetry operation, so the electronic wave function must be an eigenfunction of all the symmetry operators, and we have only nondegenerate (A and B) symmetry species. Since the three rotations, the three reflections, and the inversion operation each have their squares equal to the identity operation, these symmetry operations must have the eigenvalues ± 1. All eight symmetry operations can be expressed as the product of one, two, or three of the reflections (each reflection simply converts one coordinate to its negative):

$$\hat{E} = [\hat{\sigma}(xy)]^2 , \qquad \hat{\imath} = \hat{\sigma}(xy)\hat{\sigma}(xz)\hat{\sigma}(yz)$$

$$\hat{C}_2(x) = \hat{\sigma}(xy)\hat{\sigma}(xz) , \qquad \hat{C}_2(y) = \hat{\sigma}(xy)\hat{\sigma}(yz) , \qquad \hat{C}_2(z) = \hat{\sigma}(xz)\hat{\sigma}(yz)$$

Since the symmetry eigenvalues multiply the same way the symmetry operations do, specification of the eigenvalues of the three reflections is sufficient to fix all eight symmetry eigenvalues. There are thus $2^3 = 8$ possible symmetry species. The standard notation for these is

	$\hat{\sigma}(xy)$	$\hat{\sigma}(xz)$	$\hat{\sigma}(yz)$
A_g	$+1$	$+1$	$+1$
A_u	-1	-1	-1
B_{1g}	$+1$	-1	-1
B_{1u}	-1	$+1$	$+1$
B_{2g}	-1	$+1$	-1
B_{2u}	$+1$	-1	$+1$
B_{3g}	-1	-1	$+1$
B_{3u}	$+1$	$+1$	-1

$$(15.62)$$

The g or u subscript corresponds to eigenvalue $+1$ or -1 for $\hat{\imath}$. For $\mathscr{D}_{2h}$ the designation A is used only for symmetry species with eigenvalue $+1$ for all three $\hat{C}_2$ rotations.

There are 14 minimal-basis-set AOs. It is easy to set up symmetry orbitals by trial and error. The reader will readily verify the symmetry species of the (unnormalized) symmetry functions listed in Table 15.3.

The 14 lowest MOs thus include four a_g, four b_{1u}, two b_{2u}, two b_{3g}, one b_{3u}, and one b_{2g} MO. Eight of these MOs are occupied in the ground state. SCF calculations give as the electron configuration of the ^1A_g ground state [U. Kaldor and I. Shavitt, *J. Chem. Phys.*, **48**, 191 (1968)]

$$(1a_g)^2(1b_{1u})^2(2a_g)^2(2b_{1u})^2(1b_{2u})^2(3a_g)^2(1b_{3g})^2(1b_{3u})^2$$

Each canonical MO of a planar molecule is classified as σ or π, according to whether its eigenvalue for reflection in the molecular plane is $+1$ or -1, respectively. (This usage is somewhat inconsistent with the $\sigma, \pi, \delta, \ldots$ classification of linear-molecule MOs. For linear molecules the symbols σ and π signify an axial component of electronic orbital angular momentum of 0 and $\pm\hbar$, respectively; for nonlinear molecules we cannot specify the component of $\mathbf{L}$ along an internuclear line. For linear molecules, σ MOs are nondegenerate and π MOs are doubly degenerate; for nonlinear molecules the $\sigma-\pi$ classification is unrelated to

TABLE 15.3   Symmetry-adapted Basis Functions for C_2H_4

Symmetry Function	Symmetry Species
$H_1 1s + H_2 1s + H_3 1s + H_4 1s$	a_g
$C_1 1s + C_2 1s$	a_g
$C_1 2s + C_2 2s$	a_g
$C_1 2p_z - C_2 2p_z$	a_g
$H_1 1s + H_2 1s - H_3 1s - H_4 1s$	b_{1u}
$C_1 1s - C_2 1s$	b_{1u}
$C_1 2s - C_2 2s$	b_{1u}
$C_1 2p_z + C_2 2p_z$	b_{1u}
$H_1 1s - H_2 1s + H_3 1s - H_4 1s$	b_{2u}
$C_1 2p_y + C_2 2p_y$	b_{2u}
$H_1 1s - H_2 1s - H_3 1s + H_4 1s$	b_{3g}
$C_1 2p_y - C_2 2p_y$	b_{3g}
$C_1 2p_x + C_2 2p_x$	b_{3u}
$C_1 2p_x - C_2 2p_x$	b_{2g}

the degeneracy.) For the ground state of water, all the occupied MOs are σ MOs except the lone-pair $1b_1$ MO, which is a π MO.

For ethylene the only occupied π MO in the ground electronic state is the $1b_{3u}$ MO, the highest occupied MO. Since there is only one b_{3u} symmetry orbital in Table 15.3, the $1b_{3u}$ MO must be identical (apart from a normalization constant) to this symmetry orbital. One finds (using STOs)

$$1b_{3u} = 0.63(C_1 2p_x + C_2 2p_x)$$

This bonding MO formed by overlap of the $2p_x$ AOs of carbon resembles the π_u MO of Fig. 13.14. The $1b_{3u}$ π MO accounts for the planarity of ethylene in its ground state. As one CH_2 group is rotated relative to the other, the overlap between the two carbon $2p_x$ AOs rapidly diminishes, and the energy of the $1b_{3u}$ MO increases; hence the molecule strongly resists torsion about the carbon–carbon bond.

The lowest-lying unoccupied MO of ethylene is the $1b_{2g}$ antibonding π MO:

$$1b_{2g} = 0.82(C_1 2p_x - C_2 2p_x)$$

It resembles the π_g MO in Fig. 13.14. The excited ethylene configuration

$$(1a_g)^2 \cdots (1b_{3g})^2 (1b_{3u})(1b_{2g})$$

gives rise to two terms (a singlet and a triplet), and it is likely that these electronic states are nonplanar, with one CH_2 group rotated 90° with respect to the other.

The canonical σ MOs of ethylene each contain contributions from AOs of all six atoms and are delocalized over the whole molecule. By taking appropriate linear combinations of the canonical σ MOs, we can form localized σ MOs. We

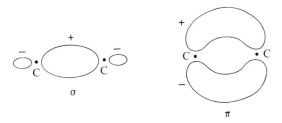

Figure 15.11 $\sigma-\pi$ description of the ethylene double bond. The cross sections are taken in a plane perpendicular to the molecular plane, which is a nodal plane for the π MO.

expect these MOs to consist of the following inner-shell and bonding MOs:

$$i(C_1), \qquad i(C_2)$$

$$b(C_1C_2), \qquad b(C_1H_1), \qquad b(C_1H_2), \qquad b(C_2H_3), \qquad b(C_2H_4) \tag{15.63}$$

The MOs $b(C_1C_2)$ and $1b_{3u}$ constitute the familiar description of the carbon–carbon double bond as one σ and one π bond (Fig. 15.11). We still do not, however, have the equivalent MOs of ethylene, since the π MO $1b_{3u}$ is not equivalent to any of the σ MOs in (15.63), nor is it equivalent to itself; it goes into its negative upon a $\hat{C}_2(z)$ rotation. By adding and subtracting the $b(C_1C_2)$ and $1b_{3u}$ MOs, we can form two equivalent localized carbon–carbon bonding MOs:

$$b_1(C_1C_2) = 2^{-1/2}[b(C_1C_2) + 1b_{3u}] \tag{15.64}$$

$$b_2(C_1C_2) = 2^{-1/2}[b(C_1C_2) - 1b_{3u}] \tag{15.65}$$

The MOs (15.64) and (15.65) lead to the description of the carbon–carbon bond as composed of two "banana" bonds (Fig. 15.12).

There has been considerable controversy as to whether the ethylene double bond is best described as two equivalent bent banana bonds or as a σ and a π bond. Kaldor calculated the energy-localized MOs of ethylene from the minimal-basis-set SCF MOs [U. Kaldor, *J. Chem. Phys.*, **46**, 1981 (1967)]. Since there is

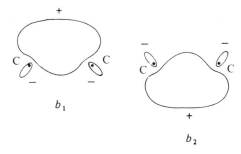

Figure 15.12 Banana-bond description of the ethylene double bond. The cross sections are the same as in Fig. 15.11.

no a priori necessity that the energy-localized MOs be equivalent orbitals, this calculation provides evidence as to which is the "better" description of the carbon–carbon double bond. Kaldor found the energy-localized carbon–carbon bond orbitals in ethylene to be the two equivalent banana bonds. For acetylene he found the energy-localized carbon–carbon bond orbitals to be three equivalent banana bonds and not one σ and two π bonds. Of course, although the electron densities in the individual MOs differ for the banana-bond versus $\sigma-\pi$ descriptions, the total probability density for the four or six electrons in the double or triple bond is the same in either picture.

The energy-localized C_1H_1 bond MO in ethylene was found by Kaldor to be

$$0.3637(C_1 2s_\perp) + 0.4143(C_1 2p_y) - 0.2574(C_1 2p_z) + 0.4939(H_1 1s) + \cdots$$

where the dots indicate small contributions from other AOs. From (15.61), the hybridization of carbon in this localized MO is $sp^{1.8}$, which is close to the sp^2 of VB theory. The angle made by the carbon hybrid AO in $b(CH_1)$ with the C–C axis is $\pi - \arctan(1.61) = 122°$, essentially the same as the experimental bond angle. Hence the localized C—H bond orbitals are not bent in planar ethylene.

Localized MOs for butadiene and benzene are discussed at the end of Section 16.3.

Localized MOs have been calculated for some of the boron hydrides. These compounds have fewer than $2(n-1)$ valence electrons, where n is the number of atoms. Thus there are not enough electrons to form $n-1$ electron-pair bonds between the atoms. From the VB viewpoint such compounds are anomalous and have been termed electron deficient. However, it is only chemical prejudices (and not quantum-mechanical requirements) that lead us to expect $n-1$ electron-pair bonds. The MO treatment of boron hydrides shows all the bonding MOs to be completely filled in the ground state, and the term electron deficient is not really appropriate. Calculation of the localized bonding MOs in B_2H_6 shows that there are two localized MOs, each of which extends over three nuclei (two borons and one hydrogen), giving two three-center bonds [E. Switkes et al., *J. Chem. Phys.*, **51**, 2085 (1969)]. These three-center localized MOs resemble the banana orbitals of the isoelectronic molecule ethylene; we can view B_2H_6 as being obtained from C_2H_4 by removing a proton from each carbon nucleus and placing it in the midst of one of the banana orbitals. The higher boron hydrides have three-center bonds involving three boron atoms. (See *Murrell, Kettle, and Tedder*, Chapter 14.)

We noted earlier in this section that each *canonical* MO of a planar molecule is classified as π or σ according to whether or not the molecular plane is a nodal plane for the MO. A *localized* bonding MO of a molecule (planar or nonplanar) is classified as σ or π or δ according to whether this MO has 0 or 1 or 2 nodal planes containing the line joining the nuclei of the two bonded atoms. The bonding localized MOs in water (Fig. 15.7) are σ MOs; the ethylene double-bond localized MOs in Fig. 15.11 consist of one σ and one π MO; the $b(CH_1)$ MO (15.60) in CH_4 is a σ MO. In certain transition-metal compounds (for example, $Re_2Cl_8^{2-}$), overlap of d AOs produces a δ localized bonding MO, and these compounds have a quadruple bond consisting of one σ, two π, and one δ MO; see F. A. Cotton, *Chem. Soc. Rev.*, **4**, 27 (1975).

15.10 MOLECULAR GEOMETRY

The equilibrium geometry of a molecule corresponds to the nuclear configuration that minimizes the molecular electronic energy including nuclear repulsion.

The molecular energy is found to be relatively insensitive to even large changes in molecular geometry. For example, a CI calculation of the H_2O potential-energy surface U as a function of molecular geometry that used a large basis set and included all single and double excitations (as many as 30380 CSFs), and an estimate of the effect of quadruple excitations yielded the following values in hartrees for the electronic energy including nuclear repulsion as a function of the bond angle θ at the equilibrium bond distance [P. Hennig, W. P. Kraemer, and G. H. F. Diercksen, *Theor. Chim. Acta*, **47**, 233 (1978)]:

θ	90°	105°	120°	180°
U	-76.342	-76.348	-76.343	-76.294

The variation in energy over the range of bond angles is only 0.07%, as compared with a correlation energy (Table 15.1) of 0.5%. Especially noteworthy is the relative flatness of the surface in the region of the equilibrium geometry; a variation of $\pm 15°$ from the equilibrium 105° angle produces a variation of only 0.008% in U. (For near-Hartree–Fock SCF calculations with the same basis set as the CI calculations, the magnitudes of the changes versus geometry are similar to those given by the above CI values.) Changes in bond lengths also produce relatively small changes in energy. For example, for H_2O at the equilibrium bond angle, CI calculations (Hennig et al., op. cit.) gave these energies in hartrees as a function of OH bond distance for bonds of equal length: -76.294 at 0.794 Å, -76.348 at 0.957 Å, -76.295 at 1.164 Å. (Note that the small bond bending of 15° has much less effect on the energy than the small bond stretching of 0.2 Å; recall from organic chemistry that infrared vibrational bands corresponding to bond bending occur at lower frequencies than bond-stretching vibrations.)

Despite the smallness of these changes in molecular energy versus molecular geometry as compared with the energy error (the correlation energy) inherent in the SCF method, SCF wave functions give good predictions (0% to 3% error) of equilibrium bond angles and distances. Some SCF 4-31G-basis-set bond angles and lengths follow (J. A. Pople in *Schaefer, Applications of Electronic Structure Theory*, Chapter 1), where the experimental values are in parentheses. NH_2: 108° (103°) and 1.02 Å (1.02 Å); H_2O: 111° ($104\frac{1}{2}°$) and 0.95 Å (0.96 Å); C_2H_6: 108° (108°) for HCH and 1.53 Å (1.53 Å) for C—C; C_2H_4: 116° ($116\frac{1}{2}°$) for HCH and 1.32 Å (1.33 Å) for C=C; H_2CO: 116° (116°) for HCH and 1.21 Å (1.20 Å) for C=O. (See also Section 17.1.)

Evidently, the correlation energy remains approximately constant for bond angle and length variations in the region of the equilibrium geometry, and the SCF potential-energy surface, although lying above the true potential-energy surface, is approximately parallel to it in this region.

A molecule of considerable interest to organic chemists is the reaction intermediate CH_2, the methylene radical. The first spectroscopic observations on CH_2 were interpreted as indicating a linear ground state. However, a DZ CI-SD calculation predicted a bond angle of 135°. Further experimental work showed that indeed the molecule is bent, with an equilibrium bond angle of 133.8° [P. R. Bunker and P. Jensen, *J. Chem. Phys.*, **79**, 1224 (1983)]. Both SCF and CI calculations work well for the CH_2 bond angle; for example, an SCF 4-31G calculation gives 132°. For the story of the interaction of theoretical and experimental CH_2 work, see H. F. Schaefer, III, *Science*, **231**, 1100 (1986).

Finding the Equilibrium Geometry.   The geometry of a nonlinear molecule with N nuclei is defined by $3N - 6$ independent nuclear coordinates, and its electronic energy U is a function of these $3N - 6$ coordinates. The number 6 is subtracted from the total number of nuclear coordinates because there are 3 translational and 3 rotational degrees of freedom, and these leave U unchanged. Many systematic mathematical procedures (algorithms) exist to find the minimum of a function U of several variables; see H. B. Schlegel, *Adv. Chem. Phys.*, **67**, 249 (1987), and *Press* et al., Chapter 10. Some of these procedures require only repeated calculation of U at various values of its variables, but these are not very efficient. More efficient procedures require repeated calculation of both U and its derivatives (Schlegel, op. cit., pages 262–268). The set of $3N - 6$ first derivatives of U with respect to each of its variables constitutes a vector (in a $3N - 6$ dimensional "space") called the ***gradient*** of U [Eq. (5.31)].

The SCF energy is (13.168), and its derivatives with respect to the nuclear coordinates would seem to involve the derivatives of the H_{rs}^{core} and $(rs|tu)$ integrals (which occur in ε_i), the derivatives of V_{NN}, and the derivatives of the SCF coefficients c_{si} (which occur in P_{rs}). However, the terms involving the derivatives of the c_{si}'s turn out to add up to zero, leaving only the derivatives of the integrals and of V_{NN}. The derivatives of the integrals are readily calculated, since the derivative of a Gaussian-type function with respect to a nuclear coordinate is another GTF, and an analytic formula for the gradient of the SCF energy is known [see *Hehre* et al., Section 3.3.3; P. Pulay, *Adv. Chem. Phys.*, **69**, 241 (1987)]. Once the SCF energy U and wave function have been found for some chosen geometry, the time needed to analytically calculate the energy gradient is roughly equal to the time needed to do the SCF wave function and energy calculation. (To calculate the gradient numerically requires varying the $3N - 6$ coordinates one at a time by a small amount, repeating the SCF calculation for each new geometry to get U for that geometry, and estimating each derivative as the ratio of the change in U to the change in the coordinate. Numerical evaluation of the gradient thus takes about $3N - 6$ times as long as analytical evaluation of the gradient.)

Methods that use both the first and second derivatives of U are even more efficient at finding a minimum than methods using only first derivatives. However, analytical calculation of the second derivatives of an SCF energy is much more time consuming than analytical calculation of the first derivatives, so use of a minimum-finding method based on second derivatives may not be more efficient than gradient-based methods. Once the equilibrium geometry has been found, however, one might well want to calculate the second derivatives of U, since the second derivatives are the molecular force constants, and from the force constants, we can find the molecular vibrational frequencies.

To find the equilibrium geometry of a molecule using the SCF MO method, one starts with an initially assumed plausible structure and calculates the SCF wave function, energy, and energy gradient for this structure. Using the calculated gradient, the computer program changes the $3N - 6$ nuclear coordinates to a new set that is likely to be closer to the equilibrium geometry than the initial set, and the SCF wave function, energy, and energy gradient are calculated at this new structure. Using the results of the new calculation, a further improved set of nuclear coordinates is calculated, and the SCF calculation is repeated at this new geometry. The process is repeated until the first derivatives of U are all less than a tiny, preset value, indicating that the minimum-energy geometry (at which the gradient of U is zero) has been found. Typically, about $3N - 6$ to $2(3N - 6)$ repetitions of the SCF and gradient calculations are needed to find the minimum-energy geometry.

The availability of analytical gradients makes possible the efficient determination of the ab initio equilibrium geometry of moderate-size molecules, and the introduction of analytical gradients into ab initio calculations has been called a "revolution" [L. Schäfer, *J. Mol. Struct.*, **100**, 51 (1983)]. According to a 1987 review, "With reasonable computational facilities, full geometry optimizations are feasible for molecules as large as 10 non-hydrogen atoms, and perhaps larger if small basis sets are used" (H. B. Schlegel, op. cit., page 250).

Actually, the "equilibrium" geometry found by the above procedure may not be the true equilibrium geometry. The above procedure locates the *local* minimum in U, which is the minimum in the neighborhood of the initially assumed geometry. A molecule with several single bonds in its framework has many possible conformations that differ in the dihedral angles of rotation about the single bonds, and one must repeat the entire search procedure for each possible conformation so as to locate the true (or *global*) minimum in U. Even so, the true equilibrium geometry might correspond to a highly unconventional structure that the researcher might not think to consider. For example, high-level calculations that include electron correlation and use large basis sets show that, for the vinyl cation, the classical structure $CH_2{=}\overset{+}{C}H$ lies about 4 kcal/mol higher in energy than the true equilibrium structure

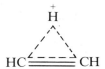

which has a three-center bond [C. Liang, T. P. Hamilton, and H. F. Schaefer, III, *J. Chem. Phys.*, **92**, 3653 (1990)]; the infrared spectrum of the vinyl cation also shows the three-center-bond structure to be more stable [M. W. Crofton et al., *J. Chem. Phys.*, **91**, 5139 (1989)]. The ethyl cation has a similar symmetrical bridged structure [B. Ruscic et al., *J. Chem. Phys.*, **91**, 114 (1989)].

The VSEPR Method.   Accurate SCF calculations for large molecules are difficult. We want a simple procedure that will give reasonably accurate predictions of bond angles. One such method is the *valence-shell electron-pair repulsion* (VSEPR) *theory*, which originated with Sidgwick and Powell in 1940 and has been extended by Gillespie [R. J. Gillespie, *J. Chem. Educ.*, **40**, 295 (1963); **47**, 18 (1970); *Can. J. Chem.*, **38**, 818 (1960); *Angew. Chem. Intern. Ed. Engl.*, **6**, 819 (1967); *Molecular Geometry*, Van Nostrand, 1972; R. J. Gillespie and I. Hargittai, *The VSEPR Model of Molecular Geometry*, Prentice Hall, Englewood Cliffs, N.J., 1991.]

We saw in Section 15.8 that the energy-localized MOs for the eight valence-shell electrons of water consist of two bonding orbitals $b(OH_1)$ and $b(OH_2)$ and two lone-pair MOs $l_1(O)$ and $l_2(O)$. Roughly speaking, these four MOs are tetrahedrally disposed about the oxygen atom. (Actually, the bonding MOs are separated by 103° and the lone-pair MOs by 114°.) Similarly, the energy-localized MOs found by Edmiston and Ruedenberg for the valence-shell electrons of NH_3 consist of three bonding MOs making angles of $104\frac{1}{2}°$ with one another and a lone-pair MO making an angle of 114° with each bonding localized MO. Again, these four MOs are approximately tetrahedrally disposed.

To apply the VSEPR method, we draw the Lewis dot structure of the molecule and count the number of valence electron pairs around the central atom. These valence electron pairs are then assumed to be arranged in space so as to minimize the repulsions between the localized MOs of the various pairs. [Recall that the condition for the energy-localized MOs is the minimization of (15.35).] In addition to the Coulombic repulsion between localized pairs is the Pauli repulsion (Section 10.3), which forces electrons with like spins to keep apart; each localized pair contains electrons with both possible spin values and thus excludes other electrons from the space it occupies.

For four pairs of electrons, the minimum repulsion occurs with the four MOs disposed tetrahedrally about the central atom. The most favorable arrangements for various numbers of electron pairs are (see Fig. 15.13)

2	3	4	5	6	8
straight line	equilateral triangle	tetrahedron	trigonal bipyramid	octahedron	square antiprism

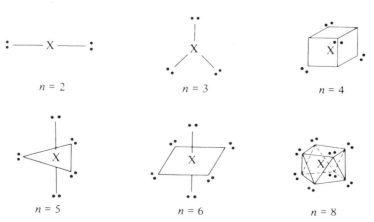

Figure 15.13 Preferred arrangements for n electron pairs about a central atom X.

Consider some examples. For water there are eight valence electrons, and we have two bonding pairs and two lone pairs arranged tetrahedrally. The bond angle should be close to the $109\frac{1}{2}°$ tetrahedral angle. To explain the deviation of the bond angle ($104\frac{1}{2}°$) from tetrahedral, the VSEPR approach points out that since lone-pair orbitals are under the influence of only one nucleus (rather than two), they are "fatter" (more spread out) than bonding orbitals; hence lone-pair electrons exert greater repulsions on neighboring pairs than bonding pairs exert on neighboring pairs. The angle between the lone pairs of water is thus greater than tetrahedral, while the angle between the bonding pairs is less than tetrahedral. This view is confirmed by the values of 114° and 103° found for the angles between the lone-pair and bonding energy-localized MOs of water. [The "fatness" of lone-pair MOs is illustrated by the greater contribution of the $2s$ oxygen AO to the lone-pair water MOs than to the bonding MOs; see Eq. (15.37).] For ammonia the four valence pairs are arranged approximately tetrahedrally; since there is only one lone pair instead of two, the bonding pairs are not pushed together as much as in water, and the observed bond angle (107°) is closer to tetrahedral than it is for water.

For IF_5 the dot formula shows six electron pairs around the central atom, I. These are arranged octahedrally. One pair is a lone pair, giving the observed square-pyramidal structure. For XeF_4 the dot formula has six pairs around Xe, which are arranged octahedrally; the two fat lone pairs are directed toward opposite vertexes of the octahedron, giving the observed square-planar structure.

ClF_3 has five pairs around Cl. The two fat lone pairs are placed so as to minimize the repulsions. A trigonal bipyramid has three equatorial and two axial positions. An axial electron pair makes an angle of 90° with three neighboring pairs; an equatorial pair makes an angle of 90° with only two neighboring pairs and an angle of 120° with two other neighboring pairs. The repulsions fall off rapidly with increasing angle, so the most favorable arrangement is with the two lone pairs of ClF_3 in equatorial positions. This gives the molecule a (planar) T structure. The observed slight distortion of $2\frac{1}{2}°$ from the T structure is explained by the fat lone pairs pushing the bonding axial pairs toward the bonding equatorial pair.

For molecules with multiple bonds, Gillespie recommends that for simplicity a double or triple bond be considered as a single electron pair. For example, for $CH_2{=}CH_2$, the expected bond angles are all 120°, since each carbon is to be considered as surrounded by three electron pairs. The observed H—C—H angle of 117° can be explained by the double-bond electrons pushing together the single-bond pairs. An alternative view (which is more consistent with the energy-localized MOs of ethylene) is to consider the double bond as two electron pairs forming two banana bonds that start out from each carbon atom at roughly the tetrahedral angle (the angle between the energy-localized banana bonds in ethylene is 112°); with four pairs about each carbon, the expected H—C—H angle is $109\frac{1}{2}°$. The observed angle results from the banana double-bond orbitals exerting less repulsion on the bonding C—H pairs than would two ordinary tetrahedrally directed orbitals.

The arrangements of Fig. 15.13 assume that the inner shells of the central atom are spherically symmetric and so do not affect the arrangement of the valence pairs. Recall (Section 11.1) that filled and half-filled subshells give spherically symmetric probability densities. Hence Fig. 15.13 applies to main-group elements (which all have filled inner subshells) and applies to those transition elements with d^0, d^5, or d^{10} subshells. In transition elements with nonspherical d subshells, distortions from the shapes of Fig. 15.13 may be observed.

The VSEPR model has a remarkably high batting average but is not infallible. Thus $TeCl_6^{2-}$ has seven electron pairs about Te, but its observed structure is octahedral. Other exceptions are some of the alkaline earth dihalides (for example, BaF_2, $SrCl_2$) that are bent, rather than linear. Gillespie has discussed the probable reasons for these exceptions in *J. Chem. Educ.*, **47**, 18 (1970).

Walsh Diagrams.   An alternative approach to predicting molecular geometry is the method of Walsh diagrams, developed by A. D. Walsh in 1953.

Consider the molecule AH_n, where A is one of the elements Li, Be, B, C, N, O, or F. We shall examine the AH_2 canonical (delocalized) MOs, guided by our experience with H_2O (Section 15.6). AH_2 might be linear or bent; for the purposes of discussion, we shall use the symmetry-species labels corresponding to a bent geometry, point group $\mathscr{C}_{2v}$; see Eq. (15.3). We choose the coordinate axes as in Fig. 15.1.

We expect the MOs to lie in the following order: inner shell, bonding, nonbonding lone pair, antibonding.

Clearly, the lowest MO is the inner-shell $1a_1$ MO, which is a nearly pure $1s$ AO on atom A; $1a_1 \approx A1s$.

To form the bonding MOs, we combine the $H_1 1s + H_2 1s$ symmetry orbital, which has a_1 symmetry [see (15.15)], with those valence AOs of A that have a_1 symmetry, and we combine the b_2 symmetry orbital $H_1 1s - H_2 1s$ with the valence AO of A that has b_2 symmetry. Overlap of $H_1 1s + H_2 1s$ with the A2s AO [and with a small contribution from $A2p_z$, as in (15.20)] forms the bonding $2a_1$ MO. Overlap of $H_1 1s - H_2 1s$ with $A2p_y$ gives the bonding $1b_2$ MO. The nodal plane in $1b_2$ (Fig. 15.4) and the fact that it is formed from a $2p$ AO as compared with the lower-energy $2s$ AO in $2a_1$ indicate that $1b_2$ lies above $2a_1$.

Next we have the lone-pair MO $3a_1$, composed of a hybrid of A2s and $A2p_z$ and $H_1 1s + H_2 1s$, all of which have a_1 symmetry. This MO contains a positive (bonding) overlap between $A2p_z$ and $H_1 1s + H_2 1s$ [see Eq. (15.20)] but is nonbonding overall, as shown by population analysis (Section 15.7). Next is the lone-pair MO $1b_1$, consisting solely of $A2p_x$, the only valence AO with b_1 symmetry.

Next is the antibonding MO $4a_1$, formed by antibonding interaction between $H_1 1s + H_2 1s$ and an $A2s–A2p_z$ hybrid, as in Eq. (15.23). Finally, we have the antibonding $2b_2$ MO, formed by antibonding interaction between the b_2 symmetry orbitals $H_1 1s - H_2 1s$ and $A2p_y$. (See Fig. 15.5.) The $2b_2$ MO lies above the $4a_1$ MO because $A2p_y$ of $2b_2$ lies above the $A2s–A2p_z$ hybrid of $4a_1$.

To decide whether AH_2 will be linear or bent, we examine the effect of a change in bond angle on each MO's orbital energy.

We expect the energy of the inner-shell $1a_1$ MO to be essentially unaffected by the bond angle.

For $2a_1$, as the bond angle decreases from 180°, the overlap between A2s and $H_1 1s + H_2 1s$ is unaffected, since this overlap depends only on the AH distances; the overlap between $H_1 1s + H_2 1s$ and A2p_z (which makes only a small contribution to $2a_1$) increases; and the overlap between $H_1 1s$ and $H_2 1s$ increases. Because H_1 and H_2 are relatively far apart, this increase in H–H overlap will have only a small energy-lowering effect. As a result of the small increases in overlap, we expect the orbital energy ε_{2a_1} to decrease slowly as the bond angle θ decreases from 180°.

For $1b_2$ a decrease of angle from 180° decreases substantially the bonding overlap between A2p_y (whose maximum probability density lies on the y axis in Fig. 15.1) and $H_1 1s - H_2 1s$; also, the antibonding interaction between $H_1 1s$ and $H_2 1s$ is increased, which is a smaller effect. Therefore, ε_{1b_2} increases rapidly as θ decreases below 180°.

For $3a_1$, a decrease in θ increases substantially the bonding overlap between A2p_z and $H_1 1s + H_2 1s$ and increases slightly the H–H bonding overlap. Therefore, ε_{3a_1} decreases rapidly as θ decreases from 180°.

The orbital energy of the pure lone pair $1b_1$ MO will be essentially unchanged as θ changes.

For $4a_1$ a decrease in θ will increase the antibonding interaction between A2p_z and $H_1 1s + H_2 1s$, causing ε_{4a_1} to increase rapidly. For $2b_2$, decreasing θ reduces the antibonding overlap between A2p_y and $H_1 1s - H_2 1s$ and decreases ε_{2b_2} rapidly.

Figure 15.14 is a rough sketch of the expected orbital energies versus bond angle θ for the AH_2 valence MOs. The important overlaps for the linear and bent MOs are sketched at each side. Also, the symmetry species of the linear-AH_2 MOs are listed.

Note that for linear AH_2 the $1\pi_{uz}$ MO, which correlates with the $3a_1$ MO of bent AH_2, cannot contain any contributions from $H_1 1s + H_2 1s$ or A2s, since these orbitals do not have π_u symmetry; the contributions of these orbitals grow in as the molecule is bent. Similarly, in linear AH_2 there is no contribution of A2p_z to σ_g MOs.

We assume that minimization of the sum of the orbital energies of the valence electrons will determine whether the molecule is linear or bent.

Although we have so far assumed A to be one of the atoms Li–F, the same sort of diagram will hold for the valence MOs of a main-group element of a later period. We shall have more inner-shell MOs, but the nature of the valence MOs will be the same; only the numbering of the valence-shell MOs will be changed.

AH_2 molecules with four valence electrons will have the ground-state valence-electron configuration $(2a_1)^2(1b_2)^2$. The rapid increase in ε_{1b_2} as the molecule is bent will outweigh the slow decrease in ε_{2a_1}, and we expect a linear geometry. The isolated BeH_2 molecule has not been observed (this compound forms a polymeric solid), but ab initio SCF and CI calculations show that the BeH_2 molecule is linear.

For five valence electrons, $(2a_1)^2(1b_2)^2(3a_1)$ is the AH_2 ground-state configuration. It turns out that the combined effect of the two $2a_1$ electrons (which

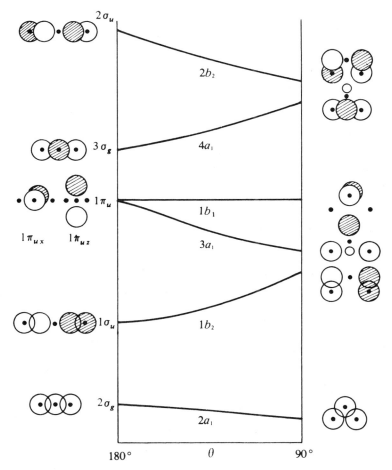

Figure 15.14 AH_2 MO energies versus bond angle. Shading indicates negative portions of AOs.

weakly favor bent geometry) and the $3a_1$ electron (which strongly favors bent geometry) outweigh the effect of the two $1b_2$ electrons, and such species are bent. Some experimentally observed angles are $131°$ for BH_2 and $119°$ for AlH_2.

For six valence electrons, the ground-state configuration $(2a_1)^2(1b_2)^2(3a_1)^2$ produces bent geometry. The double occupancy of $3a_1$ produces smaller bond angles than for five-valence-electron species. An example is the $93°$ angle observed for SiH_2. However, CH_2 (methylene) does not have the expected ground-state configuration; rather, its ground term is the triplet term of the configuration $(2a_1)^2(1b_2)^2(3a_1)(1b_1)$, which lies 9 kcal/mol (0.4 eV) below the singlet term of the $(2a_1)^2(1b_2)^2(3a_1)^2$ configuration [A. R. W. McKellar et al., *J. Chem. Phys.*, **79**, 5251 (1983)]. The CH_2 ground state has a $134°$ bond angle, similar to that of BH_2, which also has one $3a_1$ electron. The $CH_2 \ldots (3a_1)^2$ excited state has a $102°$ angle.

For seven valence electrons, the predicted ground-state configuration is $(2a_1)^2(1b_2)^2(3a_1)^2(1b_1)$. The molecule will be bent with an angle similar to those of the six-electron species, since $1b_1$ favors neither linear nor bent geometry. Examples are NH_2 (bond angle 103°), PH_2 (92°), AsH_2 (91°), H_2O^+ ($110\frac{1}{2}°$), and CH_2^- (99°).

For eight valence electrons and configuration $(2a_1)^2(1b_2)^2(3a_1)^2(1b_1)^2$, the molecule is bent with an angle similar to those of the six- and seven-valence-electron species. Examples are H_2O ($104\frac{1}{2}°$), H_2S (92°), H_2Se (91°), H_2Te (90°), and NH_2^- (104°).

The MOs $2a_1$ and $1b_2$ are bonding; $3a_1$ and $1b_1$ are nonbonding; and $4a_1$ and $2b_2$ are antibonding. Therefore, BeH_2, BH_2, CH_2, NH_2, and OH_2 each have four net bonding electrons and a total bond order of 2.

Although we considered only AH_2 molecules, similar arguments have been developed by Gimarc, Hoffmann, and others to rationalize and predict the geometry of many other species, including AH_3, AH_4, A_2H_2, A_2H_4, A_2H_6, HAB, AB_2, AB_3, AB_4, AB_5, and AB_6, and to achieve qualitative understanding of reaction mechanisms (see Section 16.7 for an example). This general approach is called *qualitative molecular-orbital theory*; see B. M. Gimarc, *Molecular Structure and Bonding*, Academic Press, New York, 1979; *Lowe*, Chapter 14.

The Walsh-diagram method works quite well, although there are a few exceptions. However, a good theoretical justification for the method is lacking. The method assumes that the equilibrium geometry is determined by minimizing the sum $\sum_i n_i \varepsilon_i$ of orbital energies, where n_i is the number of electrons in MO i. Actually, this is false, since the equilibrium geometry is determined by minimization of the molecular electronic energy including nuclear repulsion. In the SCF MO method (which is known to predict ground-state geometries accurately), this energy E_{HF} is equal to [Eq. (13.155)] $\sum_i n_i \varepsilon_i - V_{ee} + V_{NN}$, where V_{ee} is the interelectronic repulsion energy including exchange interactions (and is needed to correct for the fact that the sum of orbital energies counts each interelectronic repulsion twice) and V_{NN} is the internuclear repulsion energy. Various conflicting arguments have been advanced to explain why Walsh diagrams work so well, but the "ultimate theoretical basis of Walsh diagrams is still unclear" [I. Ferguson and N. C. Pyper, *Theor. Chim. Acta*, **55**, 283 (1980)]. Further references on this question include P. K. Mehrotra and R. Hoffmann, *Theor. Chim. Acta*, **48**, 301 (1978); R. J. Buenker and S. D. Peyerimhoff, *Chem. Rev.*, **74**, 127 (1974); B. M. Gimarc, *Molecular Structure and Bonding*, Academic Press, New York, 1979, Chapter 9; N. H. March and B. M. Deb (eds.), *The Single-Particle Density in Physics and Chemistry*, Academic Press, New York, 1987, Section 1.11.

15.11 CONFIGURATION INTERACTION

The four sources of error in ab initio molecular electronic calculations are (1) neglect of or incomplete treatment of electron correlation, (2) incompleteness of the basis set, (3) relativistic effects, and (4) deviations from the Born–Oppenheimer approximation. Deviations from the Born–Oppenheimer approximation are usually negligible for ground-state molecules. Relativistic effects will be discussed in Section 15.15. In calculations on molecules without heavy atoms, (1) and (2) are the main sources of error.

Almost all computational methods expand the MOs in a basis set of one-electron functions. The basis set has a finite number of members and hence is incomplete. The incompleteness of the basis set produces the **basis-set truncation error**. In CI treatments of correlation, one usually includes only CSFs (configuration state functions) with single and double excitations; the failure to go to full CI produces error (1).

Sections 15.11 to 15.14 discuss approaches that allow for electron correlation.

CI was discussed in Section 13.21. For a molecule with n electrons and with spin quantum number $S = 0$, the number of CSFs in a full CI calculation (with spatial symmetry restrictions ignored) is (*Wilson*, page 199)

$$\frac{b!(b + 1)!}{(\tfrac{1}{2}n)!(\tfrac{1}{2}n + 1)!(b - \tfrac{1}{2}n)!(b - \tfrac{1}{2}n + 1)!} \tag{15.66}$$

where b is the number of one-electron basis functions used to express the MOs. For a 6-31G** full CI calculation of the small molecule C_2H_6, $b = 2(15) + 6(5) = 60$. For $n = 18$ and $b = 60$, the number of CSFs given by (15.66) is 2.6×10^{19}, so a full CI 6-31G** calculation on ethane is not possible. One might therefore try a full CI calculation with the minimal STO-3G basis set, which has $b = 2(5) + 6(1) = 16$ for ethane; this gives a mere 2.8×10^7 CSFs. This calculation is feasible, since CI calculations with as many as 10^8 Slater determinants have been done [P. J. Knowles and N. C. Handy, *J. Chem. Phys.*, **91**, 2396 (1989)], but would be a waste of time. Experience shows that in order to get a substantial portion of the correlation energy, one must use a large basis set. For example, for H_2O, full CI calculations with a (relatively small) DZ basis set and 256473 CSFs gives an energy of -76.158 hartrees, compared with the Hartree–Fock limit of -76.068 hartrees and the true nonrelativistic energy of -76.440 hartrees (Table 15.1), so only a relatively small portion of the 0.372-hartree correlation energy has been obtained. The SCF energy obtained with this DZ H_2O basis set is -76.010 hartrees, and the difference of 0.148 hartree between this SCF energy and the full-CI DZ energy is called the **basis-set correlation energy**. Even for a DZ + P basis set, a full CI calculation on H_2O gives only -76.257 hartrees (Table 15.4), still quite far from -76.440 hartrees.

High-level correlation calculations may use basis sets that are triple zeta with two sets of polarization functions with different orbital exponents added on each atom; such a set is designated TZ2P. A commonly used basis set in electron correlation calculations is the 6-311G** set for first-row atoms, which is single zeta for the core and triple zeta for the valence AOs and contains five d-type Gaussian polarization functions on each nonhydrogen atom and three p-type polarization functions on each hydrogen atom. The 6-311G** set cannot really be considered to be a large basis set. An example of a large basis set for correlation calculations is the 6-311++G(3df,3pd) set; the pluses indicate diffuse functions (Section 15.4) on all atoms; the letters in parentheses indicate that three sets of five d-type Gaussian polarization functions (each set having a different orbital exponent) are added to each nonhydrogen (giving fifteen d functions on nonhydrogens), one set of seven f-type Gaussians is added to each nonhydrogen, and three sets of three p-type Gaussians and one set of five d-type Gaussians are added to each hydrogen.

Since full CI (FCI) is impossible except for small molecules and small basis sets, one resorts to **limited** CI, the most common form of which is CI-SD (Section 13.21). In addition, the **frozen-core** approximation is often used; here, excitations out of the inner-shell (core) MOs of the molecule are not included. The contribution of such excitations is not small, but their contribution changes very little with change in environment.

For calculations on a few 10-electron molecules, CI-SD gave about 94% of the basis-set correlation energy [R. J. Harrison and N. C. Handy, *Chem. Phys. Lett.*, **95**, 386 (1983)]. This result might seem to be a cause for optimism about CI-SD, but CI-SD has serious drawbacks. Calculations on a simple model system [F. Sasaki, *Int. J. Quantum Chem. Symp.*, **11**, 125 (1977)] indicate that the percentage of the basis-set correlation energy obtained by CI-SD decreases as the size of the molecule increases. For molecules that consist of first-row atoms, CI-SD is estimated to give 82% to 90% of the correlation energy of 20-electron molecules, 68% to 78% of the correlation energy of 50-electron molecules, and 55% to 67% for 100-electron molecules (Problem 15.29).

A related defect is that CI-SD calculations are not size consistent. A *size-consistent* quantum-mechanical method is one for which the energy and hence the energy error in the calculation increase in proportion to the size of the molecule. Size consistency is important whenever calculations on molecules of substantially different sizes are to be compared, as, for example, in calculation of the energy change in the dissociation reaction $A \rightarrow B + C$. A special case of size consistency is *size consistency for infinitely separated systems*, which means that the method gives the energy of a system of two or more infinitely separated molecules as equal to the sum of the energies of the individual molecules.

To see that CI-SD is not size consistent, consider two infinitely separated helium atoms He_a and He_b. If we do a CI-SD calculation of He_a using a complete set, we get the exact energy E_a of He_a, since CI-SD is the same as full CI for this two-electron atom. Likewise for He_b. Now consider a complete-basis-set CI-SD calculation for the composite system of infinitely separated He_a and He_b. This composite system has four electrons, so a CI-SD calculation is not equivalent to a full CI calculation, and the CI-SD calculation will give an energy higher than the exact energy $E_a + E_b$ of the composite system. Therefore, CI-SD is not size consistent. (Because the CI-SD wave function is a variation function, the variation theorem assures us that the CI-SD energy cannot be less than the true energy. The CI-SD method is therefore said to be **variational**.)

Full CI is size consistent, as are SCF MO calculations.

Theory indicates that after double excitations, quadruple excitations are next in importance. Calculations on some 10-electron molecules found that CI with inclusion of single, double, triple, and quadruple excitations (CI-SDTQ) gave over 99% of the basis-set correlation energy (Harrison and Handy, op. cit.). For molecules containing only first-row atoms, CI-SDTQ is estimated to give the following percentages of the basis-set correlation energy: 98% to 99% for 20-electron molecules, 90% to 96% for 50-electron molecules, and 80% to 90%

for 100-electron molecules (Sasaki, op. cit.). Provided we confine ourselves to molecules with no more than about 50 electrons, CI-SDTQ will come reasonably close to full CI and hence will be approximately size consistent. However, CI-SDTQ calculations with basis sets large enough to give good results for the correlation energy involve far too many CSFs to be practical.

An approximate formula due to Davidson [S. R. Langhoff and E. R. Davidson, *Int. J. Quantum Chem.*, **8**, 61 (1974)] is widely used to estimate the energy contribution ΔE_Q due to quadruple excitations:

$$\Delta E_Q \approx (1 - a_0^2)(E_{\text{CI-SD}} - E_{\text{SCF}}) \tag{15.67}$$

where a_0 is the coefficient of the SCF function Φ_0 in the normalized CI expansion $\psi = \Sigma_i\, a_i \Phi_i$, and $E_{\text{CI-SD}}$ and E_{SCF} are the CI-SD and SCF energies calculated with the basis set. For example, a DZ calculation on H_2O gave (in atomic units) at the equilibrium geometry $E_{\text{SCF}} = -76.009838$, $E_{\text{CI-SD}} = -76.150015$, and $a_0 = 0.97874$; from (15.67), we find $\Delta E_Q \approx -0.005897$, which when added to $E_{\text{CI-SD}}$, gives -76.155912; this result is reasonably close to the CI-SDTQ result of -76.157603 (the contribution of triple excitations is small).

Use of the Davidson correction reduces the size-consistency error. For example, CI-SD/DZP calculations on two H_2O molecules separated by 500 Å, a distance at which their interaction energy is utterly negligible, found that the CI-SD energy of this system exceeded twice the CI-SD energy of one H_2O molecule by 12.3 kcal/mol, which is the size-consistency error; use of the Davidson correction reduced this error to 3.8 kcal/mol [M. J. Frisch et al., *J. Chem. Phys.*, **84**, 2279 (1986)].

In the notation CI-SD/DZP, the method used precedes the slash and the basis-set designation follows the slash. For the SCF MO method, the abbreviation HF (for Hartree–Fock) is often used, without implying that the Hartree–Fock limiting energy has been reached. Thus, HF/3-21G signifies an SCF MO calculation with the 3-21G basis set.

Table 15.4 lists some calculations on H_2O that include electron correlation. Rosenberg, Ermler, and Shavitt's CI-SD calculations used a basis set of 39 STOs: $(5s4p2d)$ on O and $(3s1p)$ on each H; they obtained 75% of the correlation energy. Feller, Boyle, and Davidson did several CI-SD calculations with large Gaussian basis sets. To reduce the computational effort, they used perturbation theory to estimate the contribution of each doubly excited CSF to the wave function (see I. Shavitt in *Schaefer, Methods of Electronic Structure Theory*, pages 262–266) and omitted those with very slight contributions. Also, they transformed the virtual SCF orbitals to orbitals that approximated natural orbitals (Section 13.21). With a 120 CGTF basis set, their CI-SD energy of -76.378 hartrees obtained 84% of the correlation energy. With a 140 CGTF basis set and a reference function that included 53 CSFs, their MRCI-SD calculation obtained 88% of the correlation energy.

To speed up the convergence of CI calculations, multireference CI is often used (Section 13.21). Since multireference CI-SD calculations include the most important quadruple excitations, such calculations substantially reduce the lack of size consistency of CI-SD calculations.

TABLE 15.4 H_2O Calculations That Include Correlation[a]

Reference[b]	Method/Basis Set	Energy/E_h	μ/D	θ	R_{OH}/Å
Harrison, Handy	CI-SD/DZ	−76.150			
Harrison, Handy	CI-SDTQ/DZ	−76.1576			
Harrison, Handy	FCI/DZ	−76.1579			
Frisch et al.	MP2/6-31G*	−76.199	2.20	104.0°	0.969
Gauss, Cremer	MP3/6-31G*	−76.205	2.19	104.2°	0.967
Frisch et al.	MP2/6-31G**	−76.219	2.11		0.961
Bauschlicher, Taylor	CI-SD(FC)/DZP$_1$	−76.244			
Pople et al.	QCISD(FC)/DZP$_1$	−76.253			
Kucharski et al.	MP5(FC)/DZP$_1$	−76.256			
Noga, Bartlett	CCCSDT(FC)/DZP$_1$	−76.256			
Pople et al.	QCISD(T)(FC)/DZP$_1$	−76.256			
Bauschlicher, Taylor	FCI(FC)/DZP$_1$	−76.257			
Frisch et al.	MP2/DZP$_2$	−76.257	2.16	104.5°	0.962
Scuseria, Schaefer	CI-SD/DZP$_2$	−76.258	2.14	104.9°	0.958
Trucks et al.	CCD/DZP$_2$	−76.2656	2.14	104.7°	0.961
Scuseria, Schaefer	CCSD/DZP$_2$	−76.2664	2.14	104.6°	0.962
Trucks et al.	MP4/DZP$_2$	−76.268	2.13	104.5°	0.962
Scuseria, Schaefer	CCSDT-1/DZP$_2$	−76.270	2.13	104.5°	0.963
Scuseria, Schaefer	CI-SDTQ/DZP$_2$	−76.270	2.13	104.5°	0.963
Rosenberg et al.	CI-SD/39 STOs	−76.340	1.92	104.9°	0.953
Bartlett et al.	MP4/39 STOs	−76.360		104.4°	0.960
Feller et al.	CI-SD/120 CGTFs	−76.378			
Feller et al.	MRCI-SD/140 CGTFs	−76.396			
Nonrelativistic fixed-nuclei energy		−76.440			
Experimental values		−76.480	1.85	104.5°	0.958

[a] See footnotes a and c to Table 15.1 in Section 15.6. FC means frozen core.

[b] R. J. Harrison and N. C. Handy, *Chem. Phys. Lett.*, **95**, 386 (1983); M. J. Frisch, J. E. Del Bene, J. S. Binkley, and H. F. Schaefer, III, *J. Chem. Phys.*, **84**, 2279 (1986); J. Gauss and D. Cremer, *Chem. Phys. Lett.*, **138**, 131 (1987); Frisch et al., op. cit., and M. J. Frisch, J. A. Pople, and J. E. Del Bene, *J. Phys. Chem.*, **89**, 3664 (1985); C. W. Bauschlicher and P. R. Taylor, *J. Chem. Phys.*, **85**, 2779 (1986); J. A. Pople, M. Head-Gordon, and K. Raghavachari, *J. Chem. Phys.*, **87**, 5968 (1987); S. A. Kucharski, J. Noga, and R. J. Bartlett, *J. Chem. Phys.*, **90**, 7282 (1989); J. Noga and R. J. Bartlett, *J. Chem. Phys.*, **86**, 7041 (1987); G. E. Scuseria and H. F. Schaefer, III, *Chem. Phys. Lett.*, **146**, 23 (1988); G. W. Trucks, J. D. Watts, E. A. Salter, and R. J. Bartlett, *Chem. Phys. Lett.*, **153**, 490 (1988); B. J. Rosenberg and I. Shavitt, *J. Chem. Phys.*, **63**, 2162 (1975) and B. J. Rosenberg, W. C. Ermler, and I. Shavitt, *J. Chem. Phys.*, **65**, 4072 (1976); R. J. Bartlett et al., *J. Chem. Phys.*, **87**, 6579 (1987); D. Feller, C. M. Boyle, and E. R. Davidson, *J. Chem. Phys.*, **86**, 3424 (1987).

Another way to speed up convergence is to use localized SCF MOs (Section 15.8) instead of canonical SCF MOs in the CSFs. CI calculations on 1,3-butadiene using this *localized correlation* method showed speedups by factors of 20 to 40, since fewer CSFs needed to be included [S. Saebø and P. Pulay, *Chem. Phys. Lett.*, **113**, 13 (1985)].

Because of slow convergence and lack of size consistency, CI calculations have lost their former dominance in correlation calculations, and several other methods have been developed to allow for correlation.

15.12 MØLLER–PLESSET (MP) PERTURBATION THEORY

Physicists and chemists have developed various perturbation-theory methods to deal with systems of many interacting particles (nucleons in a nucleus, atoms in a solid, electrons in an atom or molecule), and these methods constitute **many-body perturbation theory** (MBPT). In 1934, Møller and Plesset proposed a perturbation treatment of atoms and molecules in which the unperturbed wave function is the Hartree–Fock function, and this form of MBPT is called **Møller–Plesset** (MP) **perturbation theory**. Actual molecular applications of MP perturbation theory began only in 1975 with the work of Pople and co-workers and Bartlett and co-workers [R. J. Bartlett, *Ann. Rev. Phys. Chem.*, **32**, 359 (1981); *Hehre* et al.].

The treatment of this section will be restricted to closed-shell, ground-state molecules. Also, the development will use spin-orbitals u_i, rather than spatial orbitals ϕ_i. For spin-orbitals, the Hartree–Fock equations (13.148) and (13.149) for electron m in an n-electron molecule have the forms (*Szabo and Ostlund*, Section 3.1)

$$\hat{f}(m)u_i(m) = \varepsilon_i u_i(m) \tag{15.68}$$

$$\hat{f}(m) \equiv -\tfrac{1}{2}\nabla_m^2 - \sum_\alpha \frac{Z_\alpha}{r_{m\alpha}} + \sum_{j=1}^{n} [\hat{j}_j(m) - \hat{k}_j(m)] \tag{15.69}$$

where $\hat{j}_j(m)$ and $\hat{k}_j(m)$ are defined by equations like (13.151) and (13.152) with spatial orbitals replaced by spin-orbitals and integrals over spatial coordinates of an electron replaced by integration over spatial coordinates and summation over the spin coordinate of that electron.

The MP unperturbed Hamiltonian is taken as the sum of the one-electron Fock operators $\hat{f}(m)$ in (15.68):

$$\hat{H}^0 \equiv \sum_{m=1}^{n} \hat{f}(m) \tag{15.70}$$

The ground-state Hartree–Fock wave function Φ_0 is the Slater determinant $|u_1 u_2 \cdots u_n|$ of spin-orbitals. This Slater determinant is an antisymmetrized product of the spin-orbitals [for example, see Eq. (10.36)] and, when expanded, is the sum of $n!$ terms, where each term involves a different permutation of the electrons among the spin-orbitals. Each term in the expansion of Φ_0 is an eigenfunction of the MP $\hat{H}^0$; for example, for a four-electron system, application of $\hat{H}^0$ to a typical term in the Φ_0 expansion gives

$$[\hat{f}(1) + \hat{f}(2) + \hat{f}(3) + \hat{f}(4)]u_1(3)u_2(2)u_3(4)u_4(1)$$

$$= (\varepsilon_4 + \varepsilon_2 + \varepsilon_1 + \varepsilon_3)u_1(3)u_2(2)u_3(4)u_4(1)$$

where $\hat{f}(m)u_i(m) = \varepsilon_i u_i(m)$ [Eq. (15.68)] was used. Similarly, each other term in the expansion of $|u_1 u_2 u_3 u_4|$ is an eigenfunction of $\hat{H}^0$ with the same eigenvalue $\varepsilon_1 + \varepsilon_2 + \varepsilon_3 + \varepsilon_4$. Since Φ_0 is a linear combination of these $n!$ terms, Φ_0 is an eigenfunction of $\hat{H}^0$ with this eigenvalue:

$$\hat{H}^0\Phi_0 = \left(\sum_{m=1}^{n} \varepsilon_m \right)\Phi_0 \tag{15.71}$$

The eigenfunctions of the unperturbed Hamiltonian $\hat{H}^0$ are the zeroth-order (unperturbed) wave functions [Eq. (9.2)], so the Hartree–Fock ground-state function Φ_0 is one of the zeroth-order wave functions. What are the other eigenfunctions of $\hat{H}^0$? The Hermitian operator $\hat{f}(m)$ has a complete set of eigenfunctions, these eigenfunctions being all the possible spin-orbitals of the molecule; the n lowest-energy spin-orbitals are occupied, and there are an infinite number of unoccupied (virtual) orbitals. The operator $\hat{H}^0 \equiv \sum_{m=1}^{n} \hat{f}(m)$ is the sum of the operators $\hat{f}(m)$, and so the eigenfunctions of $\hat{H}^0$ are all possible products of any n of the spin-orbitals. However, the wave functions must be antisymmetric, so we must antisymmetrize these zeroth-order wave functions by forming Slater determinants. Thus, the zeroth-order wave functions are all possible Slater determinants formed using any n of the infinite number of possible spin-orbitals. (Of course, the n chosen spin-orbitals must all be different or the Slater determinant would vanish.)

The perturbation $\hat{H}'$ is the difference between the true molecular electronic Hamiltonian $\hat{H}$ and $\hat{H}^0$; $\hat{H}' = \hat{H} - \hat{H}^0$. Use of (15.10) for $\hat{H}$ and (15.70) and (15.69) for $\hat{H}^0$ gives (Problem 15.31)

$$\hat{H}' = \hat{H} - \hat{H}^0 = \sum_{l} \sum_{m>l} \frac{1}{r_{lm}} - \sum_{m=1}^{n} \sum_{j=1}^{n} [\hat{j}_j(m) - \hat{k}_j(m)] \qquad (15.72)$$

The perturbation $\hat{H}'$ is the difference between the true interelectronic repulsions and the Hartree–Fock interelectronic potential (which is an average potential).

The MP first-order correction $E_0^{(1)}$ to the ground-state energy is [Eq. (9.22)] $E_0^{(1)} = \langle \psi_0^{(0)} | \hat{H}' | \psi_0^{(0)} \rangle = \langle \Phi_0 | \hat{H}' | \Phi_0 \rangle$, since $\psi_0^{(0)} = \Phi_0$. The subscript 0 denotes the ground state. We have

$$E_0^{(0)} + E_0^{(1)} = \langle \psi_0^{(0)} | \hat{H}^0 | \psi_0^{(0)} \rangle + \langle \Phi_0 | \hat{H}' | \Phi_0 \rangle = \langle \Phi_0 | \hat{H}^0 + \hat{H}' | \Phi_0 \rangle = \langle \Phi_0 | \hat{H} | \Phi_0 \rangle$$

But $\langle \Phi_0 | \hat{H} | \Phi_0 \rangle$ is the variational integral for the Hartree–Fock wave function Φ_0 and it therefore equals the Hartree–Fock energy E_{HF}. Hence (recall the beginning of Section 9.4)

$$E_0^{(0)} + E_0^{(1)} = E_{HF}$$

To improve on the Hartree–Fock energy, we must find the second-order energy correction $E_0^{(2)}$. From (9.35),

$$E_0^{(2)} = \sum_{s \neq 0} \frac{|\langle \psi_s^{(0)} | \hat{H}' | \Phi_0 \rangle|^2}{E_0^{(0)} - E_s^{(0)}} \qquad (15.73)$$

We saw above that the unperturbed functions $\psi_s^{(0)}$ are all possible Slater determinants formed from n different spin-orbitals. Let $i, j, k, l, \ldots$ denote the occupied spin-orbitals in the ground-state Hartree–Fock function Φ_0, and let $a, b, c, d, \ldots$ denote the unoccupied (virtual) spin-orbitals. Each unperturbed wave function can be classified by the number of virtual spin-orbitals it contains; this number is called the **excitation level**. Let Φ_i^a denote the singly excited determinant that differs from Φ_0 solely by replacement of u_i by the virtual spin-orbital u_a. Let

Φ_{ij}^{ab} denote the doubly excited determinant formed from Φ_0 by replacement of u_i by u_a and u_j by u_b; and so on.

Consider the matrix elements $\langle \psi_s^{(0)} | \hat{H}' | \Phi_0 \rangle$ in (15.73), where Φ_0 is a closed-shell single determinant. One finds (*Szabo and Ostlund*, Section 6.5) that this integral vanishes for all singly excited $\psi_s^{(0)}$'s; that is, $\langle \Phi_i^a | \hat{H}' | \Phi_0 \rangle = 0$ for all i and a. Also, $\langle \psi_s^{(0)} | \hat{H}' | \Phi_0 \rangle$ vanishes for all $\psi_s^{(0)}$'s whose excitation level is three or higher. This follows from the Condon–Slater rules (Table 11.3). Hence, we need consider only doubly excited $\psi_s^{(0)}$'s to find $E_0^{(2)}$. Also, the same reasoning applied to Eq. (9.27) shows that $\psi_0^{(1)}$, the first-order correction to the wave function, contains only doubly excited $\psi_s^{(0)}$'s.

The doubly excited function Φ_{ij}^{ab} is an eigenfunction of $\hat{H}^0 = \Sigma_m \hat{f}(m)$ with an eigenvalue that differs from the eigenvalue of Φ_0 solely by replacement of ε_i by ε_a and replacement of ε_j by ε_b. Hence, $E_0^{(0)} - E_s^{(0)}$ in (15.73) equals $\varepsilon_i + \varepsilon_j - \varepsilon_a - \varepsilon_b$. Use of (15.72) for $\hat{H}'$ and of the Condon–Slater rules allows the integrals involving Φ_{ij}^{ab} to be evaluated; one finds (Problem 15.32)

$$E_0^{(2)} = \sum_{b=a+1}^{\infty} \sum_{a=n+1}^{\infty} \sum_{i=j+1}^{n} \sum_{j=1}^{n-1} \frac{|\langle ab | r_{12}^{-1} | ij \rangle - \langle ab | r_{12}^{-1} | ji \rangle|^2}{\varepsilon_i + \varepsilon_j - \varepsilon_a - \varepsilon_b} \quad (15.74)$$

where

$$\langle ab | r_{12}^{-1} | ij \rangle \equiv \int \int u_a^*(1) u_b^*(2) r_{12}^{-1} u_i(1) u_j(2) \, d\tau_1 \, d\tau_2 \quad (15.75)$$

The integrals over spin-orbitals (which include a sum over spins) are readily evaluated in terms of electron-repulsion integrals. The sums over a, b, i, and j in (15.74) provide for the inclusion of all the doubly substituted $\psi_s^{(0)}$'s in (15.73).

Taking the molecular energy as $E^{(0)} + E^{(1)} + E^{(2)} = E_{HF} + E^{(2)}$ gives a calculation designated as MP2 or MBPT(2), where the 2 indicates inclusion of energy corrections through second order.

Formulas for the MP energy corrections $E^{(3)}$, $E^{(4)}$, and so on, have also been derived [for example, see R. Krishnan and J. A. Pople, *Int. J. Quantum Chem.*, **14**, 91 (1978)]. Since $\psi^{(1)}$, the first-order correction to the wave function, determines both $E^{(2)}$ and $E^{(3)}$ (Section 9.2), and since $\psi^{(1)}$ contains only doubly excited determinants, $E^{(3)}$ contains summations over only double substitutions. The MP $E^{(4)}$ involves summations over single, double, triple, and quadruple substitutions. Calculations that include energy corrections through $E^{(3)}$ are designated MP3 or MBPT(3), and those that include corrections through $E^{(4)}$ are MP4 or MBPT(4).

To do an MP electron-correlation calculation, one first chooses a basis set and carries out an SCF calculation to obtain Φ_0, E_{HF}, and virtual orbitals. One then evaluates $E^{(2)}$ (and perhaps higher corrections) by evaluating the integrals over spin-orbitals in (15.74) in terms of integrals over the basis functions. One ought to use a complete set of basis functions to expand the spin-orbitals. The SCF calculation will then produce the exact Hartree–Fock energy and will yield an infinite number of virtual orbitals. The first two sums in (15.74) will then contain an infinite number of terms. Of course, one always uses a finite, incomplete basis set, which yields a finite number of virtual orbitals, and the sums in (15.74) contain only a finite number of terms. One thus has a basis-set truncation error in addition to the error due to truncation of the MP perturbation energy at $E^{(2)}$ or $E^{(3)}$ or whatever.

In MP4 calculations, evaluation of the terms that involve triply substituted determinants is very time consuming, so these terms are sometimes neglected [even though their contribution to $E^{(4)}$ is not small], giving an approximation to MP4 that is designated MP4-SDQ or SDQ-MBPT(4), where SDQ indicates inclusion of single, double, and quadruple excitations. Evaluation of $E^{(5)}$ is extremely time consuming and so is never done except by specialists investigating the convergence of the series. To save time in MP2, MP3, and MP4 computations, the frozen-core approximation is often used; here, terms involving excitations out of core orbitals are omitted.

MP calculations are much faster than CI calculations, and the widely used ab initio programs GAUSSIAN 82, 86, 88, and 90 contain provision for doing MP2, MP3, and MP4 calculations. Relative times for single ab initio frozen-core 6-31G* calculations on CH_3NH_2 are (*Hehre*, page 91) 1 for SCF, 17 for CI-SD, 1.5 for MP2, 3.6 for MP3, and 5.8 for MP4-SDQ.

In addition to their computational efficiency, MP calculations truncated at any order can be shown to be size consistent (*Szabo and Ostlund*, Section 6.7.4). However, MP calculations are not variational and can produce an energy below the true energy. Currently, size consistency is viewed as more important than being variational.

The MP perturbation series has not been proved to converge to the true energy, but calculations on H_2O and other small molecules that go as high as MP48 show convergence and indicate that "the MP series will converge for most molecular calculations" [N. C. Handy et al., *Theor. Chim. Acta*, **68**, 87 (1985); P. J. Knowles et al., *Chem. Phys. Lett.*, **113**, 8 (1985)].

As with CI calculations, MP calculations with small basis sets are of little practical value, and MP calculations with the GAUSSIAN programs should use a 6-31G* or larger basis set for useful results. For DZ + P basis sets, MP2 calculations on closed-shell molecules typically yield 85% to 95% of the basis-set correlation energy [R. J. Bartlett, *Ann. Rev. Phys. Chem.*, **32**, 359 (1981)] and substantially improve the accuracy of equilibrium-geometry and vibrational-frequency predictions.

Experience indicates that in most electron-correlation calculations, the basis-set truncation error is larger than the error due to truncation of the correlation treatment. For example, when one goes from a 6-31G* basis set to a TZ2P basis set, the errors in MP2-predicted equilibrium single-bond lengths are reduced by a factor of 2 or 3 [E. D. Simandiras et al., *J. Chem. Phys.*, **88**, 3187 (1988)], but when one goes from MP2/TZ2P to MP3/TZ2P calculations, no improvement in geometry accuracy is obtained [I. L. Alberts and N. C. Handy, *J. Chem. Phys.*, **89**, 2107 (1988)].

The energy gradient in MP2 calculations is readily evaluated analytically [see P. Pulay, *Adv. Chem. Phys.*, **69**, 276 (1987)], and this allows MP2 geometry optimization to be done easily and allows calculation of MP2 vibrational frequencies.

Special techniques have been devised to allow efficient calculation of dipole moments and other one-electron properties at the MP2, MP3, and MP4 levels [G. W. Trucks et al., *Chem. Phys. Lett.*, **147**, 359 (1988); **150**, 37 (1988)].

Saebø and Pulay applied their local correlation method, in which the occupied canonical SCF MOs are replaced by localized MOs (Section 15.11), to MP calculations and did frozen-core MP4-SDQ/6-311G** calculations at fixed geometries on molecules as large as 1,3,5,7-octatetraene, C_8H_{10}, the first correlated calculation on a molecule this large with a basis set of this quality [S. Saebø and P. Pulay, *J. Chem. Phys.*, **88**, 1884 (1988)].

Saebø and Almlöf have developed a ***direct correlation*** method that (like the direct SCF method discussed in Section 15.5) minimizes the amount of input and output during calculations and have implemented it for MP2 calculations [S. Saebø and J. Almlöf, *Chem. Phys. Lett.*, **154**, 83 (1989)]. Direct-correlation frozen-core MP2/6-311G** calculations were done on the quite large molecule diphenylacetylene, $C_6H_5C{\equiv}CC_6H_5$, at four different torsional angles [S. Saebø, J. Almlöf et al., *THEOCHEM*, **59**, 361 (1989)]. These monumental supercomputer calculations included 312 basis functions and 40 million doubly substituted determinants. The calculations predicted coplanar phenyl rings with a 0.64-kcal/mol barrier to internal rotation, in good agreement with the experimental results, which show a planar molecule with a 0.57-kcal/mol barrier.

Table 15.4 lists results of some MP calculations on H_2O.

For species involving open-shell ground states (for example, O_2, NO_2, and OH), one bases an MP calculation on the unrestricted SCF wave function (Section 15.3), giving calculations designated UMP2, UMP3, and so on. Unrestricted SCF wave functions are not eigenfunctions of $\hat{S}^2$, and this "spin contamination" can sometimes produce serious errors in UMP-calculated quantities [K. Wolinski and P. Pulay, *J. Chem. Phys.*, **90**, 3647 (1989)].

Another limitation of MP calculations is that, although they work well near the equilibrium geometry, they do not work well at geometries far from equilibrium. For example, DZ calculations on H_2O showed that at the equilibrium geometry an MP2 calculation obtained 94% of the basis-set correlation energy, but at a geometry with the bonds at twice their equilibrium lengths an MP2 calculation obtained only 83% of the basis-set correlation energy [W. D. Laidig et al., *Chem. Phys. Lett.*, **113**, 151 (1985)]; also, the MP energy series at this stretched geometry converged erratically [N. C. Handy et al., *Theor. Chim. Acta*, **68**, 87 (1985)].

A third limitation is that MP calculations are not generally applicable to excited electronic states. [See, however, K. Wolinski and P. Pulay, *J. Chem. Phys.*, **90**, 3647 (1989).]

Because of these limitations, MP calculations have not made CI calculations obsolete, and multireference CI calculations are widely used for excited states and for geometries far from equilibrium.

Because of their computational efficiency and the good results they provide for molecular properties, MP2 calculations have become the method of choice for dealing with the effects of electron correlation on molecular ground-state equilibrium properties. MP perturbation theory "is the only correlation method routinely used by nonspecialists" [K. Wolinski and P. Pulay, *J. Chem. Phys.*, **90**, 3647 (1989)].

15.13   THE COUPLED-CLUSTER METHOD

The *coupled-cluster* (CC) *method* for dealing with a system of interacting particles was introduced around 1958 by Coester and Kümmel in the context of studying the atomic nucleus. CC methods for molecular electronic calculations were developed by Cizek, Paldus, Sinanoglu, and Nesbet in the 1960s and by Pople and co-workers and Bartlett and co-workers in the 1970s. For a review of the CC method, see R. J. Bartlett, *J. Phys. Chem.*, **93**, 1697 (1989).

The fundamental equation in CC theory is

$$\psi = e^{\hat{T}} \Phi_0 \tag{15.76}$$

where ψ is the exact nonrelativistic ground-state molecular electronic wave function, Φ_0 is the normalized ground-state Hartree–Fock wave function, the operator $e^{\hat{T}}$ is defined by the Taylor-series expansion

$$e^{\hat{T}} \equiv 1 + \hat{T} + \frac{\hat{T}^2}{2!} + \frac{\hat{T}^3}{3!} + \cdots = \sum_{k=0}^{\infty} \frac{\hat{T}^k}{k!} \tag{15.77}$$

and the *cluster operator* $\hat{T}$ (no connection with kinetic energy) is

$$\hat{T} \equiv \hat{T}_1 + \hat{T}_2 + \cdots + \hat{T}_n \tag{15.78}$$

where n is the number of electrons in the molecule and the operators $\hat{T}_1$, $\hat{T}_2, \ldots$ are defined below. Proof of (15.76) is omitted [see references in R. F. Bishop and H. G. Kümmel, *Physics Today*, March 1987, p. 52], but its plausibility will be shown below. The wave function ψ in (15.76) is not normalized (Problem 15.34) but can be normalized at the end of the calculation.

The *one-particle excitation operator* $\hat{T}_1$ and the *two-particle excitation operator* $\hat{T}_2$ are defined by

$$\hat{T}_1 \Phi_0 \equiv \sum_{b=n+1}^{\infty} \sum_{i=1}^{n} t_i^a \Phi_i^a , \qquad \hat{T}_2 \Phi_0 \equiv \sum_{b=a+1}^{\infty} \sum_{a=n+1}^{\infty} \sum_{j=i+1}^{n} \sum_{i=1}^{n-1} t_{ij}^{ab} \Phi_{ij}^{ab} \tag{15.79}$$

where Φ_i^a is a singly excited Slater determinant with the occupied spin-orbital u_i replaced by the virtual spin-orbital u_a, and t_i^a is a numerical coefficient whose value depends on i and a and will be determined by requiring that Eq. (15.76) be satisfied. The operator $\hat{T}_1$ converts the Slater determinant $|u_1 \cdots u_n| = \Phi_0$ into a linear combination of all possible singly excited Slater determinants. Φ_{ij}^{ab} is a Slater determinant with the occupied spin-orbitals u_i and u_j replaced by the virtual spin-orbitals u_a and u_b, respectively; t_{ij}^{ab} is a numerical coefficient. Similar definitions hold for $\hat{T}_3, \ldots, \hat{T}_n$. Since no more than n electrons can be excited from the n-electron function Φ_0, no operators beyond $\hat{T}_n$ appear in (15.78). The limits in (15.79) are chosen so as to include all possible single and double excitations without duplication of any excitation. By definition, when $\hat{T}_1$ or $\hat{T}_2$ or ... operates on a determinant containing both occupied and virtual spin-orbitals, the resulting sum contains only determinants with excitations from those spin-orbitals that are occupied in Φ_0 and not from virtual spin-orbitals. $\hat{T}_1^2 \Phi_0 \equiv \hat{T}_1(\hat{T}_1 \Phi_0)$ contains only doubly excited Slater determinants, and $\hat{T}_1^2 \Phi_0$ contains only quadruply excited determinants. When $\hat{T}_1$ operates on a determinant containing only virtual spin-orbitals, the result is zero, by definition.

The effect of the $e^{\hat{T}}$ operator in (15.76) is to express ψ as a linear combination of Slater determinants that include Φ_0 and all possible excitations of electrons from occupied to virtual spin-orbitals. A full CI calculation also expresses ψ as a linear combination involving all possible excitations, and we know that a full CI calculation with a complete basis set gives the exact ψ; hence, it is plausible that Eq. (15.76) is valid. The mixing into the wave function of Slater determinants with electrons excited from occupied to virtual spin-orbitals allows electrons to keep away from one another and thereby provides for electron correlation.

In the CC method, one works with individual Slater determinants, rather than with CSFs, but each CSF is a linear combination of one or a few Slater determinants, and the CC and CI methods can each be formulated either in terms of individual Slater determinants or in terms of CSFs.

The aim of a CC calculation is to find the coefficients t_i^a, t_{ij}^{ab}, t_{ijk}^{abc}, . . . for all $i, j, k, \ldots$, and all $a, b, c, \ldots$; once these coefficients (called **amplitudes**) are found, the wave function ψ in (15.76) is known.

To apply the CC method, two approximations are made. First, instead of using a complete, and hence infinite, set of basis functions, one uses a finite basis set to express the spin-orbitals in the SCF wave function; one thus has available only a finite number of virtual orbitals to use in forming excited determinants. As usual, we have a basis-set truncation error. Second, instead of including all the operators $\hat{T}_1, \hat{T}_2, \ldots, \hat{T}_n$, one approximates the operator $\hat{T}$ by including only some of these operators. Theory shows (*Wilson*, p. 222) that the most important contribution to $\hat{T}$ is made by $\hat{T}_2$. The approximation $\hat{T} \approx \hat{T}_2$ gives

$$\psi_{\text{CCD}} = e^{\hat{T}_2}\Phi_0 \qquad (15.80)$$

Inclusion of only $\hat{T}_2$ gives an approximate CC approach called the **coupled-cluster doubles** (CCD) **method**, also known as the **coupled-pair many-electron theory** (CPMET). Since $e^{\hat{T}_2} = 1 + \hat{T}_2 + \frac{1}{2}\hat{T}_2^2 + \cdots$, the wave function ψ_{CCD} contains determinants with double substitutions, quadruple substitutions, hextuple substitutions, and so on. Recall (Section 15.11) that quadruple substitutions are next in importance after double substitutions in a CI wave function. The treatment of quadruple substitutions in the CCD method is only approximate. The CCD quadruple excitations are produced by the operator $\frac{1}{2}\hat{T}_2^2$, and so the coefficients of the quadruply substituted determinants are determined as products of the coefficients of the doubly substituted determinants [see Eq. (15.79)], rather than being determined independently, as in the CI-SDTQ method. The CCD approximation of the coefficients of the quadruply substituted determinants turns out to be pretty accurate.

We need equations to find the CCD amplitudes. Substitution of $\psi = e^{\hat{T}}\Phi_0$ [Eq. (15.76)] in the Schrödinger equation $\hat{H}\psi = E\psi$ gives

$$\hat{H}e^{\hat{T}}\Phi_0 = Ee^{\hat{T}}\Phi_0 \qquad (15.81)$$

Multiplication by Φ_0^* and integration gives

$$\langle \Phi_0 | \hat{H} | e^{\hat{T}}\Phi_0 \rangle = E\langle \Phi_0 | e^{\hat{T}}\Phi_0 \rangle \qquad (15.82)$$

We have $e^{\hat{T}}\Phi_0 = (1 + \hat{T} + \cdots)\Phi_0 = \Phi_0 + \hat{T}\Phi_0 + \frac{1}{2}\hat{T}^2\Phi_0 + \cdots$. Since $\hat{T} = \hat{T}_1 + \hat{T}_2 + \cdots + \hat{T}_n$, the functions $\hat{T}\Phi_0$, $\frac{1}{2}\hat{T}^2\Phi_0$, and so on, contain only Slater determinants with at least one occupied orbital replaced by a virtual orbital. Because of the orthogonality of the spin-orbitals, all such excited Slater determinants are orthogonal to Φ_0, as can be seen by replacing $\sum_{i=1}^{n}\hat{f}_i$ with 1 in Table 11.3. Therefore, $\langle\Phi_0|e^{\hat{T}}\Phi_0\rangle = \langle\Phi_0|\Phi_0\rangle = 1$, and (15.82) becomes

$$\langle\Phi_0|\hat{H}|e^{\hat{T}}\Phi_0\rangle = E \qquad (15.83)$$

Multiplication of the Schrödinger equation (15.81) by Φ_{ij}^{ab*} and integration gives

$$\langle\Phi_{ij}^{ab}|\hat{H}|e^{\hat{T}}\Phi_0\rangle = E\langle\Phi_{ij}^{ab}|e^{\hat{T}}\Phi_0\rangle \qquad (15.84)$$

Use of (15.83) to eliminate E from (15.84) gives

$$\langle\Phi_{ij}^{ab}|\hat{H}|e^{\hat{T}}\Phi_0\rangle = \langle\Phi_0|\hat{H}|e^{\hat{T}}\Phi_0\rangle\langle\Phi_{ij}^{ab}|e^{\hat{T}}\Phi_0\rangle \qquad (15.85)$$

So far, the treatment is exact. We now invoke the CCD approximation $\hat{T} \approx \hat{T}_2$, and Eqs. (15.83) and (15.85) become

$$E_{\text{CCD}} = \langle\Phi_0|\hat{H}|e^{\hat{T}_2}\Phi_0\rangle \qquad (15.86)$$

$$\langle\Phi_{ij}^{ab}|\hat{H}|e^{\hat{T}_2}\Phi_0\rangle = \langle\Phi_0|\hat{H}|e^{\hat{T}_2}\Phi_0\rangle\langle\Phi_{ij}^{ab}|e^{\hat{T}_2}\Phi_0\rangle \qquad (15.87)$$

Since these equations are approximate, the exact energy E has been replaced by the CCD energy E_{CCD}. Also, the coefficients t_{ij}^{ab} (produced when $e^{\hat{T}_2}$ operates on Φ_0) in these equations are approximate. The first integral on the right side of (15.87) is

$$\langle\Phi_0|\hat{H}|e^{\hat{T}_2}\Phi_0\rangle = \langle\Phi_0|\hat{H}|(1 + \hat{T}_2 + \tfrac{1}{2}\hat{T}_2^2 + \cdots)\Phi_0\rangle$$

$$= \langle\Phi_0|\hat{H}|\Phi_0\rangle + \langle\Phi_0|\hat{H}|\hat{T}_2\Phi_0\rangle + 0 = E_{\text{HF}} + \langle\Phi_0|\hat{H}|\hat{T}_2\Phi_0\rangle \qquad (15.88)$$

where E_{HF} is the Hartree–Fock (or SCF) energy. The integral $\langle\Phi_0|\hat{H}|\hat{T}_2^2\Phi_0\rangle$ and similar integrals with higher powers of $\hat{T}_2$ vanish because $\hat{T}_2^2\Phi_0$ contains only quadruply excited determinants; hence, $\hat{T}_2^2\Phi_0$ differs from Φ_0 by four spin-orbitals, and the Condon–Slater rules (Table 11.3) show that the matrix elements of $\hat{H}$ between Slater determinants differing by four spin-orbitals are zero. Similar use of the Condon–Slater rules gives for the integral on the left of (15.87) (Problem 15.35)

$$\langle\Phi_{ij}^{ab}|\hat{H}|e^{\hat{T}_2}\Phi_0\rangle = \langle\Phi_{ij}^{ab}|\hat{H}|(1 + \hat{T}_2 + \tfrac{1}{2}\hat{T}_2^2)\Phi_0\rangle \qquad (15.89)$$

Also, use of the orthogonality of different Slater determinants gives (Problem 15.35)

$$\langle\Phi_{ij}^{ab}|e^{\hat{T}_2}\Phi_0\rangle = \langle\Phi_{ij}^{ab}|\hat{T}_2\Phi_0\rangle \qquad (15.90)$$

Use of (15.88) to (15.90) in (15.87) gives

$$\langle\Phi_{ij}^{ab}|\hat{H}|(1 + \hat{T}_2 + \tfrac{1}{2}\hat{T}_2^2)\Phi_0\rangle = (E_{\text{HF}} + \langle\Phi_0|\hat{H}|\hat{T}_2\Phi_0\rangle)\langle\Phi_{ij}^{ab}|\hat{T}_2\Phi_0\rangle$$

$$i = 1, \ldots, n; \quad j = i+1, \ldots, n; \quad a = n+1, \ldots; \quad b = a+1, \ldots$$

$$(15.91)$$

Next, one uses the definition (15.79) of $\hat{T}_2$ to eliminate $\hat{T}_2$ from (15.91). $\hat{T}_2 \Phi_0$ is a multiple sum involving $t_{ij}^{ab} \Phi_{ij}^{ab}$, and $\hat{T}_2^2 \Phi_0 \equiv \hat{T}_2(\hat{T}_2 \Phi_0)$ is a multiple sum involving $t_{ij}^{ab} t_{kl}^{cd} \Phi_{ijkl}^{abcd}$. For each unknown amplitude t_{ij}^{ab}, there is one equation in (15.91), so the number of equations is equal to the number of unknowns. After replacing $\hat{T}_2 \Phi_0$ and $\hat{T}_2^2 \Phi_0$ by these multiple sums, one expresses the resulting integrals involving Slater determinants in terms of integrals over the spin-orbitals by using the Condon–Slater rules (Table 11.3); the integrals over spin-orbitals are then expressed in terms of integrals over the basis functions. The net result is a set of simultaneous nonlinear equations for the unknown amplitudes t_{ij}^{ab}, whose form is (see *Carsky and Urban*, pages 96–97)

$$\sum_{s=1}^{m} a_{rs} x_s + \sum_{t=2}^{m} \sum_{s=1}^{t-1} b_{rst} x_s x_t + c_r = 0 , \qquad r = 1, 2, \ldots, m \qquad (15.92)$$

where $x_1, x_2, \ldots, x_m$ are the unknowns t_{ij}^{ab}, the quantities a_{rs}, b_{rst}, and c_r are constants that involve orbital energies and electron-repulsion integrals over the basis functions, and m is the number of unknown amplitudes t_{ij}^{ab}. The set of equations (15.92) is solved iteratively, starting with an initial estimate for the x's found by neglecting many of the terms in (15.92). Once the x's (that is, the t_{ij}^{ab}'s) are known, the wave function is known from (15.80) and the energy is found from (15.86).

The CCD method is size consistent, but not variational. It requires more computational time than a CI-D calculation, but the CCD method gives substantial improvement over CI-D calculations since the CCD method is size consistent and has a reasonably accurate treatment of quadruple excitations, without requiring the huge amounts of computational time needed for CI-SDTQ. Methods for analytical evaluation of the CCD gradient have been developed [J. D. Watts, G. W. Trucks, and R. J. Bartlett, *Chem. Phys. Lett.*, **164**, 502 (1989) and references cited therein]. The CC method has been applied mainly to ground states. CC methods for excited electronic states are under development [see references 23 and 24 in R. J. Bartlett, *J. Phys. Chem.*, **93**, 1697 (1989)].

Several approximations to the CCD method have been developed. The *linearized CCD* (LCCD) *method* (also called L-CPMET) uses the approximation $e^{\hat{T}_2} \approx 1 + \hat{T}_2$. With $\hat{T}_2^2$ eliminated from (15.91), the nonlinear terms involving $x_s x_t$ are eliminated from (15.92) in the LCCD method. The *coupled electron-pair approximation* (CEPA) neglects some (but not all) of the nonlinear terms in (15.92) (see *Carsky and Urban*, pages 97–100) and exists in the versions called CEPA-1 and CEPA-2 (see W. Kutzelnigg in *Schaefer, Methods of Electronic Structure Theory*, pages 172–174). (The LCCD method is sometimes called CEPA-0.) CEPA calculations are often done using pair natural orbitals. *Pair* (or *pseudo*) *natural orbitals* (PNOs) are approximate natural orbitals calculated for separate pairs of electrons (rather than for all electrons considered simultaneously, as should be done to obtain true natural orbitals). LCCD and CEPA calculations have given good results for correlation energies of a number of small molecules without requiring large amounts of computer time.

A method that uses more approximations than CEPA is the independent electron-pair approximation (IEPA), developed by Sinanoglu and by Nesbet in the 1960s. IEPA gives substantial errors in calculated correlation energies and is little used nowadays.

The next step in improving the CCD method is to include the operator $\hat{T}_1$ and take $\hat{T} = \hat{T}_1 + \hat{T}_2$ in $e^{\hat{T}}$; this gives the *CC singles and doubles* (CCSD) *method*. With $\hat{T} = \hat{T}_1 + \hat{T}_2 + \hat{T}_3$, one obtains the *CC singles, doubles, and triples* (CCSDT) *method* [J. Noga and R. J. Bartlett, *J. Chem. Phys.*, **86**, 7041 (1987)]. CCSDT calculations give very accurate results for correlation energies but are very demanding computationally and are only feasible for small molecules with small basis sets. Several approximate forms of CCSDT have been developed and bear such designations as CCSD(T), CCSDT-1, and CCSD + T(CCSD).

Pople and co-workers developed the nonvariational *quadratic configuration-interaction* (QCI) *method,* which is intermediate between the CC and CI methods. The QCI method exists in the size-consistent forms QCISD, which is an approximation to CCSD, and QCISD(T), which is similar to CCSD + T(CCSD). QCISD(T) has given excellent results for correlation energies in several calculations and is available as an option in the GAUSSIAN 88 and 90 programs [J. A. Pople, M. Head-Gordon, and K. Raghavachari, *J. Chem. Phys.*, **87**, 5968 (1987); **90**, 4635 (1989); K. Raghavachari and G. W. Trucks, *J. Chem. Phys.*, **91**, 1062, 2457 (1989); J. Paldus et al., *J. Chem. Phys.*, **90**, 4356 (1989)].

The *coupled-pair functional* (CPF) *method* is a size-consistent modification of the CI-SD method that is closely related to CEPA-1 and has given promising results [R. Ahlrichs et al., *J. Chem. Phys.*, **82**, 890 (1985)].

The exact basis-set correlation energy is obtained by full CI, by CC calculations with $\hat{T}$ not truncated, and by MP perturbation theory carried to infinite order. Full CI calculations with DZ + P basis sets have been done for H_2O, HF, BH, and CH_2. Comparison of the deviations from the full CI energies given by calculations with the CI-SD, CI-SDTQ, MP4, MP5, LCCD, CCD, CCSD, CCSD + T(CCSD), CCSDT-1, CCSDT, QCISD, and QCISD(T) methods showed the CI-SD method to give the poorest results, with an average absolute error of $5\frac{1}{2}$ kcal/mol at the equilibrium geometry, and showed the CI-SDTQ, QCISD(T), CCSD + T(CCSD), CCSDT-1, and CCSDT methods to give the best results, with average absolute deviations of only 0.2 to 0.3 kcal/mol at the equilibrium geometry [Table I in R. J. Bartlett, *J. Phys. Chem.*, **93**, 1697 (1989); J. A. Pople et al., *J. Chem. Phys.*, **87**, 5968 (1987)].

Table 15.4 lists some CC and CC-related calculations on H_2O.

15.14   DENSITY-FUNCTIONAL THEORY

The electronic wave function of an n-electron molecule depends on $3n$ spatial and n spin coordinates. Since the Hamiltonian (15.10) contains only one- and two-electron spatial terms, the energy can be written in terms of integrals involving only six spatial coordinates (Problem 15.37). In a sense, the wave function of a many-electron molecule contains more information than is needed and is lacking in direct physical significance. This has prompted the search for functions that involve fewer variables than the wave function and that can be used to calculate the energy and other properties. The molecular energy can be expressed in terms of the first-order spinless density matrix (which is a function of the spatial coordinates of one electron) and the second-order spinless density matrix (which is a function of the spatial coordinates of two electrons). Unfortunately, no convenient principle (analogous to the variation principle used to calculate wave functions) has been developed that would allow direct calculation of these density matrices without first requiring calculation of the wave function. (For more on density matrices, see *Pilar*, Section 10-5.)

In 1964, Hohenberg and Kohn proved that the ground-state molecular energy, wave function, and all other molecular electronic properties are uniquely determined by the electron probability density $\rho(x, y, z)$, a function of only three variables [P. Hohenberg and W. Kohn, *Phys. Rev.*, **136**, B864 (1964)]. One says that the ground-state energy E_0 is a functional of ρ and writes $E_0 = E_0[\rho]$, where the square brackets denote a functional relation.

What is a functional? Recall that a function $f(x)$ is a rule that associates a number with each value of the variable x. For example, the function $f(x) = x^2 + 1$ associates the number 10 with the value 3 of x and associates a number with each other value of x. A ***functional*** $F[f]$ is a rule that associates a number with each function f. For example, the functional $F[f] = \int_{-\infty}^{\infty} f^*(x)f(x)\,dx$ associates a number, found by integration of $|f|^2$ over all space, with each quadratically integrable function $f(x)$. The variational integral $W[\phi] = \langle \phi | \hat{H} | \phi \rangle / \langle \phi | \phi \rangle$ is a functional of the function ϕ, and gives a number for each well-behaved ϕ.

If we know the ground-state electron density $\rho(x, y, z)$, the Hohenberg–Kohn theorem tells us that it is possible in principle to calculate all the ground-state molecular properties from ρ. [Recall that in the traditional quantum-mechanical approach, one first finds the wave function ψ and then finds ρ by integration; Eq. (13.128).] The Hohenberg–Kohn theorem does not tell us *how* to calculate E_0 from ρ or *how* to find ρ without first finding ψ. A key step toward these goals was taken by Kohn and Sham [W. Kohn and L. J. Sham, *Phys. Rev.*, **140**, A1133 (1965)], who showed that the exact ground-state purely electronic energy E_0 of an n-electron molecule with ground-state electron probability density ρ is given by

$$E_0 = -\frac{1}{2} \sum_{i=1}^{n} \langle \psi_i(1) | \nabla_1^2 | \psi_i(1) \rangle - \sum_{\alpha} \int \frac{Z_\alpha \rho(1)}{r_{1\alpha}}\,dv_1$$

$$+ \frac{1}{2} \int\int \frac{\rho(1)\rho(2)}{r_{12}}\,dv_1\,dv_2 + E_{\text{xc}}[\rho] \qquad (15.93)$$

where the ***Kohn–Sham orbitals*** $\psi_i(1)$, $i = 1, 2, \ldots, n$, are found by a procedure discussed below, and the ***exchange–correlation energy*** $E_{\text{xc}}[\rho]$ is a functional of ρ and is discussed below. The notations $\psi_i(1)$ and $\rho(1)$ indicate that ψ_i and ρ are taken as functions of the spatial coordinates of electron 1; the kinetic-energy integrals in (15.93) are over the coordinates of electron 1. Kohn and Sham also showed that the exact ground-state ρ can be found from the ψ_i's, according to

$$\rho = \sum_{i=1}^{n} |\psi_i|^2 \qquad (15.94)$$

The Kohn–Sham orbitals are found by solving the one-electron equations

$$\hat{F}_{\text{KS}}(1)\psi_i(1) = \varepsilon_{i,\text{KS}}\psi_i(1) \qquad (15.95)$$

where the Kohn–Sham operator $\hat{F}_{KS}$ is

$$\hat{F}_{KS} \equiv -\tfrac{1}{2}\nabla_1^2 - \sum_\alpha \frac{Z_\alpha}{r_{1\alpha}} + \sum_{j=1}^n \hat{J}_j(1) + V_{xc}(1) \tag{15.96}$$

where the Coulomb operator $\hat{J}_j(1)$ is defined by (13.151) with $\phi_j(2)$ replaced by $\psi_j(2)$, and where the **exchange–correlation potential** V_{xc} is found as the functional derivative of E_{xc}:

$$V_{xc} = \delta E_{xc}[\rho]/\delta\rho \tag{15.97}$$

The precise definition of the functional derivative need not concern us. (See Problem 15.39 and *Parr and Yang*, Appendix A, for how to find a functional derivative.) If $E_{xc}[\rho]$ is known, its functional derivative is readily found, and so V_{xc} is known. $\hat{F}_{KS}$ is like the Hartree–Fock operator (15.69), except that the exchange operators $-\sum_{j=1}^n \hat{k}_j$ are replaced by V_{xc}, which handles the effects of both exchange (antisymmetry) and electron correlation.

For a closed-shell ground state, the electrons are paired in the Kohn–Sham orbitals, with two electrons of opposite spin having the same spatial Kohn–Sham orbital (as in the RHF method).

Substitution of (15.96) in (15.95) and use of (13.151) and (15.94) gives

$$\left(-\frac{1}{2}\nabla_1^2 - \sum_\alpha \frac{Z_\alpha}{r_{1\alpha}} + \int \frac{\rho(2)}{r_{12}}\,dv_2 + V_{xc}(1)\right)\psi_i(1) = \varepsilon_{i,KS}\psi_i(1) \tag{15.98}$$

The Kohn–Sham orbitals ψ_i have no physical significance other than in allowing the exact ρ to be calculated from (15.94). The density functional (DF) wave function is not a Slater determinant of spin-orbitals; in fact, there is no DF molecular wave function. Likewise, the Kohn–Sham orbital energies should not be confused with molecular-orbital energies.

There is only one problem in using the Kohn–Sham equations to find ρ and E: No one knows what the correct functional $E_{xc}[\rho]$ is for molecules.

Various approximate functionals $E_{xc}[\rho]$ have been used in molecular DF calculations. To investigate the accuracy of a particular approximate $E_{xc}[\rho]$, one uses it in DF calculations and compares the calculated molecular properties with their experimental values. The lack of a systematic procedure for improving $E_{xc}[\rho]$ and hence improving calculated molecular properties is the main drawback of the DF method.

If ρ varies very slowly with position, one can show that $E_{xc}(\rho)$ is given by

$$E_{xc}^{LDA}[\rho] = \int \rho(x,\,y,\,z)\varepsilon_{xc}(\rho)\,dx\,dy\,dz \tag{15.99}$$

where the integral is over all space and $\varepsilon_{xc}(\rho)$ is the exchange plus correlation energy per electron in a homogeneous electron gas with electron density ρ. A homogeneous electron gas is a hypothetical electrically neutral, infinite-volume system consisting of an infinite number of electrons moving in a space throughout which positive charge is continuously and uniformly distributed; the number of electrons per unit volume has a nonzero value ρ. An accurate expression for $\varepsilon_{xc}(\rho)$ has been found by Vosko, Wilk, and Nusair (see *Parr and Yang*, Appendix E).

In a molecule, the positive charge is not uniformly distributed, but is located only at the nuclei. Hence, ρ varies rapidly in a molecule, and (15.99) is only an approximation when applied to molecules, an approximation called the *local density approximation* (LDA). Molecular LDA calculations show only fair agreement with experimental molecular properties. Much better results are obtained with an improved version of the LDA called the *local spin-density approximation* (LSDA); the LSDA uses different orbitals and different densities ρ^α and ρ^β for electrons with different spins (see *Parr and Yang*, Chapter 8).

How does one do a DF calculation with E_{xc}^{LDA}? Starting with an initial guess for ρ, one uses (15.97) with $E_{xc} = E_{xc}^{LDA}$ to find an initial V_{xc}, which is then used in (15.98) to find an initial set of Kohn–Sham orbitals ψ_i. [In solving (15.98), one expands the ψ_i's in terms of a basis set of STOs or GTFs, or rather peculiar functions called localized muffin-tin orbitals (LMTOs). Alternatively, (15.98) can be solved numerically, without the use of a basis-set expansion.] The initially found ψ_i's are used in (15.94) to get an improved ρ, which is then used to find an improved V_{xc}, which is used to find improved ψ_i's, and so on. Once the calculation has converged, E_0 is found from (15.93) using the converged ρ and E_{xc}^{LDA}. The dipole moment can be calculated from ρ using (13.144) and similarly for other one-electron properties. For large molecules, DF calculations are faster than SCF calculations. The main source of error in DF calculations usually comes from use of an approximate E_{xc}, rather than from basis-set inadequacy.

LSDA calculations that solved for the ψ_i's numerically (to avoid basis-set truncation error) for 13 homonuclear diatomic molecules found average absolute errors of 0.02 Å in R_e, 1.0 eV in D_e, and 3.3% in vibrational frequencies [A. D. Becke, *Phys. Rev. A*, **33**, 2786 (1986)]; the results for bond lengths and vibrational frequencies are good, but the D_e values are not satisfactory in several cases (for example, F_2, where D_e is calculated as 3.4 eV versus the true value 1.7 eV).

Another commonly used version of the density-functional method is the **Xα** *method* (the X stands for exchange). Here, the correlation contribution to E_{xc} is neglected and (based on the homogeneous electron-gas model) the exchange contribution is taken as

$$E_{x\alpha} = -\frac{9}{8}\left(\frac{3}{\pi}\right)^{1/3} \alpha \int \int \int \rho^{4/3}(x, y, z) \, dx \, dy \, dz$$

where α is an adjustable parameter; values of α from 2/3 to 1 have been used. Functional differentiation of $E_{x\alpha}$ yields the Xα exchange potential as $V_{x\alpha} = -(3\alpha/2)(3\rho/\pi)^{1/3}$ (Problem 15.39). The Xα equations (15.98) with V_{xc} replaced by $V_{x\alpha}$ can be solved by expansion of the ψ_i's using a basis set, but this requires a lot of computation. A much more efficient procedure is to solve the Xα equations numerically, using the following technique.

One divides space into several regions by imagining a sphere around each atom and a larger sphere around the entire molecule. Within each atomic sphere, and in the region outside of the molecular sphere, the potential energy terms in (15.98) are averaged over the angles θ and ϕ so as to produce a spherically symmetric potential energy in each of these regions [this greatly facilitates numerical solution of (15.98)]; in the interstitial region between the atomic spheres and within the molecular sphere, the potential energy is assumed constant. These approximations for the potential energy give the **muffin-tin potential**. Within each atomic sphere, one uses an α value appropriate to that atom; these are taken as the α values needed for an Xα calculation on the isolated atom to produce an energy equal to the Hartree–Fock energy of that atom. One solves for the ψ_i's in each region of space using a numerical procedure called the *scattered wave* (SW) method and then applies the conditions of continuity for the ψ_i's and their derivatives at the boundaries between regions.

The SW-Xα method with the muffin-tin approximation provides a very rapid computational procedure for even rather large molecules. Unfortunately, the muffin-tin approximation occasionally leads to wildly incorrect results; for example, H_2O is predicted to be linear for most choices of atomic-sphere radii [Y. Takai et al., *Chem. Phys. Lett.*, **159**, 376 (1989)]. Xα calculations that use basis-set expansions for the orbitals, rather than the SW muffin-tin approach, give an accurate bond angle for H_2O. SW-Xα muffin-tin-potential calculations have been widely applied to transition-metal complexes with good results for ionization potentials and transition energies but poor results for bond energies and geometries [see the references in L. Versluis and T. Ziegler, *J. Chem. Phys.*, **88**, 322 (1988)].

Versluis and Ziegler developed a computationally efficient method to find analytical energy gradients in the Xα method using an STO basis-set expansion, rather than the somewhat unreliable SW muffin-tin approach (Versluis and Ziegler, op. cit.). When a basis-set expansion is used, the single value $\alpha = 0.7$ is used throughout the molecule. With TZ2P basis sets, they found the Xα method to give good results for molecular geometries, the bond distances and angles being at least as accurate as those given by HF/6-31G* calculations.

The Xα method was originally developed by Slater for atoms and is sometimes called the Hartree–Fock–Slater method, since it was viewed as providing an approximation to the Hartree–Fock method. However, the Xα method is best viewed as a special case of the density-functional method, and not as an approximate Hartree–Fock theory; for example, the orbital energies obtained in the Xα method do not give good approximations to the energies needed to remove an electron from each of the MOs and differ substantially from SCF orbital energies. (A formula has been developed to calculate Xα and other density-functional ionization potentials; see *Parr and Yang*, page 166.)

Although the Xα functional $E_{x\alpha}$ does not explicitly include electron correlation, nonmuffin-tin Xα calculations usually give better results than the Hartree–Fock method for molecular dissociation energies and vibrational frequencies and, unlike the Hartree–Fock method, give the correct behavior of the H_2 $U(R)$ curve in the $R \rightarrow \infty$ limit. A study of the Xα method concluded that, as applied to atoms, it does not include any electron correlation, but that when applied to molecules it has the effect of including some degree of electron correlation; whether or not this should be considered true electron correlation is not a clearly answerable question [M. Cook and M. Karplus, *J. Phys. Chem.*, **91**, 37 (1987)].

Density-functional calculations are often used to study solids, and their use in molecular calculations is increasing. Density-functional calculations are currently limited to ground states. Because density-functional calculations do not use the exact E_{xc} and V_{xc} in Eqs. (15.93) and (15.98), they are not, strictly speaking, ab initio calculations. However, they do not use parameters fitted to experimental data, as do the semiempirical methods of Chapter 16, and so are closer in spirit to ab initio calculations than to semiempirical ones.

For reviews of the density-functional method, see R. O. Jones and O. Gunnarsson, *Rev. Mod. Phys.*, **61**, 689 (1989) and S. Borman, *Chem. Eng. News*, April 9, 1990, p. 22.

15.15 RELATIVISTIC EFFECTS

For a nonrelativistic hydrogenlike atom, the root-mean-square speed of a $1s$ electron is $Zc/137$, where Z is the nuclear charge and c is the speed of light (Problem 15.40). Hence, for atoms of high atomic number, the average speed of inner-shell electrons is a significant fraction of c, and relativistic corrections to inner-shell orbitals and orbital energies are substantial for high-Z atoms. The valence electrons in an atom or molecule are well shielded from the nuclei, and their average speeds are far less than the speed of light, even in heavy atoms. Hence, it was formerly believed that one need not worry about relativistic corrections in molecules containing high-Z atoms. It is now realized that relativistic effects on properties of molecules with heavy atoms can be quite substantial.

The average radius of a hydrogenlike atom is proportional to the Bohr radius a_0, and a_0 is inversely proportional to the electron mass. Hence, the relativistic increase of mass with velocity shrinks the inner s orbitals in a heavy atom. To maintain their orthogonality to the inner s orbitals, the outer s orbitals are required to shrink also. The relativistic increase of mass also shrinks the p orbitals, but to a lesser extent than the s orbitals. Because of the relativistic shrinkage of the s and p orbitals, these orbitals screen the nucleus more effectively than in nonrelativistic atoms, and this produces an expansion of the d and f orbitals. Calculated relativistic contractions of the $6s$ orbital radius in some atoms are 4% for $_{55}$Cs, 7% for $_{70}$Yb, 12% for $_{75}$Re, 18% for $_{79}$Au, and 12% for $_{86}$Rn [P. Pyykkö, *Chem. Rev.*, **88**, 563 (1988)]. Because of relativistic contraction, the atomic radius of Fr is less than that of Cs, which lies above Fr in the periodic table.

The relativistic form of the one-electron Schrödinger equation is the Dirac equation. One can do relativistic Hartree–Fock calculations using the Dirac equation to modify the Fock operator, giving a type of calculation called Dirac–Fock. Likewise, one can use a relativistic form of the Kohn–Sham equation (15.98) to do relativistic density functional calculations. Relativistic $X\alpha$ calculations are often called Dirac–Slater calculations. To help simplify calculations, one can use all-electron Dirac–Fock calculations on heavy atoms to derive relativistic effective core potentials (Section 15.5) and then use these pseudopotentials to eliminate inner-shell electrons in relativistic calculations on molecules. After doing a Dirac–Fock molecular calculation, one can apply CI or MP perturbation theory to improve the results. A commonly used procedure is relativistic CASSCF calculations followed by CI. Still another approach is to do a nonrelativistic calculation and then use perturbation theory to correct for relativistic effects.

Some examples of relativistic effects on molecular properties follow, where each number is the value found in a relativistic calculation minus the value found in a nonrelativistic calculation (P. Pyykkö, op. cit.). Bond-length changes: -0.01 Å in I_2, -0.1 Å in Ag_2, -0.02 Å in SnH_4, -0.03 Å in CsH, -0.2 Å in AuH, -0.4 Å in Au_2, and -0.3 Å in Hg_2^{2+}. Dipole moment changes: 0.01_5 D in HCl, -2.4 D in AuH, and 0.9 D in PbH_2. For the vibrational wavenumber in Au_2, the calculated nonrelativistic value is 93 cm^{-1}, the calculated relativistic value is 201 cm^{-1}, and the experimental value is 191 cm^{-1}. Relativistic effects increase the calculated D_e of Au_2 from 1.2 to 2.5 eV and of AuH from 1.6 to 3.0 eV, and decrease the calculated D_e of AuCl from 3.0 to 2.6 eV.

Some review articles on relativistic quantum chemistry are P. Pyykkö, *Chem. Rev.*, **88**, 563 (1988); W. C. Ermler, R. B. Ross, and P. A. Christiansen, *Adv. Quantum. Chem.*, **19**, 139 (1988); P. A. Christiansen, W. C. Ermler, and K. S. Pitzer, *Ann. Rev. Phys. Chem.*, **36**, 407 (1985); K. Balasubramanian and K. S. Pitzer, *Adv. Chem. Phys.*, **67**, 287 (1987); K. Balasubramanian, *J. Phys. Chem.*, **93**, 6585 (1989).

15.16 VALENCE-BOND TREATMENT OF POLYATOMIC MOLECULES

The valence-bond treatment of polyatomic molecules is closely tied to chemical ideas of structure. One begins with the atoms that form the molecule and pairs up the unpaired electrons to form chemical bonds. There are usually several ways of pairing up the electrons. Each pairing scheme gives a VB **structure**. A Heitler–London-type function Φ_i (called a **bond eigenfunction**) is written for each structure i, and the molecular wave function ψ is taken as a linear combination $\sum_i c_i \Phi_i$ of the bond eigenfunctions. The variation principle is then applied to determine the coefficients c_i. The VB wave function is said to be a **resonance hybrid** of the various structures.

The weight of each resonance structure in the wave function is sometimes taken as proportional to the square of its coefficient in the wave function. Because the bond eigenfunctions are not mutually orthogonal, the electron probability density is not equal to the weighted sum of the probability densities of the various structures, and the $|c_i|^2$ quantities are somewhat lacking in direct physical significance. There are several other ways of assigning weights to VB structures. One procedure defines the *occupation number* n_i for the VB structure i as $n_i \equiv c_i^* \sum_j c_j S_{ij}$, where the sum goes over all the VB structures and $S_{ij} = \langle \Phi_i | \Phi_j \rangle$. The n_i's add to 1 (Problem 15.41).

Water. The oxygen-atom ground-state electron configuration is $1s^2 2s^2 2p^4$ with an unpaired electron in each of the AOs $2p_y$ and $2p_z$. We thus assume that these AOs along with the hydrogen $1s$ AOs will form electron-pair bonds. There are three possible ways of pairing these four AOs to get covalent structures:

$$
\begin{array}{ccccc}
\text{O}2p_y & \text{O}2p_z & \text{O}2p_y\text{——}\text{O}2p_z & \text{O}2p_y \quad \text{O}2p_z \\
| & | & & \diagdown\!\!\diagup \\
| & | & & \diagup\!\!\diagdown \\
\text{H}_1 1s & \text{H}_2 1s & \text{H}_1 1s\text{——}\text{H}_2 1s & \text{H}_1 1s \quad \text{H}_2 1s
\end{array}
\qquad (15.100)
$$

$$
\qquad\qquad A \qquad\qquad\qquad\quad B \qquad\qquad\qquad\quad C
$$

Let

$$
s_1 \equiv \text{H}_1 1s , \qquad s_2 \equiv \text{H}_2 1s , \qquad p_y \equiv \text{O}2p_y , \qquad p_z \equiv \text{O}2p_z
$$

The normalized VB function corresponding to structure A is

$$
\begin{aligned}
\Phi_A = N|\cdots \widehat{p_y s_1}\, \widehat{p_z s_2}| &= N|\cdots p_y \overline{s_1} p_z \overline{s_2}| - N|\cdots \overline{p_y} s_1 p_z \overline{s_2}| \\
&\quad - N|\cdots p_y \overline{s_1}\, \overline{p_z} s_2| + N|\cdots \overline{p_y} s_1 \overline{p_z} s_2|
\end{aligned}
\qquad (15.101)
$$

where the signs are determined by the rule in Section 13.12. The dots stand for $\overline{\text{O}1s}\text{O}1s\overline{\text{O}2s}\text{O}2s\overline{\text{O}2p_x}\text{O}2p_x$. Similarly,

$$
\Phi_B = N|\cdots \widehat{p_y p_z}\, \widehat{s_1 s_2}| , \qquad \Phi_C = N|\cdots \widehat{p_y s_2}\, \widehat{p_z s_1}| \qquad (15.102)
$$

We then take as a trial variation function ψ, a linear combination of the bond eigenfunctions of structures A, B, and C. However, the functions Φ_A, Φ_B, and Φ_C are not linearly independent; we have (Problem 15.42)

$$
\Phi_C = -(\Phi_A + \Phi_B) \qquad (15.103)
$$

It is wasted effort to include all three structures; we shall drop structure C, taking $\psi = c_A \Phi_A + c_B \Phi_B$.

For a molecule with n valence AOs to be paired, where n is even, Rumer showed in 1932 that the following procedure gives the linearly independent covalent structures for singlet states: The n AOs are arranged in a ring and lines are drawn between pairs of AOs; those structures where no lines cross are linearly independent. These are the VB *canonical covalent structures* of the molecule. Any structure with lines crossing is linearly dependent on the canonical structures and is omitted from the VB wave function. The number of ways of drawing $\frac{1}{2}n$ noncrossing lines between n points on a circle is

$$
\frac{n!}{(\frac{1}{2}n)!(\frac{1}{2}n + 1)!} \qquad (15.104)
$$

[For a proof of (15.104), see J. Barriol, *Elements of Quantum Mechanics with Chemical Applications*, Barnes and Noble, New York, 1971, pages 281–282.] H_2O has $4!/2!3! = 2$ canonical covalent structures, and these are A and B. (Actually, which structures are taken as the canonical ones is arbitrary, since the orbitals can be arranged in various ways on the ring.) To use Rumer's method when the number of orbitals to be paired is odd, we add a "phantom" orbital, whose contribution is subtracted at the end of the calculation.

Rumer's procedure is easily justified. Let $\Phi(|\ \ |)$, $\Phi(\underline{})$, and $\Phi(\times)$ be three bond eigenfunctions that involve any number of AOs, but that differ only in the way they pair up a certain subset of four AOs; each of these functions corresponds to one of the three different ways of pairing these four AOs [see

(15.100)]. By a slight extension of (15.103), it follows that

$$\Phi(\times) = -[\Phi(|\ \ |) + \Phi(\underline{\ \ })]\qquad(15.105)$$

Any pairing scheme involving lines that cross can be shown by repeated application of (15.105) to be a linear combination of structures with no lines crossing. For example, consider the following pairing scheme for six orbitals:

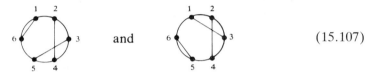

$$(15.106)$$

Application of (15.105) to the set of four AOs 1, 5, 3, 6 shows that (15.106) is a linear combination of

and

$$(15.107)$$

Use of (15.105) then shows each of the structures in (15.107) to be a linear combination of two structures with no lines crossing. Thus (15.106) has been reduced to a linear combination of canonical structures.

The structures corresponding to the pairing schemes A and B in (15.100) for water are

$$(15.108)$$

A   *B*

The separation between H_1 and H_2 is considerably greater than between O and each hydrogen; the small overlap between the hydrogen $1s$ AOs makes structure A far more significant than structure B. Since the $2p_y$ and $2p_z$ AOs are at 90° to each other, the VB structure A predicts a bond angle of 90°, since this allows maximum overlap between the bonding oxygen and hydrogen AOs. The observed angle of $104\frac{1}{2}°$ can be rationalized by considering electrostatic repulsions between the hydrogen atoms (ionic structures) and by allowing some mixing in (hybridization) of the $2s$ oxygen AO with the two bonding $2p$ AOs.

For H_2 the Heitler–London function was improved by inclusion of small contributions from ionic structures. Because of the considerable electronegativity difference between O and H, we expect ionic structures to be important in H_2O. Ionic structures for water include

D   *E*   *F*   *G*   *H*   *I*

Because oxygen is much more electronegative than hydrogen, the contributions of G, H, and I are likely to be quite small. The point group $\mathscr{C}_{2v}$ has no degenerate symmetry species, and all the electronic wave functions of H_2O must be eigenfunctions of the $\hat{O}_{C_2}$ operator (see Section 15.2). The ground electronic state belongs to the totally symmetric symmetry species. Therefore, the coefficients of Φ_D and Φ_E in the wave function must be equal. In place of the terms $c_D\Phi_D + c_E\Phi_E$ in the VB wave function, we write $c_{DE}[N(\Phi_D + \Phi_E)]$, where the normalization constant is $N = (2 + 2\langle\Phi_D|\Phi_E\rangle)^{-1/2}$. The function $N(\Phi_D + \Phi_E)$ is called a VB *symmetry function* (or symmetry structure), whereas Φ_D and Φ_E are called *individual* VB functions. Thus, a reasonable variation function for H_2O is

$$c_A\Phi_A + c_B\Phi_B + c_{DE}[N(\Phi_D + \Phi_E)] + c_F\Phi_F \qquad (15.109)$$

The ionic functions are written by analogy to the equation preceding (13.186); for example,

$$\Phi_D = |\cdots \overset{\frown}{p_y s_1} p_z \overline{p_z}| = |\cdots p_y \overline{s_1} p_z \overline{p_z}| - |\cdots \overline{p_y} s_1 p_z \overline{p_z}|$$

The linear variation function (15.109) leads to a secular equation

$$\det(H_{ij} - ES_{ij}) = 0 \qquad (15.110)$$

The AOs are expressed as STOs or as linear combinations of GTFs, and the matrix elements H_{ij} and overlap integrals between the functions $\Phi_A, \ldots, \Phi_F$ are calculated. The lowest root of the secular equation gives an approximation to the ground-state energy (higher roots correspond to excited singlet states). Evaluation of the corresponding coefficients gives the VB ground-state wave function. The calculation can be done semiempirically using experimental data to evaluate some of the integrals. Calculation of the matrix elements is involved, and approximations are often made; for example, overlap integrals between different AOs (but *not* between different structures) are neglected and exchange integrals involving interchange of the coordinates of more than two electrons are neglected. Systematic procedures for evaluating the matrix elements have been developed.

One of the relatively few ab initio all-electron VB calculations on polyatomic molecules is Peterson and Pfeiffer's calculation on H_2O [C. Peterson and G. V. Pfeiffer, *Theor. Chim. Acta*, **26**, 321 (1972)]. They included 10 covalent and 39 ionic VB structures in their wave function. The large number of structures arises because they considered structures arising from such oxygen configurations as $1s^2 2s 2p^5$ and $1s^2 2p^6$, as well as $1s^2 2s^2 2p^4$. The contribution of the ionic structures, as measured by the sum of their occupation numbers, was found to be 62%. The calculated energy and equilibrium geometry are -76.02 hartrees and 106.5°, 0.968 Å, which may be compared with the values in Table 15.1. They also did calculations on OH and O so as to calculate the first and second bond dissociation energies of H_2O. The calculated dissociation energies were not in good agreement with experiment.

Methane.   The carbon-atom ground-state electron configuration $1s^2 2s^2 2p^2$ has two unpaired electrons and would seem to indicate a valence of 2. To get the well-known tetravalence of carbon, we assume that a $2s$ electron is promoted to the vacant $2p$ orbital, giving the configuration $1s^2 2s 2p^3$. If we then assume one bond is formed with the $2s$ electron and three bonds are formed with the $2p$ electrons, the bonds are not all equivalent as they are known to be. Hence Pauling proposed that the L-shell s and p functions be linearly combined to form **hybridized sp^3** atomic orbitals, of the form

$$b_i(C2s) + d_i(C2p_x) + e_i(C2p_y) + f_i(C2p_z) , \qquad i = 1, \ldots, 4 \quad (15.111)$$

For maximum overlap we want each function to point to a vertex of a tetrahedron. From the discussion following Eq. (15.45), the constants d_i, e_i, f_i are proportional to the direction cosines in (15.55). Also, each orbital should have the same value of b_i so that the four hybrid orbitals will be equivalent. Thus the orbitals (15.111) are of the form (15.56) with $a = 0$ and $2a_1$, $1t_{2x}$, and so on, replaced by C2s, C2p$_x$, and so on. If we impose the requirement that the hybrid AOs be orthonormal, we have

$$b^2 + c^2 = 1 , \qquad b^2 - \tfrac{1}{3}c^2 = 0$$

$$c = \tfrac{1}{2}\sqrt{3} , \qquad b = \tfrac{1}{2}$$

The four equivalent sp^3 hybrid carbon AOs are then

$$te_1 = \tfrac{1}{2}[C2s + C2p_x + C2p_y + C2p_z]$$

$$te_2 = \tfrac{1}{2}[C2s + C2p_x - C2p_y - C2p_z]$$

$$te_3 = \tfrac{1}{2}[C2s - C2p_x + C2p_y - C2p_z] \qquad (15.112)$$

$$te_4 = \tfrac{1}{2}[C2s - C2p_x - C2p_y + C2p_z]$$

where "te" stands for tetrahedral. A typical sp^3 hybrid AO contour is shown in Fig. 15.15. (The three-dimensional shape is obtained by rotation about a horizontal axis through carbon.) Carbon sp^2 and sp hybrid AOs have similar shapes. [See I. Cohen and T. Bustard, *J. Chem. Educ.*, **43**, 187 (1966).]

There are many canonical covalent VB structures for methane, as well as ionic structures; however, chemical intuition suggests that the main contribution

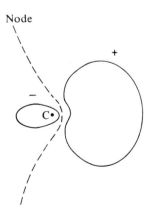

Figure 15.15 Carbon sp^3 hybrid orbital.

to the wave function is from the following covalent structure:

$$|C1s \; \overline{C1s} \; \overbrace{s_1 te_1} \overbrace{s_2 te_2} \overbrace{s_3 te_3} \overbrace{s_4 te_4}| \tag{15.113}$$

where $s_1 = H_1 1s$, and so on. The structure (15.113) is a linear combination of $2^4 = 16$ determinants. Taking the CH_4 VB wave function as the single covalent structure (15.113) or the H_2O VB wave function as the function corresponding to structure A in (15.108) gives what is called the *perfect pairing* approximation.

Raimondi, Campion, and Karplus did an ab initio all-electron VB calculation on CH_4 using a minimal AO basis set of STOs [M. Raimondi, W. Campion, and M. Karplus, *Mol. Phys.*, **34**, 1483 (1977)]. The wave function is a linear combination of 104 symmetry functions (and a much larger number of individual functions) and contained 4900 Slater determinants. Very surprisingly, the function with the largest coefficient in the wave function is not the perfect-pairing function (15.113), but is an ionic symmetry function with two covalent bonds and two ionic bonds; in this function, one of the sp^3 hybrid AOs on C has two electrons and the H atom it points to has no electrons, giving a C^-H^+ ionic bond, and a second sp^3 hybrid has no electrons and the H atom it points to has two electrons, giving a C^+H^- ionic bond. The perfect-pairing function, with four covalent bonds, has the second-highest coefficient. Two symmetry functions with three covalent bonds and one ionic bond (one symmetry function with a C^+H^- bond and one with a C^-H^+ bond) have coefficients almost as large as that of the perfect-pairing function. The perfect-pairing function has too little electron density in the overlap regions between atoms, and the great importance of the ionic structures is due to the fact that they increase the electron density in the overlap regions.

For ethylene the VB method uses sp^2-hybridized carbon AOs to form the σ bonds by overlap with the $1s$ hydrogen AOs; this leaves a p orbital on each carbon to form the π bond. The HCH bond angle is predicted to be 120°, in reasonable agreement with the observed value of $117\frac{1}{2}°$. For acetylene each carbon is sp hybridized.

Conjugated Molecules. Consider benzene. The σ bonds are formed by sp^2 carbon hybrid AOs and $1s$ hydrogen orbitals; this leaves a p orbital on each carbon to form π bonds. There are $6!/3!4! = 5$ canonical covalent structures for pairing the π orbitals, and these are

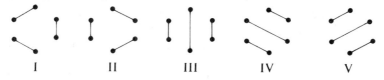

I   II   III   IV   V

Structures I and II are the Kekulé structures, and III, IV, and V are the Dewar structures. The VB Dewar structures of benzene are formal ways of pairing up electrons in AOs. Each VB Dewar structure is based on a regular-hexagonal arrangement of carbon atoms. Likewise, structures I and II correspond to a regular hexagon of carbons and differ from the hypothetical molecule cyclohexatriene with alternating bond lengths. Each VB resonance structure is based on the same internuclear distances, but a different electron-pairing scheme.

The ground-state electronic wave function of C_6H_6 belongs to the totally symmetric symmetry species and is an eigenfunction of $\hat{O}_{C_6}$ with eigenvalue $+1$. The functions I and II combine with equal coefficients to form a single symmetry function, and the functions III, IV, and V combine to form a second symmetry function.

The following types of singly polar ionic structures occur:

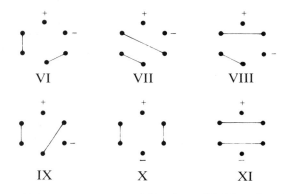

VI   VII   VIII

IX   X   XI

There are twelve individual structures of the form VI (the plus sign can be put on each of six carbons, and the minus sign can be put on the preceding or following carbon), twelve of the form VII, twelve of the form VIII, twelve of the form IX, six of the form X, and six of the form XI. There are also contributions from doubly polar and triply polar structures. Of course, we include only linearly independent ionic structures; an ionic structure that involves lines crossing is a linear combination of two ionic structures with no lines crossing and is omitted.

An all-electron ab initio VB calculation for this 42-electron molecule is a formidable task. To keep the calculation manageable, Norbeck and Gallup did an ab initio calculation using a mixed VB-MO wave function that is an antisymmetrized function in which the σ electrons occupy 18 SCF MOs that were found in an ab initio SCF MO calculation on benzene, and the six $C2p_z$ AOs are used to form VB functions for the π electrons [J. M. Norbeck and G. A. Gallup, *J. Am. Chem. Soc.*, **96**, 3386 (1974); see also G. F. Tantardini, M. Raimondi, and M. Simonetta, *J. Am. Chem. Soc.*, **99**, 2913 (1977)]. Their wave function included the following 22 symmetry functions: 2 covalent (1 Kekulé and 1 Dewar), 6 singly polar, 11 doubly polar, and 3 triply polar. The total number of individual VB functions is 175. Quite surprisingly, the symmetry function with the highest occupation number is not the Kekulé symmetry function, but is that corresponding to the linear combination of the 12 ionic individual structures of type VI. Moreover, the Kekulé symmetry function has relatively small π bond orders between adjacent atoms, and the symmetry function with the largest π bond order between adjacent atoms is that of type VI. These conclusions disagree with those of semiempirical VB calculations on benzene, which found the ionic structures to be unimportant. Norbeck and Gallup point out that semiempirical attempts to evaluate the matrix elements of VB functions "are likely to fail," since the matrix elements involve "a delicate balance of large terms."

Recall that increasing the number of functions f_i in a linear variation function $\Sigma\, c_i f_i$ improves the variation function, that is, lowers the value of the variational integral. If we were to consider structure I only, then the energy obtained would be considerably higher than when several VB structures are considered. The difference between the energy for the individual structure I and that found when all VB structures are included is the *resonance energy* of benzene. One says that benzene is "stabilized by resonance," but of course resonance is not a real phenomenon.

Such concepts as "configuration interaction," "resonance," "hybridization," and "exchange" are not real physical phenomena, but only artifacts of the approximations used in the calculations. Likewise, the concept of orbitals is but an approximation, and, strictly speaking, orbitals do not exist.

Atomic Valence States.   The *valence state* of an atom for a given molecular electronic state is the state in which the atom exists in the molecule. Since individual atoms do not really exist in molecules, the valence-state concept is an approximate one. The VB approximation constructs molecular wave functions from wave functions of the individual atoms. We use the VB wave function to define the valence state of an atom as the wave function obtained on removing all other atoms to infinity, while keeping the form of the molecular wave function invariant. This process is purely hypothetical, and the valence state is not in general a stationary atomic state.

The simplest example is the H_2 ground state. The Heitler–London function is

$$N|1s_a \overset{\frown}{1}s_b| = N[|1s_a \overline{1s_b}| - |\overline{1s_a}1s_b|]$$

Removal of hydrogen atom b leaves (at large internuclear separation, the normalization constant for each Slater determinant becomes $1/\sqrt{2}$)

$$2^{-1/2}1s_a - 2^{-1/2}\overline{1s_a}$$

The valence state is a hydrogen atom with a $1s$ spatial function and a 50% probability of having each of spin α or spin β. The valence-state ionization potential of hydrogen is thus 13.6 eV.

A less trivial example is H_2O. Although the H_2O ionic VB structures D and E are very important, it is traditional to ignore the ionic structures and find the oxygen valence state in H_2O from the perfect-pairing covalent function (15.101) corresponding to structure A. Removal of the H atoms from (15.101) gives

$$N[|\cdots 2p_y 2p_z| - |\cdots \overline{2p_y}2p_z| - |\cdots 2p_y\overline{2p_z}| + |\cdots \overline{2p_y}\,\overline{2p_z}|] \quad (15.114)$$

Each determinant in (15.114) belongs to the oxygen configuration $1s^2 2s^2 2p^4$. This configuration gives the terms ^3P, ^1D, and ^1S (Table 11.2). The first and last determinants are eigenfunctions of $\hat{S}_z$ with eigenvalues $+1\hbar$ and $-1\hbar$, respectively; hence these two determinants must correspond to states of the ^3P term (which has $S = 1$). Analysis (which we omit) of the other two determinants shows each of them to be an equal mixture (coefficients $1/\sqrt{2}$) of states belonging to the ^1D and ^3P terms. Thus the valence state is a mixture of states of the terms ^1D and ^3P and is definitely not a stationary state of the atom. The valence-state wave function ψ_{vs} is not an eigenfunction of the atomic Hamiltonian; we can, however, calculate an average energy $\langle E_{vs} \rangle = \langle \psi_{vs} | \hat{H} | \psi_{vs} \rangle$. From $\psi_{vs} = N'[c_1 \psi(^3P) + c_2 \psi(^1D)]$, we have

$$\langle E_{vs} \rangle = |N'|^2 [|c_1|^2 E(^3P) + |c_2|^2 E(^1D)]$$

The preceding discussion gives

$$|c_1|^2 = 1^2 + (2^{-1/2})^2 + (2^{-1/2})^2 + 1^2 = 3, \qquad |c_2|^2 = (2^{-1/2})^2 + (2^{-1/2})^2 = 1$$

We have $|N'|^2 = (|c_1|^2 + |c_2|^2)^{-1} = \frac{1}{4}$ and

$$\langle E \rangle_{vs} = \tfrac{3}{4} E(^3P) + \tfrac{1}{4} E(^1D)$$

$\langle E \rangle_{vs}$ can then be calculated from tables of atomic energy levels (Section 11.5).

If $2s$ hybridization is included in the $2p$ bonding oxygen orbitals of water, the oxygen valence state is found to be a linear combination involving terms of the configurations $1s^2 2s^2 2p^4$, $1s^2 2s 2p^5$, and $1s^2 2p^6$ (M. Kotani et al. in S. Flugge, ed., *Encyclopedia of Physics/Handbuch der Physik*, Springer, New York, 1961, Volume 37, pages 110–115). Hybridization gives a mixing of configurations in the valence state.

The valence state of carbon in CH_4 is important. Although the VB wave function of ground-state CH_4 is now known to have major contributions from ionic structures, the traditional way to find the C valence state is to start with the covalent perfect-pairing function (15.113). It is sometimes carelessly stated that the carbon sp^3 valence state corresponds to the ^5S term of the carbon atom $2s 2p^3$ configuration. This is not correct. The $M_S = 2$ state of the $2s 2p^3$ $^5S_2$ level has one $2s$ and three $2p$ electrons, with all four of these electrons having spin α. It is true that we can use the procedure of Section 15.8 to form linear combinations of the $2s$ and $2p$ orbitals and thereby put each of the four outer electrons of the $^5S_2$ $M_S = 2$ state into an sp^3 hybrid AO without changing the wave function. However, each such hybrid AO would still have spin α. On the other hand, when we remove the hydrogen AOs from the CH_4 VB wave function (15.113), we are left with a linear combination of sixteen determinants in which each sp^3 hybrid AO has spin α in eight determinants and spin β in eight determinants. Thus the carbon valence state differs from the $2s 2p^3$ $^5S_2$ $M_S = 2$ state and in fact differs from the other states of this term. The valence state obtained on removal of the hydrogens from (15.113) turns out to be a mixture of states of the ^5S, ^3D, and ^1D terms of the $2s 2p^3$ atomic configuration. [When other CH_4 VB structures besides (15.113) are included, we also get contributions from terms of the $2p^4$ and $2s^2 2p^2$ configurations to the C valence state.] The valence-state energy of carbon is well above the energy of the ground ^3P term of the $2s^2 2p^2$ configuration, but the energy gained by forming four bonds instead of two more than compensates for the energy needed to form the valence state.

Valence-state ionization potentials are used to estimate integrals in semiempirical calculations (Sections 16.4 and 16.5). The valence-state ionization potential for a $2p$ electron in an sp^3-hybridized carbon atom is the energy difference between the valence state of sp^3-hybridized C and the valence state of sp^2-hybridized C^+.

Status of the VB Method.   The MO method puts electrons into orthogonal MOs. The VB method puts electrons into nonorthogonal AOs. This nonorthogonality makes VB calculations a huge computational task for a molecule with many electrons. To reduce the calculations, the σ electrons in the C_6H_6 ab initio VB calculation discussed above were put into MOs, but even so this calculation did not optimize the molecular geometry, which is a relatively simple task for an SCF MO calculation. The VB method allows for the change in the AOs that occurs on molecule formation by including ionic and other resonance structures, thereby complicating the wave function. One finds quite large contributions from ionic structures even for such compounds as CH_4 and C_6H_6. In contrast to the complicated VB wave function, a single-determinant SCF MO wave function usually furnishes a good description of the molecular ground electronic state in the region of the equilibrium geometry. The VB wave function does have the advantage of dissociating properly. For quantitative calculations, the MO method has greatly overshadowed the VB method, but there is still interest in generalizations of the VB method (see the next section).

Valence-bond theory has been used to describe the electronic structure of transition-metal complex ions, with such concepts as d^2sp^3 hybridization of the metal orbitals. However, the simple VB treatment of complex ions is not fully satisfactory and has been replaced by ligand-field theory, which is MO theory applied to species whose atoms have d (or f) electrons (see *Murrell, Kettle, and Tedder*, Chapter 13; *Offenhartz*, Chapter 9).

15.17 THE GENERALIZED VALENCE-BOND AND SPIN-COUPLED VALENCE-BOND METHODS

The generalized valence-bond (GVB) method was developed about 1970 by Goddard and co-workers [W. J. Hunt, P. J. Hay, and W. A. Goddard, *J. Chem. Phys.*, **57**, 738 (1972); P. J. Hay, W. J. Hunt, and W. A. Goddard, *J. Am. Chem. Soc.*, **94**, 8293 (1972); W. A. Goddard, T. H. Dunning, W. J. Hunt, and P. J. Hay, *Acc. Chem. Res.*, **6**, 368 (1973)].

The Heitler–London VB wave function for ground-state $\bar{H}_2$ is [Eq. (13.101)] $1s_a(1)1s_b(2) + 1s_a(2)1s_b(1)$ multiplied by a normalization constant and a spin function. The GVB ground-state H_2 wave function replaces this spatial function by $f(1)g(2) + f(2)g(1)$, where the functions f and g are found by minimization of the variational integral. To find f and g, one expands each of them in terms of a basis set of AOs and finds the expansion coefficients by iteratively solving one-electron equations that resemble the equations of the SCF method.

Clearly, the GVB method will give a lower energy than the simple VB wave function. The GVB method allows for the change in the AOs that occurs on molecule formation by solving variationally for f and g. In the VB method, this change is allowed for by adding to the wave function terms that correspond to ionic and other resonance structures. The GVB wave function is thus much simpler than a VB wave function with resonance structures and the calculations are simpler.

The GVB method gives a D_e of 4.12 eV for ground-state H_2, as compared with 3.15 eV for the Heitler–London VB function, 3.78 eV for the Heitler–London–Wang function with an optimized orbital exponent, 4.03 eV for the Weinbaum function (13.110) that includes an ionic term, and 4.75 eV for the experimental value. At very large internuclear distance R, the GVB functions f and g approach the atomic orbitals $1s_a$ and $1s_b$. Thus, like the VB ψ (but unlike the MO ψ), the GVB wave function shows the correct behavior on dissociation. At intermediate distances, f is a linear combination of AOs that has its most important contribution from $1s_a$ but that has a significant contribution from $1s_b$ and lesser contributions from the other AOs (these contributions reflect the polarization of the $1s_a$ AO that occurs on molecule formation).

In the MO method, there is one orbital (an MO) for each electron pair. In the GVB method, there are two orbitals (f and g in the H_2 example) for each electron pair.

For CH_4 the VB wave function with resonance structures omitted is (15.113). For CH_4 the GVB wave function is

$$|i_a i_b \overbrace{b_{1a} b_{1b}} \overbrace{b_{2a} b_{2b}} \overbrace{b_{3a} b_{3b}} \overbrace{b_{4a} b_{4b}}| \qquad (15.115)$$

The electrons are divided into pairs, and each pair is given two orbitals. In the VB method, the inner-shell orbitals i_a and i_b are each assumed to be $1s$ AOs on carbon, the bonding orbital b_{1a} is assumed to be an sp^3-hybridized carbon AO pointing toward H_1, and b_{1b} is assumed to be a $1s$ AO on hydrogen number 1. In the GVB method, no assumptions are made about the nature of the orbitals. One simply expands each of them in terms of the chosen basis set and solves the GVB equations until self-consistency is attained.

To simplify the calculation, the GVB method restricts orbitals in different pairs to be orthogonal to one another. (For example, b_{1a} and b_{2a} are assumed orthogonal, but b_{1a} and b_{1b} are not assumed orthogonal.) This should be a good assumption since Pauli repulsion between pairs keeps them well separated spatially; moreover, a few test calculations have shown the orthogonality requirement to lead to little error.

For CH_4 one finds that i_a and i_b are essentially $C1s$ AOs; b_{1a} is an orbital centered mainly on C with some contribution from H_1; b_{1a} points toward H_1 and has a carbon AO hybridization of $sp^{2.1}$ (using a minimal basis set). The hybridization differs from the sp^3 hybridization of the VB wave function as a result of the contribution of $H_1 1s$ to b_{1a}. The orbital b_{1b} is found to be mainly an $H_1 1s$ AO with some contribution of carbon AOs mixed in, thereby polarizing the orbital toward C.

Since inner-shell electrons are little changed on molecule formation, one can simplify the GVB calculation by assuming each of i_a and i_b to be a $C1s$ AO, as is done in the VB wave function. This gives little loss in accuracy.

For C_2H_6 a GVB calculation with a minimal basis set gave a 3.1 kcal/mol rotational barrier, in good agreement with the 2.9 kcal/mol experimental value.

For C_2H_4 a GVB calculation gave a description of the double bond as composed of one σ and one π bond, in contrast to the energy-localized MOs (Section 15.8), which are two equivalent bent banana bonds.

For H_2O a GVB calculation produced an inner-shell pair on oxygen, two equivalent bonding pairs, and two equivalent lone pairs. With a DZ + P on oxygen basis set, an energy of -76.11 hartrees was obtained, which is below the Hartree–Fock limit of -76.07 hartrees (Table 15.1).

The GVB method (like the VB method) gives a description in terms of localized inner-shell, bonding, and lone pairs, whereas one must carry out a time-consuming special procedure to find localized MOs from canonical SCF MOs.

The GVB method is most applicable to molecules for which a single VB covalent structure is a good approximation.

The GVB method has been used to develop qualitative descriptions of chemical bonding; see W. A. Goddard and L. B. Harding, *Ann. Rev. Phys. Chem.*, **29**, 363 (1978).

The GVB method can be extended by writing the wave function as a linear combination of functions that correspond to different ways of coupling (combining) the electron spins to give a singlet state (just as is done in the VB method when the wave function is written as a linear combination of canonical covalent structures). This gives the *unrestricted* GVB wave function. A GVB wave function like (15.115) is sometimes called the GVB *perfect-pairing* (PP) wave function, to distinguish it from the unrestricted GVB wave function. The unmodified term GVB wave function usually refers to the GVB-PP wave function.

Gerratt, Cooper, and Raimondi developed the **spin-coupled VB method**, which (like the GVB method) uses a different, nonorthogonal, optimized orbital for each electron. The **spin-coupled wave function** is a linear combination of terms, where each term corresponds to a different way of coupling the electron spins so that they add up to the desired total spin of the molecule. [Wave functions like the GVB function (15.115) and the VB perfect-pairing function (15.113) involve just one way of coupling the electron spins.] As usual, the orbitals are expanded using a basis set. The coefficients in the orbital expansion and the coefficients that multiply each different spin-coupling term are simultaneously optimized to minimize the energy, thereby giving the spin-coupled wave function. For improved accuracy, one adds to the spin-coupled wave function CI terms that involve excitation of electrons to higher orbitals from the spin-coupled reference wave function; this gives the **spin-coupled VB wave function**. To keep the calculations manageable, one can place the inner-shell electrons (or the σ electrons in conjugated molecules) in SCF MOs. Each spin-coupled orbital found is (as in the GVB method) localized mainly on one atom with small contributions from AOs centered on adjacent atoms. Because the spin-coupled orbitals are optimized, there is little need to include ionic structures in the calculation.

The GVB PP wave function can be viewed as a special case of the spin-coupled wave function in which only the perfect-pairing spin-coupling function is included and in which orbitals in different pairs are constrained to be orthogonal.

A spin-coupled calculation on diazomethane suggests that the central nitrogen participates in five covalent bonds: CH_2=N≡N: [D. L. Cooper et al., *Chem. Phys. Lett.*, **138**, 296 (1987)]; observed bond lengths and a UHF 6-31G* calculation give the same result [S. D. Kahn et al., *J. Am. Chem. Soc.*, **109**, 1871 (1987)].

For more on the spin-coupled VB method, see D. L. Cooper, J. Gerratt, and M. Raimondi, *Int. Rev. Phys. Chem.*, **7**, 59 (1988); *Adv. Chem. Phys.*, **69**, 319 (1987).

15.18 CHEMICAL REACTIONS

The course of a chemical reaction is determined by the potential-energy function for nuclear motion $U(q_\alpha)$ (Section 13.1), where q_α indicates the coordinates of the N nuclei of the reactant molecules. For two atoms coming together to form a diatomic molecule, U is a function of one variable, the internuclear distance R, and $U(R)$ is the usual potential-energy curve. For more interesting chemical reactions, U is a function of $3N - 6$ variables; we subtract 6 because the three translational and three rotational degrees of freedom of the system leave U unchanged. (For a two-atom system, there are only two rotational degrees of freedom.) U gives what is called the ***potential-energy surface*** (PES) for the reaction. If U depends on two variables, then a plot of $U(q_1, q_2)$ in three dimensions gives a surface in ordinary three-dimensional space. Because of the large number of variables, $U(q_\alpha)$ is a "surface" in an abstract "space" of $3N - 5$ dimensions. To find $U(q_\alpha)$, we must solve the electronic Schrödinger equation at a large number of nuclear configurations, a truly formidable task.

Once U is found, we look for the path of minimum potential energy on U connecting reactants and products. The point of maximum potential energy U on the minimum-energy path is called the ***transition state***; this is a saddle point on the U surface, since it is a maximum point on a minimum-energy path. The transition state is not a stable molecule, and the transition from reactants to products is a smooth one. (However, in certain theories of reaction rates, it is convenient to ascribe properties such as entropy, free energy, vibrational frequencies, and so on, to the transition state.) The energy difference between the transition state and the reactants (omitting zero-point vibrational energies) is called the (***classical***) ***barrier height*** for the forward reaction. For the reverse reaction, the surface U is the same as for the forward reaction.

When the U surface is known, it is possible (at least in principle) to calculate the reaction rate constant k as a function of temperature. Such calculations are extremely difficult. One must allow for quantum-mechanical tunneling through the barrier; tunneling is important in reactions involving light species (e^-, H^+, H, H_2), including reactions that transfer such species between heavy molecules. Moreover, there is significant probability for molecules to traverse the reaction surface on paths that deviate somewhat from the minimum-energy path. Thus one must perform averaging over the various possible paths. (In other words, we must consider different approaches of the reactant molecules, calculate the probability of reaction occurring for each approach, and then suitably average these probabilities.) The rate constant thus depends not only on the barrier height, but on the whole shape of the reaction potential-energy surface. For qualitative discussion we can use the fact that a large barrier height means a small rate constant, and a low barrier means a fast reaction.

By measuring the experimental rate constant k as a function of temperature, one can determine an experimental **activation energy** E_a, where $k =$ $A \exp(-E_a/RT)$. The experimental quantity E_a differs slightly from the barrier height on the U surface. [See I. Shavitt, *J. Chem. Phys.*, **49**, 4048 (1968); M. Menzinger and R. Wolfgang, *Angew. Chem. Int. Ed. Engl.*, **8**, 438 (1969); *Levine, Physical Chemistry*, Section 23.4.]

To determine a complete reaction surface $U(q_\alpha)$, one needs to solve the electronic Schrödinger equation at about 10 points on the surface for each of the $3N - 6$ variables, so one needs about 10^{3N-6} calculations. For three-, four-, and five-atom systems, one needs 10^3, 10^6, and 10^9 calculations. Moreover, since the Hartree–Fock method does not usually correctly describe the process of molecular dissociation, we must include electron correlation to calculate a PES accurately.

Except for systems with very few atoms, accurate ab initio calculation of the complete potential-energy surface is out of the question. Instead, one aims to find the most important features of the surface. One first uses analytical gradients of U (Section 15.10) to locate the points on the surface where all the first derivatives $\partial U/\partial q_\alpha$ are zero. These are called *stationary points*. A stationary point may be a local minimum, a local maximum, a saddle point, or none of these. To determine the nature of a stationary point, one examines the $(3N - 6)^2$ second derivatives $\partial^2 U/\partial q_\alpha\,\partial q_\beta$. (These second derivatives constitute a matrix of order $3N - 6$ called the *force-constant matrix* or the *Hessian*. By examining the eigenvalues of the Hessian, one can tell whether a stationary point is a minimum, a saddle point, and so on. See *Hehre* et al., page 470.)

Local minima correspond to reactants, products, or reaction intermediates. A *reaction intermediate* (which should not be confused with a transition state) is a product in one elementary step and a reactant in a subsequent elementary step of a multistep mechanism. A reaction intermediate lies at a minimum in U for all nuclear displacements. Reaction intermediates are often too short lived to allow spectroscopic determination of their structure. Hence, a very significant application of quantum chemistry is the determination of the structures and relative energies of reaction intermediates; for ab initio SCF results, see *Hehre* et al., Section 7.3.

A saddle point on the U surface may correspond to a transition state. Locating transition states is harder than finding the equilibrium geometry of a stable species, since transition states do not lie at minima in U. Analytical gradients of U facilitate the location of transition states, and several techniques to find transition states are discussed in H. B. Schlegel, *Adv. Chem. Phys.*, **67**, 250 (1987). To verify that a saddle point is a transition state between particular sets of reactants and products, one must verify the existence of downhill paths from the saddle point to the reactants and to the products. Even so, a lower-energy transition state between reactants and products might exist elsewhere on the surface.

Knowledge of the minima, transition states, and barrier heights on a surface gives a good idea of the reaction mechanism.

Once the structure and vibrational frequencies of the transition state and the barrier height have been calculated, a rough estimate of the rate constant can be found using Eyring's transition-state (activated complex) theory (see any physical chemistry text). For more precise results, one must locate the minimum-energy path between reactants and products.

Although the location of the transition state is independent of the nuclear coordinates used as the variables in $U(q_\alpha)$, one finds that the location of the minimum-energy path depends on the choice of coordinates used. (For example, for H_2O, one possible choice is the $O–H_1$ and $O–H_2$ distances and the HOH angle, and another choice is the $O–H_1$, $O–H_2$, and $H_1–H_2$ distances.) The minimum-energy path (MEP) usually used is the ***intrinsic reaction coordinate*** (IRC), which is defined as the path that would be taken by a classical particle sliding downhill with infinitesimal velocity from the transition state to each of the minima (see Schlegel, op. cit.); the IRC turns out to correspond to the minimum-energy path (the path of steepest descent from the transition state) on a surface whose coordinates are the mass-weighted Cartesian coordinates $m_\alpha^{1/2}x_\alpha$, $m_\alpha^{1/2}y_\alpha$, $m_\alpha^{1/2}z_\alpha$ of the nuclei, where m_α is the mass of nucleus α. The IRC is also called the ***reaction path***. However, the reaction path is not the actual path taken by reacting molecules moving according to classical mechanics, since the molecules have translational, rotational, and vibrational kinetic energy. Techniques to determine the IRC are discussed in Schlegel, op. cit., pages 278–282.

Once the reaction path and the force-constant matrix at points along the reaction path have been found, one can use improved versions of transition-state theory (TST) called *generalized transition-state theory* to calculate rate constants more accurate than those given by TST [see references cited in D. G. Truhlar, R. Steckler, and M. S. Gordon, *Chem. Rev.*, **87**, 217 (1987)] and can construct a reaction-path Hamiltonian and use it to study such things as vibrational-energy transfer during the reaction [W. H. Miller et al., *J. Chem. Phys.*, **72**, 99 (1981); W. H. Miller, *J. Phys. Chem.*, **87**, 3811 (1983)].

PESs for gas-phase reactions are reviewed in D. G. Truhlar, R. Steckler, and M. S. Gordon, *Chem. Rev.*, **87**, 217 (1987).

Perhaps the most famous reaction surface is that for $H + H_2 \rightarrow H_2 + H$. The H_3 surface has been calculated to an accuracy of $\frac{3}{4}$ kcal/mol at 267 points using CI wave functions [B. Liu, *J. Chem. Phys.*, **58**, 1925 (1973); P. Siegbahn and B. Liu, *J. Chem. Phys.*, **68**, 2457 (1978)], and an analytical potential function has been fitted to these points [D. G. Truhlar and C. J. Horowitz, *J. Chem. Phys.*, **68**, 2466 (1978)]. From this surface, rate constants in good agreement with experiment have been calculated [M. C. Colton and G. C. Schatz, *Int. J. Chem. Kinet.*, **18**, 961 (1986); G. C. Schatz, *Chem. Phys. Lett.*, **108**, 532 (1984)]. See also J. V. Michael et al., *Science*, **249**, 269 (1990).

The $F + H_2 \rightarrow FH + H$ reaction has been intensively studied experimentally and theoretically [see H. F. Schaefer, III, *J. Phys. Chem.*, **89**, 5336 (1985)]. The initial high-level calculations with correlation included predicted the reaction path to correspond to linear geometries with the F atom approaching along the H–H internuclear line. However, the surface is very flat with respect to bending, and some calculations give a bent transition state. The experimental activation energy is 1.0 kcal/mol. The difference between the classical barrier height and the activation energy is hard to calculate, but it has been estimated that the activation energy is 0 to 1 kcal/mol less than the barrier height. The Hartree–Fock barrier height is 16 kcal/mol, and its substantial deviation from the true barrier height indicates the need for inclusion of correlation. High-level calculations with correlation done in 1976–1985 gave barrier heights in the range from 3 to 4 kcal/mol. However, high-level calculations subsequent to 1985 gave barriers in the range from 1.4 to 2.5 kcal/mol [Table IV in R. J. Bartlett, *J. Phys. Chem.*, **93**, 1697 (1989)]. The uncertainty in the barrier height and in the transition-state geometry show the need for further calculations to resolve these questions.

Comparisons of ab initio SCF calculations with calculations that include electron correlation indicate that for many reactions ab initio SCF transition-state structures are reasonably accurate, but for certain classes of reactions, SCF transition-state geometries are unreliable [F. Bernardi and M. A. Robb, *Adv. Chem. Phys.*, **67**, 155 (1987)]. SCF calculations do not generally give accurate energy differences between points on potential-energy surfaces.

PROBLEMS

15.1 Verify that (15.3) are the possible symmetry species for $\mathscr{C}_{2v}$.

15.2 Work out the possible symmetry species for $\mathscr{D}_2$.

15.3 How many independent molecular wave functions correspond to (a) a ^3E term? (b) a ^1E term?

15.4 (a) Give the number of CGTFs used in a $[4s2p/2s]$ calculation of C_3H_7OH. (b) Give the number of CGTFs used in a DZ calculation of C_4H_9OH.

15.5 For C_4H_9OH, give the number of CGTFs used in a calculation with each of the following basis sets: (a) STO-3G; (b) 3-21G; (c) 6-31G*; (d) 6-31G**; (e) 6-31+G*.

15.6 How many primitive and how many contracted GTFs are used in a calculation on $Si_{24}O_{60}H_{24}$ with each of these basis sets: (a) STO-3G; (b) 3-21G; (c) 6-31G*?

15.7 For a minimal-basis STO calculation on H_2, how many different electron-repulsion integrals $(rs|tu)$ need to be calculated, taking into account (13.170) and the symmetry of the molecule?

15.8 For formaldehyde: (a) Work out the symmetry orbitals for a minimal-basis-set calculation; give the symmetry species of each symmetry orbital. (Choose the x axis perpendicular to the molecular plane.) (b) How many σ and how many π canonical MOs will result from a minimal-basis-set calculation? (See the Section 15.9 discussion of ethylene for the definition of σ and π MOs.) How many occupied σ and occupied π MOs are there for the ground state? (c) For each of the eight energy-localized MOs, state which AOs will make significant contributions. (d) What is the maximum-size secular determinant that occurs in finding the minimal-basis-set canonical MOs?

15.9 Work out the minimal-basis symmetry orbitals and their symmetry species for *cis*-1,2-difluoroethylene. Choose the x axis perpendicular to the molecular plane.

15.10 Sketch the $1a_1$, $3a_1$, $1b_1$, $4a_1$, and $2b_2$ MOs of water.

15.11 Give the form of the normalization constants for the symmetry orbitals (15.13) and (15.14).

15.12 Suppose a ground-state calculation gives us some virtual orbitals for the molecule M. In which one of the following species would an excited electron occupy an MO that was well approximated by a virtual orbital of ground-state M? (a) M; (b) M^+; (c) M^-. Explain.

15.13 Let E_{HF} be the Hartree–Fock energy of a closed-subshell molecule; let

$E_{k,\text{HF,approx}}^{+}$ be the approximate Hartree–Fock energy of the ion formed by removal of an electron from the kth MO of this molecule, this energy being calculated using the MOs of the un-ionized molecule. For both the molecule and the ion, the Hartree–Fock wave function is a single determinant. Use Eqs. (11.80) to (11.82) [where (11.82) is modified to allow for the presence of several nuclei] and Eq. (13.153) to show that $E_{\text{HF}} - E_{k,\text{HF,approx}}^{+} = \varepsilon_k$, where ε_k is the orbital energy of MO k.

15.14 Write symmetry orbitals for a minimal-basis-set calculation of H_2.

15.15 Suppose an SCF calculation that includes $3d$ orbitals on oxygen is done on H_2O (Fig. 15.1). For each of the occupied ground-state H_2O MOs, use symmetry-species arguments to decide which of the following $3d$ oxygen AOs will contribute to that MO: $3d_{z^2}$, $3d_{xz}$, $3d_{yz}$, $3d_{xy}$, $3d_{x^2-y^2}$.

15.16 Use (15.24) to show that $1 = \sum_r c_{ri}^2 + 2 \sum_{r>s} \sum_s c_{ri}c_{si}S_{rs}$. (b) Use the result of (a) to verify that $\sum_r n_r + \sum_{r>s} \sum_s n_{r-s} = n$.

15.17 Use the tables referred to in Problem 15.20c [or one of the other available overlap-integral tables; see the reference given after (16.86)] to verify the values of the H_2O overlap integrals given in Section 15.7.

15.18 (a) Verify the net populations given at the end of the Section 15.7 example. (b) Verify the H_2O $2a_1$ and $1b_2$ interatomic overlap populations given in Section 15.7. (c) Verify the gross populations given in Section 15.7 for the H_2O basis functions.

15.19 Frequently the x and z axes of ethylene are interchanged as compared with Fig. 15.10. What change does this cause in the MO labeling? [The symmetry-species designations (15.62) are retained.]

15.20 (a) Does having the coefficient of C1s greater than 1.0 in the $1a_1$ methane MO in (15.44) violate the condition that the MO be normalized? Explain. (b) Use the results of Problem 13.49 to express the $2a_1$ methane MO of (15.44) using a nonorthogonal $2s$ STO. (c) Verify that the $2a_1$ MO found in part (b) is normalized. [The needed overlap integrals can be found by interpolating in the tables of R. S. Mulliken, C. A. Rieke, D. Orloff, and H. Orloff, *J. Chem. Phys.*, **17**, 1248 (1949).]

15.21 Call the ethylene symmetry orbitals in Table 15.3 g_1 to g_{14}, in order. For each of the eight canonical ethylene MOs occupied in the ground state, decide, as best you can, which symmetry orbitals will make major contributions; give the sign of each such symmetry orbital in the MO expression. (*Hint:* Decide how many inner-shell and how many bonding canonical MOs there are; combine the symmetry orbitals so as to build up charge density between the nuclei for the bonding MOs. *One further hint:* The third symmetry orbital in Table 15.3 makes no significant contribution to $3a_g$.) Sketch the canonical MOs. Assume an orthogonalized $2s$ orbital is used.

15.22 Predict the ground-state geometry of: (a) $SnBr_2$; (b) $HgBr_2$; (c) $TeCl_2$; (d) OF_2; (e) XeF_2; (f) H_2S; (g) I_3^-; (h) $HOCl$; (i) ClO_2^-; (j) GaI_3; (k) BrF_3; (l) PF_3; (m) BF_3; (n) H_3O^+; (o) ClO_3^-.

15.23 Predict the ground-state geometry of: (a) BrF_4^-; (b) SnH_4; (c) SeF_4; (d) XeF_4; (e) BH_4^-; (f) PCl_5; (g) $SbCl_5$; (h) BrF_5; (i) SF_6.

15.24 Give the point group of each of the molecules of Problem 15.23.

15.25 Predict the ground-state geometry of each of the following molecules, considering the electrons in a multiple bond as a single pair occupying a "large orbital": (a) ONF; (b) NO_2^+; (c) SO_2; (d) CO_2; (e) ClO_2^-; (f) NO_2^-; (g) F_2CO; (h) SO_3; (i) HNO_2; (j) XeO_3; (k) $SOBr_2$; (l) POF_3; (m) $FClO_3$; (n) $F_2IO_2^-$; (o) XeO_4; (p) $XeOF_4$; (q) SOF_4; (r) IOF_5.

15.26 Use the Walsh diagram of Fig. 15.14 to predict the ground-state geometry of (a) BH_2^+; (b) CH_2^+; (c) AsH_2; (d) CH_2^-; (e) OH_2^+; (f) NH_2^-; (g) H_2F^+.

15.27 Use Fig. 15.14 to predict the geometry of the lowest excited state of BH_2.

15.28 Find the number of CSFs in a full CI calculation of CH_2SiHF using a 6-31G** basis set.

15.29 Let γ be the fraction of the basis-set correlation energy obtained by a CI-SD calculation on a molecule with n electrons. Sasaki showed that an approximate equation satisfied by γ is $\gamma^{-1} - 1 = \frac{1}{2}\beta\gamma n - \beta$, where β is a quantity whose value is typically 0.015 to 0.03 for molecules consisting of first-row atoms. For such molecules, use this equation to estimate the percent of the basis-set correlation energy obtained by CI-SD calculations for $n = 20$, 50, 100, and 200.

15.30 Frozen-core SCF/DZP and CI-SD/DZP calculations on H_2O at its equilibrium geometry gave energies of -76.040542 and -76.243772 hartrees; application of the Davidson correction brought the energy to -76.254549 hartrees. Find the coefficient of Φ_0 in the normalized CI-SD wave function.

15.31 Verify (15.72) for the MP perturbation $\hat{H}'$.

15.32 To derive the MP $E_0^{(2)}$, we set $\psi_s^{(0)} = \Phi_{ij}^{ab}$ in (15.73); the sum over excited states $s \neq 0$ in (15.73) is replaced by a quadruple sum over i, j, a, and b that produces all possible determinants that contain two excited spin-orbitals and that represent different states. (a) For this quadruple sum, explain why we want the limits for i and j to be as in (15.74); explain why $\Phi_{ij}^{ab} = 0$ if $a = b$, and explain why we want the limits for a and b to be as in (15.74). (b) Use the Condon–Slater rules to evaluate $\langle \Phi_{ij}^{ab} | \hat{H}' | \Phi_0 \rangle$, and show that (15.74) follows.

15.33 True or false? A nonrelativistic MP2 calculation can give an energy that is less than the true nonrelativistic ground-state energy.

15.34 (a) Show that ψ in (15.76) satisfies $\langle \psi | \Phi_0 \rangle = \langle \Phi_0 | \Phi_0 \rangle = 1$, and show that $\langle \psi | \psi \rangle \neq 1$. In what earlier chapter in this book does an equation like $\langle \psi | \Phi_0 \rangle = 1$ appear?

15.35 (a) Verify the CC equation (15.89). (b) Verify (15.90).

15.36 Which of the following are functionals? (a) $\int f(x)\,dx$; (b) $\int_0^1 f(x)\,dx$; (c) $\int_0^2 [f(x) + 1]^2\,dx$; (d) $[f(x) + 1]^2$; (e) $df(x)/dx\,|_{x=0}$.

15.37 Express the electronic energy of a molecule in terms of integrals involving no more than six spatial coordinates.

15.38 Verify (15.98) for the Kohn–Sham orbitals.

15.39 For a functional defined by

$$F[\rho] \equiv \int_e^f \int_c^d \int_a^b g(x, y, z, \rho, \rho_x, \rho_y, \rho_z)\,dx\,dy\,dz$$

where ρ is a function of x, y, and z that vanishes at the limits of the integral, and where $\rho_x \equiv (\partial\rho/\partial x)_{y,z}$, and so on, the functional derivative can be shown to be

$$\delta F/\delta\rho = \partial g/\partial\rho - (\partial/\partial x)(\partial g/\partial\rho_x) - (\partial/\partial y)(\partial g/\partial\rho_y) - (\partial/\partial z)(\partial g/\partial\rho_z)$$

(a) Find $\delta E_{x\alpha}/\delta\rho$, where $E_{x\alpha}$ is given in Section 15.14. (b) If $F[\rho] = \int \rho^{-1}\nabla\rho\cdot\nabla\rho\,dv$, where the integral is over all space and ρ is a function of x, y, and z that vanishes at infinity, find $\delta F/\delta\rho$.

15.40 Use the virial-theorem result (14.18) to show that $\langle v^2 \rangle^{1/2}/c = Ze'^2/\hbar c \approx Z/137$ for the ground-state H atom.

15.41 Prove that the sum of the VB occupation numbers n_i is 1.

15.42 Verify Eq. (15.103) for the H_2O covalent VB structures.

15.43 For 1,3-butadiene: (a) How many canonical covalent VB structures are there for the π electrons? (b) Draw these structures. (c) Draw the 12 individual singly polar ionic structures for the π electrons.

15.44 (a) How many canonical covalent structures are there for the naphthalene π electrons? (b) Of these, how many are Kekulé structures (no long bonds)? (c) Considering only the Kekulé structures, which bonds in naphthalene does the resonance method predict to be the shortest?

15.45 The three equivalent sp^2 hybrid AOs of carbon point to the corners of an equilateral triangle. Derive expressions for them, assuming orthonormality.

15.46 The two equivalent sp hybrid carbon AOs make an angle of 180° with each other. Derive them.

15.47 Slater's rules for finding approximate orbital exponents of K-, L-, and M-shell Slater AOs are as follows. The orbital exponent ζ is taken as $(Z - s_{nl})/n$, where n is the principal quantum number and Z is the atomic number. The screening constant s_{nl} is calculated as follows: The AOs are divided into the following groups:

$$(1s) \quad (2s, 2p) \quad (3s, 3p) \quad (3d)$$

To find s_{nl}, the following contributions are summed: (a) 0 from electrons in groups outside the one being considered; (b) 0.35 from each other electron in the group considered, except that 0.30 is used in the $1s$ group; (c) for an s or p orbital, 0.85 from each electron whose quantum number n is one less than the orbital considered and 1.00 from each electron still further in; for a d orbital, 1.00 for each electron inside the group.

Calculate the orbital exponents according to Slater's rules for the atoms H, He, C, N, O, S, and Ar. The optimum values of ζ to use when approximating an AO as a single STO have been calculated and are given in E. Clementi and D. L. Raimondi, *J. Chem. Phys.*, **38**, 2686 (1963); E. Clementi et al., *J. Chem. Phys.*, **47**, 1300 (1967). Compare these optimum values with the above values found by Slater's rules. [For $n = 4$, Slater took ζ as $(Z - s_{nl})/3.7$; however, Slater's rules are generally unreliable for $n \geqslant 4$.]

16

Semiempirical Treatments
of Polyatomic Molecules

Because of the difficulties in applying ab initio methods to medium and large molecules, many semiempirical methods have been developed to treat such molecules. The first semiempirical methods were developed to treat the π electrons of conjugated molecules. This chapter begins with π-electron semi-empirical methods (Sections 16.1 to 16.4) and then considers more general semiempirical methods (Section 16.5).

16.1 SEMIEMPIRICAL MO TREATMENTS OF PLANAR CONJUGATED MOLECULES

The canonical MOs of a planar unsaturated organic molecule can be divided into σ and π MOs (Section 15.9). Semiempirical treatments of planar conjugated organic compounds usually make the approximation of treating the π electrons separately from the σ electrons. This is sometimes justified by arguing that since the π MOs have a node in the molecular plane, the π-electron density is well separated from the σ-electron density. Actually this is false, and the σ and π probability densities overlap appreciably. Coulson has stated that the justification for $\sigma-\pi$ separability lies in the different symmetry of the σ and π orbitals and in the greater polarizability of the π electrons, which makes them more susceptible to perturbations such as those occurring in chemical reactions.

In the π-*electron approximation*, the n_π π electrons are treated separately by incorporating the effects of the σ electrons and the nuclei into some sort of effective π-electron Hamiltonian $\hat{H}_\pi$ (recall the similar valence-electron approximation; Section 13.20):

$$\hat{H}_\pi = \sum_{i=1}^{n_\pi} \hat{H}_\pi^{core}(i) + \sum_{i=1}^{n_\pi} \sum_{j>i} \frac{1}{r_{ij}} \tag{16.1}$$

$$\hat{H}_\pi^{core}(i) = -\tfrac{1}{2}\nabla_i^2 + V_i \tag{16.2}$$

where V_i is the potential energy of the ith π electron in the field produced by the nuclei and the σ electrons. The variational principle is then applied to find a

π-electron wave function ψ_π that minimizes the variational integral $\int \psi_\pi^* \hat{H}_\pi \psi_\pi \, d\tau$ to give a π-electron energy E_π. The validity of the π-electron approximation has been discussed by Lykos and Parr (see *Parr*, pages 41–45, 211–218). Since (16.1) is not the true molecular electronic Hamiltonian, treatments that make the π-electron approximation are semiempirical. The main π-electron MO theories are the free-electron MO method (Section 16.2), the Hückel MO method (Section 16.3), and the Pariser–Parr–Pople method (Section 16.4).

16.2 THE FREE-ELECTRON MO METHOD

Perhaps the simplest semiempirical π-electron theory is the *free-electron molecular-orbital* (FE MO) *method*, developed about 1950 by Kuhn, Bayliss, Platt, and Simpson. Here the interelectronic repulsions $1/r_{ij}$ are ignored, and the effect of the σ electrons is represented by a particle-in-a-box potential-energy function: $V = 0$ in a certain region, while $V = \infty$ outside this region. With the interelectronic repulsions omitted, $\hat{H}_\pi$ in (16.1) becomes the sum of Hamiltonians for each π electron; hence (Section 6.2)

$$\hat{H}_\pi \psi_\pi = E_\pi \psi_\pi \tag{16.3}$$

$$\psi_\pi = \prod_{i=1}^{n_\pi} \phi_i \tag{16.4}$$

$$\hat{H}_\pi^{\text{core}}(i)\phi_i = e_i \phi_i \tag{16.5}$$

$$E_\pi = \sum_{i=1}^{n_\pi} e_i \tag{16.6}$$

The wave function (16.4) takes no account of spin or the Pauli principle. To do so, we must put each electron in a spin-orbital $u_i = \phi_i \sigma_i$, where σ_i is a spin function (either α or β). The wave function ψ_π is then written as an antisymmetrized product (Slater determinant) of spin-orbitals. Since $\hat{H}_\pi^{\text{core}}(i)$ does not involve spin, we have

$$\hat{H}_\pi^{\text{core}}(i)u_i = e_i u_i \tag{16.7}$$

Each term in the antisymmetrized product function ψ_π has each electron in a different spin-orbital [see, for example, (10.48)]. When $\hat{H}_\pi$, which is being approximated as the sum of the $\hat{H}_\pi^{\text{core}}(i)$'s, acts on each term in ψ_π, it gives the sum of the e_i's. Hence $\hat{H}_\pi \psi_\pi$ equals $\sum_i e_i \psi_\pi$, and (16.6) still holds when spin and the Pauli principle are allowed for.

For conjugated chain molecules, the FE MO approximation takes the box in which the π electrons move as one dimensional. We have

$$e_i = \frac{n_i^2 h^2}{8m_e l^2}, \qquad n_i = 1, 2, \ldots \tag{16.8}$$

The n_i's are restricted by the Pauli principle: no more than two electrons can be in a given spatial MO. In the ground electronic state, the π electrons fill the $\frac{1}{2}n_\pi$ lowest free-electron π MOs.

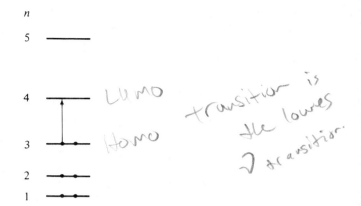

Figure 16.1 The longest-wavelength FE MO electronic transition in a conjugated molecule with six π electrons.

The π-electron transition giving the lowest-frequency electronic absorption involves an electron going from the highest occupied to the lowest vacant MO (Fig. 16.1). (The particle-in-a-box selection rule allows the quantum number of an electron in the FE MO model to change by an odd integer, as shown in Section 9.10; hence the transition of Fig. 16.1 is allowed.) For the wavelength λ of this transition, the FE MO model gives

$$\frac{1}{\lambda} = \frac{1}{hc}\left(e_{n_\pi/2+1} - e_{n_\pi/2}\right) = \frac{h}{8m_ecl^2}\left(n_\pi + 1\right) \tag{16.9}$$

We shall apply the FE MO model to the conjugated polyenes

$$CH_2=[CH-CH=]_k CH_2 \tag{16.10}$$

We take l as the zig-zag length along the carbon chain. (Another possibility is to use the direct distance between the end carbons.) Let l_1 and l_2 be the carbon–carbon single- and double-bond lengths in (16.10). For *s-trans*-1,3-butadiene these values are 1.46 Å and 1.34 Å.

The term *s-trans* refers to the conformation about the single bond:

s-trans   *s-cis*

The *s-trans* conformation is the dominant one observed for 1,3-butadiene. However, the UV, IR, and Raman spectra of 1,3-butadiene show the presence of small amounts of a second conformation. Analysis of the temperature dependence of the UV spectrum of 1,3-butadiene between 0°C and 100°C showed that $\Delta H° = 2.95$ kcal/mol and $\Delta S° = 3._3$ cal/mol-K for the reaction *s-trans* $\rightleftharpoons$ second conformation, which corresponds to $3\frac{1}{2}\%$ of butadiene present as the second conformation at room temperature [Y.-P. Sun, D. F. Sears, and J. Saltiel, *J. Am. Chem. Soc.*, **110**, 6227 (1988)].

Planarity of the second conformation is favored by conjugation of the double bonds and opposed by steric repulsion between two of the H atoms. SCF, MP2, MP4, CI-SD, and CCSD calculations with large basis sets (including polarization functions) and geometry optimization yielded the following predictions [J. E. Rice, B. Liu, T. J. Lee, and C. M. Rohlfing, *Chem. Phys. Lett.*, **161**, 277 (1989); I. L. Alberts and H. F. Schaefer, III, *Chem. Phys. Lett.*, **161**, 375 (1989)]: The planar *s-cis* structure lies at an energy maximum with respect to rotation about the C–C single bond; the equilibrium structure of the second conformation is predicted to be the nonplanar *gauche* form with a twist dihedral CCCC angle of about 38° (with an uncertainty of ±10°). The energy of the planar *s-cis* geometry is about 0.8 kcal/mol above that of the minimum corresponding to the *gauche* form. The carbon–carbon bond distances change negligibly on going from *s-cis* to *gauche* to *trans*, indicating that butadiene has only a small amount of conjugation. The experimental evidence on the nature of the second conformation is ambiguous. An analysis of the IR and UV spectra of gas-phase butadiene concluded that the second conformation is the nonplanar *gauche* form [K. B. Wiberg and R. E. Rosenberg, *J. Am. Chem. Soc.*, **112**, 1509 (1990)], but an analysis of the IR spectrum of the second conformation trapped in a solid argon matrix concluded that this conformation is the planar *s-cis* form [B. R. Arnold et al., *J. Am. Chem. Soc.*, **112**, 1808 (1990)].

Returning to the FE MO treatment of butadiene, we shall allow the π electrons to move a bit beyond the ends of the carbon chain by adding in a distance $\frac{3}{4}l_1 + \frac{1}{4}l_2$ at each end. This is about one average bond length and allows for the fact that the MOs are not confined to the regions between the bonding atoms. We have

$$l = kl_1 + (k+1)l_2 + \tfrac{3}{2}l_1 + \tfrac{1}{2}l_2 = (k+\tfrac{3}{2})(l_1+l_2) = \tfrac{1}{2}(n_C+1)(l_1+l_2) \quad (16.11)$$

where n_C is the number of carbons. Using (16.9), (16.11), and $n_\pi = n_C$, we have

$$\lambda = 2m_e ch^{-1}(l_1+l_2)^2(n_C+1) = (n_C+1)(64.6\,\text{nm}) \quad (16.12)$$

The observed longest-wavelength electronic absorption bands for the conjugated polyenes (16.10) increase monotonically from $\lambda = 162.5$ nm to 447 nm as k goes from 0 to 9 (H. Suzuki, *Electronic Absorption Spectra*, Academic Press, New York, 1967, p. 204). The FE MO equation (16.12) correctly predicts the trend of increasing wavelength with increasing chain length, but quantitative agreement with experiment is very poor; the predicted λ values have a huge 110% average absolute error (Problem 16.7a). This disagreement is hardly surprising in view of the crudity of the model. [The first excited FE MO configuration gives rise to two electronic terms, a singlet and a triplet. In the FE MO model, neglect of electronic repulsions gives the singlet and triplet terms the same energy. The observed longest wavelength electronic transition is a singlet–singlet transition, since singlet–triplet transitions are forbidden. Strictly speaking, we should compare (16.12) with the average of the singlet–singlet and singlet–triplet energy differences.]

Although the simple FE MO method fares poorly for the conjugated polyenes, it should work better for polymethine ions with the formula

$$(CH_3)_2\overset{+}{N}=CH(-CH=CH)_k-\ddot{N}(CH_3)_2 \quad (16.13)$$

Here the carbon–carbon bond lengths are all equal. (There is an equivalent VB resonance structure with carbon–carbon single and double bonds interchanged and the charge on the other nitrogen.) We shall assume the carbon–carbon bond length to be 1.40 Å as in benzene. We shall not worry about the small difference in carbon–nitrogen and carbon–carbon bond lengths. Adding in an extra bond length at each end of the chain, we set $l = (2k + 4)(1.40 \text{ Å})$ and $n_\pi = 2k + 4$ in (16.9) to get

$$\lambda = \frac{8m_e c(1.96 \text{ Å}^2)}{h} \frac{(2k + 4)^2}{2k + 5} = \frac{(2k + 4)^2}{2k + 5} (64.6 \text{ nm}) \qquad (16.14)$$

The observed λ values increase monotonically from 224 nm to 848 nm as k goes from 0 to 6 (Suzuki, *Electronic Absorption Spectra*, page 365). Values calculated from (16.14) are in pretty good agreement with experiment, the average absolute error being 12% (Problem 16.1).

16.3 THE HÜCKEL MO METHOD

The most celebrated semiempirical π-electron theory is the ***Hückel molecular-orbital*** (HMO) ***method***, developed in the 1930s. Here the π-electron Hamiltonian (16.1) is approximated by the simpler form

$$\hat{H}_\pi = \sum_{i=1}^{n_\pi} \hat{H}^{\text{eff}}(i) \qquad (16.15)$$

where $\hat{H}^{\text{eff}}(i)$ somehow incorporates the effects of the π-electron repulsions in an average way. This sounds rather vague, and in fact the Hückel method does not specify any explicit form for $\hat{H}^{\text{eff}}(i)$. Since the Hückel π-electron Hamiltonian is the sum of one-electron Hamiltonians, a separation of variables is possible, as in FE MO theory. Equations (16.4) and (16.6) hold, where the Hückel MOs satisfy

$$\hat{H}^{\text{eff}}(i)\phi_i = e_i\phi_i \qquad (16.16)$$

Since $\hat{H}^{\text{eff}}$ is not specified, there is no point in trying to solve (16.16) directly. Instead, the variational method is used.

The next assumption in the HMO method is to approximate the π MOs as LCAOs. In a minimal-basis-set calculation of a planar conjugated hydrocarbon, the only AOs of π symmetry are the carbon $2p\pi$ orbitals, where by $2p\pi$ we mean the real $2p$ AOs that are perpendicular to the molecular plane. We thus write

$$\phi_i = \sum_{r=1}^{n_C} c_{ri} f_r \qquad (16.17)$$

where f_r is a $2p\pi$ AO on the rth carbon atom and n_C is the number of carbon atoms. (16.17) is a linear variation function. The optimum values of the coefficients for the n_C lowest π MOs satisfy Eq. (8.54):

$$\sum_{s=1}^{n_C} [(H_{rs}^{\text{eff}} - S_{rs}e_i)c_{si}] = 0, \qquad r = 1, 2, \ldots, n_C \qquad (16.18)$$

where the e_i's are the roots of the secular equation (8.58)

$$\det (H_{rs}^{\text{eff}} - S_{rs}e_i) = 0 \qquad (16.19)$$

The key assumptions in the Hückel theory involve the integrals in (16.19). The integral H_{rr} is assumed to have the same value for every carbon atom in the molecule. (For benzene the six carbons are equivalent, and this is no assumption; for 1,3-butadiene one would expect H_{rr} for an end carbon and a middle carbon to differ slightly.) Moreover, H_{rr} is assumed to be the same for carbon atoms in different planar hydrocarbons. The integral H_{rs} is assumed to have the same value for any two carbon atoms bonded to each other and to vanish for two nonbonded carbons. The integral S_{rr} is equal to 1, since the AOs are normalized. The overlap integral S_{rs} is assumed to vanish for $r \neq s$. We have

$$H_{rr}^{\text{eff}} = \int f_r^*(i)\hat{H}^{\text{eff}}(i)f_r(i) \, dv_i \equiv \alpha \qquad (16.20)*$$

$$H_{rs}^{\text{eff}} = \int f_r^*(i)\hat{H}^{\text{eff}}(i)f_s(i) \, dv_i \equiv \beta \quad \text{for C}_r \text{ and C}_s \text{ bonded} \qquad (16.21)*$$

$$H_{rs}^{\text{eff}} = 0 \quad \text{for C}_r \text{ and C}_s \text{ not bonded together} \qquad (16.22)*$$

$$S_{rs} = \int f_r^*(i)f_s(i) \, dv_i = \delta_{rs} \qquad (16.23)*$$

$$f_r \equiv \text{C}_r 2p\pi \qquad (16.24)$$

where δ_{rs} is the Kronecker delta. The integral α is called the **Coulomb integral** and β is called the **bond integral** (or resonance integral). Since carbons not bonded to each other are well separated in space, the assumption (16.22) is reasonable. However, taking the overlap integral as zero for carbons bonded to each other is a poor assumption; for Slater orbitals, S_{rs} for adjacent carbons ranges between 0.2 and 0.3, depending on the bond distance. Inclusion of overlap will be considered later.

We want each π MO to be normalized. Use of (16.17) and (16.23) gives
$$1 = \langle \phi_i | \phi_i \rangle = \langle \Sigma_r c_{ri} f_r | \Sigma_s c_{si} f_s \rangle = \Sigma_r \Sigma_s c_{ri}^* c_{si} \langle f_r | f_s \rangle = \Sigma_r \Sigma_s c_{ri}^* c_{si} \delta_{rs} = \Sigma_r c_{ri}^* c_{ri},$$

$$\sum_{r=1}^{n_C} |c_{ri}|^2 = 1 \qquad (16.25)$$

This is the normalization condition for the ith Hückel MO.

In application of HMO theory to conjugated hydrocarbons, planarity is assumed. Occasionally this assumption does not hold. In gas-phase biphenyl the two rings are twisted at an angle of 40° with each other, because of steric interference between the *ortho* hydrogens.

From (16.18) it is clear that the order of the HMO secular determinant equals the number of conjugated atoms. Students sometimes make the error of assuming that this order always equals the number of π electrons.

Butadiene. To illustrate the HMO method, we consider 1,3-butadiene. The only thing of significance in the simple HMO treatment of a planar hydrocarbon is the topology of the carbon-atom framework; by this we mean which carbons are bonded together. No distinction is made between *s-cis* and *s-trans* butadiene. The numbering of the carbon atoms is

$$CH_2\!=\!CH\!-\!CH\!=\!CH_2$$
$$1 \qquad 2 \qquad 3 \qquad 4$$

The Hückel assumptions give

$$H_{11}^{\text{eff}} = H_{22}^{\text{eff}} = H_{33}^{\text{eff}} = H_{44}^{\text{eff}} = \alpha$$

$$H_{12}^{\text{eff}} = H_{23}^{\text{eff}} = H_{34}^{\text{eff}} = \beta$$

$$H_{13}^{\text{eff}} = H_{14}^{\text{eff}} = H_{24}^{\text{eff}} = 0$$

The secular equation (16.19) is

$$\begin{vmatrix} \alpha - e_k & \beta & 0 & 0 \\ \beta & \alpha - e_k & \beta & 0 \\ 0 & \beta & \alpha - e_k & \beta \\ 0 & 0 & \beta & \alpha - e_k \end{vmatrix} = 0$$

We now divide each row of the determinant by β. This divides the determinant by β^4, and since $0/\beta^4 = 0$, we get

$$\begin{vmatrix} x & 1 & 0 & 0 \\ 1 & x & 1 & 0 \\ 0 & 1 & x & 1 \\ 0 & 0 & 1 & x \end{vmatrix} = 0 \tag{16.26}$$

where

$$x \equiv \frac{\alpha - e_k}{\beta}, \qquad e_k = \alpha - \beta x \tag{16.27}$$

A determinant in which all elements are zero except the elements of the principal diagonal and the elements immediately above and below this diagonal is called a *continuant*. When the elements immediately above the principal diagonal are all equal, those on the principal diagonal are all equal, and those immediately below the principal diagonal are all equal, the nth-order continuant can be shown to have the value (T. Muir, *The Theory of Determinants in the Historical Order of Development*, Dover, New York, Volume 4, 1960, page 401)

$$\begin{vmatrix} a & b & 0 & 0 & \cdot & \cdot & \cdot & \cdot & 0 \\ c & a & b & 0 & \cdot & \cdot & \cdot & \cdot & 0 \\ 0 & c & a & b & \cdot & \cdot & \cdot & \cdot & 0 \\ \cdot & \cdot & \cdot & \cdot & \cdot & \cdot & \cdot & \cdot & \cdot \\ 0 & \cdot & \cdot & \cdot & 0 & c & a & b \\ 0 & \cdot & \cdot & \cdot & 0 & 0 & c & a \end{vmatrix}_n = \prod_{j=1}^{n} \left[a - 2(bc)^{1/2} \cos\left(\frac{j\pi}{n+1}\right) \right] \tag{16.28}$$

where the subscript on the determinant denotes its order. The definition of the product notation used in (16.28) is

$$\prod_{j=1}^{n} b_j = b_1 b_2 b_3 \cdots b_n$$

Use of (16.28) in (16.26) gives

$$\prod_{j=1}^{4}\left(x - 2\cos\frac{j\pi}{5}\right) = 0$$

$$x = 2\cos(j\pi/5), \qquad j = 1, 2, 3, 4 \tag{16.29}$$

$$x = -1.618, -0.618, 0.618, 1.618 \tag{16.30}$$

Alternatively, we can expand the secular determinant to yield the algebraic equation $x^4 - 3x^2 + 1 = 0$. The substitution $z = x^2$ yields $z^2 - 3z + 1$, and the quadratic formula gives $z = (3 \pm 5^{1/2})/2$. Since $x = \pm z^{1/2}$, we have

$$x = (3 \pm 5^{1/2})^{1/2}/2^{1/2}, \quad -(3 \pm 5^{1/2})^{1/2}/2^{1/2}$$

$$x = (5^{1/2} \pm 1)/2, \quad -(5^{1/2} \pm 1)/2 \tag{16.31}$$

Corresponding to (16.26), the equations for the HMO coefficients of butadiene are

$$
\begin{aligned}
xc_{1j} + c_{2j} &= 0 \\
c_{1j} + xc_{2j} + c_{3j} &= 0 \\
c_{2j} + xc_{3j} + c_{4j} &= 0 \\
c_{3j} + xc_{4j} &= 0
\end{aligned}
\tag{16.32}
$$

We must now substitute each of the roots (16.30) in turn into (16.32), as discussed in Section 8.5.

Consider first the root $x = -1.618$. The first equation of (16.32) reads $-1.618c_1 + c_2 = 0$ (where, for simplicity, the j subscript has been omitted). As discussed in Section 8.4, the solutions c_1, c_2, c_3, c_4 each contain an arbitrary multiplicative constant. Hence we shall solve for c_2, c_3, and c_4 in terms of c_1. We have $c_2 = 1.618c_1$. The second equation in (16.32) gives $c_3 = -c_1 - xc_2 = -c_1 + 1.618(1.618c_1) = 1.618c_1$. The fourth equation in (16.32) gives $c_4 = -c_3/x = 1.618c_1/1.618 = c_1$.

The normalization condition (16.25) is now used to fix c_1. Taking c_1 to be a real, positive number, we have $1 = c_1^2 + c_2^2 + c_3^2 + c_4^2 = c_1^2 + (1.618c_1)^2 + (1.618c_1)^2 + c_1^2 = 7.236c_1^2$, and $c_1 = 0.372$. Then $c_2 = 1.618c_1 = 0.602$, $c_3 = 1.618c_1 = 0.602$, and $c_4 = c_1 = 0.372$. The HMO corresponding to $x = -1.618$ is then $\phi = 0.372f_1 + 0.602f_2 + 0.602f_3 + 0.372f_4$. The energy of this HMO is given by (16.27) as $e = \alpha + 1.618\beta$.

Substitution of each of the three remaining roots into (16.32) yields three more HMOs. We find the following four normalized HMOs (Problem 16.8):

$$
\begin{aligned}
\phi_1 &= 0.372f_1 + 0.602f_2 + 0.602f_3 + 0.372f_4 \\
\phi_2 &= 0.602f_1 + 0.372f_2 - 0.372f_3 - 0.602f_4 \\
\phi_3 &= 0.602f_1 - 0.372f_2 - 0.372f_3 + 0.602f_4 \\
\phi_4 &= 0.372f_1 - 0.602f_2 + 0.602f_3 - 0.372f_4
\end{aligned}
\tag{16.33}
$$

The HMO energies are

$$e_1 = \alpha + 1.618\beta \, , \qquad e_2 = \alpha + 0.618\beta$$
$$e_3 = \alpha - 0.618\beta \, , \qquad e_4 = \alpha - 1.618\beta$$

(16.34)

The MO ϕ_1 in (16.33) has no nodes (other than the molecular plane) and leads to maximum charge buildup between the atoms; clearly this must be the lowest-energy π MO; its energy is $\alpha + 1.618\beta$ and, therefore, the bond integral β must be negative. [See also Eq. (13.62).] The HMOs and energy levels are sketched in Fig. 16.2. The pattern of nodes is the same as for the FE MO wave functions of butadiene; the number of vertical nodal planes is zero for the ground MO and

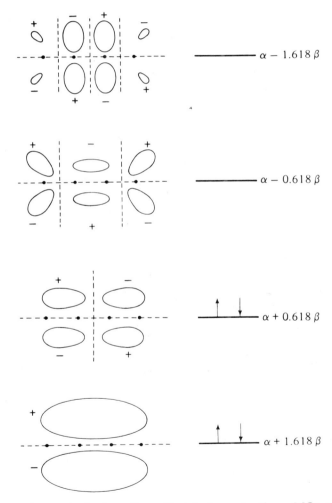

Figure 16.2 HMOs for butadiene. Nodal planes for the π MOs are indicated by dashed lines. The ground-state π-electron MO configuration is shown.

increases by one for each higher MO. From the figure, it is clear that these MOs are orthogonal. We can approximate the energy of an electron in a carbon $2p\pi$ AO in the molecule by $\int f_i^* \hat{H}^{\text{eff}} f_i \, dv = \alpha$; an HMO is classified as bonding or antibonding according to whether its energy is less than or greater than α.

Conjugated Polyenes. For the conjugated polyene (16.10) with n_C carbon atoms, the HMO secular equation involves a continuant similar to (16.26) but of order n_C; Eq. (16.28) with $b = c = 1$, $a = x$, and $n = n_C$ gives $x = 2\cos[j\pi/(n_C + 1)]$, where $j = 1, \ldots, n_C$. Since the values of x for $j = k$ and for $j = n_C + 1 - k$ are simply the negatives of each other, we can write

$$x = -2\cos \frac{j\pi}{n_C + 1}$$

$$e_j = \alpha + 2\beta \cos \frac{j\pi}{n_C + 1}, \qquad j = 1, 2, \ldots, n_C \qquad (16.35)$$

Since β is negative, e_1 is the lowest π energy level. All the π-electron levels are nondegenerate. The HMO coefficients are (Problem 16.9)

$$c_{rj} = \left(\frac{2}{n_C + 1} \right)^{1/2} \sin \frac{jr\pi}{n_C + 1} \qquad (16.36)$$

In the ground electronic state of (16.10), the highest occupied and lowest vacant π MOs have $j = \frac{1}{2}n_C$ and $(\frac{1}{2}n_C + 1)$, respectively; HMO theory predicts the longest wavelength band of the electronic absorption spectrum of a conjugated polyene to occur at

$$\frac{1}{\lambda} = -\frac{4\beta}{hc} \sin \frac{\pi}{2n_C + 2} \qquad (16.37)$$

where $\cos a - \cos b = -2\sin[\frac{1}{2}(a + b)] \sin[\frac{1}{2}(a - b)]$ was used. The bond integral β is a semiempirical parameter and is adjusted to give the best fit to experimental data. If we use the observed butadiene absorption wavelength $\lambda = 217$ nm, we find $|\beta|/hc = 37300$ cm^{-1}, $|\beta| = 4.62$ eV. With this value of β, we then calculate wavelengths for the polyenes (16.10). The predicted values do show the correct trend of increase in λ with increase in n_C, but agreement with experiment is poor; predicted λ values show an average absolute error of 44% for the first several members of the series (Problem 16.7b).

An obvious defect of the Hückel approximation for conjugated polyenes is the use of a single value of β for each pair of adjacent carbon atoms. The bond lengths in these molecules alternate, and we expect β to be larger for doubly bonded carbons than for singly bonded carbons. The refinement of using two polyene bond integrals β_1 and β_2 for C—C and C=C bonds, respectively, was introduced by Lennard-Jones [J. E. Lennard-Jones, *Proc. Roy. Soc.*, **A158**, 280 (1937)]. With two β's, the mathematics is complicated and is omitted. With $\beta_1 = -3.32$ eV, $\beta_2 = -4.20$ eV, agreement with experiment is much improved over the single-β predictions; the average absolute error of predicted λ values is reduced to 9%.

Benzene. For benzene

the HMO secular equation is

$$\begin{vmatrix} x & 1 & 0 & 0 & 0 & 1 \\ 1 & x & 1 & 0 & 0 & 0 \\ 0 & 1 & x & 1 & 0 & 0 \\ 0 & 0 & 1 & x & 1 & 0 \\ 0 & 0 & 0 & 1 & x & 1 \\ 1 & 0 & 0 & 0 & 1 & x \end{vmatrix} = 0 \qquad (16.38)$$

where x is given by (16.27).

The determinant in (16.38) is a special kind called a *circulant*. A circulant has only n independent elements; they appear in the first row, and succeeding rows are formed by successive cyclic permutations of these elements. The nth-order circulant $C(a_1, a_2, \ldots, a_n)$ is

$$C(a_1, a_2, a_3, \ldots, a_n) = \begin{vmatrix} a_1 & a_2 & a_3 & \cdots & a_n \\ a_n & a_1 & a_2 & \cdots & a_{n-1} \\ a_{n-1} & a_n & a_1 & \cdots & a_{n-2} \\ \cdot & \cdot & \cdot & \cdots & \cdot \\ a_2 & a_3 & a_4 & \cdots & a_1 \end{vmatrix}$$

The value of the nth-order circulant can be shown to be (T. Muir, *A Treatise on the Theory of Determinants*, Dover, New York, 1960, pages 442–445)

$$C(a_1, a_2, a_3, \ldots, a_n) = \prod_{k=1}^{n} (a_1 + \omega_k a_2 + \omega_k^2 a_3 + \cdots + \omega_k^{n-1} a_n) \quad (16.39)$$

where $\omega_1, \omega_2, \ldots, \omega_n$ are the n different nth roots of unity [Eq. (1.34)]:

$$\omega_k = e^{2\pi i k/n}, \qquad k = 1, 2, \ldots, n, \qquad i = \sqrt{-1} \qquad (16.40)$$

Substitution of (16.39) into (16.38), followed by the use of $\exp[2\pi i k(5/6)]$ $= \exp(-2\pi i k/6)$ and $e^{i\theta} = \cos\theta + i\sin\theta$, gives

$$\prod_{k=1}^{6} (x + e^{2\pi i k/6} + e^{-2\pi i k/6}) = 0, \qquad i = \sqrt{-1}$$

$$x = -2\cos\left(\frac{2\pi k}{6}\right), \qquad k = 1, \ldots, 6 \qquad (16.41)$$

$$x = -1, +1, +2, +1, -1, -2$$

$$e_i = \alpha + 2\beta, \quad \alpha + \beta, \quad \alpha + \beta, \quad \alpha - \beta, \quad \alpha - \beta, \quad \alpha - 2\beta \qquad (16.42)$$

There are two nondegenerate and two doubly degenerate Hückel levels. Figure 16.3 shows the HMO ground-state π-electron configuration.

The HMO coefficients can be found by solving the usual set of simultaneous equations, but it is simpler to use molecular symmetry. The $\hat{O}_{C_6}$ symmetry

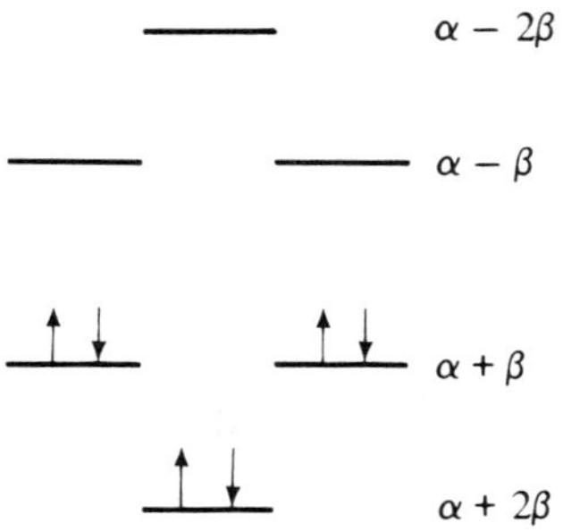

Figure 16.3 Hückel MOs for benzene.

operator commutes with the π-electron Hamiltonian, so we can choose each MO to be an eigenfunction of this 60° rotation. Since $(\hat{O}_{C_6})^6 = \hat{1}$, the eigenvalues of $\hat{O}_{C_6}$ are the six sixth roots of unity (Problem 7.18):

$$e^{2\pi ik/6}, \qquad k = 0, 1, \ldots, 5 \qquad (16.43)$$

(Since $\hat{O}_{C_6}$ has some complex eigenvalues, it is not Hermitian. Only operators representing physical quantities need be Hermitian, and $\hat{O}_{C_6}$ does not correspond to any physical property of the molecule.) Substitution of $\phi_j = \sum_{r=1}^{n_C} c_{rj} f_r$ [Eq. (16.17)] into the eigenvalue equation $\hat{O}_{C_6}\phi_j = e^{2\pi ik/6}\phi_j$ gives

$$e^{2\pi ik/6}\phi_j = \hat{O}_{C_6}\phi_j$$

$$\sum_{r=1}^{6} c_{rj} e^{2\pi ik/6} f_r = \sum_{r=1}^{6} c_{rj}\hat{O}_{C_6} f_r = \sum_{r=1}^{6} c_{rj} f_{r-1} = \sum_{r=1}^{6} c_{r+1,j} f_r$$

where $f_0 \equiv f_6$ and $c_{7j} \equiv c_{1j}$. Equating the coefficients of corresponding AOs, we have

$$c_{r+1,j} = e^{2\pi ik/6} c_{rj} \qquad (16.44)$$

The normalization condition (16.25) is $\sum_{r=1}^{6} |c_{rj}|^2 = 1$, and Eq. (16.44) shows that all the coefficients in the jth MO have the same absolute value. Hence

$$|c_{rj}| = 1/\sqrt{6}, \qquad r = 1, 2, \ldots, 6 \qquad (16.45)$$

By setting $c_{1j} = 1/\sqrt{6}$ and using (16.44), we obtain the desired coefficients. To determine which coefficients go with which energy, we evaluate the variational integral:

$$e_j = \int \phi_j^* \hat{H}^{\text{eff}} \phi_j \, dv = \sum_{r=1}^{6} \sum_{s=1}^{6} c_{rj}^* c_{sj} \int f_r^* \hat{H}^{\text{eff}} f_s \, dv$$

$$= \sum_{r=1}^{6} |c_{rj}|^2 \alpha + \sum_{r=1}^{6} c_{rj}^* c_{r+1,j} \beta + \sum_{r=1}^{6} c_{rj}^* c_{r-1,j} \beta$$

$$= \alpha + e^{2\pi ik/6} \sum_{r=1}^{6} |c_{rj}|^2 \beta + e^{-2\pi ik/6} \sum_{r=1}^{6} |c_{rj}|^2 \beta$$

$$= \alpha + 2\beta \cos(2\pi k/6), \qquad k = 0, \ldots, 5 \qquad (16.46)$$

which agrees with (16.42). From (16.44), (16.45), and (16.46), the Hückel MOs and energies for benzene are

$$\phi_1 = 6^{-1/2}(f_1 + f_2 + f_3 + f_4 + f_5 + f_6)$$

$$\phi_2 = 6^{-1/2}(f_1 + e^{\pi i/3}f_2 + e^{2\pi i/3}f_3 - f_4 + e^{4\pi i/3}f_5 + e^{5\pi i/3}f_6)$$

$$\phi_3 = 6^{-1/2}(f_1 + e^{-\pi i/3}f_2 + e^{-2\pi i/3}f_3 - f_4 + e^{-4\pi i/3}f_5 + e^{-5\pi i/3}f_6)$$

$$\phi_4 = 6^{-1/2}(f_1 + e^{2\pi i/3}f_2 + e^{4\pi i/3}f_3 + f_4 + e^{2\pi i/3}f_5 + e^{4\pi i/3}f_6) \qquad (16.47)$$

$$\phi_5 = 6^{-1/2}(f_1 + e^{-2\pi i/3}f_2 + e^{-4\pi i/3}f_3 + f_4 + e^{-2\pi i/3}f_5 + e^{-4\pi i/3}f_6)$$

$$\phi_6 = 6^{-1/2}(f_1 - f_2 + f_3 - f_4 + f_5 - f_6)$$

$$e_1 = \alpha + 2\beta, \qquad e_2 = \alpha + \beta, \qquad e_3 = \alpha + \beta$$

$$e_4 = \alpha - \beta, \qquad e_5 = \alpha - \beta, \qquad e_6 = \alpha - 2\beta$$

The condition (16.44) determining the π-MO coefficients for benzene was derived solely from symmetry considerations, without use of the Hückel approximations. Thus the MOs (16.47) are (except for normalization constants) the correct minimal-basis-set SCF π-electron MOs for benzene. (The Hückel energies $e_1, \ldots, e_6$ are, however, not the true SCF orbital energies; the Hückel method ignores electron repulsions and takes the total π-electron energy as the sum of orbital energies. The SCF method takes electron repulsions into account in an average way, and the total SCF energy is not the sum of orbital energies.) A similar situation occurs for ethylene, where the minimal-basis-set π MOs (Section 15.9) are determined solely by symmetry. (An extended-basis-set benzene calculation would mix in $3p\pi$, $3d\pi$, and so on, carbon AOs, and the contributions of these AOs must be determined by an explicit SCF calculation.)

The MOs in (16.47) for the degenerate π levels are complex. Often, real forms of these MOs are used. The two MOs of each degenerate level are complex conjugates of each other; by adding and subtracting these MOs, we get the real forms:

$$\phi_{2,\mathrm{real}} = 2^{-1/2}(\phi_2 + \phi_3), \qquad \phi_{3,\mathrm{real}} = -2^{-1/2}i(\phi_2 - \phi_3)$$

$$\phi_{2,\mathrm{real}} = 12^{-1/2}(2f_1 + f_2 - f_3 - 2f_4 - f_5 + f_6)$$

$$\phi_{3,\mathrm{real}} = \tfrac{1}{2}(f_2 + f_3 - f_5 - f_6) \qquad (16.48)$$

$$\phi_{4,\mathrm{real}} = 12^{-1/2}(2f_1 - f_2 - f_3 + 2f_4 - f_5 - f_6)$$

$$\phi_{5,\mathrm{real}} = \tfrac{1}{2}(f_2 - f_3 + f_5 - f_6)$$

Figure 16.4 shows the real benzene π-electron MOs. Note the charge buildup between the nuclei for the bonding MOs.

The symmetry species of the benzene π MOs are (*Schonland*, p. 210)

MO	ϕ_1	ϕ_2	ϕ_3	ϕ_4	ϕ_5	ϕ_6
Symmetry species	a_{2u}	e_{1g}	e_{1g}	e_{2u}	e_{2u}	b_{2g}

The ground-state π-electron configuration is $(1a_{2u})^2(1e_{1g})^4$.

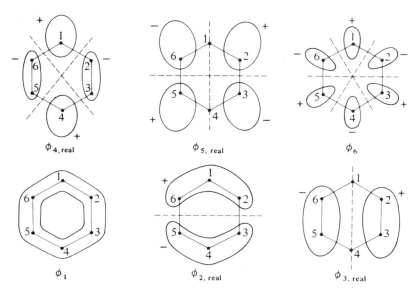

$\phi_{4,\text{real}}$ $\phi_{5,\text{real}}$ ϕ_6

ϕ_1 $\phi_{2,\text{real}}$ $\phi_{3,\text{real}}$

Figure 16.4 Benzene π MOs (real form). A top view is shown. The π MOs change sign on reflection in the molecular plane, which is a nodal plane for them. Dashed lines indicate nodal planes perpendicular to the molecular plane.

Monocyclic Conjugated Polyenes.   For the monocyclic planar conjugated polyene C_nH_n, we can use the same treatment as for benzene, C_6H_6. Replacing 6 by n_C in (16.46), (16.44), and (16.45), we find as the HMO energies and coefficients

$$e_k = \alpha + 2\beta \cos \frac{2\pi k}{n_C}, \qquad k = 0, \dots, n_C - 1 \qquad (16.49)$$

$$c_{rk} = \frac{1}{\sqrt{n_C}} \exp\left[\frac{2\pi i(r-1)k}{n_C}\right], \qquad i = \sqrt{-1} \qquad (16.50)$$

$$\phi_k = \frac{1}{\sqrt{n_C}} \sum_{r=1}^{n_C} \exp\left[\frac{2\pi i(r-1)k}{n_C}\right] f_r \qquad (16.51)$$

where $f_r = C_r 2p\pi$. Note that the index k in these equations does not correspond to the actual order of the MOs: The lowest MO has $k = 0$; next are the MOs with $k = 1$ and $k = n_C - 1$; next are the MOs with $k = 2$ and $k = n_C - 2$; and so on.

An amusing mnemonic device is available for the HMO energies (16.49) [A. A. Frost and B. Musulin, *J. Chem. Phys.*, **21**, 572 (1953)]. One inscribes a regular polygon of n_C sides in a circle of radius $2|\beta|$, putting an apex at the bottom of the circle. If a vertical scale of energy is set up with energy α coinciding with the center of the circle, then each polygon vertex is located at an HMO energy (Fig. 16.5). The method gives the correct degeneracies and spacings of the Hückel levels of the ring hydrocarbon C_nH_n (Problem 16.12).

Consider the Hückel energy levels for the monocyclic polyene C_nH_n. The lowest shell consists of a nondegenerate level and holds two electrons. Each of the

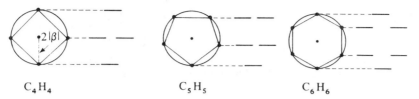

C_4H_4 C_5H_5 C_6H_6

Figure 16.5 Mnemonic device for HMO energies of monocyclic planar hydrocarbons. The levels shown are for cyclobutadiene, the cyclopentadienyl radical, and benzene.

remaining low-lying shells consists of a doubly degenerate level and holds four electrons. (If n_C is even, the highest π energy level is nondegenerate, but this level is not occupied in the ground state.) To have a stable filled-shell π-electron configuration, we see that the number of π electrons must satisfy

$$n_\pi = 4m + 2, \qquad m = 0, 1, 2, \ldots \tag{16.52}$$

This is Hückel's famous $4m + 2$ rule, which ascribes extra stability to monocyclic conjugated systems that satisfy (16.52). With $4m + 1$ or $4m - 1$ π electrons, the compound is a free radical. With $4m$ π electrons, there are two electrons in a shell that can hold four electrons, and Hund's rule predicts a triplet (diradical) ground state.

Benzene satisfies the $4m + 2$ rule. The cyclopentadienyl radical $\cdot C_5H_5$ is one electron short of satisfying (16.52); the cyclopentadienyl anion $C_5H_5^-$ satisfies the $4m + 2$ rule; the cation $C_5H_5^+$ is predicted to have a triplet ground state and be highly reactive. These predictions are borne out; $C_5H_5^-$ is found to be considerably more stable than either $C_5H_5^+$ or $\cdot C_5H_5$. Similarly, $C_7H_7^+$ should be more stable than $\cdot C_7H_7$ or $C_7H_7^-$, as is verified experimentally; for example, the salt $C_7H_7^+Br^-$ is readily prepared.

Cyclobutadiene, C_4H_4, first synthesized in 1965, is a highly reactive compound that dimerizes at temperatures above 35 K. Experimental observations and theoretical calculations are now in agreement that the ground electronic state is a singlet with a rectangular geometry and carbon–carbon bond lengths close to ordinary single- and double-bond lengths [see D. W. Whitman and B. K. Carpenter, *J. Am. Chem. Soc.*, **102**, 4272 (1980); W. T. Borden and E. R. Davidson, *Acc. Chem. Res.*, **14**, 69 (1981)]. The reason for this violation of Hund's rule is discussed in H. Kollmar and V. Staemmler, *J. Am. Chem. Soc.*, **99**, 3583 (1977).

HMO theory predicts that planar cyclooctatetraene, C_8H_8, would have a triplet ground state. Experimentally, C_8H_8 has a nonplanar "tub" structure with alternating single and double bonds. The nonplanarity results from steric strain in the planar geometry due to deviation from the 120° bond angle at the sp^2-hybridized carbons; therefore, C_8H_8 does not provide a good test of the $4m + 2$ rule. The dianion $C_8H_8^{2-}$ satisfies the $4m + 2$ rule. The salt $K_2C_8H_8$ has been prepared, and the evidence is that $C_8H_8^{2-}$ is planar. The extra stability arising from delocalization of the π electrons is sufficient to overcome the steric strain.

The monocyclic compounds C_nH_n are called *annulenes*. Benzene is

[6]annulene. The annulenes with $n = 10, 12, 14,$ and 16 suffer steric strain and substantial repulsions between nonbonded hydrogens, thereby preventing planarity and a clear-cut test of the $4m + 2$ rule. One finds that [18]annulene is nearly planar with nearly equal bond lengths and shows chemically aromatic behavior, in agreement with the $4m + 2$ rule. The stabilization gained by π-electron delocalization becomes negligible as n goes to infinity in the aromatic annulenes $C_{4m+2}H_{4m+2}$; see the discussion of delocalization energy later in this section.

In the preceding discussion, we used the same HMOs for a neutral compound and for its related ions. The Hückel method takes no account of electron repulsions, and therefore the HMOs are unchanged when π electrons are added or removed.

The $4m + 2$ rule is sometimes applied to polycyclic systems; however, the pattern (16.49) of HMO levels holds only for a monocyclic system, and use of the $4m + 2$ rule for polycyclic systems is not justified.

The $4m + 2$ rule actually does not depend on the Hückel assumptions (16.20) to (16.23). The C_nH_n π MOs (16.51) were derived solely by symmetry considerations and are the correct SCF minimal-basis-set π MOs. With $k = 0$, we have an MO with all plus signs in front of the AOs; clearly this MO has a lower energy than any of the others. For the remaining MOs, the pair with $k = j$ and $k = n_C - j$ are complex conjugates of each other and must have the same energy. (Since $\hat{H}$ is Hermitian, we have $\int \phi^* \hat{H} \phi \, dv = \int (\phi^*)^* \hat{H} \phi^* \, dv$.) Thus the excited MOs occur in pairs (except that when n_C is even, the MO with $k = \frac{1}{2}n_C$ is nondegenerate; with alternating plus and minus signs in front of the AOs, this is the highest-energy MO). The pattern of a nondegenerate lowest π level followed by doubly degenerate π levels thus holds for the minimal-basis-set SCF MOs. Of course, the energies (16.49) are not the correct SCF orbital energies.

Naphthalene. Now consider the HMO treatment of naphthalene. For butadiene and benzene, we set up the secular equation without bothering with the intermediate step of constructing symmetry orbitals from the $2p\pi$ AOs. For these molecules, the secular equation was easy enough to solve without the simplifications introduced by symmetry orbitals. For naphthalene the 10×10 secular determinant is difficult to deal with, and we first find symmetry orbitals. The point group of naphthalene (Fig. 16.6) is $\mathscr{D}_{2h}$. The possible symmetry species are (15.62). The $C2p\pi$ AOs all have eigenvalue -1 for reflection in the molecular (yz) plane. Each symmetry orbital will be some linear combination of AOs that

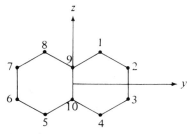

Figure 16.6 Axes for naphthalene. The x axis is perpendicular to the molecular plane.

are permuted among one another by the symmetry reflections (recall ethylene). Hence, to aid in finding the symmetry orbitals, we examine the effects of the $\hat{\sigma}(xy)$ and $\hat{\sigma}(xz)$ operations on the π AOs. We find the π AOs to fall into three sets:

$$1, 4, 5, 8 \qquad 2, 3, 6, 7 \qquad 9, 10$$

where the members of each set are permuted among one another by the symmetry reflections. (Of course, these are also the chemically equivalent carbons.) Each symmetry orbital must be some linear combination of the AOs of a given set. Naphthalene and ethylene have the same point group, and the pattern of the hydrogen symmetry orbitals in Table 15.3 gives us the symmetry orbitals of the first two above sets of naphthalene AOs. The symmetry orbitals and their readily verified symmetry species are then

$$b_{3u}: \quad g_1 = \tfrac{1}{2}(f_1 + f_4 + f_5 + f_8), \quad g_2 = \tfrac{1}{2}(f_2 + f_3 + f_6 + f_7), \quad g_3 = 2^{-1/2}(f_9 + f_{10})$$

$$a_u: \quad g_4 = \tfrac{1}{2}(f_1 - f_4 + f_5 - f_8), \quad g_5 = \tfrac{1}{2}(f_2 - f_3 + f_6 - f_7)$$

$$b_{2g}: \quad g_6 = \tfrac{1}{2}(f_1 - f_4 - f_5 + f_8), \quad g_7 = \tfrac{1}{2}(f_2 - f_3 - f_6 + f_7), \quad g_8 = 2^{-1/2}(f_9 - f_{10})$$

$$b_{1g}: \quad g_9 = \tfrac{1}{2}(f_1 + f_4 - f_5 - f_8), \quad g_{10} = \tfrac{1}{2}(f_2 + f_3 - f_6 - f_7)$$

The constants $\tfrac{1}{2}$ and $1/\sqrt{2}$ normalize the symmetry orbitals, provided the approximation $S_{rs} = \delta_{rs}$ [Eq. (16.23)] is used.

Instead of using the f_r's as basis functions, we set up the secular equation using the symmetry orbitals as basis functions:

$$\det\left[\langle g_p | \hat{H}^{\text{eff}} | g_q \rangle - \langle g_p | g_q \rangle e_k\right] = 0$$

The secular determinant in this equation is in block-diagonal form (Section 15.6), and we have two third-degree and two second-degree equations to solve. Using $S_{rs} = \delta_{rs}$, we find that

$$\langle g_p | g_q \rangle = \delta_{pq}$$

when g_p and g_q belong to the same symmetry species. Also,

$$\langle g_1 | \hat{H}^{\text{eff}} | g_1 \rangle = \tfrac{1}{4}\langle f_1 + f_4 + f_5 + f_8 | \hat{H}^{\text{eff}} | f_1 + f_4 + f_5 + f_8 \rangle = \tfrac{1}{4}(\alpha + \alpha + \alpha + \alpha) = \alpha$$

$$\langle g_1 | \hat{H}^{\text{eff}} | g_2 \rangle = \tfrac{1}{4}\langle f_1 + f_4 + f_5 + f_8 | \hat{H}^{\text{eff}} | f_2 + f_3 + f_6 + f_7 \rangle = \tfrac{1}{4}(\beta + \beta + \beta + \beta) = \beta$$

Evaluating the remaining matrix elements, we find as the secular equation for the b_{3u} MOs

$$\begin{vmatrix} \alpha - e_k & \beta & \sqrt{2}\beta \\ \beta & \alpha + \beta - e_k & 0 \\ \sqrt{2}\beta & 0 & \alpha + \beta - e_k \end{vmatrix} = 0$$

$$\begin{vmatrix} x & 1 & \sqrt{2} \\ 1 & x + 1 & 0 \\ \sqrt{2} & 0 & x + 1 \end{vmatrix} = 0$$

$$(x + 1)(x^2 + x - 3) = 0$$

$$x = -1, \qquad -\tfrac{1}{2} \pm \tfrac{1}{2}\sqrt{13}$$

Solution of the three remaining secular equations is left as an exercise. The naphthalene HMO energy levels in order of increasing energy are found to be (Problem 16.18)

$$\alpha + 2.303\beta\,,\quad \alpha + 1.618\beta\,,\quad \alpha + 1.303\beta\,,\quad \alpha + \beta\,,\quad \alpha + 0.618\beta$$

$$\alpha - 0.618\beta\,,\quad \alpha - \beta\,,\quad \alpha - 1.303\beta\,,\quad \alpha - 1.618\beta\,,\quad \alpha - 2.303\beta$$

The levels are all nondegenerate ($\mathcal{D}_{2h}$ has only A and B symmetry species).

The HMO coefficients are found by solving the appropriate sets of simultaneous equations (Problem 16.18).

Alternant Hydrocarbons. Note that the HMOs of naphthalene, like those of butadiene and benzene, are paired, meaning that for each HMO with the energy $\alpha - x\beta$ there is an HMO with energy $\alpha + x\beta$. This can be proved to be true for every alternant hydrocarbon (for the proof, see Problem 16.22). An *alternant hydrocarbon* is a planar conjugated hydrocarbon in which the carbon atoms can be divided into a starred set and an unstarred set, with starred carbons bonded only to unstarred carbons, and vice versa (Fig. 16.7). All planar conjugated hydrocarbons are alternants except those containing a ring with an odd number of carbons.

Electronic Transitions. The predicted wave number for a transition between the highest occupied and lowest vacant HMOs of a conjugated hydrocarbon is

$$\frac{1}{\lambda} = \frac{|\beta|}{hc}\,\Delta x \tag{16.53}$$

where Δx is the difference in x values [Eq. (16.27)] for the two MOs. For naphthalene, $\Delta x = 1.236$, and the observed $1/\lambda$ is $34700\ \text{cm}^{-1}$. Choosing β to fit the observed wavelength for naphthalene (because of the orbital degeneracy of its first excited term, benzene is atypical), we find

$$|\beta|/hc = 28000\ \text{cm}^{-1}\,,\qquad |\beta| = 3.48\ \text{eV}$$

Comparison of predicted and observed longest-wavelength absorptions for a large

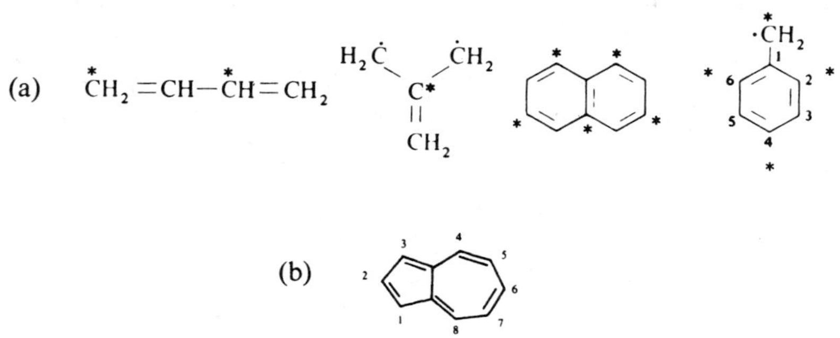

Figure 16.7 (a) Some alternant hydrocarbons. (b) Azulene, a nonalternant hydrocarbon.

number of benzenoid hydrocarbons shows only fair agreement with experiment, with many deviations of a few thousand cm^{-1}.

Improved agreement can be obtained if we fit the frequencies to a straight line that does not pass through the origin; that is, we use

$$\frac{1}{\lambda} = \frac{1}{hc} |\beta| \Delta x + a \qquad (16.54)$$

A least-squares fit of benzenoid hydrocarbon data gives [E. Heilbronner and J. N. Murrell, *J. Chem. Soc.*, **1962**, 2611]

$$a = 8200 \text{ cm}^{-1}, \qquad |\beta|/hc = 21900 \text{ cm}^{-1}, \qquad |\beta| = 2.72 \text{ eV} \qquad (16.55)$$

These constants give a good fit to the data; the standard deviation is 600 cm^{-1}. Of course, with two parameters instead of one, the agreement is bound to be improved. The fact that a semiempirical theory with several adjustable parameters gives a good fit to experimental data cannot be taken as overwhelming proof of the validity of the theory.

We can partially justify the constant term in (16.54) as follows. The Hückel method neglects electron repulsions and therefore does not distinguish between singlet and triplet terms. Hence we should compare (16.53) with the energy difference between the ground state and the average of the energies of the singlet and triplet terms of the configuration with one electron excited to the lowest vacant HMO. The experimental frequencies are for singlet–singlet transitions. If we assume that the singlet–triplet splitting of the first excited configuration is reasonably constant for aromatic hydrocarbons, then $a = 8200 \text{ cm}^{-1} = 1.0 \text{ eV}$ can be interpreted as one-half this singlet–triplet splitting. Experimental values for this splitting are available for many aromatic hydrocarbons; typically, the value of half the singlet–triplet splitting is 0.7 to 0.8 eV, in reasonable agreement with $a = 1.0 \text{ eV}$.

Delocalization Energy and Aromaticity.   There are several properties of conjugated hydrocarbons that can be defined using HMO theory. We begin with *delocalization energy*. The corresponding VB term is *resonance energy*. The energy of the occupied ethylene π HMO, the $1b_{3u}$ MO in Section 15.9, is

$$\int \frac{1}{\sqrt{2}} (f_1 + f_2)^* \hat{H}^{\text{eff}} \frac{1}{\sqrt{2}} (f_1 + f_2) \, dv = \alpha + \beta$$

There are two π electrons in this MO, and the total Hückel π-electron energy for ethylene is $2\alpha + 2\beta$. For butadiene the total Hückel π-electron energy is

$$2(\alpha + 1.618\beta) + 2(\alpha + 0.618\beta) = 4\alpha + 4.472\beta \qquad (16.56)$$

If butadiene had two isolated double bonds, its π-electron energy would be twice that of ethylene, namely $4\alpha + 4\beta$. The effect of delocalization is to change the butadiene π-electron energy by

$$4\alpha + 4.472\beta - (4\alpha + 4\beta) = 0.472\beta$$

Since β is negative, butadiene is stabilized by π-electron delocalization, and $0.472|\beta|$ is its delocalization energy.

For benzene the π-electron energy is

$$2(\alpha + 2\beta) + 4(\alpha + \beta) = 6\alpha + 8\beta \qquad (16.57)$$

as compared with $6\alpha + 6\beta$ for the π-electron energy of three isolated double bonds. The delocalization energy of benzene is $2|\beta|$. An "experimental" delocalization energy for benzene can be calculated as follows. The enthalpy of hydrogenation of cyclohexene to cyclohexane is -28.6 kcal/mol. If benzene had three isolated double bonds, its enthalpy of hydrogenation would be three times -28.6 kcal/mol which is -85.8 kcal/mol. The observed value is only -49.8 kcal/ mol, indicating that benzene is more stable by 36 kcal/mol than it would be if its double bonds were isolated. (A similar delocalization energy is arrived at by adding up the bond energies of six C—H bonds, three C—C bonds, and three C=C bonds and comparing this to the enthalpy of formation of benzene from its atoms; see Problem 16.14.) Setting $2|\beta| = 36$ kcal/mol, we get $|\beta| = 18$ kcal/ mol $= 0.8$ eV/molecule. This is far less than the value 2.72 eV/molecule $= 63$ kcal/ mol found by spectroscopic observations [(16.55)]. Part of the discrepancy is explainable as follows. In the hydrogenation of cyclohexene, the carbon–carbon double-bond length becomes a single-bond length; the figure of -85.8 kcal/mol applies to the hydrogenation of a hypothetical molecule with three isolated double bonds and three single bonds, with alternating bond lengths. We must therefore also consider the strain energy needed to compress three single bonds and stretch three double bonds to the benzene bond length. Even with this correction (Problem 16.13), the thermochemical value of β differs sharply from the spectroscopic value. The difference is to be attributed to the crudity of the HMO method. One finds that a different value of β is required for each different physical property that is being considered. Moreover, the optimum values of β differ for chain and ring conjugated hydrocarbons.

The just-discussed traditional method of calculating Hückel delocalization energies of conjugated hydrocarbons by comparing a molecule's HMO π-electron energy E_π with $n_d(2\alpha + 2\beta)$, where n_d is the number of carbon–carbon double bonds, has serious shortcomings. This method predicts a substantial delocalization energy for the linear polyenes (16.10), whereas experiment shows the delocalization stabilization in these molecules to be rather small. For example, comparison of twice the enthalpy of hydrogenation of 1-butene with the enthalpy of hydrogenation of 1,3-butadiene gives the 1,3-butadiene delocalization energy as only 4 kcal/mol. Moreover, this method predicts substantial delocalization stabilization for certain cyclic polyenes that experiment reveals to be unstable, with no aromatic character. (A cyclic conjugated polyene is said to be **aromatic** when it shows substantially more stability than a hypothetical structure in which the double bonds do not interact with one another and when it undergoes substitution, rather than addition, when treated with electrophilic reagents like Br_2.)

To produce more reliable predictions of aromaticity, Hess and Schaad (following a suggestion of Dewar) calculated delocalization (resonance) energies of cyclic hydrocarbons by comparing the compounds' Hückel-theory E_π with a value calculated for a hypothetical acyclic conjugated polyene with the same number and kinds of bonds as in a localized structure of the cyclic hydrocarbon. [B. A. Hess and L. J. Schaad, *J. Am. Chem. Soc.*, **93**, 305, 2413 (1971); **94**, 3068

(1972); **95**, 3907 (1973); B. A. Hess, L. J. Schaad, and C. W. Holyoke, *Tetrahedron*, **28**, 3657, 5299 (1972); L. J. Schaad and B. A. Hess, *J. Chem. Educ.*, **51**, 640 (1974).]

These workers found that the Hückel π-electron energies of noncyclic conjugated polyenes could be accurately calculated as $E_\pi \approx \Sigma_b n_b E_{\pi,b}$, where n_b is the number of bonds of a given type, $E_{\pi,b}$ is an empirical parameter, and the sum goes over all the types of carbon–carbon bonds. A conjugated hydrocarbon has three types of carbon–carbon single bonds and five types of double bonds, the types differing in the numbers of H atoms bonded to the carbons. In calculations of the delocalization energy, the terms involving α always cancel, so Hess and Schaad measured energies relative to α. In the following discussion, α will be omitted.

The Hess–Schaad values of $E_{\pi,b}$ for the various bond types are

$CH_2{=}CH$	$CH{=}CH$	$CH_2{=}C$	$CH{=}C$	$C{=}C$
2.0000β	2.0699β	2.0000β	2.1083β	2.1716β

$CH{-}CH$	$CH{-}C$	$C{-}C$
0.4660β	0.4362β	0.4358β

For 1,3-butadiene, $CH_2{=}CH{-}CH{=}CH_2$, these values give an E_π of $2(2.0000\beta) + 0.4660\beta = 4.466\beta$, as compared with the accurate Hückel value of 4.472β in (16.56). Thus this approach makes the delocalization energy of noncyclic conjugated polyenes essentially zero. The resonance energy of a cyclic polyene is then found as the difference between the cyclic polyene's Hückel π-electron energy E_π and the quantity $\Sigma_b n_b E_{\pi,b}$. The method thus gives the resonance stabilization of cyclic polyenes relative to noncyclic polyenes.

For example, for C_6H_6 the Hess–Schaad resonance energy is found by comparing benzene's HMO π-electron energy (with the α term omitted) of 8β [Eq. (16.57)] with $3(2.0699\beta) + 3(0.4660\beta) = 7.6077\beta$. The Hess–Schaad resonance energy of benzene is therefore $0.3923|\beta|$. More significant is the resonance energy per π electron (REPE), which is $0.3923|\beta|/6 = 0.065|\beta|$. For cyclobutadiene we compare an HMO π-electron energy of 4β [Fig. 16.5 or Eq. (16.49)] with the quantity $2(2.0699\beta) + 2(0.4660\beta) = 5.0718|\beta|$; this gives a resonance energy of $-1.0718|\beta|$ and a REPE of $-0.268|\beta|$.

A compound with a substantially positive REPE value (greater than, say, $0.01|\beta|$) is predicted to be *aromatic*. A compound with a near-zero REPE is *nonaromatic*. A compound with a substantially negative REPE is predicted to be *antiaromatic*, being less stable than if its double bonds were isolated from one another. (Some antiaromatic hydrocarbons are cyclobutadiene and fulvalene.) For the annulenes, C_nH_n, Hess–Schaad REPE versus n values are

n	4	6	8	10	12	14	16	18	20	22
REPE/$|\beta|$	-0.268	0.065	-0.061	0.026	-0.024	0.016	-0.011	0.012	-0.005	0.010

As n increases, the REPE becomes negligible, and the aromatic $4m + 2$ compounds and antiaromatic $4m$ compounds become nonaromatic.

As originally formulated, the Hess–Schaad method did not apply to conjugated ions or radicals, but by modifying their approach, Hess and Schaad were able to treat ions and radicals also. [B. A. Hess and L. J. Schaad, *Pure Appl. Chem.*, **52**, 1471 (1980).]

Comparison with experiment shows that the Hess–Schaad method is highly successful in predicting aromaticity.

π-Electron Charges and Bond Orders. Another defined quantity is the *π-electron charge*. The probability density for an electron in the HMO (16.17) is

$$|\phi_i|^2 = \sum_r \sum_s c_{ri}^* c_{si} f_r^* f_s \tag{16.58}$$

The HMO normalization condition (16.25) is $\sum_{r=1}^{n_C} |c_{ri}|^2 = 1$, and so it is natural to say that an electron in the MO ϕ_i has the probability $|c_{ri}|^2$ of being in the vicinity of the rth carbon atom. If n_i $(=0, 1, \text{or } 2)$ is the number of electrons in the MO ϕ_i, then the total π-electron charge q_r in the region of carbon-atom r is defined to be

$$q_r \equiv \sum_i n_i |c_{ri}|^2 \tag{16.59}$$

where the sum is over the π MOs. q_r is often called the "π-electron density"; this name is misleading, since q_r is neither a charge density (which has dimensions of charge/volume) nor a probability density (which has dimensions of 1/volume); rather, q_r is a pure number that gives the approximate number of π electrons in the vicinity of carbon atom r.

For butadiene

$$q_1 = 2|c_{11}|^2 + 2|c_{12}|^2 = 2(0.372)^2 + 2(0.602)^2 = 1.000$$

$$q_2 = 2|c_{21}|^2 + 2|c_{22}|^2 = 1.000$$

For the ground state of a neutral alternant hydrocarbon, all the Hückel π-electron charges q_r are 1. (For the proof, see *Lowe*, Appendix 5.)

q_r values are sometimes used to predict the positions at which substitution reactions occur most readily in conjugated cyclic hydrocarbons, but this method sometimes fails, and it cannot be used in the many cases where all the q_r's are 1. Hence, many other reactivity indexes have been devised using Hückel theory; see *Murrell, Kettle and Tedder*, Chapter 17; *Streitwieser*, Chapters 11 to 15; *Salem*, Chapter 6.

In the population-analysis discussion of Section 15.7, we saw that for a real MO ϕ_i expanded as the linear combination $\sum_r c_{ri} f_r$ of AOs f_r, the quantity $2c_{ri}c_{si}S_{rs}$ in Eq. (15.25) (where S_{rs} is the overlap integral) is a reasonable measure of the contribution of an electron in MO ϕ_i to the bonding overlap between AOs f_r and f_s. In the HMO method, overlap integrals are neglected, so we cannot use this population-analysis expression. The carbon–carbon bond distances in a conjugated compound are all reasonably similar, so we expect the $C2p\pi$–$C2p\pi$ overlap integral S_{rs} to have similar values for all pairs of bonded carbons; for nonbonded carbons, S_{rs} should be quite small. Since S_{rs} is approximately constant for bonded atoms, we can ignore the factor S_{rs} (and the constant factor 2) and take $c_{ri}c_{si}$ as the contribution of an electron in the real HMO ϕ_i to the π-electron bonding between bonded atoms r and s. If the MO ϕ_i is complex rather than real, one finds on integrating $|\phi_i|^2$ to produce the equation corresponding to (15.25) that the quantity $\frac{1}{2}(c_{ri}^* c_{si} + c_{si}^* c_{ri})$ occurs instead of $c_{ri} c_{si}$. Coulson therefore defined the *π-electron* (or **mobile**) **bond order** p_{rs} for the bond between bonded atoms r and s as

$$p_{rs} \equiv \sum_i n_i \tfrac{1}{2}(c_{ri}^* c_{si} + c_{si}^* c_{ri}) \qquad (16.60)$$

where the sum is over the π MOs. When the coefficients are all real, (16.60) reduces to $p_{rs} = \sum_i n_i c_{ri} c_{si}$. That this definition is reasonable is indicated by the fact that it gives $p = 1$ for the π bond in ethylene.

Addition of the σ electrons' single bond gives the *total bond order* p_{rs}^{tot} as

$$p_{rs}^{\text{tot}} = 1 + p_{rs}$$

For butadiene we have $p_{12} = 2(0.372)(0.602) + 2(0.602)(0.372) = 0.894$ and $p_{23} = 2(0.602)(0.602) + 2(0.372)(-0.372) = 0.447$. The total bond orders are

$$\text{CH}_2 \xrightarrow{1.894} \text{CH} \xrightarrow{1.447} \text{CH} \xrightarrow{1.894} \text{CH}_2 \qquad (16.61)$$

The central bond has some double-bond character, and the end bonds have some single-bond character. (In VB theory, this is explained by contributions from such resonance structures as $\overset{-}{\text{C}}\text{H}_2$—CH=CH—$\overset{+}{\text{C}}\text{H}_2$.) The sum of the bond orders is 5.235, exceeding 5. [This is because the π-electron energy of butadiene exceeds that of two isolated double bonds; see Eq. (16.63).] For benzene each carbon–carbon bond order is found to be $5/3 = 1.667$.

Naturally, we expect a relation between the bond order and the bond length R_{rs} for carbon–carbon bonds. The simplest assumption is a linear relation: $R_{rs} = a + b p_{rs}^{\text{tot}}$, where a and b are constants. The π-electron MO coefficients for ethylene and for benzene are determined completely by symmetry and are independent of the Hückel approximations. We therefore use the bond orders and bond lengths of ethylene (2; 1.335 Å) and benzene (5/3; 1.397 Å) to find a and b; we get

$$R_{rs} = (1.707 - 0.186 p_{rs}^{\text{tot}}) \text{ Å} = (1.521 - 0.186 p_{rs}) \text{ Å} \qquad (16.62)$$

Equation (16.62) works reasonably well. Thus the predicted naphthalene 1–2, 2–3, 1–9, and 9–10 bond lengths (in Å), as compared with the experimental values (in parentheses) are 1.386 (1.37), 1.409 (1.413), 1.418 (1.422), and 1.425 (1.420), respectively. Note that the rings are not regular hexagons. For comparison, an ab initio SCF 6-31G calculation gave the 1–2, 2–3, 1–9, and 9–10 naphthalene bond lengths as 1.362, 1.416, 1.420, and 1.413 Å, respectively [R. C. Haddon and K. Raghavachari, *J. Chem. Phys.*, **79**, 1093 (1983)].

A much-investigated question is whether the carbon–carbon bond lengths in very large conjugated linear and cyclic polyenes ($C_{2n}H_{2n+2}$ and $C_{4m+2}H_{4m+2}$ for m and n large) are equal or alternate in length. The HMO bond orders in a linear conjugated polyene become equal (except near the chain ends) in the limit of large n. The HMO bond orders in the cyclic polyene $C_{4m+2}H_{4m+2}$ are all equal. Hence it was formerly believed that in $C_{2n}H_{2n+2}$ and in $C_{4m+2}H_{4m+2}$ the bond lengths are all equal, except for those near the chain ends in $C_{2n}H_{2n+2}$. However, HMO calculations that take into account the strain energy involved in changing the σ bond lengths indicate strongly that for large n and m both linear and cyclic conjugated polyenes should have bonds that alternate in length (*Salem*, Section 8-4); because of the crudity of the HMO method, these conclusions cannot be considered definitive.

The polymer *trans*-polyacetylene, *trans*-$(CH)_x$ consists of a very long chain of alternating single and double carbon–carbon bonds; the term *trans* refers to the configurations of the two hydrogens bonded to each pair of doubly bonded carbons (Fig. 16.8). Analysis of the ^{13}C NMR spectrum of *trans*-$(CH)_x$ film gave 1.36 Å for the carbon–carbon double-bond length and 1.44 Å for the single-bond length, thereby proving the existence of bond-length alternation in long-chain polyenes [C. S. Yannoni and T. C. Clarke, *Phys. Rev. Lett.*, **51**, 1191 (1983)]. A 6-31G ab initio SCF calculation on the linear polyene *trans*-$C_{22}H_{24}$ gave 1.338 and 1.450 Å for the double- and single-bond lengths in the center of the molecule [H. O. Villar et al., *J. Chem. Phys.*, **88**, 1003 (1987)].

Semiempirical calculations on the cyclic polyenes $C_{4m+2}H_{4m+2}$ that included electron correlation by a variety of methods predict bond-length alternation for large m [M. Takahashi and J. Paldus, *Int. J. Quantum Chem.*, **28**, 459 (1985) and earlier papers]. The fact that we can write two equivalent VB resonance structures for $C_{4m+2}H_{4m+2}$ such that a given bond is single in one structure and double in the other is no guarantee of equal bond lengths.

The Hückel π-electron energy E_π is related to the π bond orders p_{rs} and π-electron densities q_r; in fact (Problem 16.19),

$$E_\pi = \alpha \sum_r q_r + 2\beta \sum_{s-r} p_{rs} \qquad (16.63)$$

where the first sum is over the carbon atoms and where the second is over the carbon–carbon bonds. For example, for butadiene

$$E_\pi = \alpha(1 + 1 + 1 + 1) + 2\beta(0.894 + 0.447 + 0.894) = 4\alpha + 4.47\beta$$

which agrees with (16.56).

For conjugated species with partly filled degenerate MOs, there is an ambiguity in the q_r and p_{rs} values if the real forms of the MOs are used. For example, for $C_6H_6^-$, if we put the unpaired electron in $\phi_{4,\text{real}}$ of (16.48), then this electron contributes $\frac{1}{3}$ to q_1, $\frac{1}{12}$ to q_2, $\frac{1}{12}$ to q_3, $\frac{1}{3}$ to q_4, $\frac{1}{12}$ to q_5, and $\frac{1}{12}$ to q_6; but if we put the odd electron in $\phi_{5,\text{real}}$, then this electron contributes 0 to q_1 and q_4 and $\frac{1}{4}$ to each of q_2, q_3, q_5, and q_6. This ambiguity can be avoided by using the complex MOs, which have the symmetry of the molecule. Putting the odd electron in either ϕ_4 or ϕ_5 of (16.47), we get a contribution of $\frac{1}{6}$ to each q. Alternatively, we

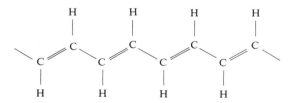

Figure 16.8 *Trans*-polyacetylene.

can average the contributions from $\phi_{4,\text{real}}$ and $\phi_{5,\text{real}}$ to give a $\frac{1}{6}$ contribution at each carbon.

The HMO quantities q_r and p_{rs} in (16.59) and (16.60) are closely related to the density matrix P_{rs} [Eq. (13.166)] of SCF theory. We see that the density matrix elements with $r = s$ (the diagonal elements) are equal to q_r; $P_{rr} = q_r$. Although HMO theory defines bond orders only for pairs of bonded atoms, if we formally define p_{rs} by (16.60) for nonbonded pairs of atoms also, then the definitions yield $p_{rs} = \frac{1}{2}(P_{rs} + P_{sr})$; for real HMOs, $p_{rs} = P_{rs}$.

Free Valence.   Another defined quantity is Coulson's *free-valence index* F_r, for carbon atom r in a planar conjugated organic compound:

$$F_r \equiv \sqrt{3} - \sum_s{}' p_{rs} \tag{16.64}$$

where the sum is over those atoms that are bonded to carbon atom r. The quantity $\sqrt{3} = 1.732$ is the value of $\sum_s' p_{rs}$ for the central carbon of the diradical $C(CH_2)_3$, whose central atom has the largest possible value for this sum of any trigonally bonded carbon. For 1,3-butadiene, $\sum_s' p_{1s} = 0.894$ and $F_1 = F_4 = 1.732 - 0.894 = 0.84$; also, $F_2 = F_3 = 1.732 - (0.894 + 0.447) = 0.39$.

F_r is a measure of the unused bonding power of atom r, and serves as a rough indicator of the susceptibility of a given atom to attack by a neutral free radical; the end carbons in butadiene are attacked preferentially by free radicals.

Heteroatomic Conjugated Molecules.   So far, we have applied the HMO method to hydrocarbons only. For planar conjugated molecules that involve π bonding to noncarbon atoms, the Coulomb and bond integrals for the heteroatoms must be modified from the carbon values. For heteroatoms X and Y, we write

$$\alpha_X = \alpha_C + h_X \beta_{CC} \tag{16.65}$$

$$\beta_{XY} = k_{XY} \beta_{CC} \tag{16.66}$$

where h_X and k_{XY} are certain constants. The best values for these constants vary, depending on which molecular property is being considered, and the values used are based on a mixture of theory and guesswork. One set of values (*Streitwieser*, Chapter 5) is

Atom	—N=	$\overset{\mid}{-}$N—	$=\overset{+}{\underset{\mid}{N}}-$	—O—	=O	—F	—Cl	—Br
h_x	0.5	1.5	2	2	1	3	2	1.5

(16.67)

Bond	$C{-}\overset{\mid}{N}{-}$	aromatic CN	$C{=}\overset{+}{\underset{\mid}{N}}{-}$	N—O—	C—O—	C=O	C—F	C—Cl	C—Br
k_{XY}	0.8	1	1	0.7	0.8	1.0	0.7	0.4	0.3

These values should not be taken too seriously; for example, different workers recommend values for $h_{=O}$ ranging from 0.15 to 2.

Consider pyridine and pyrrole as examples (Fig. 16.9). Pyridine has six π

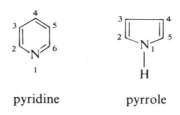

pyridine	pyrrole

Figure 16.9 Pyridine and pyrrole.

electrons, one of which is contributed by nitrogen; pyrrole has six π electrons, two of which are contributed by nitrogen. If we use the values (16.67), the HMO secular determinant for pyridine is the same as (16.38) for benzene, except that the element x in the first row and first column is replaced by $x + 0.5$. For pyrrole the HMO secular equation is

$$\begin{vmatrix} x+1.5 & 0.8 & 0 & 0 & 0.8 \\ 0.8 & x & 1 & 0 & 0 \\ 0 & 1 & x & 1 & 0 \\ 0 & 0 & 1 & x & 1 \\ 0.8 & 0 & 0 & 1 & x \end{vmatrix} = 0$$

We might also use different values of β for the carbon–carbon single and double bonds in pyrrole; Streitwieser recommends

$$\beta_{C-C} = 0.9\beta_{CC}, \qquad \beta_{C=C} = 1.1\beta_{CC} \qquad (16.68)$$

where β_{CC} is for an aromatic carbon–carbon bond (e.g., benzene).

The HMO method can be applied to the ions (16.13); however, the FE MO method is simpler to use, and we omit the HMO calculation; see Problem 16.26. The molecule (16.13) is not planar. However, the out-of-plane methyl protons are but a slight perturbation on an otherwise planar molecule, and the σ–π classification holds to a very good approximation.

For molecules with heteroatoms, Eq. (16.63) becomes

$$E_\pi = \sum_{\text{atoms}} q_r \alpha_r + 2 \sum_{\text{bonds}} p_{rs} \beta_{rs} \qquad (16.69)$$

Inclusion of Overlap. Aside from using the one-electron Hamiltonian (16.15), perhaps the most serious approximation of the simple HMO method is that of taking all overlap integrals equal to zero. Wheland proposed using a common nonzero value S for the overlap integral of carbons bonded to each other; this replaces each element β in the HMO secular equation (16.19) with $\beta - Se_i$. At the benzene carbon–carbon bond distance, S equals 0.25 for $2p\pi$ STOs with orbital exponent 1.625 (the value given by Slater's rules; Problem 15.47). Inclusion of overlap in the HMO method is easy (see Problem 16.27), but is rarely done. One finds that inclusion of overlap gives only slight changes in predicted transition frequencies of alternant hydrocarbons (Problem 16.27d) and gives no change in the π-electron charges and bond orders when these are calculated with suitably modified definitions [B. H. Chirgwin and C. A. Coulson, *Proc. Roy. Soc.*, **A201**, 196 (1950)].

Matrix Formulation.   The HMO equations (16.18) have the same form as the Roothaan equations (13.157), where H_{rs}^{eff} corresponds to F_{rs}, and e_i corresponds to ε_i. We showed the matrix form of the Roothaan equations to be $\mathbf{FC} = \mathbf{SC}\boldsymbol{\varepsilon}$ [Eq. (13.179)]. Hence the HMO equations are equivalent to the matrix equation $\mathbf{H}^{\text{eff}}\mathbf{C} = \mathbf{SC}e = \mathbf{C}e$, where $\mathbf{H}^{\text{eff}}$, $\mathbf{C}$, $\mathbf{S}$, and $\mathbf{e}$ are square matrices of order n_C with elements H_{rs}^{eff}, c_{si}, $S_{rs} = \delta_{rs}$, and $e_{mi} = \delta_{mi}e_i$. Since $\mathbf{H}^{\text{eff}}$ is real and symmetric, we have $\tilde{\mathbf{C}}\mathbf{H}^{\text{eff}}\mathbf{C} = \mathbf{e}$. Computer programs to do HMO calculations find the orthogonal matrix $\mathbf{C}$ that diagonalizes $\mathbf{H}^{\text{eff}}$. For HMO programs, see *Lowe*, Appendix 10.

Localized MOs.   For nonconjugated closed-shell molecules (for example, H_2O), we can view the electrons as residing either in the delocalized canonical MOs or in the energy-localized bond, inner-shell, and lone-pair orbitals. The latter picture allows one to accurately approximate the molecular binding energy as a sum of bond energies. For conjugated molecules (for example, 1,3-butadiene, benzene), the molecular binding energy exceeds the sum of bond energies by the delocalization energy. We therefore expect that the energy-localized bonding MOs in conjugated systems are not as well localized as those in nonconjugated systems. Newton and Switkes calculated the energy-localized MOs for *s-trans*-butadiene and benzene, starting from all-electron minimal-basis-set SCF canonical MOs. [M. D. Newton and E. Switkes, *J. Chem. Phys.*, **54**, 3179 (1971).] In butadiene they found the localized double bonds to be the bent banana bonds (as Kaldor found for ethylene), rather than $\sigma-\pi$ bonds. They defined a percentage of delocalization d of an energy-localized bond MO that measures the degree to which orbitals centered on atoms other than the bonding pair of atoms contribute to the bond MO. For the butadiene energy-localized double-bond MOs, d is $8\frac{1}{2}\%$, only slightly more than the 6% value for the ethylene localized double-bond MOs. Thus conjugation in butadiene leads to only a slight increase in delocalization of the localized double-bond orbitals.

For benzene the energy-localization procedure led to two equivalent solutions, in each of which single and double bonds alternate (as in the Kekulé structures), with each double bond composed of two bent banana bonds. (The total electron density in a localized MO wave function must have the same symmetry as the density calculated from canonical MOs; for a Kekulé-like localized benzene wave function, the tails of the localized double-bond MOs are sufficiently large to maintain $\mathscr{D}_{6h}$ symmetry.) In benzene each energy-localized double-bond MO was found to have $d = 19\%$, indicating large contributions from AOs of atoms other than the bonding pair. The high degree of delocalization is, of course, expected. Another significant point was that for benzene the bent banana double bonds were only slightly preferred (in terms of energy localization) to $\sigma-\pi$ double-bond MOs.

Summary.   Because of the simplicity of the method, carrying out HMO calculations became a favorite pastime of organic chemists, and the results of HMO calculations have been tabulated for hundreds of compounds. (See C. A. Coulson and A. Streitwieser, Jr., *Dictionary of π-Electron Calculations*, W. H.

Freeman, San Francisco, 1965; A. Streitwieser, Jr., and J. I. Brauman, *Supplemental Tables of Molecular Orbital Calculations*, Pergamon, Elmsford, N.Y., 1965; Heilbronner and Straub, *Hückel Molecular Orbitals*, Springer-Verlag, New York, 1966.) The HMO method was widely used to rationalize and predict the properties and reactivities of conjugated compounds. In view of the crudity of the HMO approximations, the fact that the method works as well as it does is surprising and not yet fully explained.

The development of semiempirical theories more sophisticated than the HMO theory led some workers to argue that "Hückel theory has largely outlived its usefulness" (*Murrell and Harget*, page v). However, these more sophisticated theories have their failings (as we shall see), and the success of the Hess–Schaad use of Hückel theory to predict aromaticity indicates that HMO theory may still be useful, "especially on a qualitative level as a guide . . . in planning and interpreting experiments" (N. Trinajstic in *Segal*, Part A, Chapter 1).

16.4   THE PARISER–PARR–POPLE METHOD

Although the Hückel theory can be used to predict the longest-wavelength bands of aromatic hydrocarbons, it would be hopeless to try to use HMO theory to predict the complete electronic spectrum of an aromatic hydrocarbon. For example, Hückel theory, which neglects interelectronic repulsions, gives no separation between singlet and triplet electronic terms arising from the same configuration. Experimentally, separations of one or two electron volts are observed between such terms. A semiempirical π-electron theory that takes electron repulsion into account and thereby improves on the Hückel method is the *Pariser–Parr–Pople* (PPP) *method*. Here, the π-electron Hamiltonian (16.1) including electron repulsions is used, and the π-electron wave function is written as an antisymmetrized product of π-electron spin-orbitals. A minimal basis set of one $2p\pi$ STO on each conjugated atom is used, and the spatial π MOs ϕ_i are taken as linear combinations of these AOs: $\phi_i = \sum_{r=1}^{b} c_{ri} f_r$. The Roothaan equations are used to find SCF π MOs within the π-electron approximation.

The π-electron Hamiltonian $\hat{H}_\pi$ in (16.1) has the same form as the all-electron operator $\hat{H}_{el} = -\frac{1}{2} \sum_i \nabla_i^2 - \sum_i \sum_\alpha Z_\alpha/r_{i\alpha} + \sum_i \sum_{j>i} 1/r_{ij}$ for a molecule except that $\hat{H}_\pi^{core}(i)$ replaces $\hat{H}^{core}(i) \equiv -\frac{1}{2} \nabla_i^2 - \sum_\alpha Z_\alpha/r_{i\alpha}$ [Eq. (13.150)], and the sums go over only the n_π π electrons rather than over all the electrons. Hence, similar to (13.148), the SCF π MOs satisfy $\hat{F}_\pi \phi_i = \varepsilon_i \phi_i$, where $\hat{F}_\pi$ is given by (13.149) with $\hat{H}_{(1)}^{core}$ replaced by $\hat{H}_\pi^{core}$ and $n/2$ replaced by $n_\pi/2$. The SCF π-electron energy is given by (13.145) with H_{ii}^{core} replaced by $H_{\pi,ii}^{core}$ and $n/2$ replaced by $n_\pi/2$.

Consider the internuclear repulsion term V_{NN}. The π electrons are viewed as moving in the field arising from the nuclei and the σ electrons. If the charges of the σ electrons that originate from atom S are added to the nuclear charge Z_S, we get a net charge equal to $n_{\pi,S}$, where $n_{\pi,S}$ is the number of π electrons contributed to the molecule by atom S. For example, a carbon atom in a conjugated molecule has atomic number 6 and contributes 5 σ electrons and one π electron, and $n_{\pi,S} = 6 - 5 = 1$ for each C atom. For the nitrogen atom in pyrrole (Fig. 16.9),

$n_{\pi,\mathrm{S}} = 2$. We can thus take

$$V_{NN} = \sum_{\mathrm{R>S}} \sum_{\mathrm{S}} \frac{n_{\pi,\mathrm{R}} n_{\pi,\mathrm{S}}}{R_{\mathrm{RS}}} \tag{16.70}$$

in the PPP method. However, the PPP method is not usually used to determine equilibrium geometries (which are assumed to be known), but is used to calculate such things as ionization energies for π electrons and the part of the electronic spectrum due to π-electron transitions. The internuclear distances are assumed not to change significantly in a π-electron transition, and so V_{NN} cancels in calculating π-electron transition energies; hence the V_{NN} term is usually omitted.

The Roothaan equations (13.157) and (13.165) become

$$\sum_s c_{si}(F_{\pi,rs} - \varepsilon_i S_{rs}) = 0, \qquad r = 1, \ldots, b \tag{16.71}$$

$$F_{\pi,rs} = H_{\pi,rs}^{\mathrm{core}} + \sum_{t=1}^{b} \sum_{u=1}^{b} P_{tu}[(rs|tu) - \tfrac{1}{2}(ru|ts)] \tag{16.72}$$

In addition to assuming $\sigma-\pi$ separability, the PPP method makes further approximations. As in Hückel theory, overlap is neglected:

$$S_{rs} \equiv \langle f_r(1)|f_s(1)\rangle = \delta_{rs} \tag{16.73}$$

where δ_{rs} is the Kronecker delta. Consistent with the neglect of overlap integrals, the PPP method makes the approximation of *zero differential overlap* (ZDO):

$$f_r^*(1)f_s(1)\,dv_1 = 0, \qquad \text{for } r \neq s \tag{16.74}$$

From (16.74) and $(rs|tu) \equiv \langle f_r(1)f_t(2)|1/r_{12}|f_s(1)f_u(2)\rangle$ [Eq. (13.162)], it follows that the electron-repulsion integrals are given by

$$(rs|tu) = \delta_{rs}\delta_{tu}(rr|tt) \equiv \delta_{rs}\delta_{tu}\gamma_{rt} \tag{16.75}$$

where $\gamma_{rt} \equiv (rr|tt)$. Thus the method ignores many (but not all) of the electron-repulsion integrals, thereby greatly simplifying the calculation. In particular, all three- and four-center electron-repulsion integrals are ignored. The integrals γ_{rt} are often treated as empirical parameters, rather than being evaluated theoretically.

The ZDO approximation is at first sight rather drastic. However, a partial theoretical justification for it can be given by reinterpreting the AOs used to express the MOs as orthogonalized AOs (rather than ordinary AOs). Each orbital in a set of orthogonalized AOs is a linear combination of ordinary AOs, the coefficients being chosen so that the members of the set are mutually orthogonal. There are many ways to choose the linear combinations to produce an orthogonal set. One approach is to make the orthogonalized AOs (OAOs) resemble the ordinary AOs as much as possible by minimizing the sum of the squares of the deviations of the OAOs from the ordinary AOs; one minimizes the sum $\Sigma_i \int |\chi_{i,\mathrm{OAO}} - \chi_i|^2\,dv$, where the sum goes over the set of AOs and where χ_i and $\chi_{i,\mathrm{OAO}}$ are the ordinary and the orthogonalized AOs. This produces what are called *symmetrically orthogonalized* (or Löwdin) AOs (see *Pilar*, Section 14-8; *Parr*, pages 51–52). One finds that in each symmetrically orthogonalized AO the coefficient of one ordinary AO is substantially greater than the coefficients of the other ordinary AOs, so the symmetrically orthogonalized AOs are not drastically different from the ordinary AOs. One finds that with symmetrically orthogonalized AOs the integrals not neglected in the ZDO approximation undergo only small changes in value as compared with their values with ordinary AOs; these changes in value can be partly allowed for by the fact that many integrals are taken as empirical parameters. Moreover, with symmetrically orthogonalized AOs, the electron-repulsion integrals neglected in the ZDO approximation are generally found to be quite small (and all overlap integrals are, of course, zero). For details, see the references cited on page 27 of *Murrell and Harget*.

With the approximation (16.75) for $(rs|tu)$, the matrix elements $F_{\pi,rs}$ in (16.72) become (Problem 16.28)

$$F_{\pi,rr} = H_{\pi,rr}^{\text{core}} + \sum_t P_{tt}\gamma_{rt} - \tfrac{1}{2}P_{rr}\gamma_{rr}$$

$$F_{\pi,rs} = H_{\pi,rs}^{\text{core}} - \tfrac{1}{2}P_{sr}\gamma_{rs} , \qquad r \neq s \tag{16.76}$$

Rather than attempting to explicitly specify $\hat{H}^{\text{core}}$ and calculate the $H_{\pi,rs}^{\text{core}}$ integrals theoretically, the PPP method takes some of them as empirical parameters and neglects the rest. For atoms R and S (on which AOs f_r and f_s are centered) not bonded to each other, $H_{\pi,rs}^{\text{core}}$ is assumed to be negligible:

$$H_{\pi,rs}^{\text{core}} \equiv \langle f_r(1)|\hat{H}_\pi^{\text{core}}(1)|f_s(1)\rangle = 0 , \quad \text{for R and S not bonded} \tag{16.77}$$

For atoms bonded to each other,

$$H_{\pi,rs}^{\text{core}} = \beta_{rs}^{\text{core}} , \qquad \text{for R and S bonded} \tag{16.78}$$

where β_{rs}^{core} is an empirical parameter whose value depends on the nature of the atoms and on the internuclear distance between them. Several methods have been proposed to assign β_{rs}^{core} values (*Murrell and Harget*, pages 17–20). One procedure takes $\beta_{rs}^{\text{core}} = k\langle f_r|f_s\rangle$, where the value of the empirical parameter k is chosen so that the predictions of the theory fit experiment as well as possible, and $\langle f_r|f_s\rangle$ is calculated from the STOs f_r and f_s [and not taken as zero as in (16.73)].

For $r = s$, we have [Eq. (16.2)]

$$H_{\pi,rr}^{\text{core}} \equiv \langle f_r|\hat{H}_\pi^{\text{core}}|f_r\rangle \equiv \alpha_r^{\text{core}} = \langle f_r(1)|-\tfrac{1}{2}\nabla_1^2 + V(1)|f_r(1)\rangle \tag{16.79}$$

where $V(1)$ is the potential energy of π-electron 1 in the field of the σ electrons and the nuclei. Let R and S be the atoms on which the $2p\pi$ AOs f_r and f_s are centered, and let the index B denote nonconjugated atoms in the molecule (for example, the H atoms in benzene). Breaking up $V(1)$ into contributions from the individual atoms in the molecule, we write

$$V(1) = V_R(1) + \sum_{S \neq R} V_S(1) + \sum_B V_B(1) \tag{16.80}$$

where the sum over S goes over all the conjugated atoms except R. If the penetration of π-electron 1 in AO f_r on atom R into the "charge clouds" of the other atoms is neglected, and if we use the fact that f_r has most of its probability density near atom R, which is at a distance R_{RS} from atom S, then the formula $V = Q_1'Q_2'/r_{12}$ gives the approximation (in atomic units) $\langle f_r|V_S|f_r\rangle = -n_{\pi,s}/R_{RS}$, where $n_{\pi,S}$ (the number of π electrons contributed by atom S) is the charge of atom S when stripped of its π electrons. Similarly, $\langle f_r | V_B | f_r \rangle = 0$, since nonconjugated atoms (such as hydrogen in benzene) have no π electrons and no charge when stripped of their π electrons. Instead of $\langle f_r | V_S | f_r \rangle = -n_{\pi,S}/R_{RS}$, the less approximate equation $\langle f_r | V_S | f_r \rangle = -n_{\pi,S}\gamma_{rs}$ [where γ_{rs} is given by (16.75)] is usually used. (For justification, see *Murrell and Harget*, pages 13–14.) The integral

$$U_{rr} \equiv \langle f_r(1) | -\tfrac{1}{2}\nabla_1^2 + V_R(1) | f_r(1) \rangle$$

looks like an average energy for a π electron in a $2p\pi$ AO centered on atom R in the molecule. Hence the PPP method takes this integral as equal to the $2p\pi$ orbital energy for an atom in its valence state. The valence state is the (hypothetical) state in which the atom exists in the molecule. The valence-state $2p\pi$ orbital energy can be readily found from atomic spectral data (see Section 15.16).

Substituting (16.80) into (16.79) and making the above approximations, we have

$$H_{\pi,rr}^{core} \equiv \alpha_r^{core} = U_{rr} - \sum_{S \neq R} n_{\pi,S}\gamma_{rs} \qquad (16.81)$$

The electron-repulsion integrals γ_{rs} in $F_{\pi,rr}$ and $F_{\pi,rs}$ [Eq. (16.76)] can be calculated from the STOs and the known separations between the atoms on which the STOs are centered. When this is done, the predicted molecular spectra do not agree with experiment. Hence, the electron-repulsion integrals are usually treated as parameters whose values are found from one of a number of proposed semiempirical formulas (see *Murrell and Harget*, pages 14–16). (Note that if the electron-repulsion integrals are all neglected, then the Fock matrix elements take on the same form as H_{rs}^{eff} in the Hückel theory.)

To do a PPP calculation, one starts with the HMO coefficients as an initial guess for the c_{si}'s, calculates the initial density matrix elements P_{rs}, calculates the initial $F_{\pi,rs}$ matrix elements, solves the equations (16.71) for π-electron orbital energies ε_i and an improved set of coefficients c_{si}, calculates improved P_{rs} values, and so on. The iteration is continued until convergence is achieved. For alternant hydrocarbons, the result $q_r = 1$ and the pairing property of MO energies hold in the PPP theory. To improve the results, CI of the π electrons may be included.

The PPP method gives a good account of the electronic spectra of many, but not all, aromatic hydrocarbons. For more on the PPP method, see *Parr*, Chapter III; *Murrell and Harget*, Chapter 2; *Offenhartz*, Chapter 11.

16.5 SEMIEMPIRICAL MO TREATMENTS OF NONPLANAR MOLECULES

Semiempirical MO theories fall into two categories: those using a Hamiltonian that is the sum of one-electron terms, and those using a Hamiltonian that includes two-electron repulsion terms, as well as one-electron terms. The Hückel method is a one-electron theory, whereas the Pariser–Parr–Pople method is a two-electron theory.

For nonplanar molecules, the simplifying approximation of treating the π

electrons separately is not available, and all the valence electrons must be considered together.

Of course, all the semiempirical treatments of this section are applicable to planar (as well as nonplanar) compounds, and they represent improvements on semiempirical π-electron theories, since they explicitly allow for the effects of the σ valence electrons.

The Extended Hückel Method. The most important one-electron semiempirical MO method for nonplanar molecules is the *extended Hückel theory.* An early version was used by Wolfsberg and Helmholz in treating inorganic complex ions; the method was further developed and widely applied by Hoffmann [R. Hoffmann, *J. Chem. Phys.*, **39**, 1397 (1963); **40**, 2745, 2474, 2480 (1964); *Tetrahedron*, **22**, 521, 539 (1966); M. Wolfsberg and L. Helmholz, *J. Chem. Phys.*, **20**, 837 (1952)].

The extended Hückel (EH) method begins with the approximation of treating the valence electrons separately from the rest (Section 13.20). The valence-electron Hamiltonian is taken as the sum of one-electron Hamiltonians:

$$\hat{H}_{val} = \sum_i \hat{H}_{eff}(i) \tag{16.82}$$

where $\hat{H}_{eff}(i)$ is not specified explicitly. The MOs are approximated as linear combinations of the valence AOs f_r of the atoms:

$$\phi_i = \sum_r c_{ri} f_r \tag{16.83}$$

In the simple Hückel theory of planar hydrocarbons, each π MO contains contributions from one $2p\pi$ AO on each carbon atom. In the extended Hückel treatment of nonplanar hydrocarbons, each valence MO contains contributions from four AOs on each carbon atom (one $2s$ and three $2p$'s) and one $1s$ AO on each hydrogen atom. The AOs used are usually Slater-type orbitals with fixed orbital exponents. For the simplified Hamiltonian (16.82), the problem separates into several one-electron problems:

$$\hat{H}_{eff}(i)\phi_i = e_i\phi_i$$

$$E_{val} = \sum_i e_i \tag{16.84}$$

Application of the variation theorem to the linear trial function (16.83) gives as the secular equation and the equations for the MO coefficients

$$\det (H_{rs}^{eff} - e_i S_{rs}) = 0 \tag{16.85}$$

$$\sum_s [(H_{rs}^{eff} - e_i S_{rs})c_{si}] = 0 , \qquad r = 1, 2, \dots \tag{16.86}$$

All this is similar to simple Hückel theory. However, the extended Hückel theory does not neglect overlap. Rather *all* overlap integrals are explicitly evaluated using the forms chosen for the AOs and the internuclear distances at which the calculation is being done. Formulas for overlap integrals of STOs are readily available. [Footnotes 12 to 18 of D. M. Bishop et al., *J. Chem. Phys.*, **45**,

1880 (1966) give references to available tabulations of overlap integrals.] Since overlap is included, off-diagonal e_i's are present in the secular determinant. A computer is used to calculate the overlap integrals and to solve (16.86).

Since $\hat{H}_{\text{eff}}(i)$ is not specified, there is the problem of what to use for the integrals H_{rs}^{eff}. For $r = s$, the one-electron integral $H_{rr}^{\text{eff}} \equiv \langle f_r | \hat{H}^{\text{eff}} | f_r \rangle$ looks like an average energy for an electron in the AO f_r centered on atom R in the molecule. Hence, the EH method takes H_{rr}^{eff} as equal to the orbital energy of the AO f_r for atom R in its valence state; see the discussion preceding (16.81). By Koopmans' theorem (Section 15.6), the valence-state orbital energy is taken as equal to minus the valence-state ionization potential (VSIP) of f_r. For hydrogen and for carbon atoms in a molecule with only single bonds to carbon (sp^3-hybridized carbon), the VSIP parametrization gives

$$\langle C2s | \hat{H}_{\text{eff}} | C2s \rangle = -20.8 \text{ eV} , \qquad \langle C2p | \hat{H}_{\text{eff}} | C2p \rangle = -11.3 \text{ eV}$$

$$\langle H1s | \hat{H}_{\text{eff}} | H1s \rangle = -13.6 \text{ eV}$$

(16.87)

The carbon VSIPs for sp^2- and sp-hybridized carbon differ from those in (16.87), but the difference is often ignored and an average set of VSIPs is used for all carbons. VSIPs are tabulated in J. Hinze and H. H. Jaffé, *J. Am. Chem. Soc.*, **84**, 540 (1962); G. Pilcher and H. A. Skinner, *J. Inorg. Nuc. Chem.*, **24**, 937 (1962); L. C. Cusachs et al., *J. Chem. Phys.*, **44**, 835 (1966).

For the off-diagonal matrix elements H_{rs}^{eff}, $r \neq s$, Wolfsberg and Helmholz took

$$H_{rs}^{\text{eff}} = \tfrac{1}{2} K (H_{rr}^{\text{eff}} + H_{ss}^{\text{eff}}) S_{rs}$$

(16.88)

where K is a numerical constant (values between 1 and 3 have been used) and the remaining quantities are evaluated as above. Since H_{rr}^{eff} and H_{ss}^{eff} are usually negative, (16.88) gives H_{rs}^{eff} as negative. Instead of (16.88), Ballhausen and Gray used

$$H_{rs}^{\text{eff}} = -K (H_{rr}^{\text{eff}} H_{ss}^{\text{eff}})^{1/2} S_{rs}$$

(16.89)

Equations (16.89) and (16.88) are the most widely used approximations, but many others have been proposed. In contrast to the simple Hückel theory, H_{rs}^{eff} is nonzero for all pairs of orbitals (unless S_{rs} vanishes for symmetry reasons).

Once the H_{rs}^{eff} and S_{rs} integrals have been evaluated, the secular equation is solved for the orbital energies and the MO coefficients are found. (Computer programs use matrix diagonalization to do this.) Since the H_{rs}^{eff} integrals do not depend on the MO coefficients, no iteration is needed. (The EH method is often used to provide an initial guess for the valence MO coefficients in ab initio SCF calculations.)

The total valence-electron energy is given by (16.84) as the sum of orbital energies. To predict molecular geometry using the EH method, one carries out a series of calculations over a range of bond distances and angles and looks for the nuclear configuration that minimizes (16.84). (This procedure is like that used in the Walsh-diagram method of Section 15.10, where one looks for the bond angle that minimizes the sum of orbital energies.) Note that (16.84) omits both electron–electron repulsions and nuclear–nuclear repulsions. It might be thought that such a theory would be useless for predicting molecular geometry, but this is not so.

Allen and Russell compared EH results with those of ab initio SCF calculations for 12 small molecules [L. C. Allen and J. D. Russell, *J. Chem. Phys.*, **46**, 1029 (1967); S. D. Peyerimhoff, R. J. Buenker, and L. C. Allen, *J. Chem. Phys.*, **45**, 734 (1966)], using (16.89) with $K = 1.5$. They found that for molecules whose bonds are not highly polar, the EH-method graph of $\Sigma_i e_i$ versus bond angle and the graph of SCF MO total energy (including nuclear repulsion) versus bond angle have minima at essentially the same angles. Since the SCF MO method yields generally accurate bond angles, so does the extended Hückel method for compounds of low polarity. The cases where the EH method failed (H_2O, Li_2O, LiOH, FOH) all involve highly polar bonds.

Allen and Russell found the order of MO energies predicted by the EH method to be usually reliable for low-polarity compounds. Hoffmann's EH calculation of benzene predicted that the lowest π MO, the $1a_{2u}$ MO, would lie below some of the occupied σ MOs, namely, the $3e_{2g}$ and $1b_{2u}$ MOs. Near-Hartree–Fock calculations of benzene show the $1a_{2u}$ π MO to have higher energy than the $1b_{2u}$ MO but very slightly lower energy than the $3e_{2g}$ pair of σ MOs [W. C. Ermler et al., *J. Am. Chem. Soc.*, **98**, 388 (1976); W. C. Ermler and C. W. Kern, *J. Chem. Phys.*, **58**, 3458 (1973)]. This MO order is confirmed experimentally by the photoelectron spectrum of C_6H_6. Thus Hoffmann's prediction is partly confirmed.

For bond lengths, Allen and Russell state that empirical formulas involving bond radii are more reliable than the extended Hückel method.

The extended Hückel MO coefficients are found by Allen and Russell to deviate significantly from those of ab initio wave functions. Thus the extended Hückel method does not give the accurate predictions of such properties as dipole moments and rotational barriers achieved by the SCF MO method.

The EH method has been widely applied to predict conformations of organic compounds, and it was once believed to be fairly reliable for such work. However, a careful EH study of benzene showed that the $\mathscr{D}_{6h}$ benzene geometry predicted by EH theory corresponds to a saddle point (rather than a true minimum) on the potential-energy surface for nuclear motion; in fact, EH theory erroneously predicts benzene to be unstable with respect to dissociation to three acetylenes. [See W. L. Bloemer and B. L. Bruner, *Chem. Phys. Lett.*, **17**, 452 (1972).] Furthermore, Pullman found EH-predicted conformations of several pharmacologically active molecules to be in serious disagreement with the results of ab initio calculations and experimental results (B. Pullman in *Advances in Quantum Chemistry*, Volume 10, Academic Press, New York, 1977, pages 251–328); Pullman concluded that these disagreements "practically disqualify [the EH method] as a valid tool for conformational studies."

Because the EH method gives poor predictions of such molecular properties as bond lengths, dipole moments, energies, and rotational barriers, Jug concluded that this method "is obsolete" [K. Jug, *Theor. Chim. Acta*, **54**, 263 (1980)]. However, this judgment is too harsh, since the EH method has been used by Hoffmann and others to provide valuable qualitative insights into chemical bonding [for example, P. J. Hay, J. C. Thibeault, and R. Hoffmann, *J. Am.*

Chem. Soc., **97**, 4884 (1975)]. Gimarc noted that "the real value of the extended Hückel method is not in its quantitative results, which have never been impressive, but rather in the qualitative nature of the results and in the interpretations those results can provide" (B. M. Gimarc, *Molecular Structure and Bonding*, Academic Press, New York, 1979, page 216).

The CNDO, INDO, and NDDO Methods. Several semiempirical two-electron MO generalizations of the PPP method have been developed for nonplanar molecules. The ***complete neglect of differential overlap*** (CNDO) ***method*** was proposed by Pople, Santry, and Segal in 1965. The ***intermediate neglect of differential overlap*** (INDO) ***method*** was proposed by Pople, Beveridge, and Dobosh in 1967. Both methods treat only the valence electrons explicitly. The valence-electron Hamiltonian has the same form as (16.1):

$$\hat{H}_{\text{val}} = \sum_{i=1}^{n_{\text{val}}} (-\tfrac{1}{2}\nabla_i^2 + V_i) + \sum_{i=1}^{n_{\text{val}}} \sum_{j>i} \frac{1}{r_{ij}} \tag{16.90}$$

where n_{val} is the number of valence electrons in the molecule and V_i is the potential energy of valence electron i in the field of the nuclei and the inner-shell electrons. The quantity in parentheses is $\hat{H}_{\text{val}}^{\text{core}}(i)$. Both methods are SCF methods that iteratively solve the Roothaan equations using certain approximations for the integrals.

The CNDO method uses a minimal basis set of valence Slater AOs f_r with fixed orbital exponents on each atom; the valence MOs ϕ_i are written as $\phi_i = \sum_{r=1}^{b} c_{ri} f_r$. The total molecular energy is given by (13.145) with H_{ii}^{core} replaced by $H_{\text{val},ii}^{\text{core}}$, $n/2$ replaced by $n_{\text{val}}/2$, and V_{NN} replaced by $\sum_{\alpha} \sum_{\beta>\alpha} n_{\text{val},\alpha} n_{\text{val},\beta}/R_{\alpha\beta}$, where $n_{\text{val},\alpha}$, the number of valence electrons contributed by atom α, is the charge on atom α due to the nucleus and the inner-shell electrons. The Roothaan equations are given by (16.71) with $F_{\pi,rs}$ replaced by $F_{\text{val},rs}$; $F_{\text{val},rs}$ is given by (16.72) with $H_{\pi,rs}^{\text{core}}$ replaced by $H_{\text{val},rs}^{\text{core}}$.

As in the PPP method, the CNDO method uses the ZDO approximation (16.74) for all pairs of AOs. Thus, $S_{rs} = \delta_{rs}$ and $(rs|tu) = \delta_{rs}\delta_{tu}(rr|tt)$. In the PPP method for conjugated hydrocarbons, there is only one basis AO per atom, the $2p\pi$ AO. In the CNDO method, there are several basis valence AOs on every atom (except hydrogens), and the ZDO approximation leads to neglect of electron-repulsion integrals containing the product $f_r(1)f_s(1)$ where f_r and f_s are different AOs centered on the *same* atom.

The $H_{\text{val},rs}^{\text{core}}$ integrals in $F_{\text{val},rs}$ are approximated by using data such as atomic ionization potentials and electron affinities and by taking some of them as semiempirical parameters adjusted so that CNDO calculations will give a good overall fit with the results of minimal basis ab initio SCF calculations. The original version of CNDO is called CNDO/1; a version using an improved parametrization is called CNDO/2.

The INDO method is an improvement on CNDO; here, differential overlap between AOs on the same atom is not neglected in one-center electron-repulsion integrals, but is still neglected in two-center electron-repulsion integrals. Thus fewer two-electron integrals are neglected, as compared with CNDO; otherwise, the two methods are the same. The INDO method gives an improvement on CNDO results, especially where electron spin distribution is important (for example, in calculating electron-spin-resonance spectra).

As to results, the CNDO and INDO methods give fairly good bond lengths and angles, somewhat erratic dipole moments, and poor dissociation energies. (For details on the CNDO and INDO methods, see *Pople and Beveridge*; G. Klopman and R. C. Evans in *Segal*, Part A, page 29; *Murrell and Harget*, Chapter 3.)

Versions of CNDO and INDO parametrized to predict electronic spectra are called CNDO/S and INDO/S. These methods include some configuration interaction; although the ground state of a closed-shell molecule is generally well represented by a single-determinant wave function, one typically requires CI for accurate representation of excited states. For details of these methods, see R. L. Ellis and H. H. Jaffé in *Segal*, Part B, page 49; J. Michl in *Segal*, Part B, page 99.

The *neglect of diatomic differential overlap* (NDDO) method (suggested by Pople, Santry, and Segal in 1965) is an improvement on INDO in which differential overlap is neglected only between AOs centered on different atoms: $f_r^*(1)f_s(1)\,dv_1 = 0$ only when AOs r and s are on different atoms. The degree of neglect of differential overlap in NDDO is more justifiable than in CNDO or INDO. A few initial attempts at parametrizing the NDDO method gave results that were rather disappointing (see G. Klopman and R. C. Evans in *Segal*, Part A, Chapter 2), and the method was little used until 1977, when Dewar and Thiel modified it to give the MNDO method, discussed later in this section.

The PRDDO Method. The *partial retention of diatomic differential overlap* (PRDDO) method was developed by Halgren and Lipscomb [T. A. Halgren and W. N. Lipscomb, *J. Chem. Phys.*, **58**, 1569 (1973); D. S. Marynick and W. N. Lipscomb, *Proc. Natl. Acad. Sci. U.S.A.*, **79**, 1341 (1982)]. This is a semiempirical minimal-basis-set SCF MO method that includes all electrons but neglects some integrals and approximates others. The method uses symmetrically orthogonalized AOs (Section 16.4) as basis functions. Whereas the CNDO method neglects all electron repulsion integrals containing the product $f_r^*(1)f_s(1)$ or $f_r^*(2)f_s(2)$, where f_r and f_s are different AOs, the PRDDO method neglects only electron-repulsion integrals that contain the product $f_r^*(1)f_s(1)f_t^*(2)f_u(2)$, where f_r and f_s are different basis functions and f_t and f_u are different basis functions. (When $r = t$ and $s = u$, the integral is not neglected.) Fewer integrals are neglected than in NDDO (which in turn neglects fewer integrals than CNDO and INDO). In PRDDO (as in CNDO and INDO) some of the integrals are taken as parameters whose values are chosen so as to reproduce results of ab initio calculations on small molecules. The PRDDO method is much more accurate than CNDO and INDO in reproducing results of ab initio calculations and gives results pretty close to those of an ab initio STO-3G calculation in most cases [T. A. Halgren et al., *J. Am. Chem. Soc.*, **100**, 6595 (1978)].

The MINDO, MNDO, AM1, MNDO-PM3, and SINDO1 Methods. The aim of Pople and co-workers in the CNDO and INDO methods and of Halgren and Lipscomb in PRDDO was to reproduce as well as possible the results of minimal-basis-set SCF MO ab initio calculations with theories that require much less computational effort than an ab initio treatment. Since these methods use approximations in the SCF calculation, we can expect their results to be similar to but less accurate than minimal-basis ab initio SCF results. Thus these methods do pretty well on molecular geometry and do poorly on binding energies. Dewar and co-workers devised several semiempirical SCF MO theories that closely resemble the INDO or NDDO theories. However, Dewar's aim was not to reproduce ab initio SCF wave functions, but to have a theory that would give molecular binding energies with chemical accuracy (that is, to within 1 kcal/mol) and that would be applicable to large molecules without a lot of computation. It might seem unlikely that one could devise an SCF theory that involves approximations to the ab initio Hartree–Fock procedure but that succeeds in an area (binding energies) where the Hartree–Fock theory fails. However, by proper choice of the parameters in the semiempirical SCF theory, one can actually get better results than ab initio SCF calculations, because the choice of suitable parameters can compensate for the partial neglect of electron correlation in ab initio SCF theory.

In 1967, Dewar and Klopman published the PNDO (partial neglect of . . .) theory. In 1969, Baird and Dewar published the MINDO/1 (first version of modified INDO) theory. The parameters in these theories were chosen so as to have the predicted molecular heats of formation fit experimental data as well as possible. Although these theories worked well for heats of formation, they were quite poor for geometry. Therefore, in 1970 Dewar and Haselbach proposed the MINDO/2 theory. In MINDO/2 a careful choice of parameters allowed molecular geometries and heats of formation to be calculated rather accurately in most cases. Since MINDO/2 gave large errors in dipole moments and in bond lengths to hydrogen atoms, an improved version, MINDO/2', was published in 1972.

Dewar's MINDO work culminated in MINDO/3, a substantially improved version published in 1975 [R. C. Bingham, M. J. S. Dewar, and D. H. Lo, *J. Am. Chem. Soc.*, **97**, 1285, 1294, 1302, 1307 (1975); Dewar et al., *J. Am. Chem. Soc.*, **97**, 1311; Dewar, *Science*, **187**, 1037 (1975)]. The MINDO/3 method has been parametrized for compounds containing C, H, O, N, B, F, Cl, Si, P, and S. Calculations have been done on molecules as large as lysergic acid diethylamide, $C_{20}H_{25}N_3O$, a 49-atom molecule for which the MINDO/3 calculation required the optimization of 141 geometrical parameters. A MINDO calculation yields the gas-phase molecular heat of atomization; one then uses experimental values of heats of formation of C, H, O, and other atoms from the elements in their reference forms (graphite, H_2, O_2, and so on) at 25°C to calculate the molecule's gas-phase heat of formation at 25°C, $\Delta H^\circ_{f,298}$. Note that $\Delta H^\circ_{f,298}$ contains rotational and vibrational contributions (Section 13.2). Rather than being calculated explicitly, these contributions are allowed for by the choice of parameters used in the calculation. For a sample of a large number of compounds, the average absolute errors in MINDO/3-calculated properties are 11 kcal/mol in heats of formation, 0.02 Å in bond lengths, 5° in bond angles, 0.4 D in dipole moments, and 0.8 eV in ionization potentials [D. N. Nanda and K. Jug, *Theor. Chim. Acta*, **57**, 95 (1980)]. Large errors in heats of formation occur for small-ring compounds, compounds containing triple bonds, aromatic compounds, compact globular molecules, boron compounds, and molecules containing adjacent atoms with lone pairs. The errors in heats of formation are larger than Dewar was aiming for, but MINDO/3 is still a significant achievement.

The MINDO/3 method has been widely applied to calculate properties of ground-state organic molecules and to calculate potential-energy surfaces of chemical reactions (Section 15.18). However, the usefulness of MINDO/3 came under attack by advocates of ab initio SCF calculations [J. A. Pople, *J. Am. Chem. Soc.*, **97**, 5306 (1975); W. J. Hehre, *J. Am. Chem. Soc.*, **97**, 5308 (1975); M. J. S. Dewar, *J. Am. Chem. Soc.*, **97**, 6591 (1975)].

Because MINDO/3 did not meet Dewar's original aims and because it proved difficult to extend MINDO/3 to include metallic elements, Dewar and co-workers developed the *MNDO* (modified neglect of diatomic overlap) method, published in 1977 [M. J. S. Dewar and W. Thiel, *J. Am. Chem. Soc.*, **99**, 4899, 4907 (1977); Dewar and H. S. Rzepa, *J. Am. Chem. Soc.*, **100**, 58, 777, 784 (1978); Dewar and M. L. McKee, *J. Am. Chem. Soc.*, **99**, 5231 (1977)]. The MNDO method is based on the NDDO approximation discussed earlier in this section. As noted, the NDDO approximation is more justifiable than the INDO approximation used in MINDO/3. Like MINDO/3, MNDO is parametrized to reproduce gas-phase $\Delta H^\circ_{f,298}$ values. The MNDO method has been parametrized for compounds containing H, Li, Be, B, C, N, O, F, Al, Si, Ge, Sn, Pb, P, S, Cl, Br, I, and Hg. For compounds containing only H, B, C, N, O, and F, the number of parameters in MNDO is 31, as compared with 61 in MINDO/3. (One criticism of MINDO/3 was the large number of adjustable parameters used.) MNDO gives substantially improved results as compared to MINDO/3; average absolute MNDO errors are 9 kcal/mol in heats of formation, 3° in bond angles, 0.025 Å in bond lengths, 0.35 D in dipole moments, and 0.5 eV in ionization potentials [D. N. Nanda and K. Jug, *Theor. Chim. Acta*, **57**, 95 (1980)]. The calculation time in MNDO is similar to that in MINDO/3.

The strengths and weaknesses of MINDO/3 and MNDO are reviewed in *Clark*, pages 144–151.

In 1985, Dewar and co-workers published an improved version of MNDO called AM1 (Austin Model 1, named for the University of Texas at Austin); M. J. S. Dewar, E. G. Zoebisch, E. F. Healy, and J. J. P. Stewart, *J. Am. Chem. Soc.*, **107**, 3902 (1985); M. J. S. Dewar and E. G. Zoebisch, *THEOCHEM.*, **49**, 1 (1988). *AM1* has been parametrized for H, B, C, Si, N, O, S, F, Cl, Br, I, Hg, and Zn. MNDO fails to reproduce intermolecular hydrogen bonds and so is unsuitable for many biological problems. AM1 partly corrects this failing.

In 1989, Stewart reparametrized MNDO to give MNDO-PM3 (MNDO parametric method 3; 1 and 2 are MNDO and AM1), which substantially reduces the errors of MNDO and AM1 in calculating heats of formation without loss of accuracy in molecular geometry and dipole moments [J. J. P. Stewart, *J. Comput. Chem.*, **10**, 209, 221 (1989); **11**, 543 (1990)]. MNDO-PM3 (also called simply *PM3*) is parametrized for H, C, Si, Ge, Sn, Pb, N, P, As, Bi, O, S, Se, Te, F, Cl, Br, I, Al, Be, Mg, Zn, Cd, and Hg. For a sample of 713 compounds, average absolute errors in $\Delta H^\circ_{f,298}$ in kcal/mol are 22.5 for MNDO, 13.8 for AM1, and 8.2 for MNDO-PM3 (Stewart, op. cit.).

The MINDO/3, MNDO, AM1, and PM3 methods are contained in the computer program package MOPAC, and MINDO/3, MNDO, and AM1 are contained in the AMPAC program. MOPAC, AMPAC, and many other semiempirical and ab initio quantum-chemistry programs can be purchased from the Quantum Chemistry Program Exchange, Chemistry Department, Indiana University, Bloomington, Indiana 47401.

In 1980, Jug and Nanda published the SINDO1 (symmetrically orthogonalized INDO) method [D. N. Nanda and K. Jug, *Theor. Chim. Acta*, **57**, 95 (1980); Jug and Nanda, *Theor. Chim. Acta*, **57**, 107, 131 (1980); K. Jug, R. Iffert, and J. Schulz, *Int. J. Quantum. Chem.*, **32**, 265 (1987)]. This semiempirical SCF MO method is based on the INDO approximation and (like Dewar's methods) uses parameters designed to reproduce molecular binding energies and molecular geometries, rather than results of ab initio calculations. Rather than using ordinary AOs as basis functions, SINDO1 calculates certain integrals using symmetrically orthogonalized AOs (Section 16.4) and uses a pseudopotential to eliminate the inner-shell electrons. SINDO1's accuracy is similar to that of MNDO, average errors being 8 kcal/mol in binding energies, $2\frac{1}{2}°$ in bond angles, 0.02 Å in bond lengths, 0.4 D in dipole moments, and 0.8 eV in ionization potentials for molecules containing only hydrogen and first-row atoms.

The PCILO Method. The PCILO (perturbative configuration interaction using localized orbitals) method was developed in 1969 by a group of French workers (see J. P. Malriu in *Segal*, Part A, Chapter 3). Unlike most other semiempirical MO methods, the PCILO method does not find SCF MOs. Instead, one chooses a reasonable set of localized bonding, lone-pair, and antibonding orbitals for the molecule and uses these localized orbitals to construct CSFs for a CI treatment. Perturbation theory is used to calculate the energy. To simplify the calculations, one uses the CNDO approximations to evaluate integrals (sometimes the INDO approximations are used). The method gives an energy at least as accurate as that of the CNDO SCF method and is much faster than CNDO. The PCILO method has been extensively applied to calculate conformations of biological molecules.

16.6   THE MOLECULAR-MECHANICS METHOD

The *molecular-mechanics* (MM) *method* [also called the *empirical force field* (EFF) *method*] is quite different from the semiempirical methods of the last section. The molecular-mechanics method is not really a quantum-mechanical method at all, since it does not deal with an electronic Hamiltonian or wave function. Instead, the method starts out with a model of a molecule as composed of atoms held together by bonds; provision is also made for interactions between nonbonded atoms. The method was developed by Westheimer, Hendrickson, Wiberg, Allinger, Warshel, and others, and is applicable to the ground electronic states of organic and organometallic compounds.

Molecular mechanics deals with the contributions to a molecule's electronic energy due to bond stretching, bond bending, van der Waals attractions and repulsions between nonbonded atoms, electrostatic interactions due to polar bonds, and energy changes accompanying internal rotation about single bonds.

Summing these contributions, we have as the potential energy V for motion of the nuclei within the molecule

$$V = V_{\text{str}} + V_\theta + V_{\text{vdW}} + V_{\text{es}} + V_\omega$$

The potential energy V_{str} of bond stretching is usually taken as a quadratic function of the displacement of each bond length l_i from its expected equilibrium length (*natural* length) $l_{i,0}$; we have $V_{\text{str}} = \frac{1}{2} \Sigma_i k_{s,i} (l_i - l_{i,0})^2$, where $k_{s,i}$ is the force constant for stretching bond i and the sum goes over all the bonds. (Recall from Section 4.3 that for small displacements of the nuclei from their equilibrium positions V_{str} is approximately a quadratic function of $l_i - l_{i,0}$.) For the bond-bending contribution, one writes $V_\theta = \frac{1}{2} \Sigma_i k_{\theta,i} (\theta_i - \theta_{i,0})^2$, where θ_i, $\theta_{i,0}$, and $k_{\theta,i}$ are the angle, the expected equilibrium angle, and the bending force constant for bond angle i. The expected lengths and angles $l_{i,0}$ and $\theta_{i,0}$ are taken from known equilibrium geometries of small unstrained molecules. (For example, carbon–carbon single-bond lengths are typically 1.53 Å.) The energy of van der Waals interactions, V_{vdW}, is written as the sum of interactions between pairs of nonbonded atoms; each pair interaction is taken as the sum of a long-range attraction and a short-range repulsion; the Lennard-Jones 6–12 potential $a/R^{12} - b/R^6$ (where a and b are constants and R is the interatomic distance) is often used. The electrostatic term V_{es} is often modeled as the sum of electrostatic interactions between empirically determined fractional point charges Q_α located at the nuclei: $V_{\text{es}} = \Sigma_\alpha \Sigma_{\beta > \alpha} Q_\alpha Q_\beta / R_{\alpha\beta}$, where $R_{\alpha\beta}$ is the distance between nuclei α and β. An alternative is to place an electric dipole moment of empirically determined magnitude (and infinitesimal separation between the dipole's charges) at the center of each polar bond, and write V_{es} as the sum of interactions between these dipoles. For saturated hydrocarbons, V_{es} is usually omitted. The term V_ω is present in molecules with internal rotation about single bonds. For ethane, which has a three-fold rotational barrier, V_ω is approximated as $\frac{1}{2} V_0 (1 - \cos 3\omega)$, where V_0 is the barrier height and ω is the angle of torsion around the carbon–carbon single bond. For increased accuracy, additional terms involving such things as interactions between bond stretching and bending are included. For molecules with small rings, a special treatment is needed to account for the steric strain.

The force constants and other parameters in V are chosen so as to give a good fit to the known geometries, energies, and vibrational spectra of small molecules. Results of high-level ab initio calculations may be used to help determine the parameters.

To do a molecular-mechanics calculation, one uses Dreiding models to find plausible conformations of the molecule. One inputs the atomic coordinates of one of these plausible conformations into a computer, which then uses a molecular-mechanics program to calculate V and its first and second derivatives for this initial guess. Using a minimization technique called the Newton–Raphson method (*Burkert and Allinger*, pages 64–72), the program varies the molecular structure until V is minimized. One then repeats the process for each other plausible conformation. The program thus gives the geometries and energies of the various conformations that correspond to local minima in V.

Very large molecules with many single bonds have a huge number of possible conformations. [For example, for a polypeptide or protein, even if one assumes that the conformation of each peptide bond is fixed and ignores the conformations of the amino acid side chains, a molecule with n amino acid residues has $2n$ adjustable dihedral angles and has 3^{2n} possible conformations with local energy minima (*Burkert and Allinger*, page 275). For $n = 50$, this is $3^{100} \approx 10^{48}$ conformations to be examined.] With too many conformations for all of them to be investigated, the true minimum-energy conformation is hard to find.

Instead of using hand-held models to find plausible conformations, one can use an interactive computer-graphics program that builds a three-dimensional picture of the molecular model on a color-monitor screen and that allows one to vary the dihedral angles to generate possible conformations. The program can then automatically input the atomic coordinates of a possible conformation into a molecular-mechanics program, which optimizes the structure.

Molecular-mechanics calculations are very fast (much faster than even semiempirical MO calculations), and molecules with as many as a thousand atoms can be handled.

For molecules with conjugated double bonds, one includes a quantum-mechanical treatment of the π electrons. In the approach of Allinger and Sprague (the MMP2 computer program), the program does a PPP calculation (Section 16.4) on the π electrons using an initially assumed geometry. The PPP MO coefficients are then used to calculate π-electron bond orders p_{rs} [Eq. (16.60)]. The bond-stretching force constants and the expected equilibrium bond lengths $l_{i,0}$ for the conjugated bonds are then calculated from the p_{rs} values with the assumption that these quantities are linearly related to the p_{rs}'s. The calculated force constants and $l_{i,0}$'s are then used in an ordinary molecular mechanics calculation. If the geometry calculated by molecular mechanics differs substantially from that initially assumed for the PPP calculation, the procedure can be repeated until convergence is obtained. A more thorough integration of the molecular-mechanics and PPP methods is the QCFF/PI (quantum-mechanical extension of the consistent force field to π-electron systems) method of Warshel and Karplus, which expresses the molecular energy as the sum of a σ-electron energy calculated from molecular mechanics and a π-electron energy calculated from the PPP equations, and minimizes this energy (see A. Warshel in *Segal*, Part A, Chapter 5).

To apply the molecular-mechanics method, one needs sufficient data to choose values for the parameters. For this reason, the method cannot be applied to novel types of compounds.

Molecular mechanics takes the zero level of energy as corresponding to all bond lengths and bond angles having their customary values l_0 and θ_0, there being no nonbonded van der Waals interactions or internal-rotation interactions. In such a hypothetical state, one can well approximate the molecular binding energy as the sum of empirical bond energies. Therefore, the equilibrium electronic energy U_{eq} of a molecule can be found by combining the molecular-mechanics-calculated equilibrium-geometry energy, called the **steric energy** V_{steric}, with bond energies. For example, for a saturated hydrocarbon with formula $C_{n_C}H_{n_H}$ (where n_C and n_H are the number of C atoms and H atoms), we have $U_{eq} = V_{steric} - n_{CH}b_{CH} - n_{CC}b_{CC}$, where n_{CH} and n_{CC} are the numbers of C—H and C—C single bonds in the molecule, b_{CH} and b_{CC} are C—H and C—C bond energies (which by convention are positive), and the zero level of energy is taken to correspond to separated (gas-phase) atoms $C(g)$ and $H(g)$. In the following discussion, all energies will be on a per mole basis.

For the formation reaction

$$n_C C(\text{graphite}) + \tfrac{1}{2} n_H H_2(g) \rightarrow C_{n_C} H_{n_H}(g)$$

the change in molar equilibrium electronic energy is $V_{\text{steric}} - n_{CH} b_{CH} - n_{CC} b_{CC} - n_C U_C - \tfrac{1}{2} n_H U_{H_2}$, where U_C and U_{H_2} are the molar equilibrium electronic energies of graphite and $H_2(g)$ with respect to the same zero level of energy as above. If we temporarily ignore zero-point vibrational energy, we can take this change in equilibrium electronic energy as equal to the change in standard-state thermodynamic internal energy for the formation reaction at absolute zero:

$$\Delta U^\circ_{f,0} = V_{\text{steric}} - n_{CH} b_{CH} - n_{CC} b_{CC} - n_C U_C - \tfrac{1}{2} n_H U_{H_2}$$

For a saturated hydrocarbon (acyclic or cyclic), it is not hard to see that the numbers of H and C atoms are related to the numbers of C—H and C—C bonds by

$$n_H = n_{CH} \quad \text{and} \quad n_C = \tfrac{1}{4} n_{CH} + \tfrac{1}{2} n_{CC}$$

At temperature T, the change in translational energy for the above formation reaction is $\tfrac{3}{2} RT - \tfrac{1}{2} n_H (\tfrac{3}{2} RT)$, and the change in rotational energy is $\tfrac{3}{2} RT - \tfrac{1}{2} n_H RT$. The relation $\Delta H^\circ_f = \Delta U^\circ_f + \Delta(PV)^\circ = \Delta U^\circ_f + \Delta n_g RT$, where Δn_g is the change in number of moles of gas in the formation reaction, gives $\Delta H^\circ_f = \Delta U^\circ_f + (1 - \tfrac{1}{2} n_H) RT$. Combining all these relations and continuing to ignore the contribution from the change in vibrational energy, we have for the standard enthalpy of formation at temperature T

$$\Delta H^\circ_{f,T} = V_{\text{steric}} - n_{CH} b_{CH} - n_{CC} b_{CC} - (\tfrac{1}{4} n_{CH} + \tfrac{1}{2} n_{CC}) U_C$$
$$- \tfrac{1}{2} n_{CH} U_{H_2} + 4RT - \tfrac{7}{4} n_{CH} RT$$

$$\Delta H^\circ_{f,T} = V_{\text{steric}} - n_{CH}(b_{CH} + \tfrac{1}{4} U_C + \tfrac{1}{2} U_{H_2} + \tfrac{7}{4} RT) - n_{CC}(b_{CC} + \tfrac{1}{2} U_C) + 4RT$$

Defining $a_{CH} \equiv -(b_{CH} + \tfrac{1}{4} U_C + \tfrac{1}{2} U_{H_2} + \tfrac{7}{4} RT)$ and $a_{CC} \equiv -(b_{CC} + \tfrac{1}{2} U_C)$, we have

$$\Delta H^\circ_{f,T} = V_{\text{steric}} + n_{CH} a_{CH} + n_{CC} a_{CC} + 4RT \qquad (16.91)$$

One uses Eq. (16.91) to determine a_{CH} and a_{CC} by a least-squares fit to 25°C experimental ΔH°_f data for several gas-phase hydrocarbons. (Typically, $a_{CH} \approx -4\tfrac{1}{2}$ kcal/mol and $a_{CC} \approx 2\tfrac{1}{2}$ kcal/mol.) Then Eq. (16.91) can be used to find the gas-phase $\Delta H^\circ_{f,298}$ for any saturated hydrocarbon from its steric energy calculated by molecular mechanics. A similar derivation gives an analogous equation for other kinds of compounds. In arriving at (16.91), the contributions of vibrational energy were ignored. It is assumed that these contributions are allowed for when a_{CC} and a_{CH} are fit to 25°C ΔH°_f data. If more than one conformation is significantly populated at 25°C, one uses $\Delta H^\circ_f = \Sigma_i x_i \Delta H^\circ_{f,i}$, where x_i is the mole fraction of conformation i as calculated using the enthalpy and entropy differences between conformations, and where $\Delta H^\circ_{f,i}$ is calculated from (16.91) for each conformation. In practice, the accuracy of (16.91) is improved by including several correction terms; see Problem 16.32.

The success of molecular mechanics in predicting ΔH_f° is extraordinary. For hydrocarbons, the errors in ΔH_f° generally run 0 to 1 kcal/mol, which is the same magnitude as the experimental errors in the ΔH_f° data. For alcohols and ethers, the molecular-mechanics ΔH_f° errors run 0 to $1\frac{1}{2}$ kcal/mol, and for carbonyl compounds they run $\frac{1}{4}$ to 2 kcal/mol (*Burkert and Allinger*, pages 182–184).

Molecular mechanics generally gives accurate molecular structures. Bond lengths are usually within 0.01 Å and bond angles within 3° of experimental values, and conformations are usually well predicted.

The molecular-mechanics programs MM2 and MMP2 of Allinger and co-workers are widely used, and an improved version MM3 has been published for hydrocarbons [N. L. Allinger, Y. H. Yuh, and J.-H. Lii, *J. Am. Chem. Soc.*, **111**, 8551 (1989); Lii and Allinger, *J. Am. Chem. Soc.*, **111**, 8566, 8576 (1989); Allinger et al., *J. Comput. Chem.*, **11**, 848, 868 (1990)]. The molecular-mechanics programs AMBER [P. K. Weiner and P. A. Kollman *J. Comput. Chem.*, **2**, 287 (1981); S. J. Weiner, P. A. Kollman et al., *J. Am. Chem. Soc.*, **106**, 765 (1984)] and CHARMM [B. R. Brooks et al., *J. Comput. Chem.*, **4**, 187 (1983)] were developed to deal with proteins and nucleic acids. YETI is a molecular-mechanics program especially for metalloproteins [A. Vedani and D. W. Huhta, *J. Am. Chem. Soc.*, **112**, 4759 (1990)]. Minimum-energy geometries obtained for the same peptide using different molecular-mechanics programs "show considerable differences" [F. A. Momany et al., *J. Comput. Chem.*, **11**, 654 (1990)]. For applications of molecular mechanics to the study of atomic motions in proteins and nucleic acids, see J. A. McCammon and S. C. Harvey, *Dynamics of Proteins and Nucleic Acids*, Cambridge University Press, New York, 1987. For a review of molecular mechanics, see *Burkert and Allinger*. MMP2 is reviewed in J. T. Sprague et al., *J. Comput. Chem.*, **8**, 581 (1987).

Many molecular-mechanics-based computer-graphics programs have been developed for molecular modeling. For example, the program MacroModel provides for assembly of structures, generation of possible conformations, molecular-mechanics calculations, and simulation of atomic motions for organic and bioorganic molecules [F. Mohamadi et al., *J. Comput. Chem.*, **11**, 440 (1990)].

16.7   CHEMICAL REACTIONS

Because of the difficulties involved in ab initio calculations of potential-energy surfaces, it would be desirable to have a semiempirical method that gave reliable reaction-surface results. An analysis of MNDO results for 24 simple organic reactions found that MNDO usually gave a realistic transition-state structure [S. Schröder and W. Thiel, *J. Am. Chem. Soc.*, **107**, 4422 (1985)], with average absolute deviations from ab initio SCF results of 0.06 Å in bond lengths, 8° in bond angles, and $11\frac{1}{2}°$ in dihedral angles. However, the calculated MNDO barrier heights were not accurate, with an average absolute deviation of 22 kcal/mol from ab initio MP4 or MP3 results. MNDO was suggested as being suitable "for fast initial scans of potential surfaces to assess their qualitative features, but it has to be kept in mind that some of these characteristics may change at higher levels" (Schröder and Thiel, op. cit.).

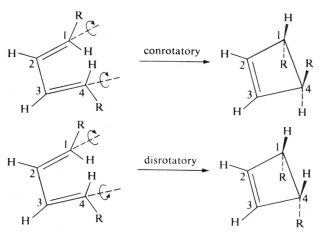

Figure 16.10 Conrotatory and disrotatory cyclizations. All atoms lie in the same plane except those shown with dashed or heavy bonds.

MO theory has been used to draw qualitative conclusions about the course of chemical reactions. The most fruitful applications have come from the **Woodward–Hoffmann rules**, which predict the preferred path and stereochemistry for many important classes of organic reactions. As an example of the application of these rules, we consider the cyclization of a substituted *s-cis*-butadiene to a substituted cyclobutene. There are two possible steric courses the reaction can take, described as *conrotatory* or *disrotatory*, depending on whether the terminal groups rotate in the same or opposite senses as the reaction proceeds. Note the difference in products in Fig. 16.10.

One finds that when the reaction is thermally induced the butadiene cyclization process is conrotatory, but when the reaction is photochemically induced, the process is disrotatory. The simplest approach that explains these facts starts with the assumption that the energy change during the reaction is determined primarily by the energy change in the highest-occupied MO (HOMO) of the substituted *s-cis*-butadiene molecule. This is the π MO ϕ_2 in Fig. 16.2. Figure 16.11 shows that for ϕ_2, a conrotatory motion of the π AOs on carbons 1

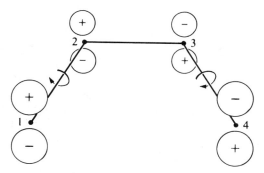

Figure 16.11 Relative phases of AOs in the highest occupied butadiene MO.

and 4 causes the positive lobe of $C_1 2p\pi$ to overlap the positive lobe of $C_4 2p\pi$; this overlap gives a bonding interaction of these AOs and leads to formation of the 1–4 σ bond of the cyclobutene. On the other hand, a disrotatory motion of these two AOs causes overlap of the positive lobe of one π AO with the negative lobe of the other π AO; this gives an antibonding interaction. Thus we expect the conrotatory process to have a lower activation energy than the disrotatory and to be preferred over the disrotatory. As a further check, we note that conrotatory motion leads to an antibonding interaction between the carbon 1 and 4 π AOs in the lowest π MO ϕ_1; this leaves the π AOs of carbons 2 and 3 in ϕ_1 to form the 2–3 π bond in the cyclobutene. (Note that if the butadiene is less symmetrically substituted than in Fig. 16.10, clockwise and counterclockwise conrotatory motions lead to a mixture of two products.)

When the above reaction is induced photochemically, an absorbed photon excites an electron from ϕ_2 to the butadiene π MO ϕ_3 (Fig. 16.2). The highest occupied MO is now ϕ_3, in which the $2p\pi$ AOs on carbons 1 and 4 have the same phases (rather than opposite phases as in ϕ_2) and in which a disrotatory motion produces positive bonding overlap between these AOs. Thus we predict disrotatory ring closure, as observed. The same reasoning correctly predicts the courses of other polyene ring closures (Problem 16.33).

Provided the reactants and products do not differ greatly in energy, a high barrier for the forward reaction implies a high barrier for the reverse reaction (in which the reaction path is traversed in the opposite direction). Thus the above reasoning also applies in determining the course of the reverse, ring-opening, reactions.

Rather than considering only the energy change in the HOMO of butadiene, one can use a less approximate approach that looks at the energy changes in all MOs (occupied or unoccupied) that are involved in the bonds being broken or formed and that uses symmetry to correlate the MOs of the reactant(s) with the MOs of the product(s); see *Lowe*, Section 14-9.

For further details of the Woodward–Hoffmann rules and their application to a wide variety of organic reactions, see R. B. Woodward and R. Hoffmann, *Angew. Chem. Intern. Ed.*, **8**, 781 (1969); *The Conservation of Orbital Symmetry*, Academic Press, New York, 1970; R. E. Lehr and A. P. Marchand, *Orbital Symmetry*, Academic Press, New York, 1972.

Pearson has applied orbital symmetry concepts to inorganic reactions. As two reactant molecules approach, electrons begin to flow from the HOMO of one to the lowest unoccupied MO (LUMO) of the other. These two MOs are called **frontier orbitals**. For a low activation energy, we require a positive overlap between these two MOs.

An example is the $H_2 + F_2 \rightarrow 2HF$ reaction. Consider a proposed mechanism in which the two molecules collide broadside to give a four-center transition state. The HOMO of H_2 is the $\sigma_g 1s$ MO; the LUMO of F_2 is the $\sigma_u^* 2p$ MO (Sections 13.6 and 13.7). Flow of electrons out of H_2 $\sigma_g 1s$ toward F_2 $\sigma_u^* 2p$ would lead to breaking of the H—H bond and formation of two H—F bonds. However, Figure 16.12a shows that these two MOs do not have a positive overlap. Hence electron flow from H_2 to F_2 is forbidden by symmetry. Figure 16.12b shows the HOMO of F_2 and LUMO of H_2. Here there is a positive overlap, but flow of

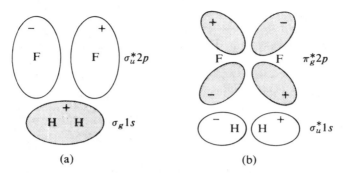

Figure 16.12 HOMOs and LUMOs for H_2 and F_2. Occupied MOs are shaded.

electrons out of the antibonding π_g^* MO would strengthen rather than weaken the F—F bond. We conclude that this bimolecular one-step mechanism has a high activation energy and is not favored. (The same reasoning applies to the famous $H_2 + I_2$ reaction.)

Further applications of orbital symmetry to inorganic reactions may be found in R. G. Pearson, *Chem. Eng. News*, Sept. 28, 1970, page 66; *Acc. Chem. Res.*, **4**, 152 (1971); *J. Am. Chem. Soc.*, **94**, 8287 (1972); *Symmetry Rules for Chemical Reactions*, Wiley, New York, 1976.

The frontier-orbital approach sometimes fails. For a scathing criticism of the frontier-orbital theory, see M. J. S. Dewar, *THEOCHEM*, **59**, 301 (1989).

PROBLEMS

16.1 For the ions (16.13), observed longest-wavelength electronic absorption bands occur at 224, 312.5, 416, 519, 625, 734.5, and 848 nm for $k = 0$, 1, 2, 3, 4, 5, and 6, respectively. Compare these values with those given by the FE MO equation (16.14) and calculate the average absolute error.

16.2 Use the FE MO method to estimate the longest-wavelength electronic transition of (a) CH_2=CH—CH=CH—CH_2^-; (b) CH_2=CH—CH=CH—CH_2^+.

16.3 The FE MO model usually takes the π electrons of a conjugated polyene as moving in a one-dimensional box. A better picture would be a long, thin, rectangular three-dimensional box. Answer the following questions for a three-dimensional-box model. (a) What restriction must be imposed on the quantum numbers of an orbital to give it the required π symmetry? (b) Show that this model still gives the expression (16.9) for the longest-wavelength transition.

16.4 Apply the FE MO model to benzene by solving the Schrödinger equation for a particle confined to move on a circle. (There is only one variable, the angle ϕ.) Draw the pattern of π-electron FE MO energies and compare to the Hückel method. Pick a reasonable value for the radius and calculate the position of the longest wavelength band of benzene; compare with the experimental value of 204 nm.

16.5 For the allyl radical $\cdot CH_2$—CH=CH_2, find (a) the HMOs and energies; (b) the mobile bond orders; (c) the π-electron charges; (d) the free valences; (e) the delocalization energy.

16.6 Calculate the quantities (a) through (e) of Problem 16.5 for the allyl cation and anion, $[CH_2CHCH_2]^+$ and $[CH_2CHCH_2]^-$. Which ion is predicted to be more stable?

16.7 For the polyenes (16.10), observed longest-wavelength electronic absorption bands are at 162.5, 217, 268, 304, 334, 364, 390, 410, and 447 nm for $k = 0$, 1, 2, 3, 4, 5, 6, 7, and 9, respectively. (a) Compare these values with those given by the FE MO equation (16.12) and calculate the average absolute percent error. (b) Do the same using the Hückel equation (16.37).

16.8 Verify the butadiene HMO coefficients for ϕ_2, ϕ_3, and ϕ_4 in (16.33).

16.9 (a) Verify that the HMO coefficients (16.36) satisfy the HMO set of simultaneous equations. [*Hint:* Use the identity $\sin a + \sin b = 2 \sin \frac{1}{2}(a + b) \cos \frac{1}{2}(a - b)$.] (b) Verify that the coefficients (16.36) give a normalized HMO. To evaluate the needed sum, express the sine function as exponentials and then use the formula for the sum of a geometric series.

16.10 Calculate the HMO total bond orders and the π-electron densities for the lowest excited state of butadiene.

16.11 Since only the topology of the carbon framework is of significance in the HMO method, the full symmetry of *s-trans*-butadiene need not be used to get the maximum simplification possible in the HMO method; instead, it is sufficient to use only the C_2 axis. (a) Write down the two possible symmetry species for the group $\mathscr{C}_2$. (b) Construct π-electron symmetry orbitals for butadiene, classifying them according to the symmetry species of $\mathscr{C}_2$. (c) Set up and solve the two Hückel secular equations for butadiene using the symmetry orbitals of (b) as basis functions.

16.12 Verify that the geometric construction of Fig. 16.5 gives the correct HMO energies of the cyclic polyene C_nH_n.

16.13 (a) Calculate the energy needed to compress three carbon–carbon single bonds and stretch three carbon–carbon double bonds to the benzene bond length 1.397 Å. Assume a harmonic-oscillator potential-energy function for bond stretching and compression. Typical carbon–carbon single- and double-bond lengths are 1.53 and 1.335 Å; typical stretching force constants for carbon–carbon single and double bonds are 5 and 9.5 millidynes/Å. (b) Use the result of (a) to calculate an improved β value for benzene from the data following Eq. (16.57).

16.14 Typical bond energies in kcal/mol are 99 for C—H, 83 for C—C, and 146 for C=C. The heat of formation of gas-phase benzene from six hydrogen and six carbon atoms is -1323 kcal/mol. Calculate the "experimental" delocalization energy of benzene using these data, first omitting the strain-energy correction and then including it. (Because bond energies vary from compound to compound, this procedure is very rough.)

16.15 (a) Calculate the quantities (a) through (e) of Problem 16.5 for the diradical trimethylenemethane $C(CH_2)_3$. (The degenerate MOs can be taken as either real or complex, similar to the benzene degenerate MOs.) (b) Do the same as in (a) for the propargyl diradical HC=C=CH. Save time by using the results of Problem 16.5. Note that this linear molecule has two sets of spatially perpendicular π MOs.

16.16 The ionization energies of the first few polyacenes are 9.4 eV for benzene, 8.3 eV for naphthalene, 7.6 eV for anthracene, and 7.0 eV for tetracene. Use these data to calculate values of α and β in the HMO method. Compare the result with the β value in (16.55). Predict the ionization energy of pentacene. The HMO x values for the highest-occupied MOs of these polyacenes are -1.00, -0.618, -0.414, -0.295, and -0.220.

16.17 Estimate the carbon–carbon bond length in (a) $C_5H_5^-$; (b) $C_7H_7^+$; (c) $C_8H_8^{2-}$.

16.18 (a) Set up and solve the a_u, b_{2g}, and b_{1g} HMO secular equations for naphthalene. (b) Find the coefficients of the lowest naphthalene HMO.

16.19 Derive Eq. (16.63) for E_π. *Hint:* Start with E_π as the sum of orbital energies and use $e_i = \langle \phi_i | \hat{H}^{\text{eff}} | \phi_i \rangle$.

16.20 (a) For P_{rs} equal to 1 and to 3, compare the bond-length predictions of Eq. (16.62) with experimental carbon–carbon single- and triple-bond lengths. (b) Look up (in one of the tabulations mentioned in Section 16.3) the Hückel bond orders of azulene and compare the predicted bond lengths with the experimental values. [The experimental data can be found in R. J. Buenker and S. D. Peyerimhoff, *Chem. Phys. Lett.*, **3**, 37 (1969).]

16.21 For each of the following, calculate the Hess–Schaad resonance energy and the REPE, and predict whether the compound will be aromatic, nonaromatic, or anti-aromatic: (a) planar [8]annulene; (b) planar [18]annulene; (c) azulene (Fig. 16.7), for which the x values in (16.27) of the occupied HMOs are -2.3103, -1.6516, -1.3557, -0.8870, and -0.4773.

16.22 (a) Show that the equations satisfied by the HMO coefficients of a conjugated hydrocarbon are $x_i c_{ri} + \sum_{s \to r} c_{si} = 0$, $r = 1, \ldots, n_C$, where the sum is over carbons bonded to carbon r. (b) If the hydrocarbon is an alternant, then we can divide the carbons into two sets such that carbons in one set are bonded only to carbons in the other set. Verify that for an alternant hydrocarbon, if we replace x_i by $-x_i$ and multiply the coefficients of one set of carbons by -1 in each HMO equation of (a), we obtain equations that are satisfied. Hence (provided $x_i \neq 0$), for each HMO of an alternant hydrocarbon with $(\alpha - e_i)/\beta = x_i$, there is an HMO with $(\alpha - e_j)/\beta = -x_i$, whose coefficients are obtained by multiplying the coefficients of one set of carbons in the first HMO by -1.

16.23 (a) What is the HMO delocalization energy of cyclobutadiene? (b) Calculate the HMO total bond orders in cyclobutadiene.

16.24 Show that $\sum_r q_r = n_\pi$ in the HMO method.

16.25 Write the simple HMO secular determinant for each of the following molecules: (a) phenol; (b) phenanthrene; (c) furan.

16.26 Find the HMO energies for the polymethine ion (16.13) with $k = 0$. Use averages of the values in (16.67); to simplify the work, use symmetry orbitals. What value of β is required to fit the observed lowest-energy transition?

16.27 (a) Show that, with inclusion of overlap between carbons bonded to each other (the Wheland method), the HMO equation for benzene is like (16.38) except that each x is replaced by w, where $w \equiv (\alpha - e_i)/(\beta - Se_i)$. Hence the values of w are the same as those found for x with overlap omitted. Provided a common value of S is used for any two bonded carbons in a molecule, the same situation holds for any planar conjugated hydrocarbon. (b) Verify that, with inclusion of overlap, $e_i = \alpha - w\gamma/(1 - Sw)$, where $\gamma \equiv \beta - S\alpha$. (c) With S taken as 0.25, find the Wheland e_i values for benzene in terms of α and γ. (d) Show that the predicted wave number of the transition between the highest-occupied (HO) and lowest-unoccupied (LU) Wheland π MOs is

$$\frac{1}{\lambda} = \frac{\gamma}{hc} \frac{w_{\text{HO}} - w_{\text{LU}}}{1 + S^2 w_{\text{LU}} w_{\text{HO}} - S(w_{\text{LU}} + w_{\text{HO}})}$$

Explain why $w_{\text{LU}} = -w_{\text{HO}}$ for an alternant hydrocarbon and use this relation to show that for an alternant hydrocarbon

$$1/\lambda = (|\gamma| \, \Delta w/hc)\{1/[1 - \tfrac{1}{4}S^2(\Delta w)^2]\}$$

where $\Delta w \equiv |w_{\text{HO}} - w_{\text{LU}}|$. Δw is identical to Δx calculated without overlap; also, $S = 0.25$ and Δw is typically about 1 or 2, so $\tfrac{1}{4}S^2(\Delta w)^2 \ll 1$. Hence the $1/\lambda$ value is nearly the same as calculated [Eq. (16.53)] without overlap, except that the empirical proportionality constant is interpreted as $|\gamma| = |\beta - S\alpha|$, rather than as $|\beta|$.

16.28 (a) Verify (16.76) for the PPP F_π matrix elements. (b) Verify (16.81).

16.29 Apply the extended Hückel method to H_2. Use $K = 1.5$; use the expression from Chapter 13 of S as a function of R; plot the valence-electron energy as a function of R; compare the predicted value of R_e with the experimental value. (To avoid solving a secular equation, use symmetry orbitals.) Use 1.0 as the orbital exponent.

16.30 (a) Set up the 8×8 extended Hückel secular determinant for methane at the equilibrium configuration; do not use symmetry orbitals. Take $K = 1.5$; use (16.88); use Slater's rules (Problem 15.47) for the orbital exponents; evaluate the overlap integrals from the reference of Problem 15.20. (b) Do the same as in (a), but now use symmetry orbitals.

16.31 Let the AOs r and s be on the same atom in a molecule and let the AO t be on a different atom. Which of the following integrals are neglected in the CNDO method? Which are neglected in the INDO method? (a) $(rr|ss)$; (b) $(rs|rs)$; (c) $(rr|tt)$; (d) $(rt|rt)$; (e) $(rr|rt)$.

16.32 MM3 parameters in kcal/mol to be used with (16.91) to calculate $\Delta H^\circ_{f,298}$ values of saturated hydrocarbons are $a_{CH} = -4.590$ and $a_{CC} = 2.447$; in addition, the following corrections (in kcal/mol) are included: 1.045 for each CH_3 group, -2.627 for each carbon bonded to three other carbons, -6.641 for each carbon bonded to four other carbons, 0.42 for each bond (except C—CH_3 bonds) with a low rotational barrier, -1.780 for each four-membered ring, and -5.508 for each five-membered ring. Given the following MM3 steric energies, calculate the gas-phase $\Delta H^\circ_{f,298}$ of each compound and compare with the experimental values given in parentheses: (a) 2.05 kcal/mol for propane (-24.82 kcal/mol); (b) 3.18 kcal/mol for isobutane (-32.15 kcal/mol); (c) 32.63 kcal/mol for cyclobutane (6.78 kcal/mol). (d) How many conformations must be considered when calculating the MM $\Delta H^\circ_{f,298}$ of butane?

16.33 Consider the thermal reaction 1,3,5-hexatriene $\rightarrow$ 1,3-cyclohexadiene. (a) Use the symmetry of the polyene HOMO to predict whether the reaction path is conrotatory or disrotatory. (The HMOs need not be found explicitly; all that is needed is the signs of the AOs in the HOMO and these can be found by noting that the HMO nodal pattern is the same as the FE MO nodal pattern.) (b) Do the same as in (a) when the reaction occurs photochemically. (c) State the general rules for the cyclization of the polyene (16.10) with n_π π electrons.

16.34 Examine the frontier orbitals and decide whether each of the following elementary reactions should have a high or low activation energy for a four-center broadside-collision reaction path. (a) $H_2 + D_2 \rightarrow 2HD$; (b) $N_2 + O_2 \rightarrow 2NO$; (c) $F_2 + Br_2 \rightarrow 2FBr$; (d) $H_2 + C_2H_4 \rightarrow C_2H_6$.

17

Comparisons of Methods

17.1 COMPARISONS OF METHODS

This section compares the accuracies of some semiempirical and ab initio methods and of molecular mechanics in calculating properties of ground-state closed-shell molecules of first- and second-row atoms (H–Ar).

Molecular Geometry. Ab initio STO-3G SCF MO calculations of molecular geometry give fairly good predictions of bond distances and quite good predictions of bond angles, but occasionally show very large bond-length errors (for example, an error of 0.72 Å for the Na_2 bond length and of 0.23 Å for NaH); STO-3G bond lengths for molecules with only first-row elements are more accurate than for second-row molecules. Successive increases in basis-set size through the series STO-3G, 3-21G, 3-21G$^{(*)}$, and 6-31G* give improved bond-length accuracy. A thorough study (*Hehre* et al., Section 6.2) gave the following average absolute errors in bond lengths and angles, where the errors are listed in the order STO-3G, 3-21G, 3-21G$^{(*)}$, 6-31G*: 0.054 Å, 0.016 Å, 0.017 Å, 0.014 Å and 2.3°, 3.8°, 3.3°, 1.5° for AH_n molecules; 0.082 Å, 0.067 Å, 0.040 Å, 0.030 Å for AB single bonds, 0.027 Å, 0.017 Å, 0.018 Å, 0.023 Å for AB multiple bonds, and 2.0°, 1.7°, 1.8°, 1.5° for bond angles in H_mABH_n molecules; 0.029 Å, 0.029 Å, 0.016 Å, 0.020 Å for bonds between nonhydrogens and 1.3°, 1.0°, 0.9° (no data for 6-31G*) for angles not involving hydrogen in molecules with first- and second-row elements only; 0.125 Å for 3-21G, 0.015 Å for 3-21G$^{(*)}$, and 0.014 Å for 6-31G* for bonds between nonhydrogens in **hypervalent** compounds (compounds for which the Lewis dot formula has more than eight electrons around one or more atoms). Hehre et al. conclude that for geometry predictions, the "relatively small 3-21G basis set (3-21G$^{(*)}$ for molecules incorporating second-row elements) appears to be the method of choice because of its wide applicability to molecules of moderate size."

Even with use of the energy gradient, ab initio SCF calculation of the equilibrium geometry of a medium or large molecule with a large basis set requires a great deal of computer time. Since the 3-21G (or 3-21G$^{(*)}$) basis set works well for geometries but not for energy differences (see below), a common procedure is to optimize the geometry of a compound using the 3-21G basis set and then use the 6-31G* basis set to calculate the energy at the 3-21G geometry.

The notation 6-31G*//3-21G denotes such a procedure, where the // lines mean "at the geometry of."

Dihedral angles are usually calculated reasonably accurately by ab initio SCF MO methods, but the number of comparisons with experiment is small (see J. A. Pople in Schaefer, *Applications of Electronic Structure Theory*, pages 11–12, 16–17). Exceptions are H_2O_2, where the 3-21G basis set predicts a 180° dihedral angle, as compared with the experimental value of 112°, and cyclobutane and cyclopentane, where STO-3G substantially underestimates the nonplanarity. In these three molecules, 6-31G* calculations predict the conformational angles well.

MP2 6-31G* geometries are significantly more accurate than SCF 6-31G* geometries. For example, average absolute errors in a sample of AB bond lengths in $H_m ABH_n$ compounds are reduced from 0.021 Å in HF/6-31G* to 0.013 Å in MP2/6-31G* (*Hehre* et al., pages 156–161).

INDO gives only fair geometries; one study (M. C. Flanigan et al., in *Segal*, Part B, Chapter 1) found average absolute errors of 0.06 Å and 4° in bond lengths and angles. Bond lengths and angles predicted by MINDO/3, MNDO, AM1, PM3, and SINDO1 are generally satisfactory. A comparison for compounds of first-row elements found average absolute errors of 0.02 Å and 5.0° for MINDO/3, 0.025 Å and 2.9° for MNDO, and 0.023 Å and 2.6° for SINDO1 [D. N. Nanda and K. Jug, *Theor. Chim. Acta*, **57**, 95 (1980)]; for compounds of second-row elements, average absolute errors are 0.067 Å and 4.0° for MNDO and 0.037 Å and 2.8° for SINDO1 [K. Jug et al., *Int. J. Quantum Chem.*, **32**, 265 (1987)]. A study of compounds of all the elements for which the methods are parametrized found the following average absolute errors in bond lengths and bond angles: 0.054 Å and 4.3° for MNDO, 0.050 Å and 3.3° for AM1, and 0.036 Å and 3.9° for PM3 [J. J. P. Stewart, *J. Comput. Chem.*, **10**, 221 (1989)].

For 16 dihedral angles, average absolute errors are 21.6° for MNDO, 12.5° for AM1, and 14.9° for PM3 (J. J. P. Stewart, op. cit.). These results are not satisfactory.

Molecular mechanics usually gives good results for bond distances, bond angles, and dihedral angles for kinds of molecules for which the method has been properly parametrized.

Energy Changes.   The failure of ab initio SCF MO calculations to predict molecular dissociation energies might seem to prevent the use of SCF energies to calculate energy changes in chemical reactions. However, for certain types of reactions, we *can* use ab initio SCF energies to estimate the reaction energy. Recall that good barriers to internal rotation can usually be obtained from SCF MO calculations, because of a near cancellation in correlation energies between different molecular conformations. As ethane goes from staggered to eclipsed, the number of chemical bonds of each type does not change. More generally, one might hope that a similar near cancellation of correlation energies might occur for an *isodesmic* chemical reaction; an isodesmic reaction (Greek *isos*, "equal"; *desm*, "bond") is one in which the number of bonds of each type does not change [W. J. Hehre et al., *J. Am. Chem. Soc.*, **92**, 4796 (1970)]. For example, the isodesmic reaction $CH_2{=}CHCH_2OH + CH_2{=}O \rightarrow CH_2{=}CHOH + CH_3CH{=}O$ has seven CH bonds, one CC double bond, one CC single bond, one CO double bond, one CO single bond, and one OH bond on each side. (Throughout this discussion, we shall consider only gas-phase reactions, since this avoids the contributions of the substantial intermolecular interactions in liquids and solids. To convert a calculated gas-phase reaction energy to one involving condensed phases, we use the experimental energies of condensation of substances.)

A special kind of isodesmic reaction is a ***bond-separation reaction***. Here, one starts with a molecule and converts it to products, each of which contains only one bond between nonhydrogen atoms. For example, starting with $CH_3{-}CH{=}C{=}O$, one would form the products $CH_3{-}CH_3$, $CH_2{=}CH_2$, and $CH_2{=}O$, in which the C—C, C=C, and C=O bonds are separated from one another. To balance the reaction, one adds an appropriate number of hydride molecules (for example, CH_4, NH_3, H_2O) to the left side. The bond-separation reaction for CH_3CHCO is then $CH_3{-}CH{=}C{=}O + 2CH_4 \rightarrow C_2H_6 + C_2H_4 + CH_2O$. The bond-separation reaction for benzene is $C_6H_6 + 6CH_4 \rightarrow 3C_2H_6 + 3C_2H_4$. (If the energy of a molecule could be represented as the sum of bond energies that were invariant from molecule to molecule, then the energy change for any isodesmic reaction would be zero. The energy change for a bond-separation reaction measures the interactions between the bonds in the molecule.)

If an SCF MO calculation could predict the energy change for the bond-separation reaction of a large molecule, then we could use the known energies of the small product molecules like C_2H_6 to get a good estimate for the energy of the large molecule from an SCF calculation, without having to use expensive correlation methods.

A study of ab initio bond-separation energies of 36 organic compounds found from SCF MO energies of reactants and products all calculated with the same basis set has been done (*Hehre* et al., pages 302–305). The STO-3G basis set gave results accurate to within 5 kcal/mol for 20 out of 36 compounds but was subject to large errors (10 to 50 kcal/mol) for compounds with small rings or several Cl atoms and for NH_2CHO. The average absolute error for STO-3G was 8.6 kcal/mol. The 3-21G basis set [3-21G$^{(*)}$ for compounds with second-row atoms] gave results accurate to 5 kcal/mol for 22 of 36 compounds, was subject to 10- to 25-kcal/mol errors for compounds with small rings or several Cl or F atoms, and had an average absolute error of 5.9 kcal/mol. 6-31G*//STO-3G calculations gave 5-kcal/mol accuracy for 26 of 36 reactions, showed errors of 4 to 13 kcal/mol for compounds with small rings or several chlorines, and had a 3.3-kcal/mol average absolute error. The minimal STO-3G set is unreliable here, but the 6-31G* set is usually reliable.

As to semiempirical SCF MO methods, a study of 15 bond-separation energies (M. C. Flanigan et al., in *Segal*, Part B, Chapter 1) found that the INDO method gives poor results, with an average absolute error of 27 kcal/mol and errors as high as 95 kcal/mol. Another study [T. A. Halgren et al., *J. Am. Chem. Soc.*, **100**, 6595 (1978)] concluded that INDO and CNDO/2 "contain serious deficiencies which frequently render them unsuitable for use in energy calculations." The MINDO/3 method is accurate to within 10 kcal/mol for two-thirds of the 15 reactions but is subject to 30 kcal/mol errors for reactions involving small rings and is in error by 57 kcal/mol for the C_6H_6 bond separation; this reaction has $6CH_4$ on the left, and MINDO/3 gives a rather poor CH_4 energy. The average absolute MINDO/3 error for the 15 reactions is $14\frac{1}{2}$ kcal/mol. The MNDO method is rather more accurate than MINDO/3 but is subject to 15- to 20-kcal/mol errors for small-ring compounds and compounds containing a CO double bond; the average absolute MNDO error for the 15 reactions is 12 kcal/mol. SINDO1 gives a 14-kcal/mol average absolute error for the 15 reactions.

An important question is the relative energies of structural isomers. The transformation between two isomers is not necessarily an isodesmic reaction (an example is cyclopropane→propene). A study of about 40 ab initio SCF MO calculated isomerization energies of organic compounds (*Hehre* et al., Section 6.5.5) found that the STO-3G basis set gives unreliable results, with errors of 10 to 20 kcal/mol being common and errors as high as 50 kcal/mol occurring. The 3-21G basis set [3-21G$^{(*)}$ for molecules with second-row atoms] did fairly well in some cases, but showed large errors (10–30 kcal/mol) for cyclic compounds and for some compounds of N or O. The 6-31G*//3-21G$^{(*)}$ calculations usually were reliable; only one energy difference was in error by more than 10 kcal/mol, but a few were in error by 5 to 10 kcal/mol. Hehre et al. note that the 6-31G*//3-21G$^{(*)}$ "level of theory is perhaps the simplest available with which to calculate relative isomer energies for a wide variety of systems to reasonable accuracy."

As to semiempirical SCF MO calculations, a study of nine relative isomer energies of small hydrocarbons (M. C. Flanigan et al. in *Segal*, Part B, Chapter 1) found INDO to be completely unreliable, with errors as high as 50 to 100 kcal/mol; MINDO/3 was unreliable, with errors running 5 to 20 kcal/mol.

For a sample of 30 isomerization reactions [those in *Hehre*, Section 6.5.5 for which semiempirical results are given in J. J. P. Stewart, *J. Comput. Chem.*, **10**, 221 (1989)], average absolute errors are 9.1 kcal/mol for MNDO, 7.4 kcal/mol for AM1, and 5.8 kcal/mol for PM3. The largest individual error was 42 kcal/mol for MNDO, 24 kcal/mol for AM1, and 23 kcal/mol for PM3. For the 26 isomerizations for which 6-31G* and PM3 results are given, the average absolute errors are 2.4 kcal/mol for 6-31G*//3-21G$^{(*)}$ and 5.4 kcal/mol for PM3.

In summary, ab initio SCF MO calculations with a 6-31G* or larger basis set usually give isodesmic and isomerization reaction energies accurate to 5 kcal/mol. The STO-3G and 3-21G basis sets are not reliable for such calculations. The INDO and CNDO/2 methods are useless for such calculations. The MINDO/3 method is not reliable here, which is disappointing, since it was parametrized to reproduce ΔH_f° data. MNDO, AM1, and especially PM3 are better than MINDO/3, but clearly inferior to ab initio 6-31G* calculations. Of course, MNDO, AM1, and PM3 can handle much larger molecules than can 6-31G* ab initio calculations.

Except for MINDO, MNDO, AM1, and PM3, which are parametrized to give gas-phase ΔH_{298}° values, a quantum-mechanical calculation of an energy change applies to all the reactants and products at their minima in potential energy. If the quantum-mechanical result is corrected for the zero-point vibrational energies of the molecules, one obtains ΔU_0°, the gas-phase thermodynamic internal energy change at absolute zero. To get the gas-phase ΔH_{298}° value, one must allow for occupation of excited translational, rotational, and vibrational levels; this is readily done if good estimates for the molecular structure and vibrational frequencies are available.

Now consider gas-phase ΔH_f° values. The following average absolute errors in kcal/mol were found for calculated $\Delta H_{f,298}^\circ$ values [J. J. P. Stewart, *J. Comput. Chem.*, **11**, 543 (1990)]. For 607 normal-valent compounds: 13.1 for MNDO, 9.6 for AM1, and 7.3 for PM3; for 106 hypervalent compounds: 75.8 for MNDO, 37.7 for AM1, and 13.6 for PM3; for all 713 compounds: 22.5 for MNDO, 13.8 for AM1, and 8.2 for PM3. For a sample of 127 normal-valent compounds containing second-row atoms, average absolute ΔH_f° errors were 11 kcal/mol for MNDO and 8.7 kcal/mol for SINDO1 (which includes d orbitals for second-row elements); errors for a sample of 15 hypervalent compounds were 129.9 kcal/mol for MNDO and 18.6 kcal/mol for SINDO1 [K. Jug et al., *Int. J. Quantum Chem.*, **32**, 265 (1987)]. The performance of MNDO and AM1 is somewhat disappointing.

If one is willing to admit some empiricism, ab initio SCF MO calculations can be used to accurately calculate gas-phase $\Delta H_{f,298}^\circ$ values without resorting to the device of bond-separation reactions. For the gas-phase reaction atoms $\longrightarrow$ compound, in which a compound is formed from its atoms, the true energy change $\Delta \varepsilon$ is $\Delta \varepsilon = \varepsilon_c - \sum_a n_a \varepsilon_a$, where ε_c is the compound's energy, n_a is the number of a atoms in the compound, and ε_a is the energy of atom a. Ordinarily, one would calculate $\Delta \varepsilon$ by using ε_c and ε_a values calculated with the same basis set: $\Delta \varepsilon_q = \varepsilon_{c,q} - \sum_a n_a \varepsilon_{a,q}$, where the q subscript indicates quantum mechanically calculated values. For ab initio SCF MO calculations, this procedure gives poor results. However, suppose we replace the quantum mechanically calculated atomic energies $\varepsilon_{a,q}$ with empirical parameters e_a, whose values are determined by a least-squares fit to known $\Delta \varepsilon$ values for a sample of typical compounds. This procedure is found to lead to pretty good agreement with experimental $\Delta \varepsilon$ values for other compounds.

Since ΔH_f° values refer to formation from reference-state elements (rather than from atoms), consider the sequence of reactions elements $\longrightarrow$ atoms $\longrightarrow$ compound. For the step elements $\longrightarrow$ atoms, the true energy change is $\sum_a n_a \Delta \varepsilon_{f,a}$, where $\Delta \varepsilon_{f,a}$ is the energy of formation of atom a from reference-state element a. [For example, for carbon, this is the energy change for C(*graphite*) $\longrightarrow$ C(g).] Thus, the estimate of the energy change for the formation reaction elements $\longrightarrow$ compound is $\Delta \varepsilon_f = \varepsilon_{c,q} - \sum_a n_a e_a + \sum_a n_a \Delta \varepsilon_{f,a} = \varepsilon_{c,q} - \sum_a n_a e_a^\#$, where $e_a^\# \equiv e_a - \Delta \varepsilon_{f,a}$. This $\Delta \varepsilon_f$ corresponds to the energy change (with zero-point vibrational energy omitted) at absolute zero. Let us suppose that the terms needed to correct $\Delta \varepsilon_f$ to a $\Delta H_{f,298}^\circ$ value can be allowed for by replacing the empirical parameters $e_a^\#$ with new parameters α_a. We then have

$$\Delta H_{f,298}^\circ = \left(\varepsilon_{c,q} - \sum_a n_a \alpha_a \right) N_A \qquad (17.1)$$

where the Avogadro constant N_A converts from per molecule to per mole energies.

The optimum values of the parameters α_a depend somewhat on the basis set used. For the 6-31G* basis set, parameter values (in hartrees) that work well are

$$\alpha_H = -0.57077, \quad \alpha_C = -37.88449, \quad \alpha_N = -54.46414, \quad \alpha_O = -74.78852$$

For C_6H_6, the ab initio SCF MO 6-31G* energy is -230.70314 hartrees. Using the above parameters, we find the quantity in parentheses in (17.1) to equal 0.02842 hartree. From (6.107) and (13.29), we find that

$$1 \text{ hartree corresponds to } 627.510 \text{ kcal/mol} \qquad (17.2)$$

Thus (17.1) predicts a 6-31G* $\Delta H^\circ_{f,298}$ of 17.8 kcal/mol for gas-phase benzene. The experimental value is 19.8 kcal/mol.

For a sample of 88 compounds, 6-31G* $\Delta H^\circ_{f,298}$ values found from (17.1) had an average absolute error of 4.3 kcal/mol, as compared with an average absolute error of 6.7 kcal/mol for AM1 [M. J. S. Dewar and B. M. O'Connor, *Chem. Phys. Lett.*, **138**, 141 (1987)]. By using different empirical parameters for atoms in different environments (for example, different parameters for H bonded to C, H bonded to O, and so on), one finds that 6-31G* $\Delta H^\circ_{f,298}$ values have an average absolute error of only 1.6 kcal/mol for a sample of 106 compounds [M. R. Ibrahim and P. von R. Schleyer, *J. Comput. Chem.*, **6**, 157 (1985)].

As noted in Section 16.6, the performance of molecular mechanics in predicting $\Delta H^\circ_{f,298}$ values and relative energies of isomers is excellent (0- to 2-kcal/mol errors) for classes of compounds for which the method has been parametrized.

Dipole Moments. The accuracy of ab initio SCF STO-3G and 3-21G dipole moments is only fair. The 3-21G$^{(*)}$ and 6-31G* basis sets give substantially improved results. For a sample of 21 small molecules, average absolute errors were 0.65 D for STO-3G, 0.49 D for 3-21G, 0.34 D for 3-21G$^{(*)}$, and 0.30 D for 6-31G* (*Hehre* et al., Section 6.6.1). CNDO/2 and INDO dipole moments often show very substantial errors, being two or three times too large for boron hydrides [T. A. Halgren et al., *J. Am. Chem. Soc.*, **100**, 6595 (1978)] and grossly in error for a number of second-row compounds [M. S. Gordon et al., *J. Am. Chem. Soc.*, **100**, 2670 (1978)]. MINDO/3, MNDO, AM1, PM3, and SINDO1 dipole moments are reasonably reliable. One comparison found average absolute errors of 0.35 D for MNDO, 0.4 D for MINDO/3, and 0.4 D for SINDO1 [D. N. Nanda and K. Jug, *Theor. Chim. Acta*, **57**, 95 (1980)]. For molecules containing second-row atoms, average absolute errors were 0.60 D for MNDO and 0.52 D for SINDO1 [K. Jug et al., *Int. J. Quantum Chem.*, **32**, 265 (1987)]. For a sample of 125 compounds, average absolute errors were 0.45 D for MNDO, 0.35 D for AM1, and 0.38 D for PM3 [J. J. P. Stewart, *J. Comput. Chem.*, **10**, 221 (1989)].

Ionization Energies. Molecular (first) ionization potentials are usually estimated as minus the orbital energy of the highest-occupied MO (Koopmans' theorem; Section 15.6). Ab initio SCF MO ionization potentials are reasonably accurate, with typical errors of 1 or $1\frac{1}{2}$ eV, provided a larger than minimal basis set is used. CNDO/2 and INDO ionization potentials are unreliable, with typical errors of 4 or 5 eV [T. A. Halgren et al., *J. Am. Chem. Soc.*, **100**, 6595 (1978)]. The MINDO/3, MNDO, AM1, PM3, and SINDO1 methods give accurate ionization potentials, with average absolute errors in the range 0.5 to 0.8 eV.

Vibrational Frequencies. As noted in the gradient discussion in Section

15.10, SCF MO calculation of molecular vibrational frequencies is readily done. The STO-3G basis set does not perform satisfactorily here. Vibrational frequencies calculated with the $3\text{-}21G^{(*)}$ or $6\text{-}31G^*$ basis sets show average absolute errors of about 10%, usually being higher than the experimental frequencies; $MP2/6\text{-}31G^*$ vibrational frequencies show average absolute errors of 5% (*Hehre et al.*, Section 6.3).

The accuracy of MINDO/3 and MNDO vibrational frequencies is only fair. Group vibrational frequencies show average absolute errors of 13% for MINDO/3 and 16% for MNDO, with frequencies for torsion about C—C and C—O single bonds in error by 40% and 25%, respectively [M. J. S. Dewar et al., *J. Mol. Struct.*, **43**, 135 (1978)].

Vibrational frequencies are readily calculated in molecular-mechanics programs. The MM2 program does not give accurate vibrational frequencies, but MM3 gives vibrational frequencies of saturated hydrocarbons with a typical accuracy of $\frac{1}{2}$% to 3%, although some frequencies are in error by as much as 12% [J. H. Lii and N. L. Allinger, *J. Am. Chem. Soc.*, **111**, 8566 (1989)].

Entropies. The entropy of a gas-phase substance is the sum of translational, rotational, vibrational, and electronic contributions. The translational contribution depends on the molecular mass; the rotational contribution depends on the moments of inertia and hence on the molecular structure; the vibrational contribution depends on the molecular vibrational frequencies; the electronic contribution depends only on the ground-state electronic degeneracy (provided no excited electronic states are significantly occupied). Ab initio SCF MO calculations with a $3\text{-}21G^{(*)}$ or larger basis set give accurate geometries; the calculated vibrational frequencies are subject to 10% errors, but only the low-frequency vibrations contribute to the room-temperature entropy. Small molecules with no heavy atoms have no or only one low-frequency vibration, so we can expect pretty accurate entropy results for such molecules. Use of the 3-21G and $6\text{-}31G^*$ basis sets to calculate the gas-phase standard-state molar entropies at 300 K of 35 small molecules containing only first-row atoms gave generally good results, with average absolute errors of 0.36 cal/mol-K for HF/3-21G, 0.27 cal/mol-K for $HF/6\text{-}31G^*$, and 0.16 cal/mol-K for $MP2/6\text{-}31G^*$ (*Hehre et al.*, Section 6.3.9). The only outright failure was the H_2O_2 HF/3-21G entropy; recall that HF/3-21G predicts the wrong H_2O_2 dihedral angle. For larger molecules with several single bonds and hence several low-frequency torsional vibrations, the errors will be larger.

MNDO and MINDO/3 are very inaccurate for torsional-vibration frequencies and so will give quite inaccurate entropies for compounds with internal rotation. For a sample of 27 molecules without internal rotations, MINDO/3 gas-phase 298-K entropies had an average absolute error of 0.4 cal/mol-K [M. J. S. Dewar and G. P. Ford, *J. Am. Chem. Soc.*, **99**, 7822 (1977)].

Rotational Barriers. Different conformers of a molecule are separated by an energy barrier to rotation about a single bond. (A *conformer* is a conformation that corresponds to an energy minimum.) A study of rotational energy barriers calculated by ab initio SCF MO calculations with full geometry optimization found the STO-3G and 3-21G basis sets to be quite unreliable, with each basis set showing an average absolute error of 35% and having errors exceeding 50% in 4 out of 13 barriers calculated; performance of the 3-21G$^{(*)}$ basis set was fair, with a 22% average absolute error and 2 out of 13 errors exceeding 50%; the 6-31G* set did fairly well, with an average absolute error of 18% and none of the 11 errors exceeding 50% (*Hehre* et al., Sec. 6.4.1). The remaining errors are probably due mainly to inadequacies of the 6-31G* basis set, rather than the effects of electron correlation. The barrier to ring inversion in cyclobutane calculated with the STO-3G basis set is grossly in error; a 6-31G* calculation gives reasonable agreement with experiment (*Hehre* et al., Section 6.4.1).

Neither MNDO nor AM1 is reliable for barrier heights. One study showed average absolute errors of 50% for MNDO and 38% for AM1 and each method in error by more than 50% in two out of five barriers calculated [M. J. S. Dewar et al., *J. Am. Chem. Soc.*, **107**, 3902 (1985)]. Another study found that MNDO erroneously predicts an energy minimum in several compounds at conformations at which energy barriers exist, and AM1 had an average absolute error of 61% in 11 barrier heights [G. Buemi et al., *THEOCHEM*, **41**, 379 (1988)]. These methods also give far too low energy barriers to ring inversion in cyclobutane.

MM3 gives reasonably accurate rotational barriers in a number of saturated hydrocarbons [N. L. Allinger, Y. H. Yuh, and J.-H. Lii, *J. Am. Chem. Soc.*, **111**, 8551 (1989)].

Energy Differences between Conformers.   A small study of five molecules indicates that ab initio SCF calculations with the STO-3G and 3-21G basis sets are very unreliable for conformer energy differences; four out of five STO-3G errors and four out of five 3-21G errors exceeded 0.5 kcal/mol, and each basis set had an average absolute error of 1.0 kcal/mol (*Hehre* et al., p. 268). Since the average energy difference between conformers is only 1.8 kcal/mol for these five molecules, the 1.0-kcal/mol average error is quite substantial. The 6-31G* set does better, with four out of five errors not exceeding 0.5 kcal/mol and an average absolute error of 0.7_4 kcal/mol. These three basis sets did correctly predict which was the more stable conformer, with the exception of the 3-21G calculation on acrolein. In certain cases, even rather large basis sets may fail. For example, for 1,2-difluoroethane, 6-31G* and 6-31G** ab initio SCF calculations incorrectly predict the *anti* conformer to be more stable than the *gauche* conformer, and basis sets larger than 6-31G** are needed to correctly predict the more stable conformer; but such large sets still substantially underestimate the energy difference [L. Radom et al., *J. Mol. Struct.*, **126**, 271 (1985)]. Radom et al. conclude that for conformational analysis, "results obtained at simple levels of theory are not always reliable. Indeed, for some problems, calculations must be performed at unexpectedly high levels before satisfactory agreement with experiment is achieved."

Despite some early successes, it is now realized (as noted in Section 16.5) that the extended Hückel method is unreliable for conformational studies.

The CNDO and INDO methods are unreliable for conformational work [A. Veillard, *Chem. Phys. Lett.*, **33**, 15 (1975); O. Gropen and H. M. Seip, *Chem. Phys. Lett.*, **11**, 445 (1971); L. L. Combs and M. Holloman, *J. Phys. Chem.*, **79**,

512 (1975); A. R. Gregory and M. N. Paddon-Row, *J. Am. Chem. Soc.*, **98**, 7521 (1976)].

The PCILO method is more reliable than CNDO for conformational work [D. Perahia and A. Pullman, *Chem. Phys. Lett.*, **19**, 73 (1973)] and has given good results in calculations of conformations of some biological molecules [B. Pullman, *Adv. Quantum Chem.*, **10**, 251 (1977)]. A study of PCILO versus ab initio STO-3G conformations of three amino acids [P. R. Lawrence and C. Thomson, *Theor. Chim. Acta*, **58**, 121 (1981)] found that all conformations corresponding to energy minima in the STO-3G calculation were also minima in the PCILO calculation; however, PCILO predicted minima for certain conformations that were not minima in the STO-3G calculation. The study concluded that "while PCILO does not reproduce exactly the results of *ab initio* calculations, a preliminary study of a molecule using PCILO may be very useful [as a starting point for more accurate calculations]."

The MINDO/3 and MNDO methods are unreliable for conformational analysis. For five conjugated molecules, MINDO/3 conformer energy differences differed greatly from experiment and MINDO/3 predicted the wrong conformer to be the most stable one in three out of five cases; also MINDO/3 predicted the existence of four conformers that actually correspond to energy maxima [H. Dodziuk, *J. Mol. Struct.*, **55**, 107 (1979)]. Likewise, MNDO frequently predicted the most stable conformation to be one that in reality corresponds to an energy maximum [G. Buemi et al., *THEOCHEM*, **41**, 379 (1988)].

The AM1 method gives mixed results for conformational analysis. For 1,3-dichloropropane conformers, AM1 (and MNDO) predicted the wrong energy differences and the wrong order of energy differences, whereas ab initio SCF 6-31G* calculations gave the correct order and the correct energy differences, while MM2 gave the correct order but the wrong energy differences [A. H. Holder and D. L. Wertz, *J. Comput. Chem.*, **9**, 684 (1988)]. For most compounds studied, AM1 (like MNDO) does poorly at predicting the magnitude of energy differences between conformers, but (unlike MNDO) usually does correctly predict which is the most stable conformer [W. M. F. Fabian, *J. Comput. Chem.*, **9**, 369 (1988)].

Hydrogen Bonding.   Hydrogen bonding is important throughout chemistry and biochemistry. Many gas-phase hydrogen-bonded dimers have been characterized spectroscopically; examples include $(H_2O)_2$, $(HCl)_2$, $HF-H_2O$, and $HF-HCN$. For $(H_2O)_2$, the structure (determined by molecular-beam microwave spectroscopy) is [T. R. Dyke, K. M. Mack, and J. S. Muenter, *J. Chem. Phys.*, **66**, 498 (1977)]

$$
\begin{array}{c}
\text{H H} \\
\diagdown \blacktriangledown \\
\text{O} \cdots \text{H} - \text{O} \\
\diagdown \\
\text{H}
\end{array}
$$

which can be interpreted as one molecule hydrogen bonding to one of the lone-pair localized orbitals of the second molecule. The oxygen–oxygen separation is 2.98 Å. For other hydrogen-bonded dimer structures, see P. Hobza and R. Zahradnik, *Chem. Rev.*, **88**, 871 (1988), Table 9.

For H-bonded dimers, ab initio SCF MO STO-3G, 3-21G, and 3-21G$^{(*)}$ calculations give equilibrium-geometry separations between the heavy atoms that are usually substantially in error (errors of 0.1 to 0.5 Å); 6-31G* calculations give heavy-atom separations in pretty good agreement with experiment (*Hehre* et al., Table 6.32).

Suppose we want to calculate the dimerization energy of H_2O using the 3-21G basis set. The natural procedure would be to calculate the energy of the dimer $(H_2O)_2$ at its 3-21G equilibrium geometry using a 3-21G basis set on each of the six atoms of $(H_2O)_2$ and to calculate the energy of each H_2O monomer at its 3-21G equilibrium geometry using a 3-21G basis set on each of the three atoms of the monomer, and to take the dimerization energy as

$$\Delta\varepsilon = \varepsilon_{AB}(\{\chi_A\} + \{\chi_B\}) - \varepsilon_A(\{\chi_A\}) - \varepsilon_B(\{\chi_B\}) \qquad (17.3)$$

Here A and B stand for the monomer molecules and AB for the dimer; in this case, A = H_2O, B = H_2O, and AB = $(H_2O)_2$. $\{\chi_A\}$ symbolizes the 3-21G basis set centered on the atoms of A and similarly for $\{\chi_B\}$. $\varepsilon_A(\{\chi_A\})$ is the equilibrium-geometry energy of monomer A calculated with the $\{\chi_A\}$ basis set, with a similar meaning for $\varepsilon_B(\{\chi_B\})$. For $(H_2O)_2$, $\varepsilon_A(\{\chi_A\}) = \varepsilon_B(\{\chi_B\})$, but this is not true for a mixed dimer such as HF–H_2O. The quantity $\varepsilon_{AB}(\{\chi_A\} + \{\chi_B\})$ is the energy of AB calculated with 3-21G basis orbitals on all atoms of AB.

However, this procedure involves an inconsistency. When the monomer energy $\varepsilon_A(\{\chi_A\})$ is calculated, the electrons of A have available to themselves only the 3-21G orbitals on the three atoms of A, whereas, when $\varepsilon_{AB}(\{\chi_A\} + \{\chi_B\})$ is calculated, the electrons of each H_2O molecule within the dimer have available not only the orbitals on their own nuclei, but also the orbitals on the nuclei of the other H_2O molecule. In effect, the dimer basis set is larger than that of each monomer, and this produces an artificial lowering of the dimer energy relative to that of the separated monomers. This artificial lowering is called the **basis-set superposition error** (BSSE). The BSSE would vanish in the limit of using a complete set for each monomer. The most often used procedure to correct for the BSSE is to calculate the dimerization energy as

$$\Delta\varepsilon = \varepsilon_{AB}(\{\chi_A\} + \{\chi_B\}) - \varepsilon_A(\{\chi_A\} + \{\chi_B\}) - \varepsilon_B(\{\chi_A\} + \{\chi_B\}) \quad (17.4)$$

where $\varepsilon_A(\{\chi_A\} + \{\chi_B\})$ is calculated with a basis set that consists of 3-21G orbitals on each nucleus of the monomer A and the appropriate 3-21G orbitals centered at the three points in space that would correspond to the equilibrium positions of the other three nuclei in the dimer. This procedure, called the **counterpoise** (CP) **correction**, has gone through a period of denunciation by many researchers, but is now recognized as probably the best way to reduce the BSSE [P. Hobza and R. Zahradnik, *Chem. Rev.*, **88**, 871 (1988)]. Because of the asymmetry of $(H_2O)_2$, $\varepsilon_A(\{\chi_A\} + \{\chi_B\}) \neq \varepsilon_B(\{\chi_A\} + \{x_B\})$.

From the temperature and pressure dependences of the thermal conductivity of water vapor, ΔH°_{373} for $2H_2O(g) \rightarrow (H_2O)_2(g)$ has been found to be -3.6 ± 0.5 kcal/mol [L. A. Curtiss et al., *J. Chem. Phys.*, **71**, 2703 (1979)]; this ΔH°_{373} is found to correspond to an energy change involving nonvibrating, nonrotating, nontranslating species of $\Delta E_{el} = -5.4 \pm 0.7$ kcal/mol, where theoretically calculated vibrational frequencies of the dimer are used to help find the electronic energy change ΔE_{el}.

Ab initio SCF MO STO-3G, 3-21G, 6-31G*, and 6-31G** calculations give H_2O dimerization electronic energies of -5.9, -11.0, -5.6, and -5.5 kcal/mol, respectively, and counterpoise-corrected dimerization energies of -0.2, -6.2, -4.6_5, and -4.5_5 kcal/mol, respectively [M. J. Frisch et al., *J. Chem. Phys.*, **84**, 2279 (1986)]. SCF calculations with huge basis sets and the CP correction give the dimerization energy in the Hartree–Fock limit as -3.73 ± 0.05 kcal/mol, and use of the MBPT and CC methods to allow for electron correlation leads to a theoretical dimerization energy of $\Delta E_{el} = -4.7 \pm 0.35$ kcal/mol [K. Szalewicz et al., *J. Chem. Phys.*, **89**, 3662 (1988)], in reasonable agreement with the experimental value of -5.4 ± 0.7 kcal/mol. CP-corrected MP2 and CEPA-1 calculations give -4.7 ± 0.3 kcal/mol [R. J. Vos et al., *J. Comput. Chem.*, **11**, 1 (1990)]. (Because of the lack of size consistency, uncorrected CI-SD calculations cannot be used here.)

The CNDO and INDO methods fail to describe hydrogen bonding, predicting that $(H_2O)_2$ has the peroxide structure $H_2O \cdots OH_2$ [M. J. S. Dewar and G. P. Ford, *J. Am. Chem. Soc.*, **101**, 5558 (1979)]. The MINDO/3, MNDO, and SINDO1 methods are unable to properly describe hydrogen bonding, giving dimerization energies of only 0 to -1 kcal/mol for $(H_2O)_2$ and giving *much* too long hydrogen bonds [*Dewar and Ford*, op. cit.; S. Scheiner, *Theor. Chim. Acta*, **57**, 71 (1980); D. N. Nanda and K. Jug, *Theor. Chim. Acta*, **57**, 95 (1980)].

The AM1 method does better than MNDO for hydrogen bonding, but still has serious failings here. One study found AM1 heavy-atom separations in H-bonded species to be 0.1 to 0.9 Å too long, with an average error of 0.44 Å, and found AM1 hydrogen-bond energies to run one-fifth to two-thirds of the experimental values, being on average 57% too low [G. Buemi et al., *THEO-CHEM*, **41**, 379 (1988)]. Another study concluded that "the AM1 method is an unreliable tool for studying systems with H-bonds" [A. A. Bliznyuk and A. A. Voityuk, *THEOCHEM*, **41**, 343 (1988)]. AM1 erroneously predicts $(H_2O)_2$ to have a structure with three hydrogen bonds, rather than the observed structure with one hydrogen bond [O. N. Ventura et al., *THEOCHEM*, **56**, 55 (1989)].

PM3 does give the correct one-H-bond structure for $(H_2O)_2$ [J. J. P. Stewart, *J. Comput. Chem.*, **10**, 221 (1989)], but underestimates the oxygen–oxygen distance by 0.23 Å. The accuracy of PM3 for hydrogen-bond energies is "a little worse than AM1" (J. J. P. Stewart, private communication).

The original version of MM2 contained no special terms in the potential to provide for hydrogen bonding, but relied on the electrostatic interaction term V_{es} to produce hydrogen bonding. This procedure worked fairly well, but gave H-bond energies that were too small by 1 to 3 kcal/mol and gave distances between the two heavy atoms involved in the H bond that were several tenths of an angstrom too long. To overcome these errors, MM2 was modified by the addition of terms specific to hydrogen bonds, and MM2 now gives a good account of hydrogen bonding [N. L. Allinger, *J. Comput. Chem.*, **9**, 591 (1988)].

Computational Times. Molecular-mechanics calculations are substantially faster than semiempirical SCF MO calculations. For example, to calculate the equilibrium geometry of propane, MM2 took only one-tenth the time of MINDO/3 or MNDO (*Clark*, p. 3).

The CNDO, INDO, MINDO/3, MNDO, AM1, and PM3 methods all require roughly the same amount of computer time, which is far less than that for ab initio SCF MO calculations. For an SCF calculation at a single geometry of a medium-size molecule (one with about 40 minimal-basis AOs, for example, B_6H_{10}), approximate relative computational times are 1 for CNDO, and so on, 100 for ab initio STO-3G, 600 for ab initio 4-31G, and 3000 for ab initio 6-31G* [T. A. Halgren et al., *J. Am. Chem. Soc.*, **100**, 6595 (1978); W. J. Hehre, *J. Am. Chem. Soc.*, **97**, 5308 (1975)]. For larger molecules, the time advantage for semiempirical methods becomes even more pronounced. The same is true if geometry optimization is done, because of the rapidity with which the energy gradient is calculated with semiempirical methods.

Conclusion. The overall reliability of the EH, CNDO, and INDO methods for calculating molecular properties is poor.

The ab initio SCF MO method is usually reliable for ground-state, closed-shell molecules, provided one uses a basis set of suitable size (at least 3-21G$^{(*)}$ for geometry and 6-31G* for energy differences) and uses appropriate procedures for finding energy differences. The STO-3G basis set is not generally reliable, and this basis set is little used nowadays.

The use of MP2 perturbation theory substantially improves calculated molecular properties as compared with SCF results. "In molecular calculations, MP2 is taking over now the position of 'the' standard method of computational quantum chemistry held by the SCF method in the late 1960s." [J. Sauer, *Chem. Rev.*, **89**, 199 (1989)].

The AM1 and PM3 semiempirical methods are substantially more reliable than the MINDO/3 and MNDO methods, but significantly less reliable than ab initio SCF calculations with basis sets of suitable size. Of course, AM1 and PM3 calculations are possible for much larger molecules than can be readily treated by ab initio methods.

The molecular mechanics method is usually reliable for those kinds of molecules for which the method has been properly parametrized.

The comparisons of this section consider only compounds of H–Ar. For compounds involving transition metals, ab initio SCF MO calculations often do not give good results; see, for example, *Hehre* et al., Section 6.2.7. The density-functional method may well be useful for transition-metal compounds; see D. R. Salahub and M. C. Zerner, eds., *The Challenge of d and f Electrons* (ACS Symp. Ser. 394), American Chemical Society, Washington, D.C., 1989.

17.2 THE FUTURE OF QUANTUM CHEMISTRY

In the 1950s there was a general belief that meaningful ab initio calculation of molecular properties for all except very small molecules was out of the question. Quantum-chemistry books written in this period contain such statements as "we cannot hope ever to make satisfactory *ab initio* calculations [for organic compounds]" and "It is wise to renounce at the outset any attempt at obtaining precise solutions of the Schrödinger equation for systems more complicated than the hydrogen molecule ion." In 1959, Mulliken and Roothaan identified the "bottleneck" holding up accurate quantum-mechanical calculations on polyatomic molecules as the difficulty in evaluating multicenter integrals. This bottleneck has now been eliminated.

Ab initio SCF calculations and geometry optimizations on moderate-size molecules have become routine, and computationally efficient procedures (for example, MP2) for inclusion of electron correlation are available. The degree of reliability of various quantum-mechanical methods and basis sets has been established by numerous calculations. The size of a molecule for which one can do an accurate ab initio calculation is limited by the speed and storage capacity of the available electronic computers. As larger and faster computers are developed, it will become feasible to treat larger molecules.

The very substantial progress in quantum chemistry in recent years has made quantum-mechanical calculations a valuable tool to help decide a wide variety of questions of real chemical interest. Whereas years ago quantum-mechanical calculations on molecules were largely confined to journals read mainly by theoretical chemists, nowadays such calculations appear in large numbers in the *Journal of the American Chemical Society*, probably the most prestigious and widely read chemistry journal in the world. Quantum chemistry is being applied to such problems as the hydration of ions in solution, surface catalysis, the structures and energies of reaction intermediates, and the conformations of biological molecules. In many cases, theoretical calculations may not give definitive answers, but they are frequently good enough to allow for a very fruitful interaction of theory and experiment. Moreover, qualitative concepts like the Woodward–Hoffmann rules and Walsh diagrams have provided considerable insight into the course of chemical reactions and into chemical bonding.

In 1929, Dirac wrote, "The underlying physical laws necessary for the mathematical theory of . . . the whole of chemistry are thus completely known, and the difficulty is only that the exact application of these laws leads to equations much too complicated to be soluble." Application of high-speed digital computers to quantum chemistry has overcome to a significant degree the difficulties referred to by Dirac. Of course, just a small fraction of chemically important problems have been successfully treated by quantum mechanics, but future prospects are bright.

Ab initio calculations are now routinely used by many chemists as a valuable guide to experimental work and are revolutionizing the way chemistry is done. The future of quantum chemistry and the future of chemistry are inextricably linked.

PROBLEMS

17.1 (a) Write the bond-separation reaction for cyclopropene. (b) Calculated 6-31G* ab initio SCF energies in hartrees are -152.91596 for CH_3CHO, -40.19517 for CH_4,

-79.22875 for C_2H_6, and -113.86633 for H_2CO. The zero-point vibrational energies in hartrees are 0.05298 for CH_3CHO, 0.04320 for CH_4, 0.07214 for C_2H_6, and 0.02567 for H_2CO. Calculate the ΔH_0° predicted by the 6-31G* basis set for the gas-phase CH_3CHO bond-separation reaction and compare with the experimental value of 11.5 kcal/mol.

17.2 Use Eq. (17.1) to estimate the gas-phase $\Delta H_{f,298}^\circ$ of each of the following compounds whose 6-31G* energies in hartrees are given. Compare with the experimental values given in parentheses. (a) Cyclopropane, -117.05887 (12.7 kcal/mol); (b) *s-trans*-1,3-butadiene, -154.91905 (26.2 kcal/mol); (c) H_2O_2, -150.76479 (-32.6 kcal/mol).

Appendix

TABLE A.1 Physical Constants[a]

Constant and Symbol[b]		SI Value	Gaussian Value
Speed of light in vacuum	c	2.99792458×10^{8} m/s	$2.99792458 \times 10^{10}$ cm/s
Proton charge	e	1.602177×10^{-19} C	
	e'		4.803207×10^{-10} statC
Permittivity of vacuum	ε_0	$8.8541878 \times 10^{-12}$ C^2/N-m^2	
Avogadro constant	N_A	6.02214×10^{23} mol^{-1}	6.02214×10^{23} mol^{-1}
Electron rest mass	m_e	9.10939×10^{-31} kg	9.10939×10^{-28} g
Proton rest mass	m_p	1.672623×10^{-27} kg	1.672623×10^{-24} g
Neutron rest mass	m_n	1.674929×10^{-27} kg	1.674929×10^{-24} g
Planck constant	h	6.62608×10^{-34} J s	6.62608×10^{-27} erg s
Faraday constant	F	96485.3 C/mol	
Permeability of vacuum	μ_0	$4\pi \times 10^{-7}$ N C^{-2} s^2	
Bohr radius	a_0	5.291772×10^{-11} m	0.5291772×10^{-8} cm
Bohr magneton	β_e	9.27402×10^{-24} J/T	
Nuclear magneton	β_N	5.05079×10^{-27} J/T	
Electron g value	g_e	2.0023193044	2.0023193044
Proton g value	g_p	5.585695	5.585695
Gas constant	R	8.3145 J/mol-K	8.3145×10^{7} erg/mol-K
Boltzmann constant	k	1.38066×10^{-23} J/K	1.38066×10^{-16} erg/K
Gravitational constant	G	6.673×10^{-11} m^3/kg-s^2	6.673×10^{-8} cm^3/g-s^2

[a]Adapted from E. R. Cohen and B. N. Taylor, *Rev. Mod. Phys.*, **59**, 1121 (1987).
[b]$F = N_A e$, $e' = e/(4\pi\varepsilon_0)^{1/2}$, $a_0 = \hbar^2/m_e e'^2 = 4\pi\varepsilon_0\hbar^2/m_e e^2$, $\beta_e = e\hbar/2m_e$, $\beta_N = e\hbar/2m_p$, $\hbar = h/2\pi$, $k = R/N_A$.

TABLE A.2 Energy Conversion Factors[a]

1 erg $= 10^{-7}$ J
1 cal $= 4.184$ J
1 eV $= 1.602177 \times 10^{-19}$ J $= 1.602177 \times 10^{-12}$ erg $\;\hat{=}\; 23.0605$ kcal/mol
1 hartree $= 4.35975 \times 10^{-18}$ J $= 27.2114$ eV $\;\hat{=}\; 627.510$ kcal/mol

[a]The symbol $\hat{=}$ means "corresponds to."

TABLE A.3   Relative Isotopic Masses

Isotope	Atomic Mass	Isotope	Atomic Mass
^{1}H	1.0078250	^{16}O	15.994915
^{2}H	2.014102	^{32}S	31.972071
^{12}C	12.000 . . .	^{35}Cl	34.968853
^{13}C	13.003355	^{37}Cl	36.965903
^{14}N	14.003074	^{127}I	126.90447

TABLE A.4   Greek Alphabet

Alpha	A	α	Iota	I	ι	Rho	P	ρ
Beta	B	β	Kappa	K	κ	Sigma	Σ	σ
Gamma	Γ	γ	Lambda	Λ	λ	Tau	T	τ
Delta	Δ	δ	Mu	M	μ	Upsilon	Y	υ
Epsilon	E	ε	Nu	N	ν	Phi	Φ	ϕ
Zeta	Z	ζ	Xi	Ξ	ξ	Chi	X	χ
Eta	H	η	Omicron	O	o	Psi	Ψ	ψ
Theta	Θ	θ	Pi	Π	π	Omega	Ω	ω

TABLE A.5   Integrals

$$\int x \sin bx \, dx = \frac{1}{b^2} \sin bx - \frac{x}{b} \cos bx \tag{A.1}$$

$$\int \sin^2 bx \, dx = \frac{x}{2} - \frac{1}{4b} \sin (2bx) \tag{A.2}$$

$$\int x \sin^2 bx \, dx = \frac{x^2}{4} - \frac{x}{4b} \sin (2bx) - \frac{1}{8b^2} \cos (2bx) \tag{A.3}$$

$$\int x^2 \sin^2 bx \, dx = \frac{x^3}{6} - \left(\frac{x^2}{4b} - \frac{1}{8b^3} \right) \sin (2bx) - \frac{x}{4b^2} \cos (2bx) \tag{A.4}$$

$$\int xe^{bx} \, dx = \frac{e^{bx}}{b^2} (bx - 1) \tag{A.5}$$

$$\int x^2 e^{bx} \, dx = e^{bx} \left(\frac{x^2}{b} - \frac{2x}{b^2} + \frac{2}{b^3} \right) \tag{A.6}$$

$$\int_0^\infty x^n e^{-qx} \, dx = \frac{n!}{q^{n+1}}, \qquad n > -1, \ q > 0 \tag{A.7}$$

$$\int_0^\infty e^{-bx^2} \, dx = \frac{1}{2} \left(\frac{\pi}{b} \right)^{1/2} \tag{A.8}$$

$$\int_0^\infty x^{2n} e^{-bx^2} \, dx = \frac{1 \cdot 3 \cdots (2n - 1)}{2^{n+1}} \left(\frac{\pi}{b^{2n+1}} \right)^{1/2}, \qquad n = 1, 2, 3, \ldots \tag{A.9}$$

$$\int_t^\infty z^n e^{-az} \, dz = \frac{n!}{a^{n+1}} e^{-at} \left(1 + at + \frac{a^2 t^2}{2!} + \cdots + \frac{a^n t^n}{n!} \right), \qquad n = 0, 1, 2, \ldots \tag{A.10}$$

Bibliography

Acton, F. S., *Numerical Methods That Work*, Harper & Row, New York, 1970.

Anderson, J. M., *Mathematics for Quantum Chemistry*, Benjamin, New York, 1966.

———, *Introduction to Quantum Chemistry*, Benjamin, New York, 1969.

Atkins, P. W., *Molecular Quantum Mechanics*, 2nd ed., Oxford University Press, New York, 1983.

Ballentine, L. E., *Quantum Mechanics*, Prentice Hall, Englewood Cliffs, N.J., 1990.

Bates, D. R., ed., *Quantum Theory*, 3 vols., Academic Press, New York, 1961.

Bethe H. A., and R. W. Jackiw, *Intermediate Quantum Mechanics*, 3rd ed., Benjamin-Cummings, Menlo Park, Calif., 1985.

———, and E. E. Salpeter, *Quantum Mechanics of One- and Two-Electron Atoms*, Academic Press, New York, 1957.

Burkert, U., and N. L. Allinger, *Molecular Mechanics* (ACS Monograph No. 177), American Chemical Society, Washington, D.C., 1982.

Carsky, P., and M. Urban, *Ab Initio Calculations*, Springer-Verlag, New York, 1980.

Christoffersen, R. E., *Basic Principles and Techniques of Molecular Quantum Mechanics*, Springer-Verlag, New York, 1989.

Clark, T., *A Handbook of Computational Chemistry*, Wiley, New York, 1985.

Cotton, F. A., *Chemical Applications of Group Theory*, 2nd ed., Wiley, New York, 1970.

Davis, J. C., Jr., *Advanced Physical Chemistry*, Ronald Press, New York, 1965.

Dicke, R. H., and J. P. Wittke, *Introduction to Quantum Mechanics*, Addison-Wesley, Reading, Mass., 1960.

Dirac, P. A. M., *The Principles of Quantum Mechanics*, 4th ed., Oxford University Press, New York, 1958.

Eyring, H., J. Walter, and G. E. Kimball, *Quantum Chemistry*, Wiley, New York, 1944.

Fong, P., *Elementary Quantum Mechanics*, Addison-Wesley, Reading, Mass., 1962.

Halliday, D., and R. Resnick, *Physics*, 3rd ed., Wiley, New York, 1978.

Hameka, H. F., *Quantum Mechanics*, Wiley, New York, 1981.

Hanna, M. W., *Quantum Mechanics in Chemistry*, 3rd ed., Benjamin, Menlo Park, Calif., 1981.

Hehre, W. J., L. Radom, P. v. R. Schleyer, and J. A. Pople, *Ab Initio Molecular Orbital Theory*, Wiley, New York, 1986.

Jammer, M., *The Conceptual Development of Quantum Mechanics*, McGraw-Hill, New York, 1966.

Johnson, C. S., and L. G. Pedersen, *Problems and Solutions in Quantum Chemistry and Physics*, Addison-Wesley, Reading, Mass., 1974.

Jørgensen, P., and J. Oddershede, *Problems in Quantum Chemistry*, Addison-Wesley, Reading, Mass., 1983.

Karplus, M., and R. N. Porter, *Atoms and Molecules*, Benjamin, New York, 1970.

Kauzmann, W., *Quantum Chemistry*, Academic Press, New York, 1957.

Kemble, E. C., *The Fundamental Principles of Quantum Mechanics*, McGraw-Hill, New York, 1937; Dover, New York, 1958.

Levine, I. N., *Molecular Spectroscopy*, Wiley, New York, 1975.

——, *Physical Chemistry*, 3rd ed., McGraw-Hill, New York, 1988.

Lowe, J. P., *Quantum Chemistry*, Academic Press, New York, 1978.

Margenau, H., and G. M. Murphy, *The Mathematics of Physics and Chemistry*, 2nd ed., Van Nostrand Reinhold, New York, 1956.

McQuarrie, D. A., *Quantum Chemistry*, University Science, Mill Valley, Calif., 1983.

Meck, H. R., *Numerical Analysis with the TI-99/4A, Commodore 64, Apple* II + /IIe, *TRS-80*, Prentice-Hall, Englewood Cliffs, N.J., 1984.

Merzbacher, E., *Quantum Mechanics*, 2nd ed., Wiley, New York, 1970.

Messiah, A., *Quantum Mechanics*, vols. 1 and 2, Halsted, New York, 1963.

Mulliken, R. S., and W. C. Ermler, *Diatomic Molecules*, Academic Press, New York, 1977.

——, and ——, *Polyatomic Molecules*, Academic Press, New York, 1981.

Murrell, J. N., and A. J. Harget, *Semi-empirical Self-Consistent-Field Molecular Orbital Theories of Molecules*, Wiley-Interscience, New York, 1971.

——, S. F. A. Kettle, and J. M. Tedder, *Valence Theory*, 2nd ed., Wiley, New York, 1970.

Offenhartz, P. O'D., *Atomic and Molecular Orbital Theory*, McGraw-Hill, New York, 1970.

Park, D., *Introduction to the Quantum Theory*, 2nd ed., McGraw-Hill, New York, 1974.

Parr, R. G., *Quantum Theory of Molecular Electronic Structure*, Benjamin, New York, 1963.

——, and W. Yang, *Density-Functional Theory of Atoms and Molecules*, Oxford University Press, New York, 1989.

Pauling, L., and E. B. Wilson, Jr., *Introduction to Quantum Mechanics*, McGraw-Hill, New York, 1935.

Pilar, F. L., *Elementary Quantum Chemistry*, 2nd ed., McGraw-Hill, New York, 1990.

Pople, J. A., and D. L. Beveridge, *Approximate Molecular Orbital Theory*, McGraw-Hill, New York, 1970.

Press, W. H., B. P. Flannery, S. A. Teukolsky and W. T. Vetterling, *Numerical Recipes*, Cambridge University Press, New York, 1986.

Salem, L., *The Molecular Orbital Theory of Conjugated Systems*, Benjamin, New York, 1966.

Schaefer, H. F., ed., *Applications of Electronic Structure Theory* (vol. 4 of *Modern Theoretical Chemistry*, W. Miller et al., eds.), Plenum, New York, 1977.

——, *The Electronic Structure of Atoms and Molecules*, Addison-Wesley, Reading, Mass., 1972.

——, ed., *Methods of Electronic Structure Theory* (vol. 3 of *Modern Theoretical Chemistry*, W. Miller et al., eds.), Plenum, New York, 1977.

Schonland, D. S., *Molecular Symmetry*, Van Nostrand Reinhold, New York, 1965.

Segal, G. A., ed., *Semiempirical Methods of Electronic Structure Calculation, Parts A and B* (vols. 7 and 8 of *Modern Theoretical Chemistry*, W. Miller et al., eds.), Plenum, New York, 1977.

Shoup, T. E., *Applied Numerical Methods for the Microcomputer*, Prentice-Hall, Englewood Cliffs, N.J., 1984.

Slater, J. C., *Quantum Theory of Atomic Structure*, vols. I and II, McGraw-Hill, New York, 1960.

———, *Quantum Theory of Molecules and Solids*, vol. I, *Electronic Structure of Molecules*, McGraw-Hill, New York, 1963.

Sokolnikoff, I. S., and R. M. Redheffer, *Mathematics of Physics and Modern Engineering*, 2nd ed., McGraw-Hill, New York, 1966.

Streitwieser, A., Jr., *Molecular Orbital Theory for Organic Chemists*, Wiley, New York, 1961.

Szabo, A., and N. S. Ostlund, *Modern Quantum Chemistry*, rev. ed., McGraw-Hill, New York, 1989.

Taylor, A. E., and W. R. Mann, *Advanced Calculus*, 2nd ed., Wiley, New York, 1972.

Wilson, S., *Electron Correlation in Molecules*, Oxford University Press, New York, 1984.

Answers to Selected Problems

1.1 3.32 Å. **1.2** (a) 3.92 eV; (b) 544 nm. **1.3** (a) 2.22×10^{-6} deg; (b) 2.22 nm.
1.5 $2mb^2x^2$. **1.6** $3c\hbar^2/m = 6.67 \times 10^{-20}$ J. *Hint:* Use the time-independent Schrödinger equation. **1.7** (a) 3.29×10^{-6}; (b) 0.0753; (c) at $x = 0$. (d) *Hint:* Use the change of variable $z = -x$ in one of the integrals and use integral (A.7) in the Appendix. **1.8** 4.978×10^{-6}. **1.9** 0.000216. **1.10** (a) The Maxwell distribution of molecular speeds. **1.11** None; the function in (c) is not normalized. **1.12** 2.24, 0.0126. **1.13** $1 - 2(13)12/(26)25 = 13/25$. **1.14** $490/1485 = 0.33$. **1.17** (a) -1; (b) $-i$; (c) 1. **1.18** (a) -4; (b) $2i$; (c) $6 - 3i$; (d) $2e^{i\pi/5}$. **1.19** (a) 1, $\pi/2$; (b) 2, $\pi/3$; (c) 2, $4\pi/3$; (d) $5^{1/2}$, 296.6°. **1.20** (a) $e^{i\pi/2}$; (b) $e^{-i\pi}$; (c) $5^{1/2}e^{5.176i}$; (d) $2^{1/2}e^{i5\pi/4}$. **1.21** (a) 1, $-\frac{1}{2} + \frac{1}{2}\sqrt{3}i$, $-\frac{1}{2} - \frac{1}{2}\sqrt{3}i$. **1.23** (a) g cm/s^2; (b) g cm^2/s^2; (c) g$^{1/2}$ cm$^{3/2}$ s^{-1}; (d) kg m s^{-2}; (e) kg m^2 s^{-2}. **1.24** 4.05×10^4 dyn $= 0.405$ N. **1.25** (a) T. (b) F. (c) F. (d) T. (e) F.

2.1 (a) $y = c_1e^{-2x} + c_2e^x$; (b) $c_1 = -1/3$, $c_2 = 1/3$. **2.2** (b) $y = ae^x + bxe^x$. **2.3** 3.0×10^{26}.
2.4 (a) $\frac{1}{4} - (2n\pi)^{-1} \sin(n\pi/2)$; (b) 3; (c) $\frac{1}{4}$; (d) correspondence principle. **2.5** (a) 1.8×10^{-10} erg; (b) 11 nm; (c) UV. **2.6** 1.8×10^{-7} cm. **2.7** 4. **2.8** The same energies and wave functions are obtained (although the mathematical expression for ψ looks different). **2.11** 323 nm.
2.12 $e^{-iEt/\hbar}$ times (2.30). **2.16** 2. **2.18** 4.02 eV, 13.6 eV. **2.19** (a) linear; (b) linear; (c) nonlinear.
2.21 (a) F. (b) F. (c) T. (d) F.

3.1 (a) $-2x \sin(x^2 + 1)$; (b) $5 \sin x$; (c) $\sin^2 x$; (d) x; (e) $-1/x^2$; (f) $36x^3 + 24x$. **3.2** (a) Operator; (b) function; (c) function; (d) operator; (e) operator; (f) function. **3.3** *Hint:* Read the definition of equality of operators. **3.6** (b) $\hat{A}$ and $\hat{B}$ linear and commute. **3.7** $\hat{1}$. **3.8** (a) Linear; (b) nonlinear; (c) linear; (d) nonlinear; (e) linear. **3.9** (a) Complex conjugation; (b) *Hint:* Consider an operator that is the product of three operators. **3.12** (a) $-\cos z$; (b) $2a + (4ax + 2b)d/dx$; (c) 0. **3.16** (a) Yes; (b) $1/p$; (c) $1/(p - a)$. **3.17** (a) Yes; (b) $1 - 2x$. **3.19** (a) Yes, 1; (b) no; (c) yes, -1; (d) yes, -1; (e) yes, -1. **3.20** The eigenfunctions are (2.30) with E replaced by the eigenvalues k, where $k \geqslant 0$. **3.22** (a) $i\hbar^3\partial^3/\partial y^3$; (b) $-i\hbar(x\partial/\partial y - y\partial/\partial x)$. **3.23** (a) $i\hbar$; (b) $2\hbar^2\partial/\partial x$; (c) 0; (d) 0; (e) $(\hbar^2/m)\partial/\partial x$; (f) $2yz\hbar^2\partial/\partial x$. **3.24** $\int_0^2 |\Psi(x, t)|^2 dx$. (b) *Hint:* Try working Problem 3.27. **3.25** (a) length$^{-1/2}$. **3.27** 7.58×10^{14} s^{-1}. **3.27** (a) 0.0108; (b) 0.306; (c) 0.306. **3.28** For (b), $n_x^2h^2/4a^2$. For (c), $n_z^2h^2/4c^2$. **3.33** (a) 17; (b) 6. **3.34** (a) Nondegenerate; (b) 6; (c) 4. **3.35** (a), (c), (d), (g). **3.36** (a) $a/2$; (b) $b/2$, $c/2$; (c) 0; (d) $(1 - 3/2n_x^2\pi^2)a^2/3$, no, yes. **3.37** (a) No; (b) yes; (c) yes; (d) yes; (e) no. **3.39** (a) T. (b) F. (c) F. (d) F. (e) F. (f) F. **3.40** (b) 12.

4.2 (a) $\sum_{n=0}^{\infty} (-1)^n x^{2n+1}/(2n + 1)!$; (b) $\sum_{n=0}^{\infty} (-1)^n x^{2n}/(2n)!$. **4.3** (a) $\sum_{n=0}^{\infty} x^n/n!$. **4.5** (a) $c_{n+2} = (n^2 + n - 3)c_n/(n + 1)(n + 2)$; (b) $c_4 = -3c_0/8$, $c_5 = -3c_1/40$. **4.6** (a) Odd; (b) even; (c) odd; (d) neither; (e) even; (f) odd; (g) neither; (h) even. **4.9** (c) 0. **4.11** $h\nu/4$, $h\nu/4$. **4.12**

$\pm(\alpha/9\pi)^{1/4}(2\alpha^{3/2}x^3 - 3\alpha^{1/2}x)e^{-\alpha x^2/2}$. **4.13** $e^{-\alpha x^2/2}(1 - 4\alpha x^2 + \frac{4}{3}\alpha^2 x^4)$. **4.15** $x = \pm\alpha^{-1/2}$. **4.17** (a)
$(v_x + \frac{1}{2})h\nu_x + (v_y + \frac{1}{2})h\nu_y + (v_z + \frac{1}{2})h\nu_z$; (b) 1, 3, 6, 10. **4.21** $(2n + \frac{3}{2})h\nu$, $n = 0, 1, 2, \ldots$.
4.22 $-(2v + 1)^{1/2} \le \alpha^{1/2}x \le (2v + 1)^{1/2}$. **4.24** (a) 4.80 mdyn/Å = 480 N/m; (b) 2.87×10^{-20} J;
(c) 6.20×10^{13} Hz. **4.26** (a) 2989.96 cm^{-1}, 52.0 cm^{-1}; (b) 8346.0 cm^{-1}. **4.27** (a) 0.00142, 0.0160;
(b) 0.159, 0.314. **4.31** (b) V is an even function with 3 nodes and minima at $x = \pm(4a)^{-1/4}$;
$V(\pm\infty) = \infty$. (c) Yes, since ψ has no interior nodes.

5.3 $(4\hbar^3/i)\partial^2/\partial x^2$. **5.6** $\Delta x = (h/8\pi^2 m\nu)^{1/2}$, $\Delta p_x = (mh\nu/2)^{1/2}$, $\Delta x \, \Delta p_x = \hbar/2$. **5.7** $\Delta x =$
$(5/192)^{1/2}l$, $\Delta p_x = (14)^{1/2}\hbar/l$. **5.8** 1, 0.707. **5.9** $|\mathbf{A}| = 7$, $|\mathbf{B}| = (33)^{1/2}$, $\mathbf{A} \cdot \mathbf{B} = 13$, $\mathbf{A} \times \mathbf{B} = -32\mathbf{i} -$
$18\mathbf{j} + 10\mathbf{k}$, $\theta = 71.1°$, $\mathbf{A} + \mathbf{B} = 2\mathbf{i} + 2\mathbf{j} + 10\mathbf{k}$, $\mathbf{A} - \mathbf{B} = 4\mathbf{i} - 6\mathbf{j} + 2\mathbf{k}$. **5.10** $\arccos(-1/3) = 109.47°$. **5.12**
$\mathbf{grad} f = (4x - 5yz)\mathbf{i} - 5xz\mathbf{j} + (2z - 5xy)\mathbf{k}$; $\nabla^2 f = 6$. **5.13** (b) 3. **5.18** (a) $r = 5^{1/2}$, $\theta = \pi/2$, $\phi =$
63.4°; (b) $r = (10)^{1/2}$, $\theta = 18.4°$, $\phi = 180°$; (c) $r = (14)^{1/2}$, $\theta = 122.3°$, $\phi = 18.4°$; (d) $r = 3^{1/2}$, $\theta =$
125.3°, $\phi = 225°$. **5.19** (a) $x = -1$, $y = 0$, $z = 0$; (b) $x = 1.414$, $y = 0$, $z = 1.414$. **5.23** 35.3°, 65.9°,
90°, 114.1°, 144.7°.

6.4 $0.31c$, $0.64c$, $0.91c$, $1.20c$, $1.24c$, and $1.80c$, where $c = 5.49 \times 10^{-12}$ erg. **6.5** (a)
1.1309 Å; (b) 230542 MHz and 345813 MHz; (c) 110189 MHz; (d) 2.945, 4.729. **6.7** 2×10^{39}.
6.10 (a) 10941.2 Å, 2.7400×10^{14} Hz; (b) 2735 Å, 1.096×10^{15} Hz. **6.11** 3971.2 Å, 3890.2 Å,
3647.1 Å. **6.12** (b) 4340.5 Å, 4101.8 Å. **6.13** -13.598 eV. **6.14** -6.8 eV. **6.15** -108.8 eV.
6.17 $5a/Z$. **6.18** $30a^2/Z^2$. **6.21** 14. **6.22** a/Z. **6.23** At the origin (nucleus). **6.24** 0.24. **6.25** 0.24.
6.26 $2.66a$. **6.27** s states. **6.30** $-e'^2/a$, $e'^2/2a$, $-\frac{1}{2}$. **6.31** 1/137.0. **6.32** *Hint:* Use the fact that all
directions of space are equivalent. **6.34** (a) All; (b) $\hat{H}$, $\hat{L}^2$; (c) all. **6.35** (a) A sphere centered at
the origin; (b) m. **6.41** (a) 16×10^{-173}; (b) 1.6×10^{-51}; (c) 2.90×10^{-5}.

7.1 Yes. **7.5** $i(d/dx)$, $4 d^2/dx^2$. **7.8** (c) and (d). **7.11** (a) Use a table of integrals to help
you. (b) $\pi^3/32 \simeq 1 - 1/3^3 + 1/5^3 - 1/7^3 + 1/9^3$; $\pi^3 \simeq 31.021$; (c) -2.70%, 0.128%, -0.022%.
7.12 (b) $-1 = (4/\pi)(-1 + 1/3 - 1/5 + 1/7 - \cdots)$; -27.32%, -10.35%, -6.31%, -4.52%.
7.13 (b) $2p_x$, $2^{1/2}2p_1 - 2p_x$ or $2p_1$, $2^{1/2}2p_x - 2p_1$; $\hat{H}$, $\hat{L}^2$. **7.15** (c) Yes; no. **7.18** The n nth roots of
1. **7.19** (b) and (c). **7.24** (a) 0, 1, 0; (b) $\frac{1}{2}$, 0, $\frac{1}{2}$; (c) 0, 0, 1. **7.26** Probability 2/3 for $2\hbar^2$;
probability 1/3 for $6\hbar^2$. **7.27** 1/4 for $h^2/8ml^2$, 3/4 for $h^2/2ml^2$. **7.28** Use a table of integrals.
Outcomes are $n^2h^2/8ml^2$, where $n = 1, 2, 3, \ldots$. Probabilities are c_n^2, where $c_n/(210)^{1/2} =$
$(6 - n^2\pi^2)(\sin n\pi)/n^4\pi^4 - (4 \cos n\pi)/n^3\pi^3 - 2/n^3\pi^3$. **7.29** $\pi^6/960$. **7.32** (b) $(2mE)^{1/2}$, $-(2mE)^{1/2}$;
(c) probabilities $|c_1|^2$ for $(2mE)^{1/2}$ and $|c_2|^2$ for $-(2mE)^{1/2}$. **7.34** (a) 1; (b) 0; (c) 1; (d) 0.
7.35 $\frac{1}{2}f(0)$. **7.39** (a) First row: 6, 2; second row: -12, -12. (b) First row: 2, 4; second row: 8,
-8. (c) First row: 3, 0; second row: 4, 1; (d) First row: 6, 3; second row: 0, -9. (e) First row:
-2, 5; second row: -16, -19. **7.40** First row of **CD**: $5i$, 10, 5; second row: 0, 0, 0; third row: $-i$,
-2, -1. **DC** is a 1×1 matrix whose sole element is $5i - 1$. **7.47** (a) 0; (b) $\hbar$. **7.48** (b) $13h^2/32ml^2$;
(c) Use product-to-sum trigonometric identities to evaluate the integral. $\langle x \rangle = \frac{1}{2}l + 8(3^{1/2}l/9\pi^2) \times$
$\cos(3h^2t/8ml^2\hbar)$; $\langle x \rangle_{\max} = 0.656l$; $\langle x \rangle_{\min} = 0.344l$, since the cosine ranges from 1 to -1.
7.54 As a further hint, see the definition of a Hermitian operator. **7.55** (a) F. (b) T. (c) F.
(d) F. (e) F. (f) T. (g) F. (h) F. (i) F. (j) F.

8.1 0% error. **8.4** 1.3% error. **8.5** (a) $\langle \phi_1|\hat{H}|\phi_1 \rangle = 5.753112\hbar^2/ml^2$; (b) $5.792969\hbar^2/ml^2$.
8.6 (a) $(5/4\pi^2)(h^2/mb^2)$. **8.8** (b) $k = 1.11237244$, 0.298%. **8.9** (a) $W = h\nu/2^{1/2}$, 41.4%.
8.11 (a) $3h^2/2\pi^2ml^2$, 21.6%. **8.12** (a) $c = 8/9\pi$, 15% error; (b) $c = 8/729\pi$.
8.13 $20.23921\hbar^2/ml^2$. **8.15** -84. **8.17** $n!$. **8.18** 1, 0, 4, -1. **8.20** -84. **8.22** (a) $x = 0$, $y = 0$, $z = 0$;
(b) $x = -5k$, $y = -2k$, $z = 3k$. **8.23** 1, -1. **8.24** $5.750518\hbar^2/ml^2$, $44.809711\hbar^2/ml^2$. **8.25** 1.3%,
6.4%. **8.35** (a) $\lambda = 3$, $c_1 = 2/5^{1/2}$, $c_2 = 1/5^{1/2}$; $\lambda = -2$, $c_1 = 1/5^{1/2}$, $c_2 = -2/5^{1/2}$. (b) Yes. Yes. (c)
Yes. Yes. (d) $\mathbf{C}^{-1} = \tilde{\mathbf{C}}$. **8.36** (a) $\lambda = 0$, $c_1 = i/2^{1/2}$, $c_2 = 1/2^{1/2}$; $\lambda = 4$, $c_1 = -i/2^{1/2}$, $c_2 = 1/2^{1/2}$. (b)
No. Yes. (c) No. Yes. (d) $\mathbf{C}^{-1} = \mathbf{C}^\dagger$. **8.38** The cubic equation for the eigenvalues can be solved by
trial and error. As a check, use the theorem that the sum of the eigenvalues of a matrix equals
the sum of the diagonal elements of the matrix. The normalized eigenvector for the lowest
eigenvalue has elements $2/5^{1/2}$, 0, $1/5^{1/2}$.

9.1 $15dh^2/64\pi^4\nu^2m^2$. **9.2** (a) $b/2 - (b/2n\pi)[\sin(\frac{3}{2}n\pi) - \sin(\frac{1}{2}n\pi)]$; (b) $5.753112\hbar^2/ml^2$, $20.23921\hbar^2/ml^2$. **9.3** $\sum_{k\neq n} H'_{kn}\psi_k^{(0)}/(E_n^{(0)} - E_k^{(0)})$, where $\psi_k^{(0)} = (2/l)^{1/2}\sin(k\pi x/l)$, $E_n^{(0)} - E_k^{(0)} = (n^2 - k^2)h^2/8ml^2$, $H'_{kn} = (b/\pi)[A_k/(n-k) - B_k/(n+k)]$ with $A_k = \sin[3(n-k)\pi/4]$ $- \sin[(n-k)\pi/4]$, $B_k = \sin[3(n+k)\pi/4] - \sin[(n+k)\pi/4]$. **9.4** (b) $E^{(2)} = -0.0027338\hbar^2/ml^2$, $E^{(0)} + E^{(1)} + E^{(2)} = 5.750378\hbar^2/ml^2$. **9.6** 1.2×10^{-8} eV. **9.7** (a) $E^{(1)} = 0$ (parity); (b) $-(30v^2 + 30v + 11)c^2/8a^3h\nu$. **9.8** -12.86 eV. **9.9** $E^{(0)} = -77.5$ eV; $E^{(1)} = 0$. **9.10** First power. **9.11** $3a/2\zeta$. **9.14** (b) $E^{(1)} = (1 + 2/\pi)^2b/4$ for the ground state. $E^{(1)} = (1 + 2/\pi)b/4$, $(1 + 2/\pi)b/4$ for the states of the first excited level. **9.15** $0, 0, \pm 3e\mathcal{E}a_0$; $2p_1, 2p_{-1}, 2^{-1/2}(2s \mp 2p_0)$. **9.16** $1s3s$, two nondegenerate levels; $1s3p$, two triply degenerate levels; $1s3d$, two fivefold-degenerate levels. **9.18** -27.2 eV. **9.20** a. **9.24** (a) T. (b) F.

10.1 $54.7°$ (note that the component in the xy plane exceeds the z component). **10.5** (1) Neither. (2) Antisymmetric. (3) Symmetric. (4) Neither. (5) Symmetric. (6) Symmetric. **10.7** Ground state: $1s(1)1s(2)1s(3)$. **10.8** (a) $2^{-1/2}(1 - \hat{P}_{12})$. **10.9** One of my students gave the answer: "A permanent is used to put the wave into the function." **10.13** $2J_{1s2s} + J_{1s1s}$. **10.14** 1.61×10^{-23} J/T. **10.15** (b) 2.80×10^{10} Hz. **10.17** (a) $\frac{1}{2}\hbar$ or $-\frac{1}{2}\hbar$; (b) $(\alpha + \beta)/2^{1/2}$, $(\alpha - \beta)/2^{1/2}$; (c) 0.5 for $\frac{1}{2}\hbar$, 0.5 for $-\frac{1}{2}\hbar$. **10.20** (d) eigenvalues $\frac{1}{2}\hbar$, $-\frac{1}{2}\hbar$.

11.1 (a) $2n^2$; (b) $4l + 2$; (c) 2; (d) 1. **11.5** 22. **11.9** 11/2, 9/2, 7/2, 5/2; (b) 11/2, 9/2, 9/2, 7/2, 7/2, 5/2, 5/2, 3/2, 3/2, 1/2. **11.11** (a) ^1F, ^1G, ^1H, ^3F, ^3G, ^3H; (c) ^2D, ^2S, ^2P, ^2D, ^2F, ^2G, ^4P, ^4D, ^4F, ^2P, ^2D, ^2F. **11.12** (a), (c), (e), (f). **11.14** $15\hbar^2/4$, $\frac{1}{2}\hbar$. **11.15** (a) $45°$; (b) $70.53°$, $70.53°$, $70.53°$, $180°$. (c) $90°$. **11.17** B, N, F. **11.18** (a) 28; (b) 1; (c) 9; (d) 10. **11.19** (a) 15; (b) 36. **11.20** (a) T. (b) T. **11.21** (a) 2; (b) 1, 3; (c) 4, 2; (d) 8, 6, 4, 2; (e) 1, 3; (f) 2. **11.22** (a) $^1S_0$, 1; (b) $^2S_{1/2}$, 2; (c) $^3F_4$, 9; $^3F_3$, 7; $^3F_2$, 5; (d) $^4D_{7/2}$, 8; $^4D_{5/2}$, 6; $^4D_{3/2}$, 4; $^4D_{1/2}$, 2. **11.23** (a) $6^{1/2}\hbar$; (b) $2^{1/2}\hbar$; (c) $(12)^{1/2}\hbar$. **11.24** $^2S_{1/2}$, $^1S_0$, $^2S_{1/2}$, $^1S_0$, $^2P_{1/2}$, $^3P_0$, $^4S_{3/2}$, $^3P_2$, $^2P_{3/2}$, $^1S_0$. **11.25** $^2D_{3/2}$, $^3F_2$, $^4F_{3/2}$, $^7S_3$, $^6S_{5/2}$, $^5D_4$, $^4F_{9/2}$, $^3F_4$, $^2S_{1/2}$, $^1S_0$. Cobalt. **11.29** 64089.8, 75254.0, 105798.7, 64073.4, 64074.5, 75237.6, 75238.9, 75239.7, 105782.3, 64043.5, 64046.4, 64047.5, 75210.6, 75211.9, 105755.3, 87685, 109685, 98230 cm^{-1}. **11.30** No. **11.32** 4.5×10^{-5} eV. **11.33** 7.7×10^{-6} eV.

12.1 The first player will win. The winning strategy is given in H. E. Dudeney, *Amusements in Mathematics*, Dover, New York, 1958. **12.2** (a) $2\sigma_v$, C_2; (b) $3\sigma_v$, C_3; (c) C_3, $3\sigma_v$; (d) σ; (g) none. **12.3** (a) $\hat{\sigma}_v$, $\hat{\sigma}'_v$, $\hat{C}_2$, $\hat{E}$; (b) $\hat{\sigma}_v$, $\hat{\sigma}'_v$, $\hat{\sigma}''_v$, $\hat{C}_3$, $\hat{C}_3^2$, $\hat{E}$; (c) same as (b); (d) $\hat{\sigma}$, $\hat{E}$; (g) $\hat{E}$. **12.4** (a) $\hat{E}$; (b) $\hat{\sigma}$; (c) $\hat{C}_2$; (d) $\hat{C}_2$; (h) $\hat{i}$. **12.5** (a) A $\hat{C}_2$ rotation about the line through F_3 and F_5. **12.6** (a) Yes. (b) No. (c) Yes. **12.7** (a) Lies along the C_2 axis; (b) lies along the C_3 axis; (e) is zero; (g) no information. **12.10** (a) The unit matrix of order 3. (b) A diagonal matrix with diagonal elements 1, 1, -1. (c) A diagonal matrix with diagonal elements -1, 1, 1. (d) A diagonal matrix with diagonal elements 1, -1, -1. **12.12** (b) No, since its eigenvalues are not all real. **12.16** $\overline{10}$, $\overline{3}$, $\overline{14}$. **12.17** (a) The regular tetrahedron. **12.18** $\mathscr{C}_1$, $\mathscr{C}_s$, $\mathscr{C}_n$, $\mathscr{C}_{nv}$. **12.19** $\mathscr{C}_1$, $\mathscr{C}_n$, $\mathscr{D}_n$, $\mathscr{T}$, $\mathscr{O}$, $\mathscr{I}$. **12.21** (a) $\mathscr{T}_d$; (b) $\mathscr{C}_{3v}$; (c) $\mathscr{C}_{2v}$; (d) $\mathscr{C}_{3v}$; (e) $\mathscr{O}_h$; (f) $\mathscr{C}_{4v}$; (g) $\mathscr{D}_{4h}$; (h) $\mathscr{C}_{3v}$. **12.22** (a) $\mathscr{D}_{2h}$. **12.23** (a) $\mathscr{D}_{6h}$; (b) $\mathscr{C}_{2v}$; (c) $\mathscr{C}_{2v}$; (d) $\mathscr{C}_{2v}$; (e) $\mathscr{D}_{2h}$. **12.24** (a) $\mathscr{C}_{\infty v}$. **12.25** (a) $\mathscr{O}_h$; (b) $\mathscr{C}_{4v}$. **12.27** The answer is not $\mathscr{D}_2$. **12.28** (a) $\mathscr{C}_{4v}$; (b) $\mathscr{C}_{\infty v}$; (c) $\mathscr{D}_{4h}$; (d) $\mathscr{C}_{4v}$; (e) $\mathscr{D}_{\infty h}$.

13.1 432.07 kJ/mol. **13.2** 15.425 eV. **13.3** (a) 4.61 eV; (b) 4.48 eV. **13.4** (b) 2.5151 eV. **13.7** (a) 1836.15; (b) -1; (c) 2π; (d) -2; (e) 4.134×10^{16} (one atomic unit of time $= \hbar a_0/e'^2$ $= 2.419 \times 10^{-17}$ s); (f) 137.036; (g) -0.49973; (h) 0.3934. **13.10** (a) Ellipsoids of revolution; (c) half-planes. **13.14** At $R = 0.5$, $k = 1.78$ and $U = 0.2682_4$. At $R = 1.0$, $k = 1.54$ and $U = -0.4410_0$. At $R = 2.0$, $k = 1.24$ and $U = -0.5865_1$. **13.15** $k = 1.238_0$ and $R = 2.003_3$. **13.16** $-(3k^3/\pi)^{1/2}e^{-kr}\cos\theta$. **13.17** The optimum k is 1.334 (use trial and error to solve the quartic equation for k); $D_e = 0.0798$ hartree. (a) 3.8%; (b) 2.1%; (c) 22.2%. **13.20** (a) Li$_2$; (b) C$_2$; (c) O$_2^+$; (d) F$_2^+$ (Actually, D_e of Li$_2^+$ is greater than that of Li$_2$). **13.21** (a) 1; (b) 3; (c) 6; (d) 2. **13.22** (a) $^1\Sigma^-$; (b) $^3\Sigma^+$; (c) $^3\Pi_2$, $^3\Pi_1$, $^3\Pi_0$; (d) $^1\Phi_3$. **13.25** (l) $1\frac{1}{2}$, 1, $^2\Pi_u$; (m) 2, 0, $^1\Sigma_g^+$; (n) $2\frac{1}{2}$, 1, $^2\Sigma_g^+$. **13.26** (a) 0, $^1\Sigma^+$; (b) 0, $^1\Sigma^+$; (c) 0, $^1\Sigma^+$; (d) 1, $^2\Sigma^+$; (e) 1, $^2\Pi$; (f) 1, $^2\Pi$. **13.27** $j + k$ must be an even number. **13.28** VB. **13.45** The calculation converges to the same final result. **13.46** (b) 4 for each, except 5 for -1. **13.47** 3.60 electrons/bohr^3, 0.0990 electron/bohr^3. **13.51** (a) F. (b) F. **13.52** 4.56 eV, 11.3 D. **13.53** (a), (c), (f), (g). **13.56** (a) 8, 8; (b) 10, 6.

14.1 (b) 0; (c) -2; (d) 3/2. **14.2** State 1. **14.4** (a) $\langle V \rangle = (5\zeta/8 - 2Z\zeta)e'^2/a_0$, $\langle T \rangle = \zeta^2 e'^2/a_0$. **14.5** $\langle V \rangle = 3\frac{1}{3}$ eV. **14.11** (a) $Z/n^2 a$. **14.12** $(v + \frac{1}{2})h\nu m$. **14.14** 1.8, 4.1, 6.9, and 10 Å. General formula: $R_e[(Z_a Z_b)^{1/2} - Z_a]$.

15.2 1, 1, 1, 1; 1, 1, -1, -1; 1, -1, 1, -1; 1, -1, -1, 1; where the symmetry eigenvalues are listed in the order $\hat{E}$, $\hat{C}_2(z)$, $\hat{C}_2(y)$, $\hat{C}_2(x)$. **15.3** (b) 2. **15.4** (a) 56. **15.5** (a) 35; (b) 65; (c) 95; (d) 125; (e) 115. **15.6** (a) 540; (b) 900; (c) 1404. **15.7** 4. **15.8** (a) a_1: $H_1 1s + H_2 1s$, C1s, C2s, C2p$_z$, O1s, O2s, O2p$_z$. b_1: C2p$_x$, O2p$_x$. b_2: C2p$_y$, O2p$_y$, $H_1 1s - H_2 1s$. (b) Ten σ, two π; seven σ, one π. (c) Partial answer: For i(C): C1s; for b(CH$_1$): $H_1 1s$, C2s, C2p$_y$, C2p$_z$; for l_1(O): O2s, O2p$_y$, O2p$_z$. (d) 7×7. **15.9** There are nine a_1, nine b_2, two b_1, and two a_2 orbitals. **15.12** (c). **15.15** For the a_1 MOs, 3d$_{z^2}$ and 3d$_{x^2-y^2}$ contribute. For the 1b$_2$ MO 3d$_{yz}$ contributes. For the 1b$_1$ MO, 3d$_{xz}$ contributes. **15.19** Interchanges the subscripts 1 and 3 for the b symmetry species. **15.20** (a) No. **15.21** Partial answer: $1a_g \approx g_2$; $1b_{1u} \approx g_6$; $2a_g \approx g_3 + g_1$; $2b_{1u} \approx g_5 + g_7 - g_8$, where the coefficients are omitted. **15.22** (a) Bond angle a bit less than 120°; (b) linear; (c) angle a bit less than 109$\frac{1}{2}$°; (j) planar with 120° angles; (k) planar, T-shaped, two bond angles a bit less than 90°. **15.23** (a) Square planar; (b) tetrahedral; (d) square planar; (i) octahedral. **15.24** (a) $\mathscr{D}_{4h}$; (b) $\mathscr{T}_d$; (c) $\mathscr{C}_{2v}$; (d) $\mathscr{D}_{4h}$; (e) $\mathscr{T}_d$. **15.25** (a) Bond angle near 120°; (b) linear; (c) angle near 120°; (d) linear; (g) planar with FCF angle a bit less than 120°. **15.26** (a) Linear; (b) bent; (c) bent. **15.27** Linear. **15.28** 1.86×10^{28}. **15.29** 82% to 89% for $n = 20$; 44% to 55% for $n = 200$. **15.30** 0.9731. **15.33** T. **15.36** (b), (c), and (e). **15.39** (a) $-\frac{3}{2}\alpha(3\rho/\pi)^{1/3}$. **15.43** (a) 2. **15.44** (a) 42; (b) 3. **15.47** 1.0 for H1s; 1.70 for He1s; 5.70 for C1s, 1.625 for C2s and C2p; 6.70 for N1s, 1.95 for N2s and N2p.

16.1 207, 332, 459, 587, 716, 844, 973 nm; 11.6%. **16.2** (a) 332 nm; (b) 465 nm. **16.4** $E = (h^2/8\pi^2 m_e R^2)J^2$, where $J = 0, \pm 1, \pm 2, \dots$, and R is the radius of the circle. **16.5** (a) $\alpha + 2^{1/2}\beta$, α, $\alpha - 2^{1/2}\beta$; $\phi_1 = \frac{1}{2}f_1 + 2^{-1/2}f_2 + \frac{1}{2}f_3$, $\phi_2 = 2^{-1/2}f_1 - 2^{-1/2}f_3$, $\phi_3 = \frac{1}{2}f_1 - 2^{-1/2}f_2 + \frac{1}{2}f_3$. (b) 0.707, 0.707; (c) 1, 1, 1; (d) 1.025, 0.318, 1.025; (e) 0.828β. **16.6** (a) Same as Problem 16.5; (b) 0.707, 0.707; 0.707, 0.707; (c) $\frac{1}{2}$, 1, $\frac{1}{2}$; 1.5, 1, 1.5; (d) 1.025, 0.318, 1.025; 1.025, 0.318, 1.025; (e) 0.828β, 0.828β. **16.10** $P_{12} = 1.448$, $P_{23} = 1.725$; $q_1 = 1.00$, $q_2 = 1.00$. **16.13** (a) 27 kcal/mol; (b) 1.4 eV. **16.14** 42 kcal/mol, 69 kcal/mol. **16.15** (a) $e_1 = \alpha + 3^{1/2}\beta$, $e_2 = \alpha$, $e_3 = \alpha$, $e_4 = \alpha - 3^{1/2}\beta$, $\phi_1 = 2^{-1/2}f_1 + 6^{-1/2}(f_2 + f_3 + f_4)$, $\phi_2 = 3^{-1/2}f_2 + 3^{-1/2}e^{2\pi i/3}f_3 + 3^{-1/2}e^{4\pi i/3}f_4$, $\phi_3 = 3^{-1/2}f_2 + 3^{-1/2}e^{-2\pi i/3}f_3 + 3^{-1/2}e^{-4\pi i/3}f_4$; $P_{12} = 0.577$; $q_1 = 1$, $q_2 = 1$, $q_3 = 1$, $q_4 = 1$; $F_1 = 0$, $F_2 = F_3 = F_4 = 1.155$; delocalization energy $= 1.464\beta$. (b) $P_{12} = P_{23} = 1.414$. *Hint:* Each set of spatially perpendicular MOs has three electrons. **16.16** $\alpha = -6.1$ eV, $\beta = -3.3$ eV; 6.9 eV. **16.17** (a) 1.40_1 Å; (b) 1.40_2 Å; (c) 1.40_9 Å. **16.18** (b) $\phi_1 = 0.301(f_1 + f_4 + f_5 + f_8) + 0.231(f_2 + f_3 + f_6 + f_7) + 0.461(f_9 + f_{10})$. **16.21** (a) $-0.061|\beta|$, antiaromatic; (c) $0.0231|\beta|$, aromatic. **16.23** (a) 0; (b) 1.5. **16.25** (a) With atom 1 being O and atom 2 the C bonded to O, the determinant has the following rows. Row 1: $x + 2$, 0.8, 0, 0, 0, 0, 0; row 2: 0.8, x, 1, 0, 0, 0, 1; row 3: 0, 1, x, 1, 0, 0, 0; row 4: 0, 0, 1, x, 1, 0, 0; row 5: 0, 0, 0, 1, x, 1, 0; row 6: 0, 0, 0, 0, 1, x, 1; row 7: 0, 1, 0, 0, 0, 1, x. **16.27** (c) $\alpha + 1.33\gamma$, $\alpha + 0.80\gamma$, $\alpha + 0.80\gamma$, $\alpha - 1.33\gamma$, $\alpha - 1.33\gamma$, $\alpha - 4.0\gamma$. **16.29** Predicted $R_e = 0$. **16.31** CNDO: (b), (d), (e); INDO: (d), (e). **16.32** (a) -25.32 kcal/mol; (b) -32.50 kcal/mol; (c) 6.29 kcal/mol. **16.33** (a) Disrotatory; (b) conrotatory. **16.34** (a) High.

17.1 (b) 11.1 kcal/mol. **17.2** (a) 12.1 kcal/mol; (b) 27.3 kcal/mol; (c) -29.0 kcal/mol.

Index